TEXTURE AND ANISOTROPY

Preferred Orientations in Polycrystals
and their Effect on Materials Properties

Many manmade and naturally occurring substances are aggregates of crystals, or polycrystals, with a non-random distribution of orientations. In such textured polycrystals, many macroscopic properties are anisotropic, i.e. they depend on direction.

This book is about the measurement and analysis of textures, the prediction of polycrystal properties from measured textures and known single-crystal properties, and the modeling of the development of texture and the ensuing anisotropic properties during plastic deformation. It also gives an overview of observed textures in metals, ceramics and rocks. There is a balance between theoretical concepts and experimental techniques. The book addresses several issues. Part I provides tools and describes methods to obtain quantitative data on textures of polycrystals; it should be of interest to experimentalists. Part II emphasizes modeling of deformation and incorporates theoretical concepts of mechanics. Part III illustrates successful applications in engineering and Earth sciences.

Graduate students and research workers in both academic and industrial environments working in materials science, mechanics, metallurgy, geology and ceramics, including high-temperature superconductivity, will find this book of value.

TEXTURE AND ANISOTROPY

Preferred Orientations in Polycrystals and their Effect on Materials Properties

U. F. Kocks, C. N. Tomé, H.-R. Wenk

Authors and Editors

with additional contributions by
A. J. Beaudoin
G. R. Canova[†]
P. R. Dawson
J. S. Kallend
A. D. Rollett
M. G. Stout
S. I. Wright
and with an Introduction by H. Mecking

CAMBRIDGE
UNIVERSITY PRESS

PUBLISHED BY THE PRESS SYNDICATE OF THE UNIVERSITY OF CAMBRIDGE
The Pitt Building, Trumpington Street, Cambridge, United Kingdom

CAMBRIDGE UNIVERSITY PRESS
The Edinburgh Building, Cambridge CB2 2RU, UK
40 West 20th Street, New York, NY 10011–4211, USA
10 Stamford Road, Oakleigh, Melbourne 3166, Australia
Ruiz de Alarcón 13, 28014 Madrid, Spain

www.cambridge.org
Information on this title: www.cambridge.org/9780521465168

The work upon which this book is based was performed under the auspices of the
US Department of Energy, under contract with the University of California, operator of
Los Alamos National Laboratory. Accordingly, the US Government retains an irrevocable,
nonexclusive, royalty-free license to reproduce the published form of this work, and to
authorize others to do so, for US Government purposes.

First published 1998
First paperback edition (with corrections) 2000

Typeset by the author [CRC]
Typeset in Adobe Utopia (11 pt) and MathType Symbol

A catalogue record for this book is available from the British Library

Library of Congress Cataloguing in Publication data
Kocks, U. F.
Texture and anisotropy: preferred orientations in polycrystals and their effect
on materials properties / U. F. Kocks, C. N. Tomé, H.-R. Wenk, authors and editors:
with additional contributions by A. J. Beaudoin ... [et al.]; and with an introduction
by H. Mecking.
p.   cm.
ISBN 0 521 46516 8
1. Polycrystals.   2. Texture (Crystallography)   3. Anisotropy. I. Tomé.
C. N. (Carlos Norberto 1951–   II. Wenk, Hans-Rudolf, 1941–   III. Title.
QD925.K63   1998
548′.84–DC21   97-29060 CIP

ISBN-13 978-0-521-46516-8 hardback
ISBN-10 0-521-46516-8 hardback

ISBN-13 978-0-521-79420-6 paperback
ISBN-10 0-521-79420-X paperback

Transferred to digital printing 2005

CONTENTS

Contributors

Prof. Armand J. Beaudoin
University of Illinois at Urbana-Champaign
140 Mechanical Engineering Building
1206 West Green Street
Urbana, IL 61801 (USA)
    abeaudoi@uiuc.edu

Dr. Gilles R. Canova
Génie Physique et Mécanique des Matériaux
Inst. Nat. Polytechnique de Grenoble
F-38402 St.Martin d'Hères Cédex (FRANCE)
    [deceased 28 July 1997]

Prof. Paul R. Dawson
Cornell University
Dept. of Mechanical & Aerospace Eng.
E.T.C., Hoy Road
Ithaca, NY 14853 (USA)
    prd5@cornell.edu

Prof. John S. Kallend
Illinois Institute of Technology
Dept. of Metallurgical & Materials Eng.
10 W 32nd Street (Rm. 208)
Chicago, IL 60616 (USA)
    kallend@charlie.iit.edu

Dr. U. Fred Kocks
Los Alamos National Laboratory
Center for Materials Science
Mail Stop K-765
Los Alamos, NM 87545 (USA)
    kocks@lanl.gov

Prof. Heinrich Mecking
Technical University Hamburg–Harburg
Section Physics and Technology of Materials
Eissendorfer Str. 42
D-21071 Hamburg (GERMANY)
    Mecking@tu-harburg.d400.de

Prof. Anthony D. Rollett
Carnegie Mellon University
Dept. of Materials Science & Engineering
3327 Wean Hall
Pittsburgh, PA 15213-3890 (USA)
rollett@andrew.cmu.edu

Dr. Michael G. Stout
Los Alamos National Laboratory
MST-8, Mail Stop G755
Los Alamos, NM 87545 (USA)
mstout@lanl.gov

Dr. Carlos N. Tomé
Los Alamos National Laboratory
MST-8, MS G755
Los Alamos, NM 87545 (USA)
tomec@lanl.gov

Prof. Hans-Rudolf Wenk
University of California
Dept. of Geology
301 McCone Hall
Berkeley, CA 94720 (USA)
wenk@seismo.berkeley.edu

Dr. Stuart I. Wright
TexSEM Laboratories, Inc.
12300 South 392 East, Suite H
Draper, UT 84020 (USA)
tsl@oim.com

GILLES CANOVA

1954-1997

This book is dedicated to him:
our friend
who so often was ahead of us all
in finding the right question
and the right answer.

FOREWORD

Most commonly used manmade materials and naturally occurring substances are aggregates of crystals or 'polycrystals'. Their crystallographic orientation is generally non-random. All non-random orientation distributions are called preferred orientations or 'texture'. In textured polycrystals, many macroscopic physical properties are anisotropic, i.e. they depend on direction. This book is about the measurement and analysis of textures, the prediction of polycrystal properties from measured textures and known single crystal properties, and the prediction of the development of texture and the ensuing anisotropic properties during elastic and plastic deformation.

The study of *texture* started in the last century. After much early research of a more qualitative nature, a quantitative description using three-dimensional orientation distributions was introduced in 1965. Over the years, various mathematical difficulties in obtaining the orientation distribution from measured pole figures have been recognized, and sophisticated methods have been devised so that it can be obtained routinely from x-ray, neutron, and most recently electron diffraction data. Since then, the attention has focused on application of these methods to a wide variety of materials.

The principles of *anisotropy* were well understood for single crystals by the 1950s, even for the grossly non-linear properties of plasticity. It was also appreciated that in polycrystals, where anisotropy additionally requires a non-randomness of the arrangements of the grains, the crystallographic texture is of prime importance. By the 1960s, one line of research explored the changes in texture caused by (anisotropic) plastic deformation. It was realized that computer simulation of polycrystal plasticity was needed for a quantitative treatment. The increasing availability of high-speed computer power led to a resurgence of activity. Advances in quantitative texture analysis could be integrated with new insights into polycrystal models, and complexities due to heterogeneities in the material could be approached. Most recently, finite-element modeling has added new possibilities, and the time of fine-tuning has begun. An important driving force for renewed interest in textures is the need for materials with small tolerances, which requires optimization of textures for industrial applications.

The range of applications has now spread to a wide variety of deformation paths and a wide variety of materials. Considerable fertilization has occurred through the interdisciplinary collaborations between materials scientists and geologists. While the goal of materials science is to use textures to develop materials with satisfactory properties, geology uses it to unravel the deformation history of the Earth.

Research results on texture have been published in a large number of important publications and some reviews and monographs, especially concerned with the theory of obtaining orientation distributions from pole figures and with polycrystal plasticity. There is no book dealing with the broad range of texture and anisotropy and it has therefore been difficult for other scientists to enter this field.

The aim of the present book is to fill this gap by capturing the state of knowledge after these major developments, and to provide a foundation for the work of the future. There will be, no doubt, much further experimentation and quantitative analysis to provide data for a wide range of materials; and these methods will be applied to increasingly complex engineering problems. We direct ourselves to *users* of texture information, for the purpose of describing the anisotropy of macroscopic properties. Readers will be mainly graduate students and researchers in academia and industry with a scientific background.

The book addresses different issues. Part I provides tools and describes methods to obtain quantitative data on textures of polycrystals. It should be of interest to experimentalists. Part II emphasizes modeling of deformation and involves theoretical concepts of mechanics. Part III illustrates successful applications in engineering and Earth sciences.

The authors have collaborated for some ten years on the topics covered here; in fact, this book contains much unpublished work that arose out of this collaboration, centered at Los Alamos. We have developed a comprehensive perspective, and have attempted to present this view in a coherent fashion. Thus, the book should not be taken as an impartial review, but rather as an exposition of our collective best judgment. We mainly use examples from our own work. However, while the subject was made homogeneous, e.g. by circulating manuscripts among all contributors, Chapters were written by individual authors, reflecting their personal styles and emphasizing different aspects, some more practical, others more theoretical.

The fourteen Chapters cover a large portion of 'Texture and Anisotropy'. But it needs to be pointed out that some aspects are missing. We deal only peripherally with the newly emerging electron diffraction measurements of individual crystals and do not go into any depth in the discussion of misorientations. Among materials, we treat metals, ceramics and rocks but leave out polymers. Among texture forming processes, emphasis is on deformation; recrystallization and growth mechanisms are not treated in any detail.

We have benefited immensely from interactions with many colleagues. We are grateful for the thoughtful comments on parts of this book provided by John Bingert, Olaf Engler, Heinz Mecking, Paul Van Houtte and many others. The work that has led to the present publication has been generously supported over many years by the Center of Materials Science at Los Alamos National Laboratory, under the guidance of Don Parkin, and with the financial means of the U.S. Department of Energy, Office of Basic Energy Sciences, Division of Materials Sciences, and the Institute of Geophysics and Planetary Physics of the University of California. Laurie Lauer and AnnMarie Dyson have provided expert help with the production, Frances Nex with copy editing.

It is a privilege of the Editors to thank all contributors for their enthusiastic collaboration and patience. Also, last but certainly not least, they acknowledge with deep gratitude the sustained and sustaining support of their spouses, Marianne, Graciela, and Julia.

Fred Kocks
Carlos Tomé
Rudy Wenk

January 1998

MOTIVATION

H. Mecking

In crystals, almost every property of interest, in the whole spectrum ranging from chemical resistance to fracture toughness, is anisotropic, and almost all materials of interest are polycrystals made up from aggregates of single crystals.

Predicting the anisotropy of polycrystals from single crystal properties involves two fundamental steps: (a) determination of texture by measuring the orientations of all crystals of a polycrystal, and data processing for presenting the texture in a suitable form; and (b) execution of a physically correct averaging scheme for combining single crystal properties into polycrystal behavior which, at least for mechanical properties, is not a trivial task.

Even the most simple case of calculating the constants in Hooke's law from single crystal elastic constants is rather complicated since tensor components are to be considered. Elastic moduli derived by the Voigt average of elastic stiffnesses can be quite different from those obtained by the Reuss average of the compliances. For a proper treatment, mechanical interaction between the grains of a polycrystal must be considered as in the model of Eshelby and Kröner where a grain is treated as an elastic inclusion in a homogeneous matrix with the average properties of the aggregate, a rather complex numerical procedure, in particular when texture effects are included.

Plastic yielding of single crystals generally occurs by shearing on selected crystallographic slip systems according to the Schmid law, so that plastic flow comes along with an extreme anisotropy of stress- and strain-components and with lattice rotations that cause orientation changes. These processes are well understood for single crystals, but to predict from these results flow behavior and texture evolution in polycrystals is still the subject of extensive research in many government and industry laboratories. The Taylor model of polycrystal deformation, assuming uniqueness of the strain field, has been applied very successfully for single phase polycrystals with a cubic lattice. It is, however, inadequate if the crystal strength is heterogeneously distributed, like in materials with a low symmetry lattice and for poly-phase materials. These cases are included in recently introduced models , which are based on the rules of visco-plastic interaction. They are increasingly applied though their predicting power is still debated.

This book is concerned with the determination and description of texture, with the mechanisms of texture evolution and with the interrelationships between texture and (mostly mechanical) anisotropy. It is an update of the level of understanding of the basic mechanisms and of the physical and mathematical tools developed in this field. It provides the background necessary for using the available experimental, analytical and computational techniques and for advancing in new directions.

The scope and practicality of the various aspects covered in this book will now be illustrated with a few examples.

Processing and material performance

The actual orientation distribution in a polycrystal is the result of the manufacturing process and thus texture contains detailed information about the production history of a work piece. On the other hand, texture has a strong effect on properties so that it contains easily accessible information on the interrelation between processing parameters and material performance.

During solidification from the melt in many materials those crystallites grow fast which have a certain lattice direction aligned with the main direction of heat flow. In these cases, texture after casting is a rather precise documentation of heat flow during solidification; conversely, the controlled cooling of a casting mold can be utilized for texture design. Not only in casting but also in chemical or physical vapor deposition, the texture depends very sensitively on the solidification conditions and thus provides key information on the interrelationship between processing parameters and product properties. Such a case is illustrated for coatings on cutting tools by the rather qualitative textures presented in Fig. 1 as incomplete pole figures. Figure 1 displays pole figures of two differently processed but in optical appearance identical titanium nitride coatings on a tungsten carbide + cobalt substrate. It is seen that the coating has a direct orientation relationship to the base material in case (b) but not so in case (c) and, most interestingly, the wear resistance of coating (b) was almost 10 times that of (c). In this specific example, texture investigation offered a quick non-destructive way for quality control and at the same time gave important hints of the effect of epitaxial growth.

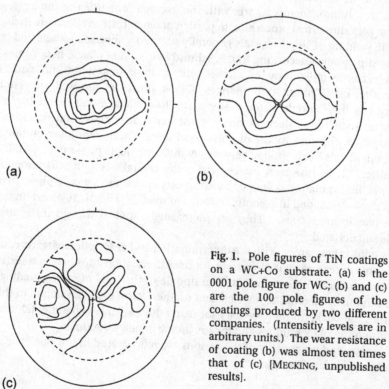

(a)

(b)

(c)

Fig. 1. Pole figures of TiN coatings on a WC+Co substrate. (a) is the 0001 pole figure for WC; (b) and (c) are the 100 pole figures of the coatings produced by two different companies. (Intensitiy levels are in arbitrary units.) The wear resistance of coating (b) was almost ten times that of (c) [MECKING, unpublished results].

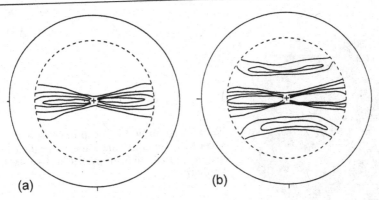

Fig. 2. 100 pole figures measured on (bundles of) Au wires with different drawing and annealing schedules. (Intensity levels are in arbitrary units.) The rupture stress was measured to be 450 MPa in case (a) and 400 MPa in case (b) [MECKING, unpublished results].

The combined effect of deformation and recrystallization on the microstructure of a work piece is due to a complex interplay between the quite different physical processes of deformation and recrystallization which is not easily treated theoretically but, nevertheless, offers a wide range of possibilities for manipulation of texture. For metallurgical applications an enormous amount of empirical knowledge has been gathered and led to standardized treatment schemes. Figure 2 is the illustration of a somewhat exotic example.

In drawn wires of face-centered cubic (fcc) metals the main component of the deformation texture is due to an alignment of ⟨111⟩ crystal directions parallel to the drawing direction. This type of texture is called fiber texture with a ⟨111⟩-fiber axis. Primary recrystallization converts this texture into a ⟨100⟩-fiber. By a suitable combination of drawing and annealing steps the relative intensity along the two fibers can be varied systematically. The 100 pole figures of Fig. 2 were measured on two gold wires, which had a standard diameter of 0.010" as they are commonly used in computer chips and which were subjected to quite different drawing and annealing schedules, thus exhibiting quite different intensities of the two texture components. The ⟨111⟩ component produces a higher tensile strength than the ⟨100⟩ component and hence by a variation of the relative intensities on the two competing fibers the tensile strength can be systematically varied and adjusted for an optimum performance in a bonding machine. Also in this case, texture inspection offered a quick method for quality assessment and process control.

As in wire drawing, also in more complex manufacturing processes crystallographic texture exhibits a marked influence on the performance of commercial products. Earing of sheet material is the most familiar and possibly the most significant example. Figure 3 shows the earing in an aluminum cup produced by deep drawing of a circular disk cut from a rolled sheet.

Enhanced formability in steel and aluminum sheet products used in stamping of automotive panels follow from control of texture through control of the production processes. Quantitative texture analysis provides the essential feedback for tuning the thermo-mechanical history on the basis of detailed insight into controlling mechanisms, thereby leading to desired properties in the finished product. Economic savings through control of texture can be significant: due to the large production volume of

Fig. 3. Cup formed by deep-drawing a rolled sheet of aluminum. The 'ears' are evidence of texture-caused anisotropy. [Courtesy Marc Verdier]

sheet material, such as aluminum can body stock, great savings can be achieved through the reduction of scrap caused by earing.

Another famous example of the commercial and technological importance of texture design are transformer cores of ferromagnetic iron-silicon alloys with the cube texture. In these, the $\langle 001 \rangle$-directions of the crystal lattice are preferentially aligned with the direction of the magnetic field in order to achieve the highest values of magnetic saturation. In this way the maximum efficiency of a transformer is obtained at a minimum of weight. The same physical effect is utilized by texture control in magnetic tapes.

Microstructure and properties

In many alloying systems the interrelationship between microstructure and properties contains the key information for establishing the strategies of material development. Texture is an essential part of the microstructure and can play a dominant role when the anisotropy of single crystals is large and crucial.

In systems with low ductility, like in many intermetallic compounds, texture can have a marked influence on fracture strain and fracture toughness. This is particularly true for intermetallic aluminides which are considered as high-performance products for high-temperature application, due to the outstanding properties of nickel based superalloys. Aluminides have become world wide the subject of extensive programs of research and development. Here texture often plays a prominent role for optimizing the microstructure. Figure 4 gives an example for texture effects in a hot rolled sheet of an alloy based on titanium aluminum. The texture shown has been inherited by transformation from the hot rolled bcc β-phase to hexagonal α with a DO_{19} ordered structure . It is seen that not only the strengths measured in the rolling and transverse directions are quite different but also that up to medium temperature the material behaves practically brittle upon loading in the rolling direction but undergoes plastic yielding combined with considerable plastic deformation when tested in the transverse direction.

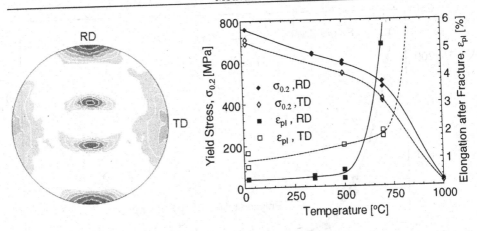

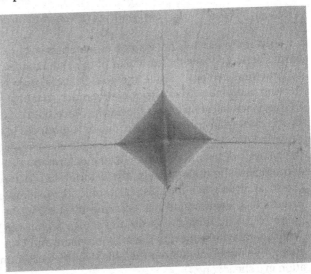

Fig. 4. (a) 0001 pole figure of the α-phase of $(Ti,Nb)_3(Al,Si)$ with a DO_{19} ordered lattice (logarithmic intensity scale; least dense shade spans the density range 0.5 to 2.0). The texture is inherited from the hot rolling texture (about 1200°C) of the bcc β-phase by transformation with a specific variant selection. (b) The anisotropy of strength and ductility are illustrated by the temperature-dependent yield stress and strain-to-failure, measured by loading in the rolling direction (RD) and the transverse direction (TD) [GERLING &AL. 1997].

In the past, ceramicists have largely ignored preferred orientations, however this is now changing. In many ceramic materials ductility and fracture toughness are properties of main concern and there texture effects can be large in particular in cleavage fracture. Figure 5 is an illustrative example for the anisotropy of preferred fracture modes and of crack propagation in silicon nitride plates which were produced by forming in plane-strain compression at high temperature.

Other advanced ceramics such as high temperature superconducting oxides are an excellent case to demonstrate dramatic effects of texture on properties. These anisotropic perovskite-like crystals all have in common a superconducting plane of Cu atoms in the crystal structure parallel to (001). The conductivity in other crystal

Fig. 5. Cracks propagated from an indentation (about 0.2 mm diagonal) on a Si_3N_4 showing anisotropy of crack resistance by the different crack lengths in different directions. The sample was produced by plane-strain compression at high temperature [LEE & BOWMAN 1992].

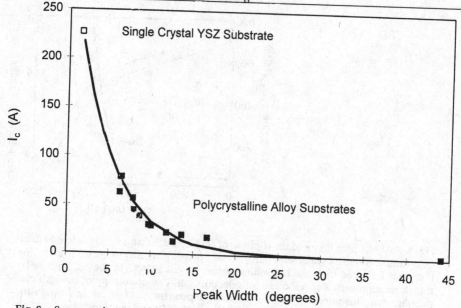

Fig. 6. Superconducting critical current for 1 μm thick YBCO films on 1 cm wide substrates. The Hastelloy substrates were coated with a textured layer of yttria-stabilized zirconia by ion-beam assisted deposition prior to laser deposition of the superconductor. In-plane alignment was characterized by the peak width obtained from an azimuthal x-ray scan on the 103 peak [FOLTYN &AL. 1995].

directions is considerably less: texture has a strong effect on electrical conductivity. In order to have satisfactory properties, crystallites need to be aligned. For example, in epitaxial thin films of YBCO deposited on a YSZ (yttria-stabilized zirconia) substrate, the strength of the texture (as evaluated by the full width at half maximum of the 103 peak) correlates with the critical current density (Fig. 6). The sharpness of the texture can be determined very quickly with an x-ray pole figure goniometer. To obtain the same information with the transmission electron microscope is a major project.

Deformation history and mechanisms

Texture research is not limited to materials science and it had in fact its origin in earth sciences. Since the early part of the 19th century it has been an important part of structural geology. While the goal of metallurgy is to use textures primarily to develop materials with satisfactory properties, that of geology is to use it to analyze the thermomechanical history of the Earth. Since plastic deformation is normally carried by slip on certain crystallographic planes it produces lattice rotations towards specific preferred positions which are characteristic of a certain deformation geometry. Thus, the analysis of texture evolution during deformation can be used to determine these slip processes and, conversely, the deformation history can be reconstructed from texture analysis when the crystallographic slip systems are known. One example is to determine if deformation has occurred by shearing or by coaxial thinning. A pole figure of calcite from the Basin and Range region of the United States (Fig. 7a) is symmetrical with respect to the foliation plane and thus indicative of coaxial deform-ation. By contrast, one from the Alps (Fig. 7b) is asymmetrical and suggests deform-ation in a shear zone.

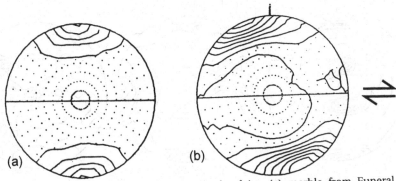

Fig. 7. 001 Pole figures of naturally deformed calcite: (a) marble from Funeral Mountains, Death Valley, California [ERSKINE &AL. 1993] (foliation plane is horizontal); (b) limestone at the base of the Morcles nappe, Swiss Alps (shear plane and direction are indicated) [RATSCHBACHER &AL. 1991]. Equal-area projection.

The final example is chosen to illustrate the usefulness and efficiency of modern texture analysis on the basis of a completely determined Orientation Distribution (OD) for a hot rolled sheet of a TiAl V alloy. In this alloying system industrial processing is usually performed by hot rolling in the two-phase field around 800°C, where the high temperature bcc β-phase is in equilibrium with the low-temperature hexagonal α-phase and thus individual texture evolution in α and β is of specific interest for optimization of the hot rolling parameters. Textures cannot be measured directly since during cooling to the test temperature (usually R.T.) the β-phase undergoes a martensitic transformation to α'' with an orthorhombic lattice structure in the case considered. The hexagonal α and the orthorhombic α'' have similar lattice spacings so that most of their x-ray diffraction peaks coincide or are so close that they cannot be resolved experimentally.

In total, seven incomplete pole figures containing 17 different diffraction peaks have been measured. From these data the ODs for the α and α'' phase could be determined separately by mathematical texture analysis combined with a deconvolution method and from that of α'' the original β-texture has been reconstructed with the help of a texture transformation scheme.

The knowledge of the ODs allows one to calculate every desired pole figure and for demonstration, Figs. 8a, b, c show for α, α'' and β the respective complete pole figures 1010, 001 and 110 obtained from this analysis. Texture inheritance by α'' from β is governed by the orientation relationships {110}$_β$∥ (001)$_{α''}$ and <111>$_β$∥ <110>$_{α''}$ which contains six equivalent variants but in the opposite direction from the orthorhombic to the bcc structure, i.e. from α'' to β the correlation is unambiguous. Thus not only the β-texture could be reconstructed unambiguously from that of α'' but also the variant selection in the martensitic transformation could be identified.

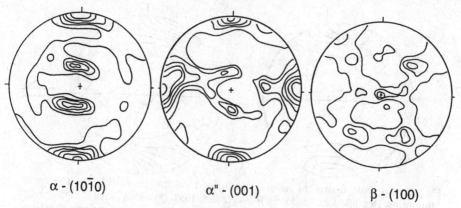

$$\alpha - (10\bar{1}0) \qquad \alpha'' - (001) \qquad \beta - (100)$$

Fig. 8. Pole figures of α and α'' TiAl obtained by a deconvolution method from a set of seven measured incomplete pole figures containing 17 different lattice reflections. From the texture of α'', that of the original β-phase was generated by a transformation scheme [DUNST & MECKING 1996].

Prospects

This short survey of various aspects of the role of texture and anisotropy in processing and application of crystalline materials was meant to create interest and attention on this important subject. The examples were chosen to illustrate the wide applicability of texture analysis in a number of disciplines. One of the attractions of this research field is the interdisciplinary nature.

Texture and anisotropy is a traditional field of materials science which has continuously expanded for about 70 years. It will continue to grow with an increasing rate due to the progress made in this area in particular during the last two decades:

- Experimental techniques for the rapid determination of texture on a macroscopic and microscopic scale have improved considerably and are widely available in many laboratories.
- Methods of texture representation and texture analysis have been perfected.
- Physical concepts for modeling texture evolution in laboratory tests and in industrial processes are developing rapidly.
- Measured texture data can be used, on a routine basis, as input for computer codes to evaluate and quantitatively predict anisotropic material behavior.

This coincides with an enormous increase of manufactured materials in number and complexity, where at the same time demands are strong to achieve optimum material behavior by adequate design of microstructure. In this context, texture evaluation and application provides many highly important aspects for the interrelation between microstructure and properties for process control and material performance and for identification of the ruling mechanisms.

Future research and application will address texture effects in ceramics, polymers, semiconductors and other functional materials. On the modeling side it is to be expected that texture and anisotropy will gain increasing interest in all sorts of transformation processes (such as recrystallization, precipitation and martensite formation) and it is assumed that the effects of interphases or of local orientation arrangements on evolution of microstructure and texture will soon be incorporated routinely in plasticity models. The progress made in quantifying the texture of polycrystals provides for improved numerical predictions in many fields, such as internal stress determination and fracture mechanics, where crystallography and anisotropy have been generally neglected. A major field is seen in implementing texture subroutines in FEM codes for simulating the mechanical behavior of materials. The tools and methods are at hand.

This book is a comprehensive update of the available tools and methods rather than a complete collection of data, and it comes at the right moment since 'Texture and Anisotropy' has reached a point where it can offer new concepts for consolidating old and advancing new areas of materials science.

Chapter 1

ANISOTROPY AND SYMMETRY

1. **Structure and Properties**

     1.1 Material state and material properties
     1.2 Anisotropy, as restricted by symmetry
     1.3 Samples and representative material elements

2. **Crystal Symmetries**

     2.1 Crystal classes and Laue groups. Lattices and crystal systems
     2.2 The rotation groups. Symmetry operators
     2.3 Lattice orientation, Laue orientation, structure orientation

3. **Sample Symmetry and Test Symmetry**

     3.1 Isotropic samples
     3.2 Fiber symmetry
     3.3 One or more planes of symmetry, orthotropy
     3.4 Heterogeneous samples
     3.5 General samples

4. **Anisotropy and Symmetry of Properties**

     4.1 Tensor field variables
     4.2 Tensor properties, scalar properties
     4.3 The kind of anisotropy of a property
     4.4 The degree of anisotropy of some properties in single crystals

ANISOTROPY AND SYMMETRY

U. F. Kocks

A material property is *isotropic* when it does not depend on how the sample is 'turned' (from the Greek root τροπ); it is *anisotropic* when it does depend on the orientation of the sample with respect to some external frame. Such an anisotropy of a property is due to the arrangement of the building blocks of the material: its *structure*.

A leading thread of the science of materials is concerned with structure/property relations. Most often, this concern is with the *level* of a certain property (or its 'strength') and its connection to the *scale* of the microstructure. In this book, the concern is not so much with the level as with the directional dependence, the *anisotropy*. The structural elements responsible for anisotropy are not so much scale as morphology; for example, not the *size* of the grains in a polycrystalline aggregate, but their *shape* and, in particular, their *orientation*. The totality of crystallite orientations is called the *texture* of the material sample. The anisotropy of properties in a polycrystalline aggregate depends both on the anisotropy of the single-crystal property and on the texture of the polycrystal.

The anisotropy of a property is generally restricted by certain *symmetry* considerations which in part follow from the symmetry elements of the underlying material structure (but also depend on the kind of property being considered). Symmetry considerations are in fact of paramount concern in the treatment of the directionality of material properties.

The words *structure* and *properties*, *sample* and *material*, *texture* and *anisotropy*, and *symmetry*, all have very specific meanings in materials science (and sometimes slightly different meanings in geology and crystallography, for example). We devote the first Section of this Chapter to clarify this framework. Subsequently, the subject of *structural symmetry* will be explored in some detail, first for individual *crystals* (Sec. 2) and then for polycrystalline *samples* (Sec. 3). Finally, the anisotropy of *properties* will be reviewed in Sec. 4. 'Neumann's Principle', which was formulated to specify the relation between symmetry and anisotropy for single crystals, will be extended to apply to the superposition of anisotropies in polycrystals.

The eventual aim of the user of this book is presumed to be an ability to describe the anisotropy of macroscopic physical properties in a quantitative way. In this Chapter, we will use particular properties only as examples. The main purpose is to summarize concepts and methods. Most topics will be presented merely as overviews, with the present perspective in mind; others will be treated in more detail, in view of the interdisciplinary audience we mean to address.

Many concepts in this Chapter are used under the assumption that the reader is at least somewhat familiar with them. Crystal symmetries are well treated in many introductory books, and will be reviewed here only insofar as they are of particular concern for textures and anisotropic properties. The level necessary is that covered by BARRETT and MASSALSKI [1980] or PHILLIPS [1963]. A most instructive review of symmetry concepts and applications, including an assessment in modern terms of the pioneering work of NEUMANN and of CURIE [1884] was presented by PATERSON & WEISS [1961], who also cover polycrystalline aggregates, after WEISSENBERG [1922]. The effect of crystal symmetry on tensor properties is exhaustively covered in NYE [1957] and has been more recently elaborated by BRANDMÜLLER & WINTER [1985]; for an introduction, see LOVETT [1989]. Many articles of relevance can be found in WENK [1985].

1. Structure and Properties

1.1 Material state and material properties

In modern materials science, all properties of a particular material sample are presumed to be fully determined by its current *state* or, in a microscopic view, by the totality of its structural elements: crystal structure, defect structure, 'microstructure' (usually meaning structure and distribution of defects of grains in a polycrystalline aggregate, and distribution of different phases), and perhaps 'mesostructure' (for composites on a more macroscopic scale).[1] Texture, the non-random distribution of the crystallographic orientation of grains, is an element of microstructure and so is the morphology of grains. The state of a material typically changes during any process: the term 'state' always refers to the *current* state, which determines the *current* properties. The current rate of evolution of the state is itself a function of the current state.

What is meant by a *property* of a material? In the context of 'structure/property relations', one would wish to reserve the term 'properties' for macroscopic quantities, and thus *not* use it also in the context of structural elements such as dislocation density and arrangement, or lattice structure. In general, we will mean by a property the *relation between a macroscopic stimulus and a macroscopic response*. Both 'stimulus' and 'response' are generally field variables, but the relation between them is provided by a *material property* or 'material behavior' [MCCLINTOCK & ARGON 1966] or 'constitutive relation'.

To illustrate some structures and some properties, Fig. 1 displays schematically (a) a heterogeneous structure of layered morphology and (b) the lattice structure of muscovite (a form of mica) which, again, has a layered appearance. It would be expected in both cases that properties perpendicular to the layer may be different from properties *in* the layer. For example, if one imagined a heat source along the y-axis, one would expect isotherms in the x-z-plane, after some time, to look like Fig. 1c: not circular.

Figure 1d describes the relevant property quantitatively. If, at a certain time and place, a unit temperature gradient ∇T is applied in a variety of directions, the ensuing heat flow $\mathbf{q}$ will, in general, be in a different direction. The property 'thermal conduct-

---

[1] This totality of microstructural elements is often called 'fabric' in geology and is commonly quoted as a translation of the German word 'Gefüge'. In materials science, Gefüge is translated as 'microstructure'. In Condensed Matter Physics, this level of structure is frequently called 'mesostructure', reserving the term 'microstructure' for the atomistic level.

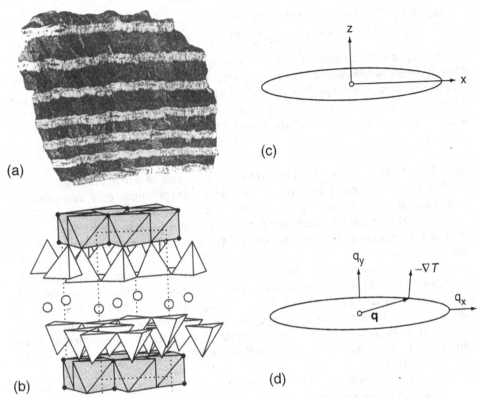

Fig. 1. Typical layered structures: (a) a heterogeneous material (asbestos), (b) an oriented single crystal of muscovite (mica). (c): a sketch of an isotherm after heat flow from a source on the y-axis. (d): an ellipse describing the thermal conductivity κ, which relates heat flow $\mathbf{q}$ to the negative temperature gradient $-\nabla T$, in both magnitude and direction. The anisotropy of κ for muscovite was used: it is 5.84 [POWELL & GRIFFITHS 1937].

ivity' (κ) specifies not only how fast the heat flows, but also in which direction (relative to the negative temperature gradient). Mathematically, this is expressed by considering ∇T and $\mathbf{q}$ *vectors* and the quantity that links them a *tensor*:

$$\mathbf{q} = -\kappa \cdot \nabla T \tag{1}$$

In general, field variables are tensorial in character (a vector being a tensor of rank one) and thus a linear relation between them requires a tensor of a rank that is the sum of the ranks of the field variables. One may write the relation in formal nomenclature, such as in Eq. 1 (see the summary "Notation" at the end of the book for details), or in components:

$$q_i = -\kappa_{ij} T_{,j} \qquad (i,j = 1,2,3) \tag{1'}$$

which, with the conventions to sum over repeated indices and to symbolize a partial derivative by a comma, is short-hand for

$$-q_1 = \kappa_{11}\, \partial T / \partial x_1 + \kappa_{12}\, \partial T / \partial x_2 + \kappa_{13}\, \partial T / \partial x_3$$
$$-q_2 = \kappa_{21}\, \partial T / \partial x_1 + \kappa_{22}\, \partial T / \partial x_2 + \kappa_{23}\, \partial T / \partial x_3$$
$$-q_3 = \kappa_{31}\, \partial T / \partial x_1 + \kappa_{32}\, \partial T / \partial x_2 + \kappa_{33}\, \partial T / \partial x_3 \qquad (1'')$$

The off-diagonal coefficients of κ need not in general be zero, and the diagonal coefficients need not be equal: this is the essence of anisotropy.[2] Only by virtue of certain *symmetry* considerations does the number of independent coefficients get reduced. These may be inherent in the material or in the property itself. For example, in many cases, tensors are 'symmetrical' about the diagonal; in the above example:

$$\kappa_{ij} = \kappa_{ji} \qquad (2)$$

Generally, this follows when the property in question can be derived from a *potential* (so that mixed second derivatives are equal). The potential may represent an equilibrium energy [NYE 1957, p.181] or a dissipation rate [KOCKS &AL. 1975, pp.10-12; NYE 1957, p.211-212]. It is in such circumstances that the quantity can be described by a representation ellipsoid (as we have used in Fig. 1d); a more detailed discussion will follow in Sec. 4.1.

Other properties that are symmetric second-rank tensors are the magnetic susceptibility (relating two vectors) and the thermal expansion (relating a second-rank tensor and a scalar). The elastic stiffness (or its inverse, the compliance) is a tensor of rank four, since it relates two field variables, stress and strain, which are themselves tensors of rank two. Tensors have certain transformation properties and are convenient to deal with; they will be used extensively throughout the more detailed parts of this book.

Not all properties are tensor properties. This is illustrated in a qualitative way in Fig. 2. Consider a plane of atoms with a rectangular pattern (Fig. 2a, open symbols) and assume that the next layer is stacked over the interstices (solid symbols); this is a representation of a body-centered tetragonal crystal. Now displace the upper layer, in a rigid manner, over the lower layer to the right: this is a form of 'shear' of the crystal. The free energy will change as sketched in Fig. 2b: in a periodic fashion (schematically represented by a sine). *Small* displacements are equivalent to a linear-elastic response; the curvature of the free energy diagram at an equilibrium position gives rise to a particular elastic constant, represented by the initial slope in a force-distance diagram, Fig. 2c. Larger displacements elicit a *non-linear* response; it could be described by a Taylor series in the displacements (with tensors of increasing rank as the coefficients). However, one property of particular interest is the *theoretical shear strength* of the material: the limit where a further increase in displacement elicits no further increase in the restoring force (the top of the diagram in Fig. 2c). This property cannot be described by any tensor at all; it is a critical value of a variable. The same situation holds for the 'Peierls stress', due to the periodic potential of a dislocation in the lattice structure. All forms of the yield strength fall in this category, and so does the phenomenon of ferromagnetic saturation.

---

[2] The word *anisotropy* will be used [NYE 1957] in connection with *properties* that depend on direction, as in Fig. 1d, rather than in relation to structural elements such as in Figs. 1a and b. In early works, the word 'anisotropy' was used for what we now call 'texture', 'anisotropy class' for what we now call 'sample symmetry' [WEISSENBERG 1922, 1924].

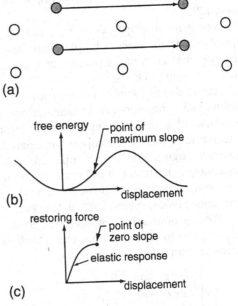

Fig. 2. a) Sketch of the atomic arrangement in a tetragonal crystal. The solid circles represent atoms one layer above the plane of the paper. When the upper layer is displaced with respect to the lower layer, the free energy (b) and the restoring force (c) change in a periodic manner. The initial response is linear elastic; when the slope in (c) becomes zero, the crystal becomes unstable.

In general, the relation between a stimulus and a response merely needs to be a *function*: only to the extent that it can be meaningfully expanded into the first few terms of a Taylor series does the response, and therefore the property, have tensor character. For example, the elastic response of stress σ to strain ε (to be discussed in detail later) can be expanded according to

$$\sigma_{ij} = \mathbb{C}_{ijkl}\varepsilon_{kl} + \mathbb{D}_{ijklmn}\varepsilon_{kl}\varepsilon_{mn} + \cdots \qquad (3)$$

where $\mathbb{C}$ and $\mathbb{D}$ are fourth and sixth-rank tensors, respectively

Finally, we must emphasize that the role of stimulus and response may be reversed. For example, the *strain* may be considered prescribed, as implied in Eq. 3, and the stress the response; or the *stress* may be prescribed, or indeed some mixture of the strain and stress components may be externally prescribed – the complementary ones follow from the material response.

1.2 Anisotropy, as restricted by symmetry

Anisotropy, as we have claimed in the last section, is in general to be assumed; it is isotropy that is special. The anisotropy of a particular property in a particular material may be quantified in two distinct ways. First, the *degree* of anisotropy – say, the ratio of the maximum to the minimum of a particular quantity – may vary widely from property to property, from material to material, and even from sample to sample; we will occasionally mention orders of magnitude: to demonstrate that it is frequently not small at all. Section 4.4 will provide some examples.

A second, more fundamental, way in which anisotropy may be quantified is by answering the question: how many independent parameters are necessary to describe the property in the actual material, as contrasted with the number needed under the assumption of isotropy? For example, six numbers are needed to describe the heat conductivity of a *general* material (Eqs. 1, 2), whereas one suffices for an isotropic one;

a general material has 21 independent elastic constants, an isotropic one has two (Sec. 4.2).

For general properties, it may not be a number of numbers that needs to be specified, but the extent of a parameter space. A property is called *anisotropic* when the dimensionality of the parameter space required to specify it completely is greater than that required for isotropy.

These distinct *kinds* of anisotropy are the same for a whole class of properties (such as all symmetric second-rank tensors) and for a whole class of materials (such as all those of a centrosymmetric cubic lattice structure). The *kinds of anisotropy* possible for a particular property in a particular sample of material are restricted by *symmetry* arguments. For example, from the definition of anisotropy (Section 1: the dependence on 'turn'), it follows that the application of any *rotation* that leaves the sample itself indistinguishable from how it was before must also leave its *properties* unchanged [WEISSENBERG 1935, p. 103; PATERSON & WEISS 1961, p. 856].

We mentioned that tensors have certain 'transformation properties'. This applies in particular to rotations from one coordinate system to another. A typical rotation operator is (in cartesian coordinates)

$$R = \begin{pmatrix} \cos\phi & \sin\phi & 0 \\ -\sin\phi & \cos\phi & 0 \\ 0 & 0 & 1 \end{pmatrix} \tag{4}$$

which achieves a counter-clockwise rotation around the z-axis by an angle ϕ. As can be derived easily (see Section 4.1), a second-rank tensor that has the form T in the 'old' coordinate system adopts the form

$$T' = R \cdot T \cdot R^T \tag{5}$$

in the 'new' one, where R^T is the transpose of R. Equation 5 is short-hand for[3]

$$T'_{ij} = R_{ik} R_{jl} T_{kl}$$

The effect of symmetry on a tensor property T is that, for the particular forms of R that are symmetry operators for the case considered, T' must be equal to T. For example, if a rotation by $\phi=180°$ around the axis is a symmetry operator, then it is required by Eqs. 4 and 5 that

$$T' \equiv \begin{pmatrix} T'_{11} & T'_{12} & T'_{13} \\ T'_{21} & T'_{22} & T'_{23} \\ T'_{31} & T'_{32} & T'_{33} \end{pmatrix} = \begin{pmatrix} -1 & 0 & 0 \\ 0 & -1 & 0 \\ 0 & 0 & 1 \end{pmatrix} \begin{pmatrix} T_{11} & T_{12} & T_{13} \\ T_{21} & T_{22} & T_{23} \\ T_{31} & T_{32} & T_{33} \end{pmatrix} \begin{pmatrix} -1 & 0 & 0 \\ 0 & -1 & 0 \\ 0 & 0 & 1 \end{pmatrix} = \begin{pmatrix} T_{11} & T_{12} & -T_{13} \\ T_{21} & T_{22} & -T_{23} \\ -T_{31} & -T_{32} & T_{33} \end{pmatrix}$$

be equal to T itself; the components 13, 23, 31, and 32 of T must therefore be zero. If H be a general symmetry operator in a certain material sample, then any second-rank tensor property T must obey

$$T = H \cdot T \cdot H^{-1} \tag{6}$$

---

[3] Note that in the component description, the sequence of indices and of symbols does not matter; in the abstract description of Eq. 4, it is assumed that the repeated indices are adjacent, and this may require the use of the 'transpose' (for which the indices are interchanged) and close attention to the sequence of symbols. See the "Notation" summary at the end of the book for more details.

For tensors of rank n, the H appears n times (see Section 4.2).[4]

In the general case when the property in question may not be a tensor, the restriction imposed by the presence of a particular symmetry element can always be ascertained by applying the symmetry operator to the 'stimulus', evaluating the response with and without the applied symmetry, and demanding that the two responses be the same. If the application of a symmetry operator to a quantity q is formally expressed by S{q}, then, for an arbitrary functional response $r = \mathcal{F}(q)$, it is to be demanded that

$$S\{r\} = \mathcal{F}(S\{q\}) \tag{7}$$

These problems will be addressed in more detail in Section 4.

A macroscopic property may have some general symmetry characteristics that do not affect its anisotropy. The most important symmetry element a property itself may possess is *centrosymmetry* or a 'center of inversion': it means that the quantity which relates the response to the stimulus remains the same when the signs of all spatial coordinates that enter the description of the stimulus and the response are changed. In matrix notation, the quantity is invariant under the inversion operator

$$\bar{I} = \begin{pmatrix} -1 & 0 & 0 \\ 0 & -1 & 0 \\ 0 & 0 & -1 \end{pmatrix} \tag{8}$$

This type of symmetry cannot be described by any rotation, and is thus not a part of 'anisotropy'. It nevertheless plays a major role in various aspects of this book. Inserting $H = \bar{I}$ into Eq. 6 shows that all second-rank tensors must be centrosymmetric – and indeed (by counting the number of times the operator appears in the equation) that *all even-rank tensors are centrosymmetric* and, conversely, that *in the presence of an inversion center in the structure, all odd-rank tensor properties must be nil*.

Let us complete this brief introduction of symmetry operators by just one more: a mirror plane. If it is perpendicular to the 3-axis, it can be expressed as:

$$M_3 = \begin{pmatrix} 1 & 0 & 0 \\ 0 & 1 & 0 \\ 0 & 0 & -1 \end{pmatrix} \tag{9}$$

In words: the third axis changes sign, everything else stays the same. If this operator were multiplied by the inversion, Eq. 8, one would obtain a rotation by $\phi=180°$ (Eq. 4). The three cases are illustrated in Fig. 3.

Note that the 'handedness' of the coordinate system changes in the operations 8 and 9. In some cases, this has an effect only on the *description*; but some crystal structures possess an innate handedness or 'chirality' (e.g., optical activity) which, in this book, plays a peripheral role only. It cannot occur in centrosymmetric materials.

Finally, properties may be inherently *isotropic*. One might think this is true for all properties deriving from scalar relations; however, caution is advised, especially in multivariate problems (Section 4.2). Isotropy is sometimes called 'spherical symmetry'. Note that

---

[4] Note that most symmetry operators are 'orthogonal', such as all rotations, and then the *inverse* (used in Eq. 6) is equal to the *transpose* (used in Eq. 5).

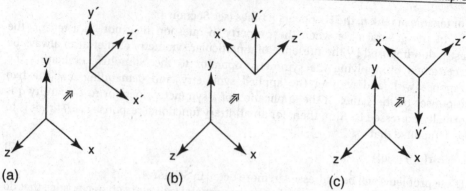

(a)   (b)   (c)

Fig. 3. Change of a set of coordinate axes under the application of (a) a mirror perpendicular to *z*, (b) a rotation by 180° around *y*; (c) a center of inversion.

isotropic samples may or may not possess centrosymmetry, and may or may not possess chirality.

In our following discussions of the kind of anisotropy to be expected in a particular case, we will need to ascertain which symmetry operators apply. The symmetry properties of materials are well known for (ideal) *single crystals* [Int. Tables 1983; NYE 1957]; they will be reviewed in Section 2. In the present context, the macroscopic properties of *polycrystalline samples* are of prime interest; their symmetry properties will be discussed in Section 3. *The anisotropy of a macroscopic property comes about by a combination of two ingredients: an anisotropy of the particular property in single crystals of the material; and the absence of spherical symmetry in the macroscopic sample.* Some specific cases as well as some generalities will be elaborated upon in Section 4.

1.3 Samples and representative material elements

The word *sample* is primarily used in this book for a *representative material element*; it is defined to be a piece of material that may be treated as having a single state and to be essentially homogeneous in terms of its structural elements. When *testing* material properties, one always chooses such samples. Occasionally, one may wish to test a piece of material in which a certain non-uniformity in structure (e.g., a gradient of texture through the diameter of an extruded wire, or an unknown conglomerate in a geological field 'sample') is known to be present, but where the interest is in its *average* properties. Such samples we refer to as quasi-homogeneous. In engineering applications, such as the design of 'structural elements' (a very different use of the word structure – which we shall call 'macro-structure') or of manufacturing processes, there is usually a significant variation from place to place not only in the field variables but also in the material properties. In this case, one introduces the notion of *material elements* (or even material 'points'); they must be sufficiently small in terms of the macro-structure, and sufficiently large in terms of the micro-structure, to qualify as 'samples' in the above sense.

Even in *test* samples, there may be a problem in comparing the size of the sample to that of the microstructural elements. For example, the number of grains in the sample may be large, and yet a large fraction of them may be surface grains subjected, perhaps, to entirely different boundary conditions than interior grains [KOCKS 1970]. For example, a cubic arrangement of 1000 grains has 488 grains on the surface! A test sample that is to be representative of material behavior must be chosen large enough to have but a small fraction of its grains in non-representative locations.

The presumed quasi-homogeneity of a 'sample' element may be used to advantage, in some applications, by describing it as a *continuum*. The concepts of anisotropy and symmetry may hold for continua as well as for discrete assemblies. While the properties we discuss will all be based on crystalline materials, their approximation, at the 'sample' level, as continua does not invalidate their properties being anisotropic and subject to symmetry requirements.

An essential element of the 'structure' of a sample is its *texture*. It may have certain symmetries associated with it – albeit only in a statistical sense [WEISSENBERG 1922,1924; PATERSON & WEISS 1961, p. 854]. These will be illuminated in Section 3.

2. Crystal Symmetries

A basic structural element common to most materials is the *crystal structure*. Thus, a most basic form of structural symmetry is that contained in the crystal structure. We find it convenient to begin our discussion with crystal symmetries that contain an inversion center (Section 2.1). All monatomic *crystals* fall into this category; and conversely, all centrosymmetric *properties* depend only on these symmetry elements.

Then, more general cases will be treated: first those that are pure rotations (Section 2.2), which are of general importance even when an inversion center exists, since rotations are the symmetry operations you can physically undertake with any actual crystal in your hand. A number of important crystals, such as quartz (α-SiO$_2$), have only pure rotation symmetry elements, and no inversion center. Crystals whose symmetry elements are neither all pure rotations nor centrosymmetric (e.g., ZnS) will be summarized only briefly (Section 2.3). Many of the advanced materials in modern materials science (such as ferro-electrics), as well as several materials of interest to the geologist fall into these less symmetric classes.

Crystal symmetry is a subject treated in depth in many books [e.g., *International Tables*, BUERGER 1956]; we only summarize aspects that are of particular relevance for defining the meaning of 'orientation' (needed for Chapter 3) and the symmetry of properties. More details concerning crystal symmetry in texture analysis can be found in MATTHIES &AL.[1987].

Our primary interest in symmetries, for the purposes of this treatise, is in their influence on the kind of anisotropy to be expected in single-crystal *properties*. Therefore, we will mention some of these as the various symmetries are introduced, to provide concrete examples. A classic book on this topic is 'Physical Properties of Crystals' written by NYE in 1957 (and re-issued many times). An at least cursory study of this book is highly recommended for a deeper understanding of this Section (and Sec. 4).

2.1 Crystal classes and Laue groups. Lattices and crystal systems.

Materials are called 'crystalline' if their structure can be idealized into having discrete translation symmetry at the atomic level. The atoms may be locally grouped into 'motifs', but the motifs are repeated periodically in three dimensions, on a *lattice*. Real materials deviate to some extent from this idealization by having 'lattice defects' (such as vacancies, dislocations, stacking faults, etc.), by being finite in extent (having surfaces and interfaces), and by their atoms being in constant vibrational motion. Nevertheless, the symmetry properties of crystals are predominantly determined by the symmetry of the lattice and the symmetry of the motif.

A lattice may be described, in various equivalent ways, by a *unit cell*. Figure 4a shows the simplest unit cell: a cube. Many lattices do not have the high symmetry implied by this unit cell. For example, one axis (commonly called z) may be different from the other two (while retaining all the right angles); this is called a *tetragonal* lattice (Fig. 4b). When all axis lengths are different, one obtains the *orthorhombic* lattice (Fig. 4c). The cube in Fig. 4a' is merely turned on a point; if it is stretched in the direction of the body diagonal (called z in this case), it becomes a *rhombohedral* unit cell. A different base is the *hexagonal* unit cell (Fig. 4d). All rhombohedral lattices can also be described by using a hexagonal unit cell (but not *vice versa*). There are two further lattice types: *monoclinic*, which is a variation on orthorhombic in which one of the angles is allowed to be non-right; and *triclinic*, in which all lengths and angles are

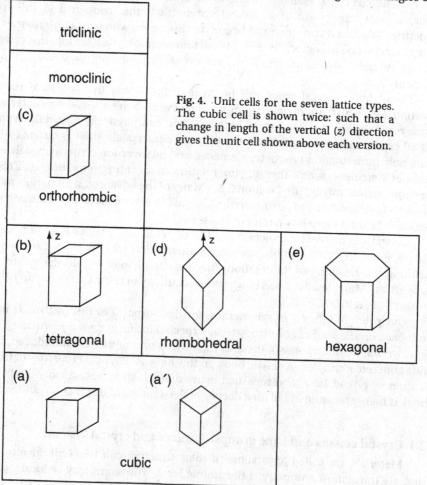

Fig. 4. Unit cells for the seven lattice types. The cubic cell is shown twice: such that a change in length of the vertical (z) direction gives the unit cell shown above each version.

arbitrary.[5]

The existence of the translation lattice limits the number of symmetry elements that can be present. There are 32 possible combinations of rotations, mirror planes, and an inversion center, collectively called the *point groups* or *crystal classes*. (The 230 space groups, which contain combinations of point group elements and translations, are generally not of interest in texture analysis.)

All *lattices* are centrosymmetric and so are all crystals with only one atom in the motif. Crystal structures with a multi-atom motif at each lattice point generally have a symmetry lower than that of the lattice; but they may nevertheless contain an inversion center. The subset of crystal classes that contain an inversion center has 11 members; these are called the 'Laue group' because an inversion center is added in a diffraction ('Laue') experiment. Their symmetry elements are shown in Fig. 5, along with crystallographers' conventions concerning the labeling of axes and the naming of the groups by 'international symbols' (or 'Hermann-Mauguin symbols'). Table I lists the Laue groups, along with other information to be discussed below.

Table I. Crystal symmetries relevant for orientations

CRYSTAL SYSTEM		crystal class Laue | rot.	no. of sym. el.	no. of centro-sym. tensor prop's. sym. 2nd-rank | sym. 4th-rank	
triclinic		$\bar{1}$ 1	1	6	21
monoclinic		2/m 2	2	4	13
orthorhombic		mmm 222	4	3	9
	†	4/m 4	4	2	7
tetragonal		4/mmm 422	8	2	6
	†	$\bar{3}$ 3	3	2	7
trigonal	†	$\bar{3}$m 3	6	2	6
	†	6/m 6	6	2	5
hexagonal		6/mmm 622	12	2	5
	†	m3 23	12	1	3
cubic		m3m 432	24	1	3

† These groups have lattice planes which have the same spacing but are not symmetrically equivalent; thus they exhibit overlapping peaks in diffraction patterns; an example is (121) and (211) in crystal class 4/m.

A frequently used subdivision of the 32 crystal classes is that into seven *crystal systems*, according to the highest rotational symmetry elements: cubic, hexagonal, trigonal, tetragonal, orthorhombic, monoclinic and triclinic. Some of these contain two Laue groups, some only one. In most cases, the *lattice* of a crystal structure has the same symmetry as the highest in its *crystal system*; an exception is the trigonal system, in which the lattice may be either rhombohedral (in which case the lattice symmetry is trigonal, $\bar{3}$m) or hexagonal (in which case the lattice has a higher symmetry than the crystal system, 6/mmm).

---

[5] The expansion of the 7 crystal systems into 14 'Bravais lattices', which may have lattice points at the center of certain faces or in the 'body center' of the unit cell, is not relevant for orientation and texture analysis, although it affects diffraction measurements and properties such as slip system selection.

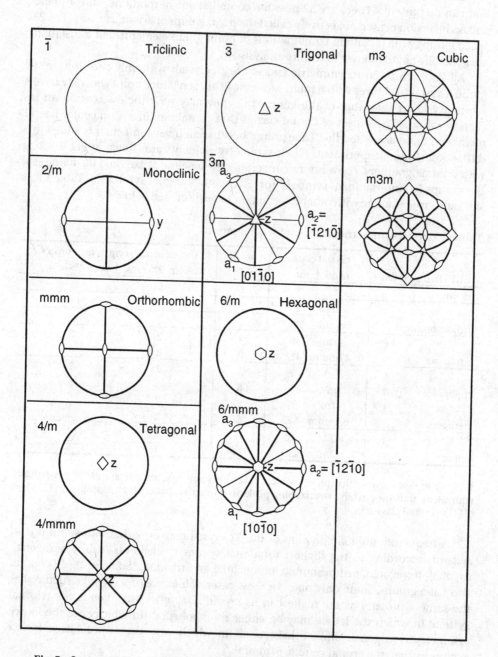

Fig. 5. Symmetry elements and crystallographers' conventions for the choice of axes in the 11 Laue groups (which are all centrosymmetric). (Note that, elsewhere in this book, x and y are usually used such that the quadrant with positive x and positive y is the upper right one.)

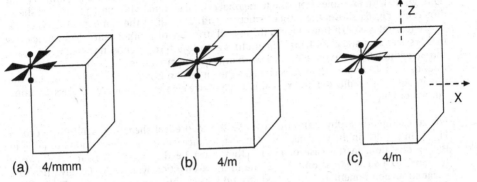

(a)   4/mmm   (b)   4/m   (c)   4/m

Fig. 6. (a) and (b): The two centrosymmetric crystal classes in the tetragonal crystal system, with a schematic motif. The four-fold axis is in the z-direction. (c): as (b), but rotated 180° around x.

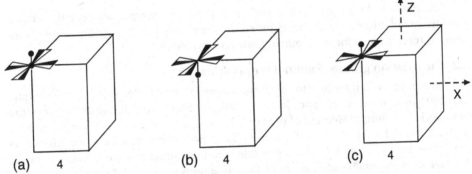

(a)   4   (b)   4   (c)   4

Fig. 7. One enantiomorphic class in the tetragonal crystal system: two forms or phases (a and b). (c): as (b), but rotated 180° around x.

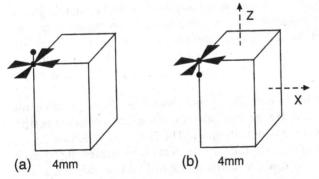

(a)   4mm   (b)   4mm

Fig. 8. One non-enantiomorphic class in the tetragonal crystal system. (b): as (a), but rotated 180° around x. They are different 'crystal orientations', but the same 'Laue orientation' (obtained by adding an inversion center). All sketches in Figs. 6, 7, 8 have the same 'lattice orientation', but the two (c) figures have a different 'Laue orientation'.

Figure 6 demonstrates the difference between the *lattice* symmetry and the symmetry of a *point group* with examples from the tetragonal crystal system (with the 4-fold axis in the z-direction): Fig. 6a shows the lattice structure with a motif of the same symmetry as the lattice (crystal class 4/mmm); Fig. 6b the crystal structure of another centrosymmetric class in this system (crystal class 4/m). The motif is entirely arbitrary, chosen so that many cases can be demonstrated in the following, with only minor variations in the motif. Note that a rotation by 180° around the x-axis changes the pattern in Fig. 6b to that of Fig. 6c, but the one in Fig. 6a is unaffected: the x-diad is a symmetry element of the crystal class 4/mmm, but not of 4/m.

For an example of a physical property, take the theoretical shear strength introduced in Fig. 2. In Fig. 6b, on the y-plane, it may be different for slip in the +x-direction from that in the −x-direction; in the case of Fig. 6a, they must be the same. Similarly, the *elastic* response to a shear stress on the y-plane in the x-direction will, in Fig. 6b, give rise to a shear strain component at 45° (rotated around z-axis); this response is zero in Fig. 6a. It is for this reason that there are seven independent elastic constants in class 4/m, but only six in 4/mmm. Table I lists these numbers of independent constants for certain properties; further details can be found in NYE [1957].

The point to be emphasized here is that the Laue groups, not only the crystal systems, are of interest for macroscopic centrosymmetric properties. The whole point group is relevant only for non-centrosymmetric properties.

2.2 The rotation groups. Symmetry operators

An example of a pure rotation element was introduced in Section 1.2: if you rotate, say, 180° around the z-axis and the new configuration is indistinguishable from the old, this is a symmetry element of the crystal. This particular element is called a 'two-fold axis around z' (also a 'z-diad'). Within the point groups, which are restricted to be applicable to *lattice* points, all the possible rotation elements are easily enumerated; they are 2-, 3-, 4-, and 6-fold axes in various directions (and a '1-fold axis' corresponds to the identity). A 'group' of symmetry elements consists of all elements that, if applied in succession (or again), give an element that is already a member of the set. Knowledge of this group is useful for enforcing all applicable symmetries in a calculation.

If you apply an inversion center to every symmetry element in the rotation group, you get precisely the symmetry elements of the Laue groups. Thus, there is a one-to-one correspondence between the Laue groups and the rotation groups; the corresponding symbols are listed next to each other in Table I. Also, all centrosymmetric properties of a certain member of the rotation group are identical to those of the corresponding Laue group.

There are 11 Laue groups and 11 rotation groups. This leaves 10 groups (out of the total of 32 point groups) which are neither pure rotations nor centrosymmetric. Examples of these, as well as of the rotation groups, will be discussed in Section 2.3.

The most elegant way of applying symmetry elements is by expressing them as symmetry operators: matrices that, when matrix-multiplied with a quantity, render the 'new' quantity. First examples were given in Section 1.2. Table II shows all rotation elements. They are grouped, for convenience, into three 'branches' (and there is some duplication between them). The identity matrix is always a formal member of the symmetry elements: you *may* leave the quantity unchanged. For triclinic crystals (symbol '1'), this is the *only* element. Monoclinic crystals (symbol '2')

Table II. Symmetry operators of rotation groups

tetragonal branch

```
| 1   0   0 |
| 0   1   0 |
| 0   0   1 |   1
```
```
| -1  0   0 |
|  0  1   0 |
|  0  0  -1 |   2
```
```
| 1   0   0 |
| 0  -1   0 |
| 0   0   1 |
```
```
| -1   0   0 |
|  0  -1   0 |
|  0   0   1 |   222
```
```
|  0   1   0 |
| -1   0   0 |
|  0   0   1 |
```
```
|  0  -1   0 |
|  1   0   0 |
|  0   0   1 |
```
```
|  0   1   0 |
|  1   0   0 |
|  0   0  -1 |
```
```
|  0  -1   0 |
| -1   0   0 |
|  0   0  -1 |   42
```

hexagonal branch

```
| 1   0   0 |
| 0   1   0 |
| 0   0   1 |
```
```
| -.5   a    0 |
| -a   -.5   0 |
|  0    0    1 |
```
```
| -.5  -a    0 |
|  a   -.5   0 |
|  0    0    1 |   3
```
```
| .5    a    0 |
| -a   .5    0 |
|  0    0    1 |
```
```
| -1   0   0 |
|  0  -1   0 |
|  0   0   1 |
```
```
| .5   -a    0 |
|  a   .5    0 |
|  0    0    1 |   6
```
```
| -.5  -a    0 |
| -a   .5    0 |
|  0    0   -1 |
```
```
|  1   0   0 |
|  0  -1   0 |
|  0   0  -1 |
```
```
| -.5   a    0 |
|  a   .5    0 |
|  0    0   -1 |
```
```
| .5    a    0 |
|  a   -.5   0 |
|  0    0   -1 |
```
```
| -1   0   0 |
|  0   1   0 |
|  0   0  -1 |
```
```
| .5   -a    0 |
| -a   -.5   0 |
|  0    0   -1 |   622
```

$$a \equiv \sqrt{3}/2$$

cubic branch

```
| 1   0   0 |     | 0   0  -1 |
| 0   1   0 |     | 0  -1   0 |
| 0   0   1 |     | -1  0   0 |
```
```
| 0   0   1 |     | 0   0   1 |
| 1   0   0 |     | 0  -1   0 |
| 0   1   0 |     | 1   0   0 |
```
```
| 0   1   0 |     | 0   0   1 |
| 0   0   1 |     | 0   1   0 |
| 1   0   0 |     | -1  0   0 |
```
```
| 0  -1   0 |     | 0   0  -1 |
| 0   0   1 |     | 0   1   0 |
| -1  0   0 |     | 1   0   0 |
```
```
| 0  -1   0 |     | -1  0   0 |
| 0   0  -1 |     | 0   0  -1 |
| 1   0   0 |     | 0  -1   0 |
```
```
| 0   1   0 |     | 1   0   0 |
| 0   0  -1 |     | 0   0  -1 |
| -1  0   0 |     | 0   1   0 |
```
```
| 0   0  -1 |     | 1   0   0 |
| 1   0   0 |     | 0   0   1 |
| 0  -1   0 |     | 0  -1   0 |
```
```
| 0   0  -1 |     | -1  0   0 |
| -1  0   0 |     | 0   0   1 |
| 0   1   0 |     | 0   1   0 |
```
```
| 0   0   1 |     | 0  -1   0 |
| -1  0   0 |     | -1  0   0 |
| 0  -1   0 |     | 0   0  -1 |
```
```
| -1  0   0 |     | 0   1   0 |
| 0   1   0 |     | -1  0   0 |
| 0   0  -1 |     | 0   0  -1 |
```
```
| -1  0   0 |     | 0   1   0 |
| 0  -1   0 |     | 1   0   0 |
| 0   0   1 |     | 0   0  -1 |
```
```
| 1   0   0 |     | 0  -1   0 |
| 0  -1   0 |     | 1   0   0 |
| 0   0  -1 |  23 | 0   0  -1 |   432
```

The dashed boxes
in this column
make up group 4.

The dashed boxes
in this column
make up group 32.

The dashed box
in this column
comprises the 3-fold axes only.

have one two-fold axis; it is conventionally taken around the y-axis. The second 3×3 matrix in the first column expresses this fact: the **x**- and **z**-directions change sign, the **y**-direction remains unchanged. Continuing down the first column, there are two more 2-fold axes – and all the four elements so far make up the orthorhombic crystal class (symbol 222).

All further crystal classes, except the cubic one, are 'uniaxial', in that one axis is preferred; it is conventionally taken in the z-direction (even though it is characterized by the *first* number in the 'international symbols'). The remaining four elements in the first column of Table II are 'permutation operators' or, equivalently, 180° rotations around axes at 45° to the original ones. All operators in this column together now make up the tetragonal crystal class (symbol 422 or 42): that's why this column was labeled the 'tetragonal branch'. One can proceed in a similar way down the second column: from triclinic (1) to trigonal (3) to hexagonal (6) to hexagonal (622, ≡ 62), and including, as a subset, trigonal (322, ≡ 32). This is the 'hexagonal branch', expressed in ortho-hexagonal coordinates.

The 'cubic branch' consists of two columns: the first alone represents the crystal class 23, all 24 elements together the class 432 (≡ 43).[6] The first nine (in the first cubic column) are a useful set: they are all the three-fold axes (plus identity), which are often hard to have explicitly present in plots and must thus be separately applied.

2.3 Lattice orientation, Laue orientation, structure orientation

It is instructive to complement the examples from the tetragonal crystal system, given in Fig. 6 for the centrosymmetric classes, to demonstrate the various types of non-centrosymmetric classes, with particular reference to what one would call an 'orientation'.

Fig. 7a shows a crystal of class 4, with a variation of the artificial motif used in Fig. 6. If you did apply an inversion center to this crystal, you would get the crystal shown in Fig. 7b: it is different. In actuality, one cannot apply an inversion center: one cannot physically make a crystal structure like Fig. 7b out of one like Fig. 7a. But both of these crystal structures may exist: they are like two different 'forms' of the same material; these are called *enantiomorphs*. You can tell the difference between the two only by methods that are sensitive to the relative atomic arrangement (such as anomalous scattering and certain physical properties such as optical activity).

If you rotate Fig. 7b 180° around the x-axis, you get Fig. 7c: it is different from both Fig. 7b (because this rotation is not a symmetry element of this crystal structure and therefore produces a different orientation) and from Fig. 7a (because it is a different enantiomorph). Thus, it would not be very meaningful to distinguish Figs. 7a and b on the basis of their 'orientation': they are different enantiomorphs. One can distinguish Fig. 7b from Fig. 7a by applying a *handedness rule* within the motif. Other crystals of the pure rotation groups have a physical, *innate handedness* in form of a screw axis: these are called 'chiral' crystals.

Finally, there are crystal classes that combine *some* elements from the rotation group with at least one that is *not* a pure rotation, but they do not contain the inversion center itself. Fig. 8a shows such an example (crystal class 4mm), and Fig. 8b shows the same crystal rotated 180° around the x-axis. This rotation is *not* a symmetry element of the crystal; it produces a different *structure orientation*.

---

[6] Some errors in Table 3.2 of reference [HANSEN &AL., 1978] have been corrected.

We now need to address the question of what is meant by the term 'orientation', in a physically meaningful sense. (Formal definitions and prescriptions for measurement will be given in Chapters 2 and 4.) All the examples in Figs. 6 to 8 have the same *lattice orientation*: all the unit cells are aligned the same way. This is the easiest orientation to measure: by 'reflection from the lattice planes', e.g. on individual crystals by electron diffraction, not considering relative diffraction intensities.

When diffraction intensities are considered, e.g. in pole figure measurements by x-ray diffraction (Chapter 4), you can distinguish *Laue orientations*: Orientations such as those in Figs. 6b and c and Figs. 7b and c can be identified. In conjunction with the meaning of the 'Laue groups' introduced in Section 2.1, the 'Laue orientations' also imply a center of symmetry. After adding a center of symmetry to each example in Fig. 7, the Laue orientation will be the same for Figs. 7a and b (and the same as Fig. 6b. However, Fig. 7c is a different Laue orientation (the same as Fig. 6c).

It must be mentioned that the word 'orientation' has been used in very different definitions over time. In one extreme [MATTHIES &AL. 1987], it has been used as synonymous with *rotation*, on the basis of an operational definition of how one may compose a polycrystal from individual grains. BUNGE [1982] has used it in the same way, only with the additional convention that two symmetrically equivalent rotations are called the same 'orientation'. At the other extreme, in BUNGE &AL. [1980], the word 'orientation' has been used as synonymous with *any* orthogonal transformation.

In this book, we shall use the term 'orientation' (or crystal orientation), when unqualified, in another operational definition: the most general orientation that can be determined by diffraction experiments alone – the Laue orientation. Not all rotations are (Laue) orientations (Fig. 8), though many are, even in non-centrosymmetric crystals (Fig. 7).

If you add an inversion center to Fig. 8, the two cases are indistinguishable, even though Fig. 8b has been obtained by a rotation from Fig. 8a, and this rotation is not a symmetry element. That is why we called them different 'structure orientations' above. Thus, structure orientations (rotations) that are different from Laue orientations cannot be determined by any centrosymmetric techniques; they can be determined only by measurement of either the atomic structure itself or a property that depends on the 'polarity'; in an assembly of crystals, such a measurement would have to be undertaken on every single crystal.

To return to the case of the pure rotation groups: here, the introduction of a term 'structure orientation' as distinct from Laue orientation is meaningless: in it, the structures are different, not the orientations. Thus, if only one structure is present in a particular sample (*either* like Fig. 7a *or* like Fig. 7b), and you know which one it is, you know the *orientation* completely from just knowing the Laue orientation; thus you could predict the macroscopic properties. If you had a mixture of the two enantiomorphs, you would have to know the *correlation* between orientations and forms to predict non-centric macroscopic properties. In summary, while the (Laue) orientations can always be mapped onto rotations, not all rotations can be mapped onto (Laue) orientations. These different definitions of 'orientation' and their implication for texture analysis have been discussed in detail by MATTHIES &AL. [1987].

3. Sample Symmetry and Test Symmetry

The shape of a sample may suggest a certain previous treatment; e.g., a flat plate may suggest that it was rolled. Then the sample might be expected to have certain planes of symmetry. This is, however, not a foregone conclusion: the history may have been different than expected; the deformation may have been non-uniform; or it may have consisted of various stages of changing symmetry, etc. The symmetry of the sample that affects subsequent properties is recorded in its current *state*, through its *microstructure* (Section 1.1). For example, all the grains in a polycrystal may be elongated in the same sample direction. Another possibility is that they may have some crystal axes aligned. The latter effect, called *texture*, is the central theme of this book. The distribution of orientations may or may not have symmetry properties. For multi-phase materials, there may be additional directional properties of the state, such as aligned fibers in a composite or 'stringers' in a casting. *The preferred axes for crystallographic and morphological features need not coincide.*

We will make extensive use of *texture* symmetries in Chapter 2. In the present Section, we discuss some of the obvious possibilities for sample symmetry that would result from certain symmetries of current, differential tests or processes. In discussions of test symmetries and texture symmetries, mirror planes are often the preferred elements, rather than rotations. We will show cause for some such cases. (It is also curious to note that in other aspects of our culture – pottery and weaving, for example – there seems to be a definite trend among different groups to prefer either mirrors or rotations.)

Many samples in practice are orthogonal parallelepipeds ('bricks'). It is then easy to define a 'sample frame' by three of the edges (and to make it a right-handed set if one so desires). An alternative that is possible even when the sample faces do *not* all form right angles is to mark just one sample plane, and one direction in it; this method actually corresponds more closely to experimental ways of mounting the sample on a stage for examination. One can always define a cartesian frame (right-handed if desired) by making the third axis perpendicular to both the marked-plane normal and the marked direction.

For a description of properties, it is appropriate to choose the sample axes in conformity with the *structural* symmetry of the sample (such as its texture), not with an *external* coordinate system (such as that in which, say, the deformations are applied). This is powerful motivation to *know* the texture of the sample. For applications to the non-uniform deformation of complex bodies (typically treated by finite-element codes), other, macroscopic coordinate systems may be required; but *only when a system based on the sample symmetry is used can it help to reduce the number of state parameters.*

3.1 Isotropic samples

In many materials, it is quite difficult to make a piece of material with a microstructure that has no directional properties at all. (We count texture as part of the microstructure.) One example is a powder sample made up of a loose assembly of equiaxed particles (i.e., particles with similar lengths in all three dimensions).

The terms 'isotropy' and 'anisotropy' were introduced, in the beginning of this Chapter, with reference to *properties* of materials, not *samples*; for example, the same material sample may be isotropic with respect to one property (e.g., the volumetric thermal expansion coefficient), anisotropic with respect to another (e.g., the elastic constants). If anisotropic grains in a sample are randomly oriented, the macroscopic sample loses any anisotropy in *any* property, and such a *sample* is called 'isotropic'.

Note that 'isotropy' of a sample does not imply that every property has but one value. For example, the elastic constants of an isotropic material consist of two independent values (e.g., the bulk and the shear moduli, or Young's modulus and Poisson's ratio).

3.2 Fiber symmetry

A common symmetry encountered in materials is uniaxial or 'fiber' symmetry. For example, a wire that was initially isotropic and then pulled in tension, or an isotropic sample subjected to compression, will have such a fiber symmetry. The properties of the resulting sample are transversely isotropic. Of course, the same holds if the compression sample had a square cross section: the *shape of the sample* is not considered part of its symmetry properties.

In a wire drawn through a set of dies, the deformation may not have been uniform throughout the cross section, and thus there may be a texture *gradient*. When one is interested in the heterogeneity, one treats the wire as a macroscopic body with (quasi-homogeneous) material *elements*. These elements may have some symmetry, such as a mirror plane containing both the particular element and the wire axis; but they will, in general, not have complete fiber symmetry. Only the material in the very vicinity of the axis (and surrounding it) has *local* fiber symmetry. Note that, if the wire was drawn in a single direction (rather than by reversing the direction of pull), this direction may be part of the information carried in the actual, heterogeneous wire; in the approximation of a homogeneous sample with a 'fiber texture' it is not.[7]

Second-rank tensor properties (such as the thermal conductivity κ) can have only two independent values in transversely isotropic samples. Thus, for example, an orthotropic material (in which there are three values, aligned with the crystal axes in each grain) will have only two values in the polycrystalline wire (parallel and perpendicular to the wire). They can be obtained by appropriate averaging procedures (Chapter 7).

An interesting case is the elastic constants. We have seen (Section 2.4) that they are transversely isotropic in hexagonal crystals (although note that they are *not* in tetragonal ones). We can conclude that a sample with fiber symmetry has five independent elastic constants: the same ones as a hexagonal material, only of course replacing the c-axis with the fiber axis.

3.3 One or more planes of symmetry, orthotropy

Many processes are inherently two-dimensional. Examples are the cases of pure and simple shear deformation. Nothing distinguishes the region above the page in which the configuration is drawn, from the region below the page. In this sense, 'the page' is a *plane of symmetry*. It is similar to a 'mirror plane' (Eq. 9). However, it is different from the mirror plane used in crystallography in two ways. First, it holds only

---

[7] An alternative description for some such cases is the 'Kegelfasertextur' proposed by SCHMID & WASSERMANN [1927] and resurrected as the 'fir tree' texture by MÜCKLICH & AL. [1981].

in a statistical sense, not at every point. Second, in this statistical sense, it holds in *every* plane parallel to the page – the sample is not subject to the restrictions of lattice translation symmetry.

Some processes have an orthogonal coordinate system associated with them. Idealized sheet rolling is an example; it is the same as 'plane-strain compression' (Fig. 9). There is a plane of symmetry perpendicular to each of the three directions: the rolling plane normal (ND), the rolling direction (RD) and the transverse direction (TD). The symmetry is similar to that of an orthorhombic crystal with inversion symmetry (class mmm); again, there is no need, in the present case of *sample* symmetry, to apply the symmetry at a *point*. We shall adopt (from common usage in Solid Mechanics) the term *orthotropy* for this type of sample or test symmetry. The three 'axes of orthotropy' are given by the intersections of the three mirror planes and are 2-fold rotation axes.

An interesting practical case of non-uniform deformation is that of non-idealized rolling of a plate (Fig. 10). The friction between the plate and the rolls introduces certain shear components at the surfaces of the workpiece and they may, in general, be different (and not the opposite of each other) above and below the center plane. If the plate is wide enough (in the TD), the problem is still two-dimensional and there will be a plane of symmetry perpendicular to TD (even though the ones perpendicular to RD and ND may be lost). The plate cannot, on the other hand, in general be transformed into itself by any *rotation* around the TD direction. Since there is a plane of symmetry without its associated two-fold axis, *this plate, as a whole, does not possess an inversion center*. Any particular element all by itself may have inversion symmetry for physical reasons, and thus also a two-fold axis; but this does not carry forward to its environment.

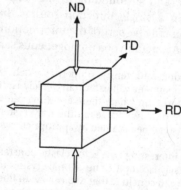

Fig. 9. Plane-strain compression (idealized sheet rolling): a case of orthotropy.

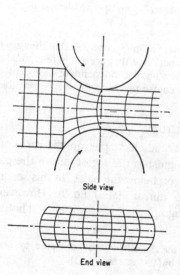

Fig. 10. Practical sheet rolling: only one mirror plane survives.

A 'sample' is often considered to be a piece of material which is (quasi-)*homogeneous*. The absence of an inversion center in the rolled plate is a direct consequence of treating a *heterogeneous* piece of material and considering, in particular, the aspects that vary from place to place. If one did smear out all differences from point to point in these samples and thus made them completely homogeneous, the inversion center and the two-fold axis would appear.

3.4 Heterogeneous samples

Sometimes, the arrangement of a second phase (rather than that of a test or process) induces a sample symmetry. Figure 11 shows an example of a fiber composite, shown in cross section. All properties that are affected by the second phase, such as plasticity, must exhibit the two-dimensionality of it; in other words, they must have a plane of symmetry perpendicular to the fiber direction.

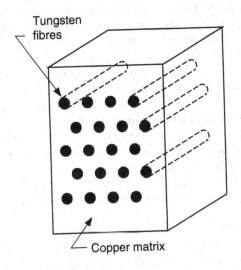

Tungsten fibres

Copper matrix

Fig.11. Morphological anisotropy due to a second phase consisting of aligned fibers.

There is a significant difference between the heterogeneity resident in the (meso-) *structure* of a material and that in, say, a non-uniformly rolled plate. The latter can be treated as a body under certain boundary conditions (for example, by the finite-element method), with the material properties at each point being the same (though not necessarily in the same current *state*). If one applies this method to a composite, one studies the composite as such: the difference between the behavior of a composite and its separate constituents; the result depends, for example, on the precise arrangement of the phases. For engineering applications, one would hope to be able to describe even a composite material as a quasi-homogeneous sample. If the second phase consists of fibers, for example, this sample will have morphological anisotropy; but the particular arrangement of the parallel fibers does not matter. This 'smearing-out' of local differences invariably leads to an effective introduction of a two-fold axis.

3.5 General samples

If a sample may have undergone a series of different types of deformation, *or if a general path is planned for the future*, it must be described without assuming any sample symmetry at all. (In *crystals*, this is called 'triclinic symmetry'.)

When the sample is homogeneous, centrosymmetry, if present in the crystals, carries forward to centrosymmetry of the sample. However, as we have seen, when the sample is heterogeneous, and this heterogeneity is not 'smeared out' for the application, centrosymmetry will, in general, not hold at all points (or even at any points) within the sample.

4. Anisotropy and Symmetry of Properties

The term 'property' was introduced in Section 1.1 as relating the response of a material to a stimulus. Both response and stimulus are generally *field variables*. When each is specified by a single number, the material property follows immediately from the ratio. In most practical situations, there are many field variables of relevance, each specified by more than one number. Typically, *some* of these are prescribed by boundary conditions (either a variation, or something held constant), and the remainder follow from the material response. In this more general sense, one means by 'property' the so-called 'constitutive' relations that specify material behavior (as distinct from the field equations and boundary conditions).

Most field variables have *tensor* character. As a result, many properties have tensor character, too. (Some exceptions were given in Section 1.2, usually associated with grossly nonlinear behavior.) In this Section, we expand upon the meaning and usefulness of tensors: first as field variables (Sec. 4.1), then as properties (Sec. 4.2). There are some significant differences in the use of tensors for these two cases [NYE 1957, DIENES 1987].

Section 4.2 also discusses some specific examples of the influence of symmetry on limiting the general anisotropy to be expected in tensors – and some subtleties that make *scalar* properties not as trivial as they might first seem. Section 4.3 will then attempt to generalize the superposition of symmetry rules, for the purpose of assessing the general *kind* of anisotropy to be expected in any given case, especially for polycrystalline samples. Finally, Section 4.4 will review a few cases of the *degree* of anisotropy observed in single crystals.

4.1 Tensor field variables

The stress is a second-rank tensor. If dK_i is the i-component of a force on an infinitesimal element of an imaginary cut whose normal has the components dA_j (proportional to the area of the element), then the stress component σ_{ij} is defined by

$$dK_i = \sigma_{ij}\, dA_j \tag{10}$$

Since the stress relates two vectors, it must transform as a second-rank tensor. For example, if a certain stress state is given by the components σ_{ij} in a chosen coordinate system, it can be 'resolved' into components $\sigma_{i'j'}$ in another coordinate system by

$$\sigma_{i'j'} = R_{i'i}R_{j'j}\sigma_{ij} \tag{11}$$

where $R_{i'i}$ is the direction cosine between the two axes i and i'. In tensor notation (see the appendix on Notation), this is expressed as

$$\sigma' = R \cdot \sigma \cdot R^T \tag{11'}$$

where R is the matrix of direction cosines and R^T its transpose (which is also its inverse).[8]

Since the stress is an even-rank tensor, it must be a centrosymmetric variable. Figure 12 shows an idealized compression test (Fig. 12a) and an idealized tension test (Fig. 12b). Algebraically, they are written, respectively, as

$$\sigma = \begin{pmatrix} 0 & 0 & 0 \\ 0 & 0 & 0 \\ 0 & 0 & -s \end{pmatrix} \quad \text{and} \quad \sigma = \begin{pmatrix} 0 & 0 & 0 \\ 0 & 0 & 0 \\ 0 & 0 & s \end{pmatrix}$$

In both, it is assumed that the only non-zero stress component is σ_{zz} (with an absolute magnitude s); this is realistic for a tension test on a long wire, but is not necessarily realistic for a typical compression test on a squat sample. If Figs. 12a and 12b represented an actual situation, it is clear that neither one of them would be changed in any way by inverting the sign of any axis (such as Z). In particular, the proven 'centrosymmetry' of the stress does not imply that the response of a sample must be the same under tension as under compression. We will later use 'stress space': a hyperspace in which the components of stress are plotted. An inversion center in *this* space would indeed mean that the behavior in tension and compression is the same. But the symmetry elements one uses to test *anisotropy* all apply to 'real', positional, space. In other words, an inversion of all spatial axes leaves the stress tensor unchanged.

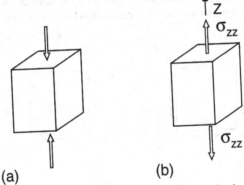

Fig.12. Compression (a) and tension (b) test. They are both centrosymmetric, but need not give the same result.

Exactly the same arguments can be made for the strain ε (treated here as infinitesimal): it is a second-rank tensor just like the stress. In plasticity, where an assumption of small strains is never meaningful, one deals with *infinitesimal* strain *increments* $d\varepsilon$ or with deformation 'rates'. The best way to start is with the velocity gradient, L; in components:

---

[8] We are here concerned with expressing a single physical quantity in different coordinate systems. In other contexts, one rotates, for example, a stress tensor to obtain a new one, all within the same reference system ('active rotation'); then, Eq. 11' is transposed.

$$L_{ij} \equiv \partial v_i / \partial x_j \equiv v_{i,j}$$

$$(12)$$

where $\mathbf{v}$ is the velocity and $\mathbf{x}$ the (*current*) coordinate. (Note this nomenclature is common in mechanics, but is a departure from our general practice in this book, in which we use $\mathbf{X,Y,Z}$ for *sample* axes and $\mathbf{x, y, z}$, with no subscripts, for *crystal* axes – *always* 'current'.)

The velocity gradient is a second-rank tensor. As distinct from the stress and the strain (rate), which are *symmetric* second-rank tensors, there is no need for $\mathbf{L}$ to be symmetric. One generally decomposes it according to

$$\mathbf{L} \equiv \mathbf{D} + \mathbf{W}$$

$$(13)$$

into a symmetric part, the 'strain rate'[9] $\mathbf{D}$ defined in components by

$$D_{ij} \equiv \left(L_{ij} + L_{ji} \right) / 2$$

$$(13')$$

and a skew-symmetric part, the 'spin' $\mathbf{W}$ defined in components by

$$W_{ij} \equiv \left(L_{ij} - L_{ji} \right) / 2$$

$$(13'')$$

Frequently, one leaves out the time derivative in the velocity gradient and talks of *displacement gradients*. Figure 13 shows two special displacement gradients, commonly referred to as 'pure shear' (Fig. 13a) and 'simple shear' (Fig. 13b).[10] The sign of the shear, in both cases, is such that the positive Z-face is displaced in the positive X-direction (and the negative Z-face in the negative X-direction). If one changed the signs of both Z and X, the test is unchanged: simple shear, too, is centrosymmetric; $\mathbf{L}$ *is centrosymmetric*. (The figures and the description so far have been two-dimensional: the third direction was of no concern. In general, this holds in three dimensions.)

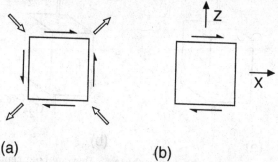

(a)   (b)

Fig. 13. Pure shear (a) and simple shear (b) test. They are both centrosymmetric, but inverting their sign need not give the same result.

Simple shear is of special concern in plasticity: both dislocation glide and deformation twinning produce simple shear. Let us look, as an example, at the *theoretical shear strength* introduced in Section 1.1. It depends on the form of the atomic

---

[9] In mechanics applications, it is emphasized that the term 'strain rate', and the symbol $\dot{\varepsilon}$ preferred by metallurgists instead of $\mathbf{D}$, is not meant to imply the time derivative of a state function (which ε is not).
[10] Since the stress is always symmetric (as defined here: the 'Cauchy stress'), the term 'pure shear' is sometimes used for the velocity gradient that ensues from a prescribed shear *stress* (for example, under an elastic response).

interaction potential and may well be different in the 'forward' and the 'backward' direction. This does not invalidate the centrosymmetric nature of the material property 'theoretical shear strength'. Thus, for example, the orientation of a crystal whose theoretical shear strength is to be measured can be unambiguously determined from diffraction experiments.

We have emphasized that even asymmetric second-rank tensors, such as the velocity gradient, are centrosymmetric. This implies that the skew-symmetric (or 'anti-symmetric') part, the 'spin' W, is centrosymmetric, too. This may seem surprising, especially when one represents the spin as an 'axial vector', with components defined by:

$$\tilde{W}_1 = \begin{pmatrix} 0 & 0 & 0 \\ 0 & 0 & W_{23} \\ 0 & -W_{23} & 0 \end{pmatrix}; \quad \tilde{W}_2 = \begin{pmatrix} 0 & 0 & -W_{31} \\ 0 & 0 & 0 \\ W_{31} & 0 & 0 \end{pmatrix}; \quad \tilde{W}_3 = \begin{pmatrix} 0 & W_{12} & 0 \\ -W_{12} & 0 & 0 \\ 0 & 0 & 0 \end{pmatrix}$$

where the W_{ij} are the components defined in Eq. 13".[11] Consider again the application of an inversion center. In Fig. 13, we saw that the situation did not change when the signs of X and Z were changed. The coordinate Y was irrelevant; but since we implied an inversion of *all* axes, we should have changed Y also, and this would have changed our presumably right-handed original frame to a left-handed one. The spin that is part of Fig. 13b does not change, physically; but if one *describes* it by an 'axial vector', one has to set a convention on 'handedness', and that convention must be subject to reversion by an inversion center. The spin W, regardless of how it is described, must remain a centrosymmetric variable. If the concept of axial vectors is extended to 'axial tensors' of all ranks, these are generally centrosymmetric when their rank is *odd* [Int. Tables]. We shall use the word 'tensor', in this book, exclusively in the non-axial ('polar') sense, and describe spins as skew-symmetric (polar) second-rank tensors.

Invariants

When the same tensor quantity is expressed in a different coordinate system, the components change. For symmetric tensors, it is always possible to find a set of axes in which all off-diagonal components are zero; this procedure is called 'diagonal-ization'. The diagonal components are then called the 'principal values' or 'eigen-values' of the tensor, and the special coordinate system for which this is true is called the set of 'principal axes' or 'eigenvectors'. (Incidentally, it is also possible to make all *diagonal* components zero, when their sum is zero, or else all equal. This is of interest, in principle, for shear deformations, but it is not used in practice.)

The principal values λ of a second-rank symmetric tensor are algebraically deter-mined by solving the so-called secular equation

$$\lambda^3 - J_1 \lambda^2 - J_2 \lambda - J_3 = 0 \tag{14}$$

where the three coefficients are *invariants* of the tensor: J_1 is the 'trace' (the sum of the diagonal components), J_3 the determinant. For example, the hydrostatic stress is an invariant of the stress tensor (one third its trace). It does not depend on the frame of reference. The second invariant has an especially simple form for *deviatoric* quantities (for which the trace is zero by definition). For example,

$$J_2 = \tfrac{1}{2}\, \mathbf{s}^2 \tag{14'}$$

---

[11] As with stress and strain, we will use the bold-face symbol W for the spin *without* implying in what representation the quantity is written.

for the stress deviator **S** (which is frequently used in plasticity).

Representation ellipsoids

The principal values of a tensor include both the maximum and the minimum value of any component. When the principal values are marked on the principal axes, they represent the extremities (and one intermediate value) of the 'magnitude' of the tensor. A symmetric second-rank tensor can be described as an *ellipsoid* with *its* principal axes aligned with the principal axes of the tensor. We made use of this, in a qualitative way, for a *property* in Fig. 1d of this Chapter. It can just as easily be done, for example, for tensor field variables such as the strain; the resulting ellipsoid is called the 'stretch ellipsoid'. For more details, the reader is referred to NYE [1957].

4.2 Tensor properties, scalar properties

Second-rank tensor properties

In the beginning of this Chapter, we used the thermal conductivity as an example of a symmetric second-rank tensor. The diffusivity is another example. Thus, in general, it has 6 independent components. Like all symmetric tensors, it can always be diagonalized by transformation into a set of principal axes, so that

$$\mathbf{D} \equiv \begin{pmatrix} D_1 & 0 & 0 \\ 0 & D_2 & 0 \\ 0 & 0 & D_3 \end{pmatrix} \tag{15}$$

Note that these are only three parameters: the information from the other three is contained in the orientation of the principal coordinate frame. This separation of the six numbers that represent 'diffusivity' in a crystal into eigenvalues and eigenvectors has no advantage *unless* the principal axes are *crystal* axes. It is indeed much more natural to describe properties in the crystal axes – and only if these crystal axes themselves have certain symmetry properties does the number of independent components of the diffusion coefficient decrease.

In particular, a crystal that has orthorhombic symmetry must have the principal axes of **D** aligned with the crystal axes; thus, only three numbers are required to specify its diffusivity. The same is true for a polycrystal with an orthotropic texture. It is immediately obvious, and can be easily verified by the tensor transformation rules outlined above, that a tetragonal crystal requires only two, a cubic crystal only one coefficient. (See also the example in Section 1.2.) Equivalently, the property ellipsoid must be aligned with any symmetry planes or axes of the material, and it degenerates into a rotational ellipsoid for tetragonal, trigonal, hexagonal, and fiber symmetries.

Fourth-rank tensor properties

Since both the stress and the strain are inherently centrosymmetric, any relation between them must be, too. In particular, the (linear) elastic constants, which are fourth-rank tensors, are centrosymmetric. But even non-linear effects are centrosymmetric: the response may be different in tension and compression (as it is when the material is 'anharmonic'), but it is independent of the signs of the axes in which it is described. This also follows formally from the fact that the elastic constants of all orders are even-rank tensors.

Fourth-rank tensors with the specific symmetry properties that follow from the relation between two symmetric second-rank tensors are often described in the 'matrix' (i.e., two-dimensional) notation suggested by VOIGT (pronounce fohgt) [NYE 1957]. For this purpose, the stress and strain are each represented by a one-dimensional array of six components:

$$\sigma = \{\sigma_1, \sigma_2, \sigma_3, \sigma_4, \sigma_5, \sigma_6\} \equiv \{\sigma_{11}, \sigma_{22}, \sigma_{33}, \sigma_{23}, \sigma_{31}, \sigma_{12}\} \tag{16}$$

$$\varepsilon = \{\varepsilon_1, \varepsilon_2, \varepsilon_3, \varepsilon_4, \varepsilon_5, \varepsilon_6\} \equiv \{\varepsilon_{11}, \varepsilon_{22}, \varepsilon_{33}, 2\varepsilon_{23}, 2\varepsilon_{31}, 2\varepsilon_{12}\} \tag{16'}$$

The multiplication of various tensor strain components by factors such as 2 and 4 [VOIGT 1928] represents one way to make these stress and strain 'vectors' work-conjugate: their scalar product does produce the work (for small strains). Although, in elasticity, the conventions in Eqs. 16 and 16' are standard, the fact that they are not analogous is somewhat awkward. We will introduce a different vectorization of stress and strain (-rate) in Chapter 7, for some special elastic problems and for later use in plasticity.

The elastic constants are then represented by a matrix of 6×6 constants; it is symmetric and therefore has, at most, 21 independent constants in it. The general form is

$$\{\mathbb{C}_{ij}\} = \begin{pmatrix} \mathbb{C}_{11} & \mathbb{C}_{12} & \mathbb{C}_{13} & \mathbb{C}_{14} & \mathbb{C}_{15} & \mathbb{C}_{16} \\ \mathbb{C}_{12} & \mathbb{C}_{22} & \mathbb{C}_{23} & \mathbb{C}_{24} & \mathbb{C}_{25} & \mathbb{C}_{26} \\ \mathbb{C}_{13} & \mathbb{C}_{23} & \mathbb{C}_{33} & \mathbb{C}_{34} & \mathbb{C}_{35} & \mathbb{C}_{36} \\ \mathbb{C}_{14} & \mathbb{C}_{24} & \mathbb{C}_{34} & \mathbb{C}_{44} & \mathbb{C}_{45} & \mathbb{C}_{46} \\ \mathbb{C}_{15} & \mathbb{C}_{25} & \mathbb{C}_{35} & \mathbb{C}_{45} & \mathbb{C}_{55} & \mathbb{C}_{56} \\ \mathbb{C}_{16} & \mathbb{C}_{26} & \mathbb{C}_{36} & \mathbb{C}_{46} & \mathbb{C}_{56} & \mathbb{C}_{66} \end{pmatrix} \tag{17}$$

These constants are written with two subscripts,[12] by contracting pairs of indices; for example, $\mathbb{C}_{12}$ represents the σ_{11} response to a strain ε_{22}. When one writes the general form of Hooke's Law as

$$\sigma = \mathbb{C} : \varepsilon \tag{18}$$

or

$$\varepsilon = \mathbb{S} : \sigma \tag{18'}$$

one concentrates on the linear relation between the physical quantities, rather than on the particular notation used ('matrix' or 'tensor'). The double contraction implied by the colon (see the appendix on Notation) refers to the *tensor* (not the matrix) rank.[13]

---

[12] We re-emphasize that this does *not* signify that they are second-rank tensors. Similarly, the single subscript on stress and strain does not signify that they are first-rank tensors. The word 'vector' is now frequently used also in the sense of a one-dimensional array, without regard to tensor properties; in this sense, one may refer Eq. 16 as representing the 'stress vector', Eq. 16' the 'strain vector'.

[13] There are various factors of 2 and 4 in the matrix notation for $\mathbb{S}$. The use of the symbols $\mathbb{C}$ and $\mathbb{S}$ is almost universal (sometimes capital, sometimes script). To avoid the use of general terms such as 'modulus' or 'constant', one commonly calls $\mathbb{C}$ the stiffness, $\mathbb{S}$ the compliance (mnemonic: the inverse of the initials!). An exception to this terminology is the general use of the terms 'shear modulus' (a stiffness) and 'Young's modulus', the inverse of a component of the compliance.

The number of elastic constants that must be zero under a given set of symmetry conditions can always be derived by making use of the *tensor* character of, for example, $\mathbb{C}$. We give one example, for a single mirror plane perpendicular to the 2-direction (permuted from Eq. 9). The tensor transformation equation

$$\mathbb{C}_{i'j'k'l'} = R_{i'i}\, R_{j'j}\, R_{k'k}\, R_{l'l}\, \mathbb{C}_{ijkl}$$

immediately demands that when 2 occurs an odd number of times as an index, the component must be zero. This is true for $\mathbb{C}_{1112} = \mathbb{C}_{16}$ and similarly $\mathbb{C}_{26}$, $\mathbb{C}_{36}$, as well as $\mathbb{C}_{14}$, $\mathbb{C}_{24}$, $\mathbb{C}_{34}$; finally for $\mathbb{C}_{1312} = \mathbb{C}_{56}$ and $\mathbb{C}_{2315} = \mathbb{C}_{45}$. Thus, eight of the 21 components in one triangle of the 6×6 matrix must be zero. It turns out that indeed there are 13 independent elastic constants in monoclinic materials (Table I). Note, however, that sometimes the derivation is not so simple: there may be *relations* between nonzero components which reduce the number of independent values [NYE 1957].

Finally, we give a very symmetric example: a cubic crystal, in which some results follow from simple inspection: the response to tension in the three cubic axes must be the same, and thus $\mathbb{C}_{11} = \mathbb{C}_{22} = \mathbb{C}_{33}$. Similarly, the three shears on the cube planes in the cube directions must behave in the same way; thus, $\mathbb{C}_{44} = \mathbb{C}_{55} = \mathbb{C}_{66}$. A detailed derivation shows that there is only one other independent combination, namely (in a common abbreviation)

$$\mathbb{C}' \equiv (\mathbb{C}_{11} - \mathbb{C}_{12})/2$$

(20)

It corresponds to shear on a {110} plane in a $\langle 1\bar{1}0 \rangle$ direction. All these simplifications apply only when the elastic constants are expressed in the cubic coordinate system (in which case they are usually written as *lower-case* c). A crystal of arbitrary orientation has 21 elastic constants; they can be derived from the three independent ones, plus the orientation information.

Isotropic properties

The case of elasticity in cubic materials provides for an easy extension into isotropic materials: the two shear constants must be equal. Yet there is no reason for any connection between tension, say, and shear. Thus an isotropic material has two independent elastic constants. They are usually quoted as the bulk modulus (or its inverse, the volume compressibility) and shear modulus, or as Young's modulus and Poisson's ratio (both applicable to tension).

In the treatment of non-tensor properties, one can rely on the fact that, in general, isotropic properties can only depend on *invariants* of the stimulus tensor. For example, the 'yield criterion' or 'yield surface' $f(\sigma)=0$ for *isotropic* materials can only depend on the three invariants of the stress. It is, therefore, usually specified in the *principal-axes* space of the stress tensor.

Note that, when yield is *anisotropic* in a particular material, there is no particular advantage to describing the yield criterion in principal axes; one will rather choose the axes of sample symmetry (see Chapter 10).

In some cases, a property is isotropic on a local basis. For example, we have seen that second-rank tensor properties are isotropic in cubic crystals. One of these is the thermal expansion coefficient. Therefore, even if the polycrystal has, for example, fiber texture (in which a symmetric second-rank tensor could, in general, have two independent components), its thermal expansion coefficient will be isotropic when the grains are cubic.

Scalars

Consider, for example, the compressibility. Since the pressure and the relative volume change are invariants of the stress and strain tensors, respectively, and are thus both scalars, the material constant that relates them may be expected to be a scalar, too. One might therefore expect it to be independent of orientation in a single crystal and of texture in a polycrystal. The former is correct, the latter not, in general. The problem is that the compressibility is properly defined by a *partial* derivative:

$$\kappa \equiv -\frac{1}{V}\left[\frac{\partial V}{\partial p}\right]_{T,\sigma'} \tag{21}$$

All the components of the deviatoric part of the stress tensor, σ', are to be held constant (just like the temperature T). It may not be actually possible to fulfill this condition in practice. For example, even when a polycrystal as a whole is subjected to a strictly hydrostatic pressure, the interactions between grains may lead to local non-zero deviatoric stresses (in non-cubic crystals), and then the macroscopic compressibility (with zero macroscopic σ') need not be the same as the local one.

This case provides a certain cautionary note to assuming that all 'scalars' should be the same in single crystals and polycrystals. The properties of polycrystals depend not only on the anisotropy and orientation distribution of the constituent grains, but also on their *interaction*. This will be treated in detail in Chapter 7.

Non-centrosymmetric properties

Quite different from the essentially even-rank tensor properties that were discussed so far are odd-rank properties. For example, it is possible for a sample to have a permanent electric polarization (an innate vector property), or it may *change* its polarization (temporarily) upon a *change* in temperature ('pyro-electricity'): a vector property. (This property is used in infra-red detectors such as tri-glycine sulfate.) Similarly, there are certain properties that have *third-rank tensor* character (such as piezoelectricity). Finally, there are some properties for which a *handedness* is part of the physical background rather than an arbitrary convention. For example, a screw axis in a space group implies an innate handedness of the structure; optical gyration is an innate handedness of a property. All these properties are *non-centrosymmetric*, and can therefore only exist in non-centrosymmetric crystals; we will not discuss them further in this book. For a systematic treatment of aspects of these properties as they are relevant to textured materials, see NYE [1960] and MATTHIES &AL.[1987].

Non-tensor properties

In Section 1.2, we described some properties that are *not* tensor properties, typically because the constitutive relation is grossly non-linear. The general rule, Eq. 6, for applying symmetry arguments holds. The yield strength and its generalization, the 'yield surface' in stress space, is of particular interest. Its symmetry properties are discussed in detail in CANOVA &AL. [1985] and will be reviewed in Chapter 10.

4.3 The kind of anisotropy of a property

Throughout this Chapter, we have seen how symmetry considerations can limit the general anisotropy to be expected in non-scalar properties. Some of these symmetry considerations refer to the property itself (e.g., it may be inherently centro-symmetric). Others refer to symmetries resident in the (polycrystalline) *sample* or in the underlying *crystals*. The overall effect on the 'kind' of anisotropy we defined in Section 1.2, namely the number of numbers needed (or the extent of a phase space) to describe the property, is subject to a complex set of superposition rules. We summarize them for single crystals, for samples without consideration of their constituents, and for polycrystals.

Single crystals

The well-known 'Neumann's Principle' (as quoted by NYE [1957, p.20]) states: "The symmetry elements of any physical property of a crystal must include the symmetry elements of the point group of the crystal." This statement, as it stands, refers to a single crystal, and an *ideal single crystal* at that. We have seen that, when microstructural features, or 'defects' in the crystal structure, are anisotropically arranged, the symmetry of a property that depends on them may be *lower* than the point group symmetry of the crystal (Section 1.1); examples are diffusion along aligned dislocations, and the magnetization of a single crystal that contains magnetic domains. Thus, we may generalize the principle to:

(1) The symmetry elements of any physical property of a single crystal must include those that are common to the point group of the crystal and the symmetry of its defect structure.

We have seen that 'anisotropy' is best defined as subject to *rotational* symmetry elements only. Thus, the anisotropy of a single-crystal property is restricted by the *rotational* symmetry elements that fall into the above category. When the property is centrosymmetric, however, a single mirror plane in the single crystal does restrict anisotropy, since it is then equivalent to a two-fold rotation axis. This is an example of the minimalist nature of statement (1), expressed by the word 'include'. In any case, we have seen that, in any particular example, the precise kind of anisotropy to be expected for a given single-crystal property can be derived by application of the invariance rule, Eq. 6.

Samples

The same principles should hold for a 'sample' of arbitrary structure (single crystal, polycrystal, even continuum), only with 'crystal' replaced by 'sample', and 'point group symmetry of the crystal' by 'sample symmetry'.

As an example, take a sample whose entire history is presumed to consist of deformation in a set of wire-drawing dies, and which thus should possess fiber symmetry. The sample as a whole will require two constants for a second-rank tensor property, five for a fourth-rank tensor property. These could be measured, and from these, the respective properties can be derived for any sample cut out of this wire at arbitrary angles. No reference to the crystal structure (if indeed it were made up of crystals) is necessary for such a macroscopic description.

Polycrystals

Now assume that the above sample was actually a polycrystal made up of cubic grains. Then, any second-rank property (such as the thermal conductivity) would be isotropic locally; thus, the two constants in the wire would have to be the same. One could even use an observation of macroscopic isotropy of a second-rank tensor property in a textured material as *evidence* that the underlying crystal structure must be cubic.

Fourth-rank tensor properties, such as elastic constants, have three independent numbers in cubic crystals, but will show five in a textured wire; the remaining information comes from the texture. This circumstance leads to the technique of deriving information on the texture from the observation of some relations between measured macroscopic constants. The information will always be *partial*, since it only uses a very limited number of relations (two in the above example).

If the wire had been a polycrystal made up of orthorhombic crystals, which have three second-rank and nine fourth-rank numbers, the anisotropy of the wire sample would still be restricted by the fiber symmetry (just two and five independent numbers, respectively).

To arrive at a rule equivalent to statement (1), but for polycrystals, we must, again, allow the sample symmetry to contain elements from morphological anisotropy, due to grain shape, inclusion shape, etc., not only from crystallographic texture. In addition, we must remember that sample symmetry holds in a statistical sense only, and thus does not have the absolute power of symmetry arguments in ideal single crystals. Thus,

(2) *the symmetry elements of any physical property of a polycrystalline sample must include the statistical symmetry elements common to its crystallographic texture and any due to morphological texture.*

As for anisotropy, it can exist in polycrystals only if the property is anisotropic in the constituent single crystals *and* if the polycrystal is textured (i.e., has a non-uniform orientation distribution). The overall 'kind of anisotropy' is determined by the anisotropy of the local property and the symmetry of the sample. Abstracting from the above examples, we attempt to formulate a general rule for anisotropies in polycrystals:

(3) *Any lack of general anisotropy of a macroscopic non-scalar property is due to restrictions imposed by the sample symmetry. A complete lack of anisotropy (i.e., isotropy) of a macroscopic property can, in addition, be due to the property being isotropic in the single crystal.*

Finally, it bears repeating that

(4) *any center of symmetry that exists everywhere locally, holds for the entire sample.*

4.4 The degree of anisotropy of some properties in single crystals

Diffusivity

We have seen Section 1.2 that the diffusivity is fully described by a second-rank tensor and is therefore isotropic in cubic crystals (Table I). In the other extreme, the diffusion of certain atomic species through the crystal structure can vary *by many orders of magnitude* in different directions; for example, in some non-cubic perovskites [RUNDE &AL. 1992]. In the kinetic relation for diffusion,

$$D = D_0 \exp\{-Q/kT\}$$ \hfill (22)

the anisotropy comes in part from D_o and in part from a directional dependence of the activation energy Q, by about a factor of two.

It is interesting to reflect on the directional dependence of the activation energy. As an 'energy', it may be presumed to be a 'scalar': independent of orientation by definition. Yet the physical reason for the dependence on direction is clear. Assume a particular atom (or vacancy) can make k different jumps to neighboring equilibrium sites. For some jumps, the activation energy may be the same, but for others it may be different. Similarly, the 'attempt frequency', ν, may depend on the direction of the jump; this can be incorporated as an activation entropy into the free enthalpy of activation, ΔG^k. With the dependence of the jump distance λ^k on the jump direction (which affects D_o), the diffusivity is

$$D_{ij} = \nu \sum_k \lambda_i^k \lambda_j^k \exp\left\{-\Delta G^k / kT\right\}$$

(23)

The activation energy ΔG^k for self-diffusion[14] is subject to the symmetry restrictions of the crystal class, not only D_o (through the λ^k), and not only D as a whole.

Elastic constants

The anisotropy of elastic properties can also be quite large [BOAS & MACKENZIE 1950; SIMMONS & WANG 1971]. As illustrative examples, Table III displays the ratio of the two shear moduli for some cubic materials. The only cubic metal that is approximately isotropic with respect to its elastic properties is tungsten at (and below) room temperature. Aluminum is marginal: most anisotropies are decidedly non-negligible, and some (as β-brass) are so extreme as to suggest the proximity of a structural instability.

The elastic anisotropy may also be expressed in terms of the variation of Young's modulus with crystal direction. Table III displays the ratio of the two extreme values for single crystals of these materials:

$$\frac{E_{100}}{E_{111}} = 1 - \frac{2}{3} \frac{\mathbb{S}_{11} - \mathbb{S}_{12} - \mathbb{S}_{44}/2}{\mathbb{S}_{11}}$$

(24)

where the $\mathbb{S}_{ij}$ are the compliances in cubic crystal coordinates. To describe the continuous variation with angle to the crystal axes, one can mold a model body such as those in Fig. 14.

In an anisotropic material, 'Poisson's ratio' will in general have two values, ν_1 and ν_2 for every tensile direction (causing ovaling of an initially circular cross-section). In highly symmetric orientations (such as <111> and <100> in cubic material), there may be only one value – but it depends on the orientation of the tensile axis. In fact, since the volume change is independent of direction, so is the combination $(1-\nu_1-\nu_2)E$; thus, the second column in Table III also gives the value of $(1-2\nu)_{100}/(1-2\nu)_{111}$.

The anisotropy in polycrystal samples is generally much reduced from that in single crystals, because of the spread in crystal orientations. For example, even in a hard-drawn copper wire, in which there is a $\langle 111 \rangle$ fiber texture component that is 36

---

[14] Note that *interstitial* diffusion is subject to the symmetry elements that are present in the neighborhood of the interstitial.

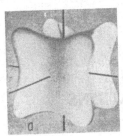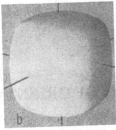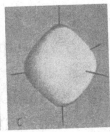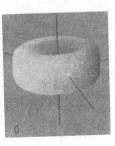

Fig.14. Three-dimensional description of the anisotropy of Young's modulus in crystals of (a) gold; (b) aluminum; (c) magnesium; and (d) zinc.

Table III. Elastic anisotropy of some cubic materials (at room temperature when not otherwise specified). Calculated from values in SIMMONS & WANG [1971].

Material	$\mathbb{C}_{44}/\mathbb{C}'$	E_{111}/E_{100}
Cu	3.21	2.87
Ni	2.45	2.18
Al	1.22	1.19
Fe	2.41	2.15
Ta	1.57	1.50
W (2000K)	1.23	1.35
W (R.T.)	1.01	1.01
V	0.78	0.72
Nb	0.55	0.57
β-CuZn	18.68	8.21
spinel	2.43	2.13
MgO	1.49	1.37
NaCl	0.69	0.74

times random, and four times as strong as the $\langle 100 \rangle$ component, the ratio of Young's moduli in the direction of the fiber and perpendicular to it is only 1.6, compared with the maximum ratio of 2.87 from Table III. A *derivation* of the polycrystal averages is usually quite difficult; Chapter 7 will discuss it in detail.

Chapter 2

THE REPRESENTATION OF ORIENTATIONS
AND TEXTURES

1. **The Representation of Directions and Planes**
   1.1 Crystal axes. Index notation for planes and directions
   1.2 The unit sphere. Signed and unsigned directions
   1.3 Stereographic and equal-area mapping
   1.4 Pole figures and inverse pole figures for sets of single crystals
   1.5 Contouring and normalization

2. **The Representation of Orientations**
   2.1 The definition of an orientation by three poles
   2.2 Description of an orientation by a vector lying in the surface the unit sphere
   2.3 Cartesian frames. Rotations.
   2.4 Matrix representation
   2.5 Symmetrically equivalent orientations
   2.6 The representation of misorientations
       Axis/angle description. Rodrigues vector description. Quaternion description
   2.7 The representation of twinning relations

3. **The Representation of Textures**
   3.1 Orthotropic sample symmetry
       Description in terms of fibers and volume fractions. Oblique sections and difference plots
   3.2 One plane of sample symmetry
   3.3 Cubic sample symmetry
   3.4 The representation of the non-textured reference state. Discrete grains files
   3.5 **Summary. Recommendations**
       Pole figures, inverse pole figures. Quantitative comparisons of similar textures
       The representation of three-dimensional ODs

4. **Continuous Distributions and Series Representation**
   4.1 The orientation distribution function
   4.2 The pole and axis distribution functions
   4.3 The misorientation distribution function

THE REPRESENTATION OF ORIENTATIONS
AND TEXTURES

U. F. Kocks

Visualization is, for many people, tantamount to 'understanding', and different methods of visualization are effective to different degrees in this task. Apart from this aid in understanding, when information on orientations and textures is to be used quantitatively, it must be represented in some way, and different methods are more appropriate for different tasks.

Any set of facts seems simple to the frequent observer. What we want to emphasize here is how you represent the facts for a case that is new to you (or to your audience); for example, where the c-axes lie in an extruded tube of Zr; what (if any) texture develops when you compress a cube of aluminum successively in all three directions; how (or whether) textures in simple shear and pure shear differ from each other. For this reason, we wish to develop, in this Chapter, a systematic way to represent the relation between the orientation of the crystallographic axes of a crystal of *arbitrary lattice structure* with respect to the coordinate system of a sample of *arbitrary symmetry*, or *vice versa*. We will emphasize the duality between describing one in terms of the other, or the other in terms of the one. In simple textures, this is the duality between (a set of) 'pole figures' and (a set of) 'inverse pole figures'.

The simplest case will be used, in Section 1, to lay the groundwork: when the interest in a particular application is just one <u>direction</u> (such as only the c-axis in a rolled sheet of Zr, or only the tensile axis in an fcc single crystal). In contrast to these (single) 'directions', we reserve the term <u>orientations</u> for a description of the relation between two complete coordinate frames. This will be the contents of Section 2. Section 3 will deal with the representation of *textures*, i.e., many discrete orientation relationships or a continuous distribution of them.

For the purposes of this Chapter, the orientation relationships are assumed known, and they are idealized. The considerable subtleties of how one obtains this knowledge are left to subsequent Chapters. Of dominating influence are the symmetries known or assumed to exist at both the crystal and the sample level. While these determine the minimum extent of orientation space that is needed, it is sometimes opportune to use more, in part to *display* these symmetries.

Finally, in Section 4, we discuss the interrelation between discrete and continuous distributions of orientations. They are important in practice: for example, one may wish to smooth out a simulated texture into a continuous distribution; or conversely, one may wish to represent a general distribution by a relatively small number of discrete grains to use in simulations of polycrystal properties. In addition, there is a fundamental difference between continuous and discrete distributions, which manifests itself in the fact that there is an inherent indeterminacy in the deconvolution of *continuous* pole figures, which does not exist for *discrete ones* in which the correlation between points can be established.

For an introductory reading into the problems of texture, we recommend the article by HATHERLY & HUTCHINSON [1979]; for details on the analysis of continuous distributions the book by BUNGE [1982].

1. The Representation of Directions and Planes

'Directions' – unit vectors, axes, plane normals – are described within a given coordinate frame. This is directly applicable to 'fiber textures' such as one might have in a drawn wire, where the *sample* symmetry is transversely isotropic, and also to cases in which only one of the *crystal* axes is of interest (e.g., the **c**-axis in hexagonal materials when considering properties such as the diffusion coefficient that have only two values: *in* the basal plane and perpendicular to it). The aim is representations that are easy to understand quantitatively and in which the aspects of primary interest are least distorted.

1.1 Crystal axes. Index notation for planes and directions

One of the coordinate frames of importance in the description of orientations and textures is tied to the crystal lattice. In crystals of cubic, tetragonal, or orthorhombic lattice structure, one chooses a cartesian frame **x**, **y**, **z** aligned with the unit cell base vectors **a**, **b**, and **c**. For descriptions of crystals of hexagonal or trigonal symmetry, it is often convenient to choose a non-orthogonal set of axes (to be described in more detail below). A cartesian frame can be associated with these; while the conversion convention is, in principle, arbitrary, in practice it is important, for example, if one wants to keep descriptions of orientations and textures comparable [MATTHIES &AL. 1988, p. 291]. We have listed the conventions we use (which are common) in Fig. 5 of Chapter 1.

Within a crystal frame, it is common to describe lattice directions and lattice planes by integer indices. A *direction* **r** in a cartesian frame is described by the vector sum

$$\mathbf{r} = u\mathbf{a} + v\mathbf{b} + w\mathbf{c} \tag{1a}$$

and concisely denoted by the symbol [*uvw*] (in brackets). The smallest integers having the same ratios are used to characterize the direction. (Specification of the *length* of the vector requires a knowledge of the lattice structure.)

A set of parallel lattice *planes* is described according to the equation

$$h\, x/a + k\, y/b + l\, z/c = 1 \tag{1b}$$

where *x*, *y*, *z* are the coordinates of any point on one specific plane of the set and *a*, *b*, *c* are the lengths of the base vectors of the unit cell. The 'Miller indices' (*hkl*) (in parentheses) denote the reciprocal multiples of the axis intercepts, again reduced to the smallest integers that have the same ratios. The lattice plane *spacing*, important for diffraction, can be easily derived from the Miller indices and the unit cell dimensions.

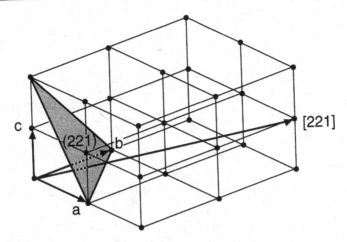

Fig. 1. Definition of crystallographic indices for planes and directions in an ortho-rhombic crystal. In non-cubic crystals, a *direction* of a certain set of indices is not, in general, perpendicular to the *plane* of the same index notation.

A (221) plane in an orthorhombic crystal is shown in Fig. 1. The normal to this plane (its pole) is also commonly labeled (221); it corresponds to a vector in 'reciprocal space'. Note that in the orthorhombic crystal displayed in Fig. 1, the *direction* [221] is not parallel to the *pole* (221): it is not perpendicular to the plane (221). Only in the cubic system do lattice plane normals (*hkl*) coincide with the *directions* of the same indices [*hkl*]; in other crystal systems, this may be true for one or more special lattice planes, but not in general.

Often, a lattice plane or lattice direction is symmetrically equivalent to some others. The generic set of all orientations that can be generated from any one of them by applying all crystal symmetry elements is called a *form*. For this, one uses angular brackets to denote directions, curly brackets for planes. For example, ⟨100⟩ in a cubic system stands for the set of directions [100], [$\bar{1}$00], [010], [0$\bar{1}$0], [001], and [00$\bar{1}$]. In a tetragonal system, {110} stands only for the set of planes (110), ($\bar{1}$10), (1$\bar{1}$0), and ($\bar{1}$$\bar{1}$0); to emphasize the difference of the z-axis, one sometimes uses mixed brackets, so that {112} stands for the set (112), ($\bar{1}$12), (1$\bar{1}$2), ($\bar{1}$$\bar{1}$2), (11$\bar{2}$), ($\bar{1}1\bar{2}$), (1$\bar{1}$$\bar{2}$), ($\bar{1}$$\bar{1}$$\bar{2}$) – but not, for example, (121). When Miller indices are used to describe diffraction from lattice planes, the parentheses are generally omitted and a common denominator indicates the order of the diffraction; e.g., 200 is the second-order diffraction on the lattice planes of form {100}. (For more information, see, e.g. BARRETT & MASSALSKI, 1980.)

Combinations such as {001}⟨110⟩ are used to specify slip systems (slip plane and slip direction) and also texture 'components' (e.g., rolling plane normal and rolling direction). The direction is always meant to be contained in the plane; this is sometimes expressed by specifying, for example, the combination of a {111}-type plane and a ⟨110⟩-type direction in cubic crystals as {111}⟨1$\bar{1}$0⟩, but is often just given as {111}⟨110⟩.

In crystals of the hexagonal and trigonal systems, a coordinate system with four axes is often used: one along the 6-fold (or 3-fold) rotation axis **c**, and three perpendicular to it; in the hexagonal basal plane, $\mathbf{a}_1$, $\mathbf{a}_2$, $\mathbf{a}_3$ are chosen in the close-packed directions, at 120° to each other (Fig. 2; see also Chap. 1 Fig. 4). (For conventions in the trigonal system, see [INT. TABLES 1983].) These axes are, of course, redundant; however, omitting one of them sacri-fices the advantage of the symmetrical description.

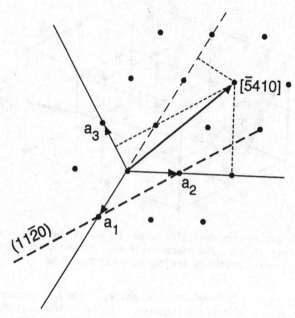

Fig. 2. The four-index ('Miller–Bravais') notation for crystals of the hexagonal lattice structure. The three redundant indices in the basal plane can be derived by the standard intercept method for planes; for directions, the *projections* are in the correct ratios.

In the hexagonal frame, lattice *planes* can still be described by their reciprocal intercepts on all four axes; this leads to the 'Miller–Bravais' indices (*hkil*), with h+k+i=0. Figure 2 shows this for the basal plane. A *vector* in three dimensions cannot be unambiguously decomposed into four components; however, it can be unambiguously *projected* onto the four axes (Fig. 2) and, in this case of a hexagonal basis, the projections are proportional to vector components so long as the sum of the three redundant indices is set to zero [OTTE & CROCKER 1965]. In this notation for hexagonal and trigonal crystals, the first three indices of a direction normal to a plane are the same as those of the plane. (Note that in conventional crystallography, directions are often described as three vector components in a non-orthogonal set, which is the least convenient of all descriptions.)

For theoretical discussions, it is easiest to retain all four indices, since symmetrically equi-valent directions (and planes) look similar: the first three indices are merely permuted or have their individual signs changed. For convenience in numerical formats, however, one often omits the third index (sometimes replacing it by a ·). In other applications, one needs an orthogonal set, and the usual convention is to set x=[2$\bar{1}\bar{1}$0], y=[01$\bar{1}$0] (Chap. 1 Fig. 5). This is a three-index set, and it is distinguishable only by inference from the abbreviated four-index set when the center dot (·) is *not* used for the latter.

1.2 The unit sphere. Signed and unsigned directions

The direction of a general unit vector can be described as a point on a unit sphere. Let us treat two particular examples of a single direction. In the first, assume you have a sample that is cut as an orthogonal parallelepiped (a 'brick'), and you have marked three faces to define what you will call the X, Y and Z axes. Figure 3a shows such a case; one of the faces is marked with a negative identifier: it is necessary to keep a clear record of the signs. For example, you may have marked the sample because you plan a simple shear test in which the positive Z-face is to be displaced in the positive X-direction, keeping the negative Z-face fixed.

Say you know that the sample is a single crystal of hexagonal lattice structure, and you are interested (for now) only in where the basal plane lies. You can get this information using an x-ray diffractometer (knowing the {c}-plane spacing and therefore the Bragg angle 2θ, see Chap. 4). You can mark the emergence of one of the stack of parallel ⟨c⟩-planes on the surface of your sample (dashed in Fig. 3a).

To describe the situation quantitatively, it is best to imagine the plane of interest to be located in the center of a unit sphere (i.e., an arbitrary point on the plane is located at the center). The plane, if extended in all directions would cut the sphere in a line (a 'great circle', defined as a circle whose center is in the origin of the sphere). A completely equivalent, and easier, description is by the plane's normal (taken at the center of the sphere): it intersects the sphere at two diametrically opposed points. You may mark one of these as 'positive' and call it, say, P (in this case, it is the crystal z-axis[1]). The positive sign may be well-defined with respect to your measurement (e.g., how the sample was mounted), but it may (depending on the crystal symmetry) be entirely arbitrary in terms of the physics: if you had chosen to mount the sample the other way around, you would have gotten exactly the opposite and would have called *that* 'positive'. When an *unsigned* normal to a set of crystallographic planes is meant, we call this a *pole*; a signed one is called a *unit vector*. With 'normal' Bragg reflection, one measures *poles*.[2]

Figure 3b shows a perspective view of the surface of a sphere ('the earth'), with some circles of constant latitude and some meridians inscribed on its surface. The location of the point P can be precisely described by these common coordinates. Let us use the pole distance[3] Θ (instead of its complement, the latitude) and an angle Ψ from a standard meridian. It is necessary, for arriving at a quantitative measure of the location, that the 'North pole' and the standard meridian be defined. We have chosen to put the North pole at the point where the Z-axis exits the sphere; and we have chosen for the standard meridian the great circle where the (X,Z)-plane intersects the sphere (arbitrary). Furthermore, we count 'longitudes' as increasing when you progress from +X toward +Y. (This is a valid definition even if you chose a left-handed coordinate system, which is sometimes advantageous. Figure 3b assumes a right-handed system.)

The angles Θ and Ψ are coordinates on the surface of a sphere. They could also have been introduced as a series of two specific rotations, such as: first by Ψ clockwise around positive Z, then by Θ clockwise around the new positive Y: these are the two 'Euler angles' for locations on the surface of a sphere.

These descriptions were based on a single point P: one of the two points of emergence of the *pole* to the plane. This single point could have been picked in two ways. If you have physical information to distinguish the sign of the two ends, you can

---

[1] We use {XYZ} for sample axes, {xyz} for crystal axes.
[2] This nomenclature is common, but not universal: WEISS & WENK [1985], as well as MATTHIES &AL. [1987], use 'pole' in the general sense (called 'direction' here) and even use the term 'true pole' for a unit vector; our *pole* is termed a 'reduced pole' by them.
[3] Since our crystal and sample axes were chosen especially to be z and Z, respectively, we used the angles {Ψ,Θ,ϕ} that will later refer to complete orientations. Pole distances and azimuths for *arbitrary* poles need more general labels (α and β in this book).

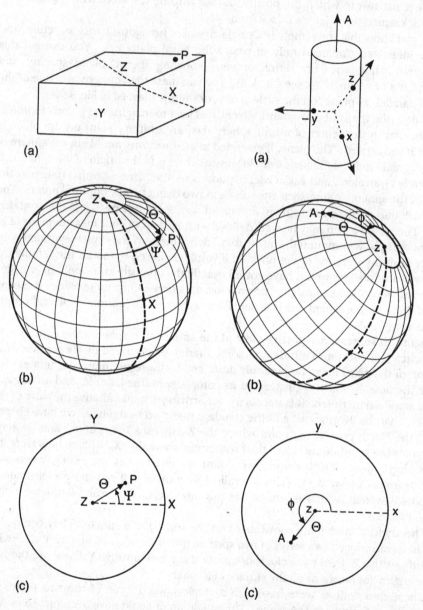

Fig. 3. Definition of a pole **P** normal to a crystallographic plane (dashed) in a sample whose faces have been labeled (with sign): (a) sketch of sample; (b) characterization of the direction **P** as a point on the unit sphere in spherical coordinates {Θ, Ψ} using the sample Z-axis as the N pole (the dashed great circle is a reference meridian); (c) the upper hemisphere of (b) mapped onto a circle.

Fig. 4. Definition of a sample axis **A** with respect to a frame {x,y,z} defined by crystal coordinates: (a) sketch of cylindrical sample with emergence of crystal directions; (b) characterization of the direction **A** as a point on the unit sphere in spherical coordinates {Θ, φ} using the crystal z-axis as the N pole; (c) the upper hemisphere of (b) mapped onto a circle.

plot the *unit vector* to **P** on the unit sphere, and **P** may lie anywhere on the sphere (i.e., Θ may go up to π). On the other hand, if you want to characterize only the (unsigned) pole of the basal planes, you pick one of the two hemispheres and plot only that end of the pole which emerges on this hemisphere. We will usually pick the 'upper' or 'North' hemisphere (geologists often pick the 'lower' or 'South' hemisphere; in either case, one must make provision for the poles intersecting the equator). A hemisphere can be mapped onto the inside of a circle: Fig. 3c describes the location of the *pole* **P** with respect to the coordinate frame **X,Y,Z**.

The representation of directions on the surface of a unit sphere is especially convenient when you have many crystals in one polycrystalline sample: their relative position is immediately evident, and the local density of points reflects exactly the number of grains in a given interval of directions: the *distribution*.

As a second example, assume you have a set of *cylindrical samples* (which you want to subject to tension, for example). In this case, again, there is only one axis of interest; label it **A**. Again, there is no physical basis to associate a *sign* with this axis. (We will use the term *axis* in the sense of an *unsigned* direction, much like pole, only that pole implies the normal to a set of planes). Of course you can always mark one side positive (for example, the one that will physically move in your laboratory, while the other one stays fixed), but this makes no sense in terms of the physics of the process.

The samples are all single crystals, and what you will want to know is the relation between the tensile axis and the crystallographic axes. Now you must determine *three* crystal axes, not just ⟨c⟩. Then you will use *them* as reference axes and plot the tensile axis **A** *with respect* to the crystal axes **x,y,z**. Again, it is convenient to think of the sample as located in the center of a (large) unit sphere, but now aligned with the *crystal* reference axes: Figs. 4a through c. To do this unambiguously, you must, in this case, assign the 'positive' points of emergence of the reference axes on each sample. (This is arbitrary, but always possible; one must preserve the handedness in the set of crystal axes in each sample.) Now the 'North pole' of the unit sphere is associated with the *crystal* **z**-direction. Figure 4b is arranged so that it coincides with the **P** pole from before, and so that the present **A** axis coincides with the previous sample **Z**-axis.

A new coordinate net was inscribed on the sphere, with its 'North pole' at the point *z* (the emergence of the axis **z**). Obviously, the angle Θ is the same as before: it describes the distance between **Z** and **z**; but the azimuth will depend on where you put the crystal **x**-axis, and therefore the standard meridian. For this reason, they were labeled differently (Ψ before, ϕ now), in preparation for the coming description of the full orientation relationship between the two frames {**x,y,z**} and {**X,Y,Z**}.

In summary, for some applications, one direction is all you need to specify, with respect to some coordinate system: one sample direction with respect to crystal coordinates, or one crystal direction with respect to sample coordinates. The direction may be *signed*, in which case it is, in standard parlance, a 'unit vector'. If it is *unsigned*, as in most applications to be dealt with in this book, it is often called an 'axis', and we shall use this term for an unsigned direction in a *sample*.[4] The word

---

[4] Again, the terminology is not without its problems: the word axis is also used commonly, and by us, in the combination 'sample axes', 'crystal axes': the three unit vectors making up a coordinate system; e.g., 'the X-axis', a signed direction.

'pole' is used, in this book, in the sense of an *unsigned* normal to a set of lattice planes (which you measure with a pole figure goniometer).

You need two numbers to specify a direction; they are most conveniently chosen as the coordinates on the surface of: a sphere, if they are signed directions; a hemisphere if they are unsigned. The latter may be mapped onto the area of a circle (the former onto *two* circles).

1.3 Stereographic and equal-area mapping

There are many methods to map the surface of a hemisphere onto a plane for representation purposes: a problem familiar to geographers. The mapping most familiar to metallurgists is the 'stereographic projection'. It is shown in Fig. 5, with meridians (great circles, rays) and latitudes (small circles) drawn. The 'North pole' is chosen to be in the center in Fig. 5a, on the periphery in Fig. 5b. The latter is the familiar Wulff net, which is typically used to determine the angle between two locations by connecting them, after an appropriate rotation of the net, along a great circle.

Another mapping is the 'equal-area mapping' or 'Lambert projection', displayed in a similar way in Figs. 5c and 5d. It is used universally in structural geology. Figure 5d is called the Schmidt net; it can be used in the same way as the Wulff net to connect data points by great circles. Note that, although the lines in the Schmidt net are not actual circles, they do represent the great circles and small circles on the surface of the sphere.

A set of points that is random on the surface of the hemisphere has been included in Figs. 5a and 5c: it appears random and of uniform density in the equal-area projection; in the stereographic projection, on the other hand, it is sparse at the periphery. This difference is a strong motivation to use equal-area projections to visualize actual orientation distributions; i.e., their departure from randomness. For this reason, it will be used in this book where we do not present the work of others. Note that the areas of the net units are not the same throughout the projection: it is only that their size represents correctly the fraction of orientation space covered by each.

On the other hand, it is true that the stereographic projection has its own advantages. The defining one, and the reason why it is so often used, is that angles *between* great circles are preserved. We demonstrate this by having drawn into one quadrant of Figs. 5b and 5d the well-known symmetry lines appropriate to cubic crystals. The 3-fold symmetry of the $\langle 111 \rangle$ direction (in the center of the quadrant) is evident by 60° angles in the stereographic projection, but not on an equal-area net. Note that, even though the angle between two great circles is not evident to the eye in the equal-area projection, it is still measurable in the standard way in either projection: find the pole for each great circle, line up the two poles on a great circle, and measure the distance on this great circle (by the number of intersections with small circles).

The fact that circles on the hemisphere remain circles in the stereographic projection is often quoted as an advantage; however, circles of the same diameter on the hemisphere (marking, e.g., a scatter radius) have different radii on the projection, depending on the distance from the pole, and do not preserve the location of their center – essentially destroying the supposed advantage. In the equal-area projection,

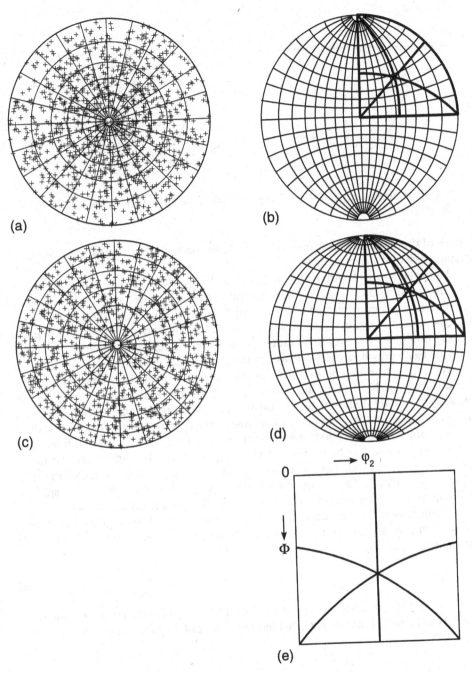

Fig. 5. Stereographic polar net (a) and equal-area polar net (c), with a set of points randomly distributed on the surface of the unit sphere. (b) and (d) display the respective nets with the N pole on top; they also contain (heavy lines) the borders of the 'crystallographic triangles' for cubic crystals in the first quadrant. Fig.4(e) shows the first quadrant projected onto a square, with the same 'crystallographic triangles' for cubic crystals.

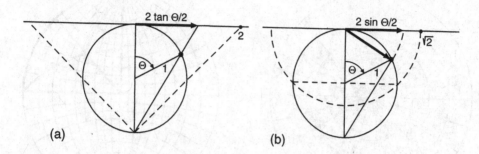

Fig. 6. Principles and formulas for stereographic (a) and equal-area mapping (b) of a hemisphere.

circles of the same diameter on the hemisphere convert into ovals of the same area (though different aspect ratios).

Figure 5e shows another way to describe the quadrant that was highlighted in Figs. 5b and d: in a cartesian, rather than polar frame, with the two angles Φ and φ_2 (equivalent to Θ and ϕ) plotted linearly along the axes. In this way, the 'North pole' of Figs. 5a through d gets extended into a line (which is therefore degenerate). The point density (not displayed) would be quite sparse in the top region. The cubic symmetry lines are again drawn in: the 3-fold axis in the center is not evident either.

We conclude that whereas for crystallographic analysis the stereographic projection will probably remain the standard, for the description of orientation distributions, the standard should be the polar equal-area projection. It also allows an immediate quantitative assessment of volume fractions 'near' a particular orientation.

To complete this section, we give the formulas for the two polar projections, within Figs. 6a and b, which demonstrate the principles. In both cases, the upper hemisphere is projected onto a plane tangent to the top.[5] The radius of the total projection is different for the two cases (both are larger than the radius of the sphere); dividing by this radius yields a projection formula that goes to unity as $\Theta=90°$.

If one calls ρ the fractional distance from the origin of the equal-area net to a point, then one can express the size of the infinitesimal area element, with the formula given in the figure, by

$$\rho\ d\phi\cdot d\rho = \tfrac{1}{2}\ d\phi\ d\left(\sqrt{2}\sin\tfrac{\Theta}{2}\right)^2 = \tfrac{1}{2}\ d\phi\ \sin\Theta\ d\Theta \tag{2}$$

Dividing this by π (the area of the unit circle), the *fractional* area element is the same as that on the sphere, $d\phi\ \sin\Theta\ d\Theta$, divided by the area of the hemisphere, 2π.

---

[5] While the stereographic mapping can be seen as an actual 'projection', with a light at the South pole, the equal-area mapping is called a 'projection' in a more abstract sense. The term 'mapping' should be preferred, also because of the frequent (and proper) use of the term 'projection' for a partial representation of truly three-dimensional information (in orientation space) in various two-dimensional views.

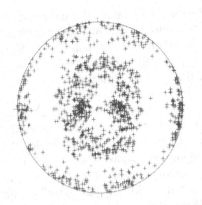

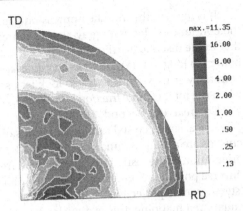

Fig. 7. Pole figure for **c**-axes of rolled Zr, simulated. The different size of symbols represents the 'weight' of the grain which represented the initial texture. RD on right.

Fig. 8. Same data as Fig. 7, but symmetrized and converted into a continuous distribution by 'binning' on a 5°×5° net. Equal-area map.

1.4 Pole figures and inverse pole figures for sets of single crystals

Let us now describe the relation between a whole set of single crystals and their crystallographic axes. First, we shall again treat the case (as in Fig. 3a) where the single crystals are 'bricks', with three faces assigned to +X, +Y, +Z, and all we want to know is the direction of the crystallographic **c**-pole in each. If you put all the crystals side by side, with their X, Y, and Z faces pointing the same way,[6] you can then plot all the **c**-poles as points on the 'pole figure'. Figure 7 shows an example. (Note that we have specialized on a pole that has no multiplicity caused by crystal symmetry.)

Incidentally, the particular points shown in Fig. 7 display a statistical symmetry: one can see that both a vertical and a horizontal mirror plane would not change the essence of the information. (We have shown in Chapter 1 that two perpendicular mirror planes imply a third, so that this is *orthotropic* sample symmetry.) If one knew nothing about the history of these samples, one could still guess that they represent grains in a rolled sheet. In fact, they came from a simulation of rolling in Zr.

The 'points' of the **c**-axes are marked by crosses. One disadvantage of this discrete representation is that, in high-density areas, the individual symbols overlap too much and then seem under-represented.

One may employ a scheme whereby the discrete points in Fig. 7 are assigned, with their respective intrinsic weights, to 'boxes' or 'bins' of the kind shown in the equal-area net of Fig. 5b. Each bin then has an 'intensity' or 'pole density' assigned to it. One may smooth these intensities and finally plot a (more-or-less) *continuous* distribution of poles. In Fig. 8, this is done with the additional assumption that the orthotropic symmetry mentioned above reflects actual physical facts, so that one can average the four quadrants. This gives rise to a smoother description, which also has higher resolution if plotted in the same size.

---

[6] You can only accomplish this feat if you chose the X, Y, Z markings of each crystal in the same *handedness*; usually one picks them in a right-handed sequence. Note, however, that in Figs. 7 and 8, we used a left-handed coordinate system: because we plotted the RD up, and still labeled it 1. (Alternatively, one may think of the positive 3-direction pointing *into* the paper.)

It is noteworthy that the impression gained by the eye of the relative intensities in the continuous plot, with gray-shades representing densities (Fig. 8) is different from that in the discrete distribution (Fig. 7): it is the continuous one that conveys the facts better (so long as it is plotted in an equal-area mapping). Even here, a smoother peak with a lower maximum tends to look less severe than a sharper narrower one – even when the total *volume fraction* is the same.

To treat our second example, let us now plot the external axes in a set of cylindrical single crystal samples, in terms of their crystal axes (as in Fig. 4). When the crystals are actually grains in a polycrystal, this procedure corresponds to plotting the orientation of the testing machine as seen from the local crystal axes. This is called an 'inverse pole figure' in texture research. Metallurgists are used to calling this merely a 'stereographic projection' of the tensile axis. Inasmuch as one may just as well use an equal-area mapping, this is a misleading term. What is meant is a 'sample axis projection', into crystal coordinates.

Since the crystal symmetry is known to start with, one need not plot the whole circle for the inverse pole figure. Figure 9 represents a typical texture in wire-drawn copper (by discrete, weighted grains, see Sec. 3.4), which has a face-centered cubic lattice structure. We plotted just one quadrant (and could, in fact, have plotted just one of the triangles in Fig. 5b).

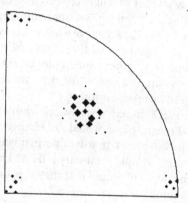

Fig. 9. A typical texture of wire-drawn Cu, described in an inverse pole figure, by weighted grains.

This texture can obviously be described as a superposition of two components: one around ⟨111⟩, one around ⟨001⟩. The ratio of the two components can be determined quantitatively; the only question is: how do you define 'the component'? Obviously, the breadth of the distribution may be different in different cases, and it may be different in different directions. For Fig. 9, the total volume fraction of what we judge to be the ⟨111⟩ component is about 71%, that of the ⟨100⟩ component about 24%.

The ratio of the components can also be quantitatively assessed on a continuous distribution: when it is plotted on an equal-area mapping, the density at each point directly represents the local volume fraction. One can add these values up over a region of interest, and one can pick the extent of this region by judgment on the particular case.

Note that this inverse pole figure is a very good way to represent the salient features. If one had chosen a pole figure, one would have merely seen rings, in accordance with the fiber symmetry of the sample.

What we have described in this example is the direction of an axis (the tensile axis) with respect to a crystallographic coordinate system. Often this is referred to as the 'orientation' of the tensile axis. We shall use the word orientation, in this book, only when it is implied that the entire relation between two coordinate frames is to be described (which requires three rather than two numbers); this will be treated in the next Section. We have also shown 'texture' information – but only for cases where a single direction was of interest. Complete textures refer to a preferred set of complete *orientations*; they will be dealt with in Section 3.

1.5 Contouring and normalization

The distribution of some parameter (such as a pole density) over an area is most effectively displayed by shading the density levels in colors or gray shades. The classical methods of contouring require only *lines* for representation, which is sometimes preferable for reproduction; however, it is imperative that the contours be marked with values or distinguished by at least three different dashing patterns.

The density values for the different contours, or density intervals for the shades, need not be linearly spaced: *logarithmic* progression is sometimes more instructive, especially for strong textures and when details in the regions of *low* intensity are also of interest. It is of considerable help in comparisons of different plots if the contour values are chosen at some standard levels (rather than, say, always 10 contours between actual maximum and minimum values).

For all these representations, it is appropriate to *normalize* the densities. (This must be done with appropriate accounting for the size of the volume elements in orientation space, see Sec. 4.1.) There are two possibilities for normalization: to a value of 1.0 for the *average*; or to a value of 1.0 for the *integral*. The former is more common. Since a *random* distribution of discrete orientations corresponds (ideally) to a uniform distribution, the normalized, continuous densities in a textured sample are then expressed as 'multiples of random distribution (m.r.d.)'. This is not an accurate term, either scientifically or grammatically, but it is long ingrained (*sic*) in the field. 'Multiples of uniform density' (or of average density) would be better.

2. The Representation of Orientations

, The word *orientation* is used, in this book, only for a complete relation between two coordinate frames; when only a single axis is of interest in either frame, we call it a *direction* and have treated it in the previous Section. In simple cases, one often specifies an orientation by the Miller indices of *two* directions: the (normal to a) plane, and a line in it (Sec. 1.1); for example, rolling texture components are specified by, e.g., {211}<111>: meaning a {211} plane is parallel to the rolling plane, and a <111> direction is parallel to the rolling direction.

In general, a relation between two coordinate frames can be represented in various ways: that is the topic of this Section. In particular, the common method of specifying a *rotation* from one frame to another is only one special description. Whether the coordinate axes must be signed or not will be discussed in some detail

(Sec. 2.1). In general, one expresses either the axes of the crystal with respect to those of the sample ('crystal orientation') or vice versa ('sample orientation'). These expressions are entirely equivalent, and we will strive for descriptions that retain this equivalence. We will deal with single orientations, or small sets of them, and leave distributions of many orientations for Sections 3 and 4. Symmetries play a major role in descriptions of orientations; they were the topic of Chapter 1 and will be taken up again in Section 3 of the present Chapter.

2.1 The definition of an orientation by three poles

In Section 1, the description of one pole (i.e., an unsigned normal to a set of crystal lattice planes) on the surface of a unit sphere was introduced. Now we add two more poles, belonging to different sets of lattice planes (Fig. 10). The reference sample coordinate system consists of the 'North pole' and the 'standard meridian'. Let us take a triclinic crystal, for which the (a), (b), and (c) poles are all non-coplanar and non-orthogonal. (The principle of what is to be discussed is similar if one takes the (001), (011), and (111) poles in a cubic lattice – without their symmetric equivalents.)

Figure 10 shows such a triplet of poles, all going through the center of a unit sphere and intersecting its surface at two, diametrically opposed, positions each. The three poles are marked differently; this reflects the fact that they really can be operationally distinguished; e.g., in the order of increasing lattice spacings. This 'pole triplet' may be used to *define* a crystal orientation. As discussed in Chap. 1 Sec. 2.3, this is a *lattice* orientation: only the orientation of lattice planes is used. By analyzing the intensities of the diffraction peaks that give rise to the locations of the poles, the *Laue* orientation may be derived; however, our use of (unsigned) *poles* precludes the description of a 'structure orientation' (i.e., a rotation) that is different from a Laue orientation (Chap. 1 Sec. 2.3).

All the orientation information is contained in just three of the intersections with the surface of the unit sphere, one for each pole, and they can be picked in an arbitrary way from among the six. (It is important for this demonstration that the poles are not orthogonal to each other.) One possibility is to pick one hemisphere, say the 'upper'; this corresponds closely to the way the poles are measured in a diffractometer. Suppose one had picked the 'lower' hemisphere: it would exhibit a different pattern. If the triangle that lies entirely on the upper hemisphere had the labels a, b, c in clockwise order, the one on the lower hemisphere has them in counter-clockwise order. If one looks out from the center of the sphere, the vectors to the upper hemisphere form a 'left-handed' set, the ones to the lower a 'right-handed' set. This is so even though nothing was assumed about the presence or absence of a handedness built into the physical structure of the crystal; in fact, to know such a physical handedness would require additional measurements (Chaps. 1 and 3).

The 'standard' (upper) hemisphere in Fig. 10 contains, in our example, the left-handed set. For many descriptions, one prefers a right-handed set. Other orientations would indeed be described by a right-handed set on the upper hemisphere. Thus, one has two choices: either take one hemisphere and keep track of the 'handedness' of each triplet; or insist on a single handedness (say, right-handed triplets only) and have such triplets *anywhere* on the surface of the sphere. The two descriptions are entirely equivalent. The first is closer to the operational measurements; the second has the advantage that all orientation relations can be described by pure rotations ('proper'

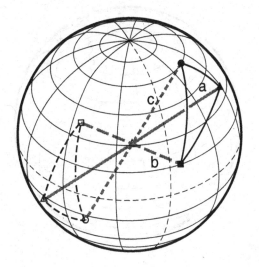

Fig. 10. Three (unsigned) poles define an orientation. The symbols and labels represent three distinguishable crystal poles. The N pole and principal meridian are determined by sample axes. The sample axes are signed and right-handed, the crystal poles are unsigned and a handedness cannot be defined. (It is the opposite on the lower hemisphere to the upper.)

rotations, without change of handedness), which is easier algebraically. The latter will be used almost exclusively in this treatise (as it is in all the literature). A single hemisphere will, nevertheless, usually suffice for the description, because of the presence of at least one symmetry element (other than the inversion center) in either the crystal or the sample, which can be strategically located on the equator; only a triclinic crystal, with no sample symmetry, requires the whole sphere.

Since the three poles in Fig. 10 are all unsigned poles, nothing changes if all directions are reversed. The fact that the descriptions of the pole triplet by a right-handed or a left-handed triangle on the surface of the unit sphere are entirely equivalent attests to the presence of an inversion center. If the crystal actually did not have an inversion center, its (Laue) *orientation* would still be completely described by the pole triplet. (See Chap. 1 Sec. 2.3)

One may assign a direction even to the unsigned poles in a triplet, e.g., by calling all those in the right-handed set positive; this is equivalent to the above recipe to always use right-handed triplets. Note that it is not possible to assign a positive direction to a *single* pole when the crystal possesses an inversion center. Thus, the notion of 'true' or 'unreduced' or 'colored' [BUNGE &AL. 1980] single poles (which we call unit vectors) is meaningful only when there is a physical distinction: an innate polarity or handedness.

Often, an orientation relationship between two frames is thought of as arising from *rotating* one frame into the other. This has led to occasional usage of the words 'orientation' and 'rotation' as synonymous. We will avoid this implication of a particular operation. When a distinction is necessary (as in crystals of the type shown in Chap. 1 Fig. 6), we will refer to the pole triplet in Fig. 10 as a 'Laue orientation' (Chapter 1). In the presence of symmetry elements, in either crystal or sample, one refers to different equivalent rotations as just one 'orientation' (Chap. 1 Sec. 2.3) [BUNGE 1982].

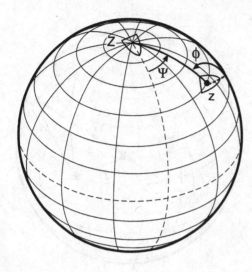

Fig. 11. Symmetric definition of Euler angles as spherical coordinates for a vector ('boat') on the surface of a unit sphere.

2.2 Description of an orientation by a vector lying in the surface of the unit sphere

If one direction is represented by a point on the surface of the unit sphere, a second direction, of fixed angular distance, can be represented in its azimuthal relation to the first, by a *vector* lying in the surface of the sphere, at the location of the first direction, and along the great circle to the second direction. A picturesque analogue is a boat on the surface of the earth (Fig. 11) [WENK & KOCKS 1987]: its *location* may represent the first direction (for which one commonly chooses the 'third' axis, Z in Fig. 11), its *heading* the azimuth toward the second direction (for which we will choose the 'first' axis, X). To complete the analogy, one could characterize the location of a third direction by a 'sail' on the boat: if it points to the left, you have a right-handed triplet, and vice versa. If one goes back to thinking of a vector instead of a sail boat, this *vector* could be, say, black or white depending on the side on which the third axis lies. In a description by right-handed systems only, there is only one kind of vector, and only 'row boats' (with no sail), as in Fig. 11.

In order to keep the crystal and sample frames in the same picture, Fig. 11 has a 'reference boat' placed at the 'North pole', heading down the standard meridian. It is apparent that either one of the boats could have been called the 'reference', the other the 'new'; correspondingly, the North pole and the grid could have been centered at either one.

Note that the equivalence of describing one frame in terms of the other, or the other in terms of the one is maintained in the angles defined in Fig. 11: ϕ or Ψ is, in each case, the angle between the heading and the connecting line – starting at the heading, and counting positive in the direction moving counter-clockwise toward the connecting line. (Θ, not shown, is the angle corresponding to the connection line between the two locations.) These are the 'symmetric Euler angles' introduced by KOCKS [1988].[7]

---

[7] Θ and Ψ are identical to those introduced by ROE [1965], but $\phi = -\Phi_{Roe}$; here lies the reason for our keeping capital Ψ but using lower-case ϕ. See Sec. 2.3.

Each of the angles φ and Ψ has a range of 2π. Either one of them can be combined with Θ to be represented by a point on the surface of the unit sphere, which then makes the range of {Θ, φ} or of {Θ,Ψ} equal an area of 4π: in either case the total volume of orientation space is $2 \times 2\pi \times 2\pi = 8\pi^2$. (If one chose only one hemisphere and kept track of the handedness of pole triangles, the total volume would still be $8\pi^2$.)

The three-dimensional orientation space can now be displayed by a set of *sections*:[8] either the pair {Θ,Ψ} or the pair {Θ, φ}, describing the crystal z-direction or the sample Z-direction, respectively, is mapped onto a circle (when a hemisphere suffices under the given symmetry conditions, or two circles when it doesn't); the third angle is then represented in the third dimension – which may be plotted linearly in a direction perpendicular to the circle(s). The average of the sections is a projection: an 001 pole figure in the first case, an inverse pole figure for the Z-direction in the second.

An early precursor to this description was that by VAN HOUTTE and AERNOUDT [1976], who plotted the rolling plane normal on the cubic unit triangle and added a flag to each point in the direction of the rolling direction: Fig. 12. Note that the 'flags' or 'boat headings' work only for individual orientations; if you want to represent a *distribution*, you may separate this figure into a number of triangles, each for a certain range of flag headings. Such a description was first introduced by WILLIAMS [1968b] and again later, with some differences in detail, by POSPIECH and LÜCKE [1975] and VADON &AL. [1980].) Representations of the kind shown in Fig. 12 are often useful in detailed modeling analyses (Chapter 9).

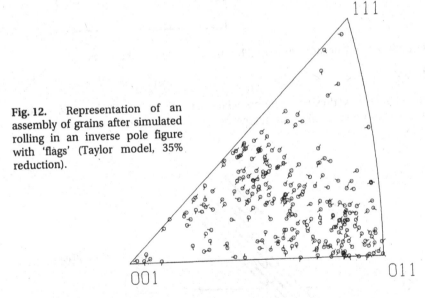

Fig. 12. Representation of an assembly of grains after simulated rolling in an inverse pole figure with 'flags' (Taylor model, 35% reduction).

The quantification of orientations by the standard polar coordinates, on the surface of the sphere, implies a singularity at the point chosen as the 'North pole': the longitude is here undefined. (This singularity was avoided in the descriptions of MORRIS & HECKLER [1968], WILLIAMS [1968b] and VADON &AL. [1980].) In the picture of

---

[8] Sections at constant Θ have been used primarily in the description of *misorientations* (see Sec. 2.6); they may reasonably be plotted in a *square* coordinate frame.

two boats (Fig. 11), it is clear that only their *relative* heading matters; in other words, the *difference* between Ψ and φ. (The parameter β introduced by WILLIAMS [1968b] goes to φ-Ψ at the origin.) Other parameters for this description were introduced by HELMING &AL.[1988] and BUNGE [1988]. In terms of the symmetric Euler angles Ψ and φ, the necessary parameter is

$$\nu \equiv (\Psi - \phi)/2$$

(3)

When ν=0, the 'reference' and 'rotated' boats are antiparallel; when ν=±π/2, they are parallel and have the 'same course' (Fig. 11). Figure 13 shows a number of parallel boats near the reference boat. When all are moved closer and closer to the reference boat, the angle ν approaches the same value for all of them.

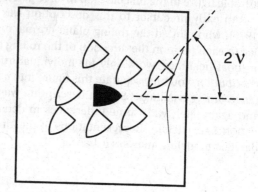

Fig. 13. A number of orientations ('boats') near the N pole (black boat).

Equation 3 may be supplemented by the introduction of an average

$$\mu \equiv (\Psi + \phi)/2$$

(4)

Figure 14 shows how these parameters relate in the Ψ-φ plane: it shows why these may be called 'oblique sections'. All angles vary between -π and +π, although ν may be

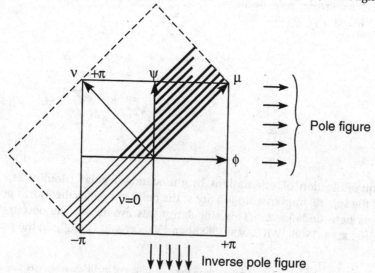

Fig. 14. Irreducible ranges in oblique sections of orientation space.

restricted, without loss of generality, to the range 0 to $+\pi$: the dashed box is obtained from the main regime by $\pm2\pi$-changes in Ψ or ϕ. (See also MATTHIES &AL. [1990a].)

There is, however, an advantage in *not* using μ for the second parameter: if one keeps Ψ, the projection remains a pole figure; if one keeps ϕ, the projection remains the inverse pole figure. Thus we find that the two representations of most direct intuitive appeal, and without any singularities or distortions, are *equal-area polar plots of Θ and Ψ or of Θ and ϕ, both in sections at constant ν*. Examples will be found in Section 3.

2.3 Cartesian frames. Rotations

So far, only the most general case of a crystal structure was discussed, where there are no natural right angles. It is, of course, easy to introduce an orthogonal coordinate system even here. Commonly this is done by defining z (and the 'North pole') as the normal to the plane containing both $\langle a \rangle$ and $\langle b \rangle$ and keeping y parallel to $\langle b \rangle$; x is on the great circle perpendicular to y (the 'zero meridian'), 90° away from z. This convention has the advantage of leaving the Euler angles defined as before and, in addition, providing a cartesian frame for the crystal as well as for the sample. For example, a description of the sample frame in terms of the crystal frame is considerably simplified when the reference frame is cartesian (and right-handed).

Once one has two orthogonal coordinate systems (of the same handedness), one could also describe their relation (not including translations) by transforming one into the other by a sequence of *rotations* (rather than by a set of spherical *coordinates* as was done above). The rotation method has been preferred in the literature: some [BUNGE 1965] rotate the sample frame into the crystal frame, others [ROE 1965] rotate the crystal frame into the sample frame. All choose a sequence of rotations about the respective x, y, and z axes, but in different sequences. All choose the successive rotations to be of the same sign – and this introduces an asymmetry into the description that was avoided by using the 'symmetric Euler angles' defined in Fig. 11 [KOCKS 1988]. Of course, all descriptions are equivalent, and Fig. 15 shows them in parallel; the algebraic relations between the symmetric angles and the two most commonly used ones can be read from Table Ia. For the case that one also has *cubic crystal symmetry and orthotropic sample symmetry*, one can keep all angles in the range 0° to 90° (and even then have a 3-fold multiplicity); for this case, one can also use Table Ib.

Table Ia. Comparison of Euler angle definitions.

	Ψ	Θ	ϕ
symmetric	Ψ	Θ	ϕ
Bunge	$\varphi_1 - \pi/2$	Φ	$\pi/2 - \varphi_2$
Roe (Matthies)	Ψ (α)	Θ (β)	$\pi - \Phi$ $(\pi - \gamma)$

Table Ib: Euler angles for cubic crystal symmetry and orthotropic sample symmetry.

	Ψ	Θ	ϕ
symmetric	Ψ	Θ	ϕ
Bunge	$\pi/2 - \varphi_1$	Φ	φ_2
Roe (Matthies)	Ψ (α)	Θ (β)	$\pi/2 - \Phi$ $(\pi/2 - \gamma)$

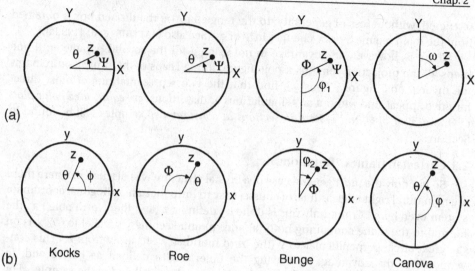

Fig. 15. Comparison of definitions of Euler angles, shown as spherical coordinates in (a) the sample system, (b) the crystal system.

Finally, the 'oblique section' parameters introduced by BUNGE [1988] and HELMING &AL.[1988] relate to the symmetric ones as:

$$\varphi^+ = \sigma = \nu + \frac{\pi}{2} \qquad \varphi^- = \delta + \frac{\pi}{2} = \mu \tag{5}$$

For a single orientation, the two basic (non-oblique) representations are shown in Fig. 16, in the complete orientation space (albeit with many empty sections replaced by dots). Note that in the symmetric definition of the Euler angles the first 90° of each occur in the quadrant limited by the *positive* first and second axes. One of these representations plots the crystal axes in terms of the sample axes, the other the sample axes in terms of the crystal axes. When there are many points, we call the former the 'crystal orientation distribution (COD)' (in term of sample axes X,Y,Z), the other the

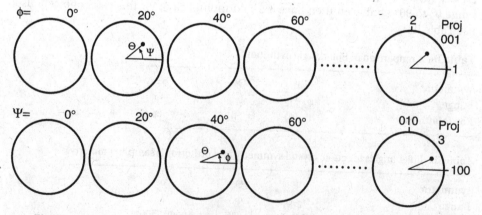

Fig. 16. One orientation shown in both COD and SOD sections of complete orientation space. {Ψ,Θ, φ} = {45°,60°,25°}. Only the first few sections are shown (in rather coarse spacing). The projections (i.e., the average of the sections) are the 001 pole figure and the Z-axis inverse pole figure, respectively.

'sample orientation distribution (SOD)' (in terms of crystal axes **x,y,z**). They are completely equivalent, but one or the other may be the more convenient representation of the same data in any particular case.

In addition, the projection is displayed for each space: the crystal orientation projection ('COP', in sample coordinates) is a 'pole figure' (namely, for the {001} poles), the sample orientation projection ('SOP', in crystal coordinates) is an 'inverse pole figure' (for the Z-axis). Thus, COD sections may be called 'partial pole figures' [WENK & KOCKS 1987] (namely only those within a certain range of ϕ), and SOD sections 'partial inverse pole figures' (namely only those within a certain range of Ψ).

2.4 Matrix representation

For practical manipulations of orientations, the most convenient representation is by an orientation matrix. For example, one can easily apply symmetry operations (Chap. 1 Sec. 2.2), or calculate the change in orientation after a finite step of plastic strain. One form of the orientation matrix is (in symmetric Euler angles)

$$\mathbf{g} = \begin{pmatrix} -\sin\phi \ \sin\Psi - \cos\phi \ \cos\Psi\cos\Theta & \sin\phi \ \cos\Psi - \cos\phi \ \sin\Psi \ \cos\Theta & \cos\phi \ \sin\Theta \\ \cos\phi \ \sin\Psi - \sin\phi \ \cos\Psi\cos\Theta & -\cos\phi \ \cos\Psi - \sin\phi \ \sin\Psi \ \cos\Theta & \sin\phi \ \sin\Theta \\ \cos\Psi \ \sin\Theta & \sin\Psi \ \sin\Theta & \cos\Theta \end{pmatrix} \quad (6)^9$$

The way Eq. 6 is written, it describes the crystal axes (*rows*) expressed in terms of the sample axes; or equivalently, a rotation *from* the sample system *to* the crystal system [BUNGE 1982, Eq. 2.10]. The reverse change is its transpose: the *columns* of **g** are the base vectors of the sample system in terms of the crystal system. An advantage of our particular, 'symmetric' choice of Euler angles {Ψ,Θ, ϕ} is that transposing **g** is equivalent to interchanging Ψ and ϕ. Using the matrix of the type of Eq. 6, rather than its transpose, to describe an orientation has the advantage that it immediately translates a property that is fixed in crystal coordinates into sample coordinates for averaging (see Chapter 7).

Note that the apparent special role of the Z-axis in Eq. 6 is due only to the preference given to Z in the Euler-angle description. The orientation matrix itself does *not* prefer any axis: it is a completely neutral description. We shall use **g** as the general symbol for (crystal) 'orientation', regardless of how it is represented.

A drawback of describing an orientation by a matrix is that it has nine components, only three of which are independent. There is no obvious way to choose three components of the matrix and have it represent 'the orientation' by a point in a three-dimensional space – so that many points correspond to a distribution. Also, a random or uniform distribution cannot be expressed in matrix form, so that deviations from randomness ('texture') cannot be visualized. In summary, the mathematically most elegant description of orientations in terms of matrices is useful in practice only for algebraic manipulations on discrete orientations.

For plotting after matrix manipulations, one wants to determine the Euler angles from the matrix. It is at this point of converting a **g**-matrix to Euler angles that it is crucial to have a single handedness convention; e.g., such that the determinant of **g** is always +1. In addition, one must find a way to maintain the correct quadrant.

---

[9] To write the orientation matrix in terms of Bunge or Roe angles, use the conversion rules for the angles, Table Ia.

2.5 Symmetrically equivalent orientations

The symmetry elements of a certain group can be expressed as operators, which transform any $\mathbf{g}$ (Eq. 6) into a symmetrically equivalent one. In Chapter 1, Table II, we listed these operators. Note that, the way $\mathbf{g}$ was written in Eq. 6, *pre*-multiplication with an operator applies a *crystal* symmetry, *post*-multiplication a *sample* symmetry. (Of course, the converse would hold, if the transpose of $\mathbf{g}$ were used to define the orientation.)

Matrix operations are the most straightforward tool to derive symmetrical equivalence, when one has discrete orientations. On the other hand, for continuous distributions (which we will deal with later) one needs to be able to transform Euler angles from one equivalent set to another. MATTHIES &AL. [1987] list all possibilities. Table II lists a few for symmetric and for Bunge angles.

Table II. The influence of symmetry elements on Euler angles.

	†	Ψ	Θ	ϕ	φ_1	Φ	φ_2
Identity (to keep $\Theta \geq 0$)		$\Psi \pm \pi$	$-\Theta$	$\phi \pm \pi$	$\varphi_1 \pm \pi$	$-\Phi$	$\varphi_2 \pm \pi$
(to keep $|\Psi| < \pi$)		$\Psi \pm 2\pi$	Θ	ϕ	$\varphi_1 \pm 2\pi$	Φ	φ_2
(to keep $|\phi| < \pi$)		Ψ	Θ	$\phi \pm \pi 2$	φ_1	Φ	$\varphi_2 \pm 2\pi$
2-fold axis on sample X		$-\Psi$	$\pi - \Theta$	$\phi \pm \pi$	$\pi - \varphi_1$	$\pi - \Phi$	$\varphi_2 \pm \pi$
2-fold axis on sample Y		$\pi - \Psi$	$\pi - \Theta$	$\phi \pm \pi$	$-\varphi_1$	$\pi - \Phi$	$\varphi_2 \pm \pi$
2-fold axis on sample Z		$\Psi \pm \pi$	Θ	ϕ	$\varphi_1 \pm \pi$	Φ	φ_2
2-fold axis on crystal x		$\Psi \pm \pi$	$\pi - \Theta$	$-\phi$	$\varphi_1 \pm \pi$	$\pi - \Phi$	$\pi - \varphi_2$
2-fold axis on crystal y		$\Psi \pm \pi$	$\pi - \Theta$	$\pi - \phi$	$\varphi_1 \pm \pi$	$\pi - \Phi$	$-\varphi_2$
2-fold axis on crystal z		Ψ	Θ	$\phi \pm \pi$	φ_1	Φ	$\varphi_2 \pm \pi$
3-fold axis on crystal z		Ψ	Θ	$\phi \pm 2\pi/3$	φ_1	Φ	$\varphi_2 \pm 2\pi/3$
4-fold axis on crystal z		Ψ	Θ	$\phi \pm \pi/2$	φ_1	Φ	$\varphi_2 \pm \pi/2$
6-fold axis on crystal z		Ψ	Θ	$\phi \pm \pi/3$	φ_1	Φ	$\varphi_2 \pm \pi/3$

† This column holds true also for Roe (or Matthies) angles, with ϕ replaced by Φ (or γ).

A question of particular interest in regard to symmetrically equivalent orientations is introduced in Fig. 17(a). Two single crystal ('sample') orientations A and B are indicated by their Z-axis (heavy dot); the direction to the sample X-axis (the 'heading of the boat') is shown as the long flag. These two orientations might seem to be symmetrically equivalent in a cubic crystal, but they are not, because of an inherent assumption made in plotting the symbol. This is demonstrated in Fig. 17(b). Given a sample A, the mirror plane perpendicular to [1$\bar{1}$0] causes an equivalent pole Z at the point provisionally labeled B', and an equivalent 'heading' toward X. However, a short flag is also shown, which is meant to point toward Y (the 'sail of the boat'). It is seen that the pole triplet at A is right-handed, the pole triplet at B' is left-handed. This is a natural consequence of our having used a mirror plane. We can get from A to B' also in another way: by first *rotating* A around [1$\bar{1}$0] into C. Its pole emergence on the 'lower' hemisphere is shown as an open symbol. The rotation, of course, did not change the handedness of the triplet. But the *same* pole triplet emerges on the 'upper' hemisphere as left-handed (Section 2.1). Thus, the sample orientations C and B' in Fig. 17(b) are *identical*, and A and B' are *equivalent* due to the presence of the two-fold axis at [1$\bar{1}$0]. However, in Fig. 17(a) the 'sails' at points A and B were not shown, and the implicit assumption would be that both were described in right-handed co-

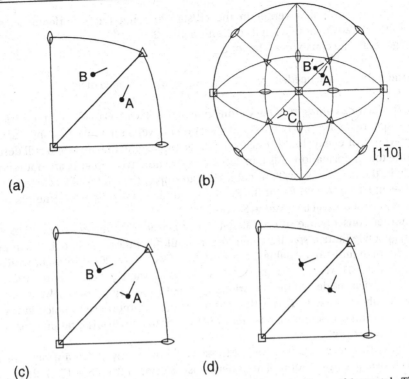

(a)   (b)

(c)   (d)

Fig. 17. Two orientations shown in a stereographic projection for a cubic crystal. The location indicates the sample axis Z, the flag the direction toward X. In (a), A and B are mirrored on the [001]-[111] line. (b) maps the whole sphere, with open symbols indicating location on the lower hemisphere; this demonstrates that A and B are equivalent only if B on the upper hemisphere is a left-handed triad (labeled B'). In (c), a short flag indicates the direction of Y; if A and B are both right-handed sample coordinate systems, these two orientations are not equivalent – unless (d) there is also a sample mirror perpendicular to the Y-direction.

ordinate systems. This is explicitly shown in Fig. 17(c): these two orientations are not equivalent by any symmetry element inherent in the crystal.

There are conditions under which A and B do become equivalent. For example, if the *sample* had a mirror plane in the Z-X plane, a 'left sail' and a 'right sail' would have to be equivalent (Fig. 17d). This is an illustrative example of how crystal symmetries and sample symmetries combine to reduce the total number of non-equivalent orientations. The 'sample' symmetry here is not meant to be that of a single crystal (under a certain test symmetry), but that of *one grain in a polycrystal*, which has an 'equivalent' partner elsewhere, under a given symmetry of the microstructure or the test.

An important effect of symmetry is that it reduces the extent of orientation space needed to represent an orientation uniquely. In general, one may divide the total volume of orientation space, $8\pi^2$ (Sec. 2.2), by the number of symmetry elements. In Chapter 1, Table I, the total number of symmetry elements was listed for all the rotation groups and for all the Laue groups (save the inversion center). Note that the presence of an inversion center in the crystal does not reduce orientation space further: it is already contained in its definition (Sec. 2.2).

In general, one can express the total

$$\text{(orientation space needed)} \ = \ 2 \ \times \ \text{(range of } \phi) \ \times \ \text{(range of } \Psi) \tag{7}$$

where the factor 2 comes from combining one of ϕ or Ψ with Θ to plot on the surface of the unit sphere (area 4π). There is an interaction between the range of ϕ and the range Ψ: sometimes, one may double one in exchange for halving the other and still describe the same total information. Also, in some cases, symmetry elements are not reflected directly in the range of ϕ or Ψ; in cubics, for example, the three-fold axes reduce the space given by Eq. 7 by a factor 3. Recommendations for different combinations of symmetry elements will be given in Section 3.5.

Another consequence of crystal symmetry is evident directly in pole figures: at one Bragg angle (lattice spacing), one obtains multiple poles; e.g. in cubic symmetry, there are three indistinguishable poles for (100), (010), and (001) – in other words, all three poles of {001} type. If one measures {111} poles, there are four. Either measurement is sufficient to determine the orientation of a cubic crystal uniquely. Moreover, they are easily visualized. This is the reason for the common use of pole figures (not three-dimensional orientation distributions) to characterize orientations (or 'components' of a texture).

Finally, the question must be addressed of how many poles are necessary to determine a single orientation. The general case is evident from Fig. 17(c): three poles are needed (or one type of pole with a multiplicity of 3). However, as is evident from a generalization of Fig. 17d, two poles suffice whenever either the crystal symmetry or the sample symmetry have a mirror plane containing both poles (or a two-fold axis perpendicular to this plane). For a general discussion, see, e.g., HELMING [1992] and Chapter 4.

We shall discuss appropriate representations of *many* orientations, or a continuous distribution of them, in Section 3.

2.6 The representation of misorientations[†]

Often the difference in orientation between two neighboring grains is of interest. Given two grains of orientation $\mathbf{g}_A$ and $\mathbf{g}_B$, the rotation required to bring the crystal lattice of grain A into coincidence with the crystal lattice of grain B is given by:

$$\Delta \mathbf{g}_{AB} = \mathbf{g}_B \, \mathbf{g}_A^{-1} \tag{8}$$

If the orthogonal matrix representation introduced in Section 2.4 is used to represent the orientations $\mathbf{g}_A$ and $\mathbf{g}_B$, then the *mis*orientation (also referred to as the *disorient*ation or orientation difference) is simply the product of $\mathbf{g}_B$ and the transpose of $\mathbf{g}_A$ (i.e. $\mathbf{g}_A^{-1} = \mathbf{g}_A^{\mathsf{T}}$). It should be noted that when $\mathbf{g}_A$ is simply the identity then $\Delta \mathbf{g}$ represents the orientation relative to the reference frame. While the representations described here can be used for orientations, they are much more appropriate for misorientations. (For an overview of such representations and their conversions, see HANSEN &AL. [1978].)

[†] Contributed by S. I. Wright.

Axis/angle description

As with orientations, the misorientation Δg can also be described in terms of three Euler angles. However, it is common to describe the misorientation in terms of a rotation axis, $\mathbf{n}$, and an angle, ω [IBE & LÜCKE 1972, POSPIECH 1972, HANSEN &AL. 1978]. This is often termed the axis/angle representation. The axis $\mathbf{n}$ represents a crystallographic direction common to both crystal lattices. The angle of rotation is the rotation about $\mathbf{n}$ required to bring the two crystal lattices into coincidence. If the axis $\mathbf{n}$ is normalized (i.e. $n_1^2 + n_2^2 + n_3^2 = 1$, where n_i represent direction cosines), then the matrix representation can be expressed as follows:

$$\Delta g = \begin{pmatrix} \left(1-n_1^2\right)\cos\omega + n_1^2 & n_1 n_2 (1-\cos\omega) + n_3 \sin\omega & n_1 n_3 (1-\cos\omega) - n_2 \sin\omega \\ n_2 n_1 (1-\cos\omega) - n_3 \sin\omega & \left(1-n_2^2\right)\cos\omega + n_2^2 & n_2 n_3 (1-\cos\omega) + n_1 \sin\omega \\ n_3 n_1 (1-\cos\omega) + n_2 \sin\omega & n_3 n_2 (1-\cos\omega) - n_1 \sin\omega & \left(1-n_3^2\right)\cos\omega + n_3^2 \end{pmatrix} \quad (9a)$$

Thus, if the misorientation is given in the form of a rotation matrix then the angle of rotation is given by:

$$\cos\omega = \tfrac{1}{2}\left(\Delta g_{11} + \Delta g_{22} + \Delta g_{33} - 1\right) \quad (9b)$$

The axis of rotation can be derived from the matrix representation as:

$$\{n_1, n_2, n_3\} = \frac{\{\Delta g_{23} - \Delta g_{32},\; \Delta g_{31} - \Delta g_{13},\; \Delta g_{12} - \Delta g_{21}\}}{\sqrt{\left(\Delta g_{23} - \Delta g_{32}\right)^2 + \left(\Delta g_{31} - \Delta g_{13}\right)^2 + \left(\Delta g_{12} - \Delta g_{21}\right)^2}} \quad (9c)$$

It should be noted that there are many axis/angle pairs due to the crystal symmetries of the lattices. For example, consider the misorientation between two lattices of cubic crystal symmetry. The cubic crystal symmetry group has 24 symmetry elements. Thus, the orientation of each of the two lattices can be described by 24 different symmetrically equivalent rotations. Therefore, 576 (=24 × 24) axis/angle pairs exist for describing the misorientation of one crystal lattice with respect to another for cubic crystal symmetry. However, since the misorientation and the inverse misorientation of two crystallites of the same symmetry are equivalent, the number of symmetrically equivalent misorientations is doubled (1152 in cubics).

Rodrigues vector description

The primary drawback of the axis/angle description of misorientation is that a degeneracy exists when the rotation angle ω approaches zero. When two crystal lattices have identical orientation, the misorientation angle is zero but the axis of rotation is not defined. However, the angle and axis of rotation can be coupled together into a single mathematical entity by multiplying the axis $\mathbf{n}$ by some function of the angle ω. If the function $f(\omega)$ is defined such that $f(\omega = 0) = 0$, then $\mathbf{n}f(0)$ is represented by a single point $(0,0,0)$, and no problems arise as the rotation angle ω approaches zero. Using this approach, the axis/angle pair can be combined into a vector representation of misorientation. FRANK [1987] and others [BECKER & PANCHA-NADEESWARAN 1989; NEUMANN 1991, 1992; RANDLE 1992; ENGLER &AL. 1994a; MORAWIEC

& FIELD 1996] have argued in favor of the Rodrigues vector representation [EULER 1775, RODRIGUES 1840]. The Rodrigues vector **R** is given by

$$\mathbf{R} = \mathbf{n}\tan(\omega/2)$$

(10a)

Note that this **R** is not a normal rotation operator: for an operation represented by $\mathbf{R}_A$ followed by a second operation similarly represented as $\mathbf{R}_B$ the resultant operation $\mathbf{R}_C$ is given by

$$\mathbf{R}_C = \frac{\mathbf{R}_A + \mathbf{R}_B - \mathbf{R}_A \times \mathbf{R}_B}{1 - \mathbf{R}_A \cdot \mathbf{R}_B}$$

(10b)

where $\times$ denotes the cross product. The Rodrigues vector representation contains some advantages over a conventional Euler representation. In Euler space, the reduced spaces arising from application of crystal symmetry generally have curved boundaries. HEINZ & NEUMANN [1991] have shown that, regardless of the type of crystallographic symmetry, all bounding surfaces of the reduced spaces are planes in Rodrigues vector space. Another advantage is that an ideal fiber texture is always represented by a straight line in Rodrigues space [NEUMANN 1990]. For some sample applications of the Rodrigues vector description see RANDLE [1992], ENGLER &AL. [1994a], FIELD [1995] and HUGHES & KUMAR [1996].

One disadvantage of the Rodrigues vector is that the space extends to infinity. However, this is not a practical disadvantage for mapping the orientation of crystals possessing symmetries more than triclinic since any remote point can be represented by a symmetrically equivalent point nearer the origin. Although the Rodrigues vector representation is mathematically more elegant, the more traditional axis/angle representation is more easily interpreted. For example, in axis/angle space the primary copper annealing twin is described as a 60° rotation about the ⟨111⟩ crystal axis. In Rodrigues space this same twin is described as $\left(\frac{1}{3}, \frac{1}{3}, \frac{1}{3}\right)$. Although it is clear from this description that the twins have ⟨111⟩ axes in common, the angular relationship is less obvious.

The distribution of misorientations measured in annealed high purity copper [HEIDELBACH &AL. 1996] is represented using Euler angles (after the manner of Bunge), axis/angle and Rodrigues vector descriptions in Fig. 18. The individual sections for the axis/angle (i.e. $\mathbf{n} - \omega$ plot are given for constant rotation angles (ω). The orientation of the axis, $\mathbf{n}$, is plotted as an equal-angle projection as shown in the $\omega = 0°$ section. The Rodrigues vector representation is plotted by sectioning Rodrigues space along the third component of the Rodrigues vector, R_z.

The main peaks in this distribution are associated with the primary copper annealing twin and geometrically necessary orientation relationship which arises when two different primary twins of the same grain share a common boundary. The locations of these peaks (and some symmetric positions that can be observed in the figure) in each of the three representations in Fig. 18 are given in Table III.

The strong peak at $\omega = 0°$ (as well as corresponding locations in the Euler angle and Rodrigues vector representations) is due to small-angle grain boundaries included in the misorientation measurements. The distribution of misorientation angle, ω, for a random set of orientations is shown in Fig. 19 [MACKENZIE 1958]. Clearly any measurements of misorientation near $\omega = 0°$ will produce a strong peak in the three dimensional misorientation distribution. This is most evident in the axis/angle plot of the mis-

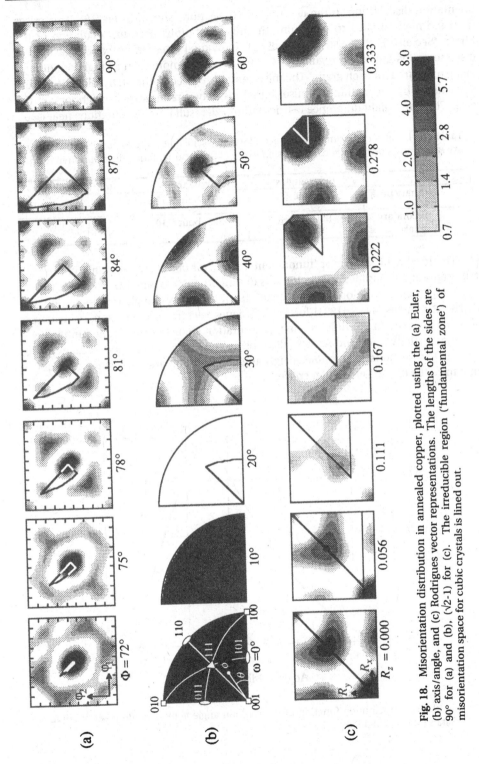

Fig. 18. Misorientation distribution in annealed copper, plotted using the (a) Euler, (b) axis/angle, and (c) Rodrigues vector representations. The lengths of the sides are 90° for (a) and (b), (√2-1) for (c). The irreducible region ('fundamental zone') of misorientation space for cubic crystals is lined out.

orientation distribution in the ω = 0° and 10° sections shown in Fig. 18. This arises from the method used to determine the misorientations used in calculation of the MDF. (See also HUTCHINSON &AL. [1996].) Individual orientations were measured at discrete points forming a regular grid. This results in more than a single orientation being measured for each grain. The misorientations used in the MDF calculation were determined by calculating the misorientations between all nearest-neighbors on the grid. Thus, any slight difference in orientation will result in misorientations near ω=0°.

Table III. Location of primary and secondary peaks in Euler angles (Bunge convention), axis/angle and Rodrigues vector representations of the misorientation distribution in Fig. 18.

	Euler angles	Axis/angle	Rodrigues vector
primary peak	(45, 70.5, 45)	60° about ⟨111⟩	$\left(\frac{1}{3},\frac{1}{3},\frac{1}{3}\right)$
secondary peak	(26, 84, 26)	38° about ⟨110⟩	$\sim\left(\frac{1}{4},\frac{1}{4},0\right)$

The irreducible region or 'fundamental zone' for cubic-cubic symmetry for each representation is outlined in Fig. 18. It is described by ZHAO and ADAMS [1988] for the Euler angle representation, by MACKENZIE [1958] for the axis/angle representation and by HEINZ and NEUMANN [1991] for the Rodrigues vector representation. Constant-Φ sections are used in the Euler angle representation so as to capture the fundamental zone set forth by ZHAO and ADAMS [1988] using the Bunge description of Euler angles with as few sections as possible. High angle Φ sections are used since in order to minimize the distortion in Euler space.

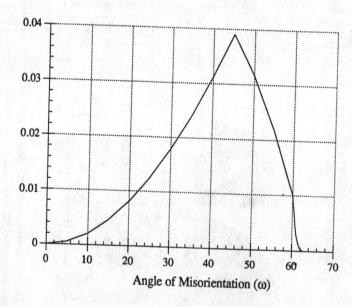

Fig. 19. The distribution function of the rotation angle ω for a random set of cubic crystals [MACKENZIE 1958].

Quaternion description

Another useful representation similar to the Rodrigues vector representation is the quarternion parameter vector representation. Each quaternion vector $\mathbf{Q} = (q_0, q_1, q_2, q_3)$ consists of a scalar q_0 (with $q_0 \geq 0$) and a vector $\mathbf{q} = (q_1, q_2, q_3)$ and represents a point on the upper half of a unit sphere in four-dimensional Euclidean space. The following normalization condition is satisfied:

$$|\mathbf{Q}| = \left[q_0^2 + q_1^2 + q_2^2 + q_3^2 \right]^{1/2} = 1 \tag{11a}$$

There is a one-to-one relationship from the quaternion parameters to the rotation angle ω and the rotation axis $\mathbf{n}$ by

$$q_0 = \cos(\omega/2) \text{ and } \mathbf{q} = \mathbf{n} \sin(\omega/2) \tag{11b}$$

The corresponding rotation matrix (see Eq. 9a) is given by:

$$\Delta g = \begin{pmatrix} q_0^2 + q_1^2 - q_2^2 - q_3^2 & 2(q_1 q_2 - q_0 q_3) & 2(q_1 q_3 + q_0 q_2) \\ 2(q_1 q_2 + q_0 q_3) & q_0^2 - q_1^2 + q_2^2 + q_3^2 & 2(q_2 q_3 - q_0 q_1) \\ 2(q_1 q_3 - q_0 q_2) & 2(q_2 q_3 + q_0 q_1) & q_0^2 - q_1^2 - q_2^2 + q_3^2 \end{pmatrix} \tag{11c}$$

The properties of quaternions and their connection to orthogonal matrices and Euler angles were reviewed by MORAWIEC & POSPIECH [1989a]. The advantage of the quaternion representation to the Rodrigues map is that the $\mathbf{Q}$ map is finite. Rotations can be combined through use of the corresponding orthogonal matrices, the $\mathbf{Q}$ parameters themselves, or as Cayley–Klein matrices. Using the quaternion vectors, two orientations can be combined as follows:

$$\mathbf{qp} = q_1 \mathbf{p} + p_1 \mathbf{q} + \mathbf{q} \times \mathbf{p} \tag{11d}$$

$$(\mathbf{qp})_0 = (q_0 p_0 - \mathbf{q} \cdot \mathbf{p}) \tag{11e}$$

Cayley–Klein matrices are 2×2 matrices with complex elements formed from the $\mathbf{Q}$ parameters as:

$$\begin{bmatrix} q_0 + i q_1 & q_2 + i q_3 \\ -q_2 + i q_3 & q_0 - i q_1 \end{bmatrix} \tag{11f}$$

It should be noted that the determinant of this matrix satisfies the normalization condition set forth in Eq. 11a. The fundamental zone for the quaternion description has been derived by ADAMS &AL. [1990].

2.7 The representation of twinning relations

Twins in crystals are a frequent occurrence especially in low-symmetry materials. They may be generated by growth accidents, by plastic deformation, or in the course of phase transformations. The crystallography of twin relations, and the kinetics of the twinning process, are quite complex in the general case and have been extensively treated and reviewed [BILBY & CROCKER 1965, CHRISTIAN 1975, BARRETT & MASSALSKI 1980]. In the context of this book, two topics are of interest: the orientation relations between twins that are possible and observed; and twinning as a deformation mode. We will illuminate some of these by example, starting from the simplest cases.

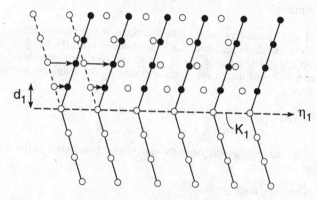

Fig. 20. Sketch of a simple compound twin, in the 'plane of shear' (S), and representation by a homogeneous shear parallel to the habit plane K_1 in the η_1 direction.

Figure 20 shows a particular plane section of a simple crystal structure (open symbols). The plane of the section will be called S. Above the plane K_1 (the so-called 'habit plane', dashed), a second set of symbols (solid) represents a 'twin' of this crystal. Conversely, the open-symbol crystal may be regarded as a twin of the closed-symbol crystal. The plane K_1 may be the physical boundary between the two twins (in which case it is called a 'coherent' boundary), or it may be merely a convenient reference plane for two infinite crystals that have the orientation relation of twins with each other.

One twin may be obtained from the other by one or more of three distinct operations: (a) by *reflection* in the plane K_1; (b) by a 180° *rotation* around the direction η_1 (the intersection of K_1 with S); and (c) by a 180° rotation around the normal to K_1. Option (c) is entirely equivalent to (a) in a centrosymmetric crystal; in non-centric crystals, it is *not* a definition of a twin. In general structures, *either* (a) *or* (b) defines a twin relation ('Type I' and 'Type II', respectively). When *both* are true, as in Fig. 20, one refers to a 'compound' twin; it can only exist in structures where S is a mirror plane.

The above descriptions can be expressed algebraically as follows. The orientation of the twinning coordinate system with respect to that of a 'matrix' is

$$R^{tw} = \begin{pmatrix} \eta_1 \\ S \\ K_1 \end{pmatrix} \tag{12a}$$

The twinning 'transformation' matrices for Type I and Type II twins are

$$t^I = \begin{pmatrix} 1 & 0 & 0 \\ 0 & 1 & 0 \\ 0 & 0 & -1 \end{pmatrix} \qquad t^{II} = \begin{pmatrix} 1 & 0 & 0 \\ 0 & -1 & 0 \\ 0 & 0 & -1 \end{pmatrix} \tag{12b}$$

Thus, the orientation g^{tw} of the twin with respect to a matrix of orientation g (itself written with respect to the 'sample' frame) is

$$g^{tw} = t \cdot R^{tw} \cdot g \tag{12c}$$

For g^{tw} to come out 'right-handed', t must have a determinant of +1; thus, one takes t^{II}, also for compound twins, which are of primary interest in this book. Then,[10]

$$g^{tw} = \begin{pmatrix} \eta_1 \\ -S \\ -K_1 \end{pmatrix} \tag{12d}$$

A different description of twin generation, which corresponds to a physical process realized in deformation twinning, is by homogeneous shear from a 'parent' (really the other twin!) parallel to the *twinning plane* (the habit plane, K_1) in the *twinning direction* (η_1): Fig. 20. 'Homogeneous shear' means that the displacement u_1 (in the direction of η_1) increases in proportion to the distance d_1 from the habit plane; the amount of shear is then

$$\gamma_t = u_1 / d_1 \tag{12e}$$

This quantity is called the *characteristic twinning shear* and is always the same for a given twin. Twinning can be an important mode of plastic deformation. (See Chaps. 8, 11).

In analogy to slip modes, one may characterize these twinning modes by giving the indices of K_1 and η_1 in the form, e.g., (112)[111]. Inasmuch as these twinning modes are uni-directional, however, it is important *not* to use generic indices such as {112}⟨111⟩ (and leave the derivation of symmetrically equivalent ones to the user). Table IV lists these twinning elements, including the twinning shear, γ_t, for some metals and minerals. Also given is S – which now, for obvious reasons, is called the 'plane of shear'.

Possible twinning operations in a given crystal structure are restricted by the obvious demand that they cannot be symmetry elements of the crystal structure itself: then, twinning would produce an identical structure, not a new orientation. For example, in a compound or a Type I twin, K_1 cannot be a mirror plane in the crystal; in a compound or a Type II twin, η_1 cannot be a two-fold axis in the crystal.

The reason why Fig. 20 represented a 'simple' case is that we assumed a structure with one atom per lattice point (which is also, *ipso facto*, centrosymmetric). Most structures in which twins occur have a multi-atomic motif on each lattice point (or, equivalently, may be represented by multiple interpenetrating lattices). In these cases, it is instructive to distinguish what happens to the lattice from what happens to the structure. The twin relation is defined in terms of the complete structure. The homogeneous shear representation refers to the *lattice*.[11] To achieve a complete transformation into the twinned structure, multi-atomic structures require homogeneous shear and, in addition, 'atomic shuffling'. For the same reason, the habit plane is, in these materials, not a mathematically sharp plane, but a 'zonal' boundary [YOO & LOH 1970], within which the atomic positions do *not* follow from a crystallographic twinning operation.

A case of special interest with respect to textures is that of 'Dauphiné twins' (so-called after the region around Grenoble, France) in α-quartz [SCHUBNIKOW & ZINSERLING 1932]. In these, the lattice does not exhibit the twin (it is hexagonal), but the motifs are in Type II

---

[10] In non-compound Type I twins in non-centric materials, one would have to replace -S by S.
[11] For this reason, even Type II twins can be obtained by homogeneous shear, though this is not so easily imagined [BEVIS & CROCKER 1969]. They are well documented in uranium and mercury [CROCKER & AL. 1966] and have recently been shown to account for a deformation mode in α-alumina [PIROUZ & AL. 1996].

Table IV. Some deformation twinning elements in metals and minerals

Lattice	pt.grp.	M,Z*	Plane S‡	K_1-Plane†	η_1-DIR	Shear γ_t	Examples
fcc	m3m	12	$(\bar{1}10)$	(111)	[112]/2	0.707	brass
bcc		12	$(1\bar{1}0)$	$(\bar{1}\bar{1}2)$	[111]/2	0.707	α-iron
hexagonal	6/mmm	6 Z	(1210)	$(10\bar{1}2)$	$[\bar{1}011]$	(-0.138:	Zn); Cd
						(0.169:	Zr); Mg, Re, Ti
						(0.199:	Be); RE#
			$(1\bar{1}00)$	(1121)	$[\bar{1}\bar{1}26]/3$	(0.628:	Zr); Re, Ti, RE
		Z	$(\bar{1}100)$	(1122)	[1123]/3	(0.224:	Zr); Re, Ti
		Z	$(\bar{1}2\bar{1}0)$	$(10\bar{1}1)$	$[10\bar{1}2]$	(0.099:	Mg); Zr
b.c.tetr.	4/mmm	4 Z	$(0\bar{1}0)$	(301)	$[\bar{1}03]:$	(0.119:	β-tin)
f.c.tetr.		4	(010)	(101)	$[10\bar{1}]/2$		In
ord.fct.		4	$(\bar{1}10)$	(111)	[112]	~0.707:	γ-TiAl)
orthorh.	mmm	2	$(00\bar{1})$	(130)	$[3\bar{1}0]/2$	(0.299:	α-U) (a,b)
		4 Z		~(197)	[512]	(0.216:	"
		4 Z		~(172)	[312]	(0.227:	"
		4 Z		(112)	~[372]	(0.227:	"
rhomboh.	3̄m	3 Z	$(2\bar{1}\bar{1}0)$	$(01\bar{1}8)$	[0441]	(0.694:	calcite) (c)
	3̄m	3 Z	$(\bar{1}2\bar{1}0)$	~(0001)	$[10\bar{1}0]$	(0.635:	α-Al$_2$O$_3$) (d)
	3̄m	3 Z	$(2\bar{1}\bar{1}0)$	$(01\bar{1}2)$	$[0\bar{1}11]$	(0.202:	α-Al$_2$O$_3$) (e)
	3̄	Z	$(2\bar{1}\bar{1}0)$	$(01\bar{1}2)$	$[0\bar{1}11]$	(0.587:	dolomite) (c)
[trigonal	32	3	--------	~$(10\bar{1}0)$	[0001]	--------	α-quartz]

† The twinning plane and twinning directions shown are one particular set of K_1 and η_1. The sign is important. Other specific representations follow from an application of the crystal symmetry elements. A ~ in front of indices signifies an irrational twinning element for non-compound twins; the ~ sign is also used in the degenerate case where the indices are in fact rational but the twin is not compound (the '~(0001)' planes).

‡ The plane of shear, S, is given such that $\{\eta_1, S, K_1\}$ form a right-handed set.

* M,Z: Multiplicity, Zonal habit plane (and shuffling)

RE: Rare Earths

REFERENCES:

(a) CAHN, R.W. (1953). *Acta Metall.* 1, 49.

(b) LLOYD, L.T., & CHISWIK, H.H. (1955). *Trans. AIME* 203, 1206.

(c) BARBER, D.J. & WENK, H.-R. (1979). *Phys. Chem. Miner.* 5, 141.

(d) PIROUZ, P. &AL. (1996). *Acta Mater.* 44, 2153.

(e) HEUER, A.H. (1966). *Philos. Mag.* 13, 379.

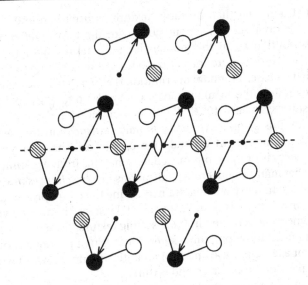

Fig. 21. Dauphiné twin in α-quartz, projection on the (0001) plane; the gray atoms are in the plane, the black and white ones are, respectively, 1/3 of a d-spacing above and below the plane of the paper. The dots show lattice points that have the point group symmetry; they are connected to an (arbitrary choice of) a motif. The dotted line marks a trace of the habit plane, (10$\bar{1}$0); the diad symbol marks a point of intersection of the rotation axis [0001], which turns the lower twin into the upper twin. Note that the gray symbols (which could be used as lattice points) form a single set above and below the plane.

twin relation (Fig. 21). The rotation axis η_1=[0001] coincides with a two-fold axis in the 6/mmm lattice, but not in the 32 structure or the 3m Laue group. Since the lattice is not twinned, these twins cannot serve as a macroscopic deformation mode. Nevertheless, an applied stress has been observed to activate such twinning (perhaps in the same way as the Snoek effect in bcc metals) – and it leads to a large change in the Laue orientation of each twinned crystal, and thus in the texture of an aggregate. Here is a case where the Laue orientations differ, but the lattice orientations are the same (see Chap. 1 Sec. 2.3).

3. The Representation of Textures

Texture is a collective term for a non-uniform distribution of crystallographic orientations in a polycrystalline aggregate. We start with the assumption that a generally large number of individual orientations are known. They correspond to a distribution of points in three-dimensional orientation space. Conversely, one may have a *continuous* distribution in orientation space – usually in the form of occupation densities in a set of contiguous (but finite) cells of orientation space. After analysis, such a continuous distribution may be expressed as a finite number of coefficients of a series expansion. We shall refer to all of these as representations of an *orientation distribution* (OD). (The additional word 'function' is not necessary and may falsely imply a truly continuous, analytic description.)

It should be noted that the term 'texture' is used here only for a *non-uniform* orientation distribution; i.e., synonymous with the term '*preferred* orientations'. A *random* distribution is, in this nomenclature, a representation of an 'untextured' material. The word 'texture' has also been used as a synonym for 'orientation distribution'; in this case, a random distribution is also a (particular kind of) texture.

Textures reflect the symmetry of the *sample* – due, e.g., to the symmetry of a previous test [WEISSENBERG 1922]. If one looks at individual grains and considers their number finite, this sample symmetry is only *statistical*: in this sense, it is different from crystal symmetry. In general, one cannot assume a priori that a particular sample symmetry is present, but must determine this from experiments in the current state. Note that many engineering applications involve *changes in straining path*; only when a certain symmetry was maintained during the entire processing (and the *initial* orientation distribution was uniform, or at least of the same symmetry) will the process symmetry be reflected in the resulting texture. Even when a certain constant path is maintained for a long time, some features of a previous texture may never be 'forgotten'; for example, empty parts of an initial distribution that would be needed as a source for parts of the expected end texture.

Inasmuch as texture symmetries represent *statistical* properties of a *sample*, we specify them by terms like 'no symmetry', 'plane symmetry', 'orthotropic' symmetry (instead of triclinic, monoclinic and orthorhombic as it would be done for *crystals*); and a most important one: 'fiber symmetry'. (See Chap. 1 Sec. 3.)

In practical cases, the texture of a sample may not be known completely or accurately. For the purposes of the present Section, we assume it known and concentrate on how to represent it.

Some textures are well described by 'components': a superposition of a small number of single crystals, with some spread (which may be quantified by Gaussians [LÜCKE &AL. 1981; JURA 1991]). Others can be idealized as 'fibers' in orientation space, in which a single angle can be used to specify an orientation within the fiber (although it may not be simply one of the Euler angles). We will discuss such cases where they are well established – but always demonstrate the power of a three-dimensional description of the OD which (in whatever form) has been the standard method of quantitative texture analysis since BUNGE [1965] and ROE [1965]; see BUNGE [1982].

The three-dimensional description is often *not* more difficult than an idealization in terms of components or fibers – especially in non-trivial or unknown cases. The following examples are not all 'typical': they were selected to demonstrate the various methods.

3.1 Orthotropic sample symmetry

The textures that have been most thoroughly investigated in metallurgy have been those in rolled sheets, or rolled and recrystallized sheets, of materials of cubic lattice structure. In face-centered cubic (fcc) metals, these have commonly been described simply by a 111 pole figure. Inasmuch as one 111 pole figure contains four {111} poles, this is sufficient. Often, it is possible to recognize a texture in this single pole figure when it is sharp, or actually consists of merely one component. Fig. 22 shows such an example: it was experimentally obtained from a rolled sheet of palladium (an fcc metal), of unknown pre-history. One can easily appreciate that this is a 'rotated cube': it is rotated by 45° around the normal direction (ND, in the center) from a fully aligned

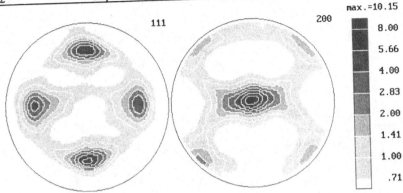

Fig. 22. Experimental texture of a cross-rolled palladium sample. In this case, one 111 pole figure suffices to identify the single-component texture; one 100 pole figure, if incomplete, might not. (The net is every 15°; the peaks in the 111 pole figure are at 55° from the origin; equal-area projection.)

cube orientation, for the 111 pole figure (Fig. 22a). This interpretation is confirmed by the 100 pole figure, Fig. 22b, although the rim region is empty in this experimental pole figure. If the weak connecting 'fibers' between the maxima are deemed unimportant, one need go no further than to describe such a texture by one component, with a certain, quantifiable spread.

The so-called *copper* component in rolling (a spread around the orientation {211}⟨111⟩), which is often the strongest, is not as obviously interpreted from a 111 pole figure. Figure 23 shows it idealized: It obviously cannot be one crystal, but consists in fact of two symmetrically disposed ones. A {111} pole in the rolling direction RD is easy to see (**a**); but if one only has this one experimental pole figure, its rim cannot be measured (in reflection) and one does not see this pole. If one is experienced, one can infer it from the symmetric disposition of the two poles **c** and the pole **b**. The latter, together with **a**, implies that the transverse direction TD must be a ⟨110⟩ crystal direction: the intersection of two {111} planes. Therefore, the ND must be a ⟨112⟩ direction, and the 'component' is labeled {112}⟨111⟩, in the sequence ND/RD. The remaining four poles follow from an assumed sample symmetry that demands a two-fold axis in the center.

When the external axes have simple crystallographic designations, *inverse* pole figures often provide the most concise description.[12] For the {112}⟨111⟩ component described above, you need one inverse pole figure for the ND, which shows a {112}, and one inverse pole figure for the RD, showing the ⟨111⟩. Figure 24 shows the irreducible region, or 'fundamental zone', for cubic inverse pole figures (for orthotropic sample symmetry), for these two sample axes. In addition to the 'copper' component discussed in detail above (Δ), a number of orientations along the 'β-fiber' (to be discussed below) are plotted, ending with the 'brass' component {110}⟨112⟩ (×).

---

[12] Inverse pole figures can, by some techniques, be measured directly (Chapter 3); but to obtain an inverse pole figure from a measured pole figure, one must first deconvolute the pole-figure projection of the three-dimensional orientation space anyway, and then re-project it in the Ψ-direction (for the ND, others for the RD and TD). It is the deconvolution of pole figures that is the essential task of texture analysis.

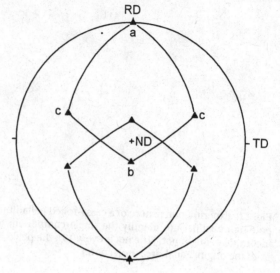

Fig. 23. A sketch of the pure 'copper component' in rolled fcc metals, 111 pole figure (stereographic projection).

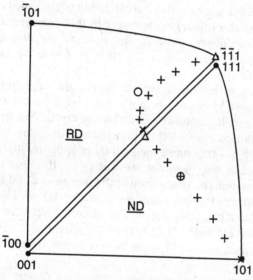

Fig. 24. Two inverse pole figures (for the ND on right, for the RD on top), showing two idealized fcc rolling components: 'copper' (Δ) and 'brass' (×), as well as an idealized 'fiber' (β).

Figure 25 shows 111 and 100 pole figures that are typical of a rolling texture in copper. They came from measurements from the center plane of a Cu sheet rolled to 90% reduction and partially recrystallized [NECKER &AL. 1993]; those displayed are recalculated pole figures.

The 'copper component' is drawn in again (Δ), and so are some other common idealizations, the 'brass' component (square), the 'S' component (o), the 'Goss' component (diamond), and the 'Taylor' component (∇). (For the indices and angles of these other components, see Table I in Chap. 5.) In addition, there is a self-evident cube component. It is seen that, while each of the components (except Goss) lies in a dense region, no individual one represents the texture well enough. In fact, as was suggested already by GREWEN and WASSERMANN [1955], it is more appropriate to describe this texture as 'fibers' of orientations, rather than 'components'. (They

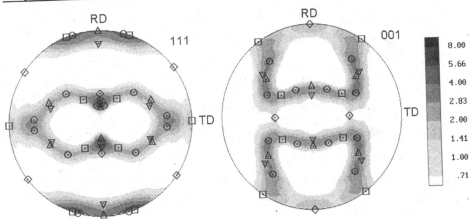

Fig. 25. Recalculated 111 and 100 pole figures of copper rolled to 90% reduction, with symbols for standard components: Δ 'Copper'; ∇ 'Dillamore'; o various 'S' components; square: 'Brass', diamond: 'Goss'.

are 'valleys' into which, and along which, orientations move; or 'ridges' of intensity, rather than 'peaks'.) In Chapter 5, some descriptions of fcc rolling textures in terms of Gaussian components or fibers will be discussed.

Figure 26 shows this same texture quantified within the framework of three-dimensional orientation space, obtained by now presuming orthotropic sample symmetry. Figure 26a shows 'square sections' of orientation space [BUNGE 1982] (albeit on a rather coarse grid); Fig. 26b shows the same crystal orientation distribution (COD) in polar sections. We show both representations in parallel for two reasons. One is to demonstrate that they can be directly compared, section by section: the periphery of each section in the polar plot is the bottom line on the square; the origin of the polar plot is the top line of the square (Fig. 27). As was seen in Table I, Bunge and symmetric angles are easy to compare in this case: the sections are at constant φ_2 or ϕ. The comparison of φ_1 and Ψ is easiest if one keeps the rolling direction (RD, axis 1) on the *right*. We do this generally for three-dimensional plots, although we retain the customary alignment of pole figures: with RD up. Note that the projection of Fig. 26b is the symmetrized 100 pole figure: one can see that this is a smoothed (and permuted) version of one quadrant of Fig. 25b (turned 90°).

The other reason for showing Fig. 26a at all is to demonstrate that it is the less desirable form of visualization: the cartesian appearance of its borders means that the space it represents is internally 'curved' [VAN HOUTTE, priv. comm.]; the polar representations, on the other hand, coupled with equal-area mapping, is 'homochoric' [WENK & KOCKS 1987].

Figure 26c, finally, shows the complementary sections: the *sample* orientation distribution (SOD). Its projection is the *inverse* pole figure for the ND: one sees the concentration at [211], but 'fibers' as well. Compare Fig. 26c with Fig. 12: apart from symmetric repetition around the quadrant, it describes the same texture (only in continuous, not discrete form); each section corresponds to a certain interval of flag directions. Figure 24 described an oversimplified idealization of this texture.

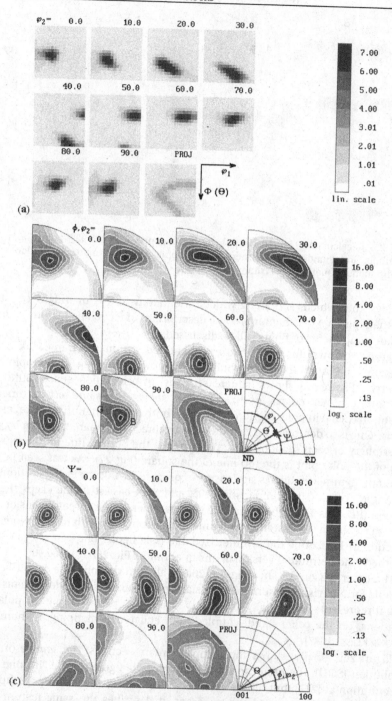

Fig. 26. Same texture as Fig. 25, assuming orthotropic sample symmetry, represented as sections of 3-D orientation space: (a) 'square' COD sections, (b) polar COD sections, (c) polar SOD sections. The last unit always shows the projection: the {001} pole figure for CODs, the inverse pole figure of NDs for SODs. Both symmetric and Bunge angles are shown in the nets. Some special 'components' are marked.

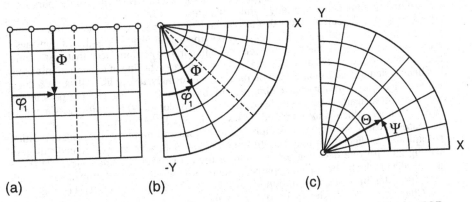

(a)   (b)   (c)

Fig. 27. Sketch showing point-by-point comparison between square and polar COD sections, with definitions of Bunge's and symmetric Euler angles. (The grid line $\varphi_1=45°$ is shown dashed for emphasis.)

Description in terms of fibers and volume fractions

Inspection of these three-dimensional distributions reveals that this texture may be idealized into fibers in orientation space. Some of the 'components' of the rolling texture are marked (C for 'copper', 'S', B for 'brass', G for 'Goss') – and they are intimately connected to each other by the fiber. The same fiber appears more than once (because the three-fold axis in the cubic crystal has not been incorporated into the analysis). The one stretching from $\phi=45°$ to $\phi=90°$ is centered on the 'β-fiber' shown in Fig. 24, and the one within the 90° section is the 'α-fiber'. Figure 28 shows this region in more detail, and for a less deformed copper (58% reduction). The peak near the X-axis in the 45° section increases in intensity and spread with increasing section angle (ϕ); it also moves to higher azimuths (Ψ) and slightly larger pole distances (Θ); finally, the *shape* of the peak becomes more and more elongated. For quick comparisons, one may condense this figure into just four sections: at 45°, 60°, 75°, and 90°.

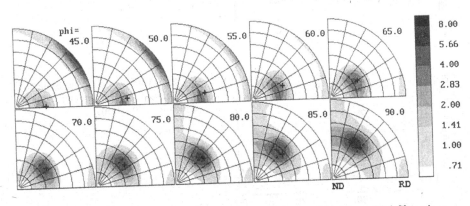

Fig. 28. Some COD sections that display a single variant of the 'β'- and 'α'-fibers in a copper polycrystal rolled to 58% reduction. The location of the 'ideal' fibers is marked by some + symbols.

Many attempts have been made to capture this 'fiber' in a concise form, for the purpose of quantitative comparisons of many similar textures, e.g., as a function of alloy content [HIRSCH & LÜCKE 1988a]. In general, it is difficult to catch the rich behavior outlined above in a few words describing Fig. 28. For example, the actual region of highest intensities is often *not* centered on a 'fiber' (in the sense of a line in orientation space that corresponds to rotations around a fixed axis [BUNGE 1982]); in that case, the line of maximum intensities has been called a 'skeleton line' [BUNGE 1982] or, confusingly, *also* a 'fiber' [HIRSCH & LÜCKE 1988a]. Figure 29 demonstrates a 'fiber plot' of the same data as Fig. 28. The peak intensities are plotted against section angle (symbol Δ). Also plotted (+) is the intensity along the ('ideal') β-fiber, which is also marked by + signs in Fig. 28. At high ϕ, the density rapidly decreases because the actual maximum departs more and more from this ideal location. To capture this effect quantitatively, HIRSCH & LÜCKE [1988] have complemented a diagram like Fig. 29 with one detailing the location of the actual skeleton line. Figure 28 shows graphically in which direction it departs: in the 90° section, the peak is moved significantly away from the 'brass' (B) orientation toward the 'Goss' (G) orientation (cf. Fig. 26a). (See also Chap. 5 for a description of fcc rolling textures.)

The most instructive quantity to derive from these data is the *volume fraction* contained in each part of the fiber. This is plotted in Fig. 29 (o) for all parts that exceed a normalized intensity of 2.0. The fact that this curve increases much more rapidly than the one for the peak intensities (Δ) is due to the increasing spread of the peaks. (The volume fraction along the 'ideal' β-fiber gives an entirely misleading abstraction.) Other definitions of the volume fraction can be obtained (as these were) from a print-out of the intensities in each pixel of the equal-area plot in Fig. 28. (To obtain volume fractions directly from intensities, one must take account of the geometry of the space [KALLEND & DAVIES 1969].)

In summary, we conclude that the three-dimensional orientation distribution, such as it is displayed in Fig. 28, is itself the most concise and quantitative representation of the texture (which is also applicable to other deformation paths). In addition, as we will see in many examples, a comparison of two textures is displayed in most instructive form as a *difference* distribution.

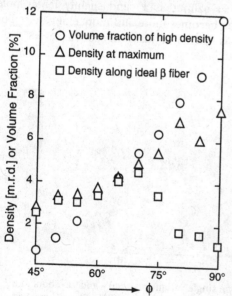

Fig. 29. Abstraction of the behavior in Fig. 28 into certain special intensities and volume fractions.

Oblique sections and difference plots

For completeness, Fig. 30a shows all the same data as Fig. 26, but now *in oblique sections*: at constant ν, with Ψ as the azimuth (so that the projection is again a quadrant of the 100 pole figure, just as in Fig. 26b). This is not an important mode of display here, because the above textures had little cube component; however, Fig. 30b shows the same sample after partial recrystallization: it has a clear cube component, which is seen at the origin, in the sections near ν=0. (In a COD or SOD polar representation it would show up at the origin in all sections, and in a cartesian representation of Euler space, it could be in every section along the entire plane Θ=0).

A good way to display the *difference* between the rolling and partially recrystallized textures is demonstrated in Fig. 30c: the actual orientation-space files were subtracted and plotted (symmetrically around zero difference). One can observe not only the appearance of a new component, but also the slight shift in some old components. We will make much use of this technique in comparisons between prediction and experiment, or different evaluations of the same experiment.

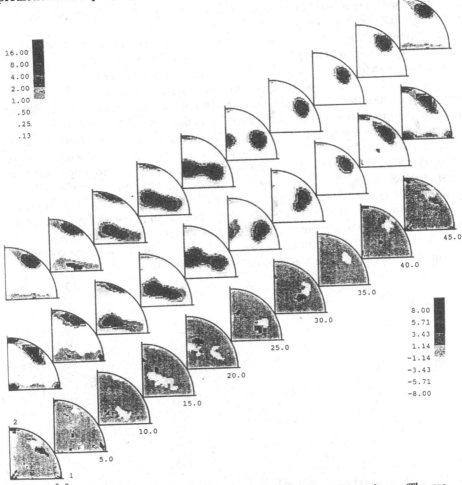

Fig. 30. (a) The same data as Figs. 26, plotted as oblique CON sections. (The projection is still the 001 pole figure). (b) the same material after partial recrystallization; the cube component makes oblique sections most appropriate. (c) A *difference* orientation distribution in oblique CON sections.

3.2 One plane of symmetry

Many forming operations of practical interest are in plane strain, or can be reasonably approximated as such. The rolling deformation idealized in the last section is in fact one of these. Only in the central plane of the sheet can one expect the higher, orthotropic sample symmetry. But even if there are shears at the surface, there is still one mirror plane, perpendicular to the transverse direction (so long as the sheet is wide enough to be approximated by plane strain). In such a case, one's theoretical inclination would be to make the TD the preferred axis, to be plotted in the center of a pole figure. This is hard to obtain by experiment directly, because of the geometry of the sample; but one can obtain it by projection of the 3-D orientation distribution (or by rotating complete pole figures). Figure 31a shows data like that in Fig. 26 one more time, in this TD-projection. The actual orthotropic symmetry is still evident; but the pole figure has been turned 45° around TD for the purposes of the following comparison. Figure 31b, on the other hand, shows the case of tube torsion (where a texture measurement from the radial direction, perpendicular to the mirror plane, is the natural one). Figures 31a and b are both from experiments to a von Mises strain of about 2.0 (not 2.5, as Fig. 26 was).

First note that Fig. 31b shows one two-fold axis only (around R): this is the equivalent of the one mirror plane present (with the inversion center present in the sample as a whole). Second, note that there is a distinct similarity between Fig. 31a and Fig. 31b: Fig. 31a is a reasonable approximation to Fig. 31b, except for the lack of orthotropy in the latter. There is a good reason for this similarity: if one thinks of both modes of deformation as 'shears', rolling corresponds to 'pure shear', on two mutually perpendicular directions, both at 45° to the ND and the RD, whereas torsion is 'simple shear', on just one plane, with the attendant rotations. The presence of a macroscopic spin may make the differences between simple and pure shear more important than the similarities, especially at large strains [GIL SEVILLANO &AL. 1977, BOLMARO & KOCKS 1992, PRANTIL &AL. 1993]. This will be discussed further in connection with simulations (Chapter 9). Here we only note that the presentation in the 'sample' coordinate system (Chapter 1), with the fixed plane (perpendicular to the axis) and the fixed line in it (in the tangential direction) changes little with strain; the rotating coordinate system of the 'stretch ellipsoid' is of no relevance to the texture.

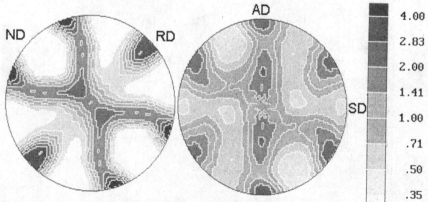

Fig. 31. Experimental comparison between 'pure' and 'simple' shear: (a) a rolled and (b) a tube-twisted copper polycrystal, plotted as recalculated 111 pole figures with the *transverse* direction in the center. (The RD and ND for the rolled crystal are at ±45°.) VonMises strain about 2 in both.

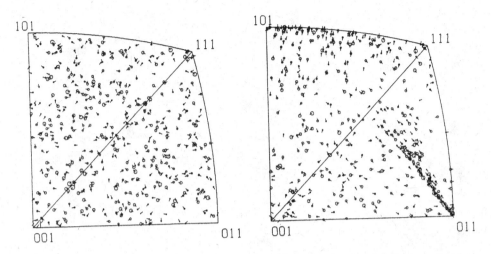

Fig. 32. (b) A shear texture similar to that used for Fig. 31, but displayed as discrete grains in an inverse pole figure (sample orientation projection) for the radial axis, with 'flags' toward the shear direction. It was obtained by Taylor simulation from the grains shown in (a).

Another representation of shear textures (just as it was for rolling textures, Fig. 12) is as inverse pole figures. Figure 32b shows two stereographic triangles with discrete grains symbolized by a circle at the location of the radial ('transverse') axis **Z** and a 'flag' in the shear direction **X** (with a right-handed coordinate system assumed). This distribution was obtained by Taylor simulation from the distributions shown in Fig. 32a; the location of the Z- and X-axes is random, the size of the circles represents the weight assumed for each grain (see Section 3.4.) (A figure much like this was first published by VAN HOUTTE and AERNOUDT [1976].)

It is, of course, possible to find descriptions of the torsion texture in terms of components and fibers [WILLIAMS 1968b, VAN HOUTTE 1978, CANOVA &AL. 1984a, MONTHEILLET &AL. 1984]; but they change with strain, material, and temperature. A *general* representation of orientation space is shown in Fig. 33, as ν-sections (like Fig. 28 – but with different sample axis assignments.) Each section must now be a semi-circle, to account for the lower sample symmetry. Since the crystal symmetry is the same, the number of sections, in this representation, remains the same.

3.3 Cubic sample symmetry

This is an unexpected case, since no testing machine for *plasticity* (at constant volume) has cubic symmetry. However, the case emerged as of interest in what may be called 'redundant forging' of a specimen: successive compression of a cube in all three directions. In the limit of infinitesimal steps, this gives rise to cubic sample symmetry. Figure 34 shows simulated results (which were in good agreement with experiments) [TAKESHITA &AL. 1989]. The first thing to notice is that the distribution is not random, as one would have expected if the test were presumed to be *isotropic*; but when there is no distinction between **X**, **Y**, and **Z**, nor with their negatives, this does *not* imply isotropy, only cubic symmetry.

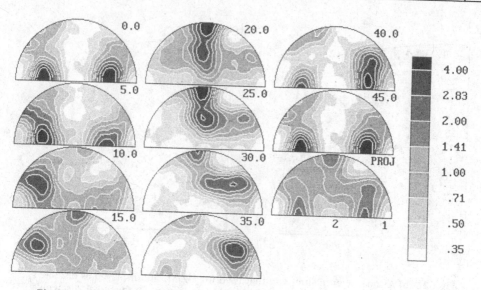

Fig. 33. The same data as Fig. 31(b) plotted as CON sections; the semicircles reflect the presence of just one two-fold axis in the sample.

Figure 34 shows both an SOD and a COD. The fact that they look so similar to each other is due to the cubic sample- as well as crystal-symmetry: the two azimuthal angles ϕ and Ψ behave the same way.

It is also worth noting that the *projections* are rather random (except for the 'hole' at $\langle 111 \rangle$). On the other hand, the *sections* are not as featureless; for example, the sections at about 45° represent an approximate cube texture, those near 0 or 90° look like a typical compression texture in the SOD – but in the COD, they look the same, and are not rings as would be demanded of a fiber texture. The lesson to be learned from this example is that projections may be deceptive: the three-dimensional description is needed unless one already knows the features.

3.4 The representation of the non-textured reference state. Discrete grains files

In a continuous distribution, the non-textured state is trivial: it has a uniform density everywhere. But for some purposes, one needs a set of individual grains; for example, for simulation of polycrystal plasticity. Also, a set of grains can, under some circumstances, be the most compact way of describing a texture. This is obvious, when a few 'components' describe the texture well enough; but it is also true for more diffuse textures – using, say, 100 grains.

The question is: how can one find the appropriate set of grains, not by inspection in every case, but in a routine way? This problem has been addressed in a number of ways [FORTUNIER & HIRSCH 1987; KOCKS &AL. 1991b]. They all start with a representation of the non-textured state by a certain number of grains, and then assign weights to these according to the density in orientation space at the location of the grain.

In a sense, the file one uses to record intensities in orientation space, *is* a set of grains: typically on a grid of 5°×5°×5°. This turns out to be about 6000 numbers for

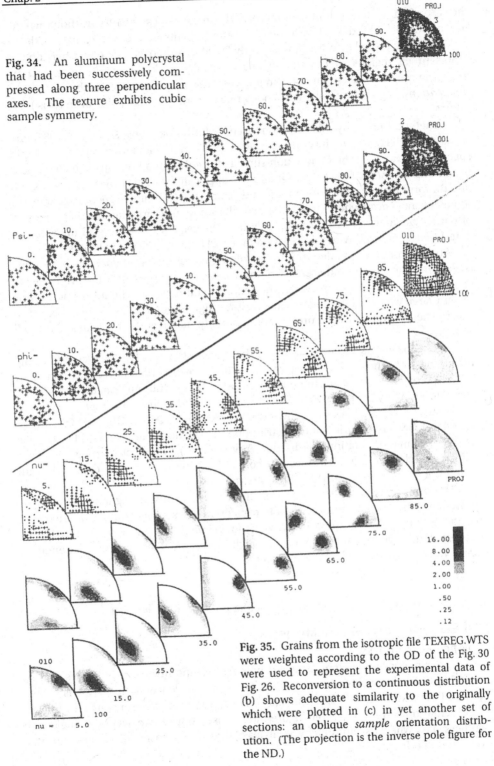

Fig. 34. An aluminum polycrystal that had been successively compressed along three perpendicular axes. The texture exhibits cubic sample symmetry.

Fig. 35. Grains from the isotropic file TEXREG.WTS were weighted according to the OD of the Fig. 30 were used to represent the experimental data of Fig. 26. Reconversion to a continuous distribution (b) shows adequate similarity to the originally which were plotted in (c) in yet another set of sections: an oblique *sample* orientation distribution. (The projection is the inverse pole figure for the ND.)

the case of orthotropic sample symmetry and cubic crystal symmetry (without making use' of the 3-fold axis). For a *general* sample symmetry coupled, say, with an orthorhombic crystal symmetry, it is nearly 50 000. Moreover, these numbers do not represent all regions of orientation space equally well; this could be solved by assigning weights to them initially. In any case, these numbers are much too large for *applications*, say, in finite-element codes that describe the behavior of macroscopic bodies.

A method to remedy this situation, used by both FORTUNIER & HIRSCH [1987] and KOCKS &AL.[1991b], is to inscribe a coarser grid onto orientation space, of a grid spacing that varies in the Θ-direction such as to represent all regions equally (as had been done by VADON &AL. [1980] for another application). This is one representation of a uniform distribution, true on a grand scale. If it is well designed, it guarantees some coverage everywhere (including near the symmetric orientations that are often of particular interest). The drawback is that the regularity of the grid may lead to 'interference patterns' with the grid used in the OD that it is meant to represent.

Another method is to determine a set of orientations by a *random* selection. This is also uniform on a grand scale. One may be able to represent a texture by fewer grains in this case, as compared to the regular grid; however, the local fluctuations remain large when the number of grains is not high enough. It is always prudent to check whether the chosen representation actually matches the particular data well enough.

Figure 35a shows a representation of the texture shown in Fig. 26; it contains only 141 grains, since those with weights less than 0.1 were discarded (a volume fraction of 0.02). Reconversion to a continuous distribution by the 'binning' method (Sec. 1.4), and smoothing by 5°, yields Fig. 35b. We have used the oblique *sample* orientation distribution here, and the same data as were shown in Fig. 26 as COD and SOD, and in Fig. 30 as CON (crystal orientation distribution in v-sections), are displayed in Fig. 35c as SON (sample orientation distribution in v-sections). The fact that Fig. 35b looks sufficiently close to Fig. 35c, especially considering the unavoidable degrading with multiple conversions, verifies the efficacy of the method.

The uniformity of a distribution of orientations of discrete grains may be improved by assigning an a priori weight to each grain – less for a grain in a dense region [KOCKS &AL. 1991b]; an example of this was shown in Fig. 32a. An actual texture is then represented by multiplying the a priori weight at a particular location in orientation space by the actual density at this location.

Finally, a different method to discretize a given OD was recently proposed by TÓTH and VAN HOUTTE [1992]. It is based on a statistical analysis, taking account of the variance considered acceptable for a particular property to be derived.

3.5 Summary. Recommendations

Pole figures, inverse pole figures

Under many simple circumstances, pole figures and inverse pole figures remain useful tools for describing texture. This is particularly true when the texture is known to be a fiber texture: here, a single inverse pole figure provides a complete and concise representation. Similarly, pole figures are most appropriate when a single one suffices; i.e., when its multiplicity is at least three. For cases one is used to, this description is frequently adequate.

Inverse pole figures are also often useful when the texture does not have fiber symmetry. In this case, one must either add a 'flag' to the point in the inverse pole figure (for discrete orientations) [VAN HOUTTE and AERNOUDT 1976], or combine it with another inverse pole figure for a second axis [WILLIAMS 1968b]. For example, rolling textures can be represented by one inverse pole figure for the rolling plane normal and one for the rolling direction. While it is true that the correlation between a point on one and the corresponding point on the other is not explicitly seen, the existence of ideal orientations or concentrations along a 'fiber' is directly seen in terms of the common Miller-indices description. This method of representation will be used to discuss model simulations (Chapter 9).

Quantitative comparisons of similar textures

Quantitative differences between textures that are qualitatively similar – such as between similar alloys, or between diverse model predictions – can often be represented concisely by specifying for each the intensities of a set of *components* or of *a fiber in orientation space*: a single bar chart of components, or a single line drawing of fiber intensities, can point up important differences graphically. When the peaks (or ridges) are not sharp, however, this compression of the information may be somewhat subjective.

When only two textures are to be compared, *density difference* plots, such as difference pole figures or difference ODs provide the most objective and accurate information. In addition, they should be resorted to when the expected differences are not known.

The comparison between discrete and continuous distributions provides another challenge. The most effective way is probably to convert the discrete to (quasi-) continuous distributions. The smoothing inherent in this process makes quantitative comparisons of peak heights difficult, but intensity *locations* should be reliable.

The representation of three-dimensional ODs.

The most complete description of a texture remains the distribution in orientation space. An enumeration of intensities on a fine grid is the least prejudicial representation. The grid needs to extend only over the part of orientation space that is irreducible (by any further application of symmetry arguments). MATTHIES &AL. [1990b] have summarized the irreducible space for all symmetry conditions and all types of OD sections.

Criteria for effective visualization may be different for different applications. Figures 36 and 37 display a set of recommendations that are based on the following rules:
- display at least the 'fundamental zone' of orientation space;
- restrict the number of sections to less than 12 if at all possible;
- make the sum of all sections exhibit the total projection (unless this requires too many sections or too coarse a section spacing).

Figure 36 shows oblique polar sections, with the azimuth the same as in a COD, so that the projection is the z-pole figure. If regular CODs are desired, the range of sections doubles which, for the same number of sections, decreases the resolution by $\sqrt{2}$ (and is also always inappropriate when there is a significant concentration near the 'North' pole). Fig. 37 shows polar SODs. The choice between *crystal* and *sample* distributions (oblique or not) depends on the case. When the sample symmetry is not completely certain, full-circle CODs will display the actual sample symmetry; SODs are most instructive when the texture is nearly uniaxial, or when ideal orientations, or fibers in orientation space, are expected.

CRYSTAL ORIENTATION DISTRIBUTION

SAMPLE SYMMETRY: Section shape

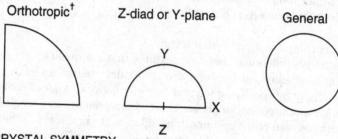

CRYSTAL SYMMETRY

		Section range		Multiplicity
Laue group		COD	Oblique sections	
Cubic	m3m	90°	45°	3
	m3	180°	90°	3
Tetragonal	4/mmm	90°	45°	1
	4/m	180°	90°	1
Orthorhombic	mmm	180°	90°	1
Hexagonal	6/mmm	60°	30°	1
	6/m	120°	60°	1
Trigonal	3̄/m	120°	60°	1
	3̄	240°	120°	1
Monoclinic	2/m	360°	180°	1
Triclinic	1̄	—	360°	1

†It is sometimes advantageous to use a semi-circle and cut the section range in half.

Fig. 36. Recommended ways to plot orientation data for the various cases of crystal and sample symmetry, as CODs.

SAMPLE ORIENTATION DISTRIBUTION

CRYSTAL SYMM.		SAMPLE SYMMETRY				
Crystal system	Laue group	Orthotropic	SOD range[†]	Other	SOD range for Z-diad or Y-diad[*]	General[†]
Cubic	m3m	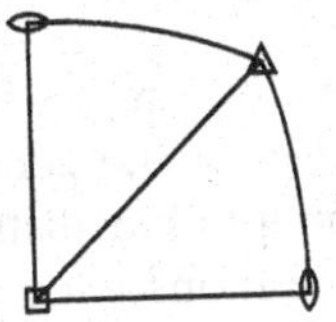	90°	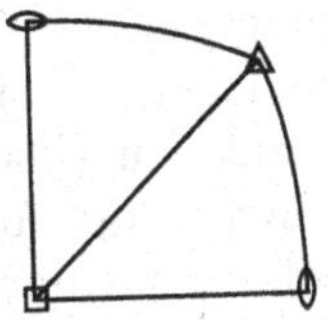		360°
	m3m		180°		360°	360°
Tetragonal	4/mmm	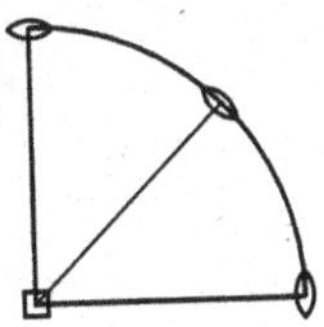	90°	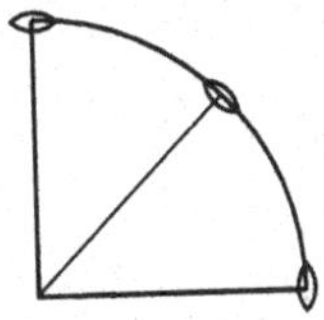	180°	360°
	4/mmm		180°		360°	360°
	4/m	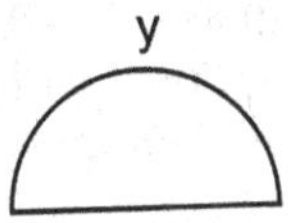	90°	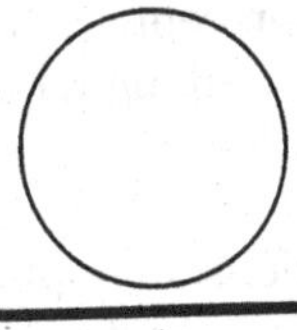	90°	180°
Orthorhombic	mmm		90°		90°	180°
Monoclinic	2/m		180°		180°	360°
Hexagonal	6/mmm	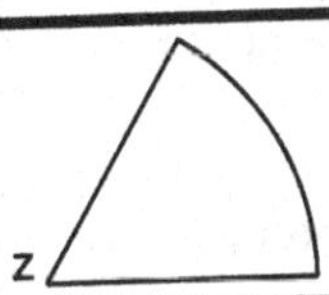	90°	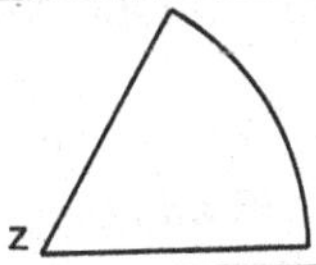	180°	360°
	6/m	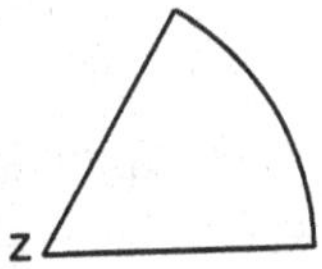	180°		180°△	360°△
Trigonal	3m		180°		180°△	360°△
	3̄	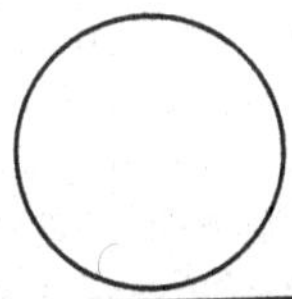	90°△	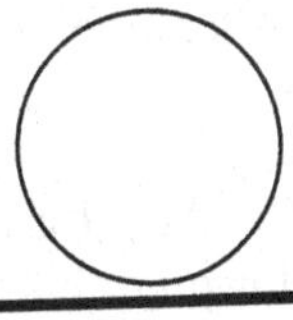	180°△	360°△
Triclinic	1̄		180°		360°	—

†Range for oblique sections is half this value
* For an X-diad, the range 90° to 270° suffices
△Multiplicity 3/2

Fig. 37. Recommended ways to plot orientation data for the various cases of crystal and sample symmetry, as SODs.

4. Continuous Distributions and Series Representation

4.1 The orientation distribution function

So far, we have described textures by an assembly of discrete orientations, or by a distribution that was represented by intensity values in cells in orientation space or pole space. This corresponds closely to how texture information is most often obtained and used in practice.

For general mathematical considerations, one represents such distributions as continuous *functions*. Retaining the symbol g for the general representation of an orientation, one then writes f(g) for the 'orientation distribution function' (ODF) [BITTER 1937]. The volume fraction of orientations within a certain region $\Delta\Omega$ of orientation space is then

$$\frac{\Delta V}{V} = \frac{\int_{\Delta\Omega} f(\mathbf{g}) \ d\mathbf{g}}{\int_{\Omega_0} f(\mathbf{g}) \ d\mathbf{g}} \tag{13}$$

where Ω_0 is the total volume of the orientation space considered: either the grand total, $8\pi^2$, or the irreducible subspace obtained by applying symmetry considerations (other than the inversion center). Note that f(g) as defined by Eq. 13 need not itself be normalized (by $8\pi^2$); common practice, however, is to set f(g) $\equiv$ 1 for uniform density. (This is often called the orientation density 'in multiples of a random distribution', m.r.d.)

When the orientations are represented by Euler angles (of any kind, with Θ being the pole distance), the infinitesimal volume element is

$$d\mathbf{g} = \sin\Theta \ d\Theta \ d\Psi \ d\phi \tag{14}$$

For *finite* increments of Θ, and especially when including regions near the origin, one should instead use the form

$$\Delta\mathbf{g} = \Delta(\cos\Theta) \ \Delta\Psi \ \Delta\phi \tag{15}[13]$$

When f(g) is plotted in an equal-area mapping {x,y}, Eq. 13 may be used to find volume fractions from f(g) (the orientation *density*), using dg=dx dy on each section. This method was used in Fig. 9.

The nature of the orientation distribution function as essentially a distribution of vectors on the surface of a sphere (Sec. 2.2) suggests that it may be represented by 'generalized spherical harmonics'. The formalism was proposed by VIGLIN [1960] and expanded into a useful tool by ROE [1965] and BUNGE [1965]; the latter has pioneered it as a general method for quantitative texture analysis, and applied it to a variety of problems [BUNGE 1982].

Because of this general use of the harmonic decomposition in the last three decades, the term ODF has sometimes been used as virtually synonymous with this representation (although there are others [WILLIAMS 1967]). Of course, a series expansion (with a finite cut-off) is merely another way to discretize a distribution. In a certain sense, the harmonic decomposition complements the representation by

---

[13] Note that the symbol Δg was used in Sec. 2.6 in a different sense: as the *misorientation*.

discrete components: it is especially appropriate for gradual variations and near-random distributions.

The harmonic decomposition proves quite generally useful when only the first few terms are needed. For example, the macroscopic average of elastic stiffnesses or compliances is fully accounted for by a series expansion to order four (in the leading index); unfortunately, these two averages provide only bounds and not actual anisotropies for the elastic constants (see Chapter 7). Low terms in the harmonic series are also useful for the discovery of certain symmetries (or lack of same); and for completing the unmeasured rim of pole figures as well as normalizing all at the same time [KALLEND &AL. 1991a].

One note of caution: while ODs are positive everywhere by definition, interpolations leading to a functional (ODF) description need not be. A general discussion of various forms of orientation distribution functions has been given by SCHAEBEN [1990]. This topic will be addressed in detail in Chapter 3.

It should also be mentioned that another decomposition into ortho-normal functions has been proposed: an expansion into a series of tensors of increasing rank [GUIDI &AL. 1992].

4.2 The pole and axis distribution functions

Similarly to the distribution of orientations $\mathbf{g}$, one may also describe the distributions of poles of family $\mathbf{h}$ (their [unsigned!] Miller indices {hkl}) at location $\mathbf{y}$, by a continuous function $p_\mathbf{h}(\mathbf{y})$. The differential element is, for a z-pole figure [$\mathbf{h}=(001)$], in symmetric Euler angles:

$$|d\mathbf{y}| = \sin\Theta \; d\Theta \; d\Psi \tag{16}$$

and the function itself may be looked upon as integration of $f(\mathbf{g})$ over the variable ϕ:

$$p_z(\mathbf{y}) = \frac{1}{2\pi}\int f(\mathbf{g}) \; d\phi \tag{17}$$

For arbitrary pole figures $\mathbf{h}$, a different set of variables must be introduced, and the integration occurs over a certain *path* in orientation space [MORAWIEC & POSPIECH 1989b]. The principle remains unchanged that continuous pole figures are generalized *projections* of the orientation distribution function. Like the ODF, they can be represented by a series of spherical harmonics (this time not 'generalized': not referring to vectors). For further discussions, see Chapter 3.

Projections contain less information than the original (3-D) distribution. When nothing further is known about the nature of the 3-D distribution (such as that it is a convex polygon, or that it consists of discrete points only), it can be shown that even infinitely many 2-D projections cannot be deconvoluted to yield a unique 3-D distribution. This indeterminacy of pole figure deconvolution is a central problem of quantitative texture determination; it will be addressed in detail in Chapter 3.

An 'inverse pole figure' is also a projection of the OD. It can be obtained for any sample direction; the most common one used is that for the sample Z-axis. At a crystallographic location $\mathbf{h}$, it is (in Roe or symmetric nomenclature)

$$a_z(\mathbf{h}) = \frac{1}{2\pi}\int f(\mathbf{g}) \; d\Psi \tag{18}$$

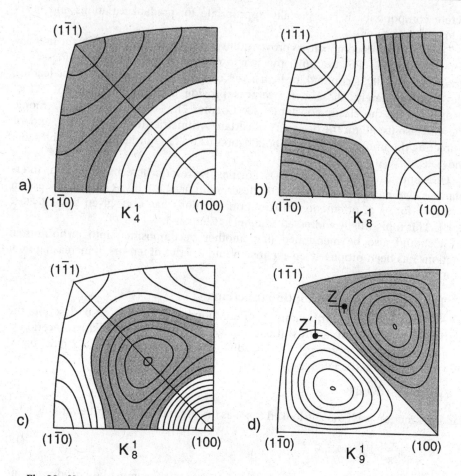

Fig. 38. Harmonic representation of the first four terms for an inverse pole figure for cubic crystal symmetry. After BUNGE [1982]. All negative intensities are shown shaded.

It, too, can be decomposed into spherical harmonics. It is interesting to look at the contours of individual harmonic coefficients of the integrand (presumably at certain value of ϕ) [BUNGE 1982, Fig. A.4.2]. Figs. 38a through c display the first three non-zero terms for the case of cubic crystal symmetry. First note how smooth the variation is: it would obviously take a large number of terms to represent a *sharp* distribution.

Figure 38d shows the first term of odd order l. It is the only one that distinguishes between the 'right-handed' and the 'left-handed' triangle. In fact, the distinction is strictly one of the *sign* of the density – and since the sign of the total density can never, in reality, be negative, the l-odd terms can never be dominating: they serve, in a sense, as correction terms to resolve the true rotational symmetry. They can actually not be determined from x-ray or neutron diffraction techniques. This is easy to see in the figure: the *axes* Z and Z' drawn in are equidistant to all the crystallographic poles; only if the *correlation* between all three poles (say, the three corners of the surrounding triangle) were measured with respect to each set of sample axes, could **Z** and **Z'** be distinguished (Sec. 2.5).

This example provides a specific illustration of the 'loss of information' in going from individual *orientation* measurements to continuous *pole* distributions. And it also shows how one may claim that the 'handedness' of orientations is the fundamental reason why they cannot be uniquely determined from continuous pole figures. (Distributions that are represented by a set of contiguous *cells* of uniform density behave, in this context, like continuous distributions.)

Finally, it is worth noting that the indeterminacy of pole figure deconvolution has nothing to do with the presence or absence of an inversion center in the crystal or sample; there is only a connection to the unsignedness of poles (for us, by definition). Assigning a sign to poles as discussed in Sec. 2.1 does not help this situation. Only if *continuous* distributions of *physically* signed poles (i.e., in non-centric materials) could be measured, would the indeterminacy disappear for these cases [MATTHIES &AL. 1987].

We have seen that ambiguities may exist in orientation distributions that are determined from pole figure deconvolution. In the harmonic representation, the differences are between odd-*l* coefficients. These are linked to a negative sign of cos Θ and thus to the 'lower hemisphere' or, equivalently, to the handedness of orientations when the upper hemisphere only is used (Secs. 2.1, 2.5).

4.3 The misorientation distribution function[†]

A misorientation can be associated with each grain boundary in a microstructure. The distribution of misorientations can be described using a continuous function, similar to the ODF. If the symbol Δg is used for the general representation of a misorientation [HAESSNER &AL. 1983], then $f(\Delta g)$ represents the 'misorientation distribution function' (MDF). The area fraction of orientations within a certain region $\Delta\Omega$ of misorientation space is then given by (compare with Sec. 4.1 Eq. 13):

$$\frac{\Delta A}{A} = \frac{\int\limits_{\Delta\Omega} f(\Delta g)\, d\Delta g}{\int\limits_{\Delta\Omega_0} f(\Delta g)\, d\Delta g} \tag{19}$$

where $\Delta\Omega_0$ is the total volume of misorientation space considered. (An area fraction is used instead of a volume fraction as grain boundaries represent planar areas in a microstructure.) Thus, just as the ODF describes the texture of a polycrystal, the MDF describes the misorientation or grain boundary texture of a polycrystal (see POSPIECH &AL. [1986] and ADAMS [1986]). As with the continuous description of orientation distributions described previously, quantitative continuous descriptions of the distribution of misorientations have usually been expressed in the form of Fourier series using generalized spherical harmonics. However, in place of the sample symmetry applied to such calculations the crystal symmetry is applied again. For example, for rolled aluminum sheet, the ODF would be calculated assuming cubic-orthotropic symmetry, whereas the MDF would be calculated using cubic-cubic symmetry. The harmonic basis functions for the series representation are defined in terms of Euler angles. Thus, although the axis/angle and Rodrigues descriptions of

---

[†] Contributed by S. I. Wright.

misorientation are generally considered superior to the Eulerian description, comparable orthogonal basis functions for these parametrizations have not yet been defined. MDFs have been presented in the literature using Euler angles, axis/angle and Rodrigues vector representations; however, the computations are generally performed in Euler space and the results then transformed to the other representations (for example, POSPIECH &AL. [1986]). With the advances in computer technology it is now practical to determine orientation distributions directly [MATTHIES & VINEL 1994] as opposed to the more traditional functional approach. Using the direct method, a misorientation distribution can be determined directly in misorientation space, employing any representation of the space.

If the spatial distribution of grain orientations were random, then the MDF could be simply derived from the ODF. For a simple example consider an ODF with two major peaks (which are not symmetrically equivalent). The MDF would contain a single peak (along with the other symmetrically equivalent peaks) at a location corresponding directly to the misorientation between the two peaks in the ODF. A method for obtaining the MDF from an ODF using the harmonic coefficients has been proposed by ZHAO &AL. [1988] and others. However, in real microstructures, a directly measured MDF generally differs from that derived from the ODF. POSPIECH &AL. [1993] have proposed dividing the measured MDF by the texture-derived MDF in order to delineate those features of the measured MDF that arise from correlated nearest-neighbor orientation relationships. This resulting MDF has been termed a texture-reduced MDF or a correlated MDF (or less aptly, a 'normalized' MDF, the term OCF or orientation correlation function has also been used). It should be noted, however, that dividing the measured MDF by the texture-derived MDF can lead to spurious results whenever zero-ranges (or near zero-ranges) exist in the texture-derived MDF (the texture-derived MDF is sometimes denoted as the orientation difference distribution function or ODDF).

While the texture-derived MDF can be obtained from conventional pole figure measurements, the measured MDF can only be obtained from individual measurements of lattice orientations. Individual misorientations can then be determined from appropriate pairs of orientation measurements. These misorientation measurements can be used to calculate the measured MDF using either the conventional series expansion method or the direct method. When single orientation measurements are used in the series expansion method, both the l-even and l-odd coefficients of the MDF can be directly obtained (in contrast to pole figures, where the odd terms can only be approximated through indirect correction techniques). Thus, a measured MDF calculated using the series expansion approach does not suffer the ambiguities related to the missing l-odd-terms. Of course, errors due to finite truncation of the series still persist. To correctly determine measured MDFs from individual measurements of misorientation, it must be remembered that each misorientation should be weighted according to the length of the corresponding boundary. (It would be more correct to weight each misorientation according to the *area* of the corresponding boundary; but currently the measurement of the *area* associated with the grain boundary plane is impractical.)

Examples of measured, texture-derived and correlated MDFs are shown in Fig. 39 and Fig. 40 for two different copper samples, one annealed at 600°C for 30 minutes [WRIGHT & HEIDELBACH 1994] and one obtained from a recovered shaped-charge jet

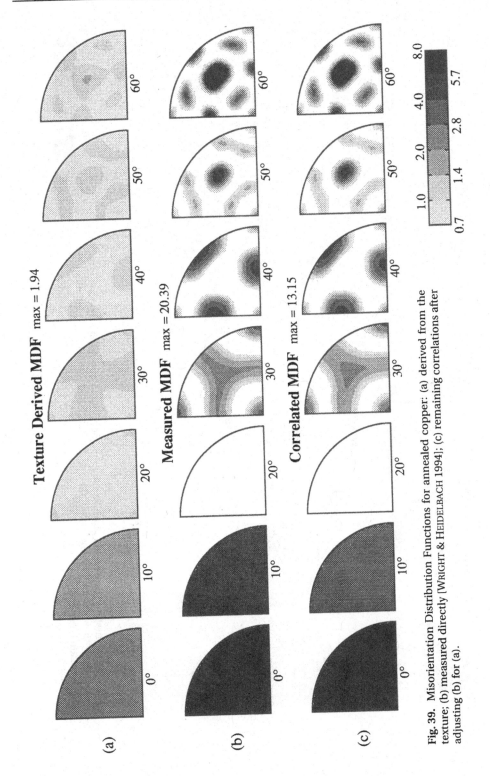

Fig. 39. Misorientation Distribution Functions for annealed copper: (a) derived from the texture; (b) measured directly [WRIGHT & HEIDELBACH 1994]; (c) remaining correlations after adjusting (b) for (a).

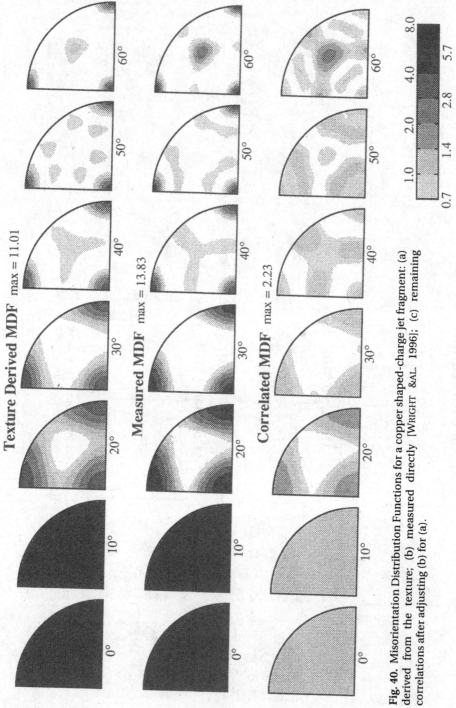

Fig. 40. Misorientation Distribution Functions for a copper shaped-charge jet fragment: (a) derived from the texture; (b) measured directly [WRIGHT &AL. 1996]; (c) remaining correlations after adjusting (b) for (a).

fragment [WRIGHT &AL. 1996]. As discussed in Section 2.6, the strong peaks at $\omega = 0°$ can be attributed to the measurement of low-angle 'grain boundaries'.

The texture in the annealed sample was nearly random. Thus, the texture-derived MDF gives very little indication that any statistically preferred grain-to-grain relationships exist in this material. However, the measured MDF shows a large number of $\langle 111 \rangle$- and $\langle 110 \rangle$- type twins. (The $\langle 110 \rangle$-type twins are geometrically necessary twins produced when two different $\langle 111 \rangle$-type twins of the same grain share a common boundary.) Since the texture-derived MDF was nearly random, the correlated MDF is simply slightly weaker than the measured MDF. This is in sharp contrast to the jet fragment sample.

The jet fragment exhibited a strong $\langle 100 \rangle$ fiber texture. Thus, we expect a large number of neighboring grains to have $\langle 100 \rangle$ axes in common. This was indeed the case, as shown in the measured MDF. However, the correlated MDF shows that the propensity for neighboring grains to have $\langle 100 \rangle$ axes in common was only slightly more than expected.

In Fig. 39 and Fig. 40, the MDFs show that the nearest-neighbor correlations are dominated by annealing twins. This result substantiates the general assumption made in texture simulations that there are no systematic grain-to-grain interactions with the exception of twinning. Until recently, it has been difficult to make statistically reliable characterizations of the local orientation relationships in polycrystals. Thus, whether the correlations observed in misorientation distributions are *generally* found to be dominated by twin relationships or not remains yet to be seen. In fact, POSPIECH &AL. [1993] observed a peak at 50° about $\langle 111 \rangle$ in the misorientation distribution of a copper sample rolled to 97% reduction; however, the number of measurements used in this study was relatively small.

Due to recent advances in the measurement of single orientations in the scanning electron microscope, it has become much easier to obtain the data necessary to determine statistically reliable misorientation textures in polycrystalline materials. The influence of the distribution of grain boundary misorientation on the properties of a polycrystal can be studied using these new techniques.

Chapter 3

DETERMINATION OF THE ORIENTATION DISTRIBUTION FROM POLE FIGURE DATA

DETERMINATION OF THE ORIENTATION DISTRIBUTION FROM POLE FIGURE DATA

J. S. Kallend

Direct measurement of the distribution of crystal orientations in a polycrystalline sample by bulk diffraction techniques is not possible. Conventional texture goniometry is only capable of determining the distribution of crystal poles or diffracting plane normals (stereographic or equal-area plots of these are the pole-figures), and further analysis is required to extract the orientation distribution from these data. This analysis is often referred to as pole-figure inversion.

In Section 1 the physical and mathematical relation between the OD and the pole figures is explained, and some of the difficulties are described. Section 2 elaborates on the earliest technique used for OD analysis, the harmonic method, and explores its advantages and disadvantages, and some of its extensions. A particular difficulty in OD analysis, the problem of ghosts, is the subject of Section 3, which also describes ways in which the harmonic method has been modified to deal with this problem.

Section 4 gives an outline of direct methods of OD analysis, describes two algorithms, WIMV and ADC in detail, and compares the advantages and disadvantages of direct methods with the harmonic method. The remaining sections of this chapter deal with operational issues in OD analysis, including data requirements, error assessment, and other practical considerations. Finally, we summarize some recommendations for both theoretical and experimental analysis.

1. Relation between Pole Figures and the OD

In a few practical cases the texture of a sample approximates a single crystal, and can be determined satisfactorily by inspection of experimental pole figures. An example would be the rotated cube texture found in palladium (Chap. 2 Fig. 22). In the majority of situations, however, this is not the case. In circumstances where the texture contains many components, or is diffuse or poorly defined, the most satisfactory way of obtaining explicit information from the pole figure data is by using those data to compute the orientation distribution, OD (or ODF), of the sample.

It has been shown in Chapter 2 that pole figures describe the distribution of the intersections of chosen crystal poles – i.e. (unsigned) plane normals – with the surface of a reference sphere (although mapped onto a plane surface for convenience). In a sample with many crystals, it is inconvenient to plot each pole, and so the *density* of poles on the surface is plotted as a continuous function (conventional x-ray techniques automatically produce this density through the averaging nature of diffraction). Two angular parameters are needed to describe the position of any pole on this surface, usually the polar angle α and the azimuthal angle β. It has also been shown that to specify the orientation, **g**, of the crystal coordinate axes **x, y, z**, with respect to an external (e.g. sample) frame **X, Y, Z**, requires three angles, such as the Euler angles Θ, ϕ, and Ψ, and specification of the handedness of the two coordinate systems. The distribution of crystal orientations over a suitable space defined by these angles comprises the OD. Once again, it is usually more convenient to plot the *density* of occupation of the orientations as a continuous function of the Euler angles. It is customary to normalize all density values so that the average is unity. Our problem, then, is to determine the OD, f(**g**), from the available experimental pole figure data, p(α,β).

In order to illustrate the relation between OD and pole figure, it is instructive to look first at the diffraction geometry of a particularly simple case. If we consider the (001) pole figure from a material with orthorhombic crystal symmetry, then the (001) pole is coincident with the z-axis. The Bragg condition specifies the direction of the pole of the diffracting planes with respect to the measuring apparatus. If the crystal satisfies the Bragg condition for (001), then rotating the crystal about its (001) pole by any arbitrary amount will not move this pole, and the crystal will still satisfy the Bragg condition in the new orientation. For this special case, the rotation about the (001) pole (or z-axis) just changes the Euler angle ϕ (Fig. 1a).

The total intensity of diffracted x-rays or neutrons that is measured at some point (α,β) on the (001) pole figure comes from all crystals in the sample satisfying the Bragg condition. In this particular case, the crystals must have the Euler angles $\Theta = \alpha$ and $\Psi = \beta$, while ϕ may have any value. The total pole density from crystals satisfying the condition is given (after normalization) by (Chap.1 Eq.17):

$$P_{(001)}(\alpha,\beta) = \frac{1}{2\pi} \int_0^{2\pi} f(\alpha,\beta,\phi) d\phi$$

(1)

In the general case for a pole of arbitrary (*hkl*), the pole figure is still obtained by integrating the OD along a path corresponding to a 2π rotation of the crystal about its (*hkl*) pole. However, in the general case this path does not necessarily correspond to a simple route in Euler space as it does for the (001) pole. The relation becomes:

$$P_{(hkl)}(\alpha,\beta) = \frac{1}{2\pi}\int_0^{2\pi} f(\Psi,\Theta,\phi)d\Gamma \qquad\qquad (2)$$

where Γ denotes the path through the OD corresponding to rotation about the (hkl) pole. Stated in words, the pole figure value at any point is obtained by integrating the OD along a path corresponding to a 2π rotation of the crystal about the diffraction vector. The pole figure may, in a general sense, be considered the *projection* of the OD along this path. This is illustrated in Fig. 1b.

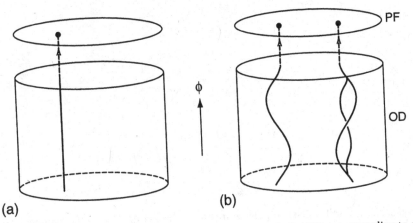

Fig. 1. The pole figure is a projection along a path through the OD, corresponding to a 2π-rotation around the diffraction vector. In simple cases, this is a straight projection along one of the Euler angles (a); in general (b) the path is curved, and symmetry may lead to multiple, equivalent paths.

It is a relatively trivial task to compute any pole figure, p, using Eq. 2, when given the OD, f. The central problem of texture analysis is the inverse of this: to compute f given p, which is very much more difficult. The task is similar to the familiar one of tomography, where a 3-D image of, for example, a human body is constructed from 2-D projections through that body made with x-rays. However, there is an important difference that makes texture analysis more challenging: there is no limit to the number of different views or projections that can be measured for tomography, whereas we are limited to relatively few pole figures that can be measured reliably. The information contained in projections is always incomplete: something else must be known about the 3-D object. For example, when it is known that it consists of discrete points, it can be unambiguously derived. However, for continuous pole figures, this indeterminacy exists; it will be addressed in Section 3.

In operational texture analysis, the continuous functions f and p are usually discretized, typically being evaluated on a 5° grid. The integral in Eq. 2 then becomes a summation. However, this does not remove the intrinsic difficulty of finding a solution. The equation is always underdetermined, and a unique solution is not available unless additional conditions can be applied. In fact, there exists a range of possible solutions to Eq. 2, which all fit the experimental pole figure data equally well.

To illustrate the problem, Fig. 2 (adapted from MATTHIES[1988]) shows a set of artificial pole figures, and three CODs (in some selected sections), each of which fit all

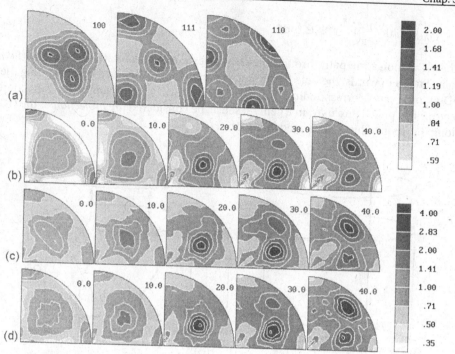

Fig. 2. (a) An artificial set of pole figures for a sample that has cubic symmetry in addition to cubic crystal symmetry (in equal-area projection, left scale). The remaining figures (right scale) show three different ODs all of which are in complete agreement with the pole figures in (a), all plotted as CODs (with φ as a parameter): (b) harmonic analysis to 22nd order (even only, positive intensities only); (c) WIMV analysis with assumption of maximal uniform background; (d) WIMV analysis without background. (After MATTHIES [1988], using popLA [KALLEND &AL. 1991a].)

the pole figure data exactly. It can be seen (for example in the 20° section) that the peak heights are different. In addition, a 'cube ghost' exists in all CODs: a component that is not reflected in the pole figure at all. (It is positive in some sections, negative in others – the latter shown as below .35 here). SCHAEBEN [1993] has analyzed in detail the nature and structure of the solution range for some typical textures. This indeterminacy manifests itself differently in different texture analysis algorithms, but is most transparent in the harmonic method. Detailed discussion of the problem will be delayed until after the harmonic method has been described. The various computational algorithms for OD analysis either implicitly or explicitly impose additional conditions in order to reach a unique solution. Examples of such additional constraints are:

(a) The OD must be non-negative everywhere.
(b) Zero values on a pole figure must correspond to zero values in all contributing orientations in the OD. (This follows from Eq. 1 and condition (a).)
(c) The OD should be the most probable, or 'maximum entropy,' solution within the range of possible solutions.
(d) The OD should be a 'smooth' function.
(e) OD peaks should have Gaussian or Lorentzian shape.
(f) The 'phon' or minimum value of the OD should be as high as possible.

Of these, only the positivity (a) and zero-range (b) constraints can physically be justified. The others are based on scientific intuition, and not all workers agree on their validity.

There are additional problems that must be handled by any method of OD analysis. Some of these will be enumerated below.

In many cases full pole figures are not available (for example, the outer $10°-15°$ range is often unmeasurable in an x-ray pole figure). To be useful, the method must accommodate this situation.

Real data are often noisy, and may contain experimental errors causing crystallographic inconsistencies within and between pole figures. An algorithm that works perfectly with model data, but becomes unstable if the data are noisy or internally inconsistent, is not suitable for routine texture analysis.

In low symmetry or polyphase materials it is frequently impossible to resolve pole figures from different forms that may have identical or very similar lattice plane spacings. In these cases a measured pole figure will contain information from two or more of these contributing forms. The relative intensity, averaged over the pole figure, contributed by each form will be proportional to its form factor, but there is no way of knowing, a priori, what the contribution is at any given location on the pole figure. Once again, it is desirable that a texture analysis method handle this situation.

2. Harmonic Method

The computational algorithms for OD analysis fall into two different categories, those in which the computations are performed in Fourier space (harmonic methods), and those in which the computations are performed directly in orientation space (direct or discrete methods). The latter will be treated in Section 4.

2.1 Fundamentals

The basic premise of the harmonic method is that the pole figure and OD are mathematically well behaved, and can therefore be fitted by a series expansion with appropriate mathematical functions. Appropriate functions to use in a spherical coordinate system are the *spherical harmonic functions* (just as sines and cosines are used in Fourier analysis of linearly periodic functions, and Bessel functions are used in cylindrical coordinate systems). Spherical harmonic functions are most familiar as the solutions to Schrödinger's equation for the hydrogen atom describing the shapes of electron orbitals. Spherical harmonic functions form an orthogonal set over the range $0 \leq \alpha < \pi, 0 \leq \beta < 2\pi$ (the surface of a sphere, which is used for pole figures).

The use of spherical harmonics to describe texture was proposed by PURSEY & COX [1954] and VIGLIN [1960], and a complete scheme for texture analysis was proposed independently by BUNGE [1965] and ROE [1965]. Although their methods are conceptually identical, the formalism used by Roe and Bunge is different. We will review Roe's formalism here, and then state the equivalent results using Bunge's formalism. Complete details of the harmonic method are given by BUNGE [1969, 1982], so it will only be outlined here.

The (hkl) pole figure, $p(\alpha,\beta)$, is expanded in a series of spherical harmonic functions:

$$p(\alpha,\beta) = \sum_{l=0}^{\infty} \sum_{m=-l}^{l} Q_{lm} P_l^m(\cos\alpha)\, e^{im\beta} \tag{3}^1$$

In this equation Q_{lm} are coefficients to be determined, $P_l^m(\cos\alpha)\, e^{im\beta}$ is a spherical harmonic function (P is an associated Legendre polynomial), and l and m are integers that govern the shape of the function (these would be the angular momentum and magnetic quantum numbers in the solution for the hydrogen atom); l is often called the 'order' of the spherical harmonic function.[2] This is an infinite series, but it is clear that truncation at some finite value of l is necessary in practice, limited by the number of data points measured. This limits the resolution of the method and leads to termination or truncation errors in the results. The coefficients Q_{lm} are, in principle, complex, but symmetries in the material often reduce the imaginary parts to zero, while the complex exponentials in Eq. 3 can be reduced to cosine functions.

Because the functions are orthogonal, the coefficients can easily be obtained from the experimental data p by integration.

$$Q_{lm} = \int_0^\pi \int_0^{2\pi} p(\alpha,\beta)\, P_l^m(\cos\alpha) e^{-im\beta} \sin\alpha\, d\beta\, d\alpha \tag{4}$$

It is assumed that the OD can be similarly expanded in a series of *generalized spherical harmonic functions*, specifically:

$$f(\Psi,\Theta,\phi) = \sum_{l=0}^{\infty} \sum_{m=-l}^{l} \sum_{n=-l}^{l} W_{lmn} Z_{lmn}(\cos\Theta)\, e^{im\Psi}\, e^{in\phi} \tag{5}$$

where W_{lmn} are the coefficients of this series, and Z_{lmn} are Jacobi polynomials. The problem becomes one of finding the relation between the unknown W_{lmn} and the experimentally accessible Q_{lm}. By substituting Eqs. 3 and 5 in Eq. 2, integrating, and making use of the Legendre addition theorem, we obtain

$$Q_{lm} = \sum_{n=-l}^{l} W_{lmn} P_l^n(\cos\xi)\, e^{-im\eta} \tag{6}$$

Here ξ and η are the polar coordinates of the (*hkl*) pole in the crystal coordinate system.

Equation 6 is a linear equation relating the pole figure coefficients with the OD coefficients. By measuring several pole figures from geometrically independent poles, we obtain a set of linear simultaneous equations that can be solved for the W_{lmn}. The W_{lmn} can then be substituted in Eq. 5 to obtain the OD.

The number of pole figures that must be measured depends on the number of independent unknowns in Eq. 6. This in turn depends on the order l to which we intend to compute the series. Symmetry in the crystal Laue group will reduce the number of independent W_{lmn} and the number of required pole figures. For example, cubic crystal symmetry dictates that $W_{2mn} = 0$ (and therefore $Q_{2m} = 0$), and in fact allows all the independent coefficients to be determined to $l = 22$ from only two pole figures [ROE 1965]. Statistical symmetry in the sample (Chap.1 Sec. 3) will further reduce both the number of independent coefficients that must be considered and the

---

[1] The summation convention is suspended for this Chapter; all sums are explicit.
[2] l is used in the literature in the generic Miller indices (*hkl*), and for the order of the spherical harmonic functions P_l^m. It is unlikely that this will confuse the reader.

irreducible volume of the OD, but does not affect the number of pole figures needed. Careful utilization of sample and crystal symmetries can considerably reduce the computational burden.

Once the W_{lmn} have been found, Eqs. 6 and 3 can be used to re-calculate pole figures, even for pole figures that were not measured or used in the analysis.

Using Bunge's formalism, the effect of symmetry is accounted for by using *symmetrized* spherical harmonic functions, and the normalization of these functions is defined differently. The overall effect is identical with Roe's formalism, but is implemented somewhat differently in the software. Bunge's equations corresponding to Eqs. 3– 6 are as follows:

$$p(\alpha,\beta) = \sum_{l=0}^{\infty} \sum_{v=1}^{N(l)} F_l^n \dot{k}_l^n(\alpha,\beta) \tag{3B}$$

where the F are pole figure coefficients, and the k are the symmetrized spherical harmonic functions.

$$F_l^v = \int_{\alpha=0}^{\pi} \int_{\beta=0}^{2\pi} p(\alpha,\beta) \dot{k}_l^{*n} \sin\alpha \, d\beta \, d\alpha \tag{4B}$$

$$f(\varphi_1,\Phi,\varphi_2) = \sum_{l=0}^{\infty} \sum_{\mu=1}^{M(l)} \sum_{v=1}^{N(l)} C_l^{\mu v} \ddot{T}_l^{\mu v}(\varphi_1,\Phi,\varphi_2) \tag{5B}$$

where T are the generalized spherical harmonics, and $C_l^{\mu v}$ are the OD coefficients, which are found from:

$$F_l^v = \frac{4\pi}{2l+1} \sum_{\mu=1}^{M(l)} C_l^{\mu v} \dot{k}_l^{*\mu}(\xi,\eta) \tag{6B}$$

(* denotes complex conjugate, and the dots over the harmonic functions denote the symmetrization of those functions. ξ an η are the same as in Eq. 6).

The relation between Bunge's and Roe's definitions of the Euler angles is given in Chapter 2. Conversion of harmonic coefficients between these two methods is straightforward, but not trivial because of the different normalization and Euler angle definitions used.

2.2 Incomplete pole figures

The simple calculation of pole figure coefficients Q_{lm} from the experimental data using Eq. 4 is a consequence of the orthogonality of the spherical harmonic functions over the surface of the sphere. Equation 4 cannot, therefore, be used as it stands with incomplete pole figures. Two ways have been used to overcome this difficulty. One of these [BUNGE 1982]) uses a least-squares analysis to find the OD coefficients that best fit the experimental data. This adds considerable complexity to the software, and has not found much application.

A second method is much easier to implement, and involves an initial extrapolation from the known to the missing part of the pole figures, usually with a third-order polynomial. For materials with cubic crystal symmetry the requirement that $Q_{2m} = 0$ can be used to properly normalize the values in this extrapolation [KALLEND &AL. 1991a]. Equations 4 and 6 are then used to calculate a first estimate of the W_{lmn}. These values are used to re-calculate the missing parts of the pole figure (from Eqs. 6 and 3). Any physically meaningless negative values are replaced with zeros. These recalculated parts contain information derived from, and crystallo-graphically consistent with, the real experimental data. The recalculated values are, therefore, much better estimates than the original extrapolated values, and are substituted for them. The process is repeated, each iteration improving the reliability of the estimate of the missing parts, until the self-consistency of the data is judged satisfactory. DAHMS & BUNGE [1989] suggest that even better results will be obtained by recalculating additional (non-measured) pole figures, correcting for any negative values, and then incorporating these into the iterative process.

2.3 Resolution of superimposed poles

This problem (which was alluded to in Sec. 1) can readily be handled by the harmonic method. Equations 3 and 6 show the linear relation between pole figure data, OD coefficients W, and the coefficients for individual poles Q. The coefficients describing a superposition pole figure will, therefore, be a linear combination of the coefficients Q^j for each of the J contributing forms, weighted by the form factor F_j for the jth form:

$$Q_{lm}^{sup} = \sum_{j=1}^{J} Q_{lm}^j F_j \tag{7}$$

With a minor modification to Eq. 6 to account for this superposition, the rest of the analysis is essentially unchanged. BAKER & WENK [1972] used this procedure in the analysis of quartz textures.

2.4 Inverse pole figures

The pole figure is a projection of the OD into sample space, whilst the inverse pole figure is a projection of the OD into crystal space, as described in Chapter 2. It is not surprising, then, that the harmonic coefficients R_{ln} describing the inverse pole figure can be obtained from the OD coefficients W_{lmn} by a variation of Eq. 6:

$$R_{ln} = \sum_{m=-l}^{l} W_{lmn} \, P_l^m(\Xi) \, e^{-im\Psi} \tag{8}$$

where Ξ and Ψ are the polar coordinates of the direction of interest with respect to the sample axes X, Y, Z. The inverse pole figure, r, is then given by:

$$r(\alpha,\beta) = \sum_{l=0}^{\infty} \sum_{n=-l}^{l} R_{ln} P_l^n(\cos\alpha) e^{in\beta} \tag{9}$$

The similarity with Eq. 3 is quite obvious, and is the result of the equivalence between the descriptions of the crystal and sample reference frames with respect to each other (Chapter 2). Software written to recalculate pole figures using Eq. 3 requires only minor modification in order to compute inverse pole figures. As a consequence, computation of inverse pole figures for any arbitrary sample direction is particularly easy in the harmonic method.

The OD may, of course, be computed from experimental inverse pole figure data by an analogous method to that described for normal pole figures. However, experimental difficulties make this a less promising method of OD determination except in special cases.

2.5 Advantages of the harmonic method

The harmonic method has a number of advantages. It is economical of computer memory, because each order of the solution can be computed independently. It is also quick, provided that a library of pre-computed spherical harmonics has been established. The inconsistencies within and between the measured pole figures are easy to evaluate, and may be used to estimate the experimental error in the OD.

Statistical symmetry in a sample is reflected in the pole figure and OD coefficients. This may be utilized in a symmetry analysis; e.g., the OD of rolled sheet metal should exhibit orthorhombic symmetry in the sample reference frame, but because of misalignment of the sample on the texture goniometer (it may have been rotated several degrees from the symmetry position), this symmetry may not be apparent. An examination of the pole figure coefficients can determine the extent of the misalignment, and the data can be corrected [KALLEND &AL. 1991a].

The set of coefficients W_{lmn} (or $C_l^{\mu\nu}$) is a compact representation of a texture, and can be used directly for very fast, simple property averaging calculations (see Chapter 7). Also, changes of coordinate system can be accommodated easily by manipulations of the coefficients. This allows texture changes during phase transformations to be analyzed easily [KALLEND &AL. 1976]. In addition, the harmonic representation is very convenient for theoretical study of texture analysis methods.

Truncation of the series expansion at a finite value of l (typically 22 for cubic crystals) has the effect of applying a noise filter. The harmonic method is stable and gives good results even when the input data are of poor quality due, for example, to large grain size in the sample, or low diffracted intensity.

2.6 Disadvantages of the harmonic method

There are several disadvantages to the harmonic method. First, an implicit assumption in the harmonic method is that the OD is a smooth function, and that truncation of the infinite series at an order l as low as (say) 22 will not degrade the solution. In practice this truncation does not seem to cause problems (in fact, for pole figures measured on a $5°$ grid, meaningful values of the pole figure coefficients cannot be obtained unless $l < 36$ anyway). The number of pole figures required to obtain a solution is modest for cubic crystals, but can become very large for the low symmetry crystal systems commonly found in non-metallic materials and minerals [BUNGE 1982]. For such cases an iterative procedure may be used to generate additional pole figures which are then re-used and refined in the next cycle (e.g. DAHMS [1992]).

Because the computations take place in Fourier space, the application of the additional conditions in Section 1 is not at all straightforward, and adds complexity to the software.

Accommodating the many possible combinations of crystal and sample symmetry also adds considerable complexity to the implementation of this method.

3. Ghosts

The distribution of poles on a sphere has an inversion center, as described in Chapter 2. The spherical harmonic functions are symmetric with respect to inversion when l is even, but are antisymmetric when l is odd.   Fitting Eq. 3 to a centrosymmetric pole distribution therefore demands that the Q_{lm} are zero when l is odd. For many years this was interpreted to imply that the W_{lmn} also had to be zero when l is odd. There is not, in fact, such a restriction [MATTHIES 1979]). All that is needed is that the W_{lmn} satisfy certain constraints so that the left-hand side of Eq. 6 is zero. As long as this is the case, it will be clear that W_{lmn} have no influence on conventional pole figures if l is odd, which means that normal diffraction pole figure data can give no information about the l-odd part of the OD (although this information is available if individual crystal orientation measurements are made).

It is convenient to consider the OD to be comprised of two parts corresponding to even and odd terms in its series expansion:[3]

$$f = \tilde{f} + \tilde{\tilde{f}}$$

$$(10)$$

where $\tilde{f}$ (the even part) can, in principle, be determined exactly from normal diffraction pole figure data. The odd part, $\tilde{\tilde{f}}$, is invisible in pole figures. It is limited by the positivity and zero range constraints, however, and its series expansion contains no even-order spherical harmonics. The decomposition of a calcite OD into even and odd parts is shown in Fig. 3 [WENK &AL. 1994b].

At first sight it may seem that fully one-half of the information is missing from the OD because of the missing odd terms. However, Eq. 6 does impose severe restrictions on the permitted values of the odd-l terms. This considerably reduces the number of independent non-zero values (e.g., for cubic crystals the first non-zero odd terms are for $l = 9$). Even so, the absence of those odd terms that should be present in the series expansion of the OD (Eq. 5) leads to underestimation of true texture maxima, and to undesirable artifacts in the function. These typically are on the order of 10% of the maximum peak height, but may be larger, especially in strong textures. Spurious peaks in a calculated OD that may result from these omissions are termed 'ghosts'. Because the integral of $\tilde{\tilde{f}}$ over the entire orientation space is zero, ghost peaks must be compensated by ghost holes elsewhere in the OD.   The physically meaningless negative regions that are often observed in ODs calculated by the harmonic method (using even terms only) are a consequence of this.

The original harmonic method outlined above does not make use of any of the additional constraints described in Section 1 above. The *implicit* assumption in its solution is that $W_{lmn} = 0$ when l is odd (or that $\tilde{\tilde{f}} = 0$). This makes such a solution particularly susceptible to ghosts. Fortunately, if the goal of the texture analysis is just to understand the anisotropic properties of a material, the absence of the odd-l terms is of little consequence, because most properties of interest are centrosymmetric and therefore are unaffected by these terms anyway (see Chap. 1 Sec. 1.2).

In Chapter 1, two categories of crystals were introduced that present special problems: enantiomorphic and polar.   Diffraction pole figures from non-centro-symmetric samples are, nevertheless, centrosymmetric (Friedel's Law). The chirality

---

[3] This decomposition applies whether or not the harmonic method is used for the texture analysis.

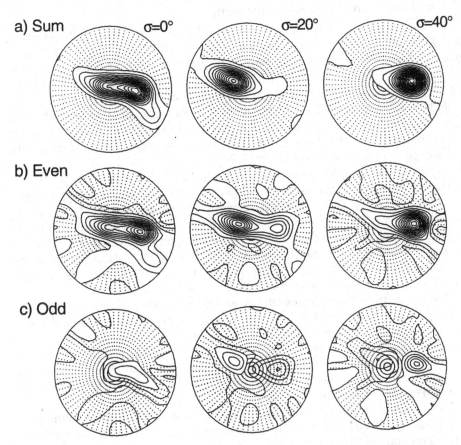

Fig. 3. Harmonic expansion of an OD that had been calculated using WIMV from seven pole figures of a quartz sample: (a) using even and odd functions (max. normalized density: 19/minimum 0); (b) even functions (14/1.8); (c) odd functions (6/-4.3). 'Ghosts' are best visible as negative regions in the odd expansion. Expansion is to order 22. Contour interval 1 m.r.d., negative regions dotted [WENK &AL. 1994b].

of enantiomorphic crystals, such as quartz, cannot be determined from pole figures. Each enantiomorph is a distinct phase and has its own independent OD, but these are indistinguishable in the analysis of normal pole figure data. All that can be computed is the superposition of these two distributions, which is the OD of the *Laue group*. Although this is different from the ghost problem described above, it does result in a similar consequence: ambiguity in the computed results.

Polar crystals present an extreme case for ghost effects. Although one end of a polar crystal is crystallographically distinct from the other end, these also cannot be distinguished in pole figures. The solution obtained in this case is also the distribution of Laue orientations. If additional information is needed the Laue OD must be supplemented with additional information that is sensitive to polarity (e.g. piezo-electric and pyroelectric response [GHOMSHEI &AL. 1988, JOHNSON & FERRARI 1988]).

3.1 Ghost correction in the harmonic method

Several methods have been suggested to deal with the problem of ghosts within the framework of the harmonic method. VAN HOUTTE [1983] suggested using a quadratic form, or an exponential form [VAN HOUTTE 1991], which guarantee no negative regions in the OD. Unfortunately these methods are computer intensive, and reduce the goodness of fit between the OD and the experimental data. LÜCKE &AL. [1986] have fitted Gaussian functions to the major OD peaks, and then used the odd-order spherical harmonics of the resulting model OD to obtain the missing terms of the experimental OD. This procedure is also computer intensive, and subjective in the choice of peaks unless the texture is quite sharp.

BUNGE & ESLING [1979] proposed the 'zero-range method', which makes use of pole figure zeros to find the parts of the OD that are known to be zero. In these regions $\tilde{\tilde{f}} = -\tilde{f}$, which is known. Odd-$l$ spherical harmonics can then be found to fit $\tilde{\tilde{f}}$ (either by least squares, or iteratively), and these are used to correct the OD computed by the standard method. Provided the texture is strong enough to produce pole figure zeros, this is the simplest of the ghost correction procedures. The disadvantage of this procedure is that the counting statistics of an x-ray (or neutron) measurement give rise to the greatest experimental errors in the low intensity range!

The positivity condition on the OD can also be achieved by an iterative procedure in which an odd correction function, $\tilde{\tilde{f}}(g)$, is added to the initially calculated (even-only) OD expansion such as to minimize the negative regions [DAHMS & BUNGE 1988, WAGNER &AL. 1990, DAHMS 1992].

These procedures generate non-negative solutions and reduce undesirable oscillations in the OD; in the literature these corrected solutions are frequently called 'true ODFs'. This should not be interpreted to imply that such a solution is either unique or correct. Nevertheless, a thorough evaluation of the different methods has shown that – provided good experimental data are available – reproducible results can be routinely obtained [WENK &AL. 1988].

4. Discrete or Direct Methods

WILLIAMS [1968a] first proposed a direct, iterative method for OD determination, but its computational demands compared unfavorably with the harmonic method and it received little use. Further development was made by RUER & BARO [1977] and IMHOF [1977], and recent improvements have been made by MATTHIES & VINEL [1982], PAWLIK &AL. [1991], LIANG &AL. [1988] and SCHAEBEN [1988]. The spur for the recent activity in this area was the discovery of the ghost problem described in Section 3. Although there are considerable differences between these methods, they all allow additional constraints on the OD solution to be incorporated far more easily than does the harmonic method. Some general features of all the direct methods will be described, and a more detailed description of the WIMV and ADC methods will be given.

In Section 1 the relationship between the OD and the pole figure was described. In the direct methods, both the pole figures and the OD are represented by discrete values, typically by dividing up their respective domains into a regular grid with a spacing of 2.5° or 5°. The projection paths relating cells in the irreducible part of the OD grid with cells on the various pole figure grids are established by consideration of the crystal geometry. Each pole figure cell corresponds to cells along one or more projection lines through the OD, depending on the crystal and sample symmetry. The value in each pole figure cell is the average of the values in the OD cells (weighted by the cell size) along the corresponding projection line(s):

$$p_h(y) = \frac{1}{N} \sum_{i=1}^{N} f(y \Leftarrow g_i) \tag{11}$$

Here we have, for brevity, adopted the convention that g represents a crystal orientation (i.e., three Euler angles), and y represents a point on the pole figure of the crystal pole h. The summation is over *only* those N orientation cells g_i in the OD that contribute to the pole density in cell y. Geometrical corrections for cell shape and size are omitted, and the data are conventionally normalized.

Equation 11 represents a system of linear equations, and its solution requires the inversion of a large, though sparse, matrix. In general, insufficient experimental data will be available to give a unique solution to Eq. 11. Additionally, the ghost problem manifests itself here in rank deficiency of Eq. 11 regardless of the amount of data available, so a solution is not possible without additional conditions. Different explicit or implied conditions lead to different solutions within the solution range (differing only in $\tilde{\tilde{f}}$). Thus the ghost problem is present in direct methods too, albeit in a (usually) different form than in the harmonic method.

One approach, the Vector Method [RUER & BARO 1977] uses OD cells that are larger than the pole figure cells in order to reduce the number of unknowns. This strategy allows direct inversion of the matrix described above. However, this method sacrifices resolution in the OD, and still does not overcome the ghost problem.

The approach of the remaining discrete methods is to make an initial estimate of the OD, $f_0(g)$, and use Eq. 11 to compute the pole figures that would result from this OD. These recalculated pole figures are compared, cell by cell, with the experimental data, and the initial OD estimate is updated based on the results of this comparison. This process is repeated until the difference between the recalculated pole figures and the experimental data falls to an acceptable level. The main differences between the methods involve the way in which the initial estimate is made, and the way in which the OD is updated on successive iterations. In all cases, calculations need only be performed over the irreducible asymmetric unit of the OD and pole figures dictated by sample and crystal symmetry.

4.1 WIMV algorithm

The WIMV (Williams-Imhof-Matthies-Vinel) algorithm [MATTHIES & VINEL 1982] makes its initial estimate (f_0) by placing in each OD cell the geometric mean of the values in the associated experimental pole figure cells:

$$f_0(g) = N_0 \prod_{j=1}^{I} \prod_{m_i=1}^{M_i} p_{h_i}^{exp}(y_{m_i})^{\frac{1}{IM_i}} \tag{12}$$

Here, I is the number of measured pole figures, M_i is the multiplicity of the ith pole, and N_0 is the normalization. (Although this may seem an arbitrary way of making an initial estimate, it is basically a time-saving step. If the initial estimate were that the OD is *random*, then the operation of the following Eq. 13 would lead to the same result as Eq. 12 after the first iteration). It is evident in Eq. 12 that pole figure zeros lead automatically to corresponding zeros in the f_o as required.

The 'inner iteration' then proceeds as follows. If p'' is the recalculated pole figure from the nth estimate of the OD, then the correction factor for each OD cell is the ratio of the geometric mean of the corresponding experimental pole figure cells ($=f_o(\mathbf{g})$) to the mean of the corresponding recalculated pole figure cells. The $(n+1)$st estimate is derived from the nth by multiplying with this correction factor:

$$f_{n+1}(\mathbf{g}) = N_n f_n(\mathbf{g}) \frac{f_o(\mathbf{g})}{\prod_{i=1}^{I} \prod_{m_i=1}^{M_i} p_{h_i}^n \left(y_{m_i} \right)^{\frac{1}{IM_i}}} \tag{13}$$

If, for example, the initial OD estimate in a cell is too large, then the corresponding recalculated pole figure values will also be too large, and the correction factor will be less than unity. The OD estimate for that cell will then be reduced in the next iteration, as required. Figure 4 is a flowchart for the inner iteration.

In practice, the algorithm converges rapidly, and a satisfactory solution is obtained after typically 10–12 iterations. Optionally, the correction factor can be raised to a power (>1) to speed convergence, but in the case of strong textures this can lead to undesirable oscillations in the solution.

An optional 'outer iteration' may be imposed, to limit the minimum value of the computed OD, thus raising the 'phon' or isotropic background.

$$f(\mathbf{g}) \geqslant f_{min} \tag{14}$$

The reasoning behind this is that any potential ghost holes or valleys will be filled, taking with them the corresponding ghost peaks. The phon-raising process is, of course, redundant in the event that the texture has a zero-range.

Examination of the algorithm makes it clear that it cannot generate negative values for $f(\mathbf{g})$, and that any zero pole figure value must give zeros in all corresponding OD cells. Both of the physically necessary constraints (a) and (b) in Section 1 are therefore satisfied automatically.

In the absence of the optional sharpening and phon-raising steps, the WIMV algorithm results in a 'maximum entropy' solution [SCHAEBEN 1988]. (Entropy is not used here in the thermodynamic sense, but because the quantity that is maximized in the formal statement of this solution has a mathematical form identical to that of entropy in statistical mechanics.) If a problem has a set of possible solutions, these solutions are concentrated about the one with maximum entropy [JAYNES 1957]; in other words, the maximum entropy solution is statistically the most probable solution that satisfies the given (insufficient) set of experimental data.

4.2 ADC algorithm

Although substantially similar to the WIMV algorithm, the ADC (Arbitrary Defined Cells) method due to PAWLIK &AL. [1991] differs in some important details. Where WIMV, and most other direct methods, consider projection *lines* in relating the pole figures to the OD cells, ADC uses projection *tubes*, whose cross section is determined by the shape of the pole figure cell concerned. Computing the intersection of these tubes with OD cells, although time consuming, allows the computation of the volume fraction of an OD cell that contributes to the pole figure, and leads to better smoothing in the final result.

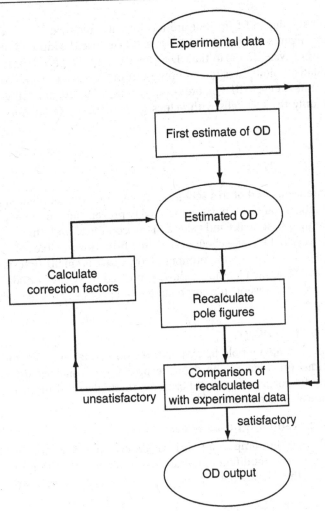

Fig. 4. Flow chart of WIMV algorithm.

The first iteration cycle is similar to the WIMV procedure described above, but is performed to obtain *recalculated* pole figures, p_o^{II}, for input to the second cycle. The purpose of this cycle is to pre-process the pole figures by averaging statistical errors over the whole set, and to provide accurate normalization of the data. Values of the OD, f_I, at the end of this cycle are saved for use in cycle III.

The second cycle starts with the allocation to each OD cell of the smallest value found in any corresponding p_o^{II} pole figure cell. These OD values are used to re-calculate pole figures as before.

Correction factors, Q_{II}^n are calculated, exactly as in WIMV, during successive iterations of this cycle. However, only if the correction factor is less than 1 is the correction applied to the current OD value:

$$f_{II}^{n+1} = \begin{bmatrix} f_{II}^n & \text{if } Q_{II}^n \geqslant 1 \\ f_{II}^n Q_{II}^n & \text{if } Q_{II}^n < 1 \end{bmatrix} \tag{15}$$

This means that values of $f_{II}(g)$ that are too high are reduced in this cycle. The re-calculated pole figures at the end of this cycle, p^{III}, are used as input to the third cycle.

At the start of cycle III, the initial OD values used are f_{II}, if $f_{II} < 1$, or f_I, if $f_{II} > 1$. Once again recalculated pole figures are computed from the current values of the OD, and correction factors for cycle III, Q_{III}, are computed as in WIMV, but using the p_{III} values. In this cycle only the OD cells with values greater than 1 (after normalization) are adjusted.

$$f_{III}^{n+1} = \begin{bmatrix} f_{III}^n & \text{if } f_{III}^n \geq 1 \\ f_{III}^n Q_{III}^n & \text{if } f_{III}^n < 1 \end{bmatrix} \tag{16}$$

The apparent purpose of this scheme is to separate the refinement of the below-random and above-random parts of the OD, preventing possible cross-coupling between potential ghost peaks and valleys. This, combined with the initial estimate of the OD used in cycle II has the effect of selecting from the possible solution range one in which the phon is large and the number of low peaks is small. In tests with a model function, the results appear similar to those produced by WIMV with the outer (phon) iteration option. At present, no compelling evidence suggests that either of these methods is superior.

4.3 Incomplete pole figures

The use of incomplete pole figures requires no special modifications to either of these algorithms, provided (i) the data are properly normalized (this may be per-formed 'on the fly', or by the use of harmonic methods), and (ii) at least three pole figure values are available for each OD cell.

4.4 Resolution of superimposed poles

The problem of superimposed poles was described in Section 1. The WIMV and ADC algorithms can be modified to handle this difficulty. Equation 11 is modified to take into account the diffracted intensity from the M contributing forms:

$$p_h^{sup}(y) = \frac{1}{M} \sum_{j=1}^{M} \frac{F_j}{N_j} \sum_{i=1}^{N_j} f(y \Leftarrow g_i) \tag{17}$$

where F_j is the relative contribution (form factor) for the jth form. This is an exact relation, and is used instead of Eq. 11 to recalculate the pole figures. This method has been used to resolve superimposed peaks in texture analysis of YCBO superconductor [KALLEND &AL. 1991b]. A further example is shown in Chap. 6, Figs. 17, where separation of the positive and negative rhombs of quartz (which have the same d-spacing) is performed. If the superposed peaks are from different phases in the sample, this method is still applicable, but the OD of each phase has to be determined independently, and the relative contribution to diffracted intensity from each phase must be incorporated in the F_j in Eq. 17.

4.5 Advantages of direct methods

Compared to the harmonic method, the addition of the positivity and zero range constraints is very straightforward in the direct methods (in most variations they form an integral part of the algorithm). The other, subjective, constraints can also be applied easily. In general, then, the problem of ghosts is handled more elegantly with these methods.

Accommodation of different crystal and sample symmetries is also straight-forward, usually involving little more than adjusting array assignments within the software. The analysis of incomplete pole figures is also accomplished easily. In general, the direct methods require fewer pole figures than the harmonic method to obtain a satisfactory solution. This is a particular advantage in the analysis of low symmetry crystal structures.

4.6 Disadvantages of direct methods

Unless explicit filtering is applied, these methods generally are more susceptible to noise in the data than the harmonic method (ADC includes an implicit filter). The robustness of most of these algorithms with respect to imperfect data has not been thoroughly analyzed. This is a particular problem because of the importance of the zero-range of the pole figures where, unfortunately, the data are least reliable.

Normalization of incomplete pole figure data is required, and may be performed 'on the fly'. However, this requires comparison of the recalculated pole figure data with the experimental data in the measurement domain, and must be repeated at each iteration step. Alternatively, the harmonic method may be used to normalize the pole figure data before performing the analysis by a direct method [KALLEND &AL. 1991a].

The output consists of a large set of numbers corresponding to OD values over a grid in Euler space. This is unwieldy (in comparison with harmonic coefficients) for use in upper and lower bounding tensor properties calculations, inverse pole figure calculations, and transformation texture computations. On the other hand, this may be the preferred form of the data for texture evolution simulations.

The geometrical information describing the relation between the OD and pole figure cells occupies a great deal of memory. For instance, cubic crystal symmetry with triclinic sample symmetry, using a 5° grid, requires nearly 480 000 pointers to relate the OD with the {200}, {220} and {111} pole figures. An efficient scheme for storing or calculating these pointers is necessary!

5. Operational Issues

5.1 Data requirements

The time required to collect pole figure data generally exceeds the time required to compute an OD by orders of magnitude, and this differential will increase as more powerful computers become available. The quantity of pole figure data needed for the analysis is clearly the limiting factor in controlling texture laboratory productivity. Use of more data than the minimum requirement is recommended for better statistics and error analysis. However, the OD solution depends also on the quality of pole figures and is thus not necessarily improved when more are added [WENK &AL. 1994].

In the harmonic method, the number of available pole figures determines the maximum order to which the harmonic expansion can be extended (and hence the resolution in the OD). Low symmetry materials require a greater number of pole figures. The reader is referred to BUNGE [1982] for an analysis of the pole figure requirements for different crystal systems.

As a general rule, three or more pole figure data points are needed that correspond to each OD cell computed by discrete methods. Each pole figure data point may, of course, be related to several OD cells. The data domain required to satisfy this condition is sensitive to the crystal symmetry, and has been investigated by HELMING [1992] for both complete and incomplete pole figures. However, for low symmetry crystals the domain depends on the crystal lattice parameters, and so no generally useful solution is available. Fortunately, texture analysis software (such as popLA) will usually inform the user if the data available are insufficient. To avoid wasting laboratory time it is recommended that a trial analysis is made with artificial data to verify that sufficient information will be available.

It is usually most satisfactory to select pole figures whose poles are geometrically independent and well distributed over the irreducible symmetry triangle of the crystal. This is especially helpful if incomplete pole figures are used, otherwise there may be some orientations whose corresponding pole figure cells are in the unmeasured range of all the available pole figures. If there are zeros in the OD, pole figures from poles with low multiplicity are more likely to exhibit a zero range. The use of these pole figures will therefore help to reduce the solution range.

5.2 Error assessment

Texture data are susceptible to errors from a number of sources: incorrectly prepared samples, misalignment of the goniometer, improper defocusing and background corrections, statistical noise in the x-ray intensity, etc. These errors are described in Chapter 4. Compared to these, errors introduced by the OD computations themselves are usually small. The various sources of error usually lead to a lack of self-consistency in the data, which can be detected during the analysis.

The method of analysis in most widespread use is the comparison of the experimental pole figure data with the pole figures recalculated from the orientation distribution. This gives a good measure of the crystallographic self-consistency within and between the pole figures for one sample. Direct comparison of these data may be made by subtraction of the experimental from the recalculated data (provided they are correctly normalized) to produce a set of 'difference pole figures.' This has the advantage of highlighting the exact location of any discrepancy in a visually striking manner; however, a large number of numerical data are involved.

More compact error measures may be obtained from the difference pole figures. The root-mean-square *absolute* error averaged over the pole figure tends to emphasize discrepancies in the strongest components of the texture; these usually dominate the anisotropic properties of a material. Another measure is the RP error (the RMS of the *relative* error, $\Delta p/p$, averaged over the pole figure) [MATTHIES &AL. 1988]. The RP error becomes indeterminate in the zero range, requiring a cut-off to be specified when evaluating this parameter; that is, the value is not evaluated for $f(\mathbf{g}) < \varepsilon$ when evaluating $RP(\varepsilon)$. The RP error emphasizes discrepancies in the weaker parts of the texture; these regions are often particularly sensitive to the mechanism of texture development. The purpose of the texture measurement should determine which of these error estimates is most appropriate.

Finally, it should be emphasized that there may be non-systematic sources of error in the data, which would not produce crystallographic inconsistencies. These will not be revealed by any of the methods described.

5.3 Severity of texture

It is sometimes useful to know how strongly a sample is textured, without regard for the details of that texture. A convenient measure, called the *texture index* [BUNGE 1982] is the mean square value of the OD (conventionally normalized to 'multiples of a random distribution', see Chap. 2). This is very easy to compute from the OD data or from harmonic coefficients. A random material has a texture index equal to unity, while textured samples have higher values. The square root of the texture index is, perhaps, more meaningfully related to the texture, since it is in the same units. It is called the *texture strength* in this book.

6. Recommendations

6.1 Comparison of algorithms

Many factors affect the selection of a method for OD analysis. All of the methods described are capable of producing a solution that is a good fit to the available experimental data, and all except the classical harmonic method can fulfill the positivity and zero-range constraints. All of the methods described can be implemented on inexpensive computers, and will perform the analysis in a reasonable time.

The investigator will need to consider the goals of the measurement, and the nature of material in making a choice of a method of analysis. If correlation of texture with the bounds or simple averages of anisotropic tensor properties in high symmetry materials is the only goal, the classical harmonic method offers many advantages and almost no disadvantages. The harmonic method loses a great deal of its elegance when modified for ghost correction and for use with incomplete pole figures, and becomes cumbersome if written in general enough form to handle many crystal and sample symmetry combinations.

If, for experimental reasons, the available data are of poor quality, a good choice will be an algorithm of proven robustness even though it may be theoretically inferior to another in tests with perfect data.

The various direct methods use different additional criteria to obtain an OD from the texture data. Of these, only the maximum entropy criterion can be said to have an objective basis, and that is only in a statistical sense. If the investigator has reason to believe that the texture of the material exhibits an isotropic background with a few strong peaks, then a method that raises the phon will be an appropriate choice. If the actual texture does not have that characteristic, then raising the phon may have the effect of suppressing some real features in the texture, such as small peaks due to the existence of twins.

The reader must be wary of tests claiming to show that one algorithm is better than another based on the analysis of model functions. A model function is a synthetic OD from which a set of pole figures is computed. These pole figures are then analyzed by the method under test to see if the results of the analysis agree with the original function. The pitfall in this approach is that a model function can easily be biased to favor one algorithm over another. For example, a model OD constructed from Gaussian functions will necessarily favor an algorithm that assumes the OD has Gaussian peaks. A model OD having a high isotropic background and a few strong peaks will necessarily favor an algorithm that raises the phon, while a model constructed only from even-l spherical harmonics will obviously favor the original harmonic method.

6.2 Analysis of experiments

The analysis of real data must be performed in a way that acknowledges the likelihood of errors in those data from a variety of sources, and minimizes their effect on the resultant OD. Factors to be considered include sample grain size, counting statistics, sample misalignment, data normalization, and number and choice of pole figures.

If the sample has large grain size, it is unlikely that x-ray pole figures will be smooth because the beam will not sample enough grains. Further, the statistical correlations within and between pole figures will be reduced because different sets of grains will respond to different Bragg reflections. Poor correlation between pole figures will become apparent during OD analysis, as it will result in poor agreement between the actual and recalculated pole figures. To rectify this problem, the investigator may (a) use neutron diffraction, which samples a much larger volume, (b) use a composite sample, or (c) prepare several samples, and average the pole figures [ENGLER &AL. 1996b].

Poor counting statistics will appear in the pole figures as 'noise' which will be readily apparent on visual inspection. The quality of the OD will be degraded as a result, but since the distribution of noise will not be crystallographically correlated, the OD will usually be less affected than the pole figures. The signal to noise ratio may be improved by using longer counting times, but if this is impractical the data may be smoothed with a suitable digital filter before OD analysis. We have found a Gaussian characteristic filter with a width of 3°– 6° to be suitable.

Visual inspection of experimental data often shows a small deviation from the expected statistical symmetry in the sample. For example, rolled sheet metal usually exhibits orthotropic symmetry, with the symmetry axes aligned with the rolling, transverse, and normal directions of the sheet. In practice, the observed symmetry axes are often misaligned slightly from these directions, due to slight misalignment when cutting the sample from the sheet or when mounting it on the goniometer stage. It is desirable to make use of symmetry in the sample during OD analysis, both to reduce computation time and to reduce the domain of the resultant OD. In order to do this, it is first necessary to 'rotate' the pole figure data to bring the symmetry axes into proper alignment. The optimum angle for this rotational correction can be found by inspection of the coefficients of the harmonic expansion of the data (the imaginary terms in the series expansion are minimized when the axes are optimally chosen). Once this angle is found, the rotation can be performed by interpolation using, for example, a cubic spline. This entire process can be performed automatically and very rapidly [KALLEND &AL. 1991a]. Figure 5 shows incomplete (111) pole figures from cold rolled copper, (a) before and (b) after rotational correction.

Incomplete pole figures are rarely normalized properly. Harmonic analysis of incomplete pole figures was described in Section 2.2. During this process, which is very rapid, the data are normalized to an accuracy that is usually better than 1%. The missing part of the pole figure can be filled in at the same time (Fig. 5c). The normal-

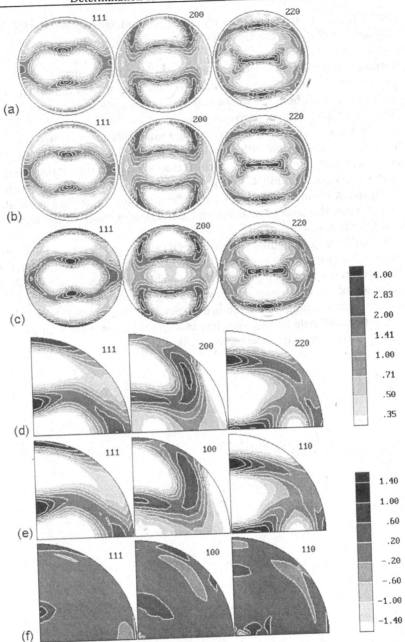

Fig. 5. (a) A set of experimental pole figures for copper, cold rolled to 80% reduction, after correction for x-ray background and defocusing, measured to an 80° tilt, and plotted in equal-are mapping (as all the following). (b) Same, after correction for a 2.6° misalignment (determined by harmonic analysis). (c) Same, after completing the unmeasured rim in a way consistent with all three pole figures, after re-normalization (performed by harmonic analysis). (d) Same, after applying orthotropic sample symmetry, and plotting only one quadrant (with greater angle resolution). (e) Recalculated pole figure, after performing WIMV analysis on the measured range of (c). (f) Difference between (d) and (e), on a sensitive scale. All operations were performed using popLA [KALLEND &AL. 1991a].

ized data may then be used as input to a direct method, such as WIMV (Fig. 5e). It is not necessary to compute the final OD by the harmonic method in order to benefit from this normalization.

Normalization may be performed during a direct OD analysis, but the process requires repeated comparison of recalculated and actual pole figure data over the measured domain. Although it works well in cubic materials, the results may be less satisfactory with low symmetry crystals.

The crystal and sample symmetry are automatically incorporated into any OD determination. Comparison with pole figures should therefore be undertaken after the latter have been symmetrized, which in turn should be undertaken after proper rotation and re-normalization: Fig. 5d provides the basis for comparison with Fig. 5e. This comparison is undertaken in a set of *difference* pole figures in Fig. 5f. It is seen that, in this case, the agreement is within 0.2 m.r.d. over much of the region; it is as much as 1 m.r.d. at just one point.

The number of pole figures required for OD analysis was covered in Section 5.1. Choice of pole figures is often dictated by the Bragg angles and relative intensities of reflections from the given sample with the available radiation. Some OD algorithms can make use of the zero range and low intensity regions of pole figures in order to narrow the solution range. If a choice is possible, these algorithms will usually perform better with pole figures from low multiplicity poles (which, in general, have more extensive regions with these characteristics).

Chapter 4

POLE FIGURE MEASUREMENTS
WITH DIFFRACTION TECHNIQUES

1. **X-ray Diffraction**

    1.1 Absorption
    1.2 Sample preparation
    1.3 Pole figure diffractometer
    1.4 Data collection in reflection geometry
    1.5 Data collection in transmission geometry
    1.6 Complete pole figures
    1.7 Intensity corrections and data reduction
    *Background correction. Empirical defocusing correction. Theoretical
    expressions for intensity corrections. Counting statistics. Normalization*

2. **Special Procedures in X-ray Diffraction**

    2.1 Energy dispersive and position-sensitive detectors
    2.2 Wires
    2.3 Thin epitaxial films: advantages of monochromatic radiation
    2.4 Determination of inverse pole figures
    2.5 Texture correction of powder data for crystal structure analysis
    2.6 Synchrotron x-rays
    Local textures in fine grained aggregates
    Individual orientations from Laue patterns
    2.7 Residual stress and dispersion

3. **Neutron Diffraction**

    3.1 Low absorption for neutrons
    3.2 Nuclear and magnetic scattering
    3.3 Conventional pole figure measurements
    3.4 Position-sensitive detectors and time-of-flight neutron diffraction
    3.5 Whole-spectrum analysis

4. **Electron Diffraction**

    4.1 Transmission electron microscope (TEM and HVEM)
    Selected area diffraction patterns (SAD). Kikuchi patterns
    4.2 Scanning electron microscope (SEM)

5. **Comparison of Methods**

POLE FIGURE MEASUREMENTS
WITH DIFFRACTION TECHNIQUES

H.-R. Wenk

Interpretation of textures in materials has to rely on a quantitative description of orientation characteristics. Two types of preferred orientations need to be distinguished: The *lattice preferred orientation* or 'texture' (also preferred crystallographic orientation) is the principal subject of this book and refers to the orientation of the crystal lattice. The *shape preferred orientation* (or preferred morphological orientation) describes the orientation of grains with anisotropic shape. Both can be correlated, such as in sheet silicates with a flaky morphology in schists, or fibers in fiber reinforced ceramics. In other cases they are not. In a rolled cubic metal the grain shape depends on the deformation rather than on the crystallography. Many methods have been used to determine preferred orientation. Geologists have extensively applied the petrographic microscope equipped with a universal stage to measure the orientation of morphological and optical directions in individual grains [see review by WENK 1985b]. Metallurgists have used a reflected light microscope to determine the orientation of cleavages and etch pits [see review by NAUER-GERHARDT & BUNGE 1986]. With advances in image analysis, shape preferred orientation can be determined quantitatively and automatically with stereological techniques [SANDLIN &AL. 1994]. In this Chapter we only deal with diffraction techniques which are most widely used to measure lattice preferred orientation. X-ray diffraction is most commonly applied and will be discussed in some detail. For some special techniques, the reader is also referred to reviews by BUNGE (pronounced boonga) [1986, 1996]. Neutron diffraction offers some distinct advantages but is not generally available. Electron diffraction using the transmission (TEM) or scanning electron microscope (SEM) is gaining interest because it permits one to correlate microstructures, neighbor relations and texture. For the latter techniques we mainly make readers aware of the possibilities and refer to the original literature.

There are two distinct ways to measure orientations. Usually one measures an average over a large volume of a polycrystalline aggregate. A pole figure collects a sum of lattice plane reflection signals from a large number of crystals. In the pole figure we have not only lost spatial information (for example what the orientations of the neighbors are), but also some orientation relations (such as how x, y, and z-axes of individual crystallites correlate). As was discussed in Chapters 2 and 3, it is not possible to determine the orientation density of crystallites in a polycrystal from densities in a pole figure without a certain ambiguity. It should be emphasized that pole figures measured with diffraction techniques are always centrosymmetric, even if crystals are non-centric. This is because diffraction averages over *volumes* and it is irrelevant if an x-ray beam impinges a lattice plane from the front or back side ('Friedel's law'). Note that this law does not hold for double diffraction in non-centric crystals which may produce a weak intensity difference ('anomalous scattering') which has so far never been applied in texture research.

The second method is to measure orientations of individual crystallites either by diffraction or optical techniques. In this case orientations and the orientation distribution can be determined unambiguously – certainly when diffraction intensities are considered. Also, if a map of the microstructure is available, the location of a grain can be determined and relationships with neighbors can be evaluated. These techniques are experimentally more involved and more limited than bulk diffraction measurements. Optical measurements have long been used to determine individual orientations of minerals. More recently electron diffraction techniques (Sec. 4) have been extensively applied.

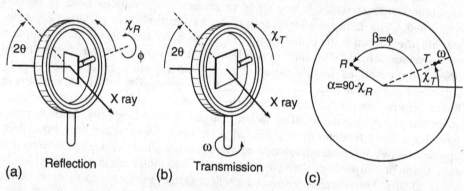

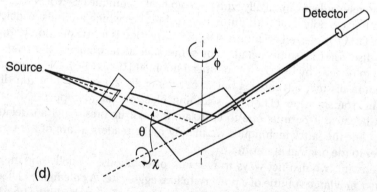

Fig. 1. Diffraction in a four-circle pole figure goniometer and definition of instrument angles. (a) Reflection geometry, (b) transmission geometry, (c) spherical projection illustrating the definition of a pole in reflection (R) by $\alpha = 90° - \chi$ and $\beta = \phi$ and in transmission (T) by $90 - \alpha = \omega$ and $\beta = \chi$. (d) In the Bragg–Brentano geometry a divergent x-ray is focused on the detector. This no longer applies when the sample is tilted about χ.

1. X-ray Diffraction

X-ray diffraction was first employed by WEVER [1924] to investigate preferred orientation in metals, but only with the introduction of the pole-figure goniometer and use of Geiger counters by DECKER &AL. [1948] and NORTON [1948] did it become a quantitative method. Many aspects of pole figure goniometry have been discussed in some detail by SCHULZ [1949a,b]. Bragg's law for monochromatic radiation is applied. It has two conditions. The first is that a lattice plane *hkl* diffracts if it is in a reflection position between incident and diffracted x-rays. The second condition is that lattice planes with a spacing d_{hkl} obey the law

$$2 d_{hkl} \sin\theta = n\lambda, \tag{1}$$

where 2θ is the angle between incident and diffracted beams, and λ is the mono-chromatic x-ray wavelength and n an integer defining the order of diffraction. The principle is simple: in order to determine the orientation of a given lattice plane, *hkl*, of a single crystallite, the detector is first set to the proper Bragg angle, 2θ, of the diffraction peak of interest, then the sample is rotated in a goniometer until the lattice plane *hkl* is in the reflection condition (i.e. the normal to the lattice plane or diffraction vector is the bisectrix between incident and diffracted beam). The gonio-meter rotations are related to the angular coordinates which define a sample orientation. In the case of a polycrystalline sample, the intensity recorded at a certain sample orientation is proportional to the volume fraction of crystallites with their lattice planes in reflection geometry.

Two methods of analysis are generally used: Determination of texture can be done on a sample of large thickness and a plane surface on which x-rays are reflected (Fig. 1a, *reflection*) or on a thin slab of thickness *t* which is penetrated by x-rays (Fig. 1b, *transmission*). Since x-rays are strongly absorbed by matter, the transmission method is only applicable to very thin foils or wires (<100μm) and to materials with relatively low absorption, except for some special experiments with tungsten radiation. Among common metals only Al, Mg, and Be can be studied with the transmission technique and most routine pole figure measurements are done with the reflection method. Depending on the compound that is analyzed details vary. Cubic metals with strong and widely spaced diffraction peaks at high Bragg angles are most easily measured (Fig. 2a). Low symmetry materials, such as many ceramics, minerals, polymers and polyphase composites, often have diffraction patterns with many closely spaced and weak peaks which often overlap (Fig. 2b). Here everything needs to be optimized: collimation, instrument alignment, choice of radiation and counting time. Other difficulties arise with very sharp textures as they occur in epitaxial films which are more similar to single crystals than to polycrystals.

For an x-ray source one generally relies on a generator and x-ray tube with a particular metal as anode material producing a wavelength spectrum with high intensity for Kα and Kβ of the anode element (Table I). Rotating anode generators produce higher intensities than ordinary vacuum tubes but require considerable maintenance. Recently some texture experiments made use of x-rays produced in a synchrotron with the advantage of high intensity, good collimation and arbitrary selection of wavelength. For x-ray tubes, the Kβ component in the spectrum can be reduced by applying absorption filters (Table I, Fig. 3). Single crystal monochromators

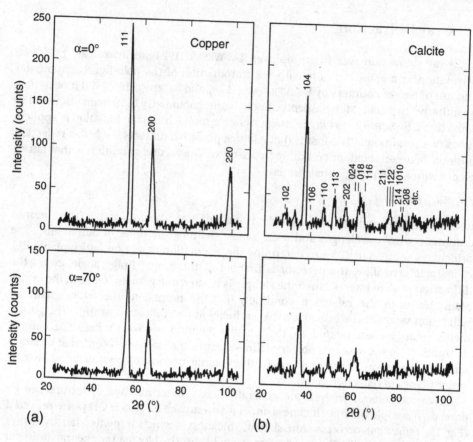

Fig. 2. X-ray 2θ-scans measured with a pole figure goniometer on samples with a random orientation distribution: (a) polycrystalline copper with cubic crystal symmetry, (b) a sample of trigonal quartz. Monochromatic FeKα radiation. The spectra on top are at a sample tilt α of 0° and those at the bottom at a tilt of 70°. Counting times for copper are 1 s, those for calcite 5 s.

or energy dispersive detectors are preferred for producing truly monochromatic radiation; their advantages for epitaxial films are discussed in Sec. 2.3.

Table I. Commonly used x-ray sources, corresponding wavelengths and β-filters and a list of elements which fluoresce strongly.

Anode	Wavelength (Å)		β-filter	Fluorescing matrices
	Kα	Kβ		
Cr	2.2909	2.0848	V	Ti, Sc, Ca
Fe	1.9373	1.7566	Mn	Cr, V, Ti
Co	1.7902	1.6207	Fe	Mn, Cr, V
Cu	1.5418	1.3922	Ni	Co, Fe, Mn
Mo	0.7107	0.6323	Zr	Y, Sr, Ru

It is assumed that the reader is familiar with basic concepts of x-ray diffraction and has consulted appropriate textbooks [e.g. AZAROFF 1968, BARRETT & MASSALSKI

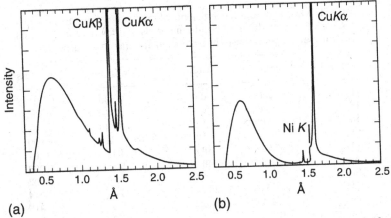

Fig. 3. X-ray spectra (a) Cu tube (operated at 40 kV) with pronounced peaks for Kα and Kβ radiation.   (b) Cu spectrum if a 0.001"-thick Ni β filter is used.   The Ni K absorption edge is indicated.

1980, CULLITY 1978]. We will first discuss some properties of x-rays which are relevant to texture analysis, briefly describe methods of sample preparation, review the geometry of reflection and transmission pole figure measurements and discuss special applications, especially for thin films and wires.

1.1 Absorption

Most x-rays used in texture analysis are highly absorbed by matter. Assume for example a slab of thickness t through which x-rays are transmitted (Fig. 4). The incident intensity I_0 is reduced to I following the relation

$$I = I_0 \exp(-\mu t) \tag{2}$$

where μ is called the linear absorption coefficient of the material (usually given in units cm^{-1}) which is a material constant that also depends on the wavelength of the x-rays. Generally, the mass absorption coefficient μ/ρ is tabulated, where ρ is the density. For materials composed of more than one element, μ/ρ can be calculated by a weighted summation over the individual mass absorption coefficients $(\mu/\rho)_i$:

$$\mu/\rho = \sum_i g_i (\mu/\rho)_i \tag{3}$$

where g_i is the mass fraction contributed by element i. Values for μ/ρ for all elements are given, for example, in Volume 3 of the INTERNATIONAL TABLES FOR X-RAY CRYSTALLOGRAPHY (Sec. 3.2). Table II lists a few examples.

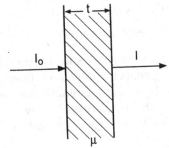

Fig. 4. Absorption in a slab of thickness t. The intensity is reduced from I_0 to I as the beam passes through material with a linear absorption coefficient μ.

Table II. Mass absorption coefficients of important elements, in units of cm^2g^{-1}, for three different anode materials and for neutrons (λ=1.08 Å). Numbers in parentheses indicate high anomalous absorption.

anode	FeKα	CuKα	MoKα	neutrons
H[1]	0.483	0.435	0.380	(0.11)
H[2]	0.483	0.435	0.380	0.0001
Be	2.80	1.50	0.298	0.0003
O	22.4	11.5	1.31	0.00001
Mg	74.8	38.6	4.11	0.001
Al	93.9	48.6	5.16	0.003
Si	117	60.6	6.44	0.002
Ti	(377)	(208)	24.2	0.044
Fe	66.4	(308)	38.5	0.015
Cu	103	52.9	50.9	0.021
Ag	391	218	25.8	0.20
Cd	412	231	27.5	(14)
Au	365	208	115	0.17
Pb	402	232	120	0.0003

In general, absorption increases with rising atomic number of the analyzed material and decreases as the wavelength becomes shorter. X-rays with short wavelengths (Mo or Cu anode) are less absorbed than those with large wavelengths (Cr or Fe anode) but there are exceptions. For example if Fe is analyzed with Cu radiation absorption is anomalously high. This anomalous absorption is caused by the fact that Cu Kα x-ray photons have a high enough energy to eject electrons from the K-shell in Fe. The released energy, due to the recombination of the electron hole, is emitted as fluorescence. A high absorption is undesirable; moreover fluorescence adds to the background of the measured signal. Therefore combinations of radiation (Kα) and material with high μ_a should be avoided. Some unfavorable combinations are indicated in Table I. For practical purposes the frequently used Cu Kα radiation should not be used to analyze titanium, steel or materials containing Co and Mn. Anomalous absorption can be used to eliminate undesirable Kβ radiation from the spectrum by filtering the primary x-ray beam with a foil of the corresponding material. Appropriate Kβ filters are also listed in Table I.

An important aspect of absorption to consider in x-ray texture analysis is the depth of x-ray penetration. For *transmission* geometry, we can calculate from Eq. 1 that the intensity of Cu Kα x-rays penetrating an Al foil 100 μm thick is reduced by 73%. In *reflection* geometry, the relationship is more complicated. According to CULLITY [1978] the fraction G of the total intensity of the measured signal which is contributed by the surface layer of thickness t is

$$G = 1 - \exp(-2\mu t / \sin\theta)$$

(4)

Sample thicknesses for G-values of 50% and 90% are shown in Table III. When analyzing copper with Fe Kα radiation, 90% of the intensity comes from a layer only 4.3 μm thick. For aluminum the corresponding layer is 15 μm thick and for silica (quartz) it is 22 μm. Textures may be heterogeneous. In a rolled metal the surface and the bulk texture are generally different and in such cases the choice of radiation may be important, depending whether the surface or the bulk texture is of interest. MORAWIEC [1992] described a method to deconvolute composite textures from reflection pole figures if the absorption coefficient is known.

Table III. Sample thicknesses t (in µm) which lead to G-values of 50% and 90% for different x-rays and materials analyzed in reflection geometry at a Bragg angle $2\theta=40°$.

Target: G:	FeKα 90%	50%	CuKα 90%	50%	MoKα 90%	50%
Al	15	4.7	30	9.0	282	85
Si	14	4.3	28	8.4	262	79
Ti	2.3	0.7	4.2	1.3	36	11
Cu	4.3	1.3	8.3	2.5	8.6	2.6
Fe	7.5	2.3	1.6	0.5	13	3.9
SiO$_2$	22	6.7	45	14	401	120

1.2 Sample preparation

Sample preparation for texture analysis is generally not difficult but nevertheless important. It requires preparation of plane surfaces of slabs for reflection measurements (Fig. 5a) or foils with plane parallel surfaces for transmission measurements

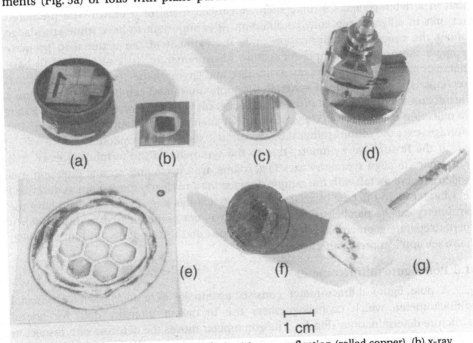

(a)   (b)   (c)   (d)

(e)   (f)   (g)

⊢———⊣
1 cm

Fig. 5. Samples used for texture analysis: (a) x-ray, reflection (rolled copper), (b) x-ray transmission (phyllosilicate thin section mounted on an aluminum frame), (c) assembly of wires for reflection measurements, (d) fragment of a bone on a glass capillary for synchrotron microbeam texture measurements, (e) TEM foil mounted on a Cu grid, (f) polished sample of recrystallized mylonite for EBSP measurements, (g) cube of a deformed quartzite for neutron diffraction, mounted on an aluminum rod and shielded with cadmium. The scale is indicated (except that the enlarged image with a hexagonal pattern in (e) has a diameter of 2 mm).

(Fig. 5b). A plane surface is essential; however, it is not necessary to have a high degree of polish which is often associated with non-planarity on the large scale. For ceramics and geological materials sawing and mechanical grinding are generally adequate; the bulk samples are usually cut with a diamond saw and then further ground with silicon carbide, alumina or diamond abrasive powders down to a grit size of 10 μm. For metals, surface layers may be deformed during specimen preparation, introducing texture artifacts. In such cases it is essential to remove the surface layer by chemical etching or by short electropolishing. Extended electropolishing produces a smooth but not plane surface.

The size of the sample is of utmost importance. It is necessary that the x-ray beam does not even partially leave the sample surface at any specimen tilt, even at oblique angles where the intersection with the x-ray beam is large as will be further discussed below. If the beam leaves the specimen in certain orientations it causes a reduction in diffracted intensity which is expressed in an apparent reduction of the texture strength. This is very difficult to correct. Keep in mind that also when the sample is translated in the goniometer for averaging it should never leave the beam. For reflection measurements it is advantageous if the sample surface has a circular shape (but clearly mark a reference sample direction, such as the rolling direction!); 2cm diameter is a good size and 1 cm is usually a minimum. It may be necessary to compose a specimen from several fragments (Fig. 5a), which is no problem as long as the fragments are strictly parallel and the sample is homogeneous. Note, however, that in a heterogeneous sample such as a rolled metal or a drawn wire measured textures in different cuts may be different. It is important to have some knowledge about the processing, deformation or heat treatment of the material to properly prepare the optimum sample for texture measurements. It is also useful to have microstructural information on grain size and grain shape to determine optimal sections. For coarse grained materials a sample translation can be applied but it may be difficult to obtain quantitative data on materials with grain sizes larger than 0.5 mm in reflection and 100 μm in transmission. For such materials neutron diffraction, that averages over the volume rather than the surface, should be considered.

If the texture has symmetry due to the symmetry of the forming process, it is customary to align symmetry axes (e.g. rolling, transverse and normal direction in a rolled sheet of metal) with the sample dimensions and clearly mark them as described in Chapter 2. In the case of axially symmetric textures it is advantageous to have the symmetry axis lie parallel to the sample surface, not perpendicular to it, in order to better separate geometrical intensity corrections and texture [WENK & PHILLIPS 1992] (also see application to wires in Sec. 2.2).

1.3 Pole figure diffractometer

A pole figure diffractometer consists essentially of a four-axis single-crystal diffractometer, which crystallographers use to collect intensity data for crystal structure determinations (Fig. 6). The goniometer moves the detector with respect to the x-ray beam (rotation 2θ); with an Eulerian cradle, the sample is positioned relative to the x-ray beam by two rotations, φ and χ (definitions in Fig. 1). The χ circle is generally symmetrical between incoming and diffracted beam (positioned at an angle θ, Bragg–Brentano focusing condition for reflection geometry, Fig. 1d) but may be different for special applications (ω is the deviation from the bisecting position, cf. Fig. 15a). The 2θ and the ω axis coincide.

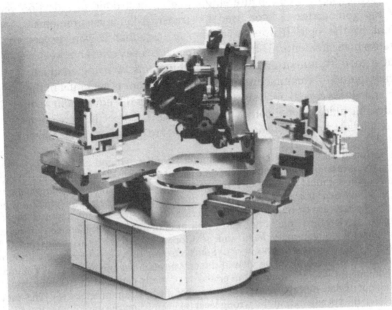

Fig. 6. Four-circle pole figure goniometer with texture attachment for sample translation (courtesy of Philips USA).

The nomenclature ϕ, χ and ω and θ is standard in single-crystal diffractometry and marked on most instruments. It is applied only in this Chapter and there should be no confusion with a different usage of ϕ and θ in other parts of the book. In texture analysis pole figure coordinates α (radial) and β (azimuthal) are more commonly used and we will apply those below. If the sample slab is mounted perpendicular to the ϕ axis (reflection geometry, Fig. 1a) the relationship is $\beta=\phi$ and $\alpha=90°-\chi$ (Fig. 1c). In a pole figure goniometer the crystallographic 'goniometer head' is replaced by a 'texture attachment' on which the sample can be mounted and oscillated. Stepper motors, controlled by a personal computer, enable one to obtain any arbitrary angular position on the four axes 2θ, ω, χ and ϕ (within a certain range to avoid mechanical collisions). For most applications only rotations on 2θ, χ and ϕ are performed and ω is kept constant at $0°$. The ω–scans (rocking curves) are sometimes used to assess the quality of crystals and exact peak shapes. For very thin films it may be advantageous to use a low incident beam angle to increase the path length of the x-ray (grazing beam technique, e.g. VAN ACKER &AL.[1994]).

For x-ray detection scintillation, proportional or energy dispersive detectors are used, the former two being most widely applied. With energy dispersive detectors the background from incoherent scattering and fluorescence can be eliminated. Recently position-sensitive detectors have been introduced and will be discussed in Section 2.1 of this Chapter. Fully functional texture systems composed of x-ray generator, goniometer, sample stage, detection system and electronics and computer control are commercially available. With some know-how it is possible to assemble a system from components at a much lower cost and in this case to tailor it to the specific needs of the user. Goniometers and electronics differ in details and instructions for installation and alignment which are supplied by manufacturers should be consulted.

The alignment of a pole figure goniometer is best done in reflection geometry even if the goniometer is later used for transmission measurements. A perfect alignment of the pole figure goniometer is essential if you expect to obtain quantitative measurements in an optimal time. It is a day or even a week well spent to arrive at a good alignment and to verify all functions by using adequate standards. Since the expertise of manufacturers of x-ray equipment is generally in the domain of powder and single-crystal analysis, the scientist using the equipment is often a more knowledgeable judge when it comes to texture analysis. Specific procedures vary from instrument to instrument but some general steps are included in the following.

- It is assumed that the four-circle goniometer is functional and aligned. Particularly the four axes of rotation must be eucentric, i.e. intersect in a single point, and the incoming beam collimator must be pointed at this geometric center. If this mechanical alignment is not achieved or if the goniometer has been misaligned by some forceful collision, it is generally necessary to return it to the manufacturer. The eucentricity can be checked optically by viewing a small crystal (0.1 mm in diameter) which has been mounted on a crystallographic goniometer head and positioned with three translations (x,y,z) in the center of the collimator viewing area. Some systems provide a telescope with crosshair, alternatively the crystal can be viewed directly through the incoming x-ray collimator. If the goniometer is aligned and if the crystal is properly centered in the eucentric point it should not deviate during rotations about any of the four goniometer axes. This procedure may fail if the specimen is at the wrong level (z) even when the goniometer is well-aligned. The eucentricity check does not require x-rays.

- The next step is to center the goniometer in front of the x-ray tube. For this at first the largest collimator which is available (e.g. 2 mm diameter) should be used. The goniometer should be shifted, raised or lowered and tilted (or alternatively the x-ray tube should be tilted if this is possible) until a beam is observed with a fluorescent screen (e.g. mounted on the sample holder instead of a specimen). Then the intensity should be maximized by further tilting and shifting. Repeat the procedure by going to smaller collimators, including the smallest one (e.g. 0.5 mm diameter), even if that is not used for later routine measurements. Special caution is required because during the goniometer alignment the fail safe system must be disabled. For the first steps of this alignment a fluorescent screen is used to detect x-rays. For the final steps the electronic detector should be used to maximize the intensity. Detectors should not be exposed to the primary beam at full power. Keep the x-ray tube at maximum voltage (to avoid shifts in the focal spot) and reduce the current to a minimum (< 1 mA). If necessary insert absorption filters in front of the detector.

- Install the collimator which is used for routine operation. It has been found that a rather small circular collimator (e.g. 1 mm) gives optimal results, at least if no Soller slits are used. Mount a standard sample in the specimen holder. A good standard sample with a random texture is an aggregate of Cu filings embedded in a matrix of epoxy and with a flat polished surface. Set χ to 90° ($\alpha=0°$, Bragg condition for sample surface) and perform a θ–2θ scan over a strong Bragg peak to determine the true value of 2θ and adjust the zero value if necessary. Go to a different value of χ (e.g. 30°) and perform another θ–2θ scan. You should obtain the same value for θ, otherwise the sample height is probably incorrect.

- Check the eucentricity with the primary x-ray beam. Either remove the detector and use the fluorescent screen or use the detector but set the beam at minimum dosage (additionally absorbing filters may have to be introduced to reduce the intensity). Set θ–2θ to 0° and rotate through χ. You should always observe half the full beam ('half moon') on the screen and the intensity should not vary during rotation of χ.

- Mount a single crystal fragment whose surface corresponds to a crystallographic plane (cleavage fragments of MgO, LiF are good materials). Find a reflection *hkl* of the sample surface by setting the detector to the appropriate 2θ and rotating ϕ in the vicinity of $\chi=90°$

which is best done manually (though this is not possible on some goniometers). Ideally the reflection should occur at $\chi=90°$ and should not change in intensity as you rotate ϕ through 360°. Condition is that the sample is exactly at the right height and with hkl parallel to its surface. Adjust χ and tilts on the goniometer slightly to achieve this condition without losing intensity.

- Attach a metal tube with a wide opening (so as not to interfere with the diffracted beam) between detector and sample, as close as possible to the sample. This reduces contributions to the signal from air scatter.

- Align detector electronics. Apply a voltage to the detector (around 1 kV) near the 'plateau' where fluctuations in voltage have least effect. For scintillation detectors an energy window is chosen by electronically discriminating pulses below and above a threshold (adjusting baseline and window, best with an oscilloscope). This improves the signal to noise ratio. The alignment can be done on the strong 111 Bragg peak of a Cu standard sample with no preferred orientation . In principle this needs only to be redone if the detector voltage or the x-ray wavelength is changed, but ought to be checked periodically.

- Choose an appropriate receiving slit. The larger the slit the smaller are the intensity corrections. But this reduces angular resolution, both for 2θ and for texture. For each application a compromise must be chosen which is different for relatively smooth textures as in deformed metals with widely spaced diffraction peaks and for very sharp textures such as Y-Ba cuprate epitaxial films with many closely spaced and overlapping peaks.

- With the standard sample inserted, count for a given amount of time and record the counts on the Bragg peak, counts for the background off the peak and the goniometer and electronics settings in the log book. It is very important to verify periodically that the measured intensities have not changed significantly. Changes would indicate misalignment of the goniometer, malfunctioning of electronics or deterioration of detector or x-ray tube.

- The choice of the x-ray tube target material depends on the material to be analyzed: Soft x-rays (Fe, Cr) penetrate the sample less and analyze the surface. Their larger wave length produces a better angular resolution. The most widely used target material, Cu, fluoresces with Fe and should not be used for steel and cobalt. Naturally, if $K\alpha$ radiation is selected with a β filter, rather than a monochromator, the proper filter element must be chosen (Ni for Cu-radiation and Mn for Fe-radiation, see Table I). For voltage and ampere settings one should consult recommendations by the tube manufacturer. They depend on anode materials and focus size. If you use ampere settings below the recommended maximum you extend the life of your x-ray tube.

1.4 Data collection in reflection geometry

The ideal specimen consists of a thick (>1 mm) slab with a flat surface and a roughly circular shape (>2 cm diameter). As discussed above, sample penetration varies with material and radiation and one should ascertain that the sample is representative over the irradiated volume.

The slab is mounted normal to the rotation axis ϕ and is rotated in its own plane. In a pole figure the angle of rotation about the ϕ axis corresponds to the azimuth β of a pole and the rotation about χ to the pole distance $\alpha = (90° - \chi)$ (Fig. 1 a, c). The third rotation axis is generally constant at $\omega = 0°$ so that the χ circle remains in a position bisecting the incident and reflected x-ray beams. For special purposes, in highly textured materials like epitaxial films, ω scans have been used to document dispersion of texture peaks [e.g. HEIDELBACH &AL. 1992]. Also, for heterogeneous textures of coatings and surface layers, equal sample penetration over the whole pole figure can be achieved by changing ω as a function of sample tilt χ but this requires availability of a position-sensitive detector [BONARSKI &AL. 1994].

If lattice planes *hkl* of all crystallites are parallel to the surface of the specimen, diffraction will only occur at $\alpha = 0°$ (i.e., when the specimen surface obeys the Bragg reflection condition). If all lattice planes are at right angles to the surface, diffraction occurs at $\alpha = 90°$ and only for that particular β which brings the lattice planes into reflection. In the case that crystallites in the sample have a random orientation distribution, as in a powder, an equal number of crystallites diffract at all angle settings, and the recorded intensity should be uniform (we will modify this statement later). In polycrystalline samples exhibiting preferred orientation, the intensity recorded at a detector setting 2θ for a Bragg peak *hkl* varies as a function of α and β, depending on the relative number of lattice planes that are in the Bragg reflection condition for each orientation. For each sample orientation the intensity diffracted on lattice planes is directly proportional to the volume of crystallites that are in that orientation and, if the grain size is uniform, to the number of crystallites.

A pole figure is scanned by measuring the diffracted intensity at different $\{\alpha,\beta\}$ settings. This can either be done in a stepwise manner by counting at discrete positions (typically 1 – 100 s) or by recording the average intensity continuously over an angular range from one setting to the next. The latter has the advantage that statistics are improved but at present the former method is almost universally used because of the ease of driving motors to specific positions. Due to the decreasing inclination of the specimen surface to the x-ray beam, the beam covers a larger area at high angles α and this causes 'defocusing' which will be discussed in Section 1.7. Therefore a pole figure can be measured in reflection reliably only to $\alpha=80°$ or at most 85° and the overall intensity decreases towards the edge of the pole figure (Fig. 7a). If proper intensity corrections are applied, this defocusing effect can be compensated for (Fig. 7b).

It has become traditional to scan the pole figure in a pattern with 5° steps, both in pole distance ($\alpha=90°-\chi$) and azimuth ($\beta=\phi$), resulting in an array of 72×19=1368 numbers that are subsequently used for processing, representation and analysis (Fig. 8a). This is statistically not optimal because the center of the pole figure gets covered in a much denser pattern in β than the periphery. An equal area coverage is recommended, particularly if measuring time is of concern. An elegant scheme, covering the pole figure uniformly with a hexagonal pattern in equal-area projection, has recently been introduced [MATTHIES & WENK 1992, Fig. 8b]. Measuring time (i.e. the number of measurements to cover the pole figure) is reduced by up to one half for

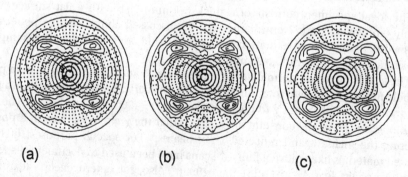

(a)   (b)   (c)

Fig. 7. 100 pole figure of recrystallized aluminum measured in reflection geometry. (a) raw data, (b) after intensity correction, measured on a 5°×5° grid, (c) measured on a hexagonal grid with 10° resolution. Equal-area projection. Contour interval 200 cps, regions with pole densities below 600 cps are dotted.

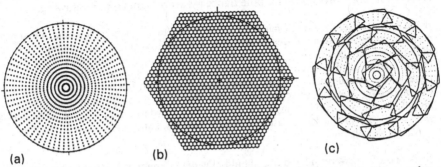

Fig. 8. Schemes for pole figure coverage by (a) an angular 5°×5°-grid, (b) a regular hexagonal grid and (c) with 27 histograms using the SCD TOF neutron diffraction area detector. Solid circles are at 30° intervals. All plots in equal-area projection.

the same 5° resolution. This scheme is particularly useful because the hexagonal grid can be easily adjusted for any desired angular resolution of the texture. With two-dimensional splines, values for the conventional angular 5°×5° grid, or for any other grid, can be interpolated and smoothing operations can be applied. Figure 7c shows the same 111 pole figure of Al represented in Fig. 7b but measured at hexagonal grid points with 10° resolution. Clearly both pole figures are very similar. The pole figure in Fig. 7c relies on measurements at only 265 $\{\alpha,\beta\}$ settings.

   Grain statistics for coarse grained materials are improved by rapidly oscillating the sample over 1 cm or more in its own plane provided that the texture is homogeneous. Even so single spikes may appear in the data due to diffraction from individual large grains. To some extent these can be eliminated by data processing. Yet grain size is often a limiting factor of x-ray texture measurements. Remember that oscillating the sample requires a larger surface area to ensure that the beam does not leave the specimen. Another method to increase statistics is to measure several samples and average pole figures [ENGLER &AL. 1996b].

   For detailed investigations of texture components in very strongly textured materials with halfwidths of texture peaks of less than 5°, one may want to scan, after a preliminary qualitative pole figure determination, certain orientation ranges with a finer grid. Figure 9 shows a scan over two texture peaks for 102 of an epitaxial Y-Ba cuprate film deposited on $LaAlO_3$ in steps of 0.5° on α and 0.2° on β.

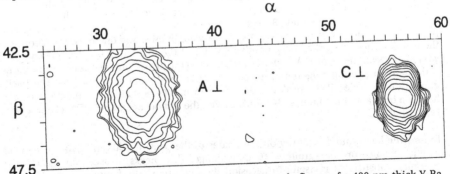

Fig. 9. Detailed α–β scan over a sector of the 102 pole figure of a 400 nm thick Y-Ba cuprate film, laser deposited on $LaAlO_3$. Increment in α is 0.5° and in β 0.2°. Monochromatic CuKα radiation. A and C denote two texture components (cf. Fig.18).

A recommended procedure for pole figure measurements is the following.

- If you are not absolutely sure what the material is, it is recommended to perform first a 2θ-scan on a powder with a powder diffractometer of higher angular (2θ) resolution than that of the pole figure goniometer. This is necessary if there is a possibility of multiple phases, of precipitates or of ordered domains which may produce superstructure reflections. All peaks of the powder pattern need to be indexed (e.g. with the help of the JCPDS (Joint Committee on Powder Diffraction Standards) file) and possible peak overlaps have to be identified. The reason for doing such a scan with a powder rather than with the textured sample is, that in the latter case, for strong textures, some peaks may be absent in a particular sample orientation.

- With a powder pattern in hand one should now determine *which* pole figures ought to be measured. This point can not be overemphasized. Several issues are at stake: (1) The quality of the OD calculation improves with the number of available pole figures [WENK &AL. 1994]. (2) However, pole figures with experimental errors are very detrimental. If possible do not use weak reflections with poor peak to background ratios, reflections that are partially or completely overlapped, or very low angle reflections. (3) It is necessary to have a minimum coverage of orientation space to resolve the OD [HELMING 1991]. This coverage is determined by symmetry, number and extent of pole figures, and conditions are different for the harmonic method and discrete methods of ODF analysis. In some cases a single incomplete pole figure is sufficient (e.g. 111 measured to 80° for cubic crystals), in other cases five incomplete pole figures may be necessary. The coverage of orientation space for discrete methods can be evaluated (e.g., with the program MIMA in the package BEARTEX [WENK &AL. 1997b]). (4) In the case of lower crystal symmetry, not all pole figures contain the same information and the symmetry of the crystal form corresponding to the pole set {hkl} needs to be evaluated. At least one pole figure belonging to the lowest symmetry of the Laue group that needs to be resolved has to be included. In the case of a tetragonal crystal a (001) pole figure does not resolve the full orientation. In the case of trigonal crystals a {11$\bar{2}$0} pole figure only resolves the hexagonal symmetry and a {h0hl}+{0hhl} pole figure needs to be included. Some of these limitations have been addressed in Chapter 3.

- Next mount the sample in the texture goniometer. Be sure that the sample surface is at the correct height (eucentric point) and at α=0° exactly in reflection condition, i.e. the sample normal is bisecting the incoming and diffracting beam. Turn on the oscillator, if you are going to use it. Perform a θ–2θ scan on the texture goniometer, for example at α=0°, over the angular range of interest and record it. A first scan may be in steps of 0.2° to find the peaks and possible overlaps and a second scan in smaller increments to determine the exact position of the peak. If the texture is very strong, and no peaks are observed, it may be necessary to repeat the scan in a sample orientation for which a texture maximum is in Bragg orientation in order to observe a peak (change α and β). Determine whether peaks are sufficiently separated at high α angles where they are much broader (for this reason it is often useful to perform a second θ–2θ scan at α=70° to evaluate the spectrum and identify possible peak shifts). Locate the center of the peak of interest and one or two positions off the peak for measuring the backgrounds (Fig. 10). If the diffraction spectrum is complex with many peaks this may not be directly adjacent to the peak of interest and in some cases with many interfering neighboring peaks it may be only possible to measure the background on one side. If the peak position changes in a θ–2θ scan as a function of α (e.g. between α=0° and α=70°) it usually indicates that the sample height is incorrect and needs to be adjusted.

- Record the background intensity on both sides of the peak at α=0° and β=0°. It is good practice to measure backgrounds 10 times longer than peaks to ascertain the needed counting statistics. It is useful to occasionally measure a pole figure on a background setting, particularly if new materials are analyzed.

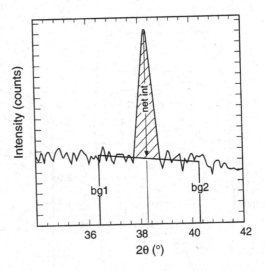

Fig. 10. Procedure to determine backgrounds in a θ–2θ scan. The net intensity is calculated by subtracting an average of background 1 (bg1) and background 2 (bg2) from the peak intensity. The dashed area outlines the integrated net intensity.

- Move the detector to the desired 2θ angle of the Bragg peak *hkl* and start the scan on α and β in a desired scanning pattern (e.g. 5°×5° or hexagonal grid). Counts are usually recorded on-line by a personal computer. In general choose the measuring time so that minimal regions yield at least 50 counts and texture maxima at least 1000. (Usually measurements are done at a constant time. Alternatively one may wish to measure the time for a pre-determined number of counts. The latter procedure may be appropriate when counting statistics are critical).

1.5 Data collection in transmission geometry

The transmission geometry is only applicable to materials with low x-ray absorption and has mainly been applied to minerals and aluminum. For a slab of 100 μm thickness the irradiated volume is small and grain statistics are poor, even if the sample is translated. Also, surface defects introduced during sample preparation are more critical. The thin transmission sample is mounted on a holder (e.g., with a thin tape) or directly glued onto a frame for better transparency (Fig. 5b) and placed normal to the χ axis. It is then rotated in its own plane about χ, which now corresponds to the azimuthal angle β in the pole figure (Fig. 1b,c). The assembly is tilted around ω, which corresponds to the complement of the pole distance (90° − α) in the pole figure. Using this geometry, the peripheral part of the pole figure is covered. There is an intensity change with tilt angle ω due to changes in absorption and the irradiated volume. Empirical correction curves on specimens with random orientation can be used. Corrections are generally minor out to ω < 30–40°. In transmission, pole figures can be measured to ω = 55° or at most 60°.

Transmission geometry is particularly attractive for samples with axial symmetry cut in such a way that the symmetry (fiber) axis lies in the thin slab. In this case, a single χ (β) scan at ω = 0° provides all the necessary texture information and intensity corrections are unnecessary. In transmission rather small sample areas can be selected (1 mm²) for studies of local texture variations, as is illustrated in Fig. 11 for a quartzite, deformed experimentally in axial compression. In this sample the type and intensity of the texture changes due to strain and temperature gradients.

The transmission method is often used to determine sheet-silicate textures in slates and schists because the basal reflection (001), which is of main interest, is at a very low diffraction angle. Owing to the low take-off angle, the broadness of the basal

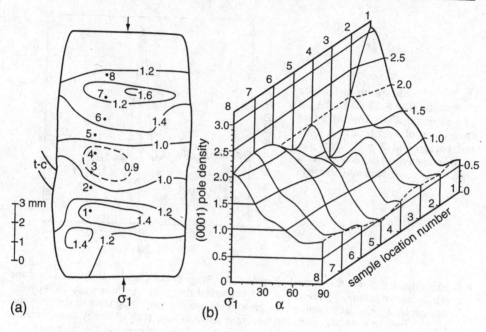

Fig. 11. Texture variation in an experimentally deformed quartzite sample due to temperature and strain gradients. Sample GB-11 compressed 35% at 780°C in center, 15 kb confining pressure and a strain rate of 10^{-6} s^{-1} (a) sample geometry with scale indicated. Contours map transmission of polarized light which is related to texture. Numbers indicate locations of texture measurements in transmission geometry; irradiated area is about 1.5 mm in diameter, t-c is location of thermocouple. (b) (0001) profiles from parallel to perpendicular to the compression axis calculated from the OD. (0001) pole densities in m.r.d. [BAKER &AL. 1969].

reflection, and streaking of the diffraction peak due to stacking disorder or inter-layering in the crystal structure, the background is difficult to determine by moving the detector off the diffraction peak. WENK [1985b, p.39] and CHEN & OERTEL [1991] describe procedures to circumvent these problems.

1.6 Complete pole figures

Because of the defocusing effect, sample geometry and absorption, x-ray diffraction only permits the measurement of incomplete pole figures on flat samples. A method that uses hemispherical samples, suggested by JETTER & BORIE [1953], never became popular, probably because of difficult sample preparation. Reflection pole figures are most accurate in the center ($\alpha = 0°$) and reasonable to 50–70°. Peripheral regions in the pole figure that are at large angles ($\alpha > 80°$) cannot be reached in a reflection scan.

In the past major efforts have been invested in constructing complete pole figures, either combining several perpendicular cuts, combining reflection and transmission scans, or using oblique cuts and assuming a sample symmetry. One way to obtain a complete pole figure is to combine scans on slabs cut parallel to the three sides of the sample cube (Fig. 12a). If a pole figure has orthorhombic symmetry, a single in-complete pole figure on a slab cut normal to the sample body diagonal ('[111]') covers the whole asymmetric (irreducible) unit of the pole figure (Fig. 12b). Reflection and

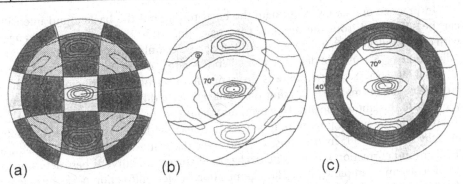

(a)   (b)   (c)

Fig. 12. Various methods for determining complete pole figures illustrated with a 111 pole figure of rolled steel. (a) Overlap of incomplete pole figures measured in reflection to 70° on three orthogonal slabs, (b) Single pole figure measured on an oblique slab and making use of the orthorhombic sample symmetry. (c) Combination of a reflection and a transmission scan. Normalization of the two is based on the annulus of overlap (dark shading).

transmission scans on the same sample can be combined to produce a complete pole figure, using an annulus of overlap to determine a scaling factor (Fig. 12c). If two or more slabs are required the specimen must be homogeneous over a large volume and this is often not the case; therefore such combined pole figures are not recommended.

In principle, complete pole figures have the advantage that they can be easily normalized but this is not necessary to calculate the orientation distribution (OD) (Chap. 3). Complete pole figures can be calculated from the OD even if only incomplete pole figures were measured. An extreme example illustrating OD calculations from minimal pole figure data was given by CHATEIGNER &AL. [1996]. Today most OD calculations from x-ray data rely on incomplete pole figures.

The pole figure region could be reduced by taking into account the sample symmetry (Chap. 1). However, this symmetry is often not strictly satisfied due to heterogeneities in the forming process or some misalignment of the specimen. It is recommended to measure as large an angular region as possible, and, if applicable, impose the symmetry later by averaging.

1.7 Intensity corrections and data reduction

Background correction

Background intensity results from incoherent scattering and fluorescence in the sample, from interaction of the beam with air molecules, and from electronic noise. Fluorescence can be reduced by selecting an x-ray tube target which does not cause anomalous absorption. A monochromator inserted between sample and detector reduces the background drastically and eliminates fluorescence. With such a monochromator it is necessary to use Soller slits to keep the x-ray beams parallel. Background can be reduced by choosing a small window for the pulse height analyzer, or even better with an energy dispersive detector, because most of the background x-ray photons have a different wavelength. Air scattering can be reduced by setting the incoming beam collimator and a receiving collimator close to the sample or by measuring in a vacuum. Background is diminished with a small receiving slit but this has an effect on the change of peak intensity with tilt.

In practice one cannot eliminate the background, but the background intensity can be subtracted from the measured intensity (Fig. 10). The background changes with α (χ in reflection, ω in transmission) but decreases only at very high tilt angles (in reflection $\alpha > 75°$) and the decrease is fairly independent of the Bragg angle and of the material, at least as long as there is no strong fluorescence. In principle the background could be measured during data collection at different α-values and some vendors do this automatically. However, due to line broadening as a function of α (as discussed in the next paragraph), it is better to measure the background only at $\alpha=0°$ in reflection or $\alpha=90°$ in transmission and infer the value at other α-values with a 'background correction curve'. The background correction curve is best determined empirically on a sample with similar composition but few diffraction peaks by placing the detector on a background region well removed from any Bragg peak. If no fluorescence occurs this measurement only needs to be redone if the collimation system has been changed.

If $I_{\alpha=0°}^{bg}$ is the measured background at $\alpha=0°$ then the background I_{α}^{bg} at α is

$$I_{\alpha}^{bg} = I_{\alpha=0°}^{bg} \frac{I_{\alpha}^{bg(stand)}}{I_{\alpha=0°}^{bg(stand)}} \tag{5}$$

where the ratio $I_{\alpha}^{bg(stand)}/I_{\alpha=0°}^{bg(stand)}$ is the empirically determined background correction curve. A first step in the data correction is to subtract the background from the measured intensity.

Empirical defocusing correction

The shape of the irradiated area depends on the collimating system and the orientation of the sample surface relative to the incident x-ray beam. If a cylindrical beam is used, the intersection with the surface is an ellipse, which becomes increasingly elongated with increasing sample tilt (increasing α in reflection, decreasing α in transmission). In this section reflection geometry is emphasized but similar arguments apply to transmission.

Figure 13 shows the change in shape and orientation of the ellipse for a 2 mm collimator and different values of α and 2θ. The beam intersection with the sample surface is more elliptical at small 2θ angles and reflection pole figures should not be measured at $2\theta<20°$. Since x-rays penetrate the sample up to about 100–200 μm, the true geometry is more complicated and the covered volume is slightly larger. The spread in area and the resulting distortion of the reflected signal causes 'defocusing' of x-rays. This is expressed by line broadening (Fig. 2, bottom) and reduced intensities at peak maxima, since only part of the peak signal passes through the detector receiving slit. At very large tilts the beam may partially leave the specimen, especially if it oscillates, and further reduce the intensity. One should keep this in mind when preparing the sample.

The line broadening also causes overlaps of closely spaced diffraction peaks which, in reflection, may be resolved at $\alpha= 0°$ ($\chi=90°$) but not at large values near the periphery of the pole figure (Fig. 2d). Since x-ray texture measurements usually record peak intensities rather than integrated intensities, the intensity decrease with tilt is severe and must be corrected. Empirical correction curves can be measured on a sample of the same material or a material with similar 2θ peak positions and the same sample geometry but with no preferred orientation. The corrections depend on the alignment of the instrument, the size and shape of collimator and receiving slit, and

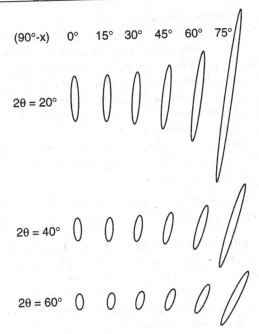

Fig. 13. Change in shape and orientation of the irradiated spot on the sample surface for different sample inclinations as a function of tilt angle α and Bragg angle 2θ. The incident beam is cylindrical with 2 mm diameter.

the diffraction angle 2θ. Corrections are smallest if a small circular collimator (~ 0.5 mm diameter) and a large receiving slit (thereby reducing the 2θ resolution) are used, and for high angle reflections. The corrections are less important if integrated intensities are recorded with a position-sensitive detector but the problem of potential peak overlaps remains. A corrected intensity I_α^{corr} is obtained from:

$$I_\alpha^{corr} = I_\alpha \frac{I_{\alpha=0°}^{rand}}{I_\alpha^{rand}} \tag{6}$$

where I_a is the measured intensity from the textured sample, and I_α^{rand} is the intensity from the random sample at the same value of α. All these intensities must first be corrected for background.

It is not trivial to produce or to identify a standard sample that has a random orientation distribution. Powder compacts, embedded in epoxy can be used ('isostatically pressed' powders often show a texture, because the stress is not purely hydrostatic). Geologists can use undeformed limestones (for calcite) and flint (for quartz). Such empirical correction curves obtained from standards can then be applied to other materials with similar absorption properties for diffraction peaks at a similar d-spacing and of a similar peak-to-background ratio. No satisfactory standards can be produced for materials with a high shape anisotropy such as sheet silicates and some high-temperature superconductors. It is necessary to ascertain the randomness of the standard by measuring a pole figure and applying empirical corrections. This should produce a uniform distribution.

Theoretical expressions for intensity corrections

Whereas empirical corrections are generally preferred, sometimes it is not possible to prepare a standard with no preferred orientation. In this case theoretical curves can be used which take into account the beam geometry, absorption and volume considerations at different angular setting and the detector receiving slit geometry. Below are expressions for the change in intensity as a function of pole distance α for a standard pole figure goniometer using symmetrical Bragg–Brentano geometry ($\alpha=0°$ is in the center of the pole figure).

For *transmission* of a thin slab of thickness t and linear absorption coefficient μ the intensity changes as a function of the Bragg angle θ and the tilt ω ($\alpha=90°-\omega$) [SCHULZ 1949b, CULLITY 1978]:

$$I_\omega / I_{\omega=0°} = \frac{\cos\theta \,[\exp\{-\mu t/\cos(\theta-\omega)\} - \exp\{-\mu t/\cos(\theta+\omega)\}]}{\mu t \exp\{-\mu t/\cos\theta\}\,[\cos(\theta-\omega)/\cos(\theta+\omega)-1]} \tag{7}$$

In *reflection on an infinitely thick sample* the absorption is exactly balanced by the volume increase and the reflected integrated intensity remains constant. However, the peak intensity ratio $I_\alpha/I_{\alpha=0°}$ (where $I_{\alpha=0°}$ is the reference peak intensity from a random sample at $\alpha=0°$ and I_α that at α, *corrected for background*), while independent of μ of the material, changes through defocusing of the beam. The beam emerging from an x-ray tube is divergent but diffraction from a flat sample surface has the effect of bringing the diffracted beam back into convergence at a focal point. Powder diffractometers are constructed so that this focal point coincides with the detector slits, to give the highest possible 2θ resolution to the diffractometer. Even a relatively wide source slit size may be used without sacrificing resolution, because of this focusing geometry (Fig. 1d). This is also the geometry of a texture diffractometer for sample tilt angles $\alpha=0°$ ($\chi=90°$) and $\omega=0°$. For other tilt angles, the focusing sphere (tangent to the sample surface) moves, and the diffracted beam is no longer focused at the receiving slits. The width of the diffracted beam increases due to this defocusing effect. Even at modest tilt angles part of the beam may fall outside the receiving slits, thereby reducing the measured intensity. TENCKHOFF [1970a] and GALE & GRIFFITHS [1960] derived expressions for this defocusing effect, assuming a Gaussian peak shape and a diffractometer geometry defined by the beam width at the sample r (in mm), the width of the receiving slit w (in mm) and the distance if the slit from the sample d (in mm). $\Delta 2\theta$ is the integral angular width of the Bragg peak (in degrees) at $\alpha=0°$:

$$I_\alpha / I_{\alpha=0°} = t_1(t_2 - t_3 + t_4 - t_5)$$

where

$$t_1 = \tau\sqrt{2}\,/\left[2a\,\mathrm{erf}\!\left(w/[2\tau\sqrt{2}]\right)\right],$$

$$t_2 = u_1\,\mathrm{erf}(u_1)$$

$$t_3 = u_2\,\mathrm{erf}(u_2)$$

$$t_4 = \frac{1}{\sqrt{\pi}}\exp\!\left(-u_1^{\,2}\right)$$

$$t_5 = \frac{1}{\sqrt{\pi}}\exp\!\left(-u_2^{\,2}\right)$$

with

$$u_1 \equiv (a+w/2)/\left(\tau\sqrt{2}\right)$$

$$u_2 \equiv (a-w/2)/\left(\tau\sqrt{2}\right)$$

$$a \equiv (r/d)\cos\theta\,\tan\alpha$$

$$\tau \equiv \Delta 2\theta/\sqrt{2\pi}$$

$$\tag{8}$$

The ratio $I_\alpha/I_{\alpha=0°}$ can be thought of as defining a correction function with tilt α. HUIJSER-GERITS & RIECK [1974] used this expression to calculate theoretical curves and to compare them with empirical curves, finding good agreement. Figure 14a shows some curves for a pole figure geometry with r=1 mm, w=2 mm and 0.5 mm, d=200 mm and $\Delta 2\theta$ =0.5° for 2θ-values of 20° and 60°. It is apparent that the corrections are less severe for larger Bragg angles and that a wider receiving slit reduces the intensity drop with tilt. It is recommended to maximize the width of the receiving slit within the range of permissible angular resolution.

As mentioned above, for very thick slabs the increase of absorption with tilt is compensated by the increase in volume. This is not the case for *thin slabs* and particularly for *thin films*. The effect of absorption and volume on the effective intensity, as a function of tilt angle α and sample thickness t is expressed by the factor [SCHULZ 1949a]:

$$F(\alpha,t) = 1 - \exp\left(-\frac{2\mu t}{\sin\theta \cos\alpha}\right) \tag{9}$$

It needs to be applied to the measured intensity, corrected for defocusing.

Figure 14b uses Eq. 9 to show the effect of slab thickness on the absorption-volume correction (dashed lines) and on the total correction, including defocusing (solid lines) for a generic material with μ= 1000 cm^{-1}. For the thicker film absorption is important and the intensity increases only gradually with sample tilt. For the ten times thinner film the volume increase is dominating and the intensity increases greatly with tilt. For an infinitely thick film (or a material with a very high absorption coefficient) the absorption-volume correction factor would stay constant at 1 (Eq. 9). Combining absorption-volume and defocusing correction we find a significant

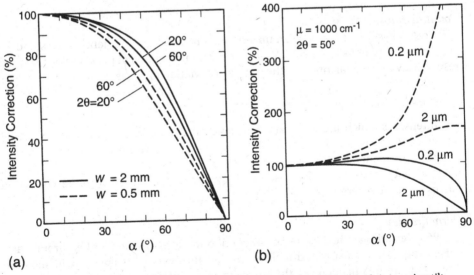

(a)   (b)

Fig. 14. Typical intensity correction curves for reflection geometry with increasing tilt angle α. (a) Theoretical correction for a thick slab with two different receiving slits (w = 2 mm solid lines, 0.5 mm dashed lines) for two values of 2θ. (b) Correction curves for a thin film of thickness 2 μm and 0.2 μm. Dashed lines give absorption-volume effect according to the Schulz formula, solid lines combine absorption-volume and defocusing correction.

difference in correction curves. At $\alpha=70°$ the correction factor for the thin film is 0.9, that for the thick film 0.5.

These corrections are fairly straight-forward and easily implemented. For thick slabs, an empirical correction based on a sample with a random orientation distribution should be used, whenever possible, because the diffractometer geometry is often not adequately modeled with Eq. 8. One may then wish to fit this curve with the GALE & GRIFFITHS [1960] expression to establish appropriate instrument parameters and then use Eq. 8 to extrapolate the curve for other 2θ-values. For thin films it is always necessary to use the theoretical absorption-volume correction (Eq.9), combined either with an empirical or a theoretical defocusing correction. [e.g. CHATEIGNER &AL. 1992, WENK &AL. 1996a].

The foregoing discussion assumed samples with a homogeneous texture. In many epitaxial films the texture near the interface with the substrate is different from that at the surface (e.g., in laser deposited Y-Ba cuprate films). Consider again Fig. 14b. If the sample is 2 μm thick and homogeneous, the corresponding correction curve applies (e.g. a factor 0.5 at $\alpha=70°$). If the sample has a different texture in a 0.2 μm layer near the surface then that correction curve should be applied to the surface texture components (e.g. a factor 0.9 at $\alpha=70°$). In the case of Y-Ba cuprate there is indeed a texture peak at 70°; this imposes a large uncertainty in the interpretation of orientation densities when the internal microstructure is not known. There is still no satisfactory algorithm to apply to textures of thick films in which the texture varies with thickness. In principle it is possible to evaluate the texture distribution within the sample by analyzing different texture peaks hkl with different radiations (e.g. FeKα and CuKα). The surface layer dominates at low θ and high α (shallow incidence angle of the x-ray beam), and more of the bulk sample is analyzed at high θ and low α unless an asymmetric geometry is used [BONARSKI &AL. 1994].

Counting statistics

Counting statistics impose an uncertainty on each measurement with a standard error of the net intensity. If $N_p + N_{bg}$ are count contributions of peak and background respectively to the total intensity the standard deviation $\sigma_{(p-bg)}$ is

$$\sigma_{(p-bg)} = \left(N_p + N_{bg}\right)^{1/2}$$

$$(10)$$

Intensities which are less than 2–3 standard deviations are not significant and longer counting times should be used. For x-ray texture measurements counting statistics are generally not a limiting factor. Errors introduced through sample heterogeneity, crystallite effects due to large grain size, and inaccurate correction curves are more severe.

Normalization

Pole densities are first expressed as count numbers corrected for background and defocusing. They must be normalized so that the texture is expressed in standard units that are independent of the intensity of a particular diffraction peak and of experimental parameters. When normalized, the data are commonly referred to as being in 'multiples of a random distribution (m.r.d.)'. These values can be calculated by summing all measured and corrected intensities over a complete pole figure and weighting them with respect to their areal contribution in the pole figure:

$$I_{\alpha\beta}^{\mathrm{norm}} = I_{\alpha\beta} \frac{\sum\limits_{i} \sin\alpha_i'}{\sum\limits_{i} I_{\alpha'\beta'} \sin\alpha_i'} \qquad (11)$$

The sum over the pole figure becomes unity. Orientations with densities above 1 m.r.d. have more lattice planes aligned than a sample with a random orientation distribution and those less than 1 m.r.d. are depleted. Densities, expressed as multiples of a random distribution, are equivalent to normalized pole densities on the sphere per 1% area units used in the older geological literature. In the case of incomplete pole figures it is not possible to achieve a true normalization but it is recommended to apply at least a preliminary normalization using Eq. 11 and performing the summation only over the measured part. A final normalization is obtained in the OD analysis (Chap. 3).

2. Special Procedures in X-ray Diffraction

Routine procedures of x-ray pole figure measurements have been discussed in some detail in the previous section. It should be mentioned that, in principle, these methods apply to both single and polyphase materials. However, it requires that diffraction peaks of the different phases are separated, and that the phases are not aligned in planes [BUNGE 1985]. For a quantitative analysis of textures and a determination of phase proportions in polyphase composites, neutron diffraction with profile analysis and electron diffraction are recommended. Some special procedures have been developed and applied to samples such as thin films and wires. They will be only briefly mentioned in the following sections and the reader is referred to the original literature.

2.1 Energy dispersive and position-sensitive detectors

In most pole figure measurements monochromatic radiation is used and intensity is recorded with a point detector. Recently energy dispersive detectors have been introduced and applied to textures of metals [SZPUNAR 1990]. For a given sample orientation, and with polychromatic radiation, a whole range of lattice planes *hkl* can be measured simultaneously.

Another advance has been the application of one dimensional (1-D) and two dimensional position-sensitive detectors (e.g., BESSIÈRES &AL. [1991], HEIZMANN & LARUELLE [1986]). Such detectors are commonly employed in neutron diffraction. With position-sensitive detectors the correlation of goniometer coordinates and pole figure coordinates is no longer trivial. In the following section we elaborate on this relationship for 1-D position-sensitive detectors mounted in the diffraction plane [Fig. 15a, BUNGE &AL. 1982].

In normal diffraction geometry the Eulerian cradle is set up in such a way that the χ circle coincides with the bisectrix of incident and diffracted beam (Bragg–Brentano condition). Any direction on the pole figure can be brought parallel to the bisectrix by two rotations $(90°-\chi)$ and ϕ, which correspond to spherical coordinates α and β of the diffracting lattice plane (Fig. 15b). With a 1-D position-sensitive detector this is true only for one 2θ angle (x_0 in Fig. 15a). For all others, the χ circle is inclined by ω to the bisectrix (x_1 in Fig. 15a). (ω in the position-sensitive detector geometry is equal to $\Delta\theta$, the deviation of the χ circle from the bisecting position.) A pole P_1 with coordinates α,

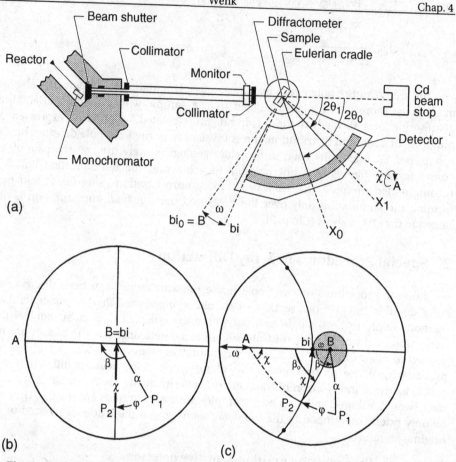

Fig. 15. Diffraction geometry for a position-sensitive detector. (a) Experimental setup for neutron diffraction at ILL Grenoble with the 1-D position-sensitive 'banana' detector extending over a 2θ range of 80°. (b) Angular relations for symmetrical geometry in which the χ circle bisects the incident and diffracted beam (position X_0 in (a)). (c) Angular relations for an asymmetric detector position X_1 with a blind area (shaded). The point bi is the bisectrix of incident and diffracted beams. For explanations of other symbols see text.

β is first rotated about B by ϕ to P_2 (Fig. 15c), and then about A by χ to the bisectrix bi, where it will diffract. In this case χ no longer coincides with α, nor ϕ with β. In the triangles ABP_2 and $biBP_2$ we obtain

$$\cos\alpha = \cos\omega - \cos\chi$$
$$\sin\beta_0 = \cos\omega - \sin\chi / \sin\alpha \qquad (12a)$$

and

$$\beta = \beta_0 + \phi \qquad (12b)$$

Note that a pole with $\alpha < \omega$ cannot be brought to coincidence with the bisectrix by a $\chi-\phi$ rotation, resulting in a blind area in the center of the pole figure (shaded in Fig. 15c).

2.2 Wires

Extrusion of wires produces heterogeneous textures with the mantle of the wire differing from the core. These differences are sometimes important for mechanical properties and need to be quantitatively characterized. With soft x-rays (such as Fe or Cr) it is possible to study the surface texture of the sample; it is determined on the wire as is. Subsequently the surface can be removed by chemical etching and the core texture can be measured [e.g. MONTESIN & HEIZMANN 1991a,b]. Wires are usually measured in reflection with parallel pieces mounted on a flat holder (Fig. 5c). Because of the uneven surface and thickness, absorption is highly anisotropic and rather complicated corrections are required [LANGOUCHE &AL. 1989, MONTESIN &AL. 1991]. Figure 16a shows a measured pole figure of a wire core which displays large deviations from axial symmetry. After an anisotropic absorption correction (Fig. 16b) has been applied, the core texture is an axially symmetric fiber (Fig. 16c); however, the mantle texture is not (Fig. 16d-f): it is a 'circular texture' – also called 'cylindrical texture' [LEBER 1961] and 'cyclic texture' [STÜWE 1961].

When only the average texture is of interest, texture measurements can be undertaken, without further corrections, on the cross-section of a bundle of wires.

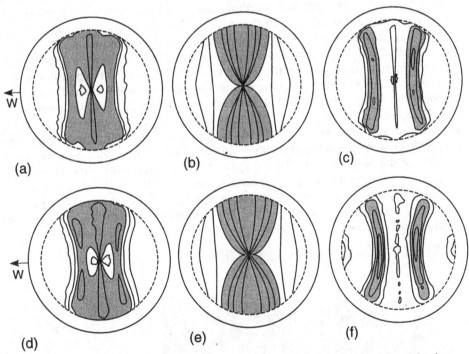

Fig. 16. Incomplete pole figures of Fe wires measured in reflection geometry, W: wire direction [MONTESIN & HEIZMANN 1991a,b]. (a-c) Core of wire, (d-f) mantle of wire, (a,d) uncorrected pole figures as measured, (b,e) correction function, (c,f) corrected pole figures. Notice that the pole figure of the core is essentially axially symmetric (c), that of the mantle is orthorhombic (f). Stereographic projection.

2.3 Thin epitaxial films: advantages of monochromatic radiation

In the case of epitaxial films on single-crystal substrates, textures are extremely strong and this requires special procedures, some of which are also advantageous for texture analysis in general [WENK 1992].

Almost universally x-ray pole figure goniometers use Cu-radiation (Fig. 3a) and the intensity of the Kβ (λ=1.392 Å) component is reduced by filtering through a Ni foil with an absorption edge at 1.488 Å (Fig. 3b). Kβ is reduced to 1% if radiation passes through a Ni filter, whereas Kα is only attenuated to 50%. This provides a relatively strong Kα component (λ=1.542 Å) with a moderate continuous spectrum. A continuous contribution remains at wavelengths larger than Kα and also at very short wavelengths. The role of this continuous radiation is best illustrated with x-ray diffraction photographs of single crystals. When 'strictly' monochromatic radiation is used, produced with a single-crystal monochromator such as graphite, diffractions appear as sharp spots (Fig. 17a). For unfiltered radiation each spot is drawn out into a spectrum which also includes Kβ and a streak, mainly towards smaller diffraction angles (Fig. 17b). For filtered radiation (Fig. 17c) the streak still exists but Kβ is reduced. The intensity of the streak depends on the peak intensity.

The intensity of the continuous spectrum is relatively small compared to the Kα intensity, yet becomes significant if diffractions from a thin film are superposed on those from a substrate crystal. In this case the use of a monochromator can be paramount. This is illustrated for a high-temperature superconductor Y-Ba cuprate film deposited on a lanthanum aluminate substrate. Figure 18 shows two (102) pole figures, projected on the plane of the substrate surface. On the left is a pole figure with truly monochromatic CuKα radiation, on the right one with Ni-filtered radiation. The monochromatic pole figure (Fig. 18a) has eight peaks labeled A⊥ and C⊥ respectively (see Chap. 6 for a discussion). The Ni-filtered pole figure (Fig. 18b) contains many additional peaks which are artifacts and not part of the 102 pole figure. They arise from diffraction on the substrate, with some wavelength of the continuous spectrum. Because of the larger volume the substrate diffracts much more strongly than the 400nm film and produces a significant signal, even for weak radiation of the continuous spectrum. For example the peaks at α =45° between the two (102) peaks in the 2θ = 27.5° (for Cu Kα) pole figures are due to diffraction on (220) (d = 2.68 Å) of the substrate with a wavelength of 1.35 Å.

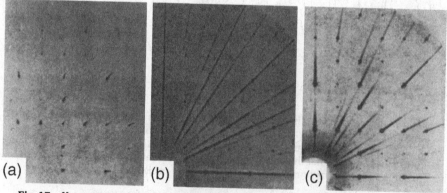

(a)   (b)   (c)

Fig. 17. X-ray precession photographs of an *hk*0 reciprocal lattice layer of a single crystal of garnet. Mo radiation. (a) with graphite monochromator, (b) unfiltered radiation, (c) Zr β-filtered radiation. Each reciprocal lattice point *hkl* is extended into a spectrum along 2θ (1/*d*) [WENK 1992].

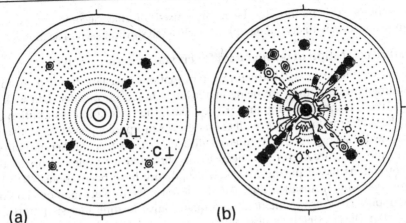

Fig. 18. 102 pole figures of a 400 nm Y-Ba cuprate film on a substrate crystal of LaAlO₃. at 2θ=27.5°(for Cu Kα). Dot pattern gives positions of actual measurements (5°×5° grid). The main texture peaks are for Y-Ba cuprate with the A ⊥ orientation at α=34° and the C ⊥ orientation at α=56°. (a) Monochromatic radiation, (b) β-filtered radiation with additional peaks due to diffractions with a different wavelength on the substrate. Equal-area projection. High intensities are contoured [WENK 1992].

For epitaxial films the use of a monochromator on the incident beam is necessary to eliminate ambiguity about peak identification and uncertainty about background determination. With a mosaic graphite monochromator, intensities of the effective beam are only slightly lower than for β-filtered x-rays. A monochromator also eliminates impurities in the spectrum in old x-ray tubes, mainly due to W-contamin-ation and reduces overall background. There are no real disadvantages and its application is recommended for general use. The slight reduction in intensity is easily compensated by the improved peak-to-background ratio. The monochromator is inserted between the x-ray tube and the collimator. Such a geometry causes no problems with defocusing during pole figure measurements. If the mocochromator is inserted between sample and detector, the ray path must be controlled by Soller slits which reduce the intensity.

2.4 Determination of inverse pole figures

In metallurgy, information is sometimes obtained from representations of the distribution of a particular specimen direction (e.g., tensile or compression direction in axial deformation, and rolling or normal direction in a rolled metal sheet) with respect to crystal coordinates (Chap. 2). This can be measured with a powder diffractometer in a θ – 2θ scan on slabs cut perpendicular to the directions of interest. The sample surface is mounted in reflection geometry, therefore not requiring a texture attachment. Peak intensities must be normalized to account for differences in diffraction intensities (structure factor) for individual reflections *hkl*. This can be done by comparing the textured sample with a standard that has no preferred orientation [e.g. MORRIS 1959]. From these normalized intensities an 'inverse pole figure' can be constructed. The coverage is restricted to directions *hkl* with diffraction peaks in the experimentally accessible 2θ range, which is inferior to the much denser coverage in a pole figure. From several inverse pole figures, measured on different slabs, the full ODF can, in principle, be calculated subject to limitations of 'ambiguity' discussed in earlier Chapters. This method has been rarely used but may become revived in the future in connection with crystallographic profile analysis discussed in Section 3.5.

2.5 Texture correction of powder data for crystal structure analysis

Whereas in the past crystal structure determinations have largely relied on single-crystal diffraction data, they are being increasingly based on powder data with the RIETVELD [1969] technique. If grains are equiaxed, the orientation distribution of a powder should be random; however, if grains show a morphology, such as platelets or needles, powder aggregates are textured. Diffraction intensities are then not directly related to the structure factor and need to be corrected for preferred orientation. Radial corrections based on the MARCH [1932] theory have been applied [DOLLASE 1986] but this is often unsatisfactory [e.g. CHOI &AL. 1993]. New methods combine crystallographic structure refinement with quantitative texture analysis [LUTTEROTTI &AL. 1997, VON DREELE 1997]; see also Sec. 3.5.

2.6 Synchrotron x-rays

Conventional x-ray tubes produce a broad beam of relatively low intensity. In a synchrotron a very fine-focused high-intensity beam with monochromatic, polychromatic or continuous wavelengths can be produced. The unique advantages of high intensity, small beam size (<5 μm) and free choice of wavelength opens a wide range of possibilities for texture analysis and even though this field is barely explored, a short discussion is warranted.

Local textures in fine grained aggregates

In a synchrotron a narrow beam of monochromatic x-ray photons can be focused on a small fine-grained polycrystalline sample (Fig. 5d). The diffraction pattern is recorded with a CCD camera that provides digitized intensities with a high spatial resolution and a wide dynamic range, within a few seconds (Fig. 19). Diffraction from a powder is recorded as a set of concentric Debye rings. If the sample has texture, the rings, corresponding to a lattice planes *hkl*, have intensity variations.

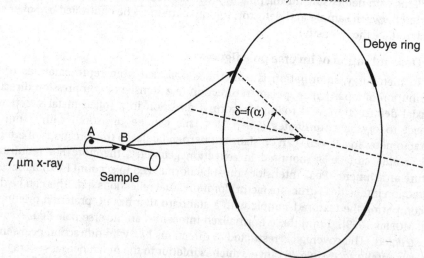

Fig. 19. Local texture measurement with a synchrotron microbeam. A small sample is irradiated with high intensity monochromatic x-rays and produces Debye rings with intensity variations that are recorded with a 2-D position-sensitive detector.

Figure 20 compares the diffraction pattern from an Al wire recorded with the CCD camera (Fig. 20b) with a historical photograph using conventional x-rays (Fig. 20a [GUINIER & GRAF 1952]). In contrast to the old picture, where gray shades represent densities on a photographic film that was exposed to x-rays for many hours, the CCD pattern was recorded in one second with each of the 1024×1024 cells containing a digital intensity value. This can be represented in a 3-D plot with the intensity variations in the third dimension displaying the texture (Fig. 20c).

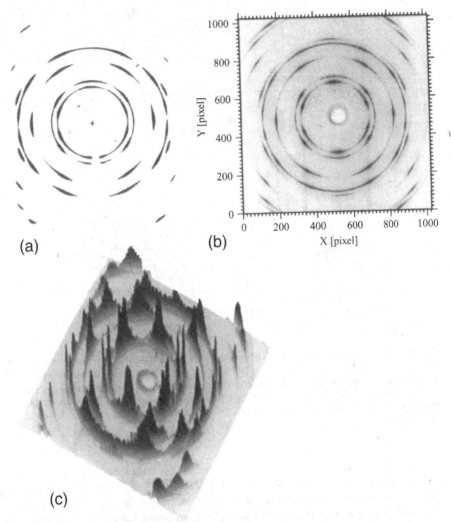

Fig. 20. (a) Historical photograph of Debye rings of an aluminum wire [GUINIER & GRAF 1952]. (b,c) Aluminum wire of 90 μm thickness analyzed with synchrotron x-rays (beam 30 μm) and digitally recorded with a CCD camera. (c) A 3-D representation of intensity variations on Debye rings that are due to texture (courtesy of F. Heidelbach and C. Riekel).

From the quantitative analysis of intensity variations on Debye rings (Fig. 21a) and proper intensity corrections, the texture can be determined such as for a sample of a small rod, 150 μm in diameter and 500 μm long, of Ni-Fe alloys prepared by

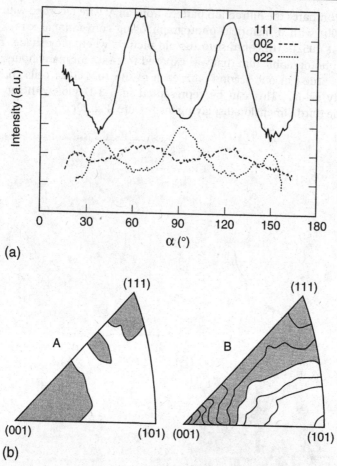

Fig. 21. Texture variations in a small rod of FeNi alloy, prepared by micro-electro-plating. (a) Intensity variations along Debye rings for 111, 002 and 022 expressed as function of pole figure angle α. (b) Inverse pole figures derived from measurements at two positions on the sample (A and B in (a)) on areas less than 10 μm in diameter. Equal-area projection, contour interval is 0.1 m.r.d., shaded above 1 m.r.d. [BACKSTRØM &AL. 1995].

electroforming techniques. Local texture variations could be documented and represented in inverse pole figures (Fig. 21b) [BACKSTRØM &AL. 1996]. Understanding and control of texture in these materials, used as magnetic microactuators, is of considerable practical importance. With synchrotron x-rays the circular texture in Fe-wires can be directly verified without complicated sample preparation and absorption corrections. A thin (50 μm) slice of a steel wire was probed in the core where the texture displays axial symmetry (Fig. 22a) and near the surface where diffraction rings show large deviations from axial symmetry, with a horizontal mirror plane (Fig. 22b). A beauty of synchrotron radiation with CCD cameras is that one can immediately view diffraction effects and qualitatively identify the presence of texture (e.g. in thin films [PLAYER &AL. 1992]). The method is ideally suited to document local variations in texture in fine grained materials.

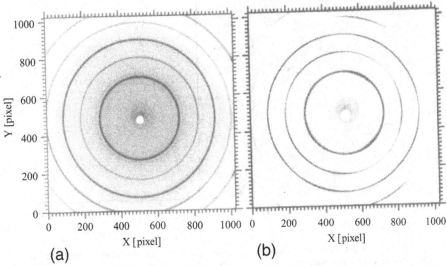

(a)   (b)

Fig. 22. Texture variations observed in a thin slice of steel wire of 1mm diameter, cut perpendicular to the wire axis. The Debye rings, recorded with a CCD camera, document axial symmetry in the center (a) and a circular texture at the periphery of the wire (b). Compare these synchrotron measurements with conventional x-ray results in Fig. 16.

The procedure for the quantitative interpretation is in many ways analogous to the much older photographic methods of texture analysis [e.g. WEVER 1924, KRATKY 1930, STARKEY 1964, WENK 1965] and similar data processing can be used.

For axially symmetric textures, analyzed in transmission with the fiber axis normal to the incident x-ray, the angle δ on the Debye ring (Fig. 19), measured from the horizontal plane, is related to the tilt of the lattice plane, α (or pole distance) [CULLITY 1978]

$$\cos\alpha = \cos\theta\cos\delta \qquad (13)$$

There is a blind region in the center of the pole figure ($\alpha < \theta$).

For non-axial textures a single set of Debye rings is not sufficient. The sample needs to be rotated (ϕ) and in each setting a pattern has to be recorded. In this case geometric relations are more complicated, particularly for reflection geometry where the sample surface is inclined by an angle v to the incident x-ray. Explicit expressions for the relationship between pole figure coordinates α and β, and δ, θ, tilt v and sample rotation ϕ have been given by WENK [1963].

Individual orientations from Laue patterns

In the previous experiment a wavelength was selected with a monochromator. One can also use a white spectrum with a large wavelength range and such radiation is employed to record single-crystal Laue patterns. GOTTSTEIN [1988] and BRODESSER &AL. [1990] suggested the use of synchrotron x-rays for Laue orientation determination. With the advent of accurate stages, CCD cameras and highly focused beams it is now possible to record Laue patterns, interpret them on-line and

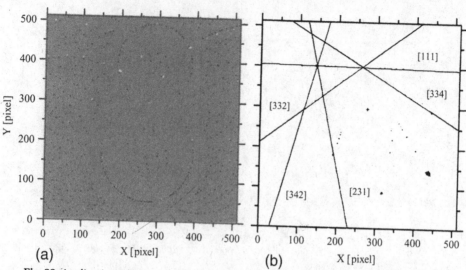

Fig. 23. Application of white synchrotron x-rays to obtain crystal orientation from the interpretation of Laue patterns. (a) Laue pattern of an olivine crystal from a peridotite, recorded with a CCD camera. White dots are measured reflections, dark dots represent a recalculated pattern. (b) Gnomonic projection of the Laue pattern in (a); it is used to index the diffraction spots and determine the orientation [WENK &AL. 1996a].

determine the local crystal orientation in a similar way as EBSP patterns with a scanning electron microscope [WENK &AL. 1997a]. The procedure is to convert Laue patterns (Fig. 23a) to gnomonic projection where co-zonal poles hkl lie on straight lines (Fig. 23b). Next the Hough transform is applied, converting lines to points (see Sec. 4.2). The computer identifies the points and calculates interzonal angles [BARRETT & MASSALSKI 1980, p.616]. Indexing and determination of the orientation is based on a comparison of the experimental interzonal angles with a reference table.

Because of the possibility to tune the synchrotron x-ray wavelength arbitrarily, without restrictions of the composition of the anode material in an x-ray tube, one can make observations near the absorption edge with high anomalous scattering. Though this has not been tested yet, it may be possible to determine the absolute orientation of non-centric crystals, making use of anomalous scattering, e.g. for ferroelectric materials.

2.7 Residual stress and dispersion

Due to internal stresses, lattice parameters of deformed crystals may deviate from those of ideal single crystals. This causes shifts in the 2θ spectrum. In a textured material the distortions ('residual strains') may be different for different crystallite orientations and in such cases 2θ-values change slightly as a function of sample orientation. This may necessitate adjustment of 2θ angle positions during pole figure data collection. The anisotropic shifts can be used to extract additional information about residual stress [VAN HOUTTE & DE BUYSER 1993, VAN ACKER &AL. 1994]. Since lattice planes of crystallites which are in diffraction position in a given sample orientation may have different orientations and a different next neighbor environment, one observes generally not only a shift in d-spacing (θ) but a more complicated peak shape (asymmetric broadening) due to superpositions of differently stressed

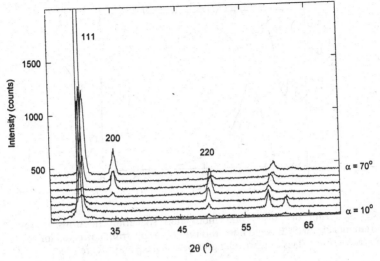

Fig. 24. X-ray diffraction spectra of a film of Y stabilized zirconia on a Si-waver. Bottom spectrum is at a sample tilt of α=10°, next one at α=20° and so on to 70°. Changes in peak heights are due to texture, changes in peak position are due to internal stresses. CuKα radiation [LUTTEROTTI &AL. 1994].

crystallites. The interpretation of such data is still unsatisfactory and not combined with ODF analysis.

Peak position and peak shape need to be extracted from θ–2θ profiles and this is presently best done with high resolution neutron diffraction [HUTCHINGS & KRAWITZ 1993, PRIESMEYER &AL. 1994]. With x-rays, the peak shape is highly dependent on sample geometry because of the defocusing effect and θ resolution is limited, but there are some applications for thin films where residual stresses can be quite significant. Figure 24 shows x-ray spectra for a film of zirconia deposited on a single-crystal silicon substrate, measured in seven different sample orientations with significant peak shifts due to internal stresses [LUTTEROTTI &AL. 1994, FERRARI & LUTTEROTTI 1994].

3. Neutron Diffraction

Instead of x-rays, neutron radiation can be used for diffraction experiments [BACON 1975]. Neutron diffraction was first applied to textures in 1953 by Brockhouse and offers several advantages, mainly low absorption, high angular resolution and a magnetic signal. Most neutron diffraction texture studies are done at reactors which produce a constant beam of thermal neutrons. A few facilities use pulsed neutrons, e.g. at spallation sources. The wavelength distribution of thermal neutrons generated in a reactor or by spallation is a broad spectrum with a peak at 1–2 Å (Fig. 25). A disadvantage of neutrons is that the flux is very low, compared even to conventional x-ray generators, and long counting times are required.

3.1 Low absorption for neutrons

For most materials, absorption is negligible (Table II). Large samples, 1–10 cm in diameter, of roughly spherical shape (Fig. 5e) can be measured in transmission on the same goniometers that are used for single crystals. Contrary to conventional x-ray

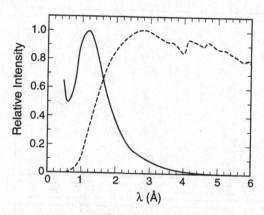

Fig. 25. Spectrum of pulsed and moderated spallation neutrons in the TOF measurements at LANSCE (solid line) and the same spectrum corrected for scattering power (dashed line). Typically wavelengths between 0.5 and 4 Å are used.

techniques where the beam must not leave the specimen, for neutrons the specimen must not leave the beam for most applications. Because the diffraction signal averages over large volumes rather than surfaces, grain statistics are better and neutron diffraction is often necessary to analyze textures of coarse grained materials [WENK &AL. 1984]. Also, intensity corrections are generally unnecessary and complete pole figures can be determined in a single scan. Because of the low absorption, environmental stages (heating, cooling, straining) can be used for in situ observation of texture changes [JUUL JENSEN & KJEMS 1983]. Figure 26 shows a diffraction spectrum and a pole figure of experimentally deformed ice–II measured in a cryostat at 70 K. Ice–II is a high pressure low-temperature form of ice. Because of unfavorable interaction of neutrons with hydrogen, the ice sample had to be prepared with deuterium [BENNETT &AL. 1997].

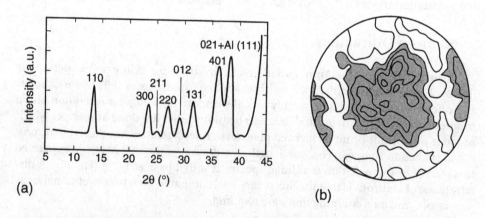

(a)   (b)

Fig. 26. Neutron diffraction of experimentally deformed low-temperature/high-pressure D_2O ice–II, measured in a cryostat at 70K. (a) 2θ spectrum with peaks indexed (Al is due to the sample holder), (b) 110 pole figure. Compression axis is in center of pole figure. Equal-area projection. Contour interval is 0.2 m.r.d., shaded above 1 m.r.d. [BENNETT &AL. 1997].

3.2 Nuclear and magnetic scattering

The elastic scattering of thermal neutrons by a crystal consists of two components, nuclear and magnetic scattering. *Nuclear scattering* is due to interactions between the neutron and the atomic nuclei and yields diffraction effects with equivalent information to x-ray scattering on electrons, but magnitudes of the scattering lengths for elements are different and therefore diffraction peaks have different relative intensities. With neutrons, signals from light elements are of similar magnitude to those from heavy ones. Different isotopes have different scattering power. Scattering amplitudes of x-rays decrease with diffraction angle, whereas those of neutrons do not. This improves the capability of measuring high angle reflections; but the overall intensity still decreases with diffraction angle because of thermal vibration and Lorentz-polarization effects. Most neutron diffraction instrumentation provides better *spectral resolution* than x-rays, which is advantageous for deconvoluting complex diffraction patterns with overlapping peaks such as in low-symmetry materials and multiphase composites.

Magnetic scattering, due to a classic dipole interaction between the magnetic moments of nucleus and shell electrons, is much weaker than nuclear scattering. In materials with magnetic elements (Mn, Fe) peaks may occur in the diffraction pattern that are solely due to magnetic scattering, i.e. the alignment of magnetic dipoles, and with those one can measure magnetic pole figures. MnO is an example of a material with a magnetic superstructure. Magnetic pole figures display preferred orientation of magnetic dipoles in component crystals. If no magnetic superstructures are present, the magnetic contribution is, with presently available instrumentation, very difficult to separate from the nuclear scattering and crystallographic and magnetic pole figures are superposed. Models have to be used to interpret them and these are only applicable to simple cases [MÜCKLICH &AL. 1984]. Interestingly the first neutron diffraction texture experiment by BROCKHOUSE [1953] aimed at determining the magnetic structure of steel.

3.3 Conventional pole figure measurements

A conventional neutron texture experiment uses monochromatic radiation. With a Cu (111) or graphite (0002) monochromator wavelengths $\lambda = 1.289$ Å and $\lambda = 2.522$ Å are often selected. The detector is set at the Bragg angle 2θ for a selected lattice plane *hkl*. The pole densities of that lattice plane in different sample directions are scanned with a goniometer by rotating the sample cube or sphere around two axes, ϕ and χ, to cover the entire orientation range. This method is analogous to that described for x-rays. An efficient equal area coverage (Fig. 8b) is particularly important because of the limited access to neutron sources.

3.4 Position-sensitive detectors and time-of-flight neutron diffraction

It is also possible to use *position-sensitive detectors* which record intensities along a ring (1-D), or over an area (2-D), rather than at a point. In the case of 1-D position-sensitive detectors, the ring can be mounted on a diffractometer so that it covers the whole 90° range of diffractometer coordinates χ [JUUL JENSEN & KJEMS 1983] or that it records a continuous 2θ range [BUNGE &AL. 1982, Fig. 15a]. The former is useful for *in situ* experiments. The latter is interesting because it permits one to determine continuous diffraction profiles and from those integrated intensities can be extracted, rather than peak counts, which are subject to large errors for slight instrument mis-

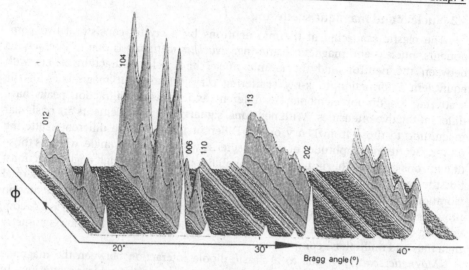

Fig. 27. 72 neutron diffraction spectra measured on an experimentally deformed calcite polycrystal (limestone) with a position-sensitive detector at ILL with monochromatic neutrons; changes in intensity as a function of ϕ are due to texture [WENK 1991].

alignments. Figure 27 is a graphic representation of 72 spectra measured with a position-sensitive detector at different sample orientations for an experimentally deformed calcite limestone [WENK 1991]. The variation in peak intensities is due to texture. As was discussed in Sec. 2.1, a diffraction pattern collected with a position-sensitive detector collects different crystal orientations and construction of pole figures from such spectra requires corresponding angle transformations.

Another method to measure a spectrum simultaneously is at a fixed detector position but with *polychromatic* neutrons and a detector system that can identify the energy of neutrons, e.g. by measuring the time of flight (TOF). (For a review of TOF texture measurements see FELDMANN &AL. [1991] and WENK [1994a].) Figure 28 is a

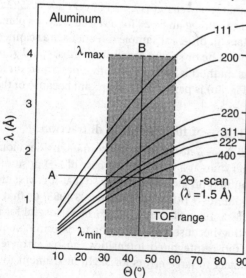

Fig. 28. Bragg's law for low order diffractions of aluminum. A illustrates a 2θ scan with a conventional goniometer and monochromatic radiation. B gives the coverage of a continuous spectrum at a fixed detector position. With a position-sensitive detector a large 2θ range is measured simultaneously (A) and with an energy sensitive or TOF detector a large wavelength range is covered (B).

representation of Bragg's law for aluminum with monochromatic and polychromatic neutrons. For monochromatic neutrons with a wavelength λ=1.5Å (A) a θ-scan can be performed over the θ-range of interest; alternatively the θ-scan can be measured simultaneously with a 1-D position-sensitive detector. There are peaks at each intersection of line A with the Bragg curves for lattice planes *hkl*.

For pulsed polychromatic neutrons, a whole range of wavelengths is available (line B in Fig. 28). A detector at a fixed scattering angle (B) can again record different Bragg peaks but at a single diffraction angle. Position-sensitive detectors and TOF have been combined [WENK &AL. 1988]. With a 2-D position-sensitive detector (Fig. 29), as it is available at Argonne and Los Alamos National Laboratories, a whole pole figure range can be measured simultaneously and the complete pole figure can be covered with 20–30 sample orientations [WENK &AL. 1991a] (Fig. 8c).

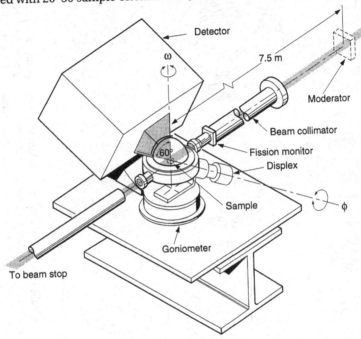

Fig. 29. Sketch of the single-crystal diffractometer with sample and 2-D position-sensitive detector at the TOF neutron diffraction facilities at Los Alamos National Laboratory (LANSCE) and at Argonne National Laboratory (IPNS).

The reliability of neutron pole figure measurements has been evaluated by circulating a textured polycrystalline calcite standard sample among thirteen different neutron diffraction facilities [WENK 1991, WALTHER &AL. 1995, VON DREELE 1997]. They include conventional reactors with monochromatic radiation and point detectors, reactors with position-sensitive detectors, pulsed reactors and spallation sources. In general textures measured on the same sample at different facilities agree very closely (some examples are shown in Fig. 30). For pole figures with strong diffraction intensities, standard deviations from the mean are 0.04–0.06 m.r.d. with a spread of maxima values of 0.2 m.r.d. The spread is considerably larger when the pole figure intensities are weak (0.4 m.r.d.). For weak diffraction peaks, position-sensitive detectors have an

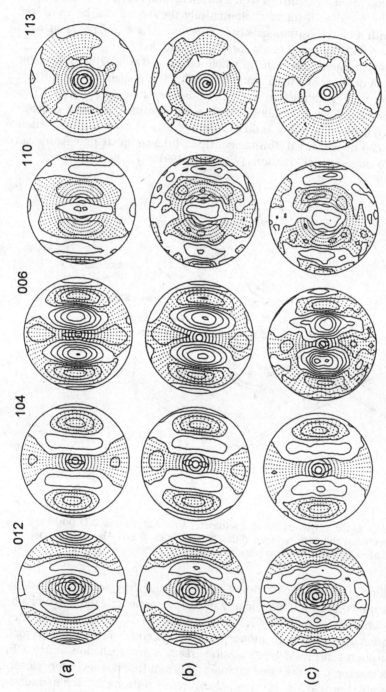

Fig. 30. Selected pole figures of an experimentally deformed limestone standard sample measured at three neutron diffraction facilities in the course of a standardization project [WENK 1991]. (a) Reactor with position-sensitive detector (ILL Grenoble), (b) Pulsed reactor with TOF measurements (Dubna, Russia), (c) Spallation neutrons with a 2-D position-sensitive detector (IPNS, Argonne). Pole figures are normalized so that densities are expressed in multiples of a uniform distribution (m.r.d.). Contour interval is 0.2 m.r.d., 5°x5° dot grid below 1 m.r.d. Equal-area projection.

advantage over single tube detectors because integrated rather than peak intensities can be determined, which yields better counting statistics. Also, the method can be used to deconvolute spectra and separate closely situated and partially overlapping peaks [e.g. ANTONIADIS &AL. 1990, SCHÄFER &AL. 1991, WILL &AL. 1989]. Position-sensitive detectors permit measurement of the whole spectrum simultaneously, not just of the six diffraction peaks considered in this study. They are therefore especially adapted to problems of low crystal symmetry and composites with many diffraction peaks. Similar advantages are provided by TOF experiments. The neutron round-robin experiment documented that pole figure measurements with neutron diffraction of the same sample by different laboratories are much more reproducible than those with x-ray diffraction [VATNE &AL. 1996].

3.5 Whole-spectrum analysis

Even with the high resolution of neutron diffraction, deconvolution of diffraction spectra by profile analysis remains a major obstacle for texture analysis of materials with complex diffraction spectra. Conventionally peaks *hkl* have been fitted with Gauss functions to obtain integrated intensities [WENK & PANNETIER 1990]. From such deconvoluted pole figures the 3-D OD can be determined. However, rather than extracting information from single diffraction peaks, it would be more efficient to use the whole diffraction spectrum and extract texture information in a similar way as crystallographers extract structural information from a powder pattern with the Rietveld technique [1969].

A method proposed by WENK &AL. [1994a] relies on an iterative combination of crystallographic Rietveld profile analysis and OD calculation (RITA), Rietveld texture analysis). Figure 31 displays a summation of TOF powder diffractometer spectra of textured limestone measured in different sample orientations. Due to the averaging, texture effects are largely removed. In the high-*d* range, diffraction peaks are well separated; in the low-*d* range, they overlap. Therefore, deviations between observed data (dots) and the refined profile (line) are minimal. This is not the case for individual spectra where the deviations in peak intensity are due to texture (Fig. 32). The Rietveld refinement proposes weights for each peak to improve the fit, these are then used for the OD calculation with the discrete WIMV method that refines the weights based on texture correlations. The iterative procedure converges very rapidly and provides quantitative information about texture, crystal structure, elastic strains, and microstructural parameters (such as grain size). The method has so far been applied to a synthetic test case [MATTHIES &AL. 1997] and a sample of calcite [LUTTEROTTI &AL. 1997]. In both examples resolution was excellent, even for the highly overlapped region of the diffraction spectrum, as is illustrated in Fig. 33.

The RITA method is expected to improve quantitative texture analysis of low symmetry compounds and polyphase materials with overlaps in the diffraction spectrum. It enables efficient data collection by requiring only a small pole figure range and making maximal use of data at TOF facilities that record complete spectra with many diffraction peaks. In the traditional approach, ODFs are determined from measurements of a *few* diffraction peaks (e.g. 3–5) in *many* sample orientations (e.g. 1000). The new method uses *many* diffraction peaks (a continuous spectrum with >50 *hkl* sets) and only a *small* pole figure region (e.g. 15 positions). An analogous scheme that relies on the series expansion method o the OD has been introduced by VON DREELE [1997]. It is encumbered by the well-known problems of the harmonic analysis.

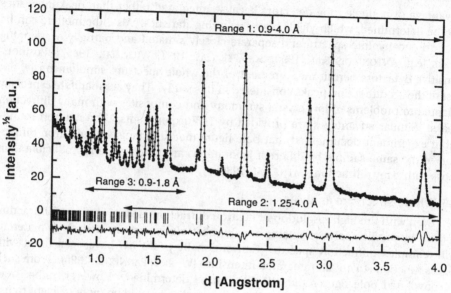

Fig. 31. Sum of 17 spectra of polycrystalline calcite recorded in different sample orientations by TOF measurements of thermal neutrons at the IPNS spallation source with a powder diffractometer. Some d-ranges are indicated that served to obtain texture information [LUTTEROTTI &AL. 1997].

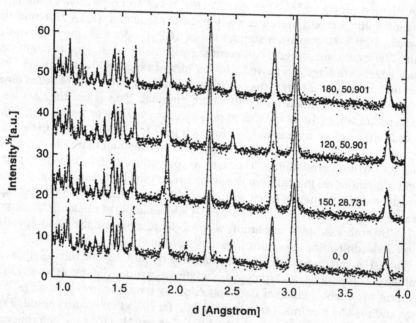

Fig. 32. Four individual spectra in different sample orientations. Dots represent measured data. The solid line is calculated based on the assumption of a random orientation distribution. The deviations are due to texture and can be used to determine the OD.

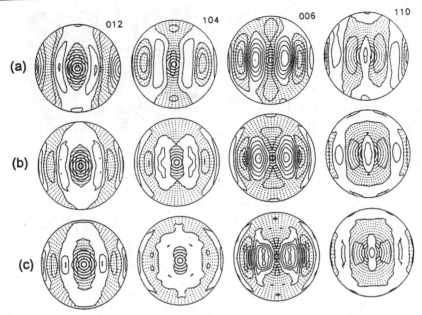

Fig. 33. Pole figures recalculated from the OD. (a) Conventional method, using peak intensities for five complete pole figures. (b,c) Results of the RITA method using only 17 spectra covering the central portion of the pole figure to 40° and using data for two resolution ranges: (b) Spectrum from 0.9–4 Å, (c) only highly overlapped region from 0.9–1.8 Å (compare Fig. 31). Equal-area projection [LUTTEROTTI &AL. 1997].

4. Electron Diffraction

4.1 Transmission electron microscope (TEM & HVEM)

The transmission electron microscope (TEM) offers excellent opportunities to study textural details in fine-grained aggregates. Like light microscopy, the TEM not only provides information about orientation but also about grain shape and, more importantly, about dislocation microstructures indicative of active deformation mechanisms, subgrains, particles, grain boundaries etc. Electrons are very strongly absorbed by matter and only very thin foils can be observed. High voltage electron microscopes (HVEM) with accelerating voltages > 1 MV are particularly useful for texture investigations because of the higher penetration which allows one to study larger sample volumes.

Selected area diffraction patterns (SAD)

One procedure for obtaining orientation information is with selected area diffraction patterns (SAD). It has been particularly used for minerals and ceramics, where Kikuchi patterns (discussed below) are not observed.

Specimens are prepared from thin sections by ion beam thinning, by chemical etching or by electropolishing (Fig. 5e), keeping track of the orientation of the foil relative to the macroscopic sample. The foil is then transferred into the electron microscope, again recording the orientation. The beam is centered on a crystallite of interest and, after observing and documenting microstructures, an SAD pattern is obtained which can be

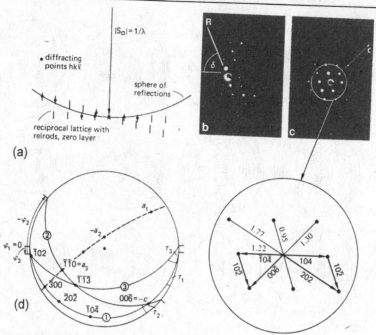

Fig. 34. Diffraction in the TEM. (a) Sphere of reflection and intersecting reciprocal lattice plane with relrods in tilted (b) and symmetric (zone axis) (c) orientation. (b) and (c) are diffraction patterns of calcite mounted on a tilt-rotation holder with R indicating the tilt axis. The enlarged portion of (c) illustrates the indexing procedure from three vectors. (d) is a stereographic projection of three SADs with respect to the foil to determine the orientation of the crystallite.

interpreted by standard diffraction theory [e.g. HIRSCH &AL. 1977]. Electrons that are accelerated to 100kV have an effective wavelength of 0.037 Å, resulting in a very large sphere of reflections in reciprocal space (Fig. 34a). These electrons interact with the specimen and diffraction occurs if reciprocal lattice points lie on the surface of the sphere. Because of the extremely thin TEM foil, reciprocal lattice points are extended to rods perpendicular to the foil, permitting a wider range of lattice planes to diffract. At first an SAD pattern is generally asymmetric (Fig. 34b). The specimen is then rotated on a goniometer stage (tilt-rotation and double-tilt stages are available) until a symmetrical zone axis pattern is obtained (Fig. 34c), keeping track of tilt τ and rotation angles ϕ (Fig. 34d). The tilt axis R is indicated on the pattern. This pattern needs now to be indexed. The vector from the origin of the reciprocal lattice to any lattice point $r^{*}_{hkl} = ha^{*} + kb^{*} + lc^{*}$ has a length of $1/d_{hkl}$, which corresponds to the spacing d of lattice planes. Because of the large radius of the sphere, the projected diffraction pattern is an almost undistorted enlarged image of the reciprocal lattice with distance F on the film

$$F_{hkl} = \lambda C r^{*}_{hkl} = \lambda C / d_{hkl}$$

$$(14)$$

where C is the camera length. To index the diffraction pattern, we prepare a list of d-spacings and F-values (Table IV) and find those which correspond to the length of vectors of the smallest triangle in the pattern (enlarged central portion of Fig. 34c). If indexing is correct, the sum of two vectors $h_3 = h_1 + h_2$, $k_3 = k_1 + k_2$, $l_3 = l_2 + l_1$ has to correspond to the third. Depending on the symmetry of the crystal and the particular lattice section, the orientation may not be unequivocally determined with one pattern. In this case the crystal is tilted into another symmetrical orientation and the procedure is repeated. Figure 34d shows the orientation of a calcite crystal with respect to the sample coordinates as determined from three SADs with a tilt-rotation stage.

Table IV. d-spacings and corresponding distances of diffraction spots on film for calcite with a wavelength $\lambda = 0.037\text{Å}$ (100kV) and a camera length C of 100cm (Fig. 34c)

hkl	d [Å]	$F = \lambda\, C / d$ [cm]
012,$\bar{1}$02,1$\bar{1}$2	3.855	0.950
104, 0$\bar{1}$4,$\bar{1}$14	3.036	1.219
006	2.845	1.301
110,1$\bar{2}$0,2$\bar{1}$0	2.585	1.431
113,1$\bar{2}$3,2$\bar{1}$3	2.285	1.619
202,0$\bar{2}$2,$\bar{2}$22	2.095	1.766

The TEM, combining imaging and SAD analysis, has recently been used to correlate microstructural observations with crystal orientations in experimentally deformed dolomite [BARBER &AL. 1994]: Fig. 35a illustrates the dislocation micro-structure in a particular grain and Fig. 35b shows an inverse pole figure determined from SAD pattern measurements on 108 grains. This technique is extremely laborious, relies on many manual operations and is difficult to automate.

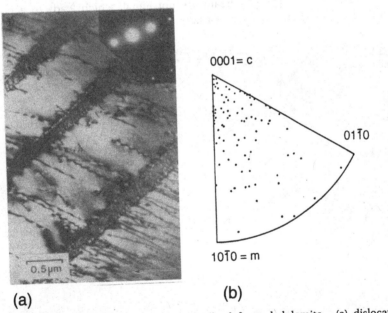

(a)   (b)

Fig. 35. TEM investigation of experimentally deformed dolomite. (a) dislocation microstructure in a grain, illustrating mechanical twinning and activity of the (0001) slip system. Arrow indicates compression direction. (b) Inverse pole figure of the compression directions of 108 grains determined by SAD analysis with the TEM [BARBER &AL. 1994].

Kikuchi patterns

If one directs the electron beam in a TEM to thicker but still transparent regions of the sample, the typical SAD spot attenuates and darker and brighter bands appear that are superposed on the diffraction pattern (Fig. 36). This so-called Kikuchi pattern is due to inelastic scattering of electrons. For a good spatial resolution of orientation determination, the illuminated area must be small which is best achieved with a convergent beam. Kikuchi lines are directly related to the orientations of reflecting lattice planes and are therefore well suited for crystal orientation determination. Elastic scattering of electrons gives rise to sharp diffraction spots. There are also electrons which arrive on lattice planes from many directions and there are always some that obey Bragg's law. These inelastic scattered electrons emerge along a cone about the normal to the reflecting lattice plane with an apex angle (180°–2θ) (Fig. 37a). Since the wavelength of electrons is very small, θ is small and the apical angle is close to 180° and consequently the intersection of the cone with the film is essentially a straight line (Fig. 36).

For each set of lattice planes two parallel lines of different intensity and with an angular distance of 2θ form. These bands are the result of a complicated diffraction and absorption process which can approximately be described by using the dynamical theory of electron contrast [e.g. REIMER 1985, HIRSCH &AL. 1977]. In a simplified way they can be understood as follows. Generally the angles between the inelastically scattered beam and the primary beam are not equal, which leads to a non-equal transfer of electrons along each cone surface. In the direction of the incident electron

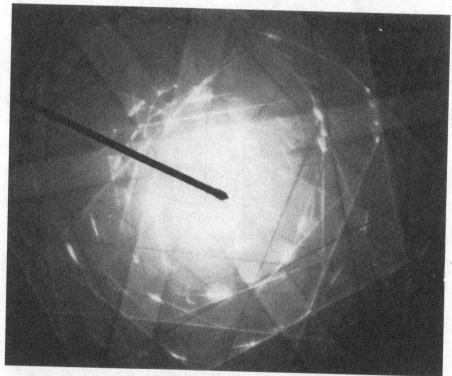

Fig. 36. Selected area diffraction pattern of Ni₃Al with a superposed Kikuchi pattern with dark and bright bands from which the orientation of the crystal can be determined (courtesy of O. Engler).

beam, where Bragg's law is satisfied, low intensity lines occur, since electrons are scattered away and do not contribute to the transmitted intensity. Lines of higher intensity are obtained in the direction into which the electron beam is diffracted since there the electrons add to the background intensity. The width of the parallel bands corresponds to the d-spacing, a band to a lattice plane, and an intersection of bands to a zone axis. By measuring the angles (distance on film) between several zone axes, the crystallographic orientation of the irradiated volume can be determined. In contrast to SAD patterns, orientations are determined without having to adjust tilts on the specimen stage and the procedure is therefore easily automated. For materials with good Kikuchi patterns this method is generally preferred. With convergent beam microdiffraction, orientations can be determined in areas smaller than 50 nm.

There are many applications of Kikuchi patterns for orientation analysis. Usually an operator manually locates a few lines. Their position is then entered into a computer which determines the orientation [e.g. HEILMANN &AL. 1982, HUMPHREYS 1988, SCHWARZER & WEILAND 1988, WEILAND & PANCHANADEESWARAN 1993, HØIER &AL. 1994, ZAEFFERER & SCHWARZER 1994, ENGLER &AL. 1996a]. HUMPHREYS [1983] and WEILAND &AL. [1990] have documented local orientation variations in heterogeneously deformed crystals with an angular resolution of <3°. HAESSNER &AL. [1983] and SZTWIERTNIA & HAESSNER [1990] described orientation correlations between neighboring grains during recrystallization. Readers interested in these detailed studies should consult the literature for more information.

4.2 Scanning electron microscope (SEM)

Local orientations can also be measured with the scanning electron microscope (SEM). Unlike the TEM, the SEM it is not restricted to thin areas located along the edge of a hole in the specimen, but enables crystal orientations to be determined on surfaces of considerable extent. Interaction of the electron beam with the uppermost surface layer of the sample produces two types of diffraction patterns from which the orientation can be determined, electron channeling patterns and electron back-scatter patterns (EBSP) [DINGLEY 1981, JOY &AL. 1982]. EBSPs are more commonly used for texture analysis and will be discussed in some detail

EBSPs are generated if a stationary beam interacts with the surface of a crystal [ALAM &AL. 1954, VENABLES & HARLAND 1973]. They are produced in an analogous way to Kikuchi patterns in the TEM with the only difference that in EBSPs electrons leave the crystal in a direction opposite to the incoming beam. EBSPs are captured on a phosphor screen (Fig. 37c) and recorded with a low intensity video camera or a CCD device [DINGLEY & RANDLE 1992]. Highest intensities are obtained if the sample surface is inclined about 20° to the incoming beam (Fig. 37c). The advantages of EBSP include the high opening angle of the pattern (50–60°) with many lines, the high spatial resolution, which is equivalent to the spot size (<0.5 μm) and the ease of generation.

Analogous to Kikuchi patterns observed in the TEM, excess and deficiency lines, as well as Kikuchi bands, can be recognized in EBSPs (Fig. 38a). The lines can be explained in the same way as described above way by applying Bragg's law to the electrons scattered back into the direction of the incoming beam (Fig. 37a).

Since EBSPs emerge from a very thin surface layer of the sample (on the order of a few nm), they require completely flat surfaces that are as free of defects as possible (Fig. 5f). It is essential to remove the damaged surface layer created during mechanical polishing. In metals electropolishing has proven to be useful [e.g. RAMAN &AL. 1989, WRIGHT &AL. 1993a],

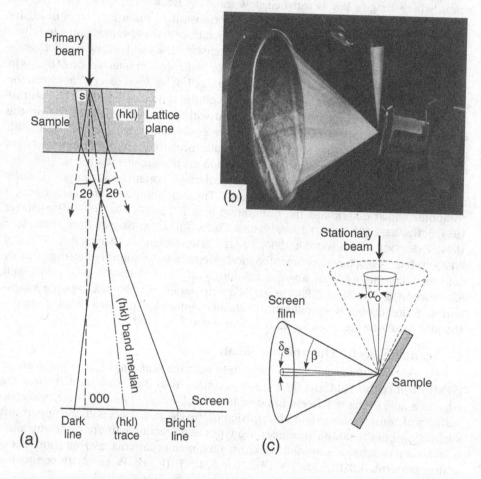

Fig. 37. (a) Sketch illustrating the origin of Kikuchi lines during electron diffraction in transmission geometry. (b) Generation of an EBSP pattern in a SEM, illustrating incident beam, sample holder and phosphor screen [HJELEN &AL. 1993]. (c) Diffraction in the SEM with the sample inclined about 70° to the beam [SCHWARZER & WEILAND 1988].

whereas for ceramics and minerals chemical polishing with a high pH colloidal silica solution improved the pattern quality [LLOYD 1987]. Although the amount of strain in the sample (i.e. high dislocation densities) limits the applicability. WRIGHT &AL. [1993a] report patterns in Al, deformed to 40% strain, ENGLER &AL. [1996a] found them in Al deformed over 90%, and HEIDELBACH &AL. [1996] were able to obtain patterns in 50% cold rolled copper. The decay of the electron diffraction pattern quality with defect density can be used as a measure to estimate the lattice strain [WILKINSON & DINGLEY 1991]. For highly strained metals, annealing at moderate temperatures may release some of the lattice strain through recovery without changing orientation characteristics. For insulators electrostatic charging may occur. A carbon coat, the standard practice for TEM and SEM imaging, can often not be applied because of reduction in pattern quality. Observation at low accelerating voltages (e.g. <10 kV) and low beam current has proven to minimize charging in minerals [KUNZE &AL. 1994].

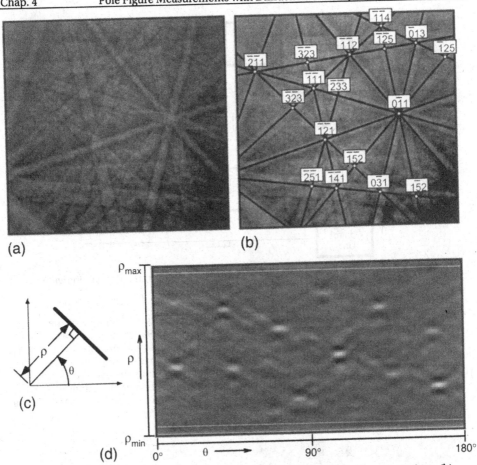

Fig. 38. EBSP of nickel. (a) Experimental EBSP in an arbitrary orientation, (b) identification and indexing of bands, (c) definition of the location of a band by an azimuth θ and a distance ρ, (d) Hough transform of the EBSP in (a) represented in θ–ρ space. A bright spot defines a band (courtesy of S. Wright).

First a map of the sample is prepared which shows the grain boundaries. Then the operator locates the grain of interest and obtains a diffraction pattern. The interpretation of EBSPs can be done on a photographic film, on an analog video screen, or on a digital image. Several methods have evolved for calibrating the sample position and geometry relative to the detector [e.g. KRIEGER LASSEN & BILDE-SØRENSEN 1993]. The angle between bands – and sometimes their width (which is inversely proportional to d) and the intensity– are used to index the pattern and determine the crystal orientation which is, for example, provided as Euler angles. Figure 38a illustrates an experimental pattern of nickel and the indexed pattern (Fig. 38b) based on parameters defining the crystal structure [YOUNG & LYTTON 1972, SCHMIDT & OLESEN 1989]. The angular accuracy of EBSPs is generally <0.5° for relative orientations (misorientations between two grains) and about 1° for absolute orientations (i.e. relative to the sample framework).

The entire procedure has been automated and a schematic illustration of the system is shown in Fig. 39 [WRIGHT & ADAMS 1992]). The sample is translated with a high precision mechanical stage in increments as small as 1 μm, or, for smaller regions, different sample locations are reached by beam deflection. At each position an EBSP is recorded. With a phosphor screen, back-scattered electrons are converted to light, this signal is transferred by a lens system or fiber optics into a CCD or analog camera. The digital EBSP is entered

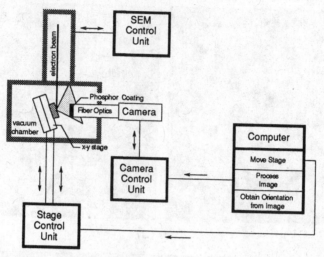

Fig. 39. Hardware configuration for automated determination of individual orientations from digitized EBSPs [WRIGHT 1993].

into a computer and indexed on-line in a matter of seconds, making use of the Hough transform [ILLINGWORTH & KITTLER 1988, KRIEGER LASSEN &AL 1992, KUNZE &AL. 1993]. In the Hough transform (Fig. 38d) diffraction bands appear as points and their position is easily identified (Fig. 38c defines coordinates θ and ρ to specify a band and point respectively). Specimen coordinates, crystal orientation, a parameter describing the pattern quality and a parameter evaluating the pattern match are recorded. Then the sample is translated to the next position and the procedure is repeated. The sample surface is covered with a hexagonal or square grid with about one measurement per second.

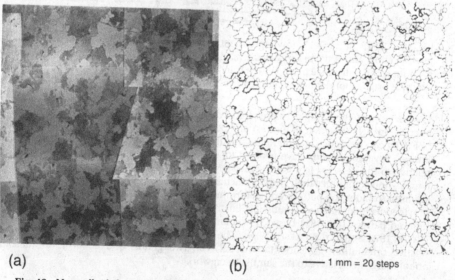

(a)   (b)   ——— 1 mm = 20 steps

Fig. 40. Naturally deformed mylonitic quartzite. (a) Microstructure observed with an optical microscope in reflected light with crossed polarizers. (b) EBSP measurements in a hexagonal grid with 50 μm increments. Grain boundaries (thin lines) were identified if misorientation angles between adjacent cells exceed 10°. Heavy lines are Dauphiné twin boundaries corresponding to a 60° rotation about the c-axis. [KUNZE &AL. 1994].

Figure 40b displays results of such a scan for a naturally deformed quartzite. Grain boundaries (thin lines) have been inserted whenever the orientation difference between adjacent points is more than 10°. The microstructure obtained by orientation imaging microscopy [OIM, ADAMS &AL. 1993, WRIGHT 1993] (Fig. 40b) is in excellent agreement with that observed by direct imaging (Fig. 40a). More importantly, pole figures generated from these individual EBSP measurements are practically identical with those obtained on a different specimen of the same quartzite by neutron diffraction (Fig. 41, KUNZE &AL. 1994). Figure 42 a,b compares 111 pole figures measured on the same rolled copper sample (though not on the same volume) by x-ray diffraction and EBSP [WRIGHT & KOCKS 1996]. The two measurements are similar but do display some significant differences which may be attributed to the fact that automatic EBSP measurements were only capable of indexing about half of the orientations due to poor image quality. The image quality is systematically linked to orientation. The differences are up to 1.5 m.r.d. and best seen in a difference pole figure (Fig. 42c). ENGLER &AL. [1994d] compare EBSP and x-ray measurements for aluminum and discuss statistical requirements. When individual orientations are measured, the determined texture strength depends on the number of grains [MATTHIES & WAGNER 1997].

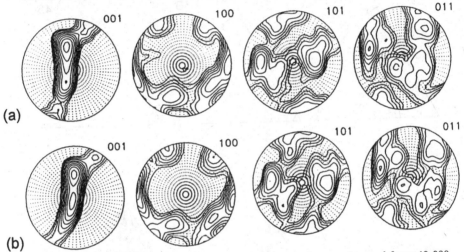

Fig. 41. Naturally deformed mylonitic quartzite. Pole figures obtained from 40 000 EBSP measurements in the SEM, spreading each orientation with a 15° Gauss (a). (b) Corresponding pole figures determined by neutron diffraction and recalculated from the OD (b) [KUNZE &AL. 1994].

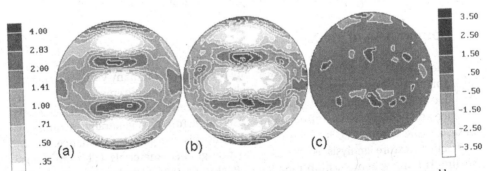

Fig. 42. Comparison of 111 pole figures of moderately deformed copper, measured by x-ray diffraction and completed with the harmonic method (a), and the automatic EBSP technique (b); (c) is the corresponding difference pole figure (scale on right).

The information on microstructure (Fig. 40b) and texture (Fig. 41) is not terribly interesting. Microstructures can be more easily imaged with an optical microscope and pole figures can be measured with an x-ray diffractometer. However, the individual orientation data contain information that cannot be obtained from bulk measurements, particularly about orientation relationships between grains. OIM is a modern extension of earlier optical measurements on geological materials (axis distribution analysis or 'AVA' introduced by SANDER [1950] and elaborated by WENK & TROMMSDORFF [1965]). POSPIECH &AL. [1993] have introduced a formalism to analyze orientation characteristics such as misorientations quantitatively. Misorientations can be represented in an axis/angle space (Chap. 2 Sec. 2.6) and collect all pairs of grains which can be brought to coincidence by a rotation about a particular axis. In the case of quartzite a misorientation distribution map (MOD) shows a strong maximum at a rotation of 60° about the c-axis (Fig. 43). This corresponds to the Dauphiné twin law which is optically invisible. In the orientation map of Fig. 40b those grain boundaries which are twin boundaries are marked as heavy lines.

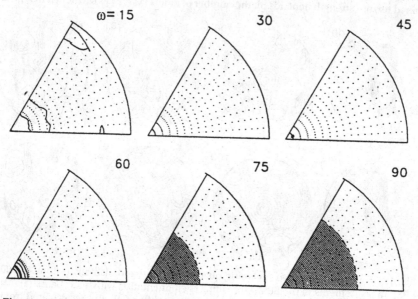

Fig. 43. Naturally deformed mylonitic quartzite. Misorientation distribution function (MOD) between adjacent grains, represented in an axis rotation space. The strong concentration at 0001, ω=60° corresponds to grains which are in Dauphiné twin relation. Areas which are outside the basic region are shaded [KUNZE &AL. 1994].

5. Comparison of Methods

The optimal choice of texture measurements depends on many variables, such as availability of equipment, material to be analyzed, and data requirements.

For routine metallurgical practice and many other applications in materials science and geology, *x-ray diffraction* in reflection geometry is generally adequate. It is fast, easily automated, inexpensive both in acquisition and maintenance. X-ray diffraction texture analysis is restricted to fine-grained materials (<1 mm) whose texture is homogeneous within the plane. Back-reflection provides incomplete pole figures, but this drawback can be overcome by data analysis (see Chap. 3). Pole figures can only be measured adequately if diffraction peaks are sufficiently separated. In geological samples and ceramics, x-ray diffraction is therefore generally limited to single phase aggregates of orthorhombic or higher crystal symmetry. Particularly for thin epitaxial films it is advantageous to produce strictly monochromatic radiation with a single-crystal monochromator or an energy dispersive detector.

Neutron diffraction is advantageous for determination of complete pole figures in coarse-grained aggregates and allows determination of magnetic pole figures. The characteristic of neutrons is that bulk samples rather than surfaces are measured, that coarse grained materials can be characterized, that environmental cells (heating, cooling, straining) are available and that angular resolution is much better than for x-rays. It is possible to measure complex polyphase composites with many closely spaced diffraction peaks. With position-sensitive detectors, and particularly with TOF, continuous spectra can be recorded. Shifts in peak positions can be used for residual stress determination and intensities to extract simultaneously texture and crystal structure information, e.g. with the RITA technique. With neutrons, texture measurements average over large volumes and homogeneity must be assumed. Whereas pole figure determinations with x-rays are generally straight-forward and results are obtained and processed within a few hours, neutron diffraction is more cumbersome. It requires the user to write proposals for a specific experiment, months in advance, and access is limited to a few days of beam time, generally only sufficient for a few selected examples. Neutron diffraction should only be used to measure samples which cannot be measured with x-rays, which is often not done at present.

Electron diffraction with a TEM is most time-consuming but provides, in addition to crystal orientation, valuable information about microstructures and, at least two-dimensionally, about interaction between neighbors and about heterogeneities within grains. These are important data to interpret deformation processes.

Recently *EBSPs*, measured with the *SEM* on polished surfaces, have become a new addition to the techniques for texture measurements. They allow for determination of local orientation correlations which are important in the study of recrystallization, or of stress concentrations on which failure may occur. Determination of individual grain orientations has advantages for OD calculations because of the absence of ambiguity which is inherent in pole figures. With the possibility of automation this technique has become comparable in expense and effort to x-ray diffraction analysis. So far, it remains confined to samples with few defects and moderate deformation. The method lends itself to texture determination in polyphase materials, for example by using in addition to the EBSP image a x-ray signal for chemical fingerprinting.

X-rays from synchrotron sources provide a highly collimated intense beam which can be used for local texture analysis of fine grained polycrystalline samples and for determination of the orientation of individual crystallites with the Laue technique. The latter is particularly useful for highly deformed materials and ceramics where EBSP measurements are not possible. As with neutrons access to synchrotron facilities is very limited.

Chapter 5

TYPICAL TEXTURES IN METALS

1. **Deformation Textures in Face-centered Cubic (fcc) Metals**
   1.1 Fiber textures: compression of aluminum, silver, copper and brass
   1.2 Rolling textures.
       Variation with strain. Variation with material and temperature.
       Fiber and component analysis
   1.3 Torsion textures (simple shear)
       Variation with material. Reversibility

2. **Deformation Textures in Body-centered Cubic (bcc) Metals**
   2.1 Basis for comparing textures in fcc and bcc metals
   2.2 Fiber textures: tension of tantalum
   2.3 Rolling textures: tantalum, steel
   2.4 Torsion textures: iron

3. **Deformation Textures in Hexagonal Close-packed (hcp) Metals**
   3.1 Fiber textures: extruded titanium and upset (compressed) zirconium
   3.2 Rolling textures
   3.3 Tube textures: zirconium

4. **Deformation Textures in 'Other' Materials**
   4.1 Orthotropic uranium
   4.2 Textures in intermetallics: NiAl and TiAl

5. **Composites**
   5.1 Wire drawn Cu-Nb composites
   5.2 Extruded Al-SiC composites
   5.3 Rolled Be-Al composites

6. **Transformation Textures**
   6.1 Phase transformations
   6.2 Recrystallization in rolled fcc metals

7. **Texture Heterogeneities**
   7.1 Local heterogeneities in deformation
       Grain-scale heterogeneity. Multi-grain heterogeneity
   7.2 Texture gradients after deformation
   7.3 Heterogeneity in recrystallization

8. **Solidification and Thin Film Textures**
   8.1 Solidification
   8.2 Vapor deposition
   8.3 Electrodeposition

9. **Summary**

TYPICAL TEXTURES IN METALS

A. D. Rollett and S. I. Wright

The purpose of this chapter is to present a selection of experimental results for textures in metallic materials. The intention is to provide a guide to what an experimenter might expect to observe for a given metal and thermo-mechanical history. The main emphasis is placed on the development of texture as a result of mechanical deformation. Despite the knowledge that almost all processes impart some texture to crystalline materials, a common omission in discussions of texture is information on the initial texture of the material under investigation. An early example of the effect of initial texture on the final deformation texture was given by FREDERICK & LENNING [1975] for cold rolled Ti-6Al-4V. Since the initial texture has an effect on the evolution of the texture during deformation, other texture forming processes such as recrystallization and solidification will also be briefly reviewed.

Since the sample symmetries (or, more accurately, sample asymmetries) tend to persist in textured materials through deformation, a fascinating and little exercised use for texture analysis is as a diagnostic of material history. There is not space in this brief review to treat this aspect. We note, however, that most types of deformation do not produce a single orientation but rather a range of orientations. More subtly, individual orientations may rotate quickly into the range of characteristic orientations, such as the partial fibers observed in plane-strain deformation, but then rotate slowly within that range. Thus the initial texture may not have much effect on the overall range of preferred orientations observed but it *can* have a strong effect on the distribution *within* that range. An intuitive way to consider this problem is to imagine that textures consist of ridges (rather than peaks) and valleys: the type of plastic deformation determines the position of the ridges; the process of deformation sweeps up grains from the valleys into the ridges (Chap. 9).

Plastic deformation and the study of plasticity as it affects texture development is dealt with in Part Two of this book. The reader should be aware, however, that an appreciation of the Taylor model of polycrystal plasticity is presumed. It treats compatibility constraints as paramount. Some textures are regarded as compatible with this basic model whereas others signify deviations, some of which can be accommodated under the 'relaxed constraints model'. Taking the Taylor model as a basis for review marks the essential difference between this chapter and the early texture literature for which no substantive theory for texture development was available. In anticipation of Chapter 9, we assert that the Taylor model provides an adequate first-order explanation of texture development during plastic deformation in many materials. Preferred orientation arising from plastic deformation, then, comprises the bulk of material in this chapter whereas other processes which lead to the formation of preferred orientation, such as recrystallization, vapor deposition and solidification are discussed in less detail.

We do not provide an exhaustive survey of all possible textures or of the extant literature; see, e.g., the introductory work by HATHERLY & HUTCHINSON [1979], and the extensive review by DILLAMORE & ROBERTS [1965]. Instead we present a broad overview of metal textures using pole figures taken both from our own work and from the literature. A consequence of this is that some plots are presented as equal-area projections whereas others are in stereographic mapping. A few orientation distributions are presented where they help to elucidate the fine points of differences between materials. Readers not familiar with the field may wish to acquaint themselves with the considerable body of work in the proceedings of the ICOTOM series, that is, the International Conferences on Textures of Materials. There have now been 11 of these conferences, held at three–year intervals.

Presenting a survey of this sort forces a choice of whether to organize the contents by type of deformation, or by type of material. For plastic deformation this article is organized by *crystal symmetry* (i.e. material type) with sample symmetry (i.e. deformation type) as an 'inner' loop. We are particularly conscious of the fact that a large fraction of the texture literature deals with rolling of fcc materials (not to say, aluminum): we *de-emphasize* these textures here. They are dealt with as examples in the chapter on Representation (Chap. 2) and on modeling (Chap. 9). Instead we endeavor to present a wide range of materials and deformation modes – albeit on scattered examples with which we are familiar.

It will be noted that variations between fcc materials are *correlated* with the variation in stacking-fault energy (SFE): this does not, however, mean that there exists a clear consensus on the micromechanical mechanisms that produce and/or explain this correlation. In addition, transitions between texture types can be obtained by variations in deformation temperature, as well as alloy content.

We will first discuss textures commonly observed in face-centered cubic (fcc) and body-centered cubic (bcc) single-phase metals. Then we will discuss the textures of hexagonal close-packed (hcp) metals, which display stronger variations with material. These variations are explained by accounting for the relative ease of slip or twinning on the various deformation systems, for example, basal slip versus prismatic slip. Following this section, other metals are examined. Certain metallic systems have even lower symmetry and yet have adequate ductility for the thermo-mechanical processing that develops strong textures. The best known example is uranium in its low-temperature, orthorhombic allotrope. The plastic deformation portion of this chapter ends with reference to the variations that arise when composites are examined, especially the characteristic weakening of texture as the volume fraction of second phase is increased.

Following the discussion of deformation textures, recrystallization and transformation textures are discussed as a separate section. While measured textures are often assumed to be representative of the entire material volume, most forming processes do not produce materials with uniform spatial distributions of texture and therefore we include a short discussion on texture inhomogeneities. This is significant, for example, when considering the fact that experimental textures are characteristically weaker (more dispersed) than the theoretical prediction for any given strain and deformation type. In the last Section, solidification and thin film textures are discussed.

1. Deformation Textures in Face-centered Cubic (fcc) Metals[†]

1.1. Fiber textures: compression of aluminum, silver, copper and brass.

The two axisymmetric deformations, tension and compression, are such that only one direction is needed to fully characterize the preferred orientation. Of course, if the material has a non–axisymmetric texture before deformation in uniaxial tension, or compression, or an axisymmetric texture where the axis of symmetry is not aligned with the tension or compression axis, then the material will have a non–axisymmetric texture after deformation. Thus, when using single directions to characterize textures, it must be kept in mind that textures are rarely perfectly axisymmetric and usually possess at least some radial component. With this *caveat* in mind, Fig. 1 [STOUT &AL. 1988] shows an array of experimentally measured inverse pole figures for four different materials, 1100 aluminum, copper, silver and brass, arranged in columns. Each inverse pole figure shows the distribution of compression axes with respect to the crystal axes. The first row shows the initial texture, which consists of the expected mixture of $\langle 111 \rangle$ and $\langle 100 \rangle$ fiber components for round bar stock. The second row shows the result of Taylor-type simulations taking the initial textures as input. The bottom row displays experimental textures after uniaxial compression to a strain of 2.0. An inspection of the first three materials (columns) shows that the slight variations in final texture can be at least qualitatively explained by variations in the initial texture alone. (For details of the simulation, see the original paper by STOUT &AL. [1988].) The lack of variation of compression texture with stacking-fault energy for these three materials runs counter to older work [ENGLISH & CHIN 1965] – which, however, had not taken account of any prior history or initial texture. In addition, results may differ because solutes can affect texture development for reasons other than their effect on stacking-fault energy [ENGLER &AL. 1994c].

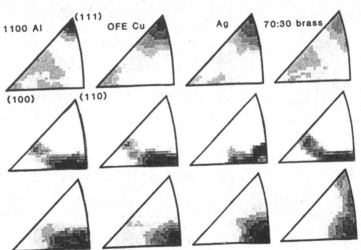

Fig. 1. Comparison of inverse pole figures for compression of fcc metals. Top row: initial texture; middle row: simulated texture; bottom row: experimental texture. [STOUT &AL. 1988].

[†] Carl Necker is acknowledged for his help with this Section.

An exception from this general behavior is displayed by the deformed brass texture, which is not well explained at all. One possible reason is that brass is known to exhibit copious twinning in addition to the usual deformation modes.

1.2 Rolling textures

Variation with strain

It is often assumed that the texture of a polycrystal rolled to a high reduction will give essentially the same texture as a polycrystal rolled to a lower reduction only stronger. While a higher reduction will generally produce a sharper texture, the character of the texture may change significantly. Figure 2 shows 111 pole figures for copper rolled to increasing reductions [NECKER 1997]. In this figure, the texture sharpens with increasing strain, as would be expected, and the general features of the texture remain relatively constant as well. However, the locations of the peaks in the pole figures shift considerably in orientation space. As discussed below, this means that caution is required when analyzing textures in terms of fibers that are fixed in orientation.

When considering rolling textures it should be remembered that the rolling reduction, r, does not accurately represent the amount of deformation the material has undergone. One often quotes the true compressive strain $\varepsilon = \ln(1-r)$, which is negative. A better metric for multi-component straining (like rolling), and for comparison with other strain paths, is the von Mises equivalent strain which for rolling is

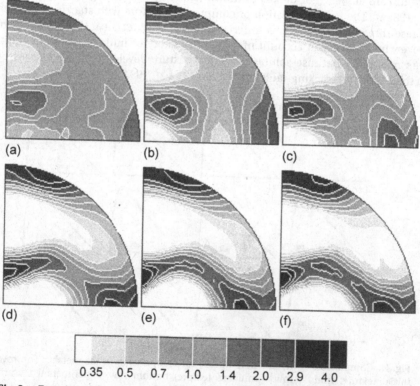

(a)   (b)   (c)

(d)   (e)   (f)

| 0.35 | 0.5 | 0.7 | 1.0 | 1.4 | 2.0 | 2.9 | 4.0 |

Fig. 2. Experimental 111 pole figures (stereographic projection, rolling direction vertical) for rolled OFE copper, recalculated by a combined harmonic and WIMV method (using popLA) at reductions of (a) 0% (initial texture), (b) 35% ($\varepsilon_{vM} = 0.5$), (c) 58% ($\varepsilon_{vM} = 1.0$), (d) 82% ($\varepsilon_{vM} = 2.0$), (e) 90% ($\varepsilon_{vM} = 2.7$), and (f) 95% ($\varepsilon_{vM} = 3.5$).

$$\varepsilon_{vM} = \left| \frac{2}{\sqrt{3}} \ln(1-r) \right|$$

The consequence of using true strains rather than reductions is that at higher rolling reductions, small increments in the rolling reduction result in relatively large increments in von Mises strain. Similarly in torsion, the von Mises equivalent strain is related to the shear strain by:

$$\varepsilon_{vM} = \left| \frac{\gamma}{\sqrt{3}} \right|$$

Variation with material and temperature

The variation in rolling texture with material has been well documented by previous authors. It is convenient to regard the 'copper' texture and the 'brass' texture as the opposite extremes of 'pure metal' and 'alloy' type textures (see Table I for definitions). Some materials consistently yield the alloy-type texture, whereas others, such as copper itself, can give either, depending on strain rate, temperature and composition. The effect of increasing the amount of zinc in copper is shown in Fig. 3 [STEPHENS 1968]. This set of 111 pole figures shows the transition from pure metal to alloy type texture over a range of Zn content in which twinning is not generally observed. This figure shows that there is a transition to the alloy texture at about 15% zinc. However, this transition can also be achieved with the addition of only a few percent of phosphorus [DILLAMORE & ROBERTS 1965].

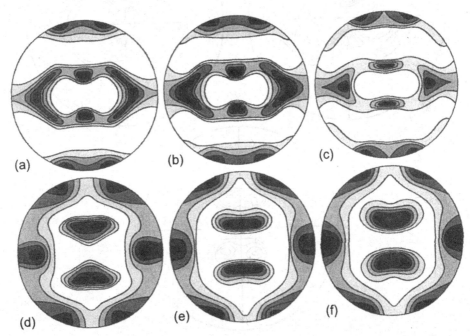

Fig. 3. 111 pole figures for Cu-Zn alloys cold-rolled to 96% reduction with varying Zn content [STEPHENS 1968] (stereographic projection, rolling direction vertical).

Similarly, elevated deformation temperatures tend to lead to the pure metal texture, whereas low temperatures have the opposite effect: for example, rolling copper at liquid nitrogen temperatures yields a brass texture [HU & GOODMAN 1963], and rolling silver at high temperature shifts the texture toward that of Al [HU & CLINE 1961, HU & AL. 1961]. Figure 4 [MOMENT 1972] shows the effect of temperature on texture evolution in 95% rolled plutonium with-9.6 atomic percent gallium. These pole figures show that the transition from the pure metal to the alloy type texture can be achieved by decreasing the deformation temperature. A rather complete table is available in DILLAMORE & ROBERTS [1965] (their Table II) summarizing the texture transition data for most fcc alloy systems. We discuss similar texture transitions for hcp metals in a later section.

It is also worth noting the continuing controversy in the literature concerning the role of microstructure in the formation of the alloy or brass texture. Although brass itself twins readily at small strains, it is not clear that twinning occurs at the higher strains (greater reductions in thickness) during rolling at which the brass component becomes strong. (See for example LEFFERS &AL. [1988] and, for very fine-grained material, YEUNG &AL. [1988]). In fact, the brass component appears without twinning in many cases; for example, this component is often dominant in aluminum–lithium alloys [VASUDEVAN &AL. 1988], obviously for different reasons.

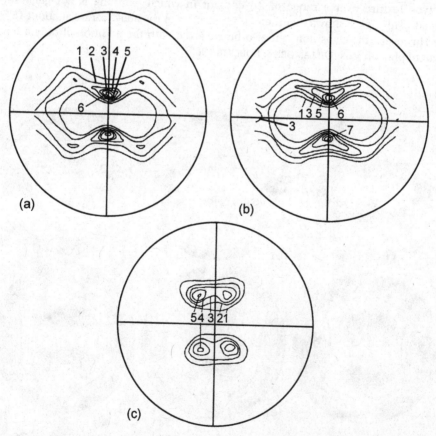

Fig. 4. 111 pole figures for Pu-9.6 atomic percent Ga rolled 95% at (a) -80°C, (b) -150°C and (c) -196°C [MOMENT 1972] (stereographic projection, rolling direction vertical).

Fiber and component analysis

There are two common approaches to analyzing rolling textures in a simplified manner: (1) using texture components and (2) using fibers. The aim of using texture components is to reduce the representation of the orientation distribution into a small set of specific orientations or texture components. The following table lists the Euler angles and Miller indices. See Chapter 2 for the Euler angle nomenclature and further discussion of the component analysis approach.

Table I. Rolling texture components: Indices and Euler angles

Name	Indices	Bunge $(\varphi_1, \Phi, \varphi_2)$	Kocks (Ψ, Θ, ϕ)
copper	$\{112\}\langle 11\bar{1}\rangle$	90, 35, 45	0, 35, 45
S1	$\{124\}\langle 21\bar{1}\rangle$	59, 29, 63	149, 29, 27
S2	$\{123\}\langle 41\bar{2}\rangle$	47, 37, 63	137, 37, 27
S3*	$\{123\}\langle 63\bar{4}\rangle$	59, 37, 63	149, 37, 27
brass	$\{110\}\langle \bar{1}12\rangle$	35, 45, 0	55, 45, 0
Taylor	$\{4\,4\,11\}\langle 11\,11\,\bar{8}\rangle$	90, 27, 45	0, 27, 45
Goss	$\{110\}\langle 001\rangle$	0, 45, 0	90, 45, 0

* This particular orientation is often quoted as 'the' characteristic S orientation. Nevertheless, there is significant variation in the literature as to which precise orientations are labeled as 'S'. *Caveat Lector!*

The other approach for studying rolling textures uses the concept of partial fibers, where a fiber is a range of orientations limited to a single degree of (rotational) freedom about a fixed axis [BUNGE 1982]; some fibers are defined in terms of an axis in sample space, and some in terms of crystal coordinates. The appearance of a fiber in OD plots is that of a line, which may or may not lie entirely in one section. There are two fibers typically used for fcc metals: the β and α fibers. The β fiber runs (in orientation space) from the 'copper' to the 'brass' component. The 'S' orientations are at intermediate locations along this fiber. The α fiber runs from 'brass' to 'Goss'. The β fiber is generally prominent in rolling textures. Ideally, these fibers should always correspond to regions of high intensity in the orientation distribution. However, the location of the maximal values in any given OD may vary significantly from the ideal. This has led to definitions of fibers that are *not* fixed in position ('skeleton lines' [BUNGE 1982]); for example, the term 'β fiber' is used by some as a generic term for all proximate locations of the β fiber defined above, and even the term 'copper orientation' is then used for all orientations *near* $\{112\}\langle 11\bar{1}\rangle$ [HIRSCH &AL. 1984]. When the skeleton line is not an ideal fiber, one needs *two* diagrams to represent it, and the advantage over using the orientation distribution itself becomes tenuous. (See also the discussion in Chapter 2.)

Nevertheless, the skeleton line analysis technique has been used extensively by many workers; we give the following example from work by HIRSCH & LÜCKE [1988a], In Fig. 5a the evolution of the (generalized) β fiber with rolling reduction is shown for copper. The results show a gradual strengthening of the fiber with strain, as expected. The fiber analysis approach is particularly well suited to quantitative comparisons of texture in large sets of related samples. The 'fibers' were defined by locating maxima in constant-φ_2 sections of the corresponding CODs[1]. The variation in the location of

---

[1] COD and SOD signify crystallite orientation distribution and sample orientation distribution, respectively. See Chapter 2 on OD representations for explanation.

these maxima (in terms of φ_1 and Φ) is shown in Fig. 5b. This figure clearly shows that the definition of the β fiber tends to differ from one COD to another. (In practice, the β fibers shown in Fig. 5 are relatively well behaved.)

The fibers described here for rolled fcc textures have an exact correspondence in bcc plane-strain textures and so the reader may use the table given above to ascertain the nomenclature, needing only to transpose plane with direction (e.g. {110}⟨$\bar{1}$12⟩ in

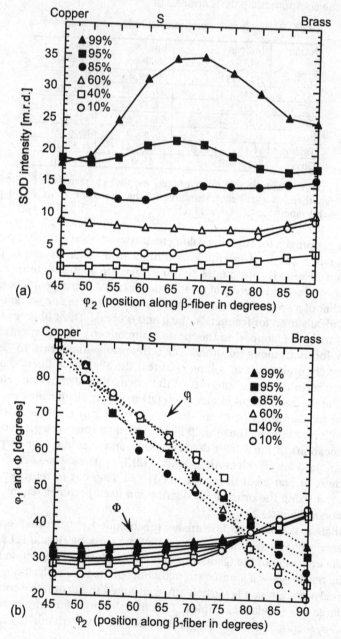

Fig. 5. Plot of (a) normalized intensities, and (b) locations of generalized β fibers in copper rolled to various reductions in thickness. After HIRSCH & LÜCKE [1988a], Fig. 26.

fcc becomes $\{\bar{1}12\}\langle 110\rangle$ in bcc). The exception (but only in nomenclature and not in a physical sense) to this rule is the Goss orientation, which is defined as $\{110\}\langle 001\rangle$, regardless of material. Another *caveat* is that some authors have resisted the use of Greek letters for names of fibers because of potential confusion in steel with the names of the various allotropes of iron; some workers [RAY &AL. 1993] favor the use of 'RD' and 'ND' fiber names (to signify the bcc α and γ fibers, to be defined in Table II).

1.3 Torsion textures (simple shear)

Torsion tests represent the simplest experimental method of obtaining simple shear. Such tests are very useful for obtaining strain hardening data for large plastic strains without the complications of friction effects at interfaces between specimens and platens that occur in compression testing. Texture development is important, not just for understanding of the underlying mechanisms, but also in order to be able to make quantitative comparisons with other deformation modes. For example, if one wishes to simulate stress/strain response in rolling, it is important that the data be corrected for differences in Taylor factor between torsion and rolling. Most torsion testing has been performed on solid cylinders with the consequence that there is a strain gradient from the surface (maximum shear) to the center (zero shear). Texture measurement requires that the surface layer be separated and flattened. The flattening operation introduces a small additional strain which may be neglected. Tubular specimens have been employed more recently in order to obtain more nearly uniform stress (and strain) states in the gauge length; the same procedure is used to measure texture. We will discuss ideal torsion textures and the dependence on material. Lastly, torsion followed by reverse torsion is used to illustrate the point that texture development is not generally reversible.

Ideal orientations are shown in Fig. 6, taken from CANOVA &AL. [1984a]. The 'A' partial fiber can be indexed (in the form plane-of-shear/direction-of-shear) as $\{111\}\langle uvw\rangle$, the 'B' partial fiber as $\{hkl\}\langle 110\rangle$, and the 'C' orientation as $\{001\}\langle 110\rangle$; the C position is just a special position on the B fiber. These designations are based on the theory of torsion textures as implemented in the Taylor model. Recent work has revealed the existence of another partial fiber, referred to in this volume as the 'D' fiber. It appears in the 111 pole figure as being approximately equivalent to the A fiber but rotated 90° about the torsion axis. When it is strong, it often has the appearance of a diagonal band joining the center of the 111 pole figure to the 2 o'clock position. One particular component that is sometimes dominant in the D fiber is $\{112\}\langle 110\rangle$. Note that care is needed when inspecting torsion textures: some publish their pole figures with the torsion axis horizontal and the shear direction pointing up; in our figures, the torsion axis is vertical and the upper shear direction points to the left. The sense of shear can be either negative, as for a left-handed screw axis (as in our figures), or positive, like a right-handed screw.

Torsion textures are rarely strong. This can be explained by considering the results of computer simulation of torsion texture development. There are no stable orientations because any particular grain is constantly rotating; the presence of a texture is simply the result of there being quasi-stationary positions in the orientation distribution where grains rotate very slowly (with respect to the specimen axes). This may be contrasted with plane-strain or axisymmetric deformation, for example, in which definite stationary end-points exist in orientation space. Further discussion of torsion textures, also at larger strains, can be found in Chap. 9 Sec. 2.3.

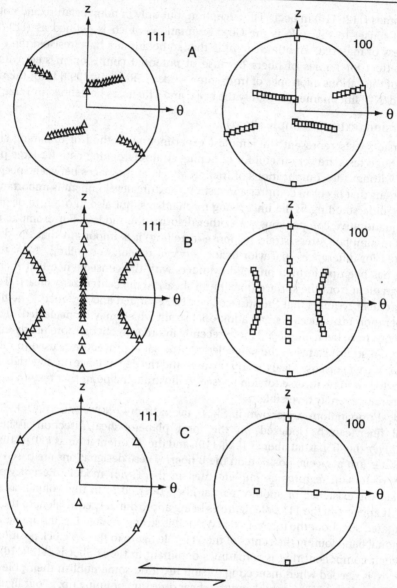

Fig. 6. 111 and 100 pole figures showing the ideal torsion texture components for the fcc crystal structure; the shear direction points to the right on top. Note that the A (partial) fiber is asymmetric with respect to the torsion axis, whereas the B fiber and the C component are apparently symmetric. After CANOVA &AL. [1984a].

Variation with material

STOUT &AL. [1988] studied the variation in torsion texture formation as a function of material. Also, HUGHES & WENK [1988][2] studied the formation of torsion textures in solid-solution Ni-Co alloys, using a short tubular specimen to reach large strains in Ni, Ni-30Co and Ni-60Co. Adding cobalt to nickel lowers the stacking-fault energy and

---

[2] The pole figures shown in this reference were plotted with an incorrect sign of shear.

the 60% Co alloy is nearly the maximum addition for which the fcc structure is stable. From these data, we have chosen to present pairs of 111 and 100 pole figures for Ni, Cu, Ag, Brass and Ni-60Co at a von Mises equivalent strain of approximately 2, Fig. 7. The purpose is to show how torsion textures vary with material. Note that the shear direction at the top of the pole figures points to the left in this case.

The high stacking-fault energy materials, Ni and Cu, show a strong B fiber and a weak, but non-zero intensity at the A fiber position. The results for the silver show a similar B fiber, but the A fiber is absent: in its place is the D fiber (see above), on the opposite side of the torsion axis. Looking next at the brass, the strength of the D fiber is accentuated, and in the Ni-60Co the D fiber dominates the texture. In all of these shear textures, the intensity is low; the maximum intensity in any of the pole figures is less than four times random. These results reproduce those of WILLIAMS [1962] for copper (at the largest shear strain) and for brass.

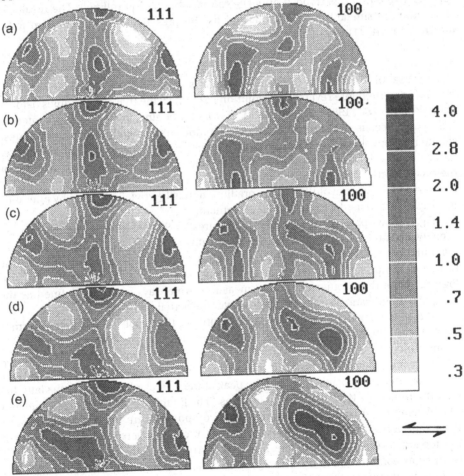

Fig. 7. Plots of 111 and 200 pole figures (with the sense of shear being left on top) for the following materials deformed in torsion: (a) Nickel at γ=3.6, (b) copper at γ=3.5, (c) Silver at γ=3.5, (d) Cu-30Zn at γ=3.5, (e) Ni-60Co at γ=3.2. Note that the partial 'A' fiber is present in Ni and Cu, but is absent in the other materials. Silver, brass and Ni-60Co show instead a 'D' fiber which is similar to the A fiber but rotated approximately 90° about the torsion axis. The B fiber is present to varying degrees in all the materials.

This texture transition has also been observed at different strain levels in Cu-3Zn by SEKINE &AL. [1981] and VAN HOUTTE &AL. [1981] where small strains showed a copper-like texture, whereas larger strains generate the alloy texture. Also, deformation at 77K induces the same transition to an alloy texture in pure copper, i.e. the A fiber is replaced by the D fiber. MONTHEILLET &AL. [1984] published an extensive investigation of torsion textures in Al, Cu and Fe. The results for copper twisted at room temperature all indicate a strong B fiber as discussed above. The B fiber is displaced with respect to the specimen axes in a fashion that depends on strain: at low strains, it is displaced counter to the sense of shear, but at large strains it is displaced with the sense of shear. The results of varying the temperature show that at the highest temperature (400°C), the texture is dominated by a strong $\{112\}\langle110\rangle$ component, which is, curiously enough, identical to the dominant component of the texture observed in brass at low temperatures and in other low stacking-fault alloys of copper. (Note that this trend is opposite to that observed in rolling.) The results for aluminum are similar, albeit complicated by the presence of a non-random initial texture [MONTHEILLET &AL. 1984].

Reversibility

Reversal of torsion is of interest because it is very simple to reverse the sense of straining and therefore reverse the evolution in grain shape that accompanies deformation [ROLLETT &AL. 1988]. Irreversibility of work hardening, in single-phase materials, is accepted because it is clear that dislocation motion through the dislocation 'forest' should not depend on the sense of straining and, therefore, the accumulation of stored dislocations is sign-insensitive. Texture development, on the other hand, depends on the accumulated change in shape of grains so reversing the strain might be expected to reverse the texture evolution. This prediction may be trivially verified by operation of a Taylor-based model for texture development.

Fig. 8 shows pairs of 111 and 100 pole figures for 99.99% pure aluminum that has been twisted and then untwisted to zero net strain. The three sets of pole figures show the initial texture, Fig. 8a, the texture after a forward strain of $\varepsilon_{vM}=2$, Fig. 8b, and the texture at a net strain of zero, corresponding to an accumulated strain of $\varepsilon_{vM}=4$, Fig. 8c. (Note that 'equivalent strain' measures are always implicitly defined as the integral of infinitesimal strain increments over the actual path.)

The initial texture is far from random and is the result of the prior thermomechanical processing; the ensuing torsion textures do not, however, suggest any strong influence of the initial texture. The intermediate state shows a typical fcc torsion texture dominated by the B partial fiber, but with an A (partial) fiber present also. Note the slight displacement of the peak at the 12 o'clock position to the left; this lack of symmetry is typical of torsion textures. The final state (zero net strain) shows a similar texture but slightly weaker; the same components are present but are not in the standard positions. The texture is different after the reverse twisting from any of the torsion textures shown in the previous figure. Nevertheless, the fact that the texture has *not* returned to the initial state shows unambiguously the irreversibility of texture development. These experiments confirmed the earlier results of BACKOFEN [1950] for twisted and untwisted copper, using a different specimen design. The dominance of the C component at large strains was also observed in simulations by CANOVA &AL. [1984a] in which the relaxed constraints variant of the Taylor model was used.

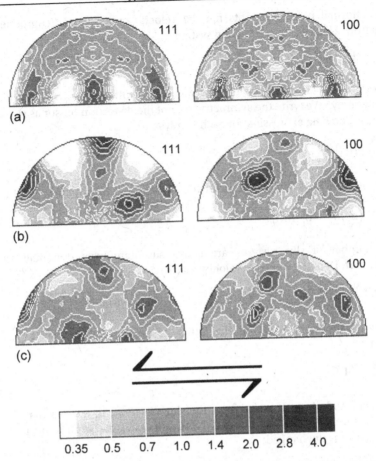

Fig. 8. Plot of 111 and 200 pole figures for aluminum (99.99%) with the torsion axis up and the shear direction pointing to the left, showing (a) the initial texture, with a remnant rolling texture (RD is horizontal in this figure); (b) the texture for a shear strain of 3.5 (von Mises equivalent strain of 2); and (c) the texture for the same forward strain but reverse-twisted to zero net strain.

2. Deformation Textures in Body-centered Cubic (bcc) Metals

2.1 Basis for comparing textures in fcc and bcc metals

The most common mode of deformation in (bcc) metals is {110}⟨111⟩ slip. Since this is merely a transposition of slip direction and slip plane with respect to fcc slip, much can be learned by a comparison of textures in the two structures. Bcc metals also slip on other planes, under various conditions, and sometimes the behavior can be adequately described by 'pencil glide' with a ⟨111⟩ slip direction. Pencil glide has no ready equivalent in fcc metals, although there is good evidence for slip on non-close-packed planes and directions in fcc metals (see e.g. CARRARD & MARTIN [1987]).

For so-called 'restricted glide', i.e., on {110} planes only, the comparison with fcc deformation is made simple by the following derivation in which we anticipate in abbreviated form the full development to be found in Chapter 8. If the slip direction of a particular slip system in an fcc material is called $\hat{\mathbf{b}}$ (components $\hat{b}_i$) and a particular slip plane normal $\hat{n}$ (components $\hat{n}_i$), the velocity gradient, L (components L_{ij}) following from a shear rate on this slip system is

$$L_{ij}^{(s)} = \dot{\gamma}^{(s)}\, \hat{b}_i^{(s)} \hat{n}_j^{(s)}$$

(1)

It is important, however, to consider the symmetric and skew parts of the distortion tensor separately. We write the symmetric part of the distortion tensor as m_{ij}, which is given by the following expression for each slip system:

$$m_{ij}^{(s)} = \frac{1}{2}\left(\hat{b}_i^{(s)}\hat{n}_j^{(s)} + \hat{b}_j^{(s)}\hat{n}_i^{(s)}\right)$$

(2)

Similarly, we write the skew part of the distortion tensor as q_{ij} which is given by:

$$q_{ij}^{(s)} = \frac{1}{2}\left(\hat{b}_i^{(s)}\hat{n}_j^{(s)} - \hat{b}_j^{(s)}\hat{n}_i^{(s)}\right)$$

(3)

When a number of slip systems are active simultaneously one may write the symmetric part, **D**, of the imposed velocity gradient, **L**, as:

$$D = \sum_s \dot{\gamma}^{(s)} m^{(s)}$$

(4)

The rotation rate, or spin, **W**, of the lattice of each grain is given by a similar expression:

$$W = \sum_s \dot{\gamma}^{(s)} q^{(s)}$$

(5)

On the other hand, if we now use $\hat{n}$ for the slip *direction* in bcc (so that the Miller indices are still 111), and $\hat{b}$ for the slip *plane* in bcc, the symmetric part of the distortion tensor is unchanged:

$$^{bcc}m_{ij}^{(s)} = \frac{1}{2}\left(\hat{n}_i^{(s)}\hat{b}_j^{(s)} + \hat{n}_j^{(s)}\hat{b}_i^{(s)}\right) = {}^{fcc}m_{ij}^{(s)}$$

(6)

The skew part however, changes sign so that the plastic spin (**W**) changes sign:

$$^{bcc}q_{ij}^{(s)} = \frac{1}{2}\left(\hat{n}_i^{(s)}\hat{b}_j^{(s)} - \hat{n}_j^{(s)}\hat{b}_i^{(s)}\right) = -\,{}^{fcc}q_{ij}^{(s)}$$

(7)

Thus, the effect of replacing fcc slip geometry by bcc slip geometry (i.e. exchanging $\hat{b}$ and $\hat{n}$) is that the rotation rates or *plastic spins* are the opposite when the strain rates are the same. Any rigid body rotation of the grains (as in torsion – see Chap. 8 Sec. 4 for detailed discussion), however, is *unchanged* by the change in slip geometry. Therefore to keep all rotations of each grain lattice the same (and obtain the same texture in bcc material as compared with fcc), one must prescribe the *opposite* strain, i.e. one must take the *negative transpose* of the velocity gradient. We will demonstrate the practical effect of this relation between fcc and bcc textures (when it applies) in a few examples.

• In uniaxial deformation, the texture developed in *tension* of fcc metals should be the same as that in *compression* of bcc metals. Indeed, the ⟨110⟩ fiber developed in compression of fcc is obtained in tensile deformation of bcc materials. Similarly, the formation of mixed ⟨100⟩ and ⟨111⟩ fibers in fcc tensile deformation corresponds to compression in bcc materials.

- In rolling, a change of sign of the strain corresponds to an interchange of rolling plane normal (ND) and rolling (RD) directions; an example is given below for an fcc to bcc comparison.
- In torsion, if one compares the same pole figure for torsion textures of fcc and bcc materials, the same texture should be observed but rotated through 90° about the radial direction.

The simplest level at which to compare results of the same deformation applied to the different crystal structures is to compare pole figures; the comparison, however, is one in which the *compression* and *extension* directions have been transposed for one of the materials. That is to say, the pole figures are plotted with the ND and RD directions transposed for the bcc case. Figure 9 shows 111 and 200 pole figures (recalculated from the OD – the {111} is not measured in bcc) for rolled copper and iron, both deformed to an 80% reduction in thickness. The comparison reveals some similarities and some differences.

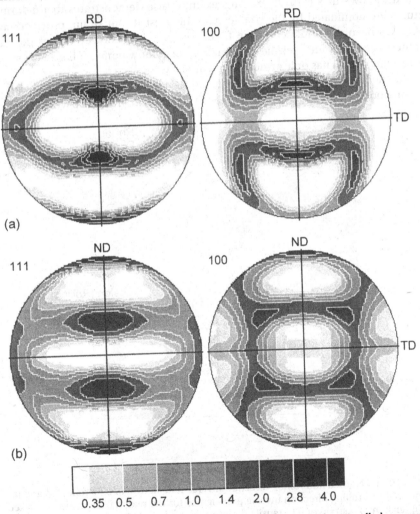

Fig. 9. 111 and 200 pole figures (recalculated, equal-area map) for (a) rolled copper (80% reduction) and (b) an Fe-3Si alloy (also 80% reduction), but with the ND and RD transposed for the bcc pole figures to show the correspondence between the textures.

There are at least three reasons why there may be differences between bcc and fcc textures compared in this way. The first was mentioned at the outset: there may be additional slip planes in the bcc material. The second is that, when grain *shape* effects matter, the morphological anisotropy is associated with the macroscopic strain, not with the grain behavior. In rolling, for example, the flatness is always in the rolling plane and is *not* transposed. Finally, we will encounter the interesting case of 'grain curling' in bcc tension, in the next Section.

2.2 Fiber textures: tension of tantalum

Bcc metals tend to form a ⟨110⟩ fiber texture under tensile deformation and, equivalently, wire-drawing. By way of example [MICHALUK &AL. 1993], Fig. 10 shows inverse pole figures for tantalum that was rolled to a 92% reduction in thickness (Fig. 10a) and then tested in tension to a strain of 0.6 (Fig. 10b). In this case, the pole figures were measured by neutron diffraction. The results show, as expected, a shift in the texture towards a ⟨110⟩ fiber. This example also demonstrates that deformation textures are dominated by the character of the most recent strain, provided enough strain has been accumulated. Pole figures would show, however, that the sample symmetry after the final tensile deformation was not a perfect fiber because of the prior rolling, which has only orthotropic symmetry.

One notable difference between bcc and fcc tensile deformation, however, is the phenomenon of grain curling, which is easily observed in drawn bcc wires, as analyzed

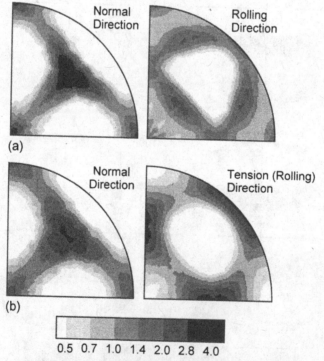

Fig. 10. (a) Normal and rolling direction inverse pole figures of 92% rolled Ta and (b) normal and rolling direction inverse pole figures for the material shown in (a), tested in tension to a strain of 0.6 (tensile direction coincident with prior rolling direction). Equal-area projection.

by HOSFORD [1964]. That is, grains tend to form ribbons and wrap around one another, instead of deforming uniaxially. This is because a $\langle 110 \rangle$ texture develops and subsequently the $\langle 110 \rangle$ orientation favors slip on two slip planes (4 systems) only, which are oriented such that plane-strain results, instead of the macroscopic uniaxial tension. The significance of the microstructural evolution is that, although the microscopic shape change of individual grains is quite different from the macroscopic average, nevertheless the texture seems largely unaffected. This point is made to show that deviations from compatibility do not always have an effect on texture development: the Taylor model often gives better results than one might expect. Note that grain curling is also expected in fcc materials after compression, but is much harder to observe because of the pancake grain shape morphology.

2.3 Rolling textures: tantalum, steel

Texture development during the rolling of steel has been studied in exhaustive detail over the years, and is still the subject of current research thanks to the technological significance of both mechanical and magnetic anisotropy in steels. Before discussing this class of textures, we should make the reader aware that the typical OD presentation for fcc rolling textures uses the COD (Φ-φ_1 sections) whereas bcc texture presentation sometimes uses the SOD convention (Φ-φ_2 sections). The most common presentation technique, however, is to show only the $\varphi_2=45°$ section (of the COD), which shows enough of the orientation distribution that many authors have used this exclusively. The textures can be described in terms of partial fibers as listed in Table II below and illustrated in Fig. 11.

Table II.   Partial fibers in rolled bcc materials. (See also Fig. 11.)

Designation	Orientation	Comments
α $\langle 110 \rangle$	$\{001\}\langle 110 \rangle$ to $\{111\}\langle 1\bar{1}0 \rangle$	110 parallel to the RD; prominent in bcc rolling textures
γ $\{111\}$	$\{111\}\langle 1\bar{1}0 \rangle$ to $\{111\}\langle 112 \rangle$	111 parallel to the ND; also prominent in bcc rolling textures
η	$\{001\}\langle 100 \rangle$ to $\{011\}\langle 100 \rangle$	100 parallel to the RD; near-relative of the β fiber
ε	$\{001\}\langle 110 \rangle$ to $\{111\}\langle 112 \rangle$	110 parallel to the TD
β	$\{112\}\langle 1\bar{1}0 \rangle$† to $\{\bar{1}1\,11\,\bar{8}\}\langle 4\,4\,\bar{1}1 \rangle$	present in simulated bcc textures; prominent in fcc rolling textures

† This orientation is known as the 'Brass' component in fcc rolling textures.

It is also useful to have maps that locate the ideal texture components and Fig. 12 shows three different diagrams of the $\varphi_2=45°$ section: diagram (a) shows a quadrant with Kocks angles; (b) and (c) show squares with Roe and Bunge angles, respectively. Note that the $\Theta=0°$ line in the square sections (corresponding to the origin in the quadrant) shows different rotations of the cube component with $\langle 100 \rangle$ parallel to the normal direction. Although ϕ and Ψ are degenerate at $\Theta=0°$ (Kocks angles), fixing one of the two angles (at a value other than 0°) means that the other one still has significance. Therefore a caution with respect to quadrant sections in SODs or CODs is that the origin of each section represents more than one orientation.

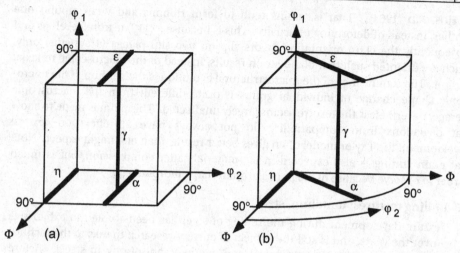

Fig. 11. Diagram showing positions of idealized (partial) fibers for rolled bcc materials in (a) Cartesian coordinates and (b) polar coordinates. The axes are labeled in the Bunge convention; note that equivalent diagrams exist for other angle conventions. The β fiber is not shown for simplicity.

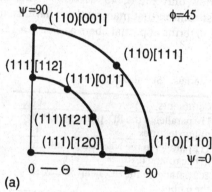

Fig. 12. Plot of the ideal rolling orientations for rolled bcc metals in the φ=45° section of the COD; (a) Kocks angles, (b) Roe angles, (c) Bunge angles. Note that the γ fiber is a straight (vertical) line in the two Cartesian plots (b & c) whereas it is a small circular arc in the quadrant (a).

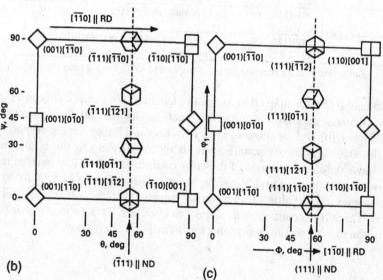

Several reviews are available for steel textures. Hutchinson, for example, recently reviewed the relationship between steel making practice and mechanical anisotropy and made the interesting point that high purity steels (so called IF or interstitial-free steels) offer the best thinning resistance, as measured by r-value [HUTCHINSON 1994]. This observation is correlated with the strong ⟨111⟩‖ND fiber that can be obtained in the IF steels; for an explanation in terms of crystal plasticity, see Chap. 9. These overviews, however, apply to the net result of several stages of thermo-mechanical treatment and therefore this aspect of bcc textures will be reserved for the section on recrystallization textures below.

The first stage of hot rolling is now known to yield significant variations in texture, despite the expectation that the γ-α transformation (for plain carbon steels) will tend to randomize the texture [RAY &AL. 1993]. This variability in 'hot band' texture often has an effect on the subsequent cold rolling texture. The presence of carbide formers (e.g. Ti, Nb) can lead to precipitation during hot rolling which restrains recrystall-ization and therefore tends to promote stronger hot band textures; see below for a discussion of transformation textures. Nonetheless, cold rolling typically generates textures which contain strong γ and α partial fibers. Increasing deformation levels shift the peak intensity in the α fiber from {001}⟨110⟩ towards {112}⟨110⟩; similarly the shift in the γ fiber is from {111}⟨112⟩ towards {111}⟨110⟩ [RAABE & LÜCKE 1994].

We give next some examples of experimental rolling textures for bcc metals. Figure 13 show pairs of 110 and 100 pole figures for a low carbon steel rolled to an 80% reduction in thickness, Fig. 13a, and for tantalum cold rolled to 92% reduction in thickness, Fig. 13b. The pole figures show textures typical of rolled bcc metals with intensity on the ⟨111⟩‖ND and ⟨110⟩‖RD fibers. The tantalum shows only a weak α fiber by contrast to the steel. The initial texture of the tantalum was not known; the steel had a weak rotated cube ({001}⟨110⟩) initial texture.

Analyzing the texture in terms of three–dimensional orientation distributions leads to the example of Fig. 14, which shows a plot of the sample orientation distribution; use of the SOD is conventional in the literature. In this plot, the γ fiber is present as a maximum at the center of each section. The α fiber is present as a line of intensity in the first quadrant (Ψ=0°) at φ=45°. Many investigators find that one section of the OD suffices for comparing one texture with another. Figure 15 compares the textures of the two materials, together with the initial texture of the steel using φ_2=45° sections of the crystallite orientation distribution (COD) for comparison with literature data[3]. This figure shows that tantalum exhibits both lower intensities generally and a weaker ⟨110⟩‖RD (α) fiber.

An excellent general survey of rolling and annealing textures in steels is available by RAY &AL. [1993] in which the texture of a wide variety of typical sheet steels is surveyed. The article covers both the older aluminum-killed steels, no longer in use because of the switch to strand casting, and the modern interstitial-free steels in which the carbon content has been balanced (i.e. precipitated) by additions of carbide-forming elements such as Ti and Nb. A useful example of cold rolling texture development is that of SUDO &AL. [1981] who studied a range of sheet steel alloys with varying amounts of silicon. The other major constituents were carbon (~0.05 w%) and

---

[3] Note that typical figures using the Bunge convention plot Φ down and ϕ_1 to the right, i.e. ϕ_2 sections, whereas those that use Roe angles plot Θ to the right and ψ down. A further variation places the origin at the lower left corner, instead of the upper left corner as just discussed: *Caveat Lector!*

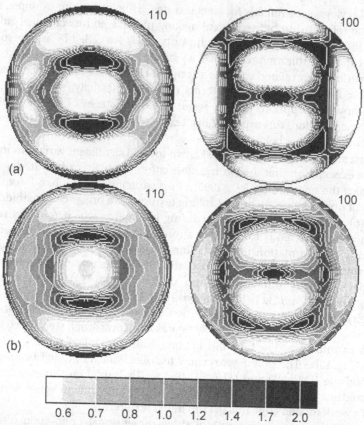

Fig. 13. 110 and 100 pole figures for (a) low-carbon steel cold rolled to a reduction in thickness of 80% (approximate equivalent strain of 2); (b) tantalum, unidirectionally rolled at room temperature to a reduction in thickness of 92%. (Equal-area projection, rolling direction vertical, transverse direction horizontal.)

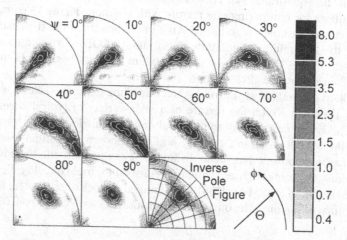

Fig. 14. SOD for low-carbon steel shown in Fig. 13, with sections at constant values of φ. The γ fiber is present as a maximum at the center of each section. The α fiber is present as a line of intensity in the first quadrant (Ψ=0°) at φ=45°.

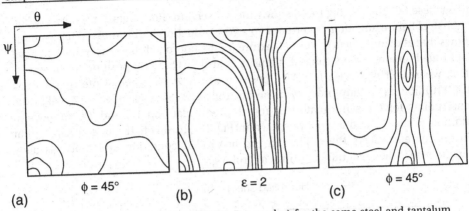

θ

ψ

(a) $\phi = 45°$ (b) $\epsilon = 2$ (c) $\phi = 45°$

Fig. 15. Plot of the 45° sections (ϕ_2=45°, Roe angles) for the same steel and tantalum textures shown in Fig. 13: (a) low-carbon steel prior to cold rolling; (b) low-carbon steel cold rolled to a reduction in thickness of 80% (approximate equivalent strain of 2); (c) tantalum, unidirectionally rolled at room temperature to a reduction of 91%. The contours are drawn at multiples of the random intensity of 1,2,3...7. Note the weaker intensities in the tantalum, and the stronger α fiber in the steel.

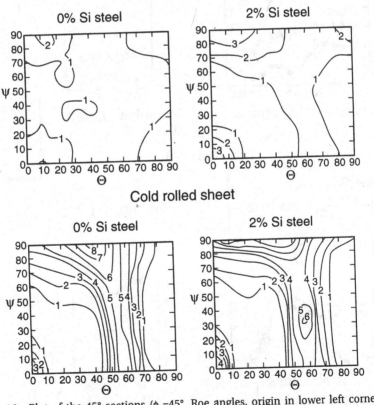

Hot rolled sheet

0% Si steel   2% Si steel

Cold rolled sheet

0% Si steel   2% Si steel

Fig. 16. Plot of the 45° sections (ϕ_2=45°, Roe angles, origin in lower left corner) for steels with 0% and 2% Si, both as hot-rolled (initial condition) and after 75% reduction cold rolling. The strongest intensity is at the {112}⟨110⟩ position in the 0% Si-steel, whereas it is at the {111}⟨110⟩ position in the 2% Si-steel. Note that a weak RD∥⟨110⟩ fiber is already present in the hot rolled 2% Si-steel.

manganese (~0.2 w%). Figure 16 shows the 45° section (Roe angles) for both 0% and 2%Si-steel before and after cold rolling to a 75% reduction in thickness. This example shows that alloying affects the initial texture (of the hot rolled sheet) as well as the final cold rolled texture. The strongest intensity is at the {112}⟨110⟩ position in the 0% Si-steel, whereas it is at the {111}⟨110⟩ position in the 2% Si-steel. Note that a weak RD‖⟨110⟩ fiber is already present in the hot rolled 2% Si-steel. Figure 17 makes the same points for the same materials, together with data on 1% and 3% Si alloys in addition to the 0% and 2% alloys shown in Fig. 16. In this figure it is apparent that silicon content affects the α fiber in the hot rolled condition and that this alloy dependence is then magnified during cold rolling.

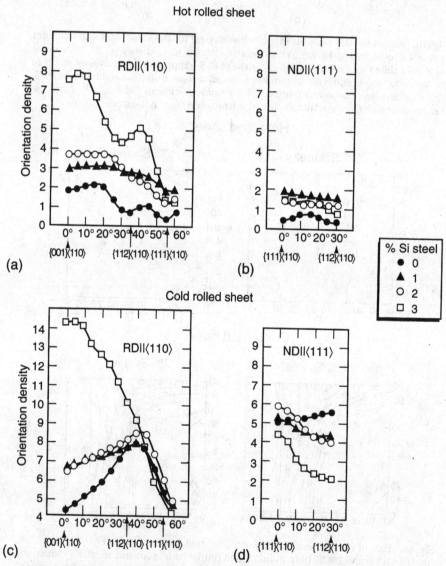

Fig. 17. Plot of the α and γ fibers for a range of iron-Si alloys, including 0, 1, 2, & 3% Si. Increasing silicon leads to stronger α fibers in both the hot-rolled (initial) condition and the cold-rolled condition.

Texture development in rolled tantalum has been studied in some detail by Garrett and co-workers, see for example [CLARK &AL. 1991]. The reader is cautioned, however, that tantalum is notorious for gradients in texture through the thickness of rolled sheet, see [WRIGHT &AL. 1994a, 1994b]. Tantalum is also known for its propensity to show a fine grain size in metallographic specimens and yet to behave as if it is coarse-grained by 'orange-peeling' during forming. This suggests that colonies of similarly oriented grains form during recrystallization [RAABE & LÜCKE 1994], which is also a problem encountered in stainless steel sheet, see e.g. SHINDO & FURUKAWA [1981].

2.4 Torsion textures: iron and tantalum

Fewer torsion texture results have been published for bcc metals than for fcc. The results by MONTHEILLET &AL. [1984] for Armco iron (0.02% carbon) show that the texture development is similar to that of Cu-Al [SEKINE &AL. 1981] discussed above. The texture is dominated by the D fiber (transposed to bcc), and at the high test temperature (800°C), only the $\{\bar{1}\bar{1}2\}\langle110\rangle$ component is present. WILLIAMS [1962] also tested Armco iron but at room temperature. At a shear strain of 2.1, the pair of 200 and 110 pole figures, Fig. 18, show that texture is dominated by the $\{11\bar{2}\}\langle111\rangle$ and $\{110\}\langle001\rangle$ components. The first component would be considered part of the A fiber in the fcc case whereas the second component is equivalent to the C component.

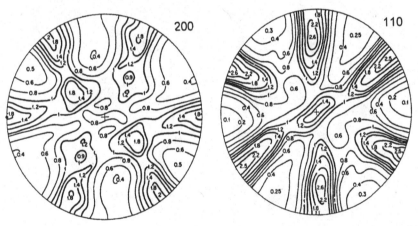

Fig. 18. Experimental 200 and 110 pole figures for Armco iron sheared to $\gamma=2.1$ ($\varepsilon_{vM}=1.2$) [WILLIAMS 1962] (Stereographic projection.) The shear direction points right on top.

Figure 19 shows pole figures for tantalum tested in torsion to a von Mises equivalent strain of 1.4, together with the initial texture. Recalculated pole figures are shown to facilitate comparison with the fcc torsion textures, keeping in mind that the torsion axis and the shear direction have to be interchanged (imagine a reflection in a line inclined at 45° between the two axes). The torsion texture is clearly different from the initial texture, albeit not very strong, with a maximum of twice random in the 200 pole figure. The texture is more similar to that of copper or nickel than to brass, as might be expected for a material that does not exhibit planar slip. The predominant component is close to $\{110\}\langle001\rangle$ which is the (transposed) equivalent of the C component in fcc metals, but tilted counterclockwise by approximately 20°. In summary, shear textures in bcc materials correspond through the appropriate transposition to those observed in fcc materials.

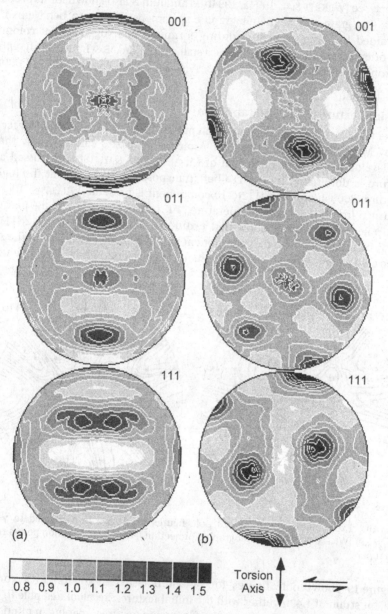

Fig. 19. Recalculated 111, 100 and 110 pole figures for tantalum: (a) initial texture; (b) tested in torsion to $\varepsilon_{vM}=1.4$. Equal-area projection.

3. Deformation Textures in Hexagonal Close-packed (hcp) Metals

Textures in hexagonal metals have attracted significant interest over the years because of the use of zirconium alloys, e.g. zircaloy, for cladding of nuclear reactor fuels, and titanium alloys as structural materials in aerospace. Useful reviews of experimental textures for both deformed and annealed materials can be found in GREWEN'S [1973] review in the proceedings of the third ICOTOM conference and from PHILIPPE'S [1994] review in the proceedings of the tenth ICOTOM. A review of deformation and texture in hexagonal materials with a particular focus on zirconium and zircaloy is given by TENCKHOFF [1988]. From a mechanistic point of view, hcp materials are different from cubic metals in that twinning almost always provides significant, if not dominant, deformation modes.

3.1 Fiber textures: extruded titanium and upset (compressed) zirconium

Fig. 20 shows inverse pole figures for a high purity, electrorefined titanium[4] that has been deformed in two different ways. The first example, Fig. 20a, shows the texture of material that had been uniaxially extruded to a von Mises equivalent strain of 1.75. A fiber texture is present with a maximum at the $\langle 10\bar{1}0 \rangle$ position such that all the basal plane normals are perpendicular to the extrusion axis in a so-called 'cylindrical' texture. As noted by DILLAMORE & ROBERTS [1965] this texture is commonly observed in extruded rods of hexagonal materials. The second example, Fig. 20b, is for the same material which has been upset and then cross-rolled to a von Mises equi-

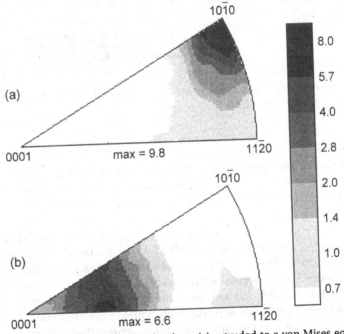

Fig. 20. Inverse pole figures of pure titanium: (a) extruded to a von Mises equivalent strain of 1.75 (extrusion-axis inverse pole figure), (b) forged and cross-rolled to a von Mises equivalent strain of 1.98 (plate normal inverse pole figure).

---

[4] Major impurities (in weight ppm) were Fe 10, Cr 5, Ni 3, V 2, Na 2, Al 3, and Mg 2.

valent strain of 1.98, which gives rise to a fiber texture that is about 25° away from $\langle 0001 \rangle$ which is approximately equivalent to the $11\bar{2}4$ position.

Electron beam melted zirconium of high purity was subjected to a similar deformation, which was a combination of upsetting to a reduction in thickness of 55%, followed by clock rolling to an overall reduction in thickness of 89%, with two intermediate anneals. The results show a similar texture but with a fiber that is very close to $\langle 0001 \rangle$, Fig. 21. Here we introduce a simplified presentation of texture based on 1-D plots of the variation of intensity with tilt angle which are useful for quantifying fiber textures. This topic will be taken up again in the discussion of texture development in hcp metals in Chap. 11, but see, e.g., LEBENSOHN & TOMÉ [1991, 1993a].

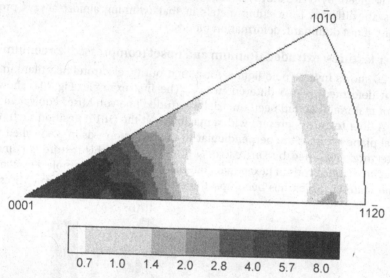

Fig. 21. Inverse pole figure (plate normals) for forged and cross-rolled zirconium, showing fiber texture near 0001.

3.2 Rolling textures

The textures of cold-rolled hcp metals and alloys can be categorized into three groups according to their c/a ratios: namely materials with c/a ratios greater than, approximately equal to, and less than the ideal value of 1.633. Schematics of the types of cold-rolled sheet textures expected for these categories are shown in Fig. 22 [TENCKHOFF 1988]. Experimental examples of 0002 pole figures for rolled magnesium, zinc and titanium are shown in Fig. 23 [GREWEN 1973]. Metals or alloys with c/a ratios above the ideal ($c/a > 1.633$) such as zinc and cadmium, tend to exhibit textures with basal poles tilted ±15 to 25 degrees away from the normal direction toward the rolling direction and $[11\bar{2}0]$ poles aligned with the rolling direction (Figs. 22a and 23b). Metals or alloys with c/a ratios approximately equal to the ideal ($c/a \approx 1.633$) such as magnesium and cobalt, tend to form [0001] fiber textures (Figs. 22b and 23a). Metals or alloys with below–ideal c/a ratios ($c/a < 1.633$) such as zirconium, titanium and hafnium, tend to form textures with basal poles tilted ±20 to 40 degrees away from the normal direction toward the transverse direction and $[10\bar{1}0]$ poles aligned with the rolling direction (Figs. 22c and 23c). See also the recent review by PHILIPPE [1994].

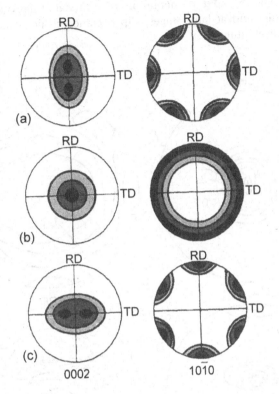

Fig. 22. Schematic rolling textures in hcp metals with *c/a* ratios of (a) greater than 1.633, (b) approximately equal to 1.633 and (c) less than 1.633. 0002 and 10$\bar{1}$0 pole figures. [TENCKHOFF 1988].

This categorization by *c/a* ratio is, however, a rather crude approach because other factors influence texture development. Strain rate, temperature [SALINAS &AL. 1992] and chemical composition all play an important role. Intentional alloying or variation in impurity content can affect texture; oxygen content for example, often has a strong effect on the properties of hexagonal metals. The variation of textures observed in hcp metals is related to the relative ease of slip and twinning. A detailed analysis of the dependence of texture development on deformation systems will be given in Chapter 8. In contrast to slip, where lattice rotations tend to proceed gradually with increasing deformation, twinning causes spontaneous, large-scale lattice rotations, even for low degrees of deformation [TENCKHOFF 1988].

As an example of the roles slip and twinning play during deformation of hexagonal materials we cite experiments of BLICHARSKI &AL. [1979] where commercially pure α-titanium was rolled at room temperature to reductions of 20, 30, 55 and 97% (von Mises strains of 0.26, 0.41, 0.92 and 1.39). The initial material was β-forged and exhibited a nearly random texture. 0002 pole figures for the materials after the rolling reductions are shown in Fig. 24. The transition from the RD split to the TD split of the peaks is evident in these pole figures. Based on these pole figures (and other microstructural evidence) BLICHARSKI &AL. concluded for their material that at the early stages of deformation (below 40% rolling reduction), twinning was the major mode of deformation leading to rapid reorientation of material unfavorably oriented for deformation by slip; and at later stages, slip became the dominating contributing factor to texture development.

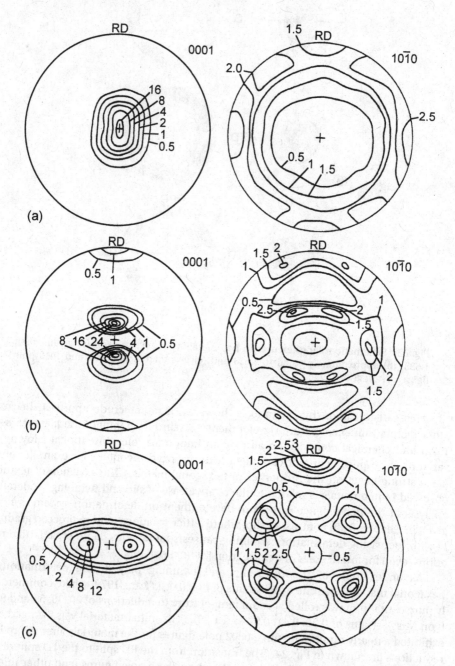

Fig. 23. 0002 pole figures for rolled (a) magnesium, (b) zinc, and (c) titanium, showing ⟨0001⟩ fiber for Mg, RD split for Zn, and TD split for Ti [GREWEN 1973] (stereographic projection).

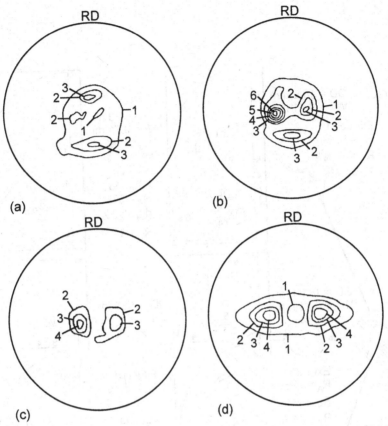

Fig. 24. 0002 pole figures for α-Ti sheet cold–rolled to thickness reductions of (a) 20%, (b) 30%, (c) 55% and (d) 97%. [BLICHARSKI &AL. 1979]. Stereographic projection.

3.3 Tube textures: zirconium

As previously mentioned, textures in zirconium and zirconium alloys (particularly zircaloy) have attracted significant interest over the years because of their use in the cladding of nuclear reactor fuels. Because of this application, a considerable amount of texture analysis of zirconium and zirconium alloys has focused on tube materials. Figure 25 shows textures from zircaloy tubes formed by extrusion and by pilgering. A schematic summary of these results is given in Fig. 26 [TENCKHOFF 1970b, 1988]. It should be noted that the schematic textures in the summary are given with the assumption of a random initial texture and are for extreme reductions. It should also be noted that the texture expected for $R_W/R_D = 1$ is similar to that expected for wire drawing and the texture expected for $R_W/R_D > 1$ is quite similar to that expected for rolled sheet (R_W: reduction in wall thickness; R_D: reduction in diameter).

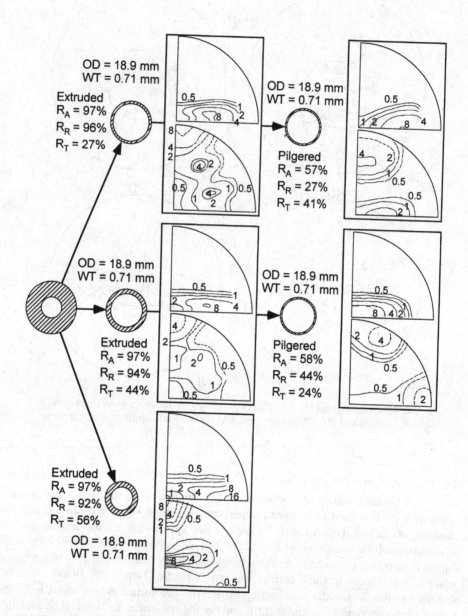

Fig. 25. 0002 and 10$\bar{1}$0 pole figures for extruded and pilgered zircaloy tubes. R_A is the axial reduction, R_W the wall thickness reduction, R_D the reduction in diameter, OD the outer diameter and WT the wall thickness. In the pole figures, the axial direction is up, the tangential direction to the right, in stereographic mapping. The dashed lines represent data obtained in transmission [TENCKHOFF 1978].

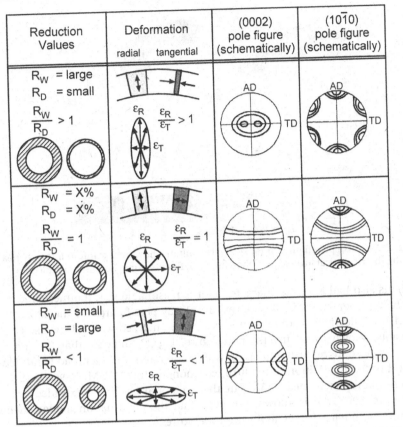

Fig. 26. Schematic summarizing tube textures for zircaloy for various reduction values. ε_R is the radial strain and ε_T is the tangential (or hoop) strain.

4. Deformation Textures in 'Other' Materials

4.1 Orthotropic uranium

The study of texture development in metals has generally been confined to the more symmetric crystal symmetries, such as hexagonal and cubic. A few examples exist, however, of metals that can be deformed in a conventional way to large strains and yet have low symmetry structures. Tin, for example, exists in a tetragonal allotrope at ambient conditions, and its texture development in cold rolling was studied by DAHMS &AL. [1988]. We present here the example of uranium, which has an orthorhombic unit cell in its low-temperature allotrope, but behaves as a ductile metal for most purposes and deforms primarily by twinning. A more complete treatment is given here, including the use of simulation to predict texture formation, because of the paucity of literature in this area.

In order to study the texture developed in tension, a specimen was taken from a uranium rod that had been swaged at room temperature to a reduction of 20% in diameter, equivalent to a strain of 45% [HOCKETT 1959]. This deformation was the last stage in a series of deformation and recrystallization steps and the texture prior to the final swaging just described was not known. Nevertheless, the previous deformation

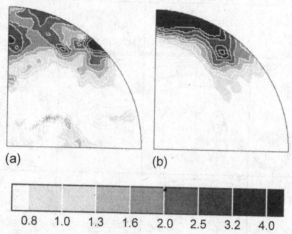

(a)   (b)

| 0.8 | 1.0 | 1.3 | 1.6 | 2.0 | 2.5 | 3.2 | 4.0 |

Fig. 27. Inverse pole figure for (a) swaged uranium showing maximum intensity near [010]; (b) theoretical inverse pole showing [010] fiber texture. Equal-area projection.

treatments also had a tensile character so it is reasonable that the measured texture would be representative of a tensile texture. X-ray pole figures for the 020, 021, 111, 112 and 131 reflections were measured and were analyzed with WIMV (see Chap. 3). The results, shown as an inverse pole figure, Fig. 27, indicate that the maximum intensity is close to the [010] direction, with some intensity in a broad band between [001] and [100]. In addition to the [010] component, JETTER & MCHARGUE [1957] also observed a component near (331) but their material had been extruded at a high temperature, 500°C. It is known, however, that the deformation modes change to being more dominated by slip at these temperatures.

The technologically important application of fiber textures in uranium is in the prediction of growth during thermal cycling, i.e. the change in length that occurs in a uranium rod because of the anisotropy between the [010] or growth direction and the [100] shrinkage direction. See MUELLER &AL. [1988] for a recent article on this subject.

Since it is not self-evident what texture is expected for a low symmetry material

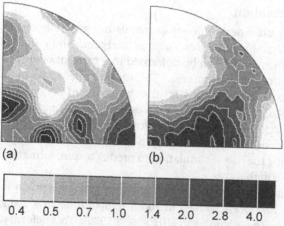

(a)   (b)

| 0.4 | 0.5 | 0.7 | 1.0 | 1.4 | 2.0 | 2.8 | 4.0 |

Fig. 28. (a) Inverse pole figure for uranium compressed 90%, showing intensity in a band between 001 and 100; (b) theoretical inverse pole figure for compression to 100% strain. Equal-area projection.

such as uranium, we break with the ground rule in this chapter of discussing only experimental textures, and show the results of theoretical calculations. These calculations have been detailed elsewhere [ROLLETT 1991] and the Volume Fraction Transfer method used is discussed later in this book in Chapter 11. The corresponding inverse pole figure for the theoretical texture development, for a tensile strain of 50%, is shown alongside in 27b. This shows a similar maximum near [010] with a gradual fall-off in intensity but no intensity along the [001]-[100] axis.

The case of uniaxial compression was studied by performing a compression test on a cylindrical specimen of polycrystalline uranium to a strain of 0.99, Fig. 28a. Again, we break with the emphasis on experimental textures, and show the corresponding inverse pole figure for the theoretical prediction for a strain of 1. The experimental result, Fig. 28a, shows several maxima in a band between [001] and [100]. The theoretical result, Fig. 28b, shows a similar band starting at [001], except that there is no intensity at [100] and instead the band ends near [310].

In the case of rolling both the sample and crystal symmetries are orthorhombic. This means that the 001, 010 and 100 pole figures can be used to give a relatively complete picture of the texture. The example of experimental pole figures is taken from MORRIS'S [1971] measurements on uranium reduced 80% by rolling, Fig. 29a. The theoretical pole figures for a rolling reduction to a von Mises equivalent strain of 200% (equivalent to an 82% reduction in thickness) are shown in Fig. 29b. The experimental

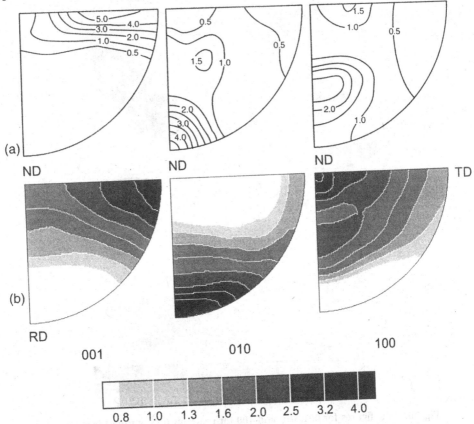

Fig. 29. Pole figures for 001, 010 and 100 for (a) cold rolled uranium [MORRIS 1971], compared with (b) theoretical pole figures for plane-strain deformation. Stereographic projection.

pole figures show maxima corresponding to (201)[010] and (263)[310]. The theoretical pole figures have a similar appearance but the dominant component is (100)[010].

The case of torsion is interesting because of the paucity of experimental results for torsion texture development of an orthorhombic metal. The experimental pole figures shown in Fig. 30a are for a specimen deformed in torsion to a von Mises equivalent strain of 60% at a strain rate of 4×10^{-3} s^{-1}. The sense of shear is noted on the figures. Note that the pole figures shown in Fig. 30a were reconstructed from the OD derived from the experimental pole figures. This illustrates one of the powerful features of the

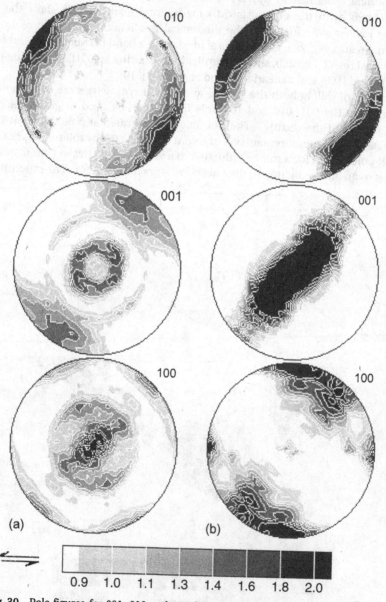

Fig. 30. Pole figures for 001, 010 and 100 for uranium (a) deformed in torsion (60% strain), compared with (b) theoretical pole figures for torsion to the same strain. Equal-area projection, shear direction left when torsion axis points up.

OD analysis, i.e. that it is possible to present the preferred orientation of a material in any convenient form. The corresponding theoretical pole figures for a strain of 50% are shown in Fig. 30b, which is plotted on a different scale to accommodate the more intense theoretical textures. The comparison of the two data sets shows again good agreement for the [010] pole and moderate agreement for the other two poles.

4.2 Textures in intermetallics: NiAl and TiAl

Intermetallics have received considerable attention in recent years due to their high strength at high temperatures, good corrosion resistance and low density. However, to date, very little texture analysis has been done on intermetallics. Therefore, an overall picture of the texture in intermetallic materials can not yet be constructed. Nonetheless, some texture results have been reported for NiAl and TiAl. These studies have primarily been done to help identify the physical mechanisms underlying plastic deformation in these materials. We will briefly review some of these results in order to give a flavor of the work being done on intermetallics where texture analysis has been used.[†]

The primary slip system at room temperature in cubic NiAl is $\langle 100 \rangle \{011\}$. This slip system contributes only three independent deformation modes of the five required to satisfy the von Mises criterion. This lack of adequate slip systems to satisfy compatibility manifests itself in NiAl by low room temperature tensile ductility and extensive grain boundary cracking in compression at relatively small strains (~ 10%). NiAl undergoes a brittle-to-ductile transition (BDT) in the 300–500°C temperature regime, where tensile ductility can increase substantially and grain boundary cracking does not occur. Figure 31 shows three inverse pole figures obtained from NiAl, one of the starting material and the other two obtained from material extruded at two different temperatures: room temperature and 300°C (which is in the BDT temperature range but below the recrystallization temperature) [MARGEVICIUS & LEWANDOWSKI 1993]. The room temperature sample exhibited considerable cracking and showed very little change in texture after extrusion which is consistent with cube type slip [MARGEVICIUS & COTTON 1995]. No cracking was observed in the sample extruded at 300°C. This

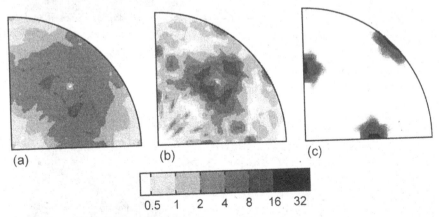

Fig. 31. Experimental inverse pole figures for NiAl: (a) starting material, (b) hydrostatically extruded at room temperature and (c) hydrostatically extruded at 300°C. Equal-area projection.

Bob Margevicius is acknowledged for his help regarding NiAl.

sample exhibited a strong ⟨110⟩ fiber which is indicative of ⟨111⟩{uvw} (pencil glide) type slip found in some bcc materials. Similar results have been reported in the literature [BIELER &AL. 1992, DOLLAR &AL. 1993]. These textures give some insight to the underlying processes involved in plastic deformation of NiAl.

TiAl has a tetragonal structure similar to the fcc structure. Slip occurs exclusively on {111} planes either by dislocation or twinning. Four modes of deformation have been identified in TiAl, namely slip by ⟨110] ordinary dislocations, by ⟨011] and ⟨112] superdislocations or by so-called order twins. (The ⟨hkl] nomenclature denotes that h and k may be interchanged while l must remain in place, i.e. [hkl] is symmetrically equivalent to [khl].) While slip by ordinary dislocations is generally accepted, the contributions of the other modes are less certain and appear to depend on the temperature and strain rate as well as alloy composition and microstructure. In order to delineate the contributions of these different modes, textures for 30% uniaxially compressed TiAl were measured and compared with textures generated by simulation [MECKING &AL. 1996a]. The comparison suggested that the contributions for ordinary, super-slip (assuming ⟨011] type) and twinning modes is 49%, 48% and 3%, respectively. Inverse pole figures of textures measured before and after the uniaxial compression are shown in Fig. 32.

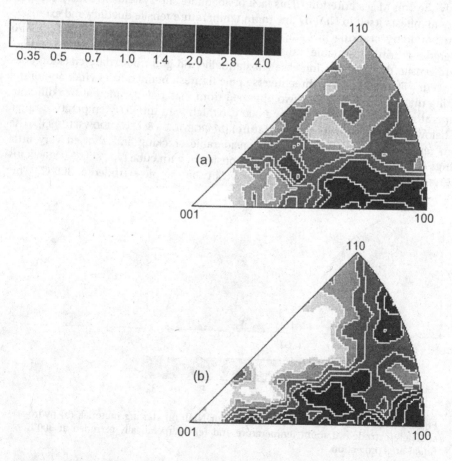

Fig. 32. Experimentally measured inverse pole figures (stereographic projection) for TiAl (a) prior to deformation and (b) after 30% uniaxial compression.

5. Composites

The development of texture in composites might at first glance be expected to occur in the same way as for the separate single-phase materials, suggesting a 'rule of mixtures' approach. That is, each phase will form the texture expected for its crystal structure and character of deformation, with allowance for hard phases undergoing less deformation than the average for the composite, and vice versa for soft phases. The first example for Cu-Nb follows this simple approach very well. Other examples demonstrate different behavior, however, as for the example of Al-SiC composites, for which the presence of the hard second phase has a strong dispersive ('randomizing') effect on the matrix texture. WASSERMANN &AL. [1978], for example, showed that fiber textures in drawn composite wires containing fcc metals exhibit strong ⟨111⟩ fibers and weak or non-existent ⟨100⟩ fibers.

The role of hard, undeforming particles on texture formation is under investigation by several authors [BOLMARO &AL. 1993, POOLE &AL. 1991]. These studies focused on tungsten wires in copper matrices, with complementary finite element studies of the deformation patterns. The results show that the heterogeneity that occurs when a soft matrix accommodates the imposed (macroscopic) strain by flowing around hard particles, leads to not just varying magnitudes of strain, but also varying local rotations, all of which tend to disperse the texture. However, to date, the role that crystallography plays at the particle/matrix interfaces has not yet been addressed to much extent (but see the work of HUMPHREYS (e.g. 1977] on small particles).

5.1 Wire drawn Cu-Nb composites

The data in Fig. 33 were taken from heavily drawn composite wire, consisting of a copper matrix (oxygen-free electrical) and a niobium reinforcement. The initial material was an ingot which was thermo-mechanically processed into its final form. The lack of mutual solubility in the Cu-Nb system means that the conductivity of the copper is nearly unaffected by the presence of Nb, and the Nb acts as a high modulus

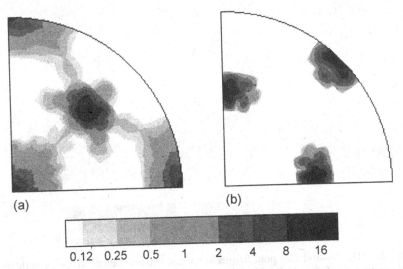

(a)   (b)

0.12   0.25   0.5   1   2   4   8   16

Fig. 33. Inverse pole figures for wire drawn copper-niobium composite, (a) copper matrix and (b) niobium reinforcement. Equal-area projection.

filamentary reinforcement of the copper. The results of the texture analysis are presented in the form of separate inverse pole figures for the two phases. These show that the textures of the two components correspond to those expected for the same deformation processing applied to each component separately. This reinforces the point that fiber textures are rather insensitive to deformation mode. Interestingly, the niobium filaments tend to assume a flat tape morphology, rather than the round wire expected from the macroscopic shape of the material [VERHOEVEN &AL. 1991, RAABE &AL. 1995; BISELLI & MORRIS 1996]. This is the result of so-called 'grain curling' whereby the ⟨110⟩ orientation prefers to deform in plane strain rather than axisymmetric elongation, as discussed in Sec. 2.2. In a single-phase bcc wire [HOSFORD 1964], compatibility requires the individual grains to curl around one another, whereas in this composite material, the lesser constraint of the copper matrix allows flat tapes to form.

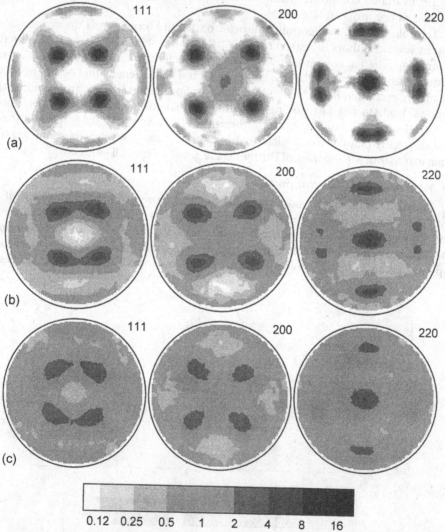

Fig. 34. 111, 200 and 220 pole figures for three extruded 8090 Al alloys with the following levels of SiC whisker reinforcement: (a) 0%, (b) 15%, and (c) 30%. Extrusion axis vertical, equal-area projection.

5.2 Extruded Al-SiC composites

Although data exist on texture formation in composites of various kinds, it has not been well developed when compared to, for example, the amount of attention that has been paid to plasticity and stress/strain behavior. We quote two examples of the effect of second phases on texture formation in an advanced aluminum–lithium alloy, 8090, that was made by high rate spray forming in an Osprey machine [PRZYSTUPA &AL. 1994]. The resulting billet was then hot extruded at ~500°C with a reduction ratio of approximately 10:1. (The initial spray-forming texture was not measured.) The first set of pole figures, Fig. 34, shows the variation in texture in unreinforced (a), 15 vol% (b), and 30 vol% of SiC whiskers (c). The qualitative texture changes only slightly, whereas the intensity of the texture decreases markedly as reinforcement is added. A similar result was obtained previously by BOWEN [1990] in Al-Li alloys reinforced by particulate silicon carbide in which the textures became progressively weaker with increasing volume fraction of reinforcement. Under certain conditions, however, textures in composites may be *stronger* [ENGLER &AL. 1989]. For example, Fig. 35 compares the texture for the same material as for Fig. 34b, but containing 15 volume percent of SiC particles instead of whiskers. The two textures are very nearly the same, except that the particle-containing material has the more intense texture, which suggests that particles are not as effective as whiskers at repressing the texture development.

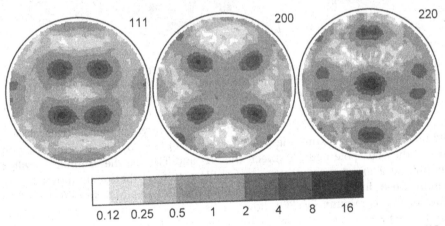

Fig. 35. 111, 200 and 220 pole figure for extruded 8090 aluminum with 15% SiC particles. Compare with Fig. 34b. Extrusion axis vertical, equal-area projection.

5.3 Rolled Be-Al composites

As an example of a combination of two ductile phases with different crystal structures and strengths, we show rolling textures for a beryllium–aluminum composite. The composite was made by centrifugally atomizing an alloy of Be with 47.5 weight-% Al and 2.5 w-% Ag to produce a fine, spherical powder. The powder was then consolidated into a sheet by hot isostatic pressing, followed by cross-rolling at 500°C to a 90% reduction in thickness. The pole figures obtained from the Be reflections, Fig. 36a, show a texture that is close to a $\langle 0001 \rangle$ fiber but with features that

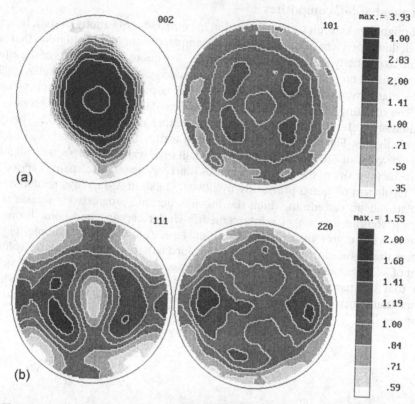

Fig. 36. Be-Al composite made by pressing powders and warm cross-rolling to 90% reduction in thickness: (a) recalculated 0002 and 10$\bar{1}$0 pole figures for the Be phase; (b) recalculated 111 and 200 pole figures for the Al phase. Equal-area projection.

reflect the lower symmetry of the actual deformation. Beryllium normally develops a (unidirectional) rolling texture that is characteristic of low-c/a-ratio hcp metals, with peaks in the 0002 pole figure between the ND and TD (see the example of rolled titanium, Fig. 22c). A more nearly comparable texture is that of the upset forged titanium shown in Fig. 19, which shows a fiber that lies about 25° from ⟨0001⟩. The texture of the aluminum, Fig. 36b, however, is quite different from the usual low-temperature fcc result, which may be due to the kind of randomization shown in the Al-SiC example above. Since the deformation temperature was a high fraction of the melting point of aluminum, however, it is also possible that recrystallization occurred, either dynamically or in short times after each rolling pass. These textures reiterate the point that the individual phases in deformed composites often exhibit a different texture than they do when not combined in a composite structure.

6. Transformation Textures

6.1 Phase transformations

The effect of a phase transformation on the texture of a material is simple in principle but complex in practice [CAHN 1991]. A certain orientation relationship exists between the parent and product phases, generally based on alignment of close-packed planes and directions. The Kurdjumov–Sachs (KS) relationship, for example, predicts that {111} planes in austenitic (fcc, γ) iron align with {110} planes in ferrite (bcc, α); correspondingly, $\langle 110 \rangle_\gamma$ directions align with $\langle 111 \rangle \alpha$ directions. Therefore each volume element of the new phase will have a definite relationship to the corresponding volume element of the parent phase but more than one *variant* is possible. The number of variants depends on the extent of coincidence between the alignment elements and the symmetry elements of the crystal structures. The {111} plane in the KS relationship (transforming from γ-Fe to α-Fe), for example, means that there are four variants of the transformed texture volume element, just based on choice of plane alone. Therefore one key conclusion is that a transformation texture is likely to be significantly weaker than the parent phase texture. The lack of a one-on-one mapping between parent and product orientations also means that no straightforward 'backwards' calculation is available that might allow one to infer the parent phase texture from the product. A more subtle point is that the experimental evidence points very strongly towards the selection of certain variants, which accounts for the complexity of transformation textures. Thus, knowledge of the orientation relationship is not sufficient for prediction of transformation textures and additional microstructural information is required.

The transformation textures of many metals have been studied and many articles may be found in the ICOTOM proceedings, see for example recent work on variant selection [MIYAJI & FURUBAYASHI 1991]. We have chosen, however, to illustrate the topic of transformation textures with some examples from a very complete review on transformation textures in steels [RAY &AL. 1994]. Steel is commonly hot-rolled in the austenitic state and, as with many other fcc metals, the dominant texture component is the cube or {001}⟨100⟩, from recrystallization. Transformation of this component should give equal intensities in ferrite at {001}⟨110⟩, {110}⟨001⟩ and {110}⟨110⟩, but in fact the {001}⟨110⟩ variant is always dominant experimentally. Examination of other major texture components also leads to the conclusion that variant selection occurs; the brass orientation in austenite, {110}⟨112⟩, leads to {332}⟨113⟩, the intensity of which is also affected by variant selection. As with most aspects of texture development, alloying affects transformation textures. In steels, however, the effect is primarily on the parent phase because the common alloy additions tend to raise the temperature at which recrystallization can occur. Thus, whereas plain carbon steels will recrystallize at temperatures just above the transus, adding Nb or Ti means that strong rolling textures can develop in the austenite. Figure 37 contrasts the textures of plain carbon steel with those of a niobium alloyed steel, for a series of finishing temperatures. The textures are presented as $\varphi_2 = 45°$ sections of the orientation distribution. The results for the plain carbon steel show that a weak texture dominated by the {001}⟨110⟩ occurs at all finishing temperatures except the lowest. Conversely, addition of niobium leads to significant intensity on the ⟨111⟩ fiber (γ fiber) at all temperatures except the highest. The cooling rate achieved during transformation can also affect the texture quite markedly; higher cooling rates, leading to non-pearlitic structures, give sharper textures.

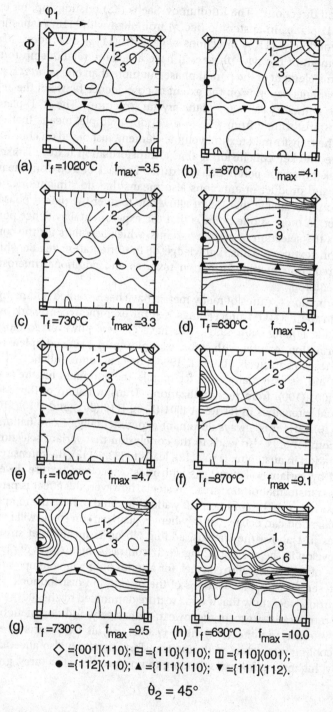

(a) T_f =1020°C f_{max} =3.5

(b) T_f =870°C f_{max} =4.1

(c) T_f =730°C f_{max} =3.3

(d) T_f =630°C f_{max} =9.1

(e) T_f =1020°C f_{max} =4.7

(f) T_f =870°C f_{max} =9.1

(g) T_f =730°C f_{max} =9.5

(h) T_f =630°C f_{max} =10.0

◇ ={001}⟨110⟩; ⊟ ={110}⟨110⟩; ▢ ={110}⟨001⟩;
● ={112}⟨110⟩; ▲ ={111}⟨110⟩; ▼ ={111}⟨112⟩.

$$\theta_2 = 45°$$

Fig. 37. Sections drawn for ϕ_2=45° of experimental orientation distribution for (a-d) a plain carbon steel finished at 1020, 870, 730, and 630°C, respectively; (e-h) for a Nb microalloyed steel finished under the same set of conditions [RAY &AL. 1994]. The symbols represent texture component positions as indicated. Note that the γ fiber is horizontal in these figures (at Φ=55°).

6.2 Recrystallization in rolled fcc metals[†]

Recrystallization shares several of the characteristics of phase transformations in that the replacement of deformed material by the nucleation and growth of recrystallized grains can lead to drastic changes in texture. The principal difference is that, in recrystallization, the 'nuclei' are regions that exist already in the deformed microstructure (the word 'nucleation' thus being used not in the thermodynamic sense) [DOHERTY 1978, HUMPHREYS & HATHERLY 1995]. Another important difference is that recrystallization does not lead to precise orientation relationships between deformed and recrystallized material in contrast to phase transformation. Approximate orientation relationships have nonetheless been used to make quantitative predictions of recrystallization textures. It should be stated that recrystallization does not always lead to changes in texture, particularly when large volume fractions of second-phase particles are present [HORNBOGEN 1979] (or when the deformation was axisymmetric). Also it should be noted that recrystallization textures have been little studied in torsion and they are not discussed here.

This topic is not merely of academic interest since the properties of many technological alloys depend critically on the control of recrystallization texture. Examples abound in the literature on silicon-bearing steels, used for their magnetic properties, and aluminum sheet for beverage cans. On the other hand, recrystallization textures in rolled copper have been investigated most thoroughly from a fundamental point of view. Copper is more easily obtained in high purity whereas aluminum very often has significant levels of both insoluble and soluble impurities.

As an illustration of the wide variety of recrystallization textures that can be obtained in fcc materials, Fig. 38 shows the 111 pole figures of the recrystallization textures of Al, Cu, brass, and an aluminum alloy. Recrystallization textures (just like rolling textures) show significant material dependence, especially on stacking-fault energy (SFE). In materials of a relatively high SFE (Al, Ni, Cu), a cube component often plays a major role. In addition to the cube component, Al and Al alloys often exhibit a component close to the typical rolling texture S-orientation, which after recrystallization is generally referred to as the R-orientation (e.g. Fig. 38a) [ITO &AL. 1983]. In Ni and Cu, recrystallization twinning leads to minor intensities of the cube orientation's first-generation twin (Fig. 38b). In contrast to these examples, low-SFE materials (Ag, brass, austenitic steels) generally exhibit much weaker recrystallization textures, which are characterized by the so-called brass recrystallization (BR) orientation {236}⟨385⟩ (Fig. 38c).

In comparison to rolling textures, recrystallization textures are much more complicated: in addition to the dependence on SFE, the deformed state, the annealing temperature, and particularly the precipitation state are known to exert strong influences on the recrystallization behavior and, consequently, on the recrystallization textures (ENGLER 1996). As an example, the recrystallization texture of an Al-Mn alloy (AA3103), which was pre-treated to contain large (>1μm) constituent particles, is shown in Fig. 38d. In comparison to pure Al (Fig. 38a), the texture of the two-phase material is much less pronounced, which is caused by the additional, nearly random nucleation in the deformation zones around the second-phase particles ('particle stimulated nucleation') [HUMPHREYS 1977].

[†] Olaf Engler and Carl Necker are acknowledged for their help with this Section.

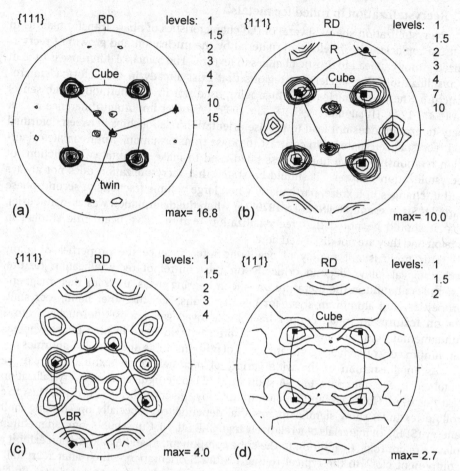

Fig. 38: Typical recrystallization textures of cold rolled (a) aluminum, (b) copper, (c) brass Cu-37%Zn, and (d) two-phase Al-Mn AA3103.

In comparison to the fcc metals, the bcc metals exhibit a narrower range of recrystallization textures. The sensitivity to dispersions of second-phase particles is similar to that in the fcc metals in that coarse particles promote weaker recrystallization textures through particle stimulated nucleation. The process of recrystallization typically increases the volume fraction of the $\langle 111 \rangle \| ND$ (partial) fiber and decreases the intensity of the $\langle 110 \rangle \| RD$ fiber [RAABE & LÜCKE 1992]. Figure 39 shows an example of rolling and recrystallization textures in a low-carbon steel after 90% cold rolling, followed by annealing at 750°C, respectively, plotted as an SOD (sections at constant φ_1). A strong dependence on solute has been repeatedly demonstrated as in, for example, the effect of Mn in low-carbon steels. For a given carbon content, increasing the Mn content decreases the $\langle 111 \rangle$ component and, at high enough levels, can lead to nearly random textures. A review of annealing textures in steels is available in [RAY &AL. 1993] which cites many examples of these effects.

One over-riding feature of recrystallization textures that has yet to be quantitatively understood is the development of high symmetry components such as cube ($\{100\}\langle 001 \rangle$) and Goss ($\{110\}\langle 001 \rangle$) after rolling [DOHERTY &AL. 1988]. As an example of

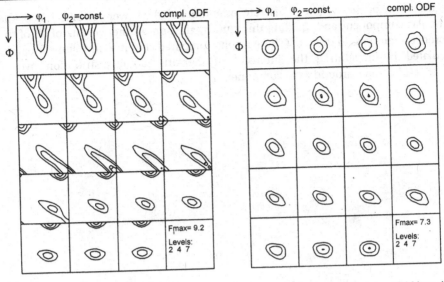

(a) rolling texture (80% red.)

(b) recrystallization texture (80% red., annealed for 150s at 700°C)

Fig. 39. SOD of (a) rolling and (b) recrystallization texture of a low-carbon steel. Courtesy D. RAABE.

current research on recrystallization in copper [NECKER 1997], Fig. 40 shows the evolution of the texture from standard fcc rolling texture, after 90% reduction in thickness by rolling, to fully recrystallized. The initial texture of the oxygen-free-electrical (OFE) copper before rolling had a weak cube component. The final texture after rolling and annealing is an extremely strong cube texture with a very small volume fraction of a twin-related derivative of cube. This evolution is of considerable practical interest to the suppliers of aluminum sheet to the beverage industry because the balance between deformation textures and annealing textures is crucial to controlling earing in beverage can stock. It happens that the anisotropy of interest, most easily characterized in terms of the variation of R-value (see Chap. 10) with direction in the rolling plane, has the same symmetry (four ears), but turned by 45°, for the rolling and the cube textures. Thus, it can happen that the texture in the partially recrystallized state (Fig. 40 b-e) is far from random, yet the anisotropy of interest is nearly absent. This point is a caution against using any particular test of mechanical isotropy as a diagnostic for texture.

A more detailed analysis of the evolution of texture during recrystallization is shown in Fig. 41. This figure shows the variation in volume fraction with fraction recrystallized for both the deformation and the recrystallization components. The first thing to note in this figure is that the volume fractions of the deformation components decrease uniformly. This indicates that no particular rolling component is depleted faster than any other during recrystallization. The steady decrease in the deformation components is in contrast to the non-linear increase observed for the cube component. The volume fractions in this case were calculated by first representing the orientation distribution by sets of weighted discrete orientations [KOCKS &AL. 1991b]. The volume fraction for a specific component was then calculated by adding up the weights of all of the weighted discrete orientations which were within 7.5° of the particular component and dividing this sum by the sum of the weights of all of the discrete orientations in the set. (These results compared closely with volume fractions calculated by integrating the intensity of the orientation distribution over a 15°×15°×15° volume around each component.)

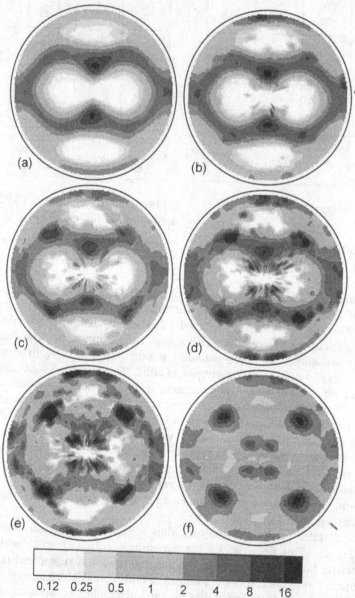

Fig. 40. 111 pole figures for a series of samples of rolled copper at successively higher fractions of recrystallization, showing the gradual strengthening of the cube component at the expense of the rolling texture. (a) 0, (b) 11, (c) 45, (d) 71, (e) 91 and (f) 100% recrystallized [NECKER 1997]. Equal-area projection, rolling direction vertical.

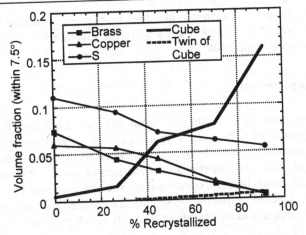

Fig. 41. Plot of the volume fraction of the major deformation (Brass, Copper and S) and recrystallization (Cube and Twin-of-Cube) components, as a function of percent recrystallized for rolled copper (90% rolling reduction) [NECKER 1997].

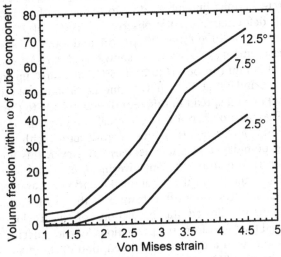

Fig. 42. Plot of volume fraction of the cube component in recrystallized copper as a function of the prior strain (in rolling) [NECKER 1997].

Another interesting aspect of texture evolution during recrystallization is the dependence on the deformed state. Since much is still being published in this area, we will confine our comments to a general observation. Figure 42 shows the variation in cube component volume fraction with prior rolling strain, expressed here in terms of the von Mises equivalent strain. Note that four different criteria have been used for the spread in the cube component and that, at high pre-strains, most of the cube component oriented material is within 7.5° of the ideal position. The results indicate that there is a marked transition from very little cube at pre-strains less than 2.0, to cube-dominated at strains over about 3.0. Many other scalar measures of texture are available for these sorts of investigations but if textures are to be compared with microscopy then volume fractions of components are more appropriate than intensities.

Lastly we note that grain growth can also lead to changes in texture because of variations of grain boundary energy and mobility associated with specific texture components [EICHELKRAUT &AL. 1988].

7. Texture Heterogeneities

In the examples given in this Chapter, it has been assumed that the textures presented are representative of the entire volume. However, while the textures shown may have been representative of the bulk of the material, most forming processes do not produce materials with uniform spatial distributions of texture. Indeed, without sufficient care in the forming process, significant texture gradients develop. Examples given in the literature include surface-to-midplane texture gradients in rolled materials, surface-to-core gradients in wires and outer-to-inner-wall gradients in tubes.

7.1 Local heterogeneities in deformation

Grain-scale heterogeneity

BARRETT & LEVENSON [1940] noted that individual grains in polycrystalline aluminum do not rotate as units but subdivide into regions with a range of orientations increasing with deformation. They observed a spread of 7° to 10° at 10% reduction in thickness, two or three times this at 30% and four or five times this at 60%. More recent measurements [HANSEN 1992] show this orientation scatter in cold-rolled polycrystalline aluminum to be up to 3° at 5% reductions, up to 15° at 20% reduction and between 5 and 20° at 30%. Barrett and Levenson noted that grains of identical orientation rotate and spread in different directions due to the influence of neighboring grains. They also observed that the break–up of grains into regions of differing orientation often forms bands in the microstructure with an orientation difference across band boundaries of a few degrees. However, this misorientation generally increases with deformation. Recent investigations of the reorientation of individual grains during plane-strain deformation [PANCHANADEESWARAN &AL. 1994] have shown similar results. The interesting feature of this work is that while the overall texture development followed the predictions of the Taylor model well, individual grains deviated markedly from the predictions for each specific orientation. Finite element calculations with a constitutive description of the material behavior that includes crystal plasticity [BECKER & PANCHANADEESWARAN 1995] show that the agreement between the observed rotations and the predicted rotations is much improved by taking grain-to-grain interactions into account.

Multi-grain heterogeneity

While several different types of deformation bands are discussed in the literature, shear bands have received the most attention within the texture community. At small strains, shear bands are observed at the grain scale where they extend through one or a few grains. As the volume fraction of microscopic shear bands increases they encompass larger regions of the microstructure. Eventually they may get organized to form sample-scale or macroscopic shear bands [CANOVA &AL. 1984b]. It should be noted that both microscopic and macroscopic shear bands are a continuum feature in the sense that they are non-crystallographic and not caused by specific slip patterns or features of the initial microstructure. This caution is reinforced by the results of recent simulations of rolling [BEAUDOIN &AL. 1995b, 1996] which show that any given grain may readily break up into regions of different orientation. These simulations were performed with finite elements (many elements per grain) and a constitutive model that included crystal plasticity. The results also showed that thin bands exist between the regions of similarly oriented material and that these transition bands can adopt special orientations such as near-cube. Lastly we note that shear bands can initiate at stress risers in a sample as, for example, at a sharp corner of the edge of a rolled sheet.

While the local textures of shear bands (or other microstructural inhomogeneities) may be quite different from the global texture, in many cases the shear bands simply contribute to the scatter around the main components of the global texture [SZTWIERTNIA & HAESSNER 1994]. Thus, even with significant volume fractions of shear banding the effect on the global texture may be imperceptible. Although the local textures of shear bands may not contribute significantly to the global texture, shear bands have long been known to affect subsequent recrystallization textures due to the large local lattice rotations present at their boundaries which generate high densities of nuclei [ADCOCK 1922, ENGLER 1996]. See the following section for a discussion of heterogeneity in recrystallization.

There are cases in the literature where the role of shear bands on texture development has been investigated [JASIENSKI &AL. 1994]. Also, LEE &AL. [1993] found that deformation banding played a significant role in the formation of the cold-rolling deformation texture, in particular in the production of {110}⟨112⟩ and {110}⟨001⟩ oriented material.

7.2 Texture gradients after deformation

Heterogeneities on the scale of the sample dimensions also exist and are highly significant for texture and texture-dependent properties. Friction at the roll/material interface, for example, results in significant texture differences between the surface and midplane of rolled materials [HANSEN & MECKING 1976, REGENET & STÜWE 1963, TRUSZKOWSKI &AL. 1980, ENGLER &AL. 1996b]. Figure 43 shows the variation in texture for rolled Fe-3%Si [CHASTEL 1993]. The surface texture exhibits a shear character. Although it is often assumed that texture gradients in rolled materials are entirely due to friction, friction effects are generally limited to a thin layer at the surface of the rolled sheet. ASBECK & MECKING [1978] have shown through rolling experiments on copper single crystals that whereas the friction can have an effect on the surface texture, the roll geometry will affect intermediate layers as well. In particular they showed that severe heterogeneities can occur at intermediate layers through the thickness of rolled samples when the ratio of the contact length between roll and sample to the sample thickness is smaller than a critical value of 0.5. More detailed analysis of the conditions required for homogeneous deformation in rolling can be found in any standard text. Texture gradients also exist in hot rolled materials and are influenced by the temperature distribution through the thickness of the sheet [ENGLER &AL. 1994b].

Wire drawing often produces significant texture gradients. An example from aluminum wire is show in Fig. 44. As in rolling, the surface textures often exhibit a shear component. One aspect of texture heterogeneities in wires is the existence of so-called 'cyclic textures', Fig. 45. It should be noted from this figure that if the textures were measured on a plane normal to the longitudinal direction of the wire then both wires would exhibit a fiber texture. Often a wire may exhibit a true fiber

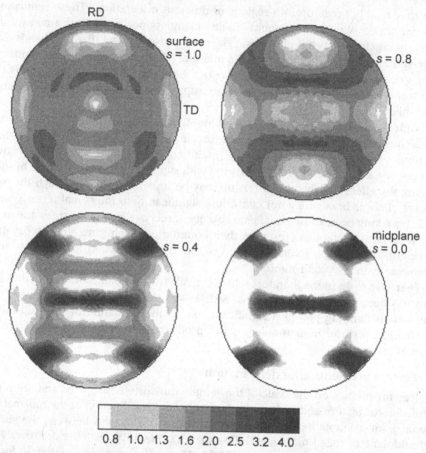

Fig. 43. 111 pole figures from rolled iron-3% silicon sheet measured at different through-thickness depths in the sheet [CHASTEL 1993]. Equal-area map.

texture at the wire core, but the texture often departs from an axisymmetric fiber texture with increasing distance from the wire core. Swaging, for example, often deforms a wire in a spiral pattern.

Drawn tubes often exhibit significant differences in textures measured at the inner and outer walls of the tubes. As with the effect of the rolling geometry on rolled materials, the texture in drawn tubes is extremely sensitive to the drawing geometry [BERTHELOOT &AL. 1976].

Significant variations in texture have also been observed in extruded materials both from surface to center [HUSSIEN &AL. 1988] and from the front end to the back end of an extrusion [MCHARGUE &AL. 1959]. As an example of textures in extruded materials we cite the work of Tenckhoff on zircaloy tubes [TENCKHOFF 1978, 1988]. Figure 46 shows the variation in texture from the inner diameter (ID) of an extruded tube to the outer diameter (OD) for three different types of deformation. It should be noted that these results are for tubes with extreme reduction values and do not represent conventionally produced tube.

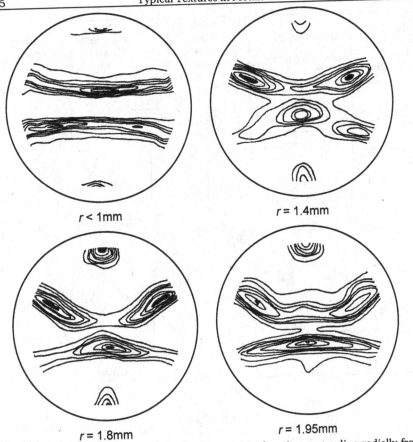

r < 1mm   r = 1.4mm

r = 1.8mm   r = 1.95mm

Fig. 44. 111 pole figures from drawn aluminum wires at locations extending radially from the wire core to the wire surface [LINSSEN &AL. 1964]. Stereographic map, wire direction vertical.

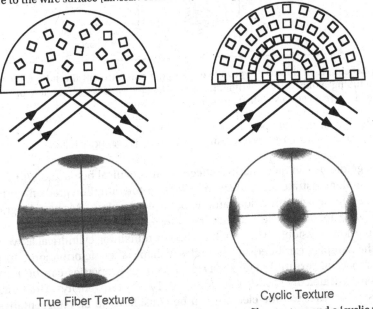

True Fiber Texture   Cyclic Texture

Fig. 45. Schematic showing the difference between a true fiber texture and a 'cyclic texture'.

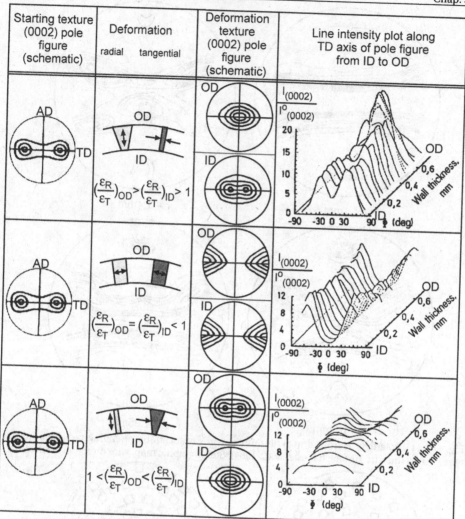

Fig. 46. Schematic of different types of texture gradients through the tube walls for zircaloy tubing along with experimental line intensity plots as a function of wall thickness for zircaloy tubes with extreme reduction values. ε_R is the radial strain and ε_T is the tangential (or hoop) strain. The line intensities were taken along the horizontal axis of the 0002 pole figure where the radial direction is normal to the plotting page and the transverse direction is to the right [TENCKHOFF 1988].

Texture gradients can have marked effects on mechanical behavior. WRIGHT &AL. [1994a, 1994b] demonstrated the effects of a surface-to-centerline gradient in tantalum plate on cylindrical compression samples cut from the plate. Because the grains at the surface were in softer orientations·than those at the center of the plate (with respect to free compression), instead of barreling or remaining cylindrical as would be expected, the samples exhibited an 'hourglass' shape after deformation. In some cases, the impact of texture heterogeneities on certain properties can be relatively small as demonstrated by BLANDFORD &AL. [1991]. Heterogeneity of deformation raises the issue of what volume element can be considered to be representative of a material, which is discussed in more detail in Chapter 12.

7.3 Heterogeneity in recrystallization

As discussed in Section 7.1, it is usually desirable to process materials in such a way as to avoid through–thickness variations in texture. However, such variations in texture play an interesting role (from both a fundamental and technological view-point) in the formation of strong Goss ($\{011\}\langle100\rangle$) textures in silicon-steel sheets for use in transformers. In this case, the through-thickness variation in texture is bene-ficial to formation of the sharp Goss texture in the final sheet product. A good example of the effect of the texture gradient on the final recrystallization texture of Fe-3%Si is work by MISHRA &AL. [1984]. They studied commercial hot-band steel which possessed the usual bcc rolling texture with a $\{112\}\langle110\rangle$-$\{001\}\langle110\rangle$ fiber at the center layer. However, the surface texture exhibited a notabble fraction of Goss oriented material. The Goss component is formed at the surface by shear during rolling at elevated temperatures. During cold rolling, the Goss grains rotated towards the $\{111\}\langle211\rangle$ orientation, eventually forming a fiber texture at the surface with $\langle111\rangle$ axes parallel to the normal direction. After subsequent annealing the Goss grains reappear, predominately in the surface layers and obtain a size advantage. During subsequent secondary recrystallization the Goss grains continue to grow primarily at the expense of the $\{111\}\langle211\rangle$ oriented grains before consuming the other grains in the matrix as well. The Goss grains grow towards the center layer of the sheet producing a sheet material with an overall bulk texture that is dominated by the Goss component. Thus, the heterogeneous (with respect to the bulk of the sheet) surface texture is critical for producing the desired sharply Goss textured sheet material. While the texture gradients in this material lead to a desired overall bulk texture, very often spatial variations in deformation textures persist after recrystallization. CLARK &AL. [1991] have shown that heterogeneities in rolled tantalum persist after annealing.

A short diversion from heterogeneities is warranted at this point in order to illustrate the importance of external surfaces in texture control of thin films or sheets. WALTER & DUNN [1959, 1960] demonstrated that thin sheets (<1mm) of Fe-3%Si can be made to undergo grain growth of the {110}<001> (Goss) component at the expense of the {100}<100> (cube) component, and vice versa, depending on the furnace atmo-sphere. The key factor was the oxygen potential: vacuum annealing of low oxygen content favors the Goss component whereas the presence of oxygen leads to growth of the cube grains. The competition between the two components is controlled by changes in the surface energy of the (110) versus the (001) planes exposed on the surface of the sheet such that in the latter case, for example, $\gamma_{100} < \gamma_{110}$. Unfortunately any improvements in performance of silicon steels obtained via surface texture control were apparently outweighed by cost and little commercial application of this elegant approach has occurred.

Heterogeneity of deformation exists at several different levels. Shear bands are (lamellar) regions of intense local shear that occur in rolling under a variety of circum-stances, see for example the brief summary discussion in HUMPHREYS & HATHERLY [1995]. At large reductions, shear bands characteristically span the entire sheet thick-ness and are inclined at an angle of approximately 35° to the rolling plane. Their effect on texture is generally to decrease the sharpness of the recrystallization texture because they are potent sources of nuclei [ADCOCK 1922, RIDHA & HUTCHINSON 1982]. Shear bands are most commonly observed in materials that exhibit planar slip, such as brass, whose deformation textures are often of the 'alloy' type (see earlier discussion, also HUTCHINSON &AL. [1979]).

Descending down the length scale of heterogeneity, we consider transition bands which arise from the tendency of grains to break up at large strains. Transition bands are significant to texture development in recrystallization because regions of large misorientations can serve as nuclei for new grains. In some cases re-orientation during slip of a symmetrically oriented volume element can occur in opposite directions (in orientation space); volume elements at the transition band sometimes remain on the symmetrical position. This phenomenon was extensively explored by DILLAMORE & KATOH [1974]. Thus in bcc metals that have undergone uniaxial compression (leading to a mixture of <111> and <100> fibers), recrystallization produces a fiber near <112> because this is the orientation most likely to be found in the transition bands [INOKUTI & DOHERTY 1978]. Similar considerations apply to rolled bcc metals [DILLAMORE & AL. 1972]. Transition bands are equally important in fcc metals for the development of recrystallization nuclei [BELLIER & DOHERTY 1977]. Here again, transition bands appear to be critical to the preservation of cube oriented volume elements during deformation, since the cube is unstable with respect to plane-strain deformation; the simulations by BEAUDOIN &AL. [1995, 1996] demonstrate how this can happen. The reasons for the development of cube oriented nuclei are still disputed and beyond the scope of this chapter (see DOHERTY &AL. [1997]). Factors that appear to favor the cube oriented nuclei include high recovery rates and proximity to orientations that give rise to boundaries with high mobility [DUGGAN &AL. 1993, VATNE &AL. 1996].

While the bulk textures of two materials may be quite similar, the two materials may have very different 'microtextures'. These local heterogeneities may then account for materials with similar deformation textures producing very different recrystallization textures. This point was demonstrated very clearly by RIDHA and HUTCHINSON [1982], for example, where they demonstrated that copper can exhibit similar rolling textures but widely varying recrystallization textures depending on minor spatial variations in purity. In general, the reader should be aware that deformation leads to misorientation development of all types and magnitudes: we have simply focused on the aspects that are most relevant to texture development, i.e. the development of large misorientations. Additional detail on texture heterogeneities on the scale of cells and subgrains is beyond the scope of this book but represents a new and promising arena for research in materials science.

8. Solidification and Thin Film Textures

8.1 Solidification

To this point, we have mainly focused on deformation textures. However, it is instructive to consider a different source of texture, namely solidification. This is important because, too often, textures are carefully measured and then rationalized on the basis of the immediate history without considering the complete thermo-mechanical history, for which the original solidification texture may have had a profound effect [CLARK &AL. 1991]. The preferred growth direction of dendrites in cubic metals is ⟨100⟩ and has been well documented to influence deformation and recrystallization textures even after extensive plastic deformation, e.g. NES &AL. [1984].

As an example we cite recent work by GANDIN &AL. [1995] on an Inconel X750 alloy that was directionally solidified by casting in a ceramic mold over a copper chill plate. As shown in Fig. 47, the texture was found to vary from nearly random at the base of

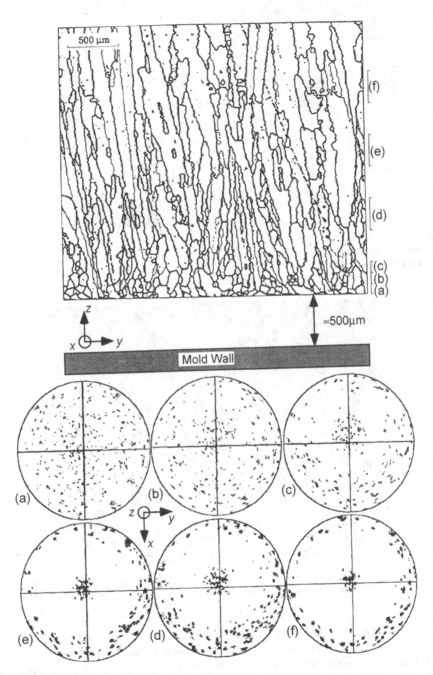

Fig. 47. Grain boundary map and corresponding discrete 100 pole figures for an Inconel X750 alloy cast on a copper chill plate in a ceramic mold. Pole figures are given for measurements in bands in the measurement domain at (a) 0 to 0.1 mm; (b) 0.1 to 0.2mm; (c) 0.2 to 0.3 mm; (d) 0.6 to 0.9 mm; (e) 1.2 to 1.5mm and (f) 1.8 to 2.1mm [GANDIN &AL. 1995].

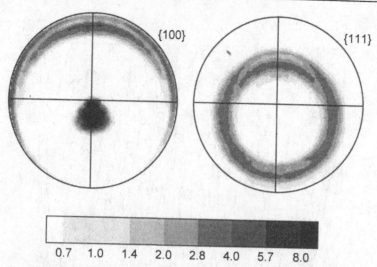

Fig. 48. 100 and 111 pole figures for melt-spun palladium, showing a strong ⟨100⟩ fiber corresponding to a growth direction from the melt that is tilted with respect to the plane of the ribbon. Equal-area projection.

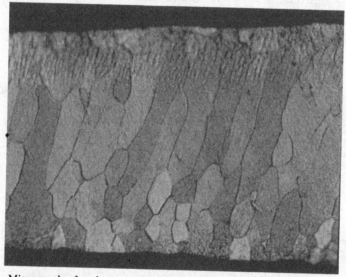

Fig. 49. Micrograph of melt-spun palladium ribbon, see previous figure for texture, showing columnar structure with tilted growth direction corresponding to texture result.

the casting to a ⟨100⟩ fiber texture away from the base. The variation of a nearly random distribution of orientations near the chilled surface to a textured structure away from the chilled surface suggests that nucleation is a random process with respect to orientation; whereas growth selection is strongly orientation dependent. (The measurements in this study were made using automatic indexing of electron diffraction patterns as described in Section 4.2 of Chapter 4.)

A second example is taken from work on rapid solidification of an unusual material. Figure 48 shows pole figures for the top surface of a ribbon of a Pd-Ni alloy that has been melt-spun, i.e. a thin strip has been made by impingement of a molten jet of the alloy onto a rapidly spinning copper wheel [KORZEKWA &AL. 1992]. The pole figures show the expected ⟨100⟩ fiber. However, the fiber is clearly offset from the pole figure normal (which corresponds to the ribbon normal). This is due to the fact that the growth direction in the ribbon is not aligned with the ribbon surface normal. The micrograph shown in Fig. 49 shows a columnar structure that has grown at an angle to the ribbon normal. An orientation distribution[5] was calculated from the pole figures and a 100 pole figure was then recalculated. Figure 50 shows a plot of intensity in the recalculated 100 pole figure along a radial line in the pole figure parallel to the long direction of the ribbon. The maximum intensity along this line is located at a tilt angle of about 20° from the ribbon normal which correlates well with the angle of tilt of the columnar structure observed in the micrograph. The texture of the bottom surface of the ribbon was also measured and was found to be much weaker. This corresponds to the previous example where a weaker texture was observed at a location where less growth selection has occurred. (See TEWARI [1988] and HUANG &AL. [1985] for additional information on texture development in melt-spinning.)

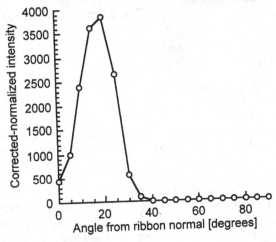

Fig. 50. Plot of intensity in 200 pole figure showing that the tilt from the ribbon normal is about 20°.

8.2 Vapor deposition

Strong fiber textures can also be achieved using vapor deposition methods. For example, Fig. 51 shows a 100 pole figure measured from a tube that was formed by chemical vapor deposition of tantalum onto a mandrel. The pole figure clearly shows that the tube has strong ⟨100⟩ fiber and that the fiber is oriented in the radial direction. This is a good example of the fact that the effect of the forming process and geometry of the sample must be considered when selecting planes to measure textures on. In this case, a weak ⟨100⟩ fiber texture was initially measured as opposed to the strong

---

[5] This example illustrates the usefulness of being able to perform a calculation of the orientation distribution without being required to assume any sample symmetry. Fiber symmetry could only be assumed if the fiber were aligned with one of the principal pole figure axes.

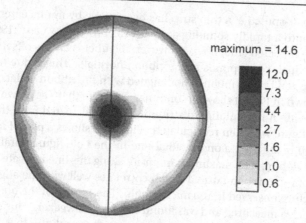

Fig. 51. 100 Pole figure measured on chemical vapor deposited tantalum. Equal-area projection.

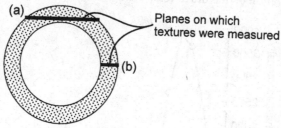

Fig. 52. Schematic of texture measurements on a tantalum tube formed by chemical vapor deposition: (a) first measurement; (b) second measurement.

⟨100⟩ fiber texture shown in Fig. 51. However, the initial plane on which the initial texture measurement was made, Fig. 52, leads to an artificial spreading of the fiber.

Knorr has reviewed textures in thin aluminum films for interconnects in microelectronic devices [KNORR 1993]. Generally speaking a ⟨111⟩ fiber texture is present; sputtering produces stronger textures than evaporation. It is also possible to obtain ⟨110⟩ fiber textures, however, and Knorr makes a strong case for applying orientation distribution analysis to pole figures (or microtexture measurements) instead of relying on diffraction intensities of the Bragg peaks. Texture has a significant effect on performance, as seen, for example, in studies of electromigration [KNORR 1993] in aluminum films deposited on oxidized silicon wafers. The results indicated that strong ⟨111⟩ fiber textures were associated with longer times to failure, where failure occurs by voiding due to electromigration. This result was reinforced by recent work on narrow interconnects by JOO & THOMPSON [1994] in which the lifetimes varied by a factor of ten between ⟨111⟩ and ⟨110⟩ oriented films (111 orientations gave the longest lifetimes).

Thin metallic films are also of technological significance for magnetic data recording. FENG &AL. [1994] have studied the formation of thin sputtered films of chromium which are commonly used as underlayers for Co films in magnetic recording media. They proposed that the variation in texture between the ⟨110⟩ and ⟨001⟩ fibers, as a function of substrate temperature (and other parameters), is caused by changes in relative interfacial energies of the substrate-Cr, (110)-vapor and (001)-vapor facets. In this case, simple x-ray spectra (2Θ scans) were used to measure texture, which is adequate in cases where the preferred orientations are limited to a small, known set. For films thinner than about 0.1 μm, special techniques of transmission electron microscopy must be employed to obtain electron diffraction data and thereby determine textures. We should also draw attention to the discussion of methods of analyzing pole figure data obtained in reflection when epitaxial films are thin enough (of order 0.5 μm, or less) that spurious peaks can arise by diffraction from the substrate (Chapter 6: typical textures in geological and ceramic materials).

8.3 Electrodeposition

Electrodeposition can generate a variety of fiber textures and there is a branch of electrochemistry that deals with the control of preferred orientation as a function of deposition conditions such as current density, overvoltage, and electrolyte composition [DAMJANOVIC 1965, KRUSHEV &AL. 1968]. PANGAROV [1991] has given a review on the texture in electrodeposited metals. The texture of electrodeposits depends primarily on two factors: (1) the crystal structure of the substrate and the deposit and (2) the bath conditions. When deposition is made on a substrate where the differences in the lattice parameters of the substrate and the deposit exceed 15% then the texture of the deposit depends only on the deposition conditions. However, when the lattice parameters of the substrate and deposit are similar then the initial layers of the deposited metals grow epitaxially, independent of bath conditions. The epitaxial influence gradually diminishes and the preferred orientation of the deposits depends solely on the bath conditions. The bath conditions that affect deposition texture include bath composition, current density, voltage, temperature and p_H. Several points noted in Pangarov's review about the effect of certain bath conditions on texture are briefly reviewed here.

(1) If deposition is done at low current densities and high temperatures, the crystallites are generally oriented so that the most densely populated atom planes lie parallel to the substrate. Thus, the axis of preferred orientation for fcc metals would be $\langle 111 \rangle$, for bcc $\langle 110 \rangle$ and for hcp $\langle 0001 \rangle$.

(2) If deposition is done at low temperatures and high current densities, the crystallites are generally oriented so that the most densely populated atom rows in the most densely populated atom planes lie perpendicular to the substrate. For the fcc metals the axes of the preferred orientation would be $\langle 110 \rangle$, for bcc $\langle 111 \rangle$ and for hcp $\langle 11\bar{2}0 \rangle$.

(3) In many cases there is a strong preferred orientation which does not belong to either type (1) or type (2). For fcc metals these preferred orientations are $\langle 100 \rangle$, $\langle 112 \rangle$, $\langle 113 \rangle$ and $\langle 210 \rangle$, for bcc $\langle 112 \rangle$ and $\langle 310 \rangle$, and for hcp $\langle 10\bar{1}0 \rangle$, $\langle 10\bar{1}1 \rangle$ and $\langle 11\bar{2}2 \rangle$.

(4) If substances are present which can be adsorbed on the electrode, the axis of preferred orientation can change.

(5) The effects of overvoltage on preferred orientation are summarized in Table III.

Recent work has shown a strong effect from pulse-reversed current, which perturbs adsorption at the metal/electrolyte interface [KOLLIA & SPYRELLIS 1993].

Table III. Variation of electrodeposition textures with overvoltage for thick deposits [PANGAROV 1965].

Metal	Crystal lattice	Low overvoltage	Intermediate overvoltage	High overvoltage
Fe	bcc	110	112 310	111
Ag	fcc	111	100	110 210
Cu	fcc	111	100	110 112 210
Zn	hcp	0001 11$\bar{1}$1	11$\bar{2}$0	11$\bar{1}$0 11$\bar{2}$2
Co	hcp	0001	11$\bar{2}$0	11$\bar{1}$0 11$\bar{2}$2
Sn	tetragonal	100	110	101

9. Summary

The textures of metals have been studied in considerable detail over the years and it is usually possible to find an example of texture to be found in any combination of alloy and processing mode. Despite the progress that has been made in developing models for texture development, however, the reader is urged to remember that texture is sensitive to almost the entire spectrum of metallurgical variables. The aim of this chapter is, in the main, to illustrate the complexity and richness of this field of enquiry.

Chapter 6

TYPICAL TEXTURES IN GEOLOGICAL MATERIALS AND CERAMICS

1. **Geological Materials**
   1.1 Calcite and aragonite ($CaCO_3$)
   1.2 Dolomite ($MgCa[CO_3]_2$)
   1.3 Quartz (SiO_2)
   1.4 Olivine ($MgSiO_4$) and perovskite ($CaTiO_3$)
   1.5 Sheet silicates
   1.6 Ice (H_2O)
   1.7 Polymineralic rocks

2. **Bulk Ceramics**
   2.1 α-Alumina (Al_2O_3)
   2.2 Zirconia (ZrO_2) transformation textures
   2.3 Hematite (Fe_2O_3) and magnetite (Fe_3O_4)
   2.4 Silicon nitride (Si_3N_4)
   2.5 Alkali halides and related structures
   2.6 Ceramic matrix composites
   2.7 Bulk high-temperature superconductors
   2.8 Cement minerals

3. **Thin Films and Coatings**
   3.1 Silicon and diamond
   3.2 Nitride, carbide and oxide coatings
   3.3 Epitaxial films

TYPICAL TEXTURES IN GEOLOGICAL MATERIALS AND CERAMICS

H.-R. Wenk

Naturally deformed rocks were some of the first materials in which preferred orientation has been studied quantitatively. SCHMIDT [1932] introduced the pole figure and equal-area projection for its representation. The state of the art in 1950 was reviewed in SANDER's monumental book on fabric analysis which includes an appendix with many of the then known fabric types. Today structural geologists use the terms texture and preferred orientation interchangeably; however, fabric implies more, including intergrowths, grain shape and correlations. Again, in 1985 the variety of geological textures was reviewed [WENK 1985a] and there is no need to repeat this here. In this Chapter we would like to present a few of the most important texture types of rock-forming minerals, using some new experimental results and emphasizing texture representations for low symmetry materials. In order to facilitate the understanding, measured textures will be now and then compared with simulated textures and reference is made to such polycrystal plasticity theories as the Taylor model or a viscoplastic self-consistent model which will be explained in later chapters. Note that in all these geological examples the rate sensitivity of plastic flow is substantial. Also, Chapter 14 will illustrate the usefulness of textures and plasticity theory for geological applications with a few examples.

Geological polycrystals are referred to as rocks; they can be monomineralic such as marble and quartzite or polymineralic such as granite and peridotite. Trigonal *calcite* is one of the best studied minerals with strong textures developing by slip, mechanical twinning and recrystallization. *Dolomite*, with a structure similar to calcite, is an example where observation of texture patterns led to the discovery of unknown slip systems. Trigonal *quartz* is the most widely studied mineral, but texture development is still very enigmatic. It is interesting from the point of view of texture representation because quartz is enantiomorphic and exists in left- and right-handed forms, recognized, for example by optical activity and a piezoelectric effect. In addition, symmetrically non-equivalent peaks are overlapped in diffraction patterns. The three trigonal minerals calcite, dolomite and quartz are of some relevance to metallurgists since they deform on similar slip systems as do hexagonal metals. Orthorhombic *olivine* has been extensively investigated because of its significance for convective flow in the Earth's mantle. *Sheet silicates* are a group of minerals with a distinct platy morphology and their alignment during deformation or compaction has been modeled as rigid particle rotations in a viscous medium [e.g. MARCH 1932]. Some aspects of the deformation and texture development of halite (NaCl) will be discussed in Section 2.5 and in Chapter 14.

Compared to metals and to geological materials where one had to select a few examples out of many, the literature on texture development in ceramics is surprisingly sparse. Some of the pioneering work was that of PENTECOST & WRIGHT [1964] who demonstrated, with pole figures, an alignment of crystallites in pressed powders of plate-shaped Al_2O_3 and needle-shaped BeO. Textures in ceramics were recently reviewed by BUNGE [1991] documenting that texture research in this field is in its infancy but expanding rapidly as new materials are being manufactured with critical properties, brittle, ductile and electrical. We report on some systems where quantitative or at least qualitative texture data exist. Since single crystal properties of ceramics are generally highly anisotropic compared to metals, even moderate textures can introduce a strong anisotropy into the polycrystal and affect the performance of ceramic materials. Whereas in metals most textures develop by casting (growth), slip, twinning and recrystallization, more mechanisms are active in ceramics. Processes such as cold pressing ('green body pressing'), hot pressing (densification and phase transformation), surface grinding, epitaxial and topotaxial growth can all lead to textures in ceramics. The discussion on ceramics is divided into two sections: bulk ceramics and films.

Whereas ODs of cubic metals can be represented in a small orientation volume and share many common features which can be catalogued, ODs of low symmetry materials require large volumes, are specific in each case, with few common features. The representation with a few ideal orientations (components such as 'cube', 'brass', 'Goss' etc.), which is common in rolled cubic metals, is often not a very efficient method for minerals and ceramics because of the generally low sample symmetries and the often unknown kinematic sample reference frame. The sample coordinates are generally not a priori defined and it is impossible to visualize an OD in a different orientation. By contrast pole figures can be mentally rotated on the sphere and one can recognize common textural features between two samples, even if these samples were measured in a different orientation. For these reasons ODs, though essential for interpretations, calculations of properties and texture simulations, are often difficult to interpret and pole figures remain an important representation. For axially symmetric textures inverse pole figures are used. Many illustrations in this Chapter are taken from the literature and could not be converted to a uniform convention.

1. Geological Materials

1.1 Calcite and aragonite ($CaCO_3$)

The trigonal mineral calcite crystallizes in the centrosymmetric point group $\bar{3}m$ and symmetry elements are shown in Fig. 1a. The asymmetric unit consists of a 60° sector which is used for inverse pole figure representations (in ODs the sector is 120° because of the lack of inversions in orientation space). In the trigonal system, lattice planes such as $10\bar{1}4$ (positive rhombs) and $01\bar{1}4$ (negative rhombs) are non-equivalent but diffract at the same Bragg angle.[1] In the case of calcite (not for quartz which will be discussed later), due to systematic extinctions for rhombohedral lattices (condition

---

[1] Note that in the text we use mainly four-index Miller-Bravais symbols $hkil$, with $i=-(h+k)$; pole figure labels are often given in three-index diffraction indices, omitting i.

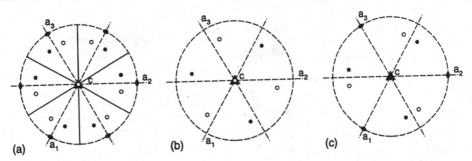

Fig. 1. Stereographic projection of symmetry elements in the trigonal point groups: (a) $\bar{3}2/m$ (calcite), (b) $\bar{3}$ (dolomite) and (c) 32 (quartz). Circles are symmetrically equivalent poles to lattice planes. Solid circles are on upper hemisphere, open circles on lower hemisphere.

for presence: $-h+k+l=3n$) these overlaps do not exist in diffraction patterns. Therefore OD calculations are straightforward.

Calcite rocks (limestone and marble) have been a model system for experimental rock deformation. The classic Yule marble studies [e.g. TURNER &AL. 1956] were the basis for a quantitative interpretation of texture development by slip and twinning. Due to the low ductility of geological materials it is necessary to perform experiments at high confining pressures which is usually only possible in compression geometry. In axial compression experiments, textures with axial symmetry are produced and for these, inverse pole figures of the compression axis are the most efficient way of representation. Texture variations were observed with temperature, pressure, strain rate and grain size [WENK &AL. 1973, SCHMID &AL. 1977, SPIERS 1979]. The observed deformation textures with a maximum near $c=(0001)$, and a shoulder towards the negative rhomb $e=(01\bar{1}2)$, are attributed mainly to slip on the rhomb $r=\{10\bar{1}4\}$ and mechanical twinning on $e=\{01\bar{1}2\}$ (Fig. 2b-d). They are distinctly different from the recrystallization texture with a maximum at high-angle positive rhombs (Fig. 2e). Under conditions where recrystallization is inhibited, SCHMID &AL. [1977] observed minimal texture development which they attributed to superplastic flow (Fig. 2f).

Since calcite rocks such as limestone are mechanically relatively ductile compared to other minerals, high confining pressure is not necessary and therefore deformation experiments other than uniaxial compression can be performed more easily. Noteworthy are plane-strain compression experiments (referred to as 'pure shear' in the geological literature) with three mirror planes in the pole figure (Fig. 3b) and simple shear with only one mirror plane and a 2-fold axis in the pole figure (Fig. 3c) as compared to axial compression experiments which yield a fiber texture (Fig. 3a) [KERN & WENK 1983]. These experimental pole figures are good examples to illustrate that the symmetry of pole figures reflects the symmetry of the deformation history (strain path) which has been a main criterion for interpreting textures in naturally deformed rocks [PATERSON & WEISS 1961].

In limestone, experimentally deformed in plane-strain compression, a transition from a low-temperature (LT) to a high-temperature (HT) texture was observed as illustrated in ODs in Fig. 4 [WENK &AL. 1987]. TAKESHITA &AL. [1987] were able to simulate similar textures based on polycrystal plasticity theory (Taylor in Fig. 4). Changes in texture were associated with texture transitions due to changing activities of slip and twinning systems. At low temperature, where e-twinning dominates, c-

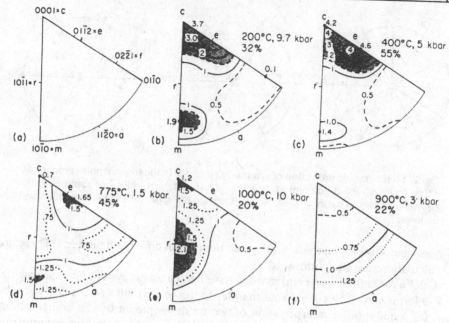

Fig. 2. Calcite textures produced in axial compression experiments at different conditions represented in inverse pole figures of the compression direction. Starting material is fine-grained limestone. Temperature, confining pressure and compressive strain are indicated. Contours give orientation densities in m.r.d. (a) Disposition of crystallographic directions. (b–e) from WENK &AL. [1973], (f) from SCHMID &AL. [1977].

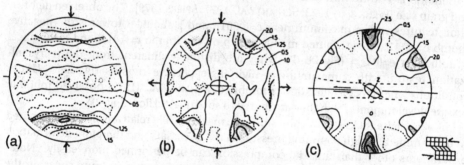

Fig. 3. Calcite 0001 pole figures of experimentally deformed calcite limestone illustrating how the symmetry of the pole figure relates to the symmetry of the strain path: (a) axisymmetric compression [WENK &AL. 1986], (b) plane-strain compression [WAGNER &AL. 1981], (c) simple shear [KERN & WENK 1983]. Some contour levels (in m.r.d.) are marked. Compression, extension and shear directions are indicated by arrows.

axes are parallel to the compression direction. At higher temperature, when r- and f-slip dominate, c-axes concentrate in symmetrically disposed maxima. (These texture transitions in calcite, which can be studied with deformation mode maps, will be discussed in more detail in Chapter 11.) KERN [1971] pointed out that textures of carbonate rocks with larger grain size correspond to those in fine grained materials deformed at higher temperature. This is probably due to the enhanced activity of mechanical twinning in coarse grained polycrystals and at low temperature.

Naturally deformed calcite rocks, limestones and marbles, often display strong preferred orientation. The texture of a marble from Palm Canyon compares well with the low-temperature experimental and theoretical textures (Fig. 4). Examples and interpretations of the deformation histories based on the observed textures, will be given in Chapter 14.

Texture development of the high pressure polymorph aragonite during deformation and recrystallization was investigated experimentally by WENK &AL. [1973].

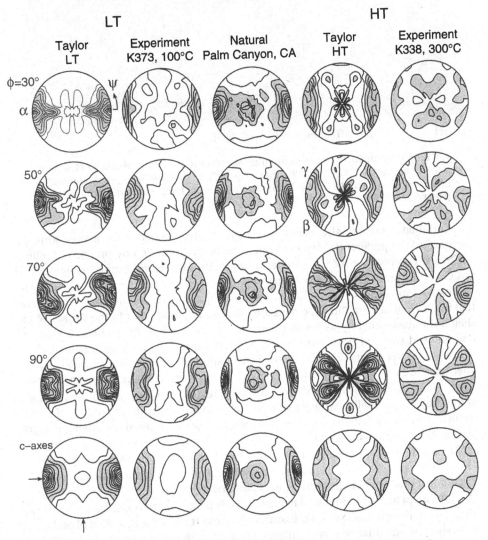

Fig. 4 Comparison of some COD Φ-sections (Roe angles) and 0001 pole figures (bottom) of experimentally deformed calcite limestone (plane-strain compression at 100°C and 300°C) and corresponding Taylor simulations, for LT assuming that e-twinning and r-slip dominate and for HT that r- and f-slip dominate and e-twinning is subordinate. Texture components α, β, and γ are shown. For comparison is a naturally deformed marble from Palm Canyon which compares well with the low-temperature texture. Equal-area projection. Arrows in bottom left pole figure indicate compression and tension directions [WENK &AL. 1987].

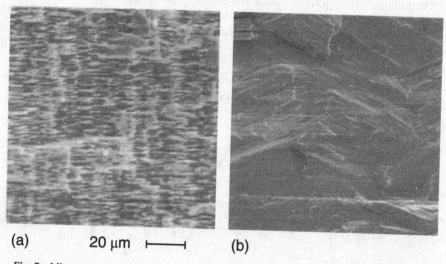

(a)   20 µm ⊢———┤   (b)

Fig. 5. Microstructures of aragonite in cross sections of mollusc shells, SEM images, scale is indicated. (a) Nacre ('mother of pearl') of Nautilus macromphalus with a brickwork structure. (b) Crossed-lamellar structure in the gastropod Patella tabularis [HEDEGAARD & WENK 1997].

Carbonate textures also occur in skeletons of organisms where they form by growth and the preferential crystallite orientation is controlled by proteins. Mollusc shells are composed either of calcite or aragonite (an orthorhombic polymorph of $CaCO_3$) and show a wide variety of textures. In aragonite nacre ('mother of pearl') the microstructure displays a brick-like framework (Fig. 5a). The texture analysis indicates that in nacre of most mollusks, **c**-axes are oriented perpendicular to the surface of the shell but **a**-axes either display a single crystal pattern such as in the bivalve Pinctada (Fig. 6a) and shells of most land snails, or they display a texture pattern with {110} twinning as in Nautilus (Fig. 6b), or they spin randomly about the **c**-axis with a [001] fiber texture as in Haliotis (Abalone, Fig. 6c) [HEDEGAARD & WENK 1997, WENK 1965]. Since [100] is the stiffest crystal direction, such a texture produces an optimal strength of the shell. Other molluscs have a cross lamellar microstructure (Fig. 5b). In this case texture analysis reveals two **c**-axis components, both of them twinned on {110}, producing a strong concentration of (110) in the growth direction (Fig. 6d). Mollusc shells have received considerable attention because of their bio-mimetic properties. Clearly, textures will be an important aspect and ought to be quantitatively related to mechanical properties, to protein type and to phylogeny.

Texture patterns in biological materials, including carbonates, silica minerals and phosphates, composing shells, skeletons, bones and teeth are a new field of endeavor, with very few quantitative investigations [LOWENSTAM & WEINER 1989]. Figure 7 illustrates a hydroxyapatite texture in a bovine ankle bone investigated with synchrotron x-rays. It displays a strong alignment of the [0001] directions of this hexagonal mineral parallel to the surface of the bone (with axial symmetry), as documented in the inverse pole figure for the direction normal to the bone (Fig. 7c). Single-cell organisms, called Foraminifera, exploit in their calcite skeleton all possible preferred orientation patterns, and also a random arrangement of calcite laths [HAYNES 1981, P.58].

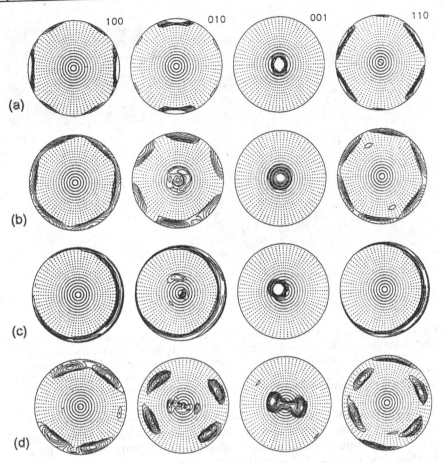

Fig. 6. Growth textures of aragonite which are correlated with the microstructures in Fig. 5. 100, 010, 001 and 110 pole figures, projected on the shell surface, recalculated from the OD. Equal-area projection, logarithmic contours 0.5, 0.7, 1, 1.4 m.r.d. etc., dot pattern below 1 m.r.d. (a) Nacre of the bivalve Pinctada maxima (Oyster) with more or less a single component texture, (b) nacre of Nautilus macromphalus with a pseudohexagonal pattern due to twinning on 110, (c) nacre of Haliotis cracherodis (Abalone) with an axial texture, (d) texture of the crossed-lamellar structure in Patella tabularis with two components, both of them twinned.

1.2 Dolomite (MgCa[CO$_3$]$_2$)

Dolomite is another carbonate, with a crystal structure similar to calcite, but a lower symmetry (point group $\bar{3}$). This would require a 120° sector for representation (Fig. 1b) and one would have to rely on minor intensity differences of general diffractions to resolve the true symmetry. This has never been attempted. For texture representations, dolomite is traditionally treated as if it had the same symmetry as calcite.

Slip systems are different from those in calcite; fewer are available, and dolomite is therefore plastically more anisotropic and much less ductile. Consequently texture development is also different which is again documented by compression experiments. WENK & SHORE [1975] described two texture types. At low and medium

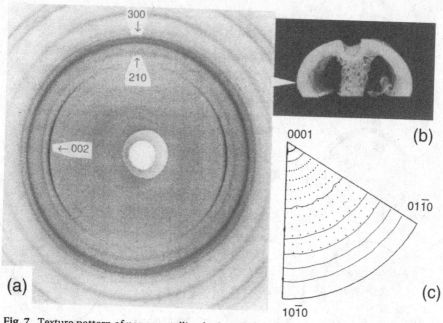

Fig. 7. Texture pattern of nanocrystalline hydroxy-apatite in a bovine bone. (a) Debye rings, recorded with a CCD camera, and using a 10μm synchrotron x-ray beam. (b) Cross section of the sample, in the same orientation as (a). The arrow points to the location of the analysis. (c) An inverse pole figure for the surface normal marked by the arrow in (b). Contour interval is 0.2 m.r.d.; dot pattern below 1 m.r.d.

temperature inverse pole figures show concentrations at low angle negative rhombs and high angle positive rhombs (Fig. 8a). At high temperature the material recrystallizes and the texture displays a single maximum at $c=(0001)$ (Fig. 8b). From the texture data they concluded that deformation systems other than (0001) slip and $f=\{01\bar{1}2\}$ twinning, which were known at the time, must be active. Dislocations for rhombohedral slip on $f=\{01\bar{1}2\}$ and $r=\{10\bar{1}4\}$ were later found by electron microscopy [BARBER &AL. 1981].

BARBER &AL. [1994] modeled the plastic behavior of polycrystalline dolomite assuming basal and rhombohedral slip and twinning. Figure 8c shows inverse pole figures of the compression direction for 200 grains after 50% deformation. The texture prediction agree very well with the experimentally observed patterns (Fig. 8a). An investigation of microstructures in individual grains in experimentally deformed coarse grained dolomite with the transmission electron microscope confirmed that those slip systems are active which have the highest resolved shear stress and grains rotate in the senses which are predicted (see Chap. 4 Fig. 35). In that study the observed number of activated slip systems in individual grains was compared with the number predicted by polycrystal plasticity models. Histograms illustrate that the Taylor model which assumes homogeneous deformation predicts four to five active systems (Fig. 9a), a viscoplastic self-consistent model which emphasizes deformation of plastically weak grains predicts one to two active systems (Fig. 9b). The latter is in better agreement with observations (Fig. 9c), suggesting heterogeneous deformation of this plastically very anisotropic material. This is supported by microstructural evidence with local brittle fracture and grain boundary effects and sub-grain scale

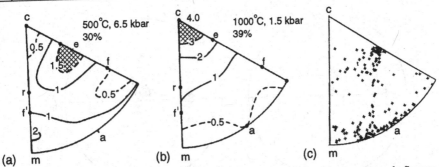

Fig. 8. (a,b) Experimentally compressed polycrystalline dolomite. Inverse pole figures of the compression direction, conditions are indicated, for symbols see Fig. 2 [WENK & SHORE 1975]. (c) Simulated texture pattern for dolomite [BARBER &AL. 1994].

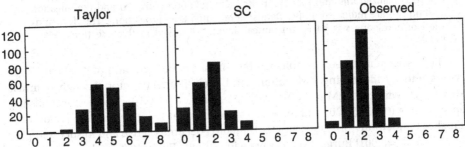

Fig. 9. Deformation of dolomite. Histograms indicating number of active slip systems per grain. (a) Predictions based on a viscoplastic Taylor theory, counting systems which contribute 10% or more to the total shear, (b) predictions based on a visco-plastic self-consistent theory (SC), (c) slip systems observed by TEM investigation of experimentally deformed dolomite [BARBER &AL. 1994].

variability of slip activity and dislocation density. The example illustrates that quantitative microstructural data can complement texture characterization and assist in interpretations of the deformation behavior.

1.3 Quartz (SiO$_2$)

Low quartz (α) is trigonal, point group 32 (Fig. 1c). At high temperature α-quartz transforms to hexagonal β–quartz (point group 622). Both low and high quartz lack a center of symmetry and exist in a right- and a left-handed form. The absence of a center of symmetry is the reason for properties such as piezoelectricity and several investigations confirm the presence of a piezoelectric effect in polycrystalline quartz [e.g. BISHOP 1981, PARKHOMENKO 1971, GHOMSHEI & TEMPLETON 1989, IVANKINA &AL. 1991]. The few investigations done so far are based on assessment of optical activity and suggest that in most quartz rocks left- and right-handed crystals occur in equal numbers and have a similar orientation distribution. An exception is a recrystallized quartzite with a dominance of left-handed crystals (Fig. 10) [WENK 1985b]. In many deformation studies and particularly investigations of preferred orientation by diffraction methods, enantiomorphism can not be identified and a higher crystal symmetry is used by adding a center of symmetry to point group 32. This produces point group $\bar{3}2/m$ which is that of calcite and is generally used in representations (Fig. 1a).

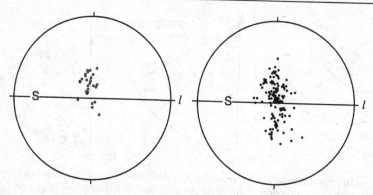

Fig. 10. Equal-area projection of c-directions of (a) right- and (b) left-handed quartz in a recrystallized metamorphic quartzite. The foliation plane (S) and the lineation direction (*l*) are indicated, upper hemisphere, equal-area projection. The determination of handedness is based on optical activity observed in thick sections [WENK 1985b].

Slip systems have been established by laboratory deformation of oriented single crystals and by identifying dislocations and their Burgers vector through contrast experiments with the TEM (for a review see e.g. BLACIC [1975]). The best established slip system is $(0001) \langle 11\bar{2}0 \rangle$ and appears to dominate at low-temperature conditions. At higher temperature prismatic slip $\{10\bar{1}0\}$ [0001] becomes active and in such samples $a = \langle 11\bar{2}0 \rangle$ and $c = [0001]$ Burgers vectors were identified. Evidence for other slip systems such as prismatic $\{10\bar{1}0\} \langle 11\bar{2}0 \rangle$, rhombohedral $\{10\bar{1}1\}$ or $\{01\bar{1}1\} \langle 11\bar{2}0 \rangle$ or $\langle \bar{1}\bar{1}23 \rangle$ and pyramidal $\{\bar{1}\bar{1}21\} \langle \bar{1}\bar{1}23 \rangle$ is more circumstantial and largely based on the fact that quartz crystals do deform if they are compressed parallel to the c-axis. These slip systems are restricted to high-temperature deformation. There is no reliable information on critical shear stresses and how they change with strain rate and temperature. It is unknown if slip on positive and negative rhombs is equivalent or if there is a preference for the sense of shear as in rhombohedral carbonates. The well established slip systems have a hexagonal symmetry, and if they alone were active, the orientation distribution in polycrystals should also display hexagonal symmetry.

Traces of water have a drastic effect on the strength of quartz and therefore critical shear stresses are not well defined [GRIGGS & BLACIC 1965]. Addition of only 0.1–1% of H_2O causes a decrease in strength by an order of magnitude [ORD & HOBBS 1986]. Water weakening and different models for its explanation have been reviewed in depth by PATERSON [1989], but there is still no agreement about the mechanism by which hydrogen affects dislocation movements. (Is it an intrinsic structural effect in which Si-O-Si bonds are replaced by weaker hydrolyzed silicon-oxygen bridges Si-OH:HO-Si [GRIGGS 1967]; or does molecular water produce mechanical instabilities [MCLAREN &AL. 1983, DEN BROK & SPIERS 1991]).

It has long been known that high stresses induce mechanical Dauphiné twinning [ZINSERLING & SHUBNIKOV 1933]. Dauphiné twinning is geometrically a 2-fold rotation about the c-axis but can be produced by a slight distortion of the lattice, either during the β–α phase transformation or by shear. Dauphiné twinning is a mechanism to produce 'trigonal textures'.

Texture development and deformation in naturally and experimentally deformed quartzites have recently been reviewed [WENK 1994b] and we only summarize the most important aspects.

Polycrystalline quartz has been the subject of numerous experimental studies, mainly in axial compression. Figure 11 shows, in inverse pole figures and c-axis profiles, typical texture types developing during plastic deformation of fine grained quartz ('flint') [GREEN &AL. 1970]. At lower temperatures and high strain rates, c-axes concentrate near the compression axis, for samples deformed in the α- as well as in the β-field (Fig. 11a,b,e,f). At higher temperature and slower strain rates a distinct maximum develops near r={10$\bar{1}$1} (Fig. 11c), and correspondingly a small-circle girdle of c-axes develops (Fig. 11g). It is generally agreed that at low temperature basal c=(0001) slip is the dominant mechanism, whereas at higher temperature prismatic and rhombohedral slip becomes active. The distinct difference between positive rhombs ($h0\bar{h}l$) and negative rhombs ($0h\bar{h}l$), e.g. in Fig. 11c, can be partially attributed to mechanical Dauphiné twinning. TULLIS & TULLIS [1972] documented strong preferred orientation in flint and quartzite samples subjected to stress but with almost no macroscopic strain. Dauphiné twinning does not change the c-axis orientation (which is almost random in the case of Fig. 11h) but reverses positive and negative rhombs (Fig. 11d). Since it switches positive and negative a-axes, Dauphiné twinning can conceivably induce a piezoelectric effect in a quartz aggregate.

TULLIS &AL. [1973] observed similar texture patterns in axially compressed coarser quartzite, documenting a transition from a 'c'- (Fig. 12a,d) to an 'r'-fabric (Fig. 12c,f) with an increasing angle of the c-axis small-circle girdle as temperature becomes higher and the strain rate decreases (Fig. 13).

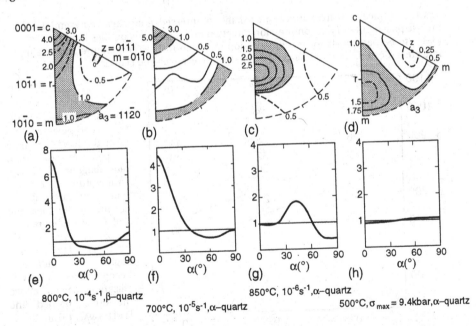

Fig. 11. Typical textures developing during axial compression of flint, a form of fine-grained polycrystalline quartz. (a-d) inverse pole figures of the compression direction. (e-h) corresponding c-axis profiles from parallel (α=0°) to perpendicular to the compression direction. Deformation conditions and contour levels (in m.r.d.) are indicated. (a-c) from GREEN &AL. [1970], (d) from TULLIS & TULLIS [1972]. The example (d) was stressed but has no permanent strain and a random c-axis orientation. The difference between positive and negative rhombs was caused by stress-induced Dauphiné twinning.

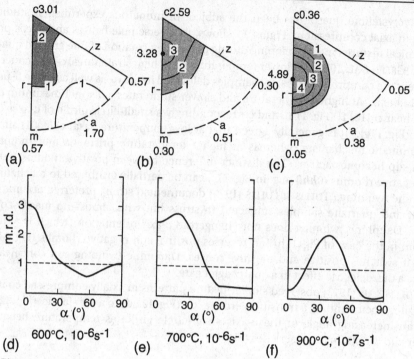

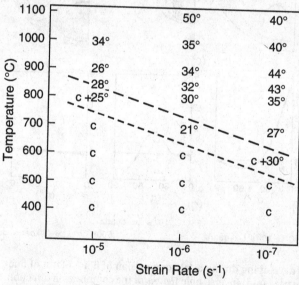

Fig. 12. Typical texture patterns in axially compressed quartzites represented as inverse pole figures of the compression direction (a-c) and c-axis profiles from parallel ($\alpha=0°$) to perpendicular to the compression direction (d-f). Temperature and strain rate are indicated. The sample in (c) is about 80% recrystallized [TULLIS &AL. 1973].

Fig. 13. Texture patterns in experimentally deformed quartzite in a temperature vs. strain rate diagram. At low temperature a **c**-maximum type fabric forms, at higher temperature and lower strain rates a **c**-small-circle girdle develops. The opening angle is indicated. Recognizable recrystallization occurs above the upper dashed line [TULLIS &AL. 1973].

Experimental investigations also addressed the behavior of quartz during dynamic recrystallization. Mainly based on microstructures, HIRTH & TULLIS [1992] identified three different mechanisms with increasing temperature, decreasing strain rate and thus decreasing flow stress: (1) Strain induced grain boundary migration; (2) progressive subgrain rotation and (3) progressive subgrain rotation with rapid boundary migration at the highest temperatures. GLEASON &AL. [1993] document that textures in experimentally recrystallized quartzites are characteristic of these mechanisms. When grain boundary migration dominates, c-axes in recrystallized grains are oriented parallel to the compression direction (Fig. 14a). When subgrain rotation is important the maximum in the inverse pole figure is near (10$\bar{1}$1) and a small-circle girdle of c-axes develops (Fig. 14b). These are similar patterns to those observed in non-recrystallized quartzites deformed under similar conditions.

There are only a few experimental fabric studies in plane strain, and thus more similar to geological conditions. In simple shear experiments DELL' ANGELO & TULLIS [1989] observed a c-axes maximum displaced from the shear plane normal against the sense of shear in samples with small deformation and without appreciable recrystallization (Fig. 15a) and a maximum of c-axes of unrecrystallized porphyroclasts displaced with the sense of shear when a highly recrystallized matrix is present (Fig. 15b). In other quartzites c-axes are more or less aligned perpendicular to the shear plane.

Many naturally deformed quartzite specimens have been examined. Most of these investigations relied on measurements of [0001] axes with a petrographic microscope, equipped with a universal rotation stage. They have been classified by SANDER [1950] into c-axis maximum fabrics, small-circle girdles, great circle girdles and crossed girdles (Fig. 16, in all the representations the schistosity plane is horizontal).

These c-axis pole figures are incomplete texture representations. There is no information on the orientation of a-axes and on differences between positive and nega-

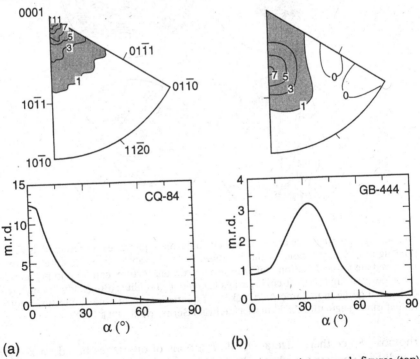

Fig. 14. Textures in dynamically recrystallized quartzites as inverse pole figures (top) and c-axis profiles from parallel ($\alpha=0°$) to perpendicular to the compression direction (bottom). (a) Texture typical of grain boundary migration. Novaculite (a form of fine-grained quartz) shortened 57% at 900°C. (b) Texture attributed to subgrain rotation. Quartzite shortened 75% at 900°C, 60% recrystallized. [GLEASON &AL. 1993].

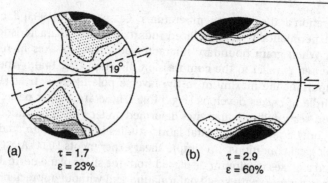

(a) $\tau = 1.7$ $\varepsilon = 23\%$ (b) $\tau = 2.9$ $\varepsilon = 60\%$

Fig. 15. Two c-axis pole figures of experimentally sheared quartzite. At low strain the c-axis maximum at high angles to the shear plane is rotated against the sense of shear (a). At higher strain where recrystallization occurs, c-axes of old non-recrystallized grains are rotated with the sense of shear (b). The implied shear plane and sense of shear are indicated [DELL' ANGELO & TULLIS 1989].

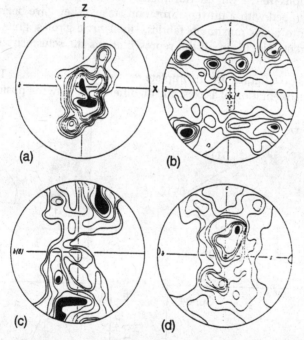

Fig. 16. Typical quartz textures of naturally deformed quartzites represented as 0001 pole figures; Z is the pole to the foliation, X the lineation, and Y (center) is the intermediate fabric direction. (a) Typical Y-maximum texture, common in mylonites (Saualpe, Austria); (b) small-circle girdle observed in granulites (Burgstädt, Saxony); (c, d) crossed girdle in metamorphic rocks. (c) quartz lens in marble (Hintertux, Tirol); (d) deformed pegmatite (Melibokus, Odenwald Germany) [SANDER 1950].

tive rhombs. Since then, several OD descriptions of quartzites based on x-ray and neutron diffraction data have been obtained (see WENK [1994b] for a review). Natural quartz textures, in contrast to calcite textures, display a wide variation and low symmetry. This may be due to changes in mechanisms such as dislocation glide, climb, and recrystallization, but also due to differences in the deformation history. Figure 17

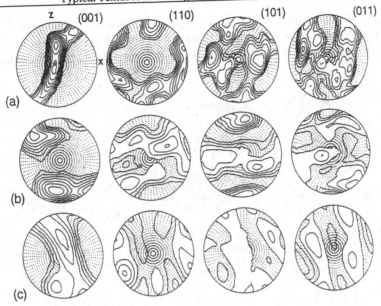

Fig. 17. Typical natural quartz textures represented as pole figures which were recalculated from the ODF. Notice the asymmetry in all patterns (X=lineation and Z=pole to the foliation are indicated). (a) Y maximum in a recrystallized quartzite Brg 420 from the Bergell Alps; (b) asymmetric small-circle girdle in quartzite from a granite contact in the Bergell Alps; (c) crossed girdle in quartzite from Wildrose Canyon, Death Valley, California. Equal-area projection. Logarithmic contour intervals, 0.5, 0.71, 1 m.r.d., etc., dot pattern below 1 m.r.d. The 101 and 011 peaks were deconvoluted based on different structure factor contributions to the overlapped diffraction peak [WENK 1994b].

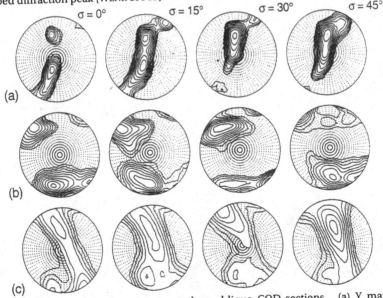

Fig. 18. Typical quartz textures represented as oblique COD-sections. (a) Y maximum in recrystallized quartzite Brg 420 from the Bergell Alps (b) asymmetric small-circle girdle in quartzite from a granite contact in the Bergell Alps, (c) crossed girdle in quartzite from metamorphic conglomerates in Wildrose Canyon, Death Valley, California. An average over all σ-sections of the COD is a 0001 pole figure. Equal-area projection. Logarithmic contour intervals, 0.5, 0.71, 1 m.r.d., etc., dot pattern below 1 m.r.d.

illustrates pole figures measured by neutron diffraction for some typical quartz textures, **c**-maximum (a), small-circle girdle (b), and crossed girdle (c); Fig. 18 displays the corresponding CODs in oblique σ-sections. A σ–rotation of 30° corresponds to a

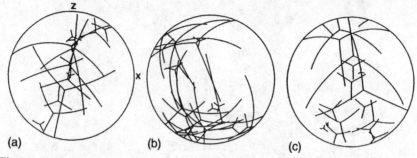

Fig. 19. Tripod representation of texture components in the three quartzites of Figs. 15 and 16. Center indicates the **c**-axes and 'legs' point to the **a**-axes. The length of the legs is proportional to the 'orientation volume' of the component. Note the components related by the Dauphiné twin law (180° rotation about **c**) [HELMING &AL. 1994].

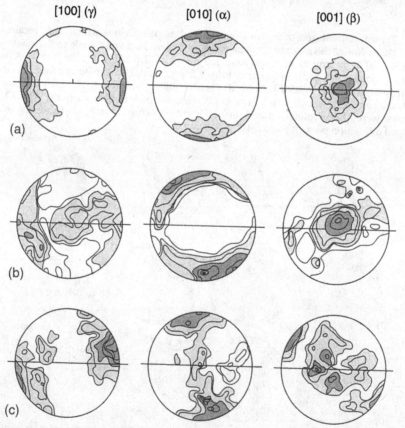

Fig. 20. Naturally deformed olivine. [100] (optical indicatrix axis nγ), [010] (nα), and [001] (nβ) pole figures. (a) Typical olivine texture in equigranular lherzolite from the Balmuccia massif, Western Alps. Contours are 1, 3, 5, 7 m.r.d. [SKROTZKI &AL. 1990]. (b) Olivine texture of matrix in a porphyroclastic chlorite peridotite from Alpe Arami, Switzerland. Contours are 1, 3, 5, 7, 11 m.r.d. [BUISKOOL 1977]. (c) Peridotite from the Bay of Islands ophiolite complex, Newfoundland [MERCIER 1985].

crystal rotation of 60° (or 180°) about the **c**-axis and thus to a Dauphiné twin or a hexagonal character. The quartzite sample in Fig. 18b is more closely 'hexagonal' than the other two since sections σ and σ+30° are similar. This quartzite was probably deformed in the stability field of hexagonal β-quartz near a granite contact.

An elegant method of representation uses 'tripods' [SCHMID &AL. 1981, SCHMID & CASEY 1986]. We apply it in a modified version to the quartz textures (Fig. 19). The complex quartz texture is divided into components which collect concentrations of orientation density [HELMING &AL. 1994]. Each component is then expressed by a tripod where the intersection illustrates the **c**-axis orientation and the legs point towards **a**-axes. The size of the tripod is proportional to the volume of the component in orientation space. The diagram for the single maximum fabric displays components which have similar **c**-axis orientations but opposite **a**-axis orientations. Such components are related by Dauphiné twinning (see also BUNGE & WENK [1977]). (Caution: a representation by components is compact and may be easy to visualize but it is purely descriptive and should not be extended to imply deformation mechanisms such as that **a**-axes are slip directions and are aligned in the macroscopic shear direction! [WENK & CHRISTIE 1991])

The actual occurrence of Dauphiné twins cannot be verified with orientation distribution data because x-ray as well as neutron data collection averages over the whole sample and local correlation information is lost. With EBSP orientation imaging analysis, obtained in a SEM, one can establish that in such naturally deformed quartzites twin boundaries do actually exist (Chap. 4 Fig. 40) and in a misorientation map in rotation-axis/rotation-angle space (to bring neighboring grains to coincidence) a strong concentration for a 60° rotation around the **c**-axis establishes the Dauphiné twin law (See Chap. 4 Fig. 42, KUNZE &AL. [1994]).

The above discussion centered on quartz deformation textures. Quantitative texture analysis has also been applied to growth textures in fibrous quartz, such as chalcedony [CADY &AL. 1997].

1.4 Olivine ($MgSiO_4$) and perovskite ($CaTiO_3$)

The orthorhombic mineral olivine is the major constituent of the upper mantle of the Earth and has therefore been of long standing importance for understanding geodynamic processes. There is a wealth of information on naturally deformed olivine-bearing rocks which was most recently reviewed by MERCIER [1985]. In a lherzolite from the Ivrea zone SKROTZKI &AL. [1990] described symmetrical pole figures with olivine [010] axes perpendicular to the foliation (Fig. 20a). Peridotite fabrics from Alpine mylonites (Alpe Arami, BUISKOOL TOXOPAEUS [1977]) differ in that olivine [010] axes are spread out along a great circle (Fig. 20b). Olivine from peridotites of the Bay of Islands ophiolite complex have [010] axes (Fig. 20c) at high angles to the macroscopic foliation and a slight asymmetry which was attributed to a component of simple shear. Mostly preferred orientation in these rocks is due to intracrystalline slip and some recrystallization [MERCIER 1985].

In experimentally compressed olivine aggregates [010] axes align with the compression direction [NICOLAS &AL. 1973, Fig. 21a]. This texture is similar to one produced during migration recrystallization [AVÉ LALLEMANT & CARTER 1970, Fig. 21b] although recent investigations by KARATO [1987, 1988] point out some differences. ZHANG & KARATO [1995] produced simple shear textures and observed at moderate strain a [100] maximum displaced against the sense of shear and, at higher strain with pervasive recrystallization, a maximum closer to the shear direction. This is somewhat similar to the observations of DELL' ANGELO & TULLIS [1989] on quartz (Section 1.3).

Simulations of olivine textures have long been difficult because of the few available slip systems such as (010)[001] and (001)[100] [e.g. BAI &AL. 1991]. TAKESHITA &AL. [1990] used the relaxed Taylor theory, and WENK &AL. [1991b] applied a viscoplastic self-consistent theory and obtained results which compare well with natural and experimental olivine textures (e.g. Fig. 21c and Chaps. 11 and 14).

Whereas olivine, composing the Earth's upper mantle, has been studied in some detail, texture development of silicate perovskite $MgSiO_3$, the main constituent of the lower mantle, is not very well known. Experiments by KARATO &AL. [1995] suggest strong texture development in the regime of dislocation creep (large grain size) and random texture in the regime of superplasticity (small grain size) and the authors imply that the apparent lack of seismic anisotropy in the lower mantle may be due to

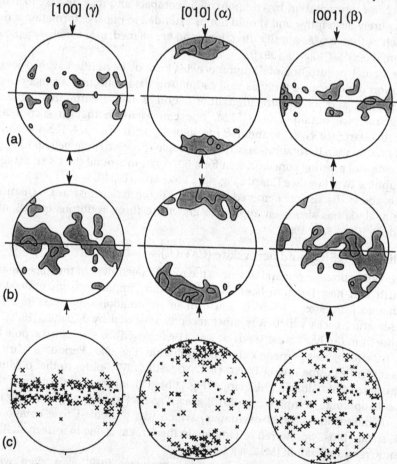

Fig. 21. Olivine deformed in axial compression. [100] (optical indicatrix axis nγ), [010] (nα), and [001] (nβ) pole figures, (a,b) experimental deformation: (a) deformation by slip at 1300°C, 1.4 GPa and 10^{-4} s⁻¹. Contours are 4, 8 m.r.d. [NICOLAS &AL. 1973]. (b) Texture developing by migration recrystallization at 1100°C, 10^{-6} s⁻¹ and 1.5 GPa. Contours are at 2, 3, 4, 5 m.r.d. [AVÉ LALLEMANT & CARTER 1970]. Arrows indicate compression direction. (c) Simulations of the olivine compression texture using a visco-plastic self-consistent theory. Symbol size is proportional to the overall deformation of individual grains [WENK &AL. 1991b].

deformation by superplasticity, in contrast to dislocation-controlled deformation in the upper mantle.

1.5 Sheet silicates

Sheet silicates such as micas (muscovite, biotite, chlorite etc.) and clay minerals (illite, smectite, kaolinite etc.) are the main component in argillaceous and pelitic rocks and occur commonly in many deformed sedimentary (mudstone, shale) and metamorphic rocks (slate, schists). Sheet silicates are characterized by a platy morphology, parallel to the basal (001) lattice plane. This morphological anisotropy largely controls the orientation changes of sheet silicates during compaction and subsequent deformation. Most pole figure measurements are limited to the (001) basal plane (often recorded by x-rays in transmission geometry) and such pole figures not only represent lattice preferred orientation but also shape preferred orientation.

The orientation behavior of non-equiaxed, rigid particles in a viscous medium during deformation has been modeled by JEFFERY [1923]. In the extension of MARCH [1932] in which particles are considered infinitely oblate (or infinite needles) the texture, as displayed on (001) pole figures, has been directly correlated with the finite strain. The theory will be briefly reviewed in Section 2.1 of Chapter 14.

In spite of the severe constraints of the model which rarely apply to real systems, the predicted strain/texture relationship is in good agreement with independent originally spherical or cylindrical strain markers, measured in naturally deformed

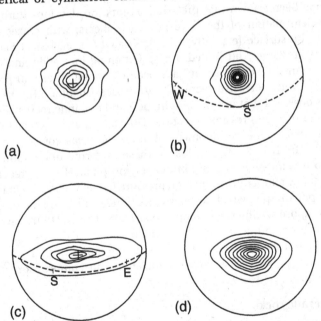

Fig. 22. Examples of (001) sheet silicate pole figures. (a-c) are illite/muscovite pole figures from shales and slates in the Variscan fold-and-thrust belt in Belgium [SINTUBIN 1994]. Dashed great circle is horizontal plane with geographic coordinates E, W, S indicated. Upper hemisphere equal-area projection. (a) axisymmetric compaction texture (maximum is 6.4 m.r.d.), (b) axisymmetric deformation textures (maximum is 20 m.r.d.), (c) orthorhombic deformation texture (maximum is 10 m.r.d.). (d) is an asymmetric biotite (00l) pole figure from mylonites in Southern California which were partially deformed in simple shear [O'BRIEN &AL. 1987] (maximum is 14.5 m.r.d.).

rocks, such as volcanic lapilli [OERTEL 1970], reduction spots around inclusions [TULLIS & WOOD 1975], or fragments of fossils [OERTEL &AL. 1989].

TULLIS [1976] tested the March strain/texture relationship experimentally. At low temperature the measured texture increasingly deviates from the predicted one with increasing strain due to particle interaction (see also Section 2.7). At high temperature, where recrystallization occurs and the March theory should not apply, there is nevertheless good agreement between predicted and observed pole densities.

The interpretation of (001) sheet silicate pole figures is based on their symmetry, their intensity and their geometrical relationship with regard to the mesoscopic structural elements such as bedding, cleavage and lineation [WEBER 1981, SINTUBIN &AL. 1995]. In pure compaction, pole figures are axially symmetric with a maximum parallel to the bedding pole (Fig. 22a). Deformation textures can be related to the development of a slaty cleavage. In this case axially symmetric pole figures, centered around the cleavage pole, indicate flattening (Fig. 22b), whereas orthorhombic textures imply a plane-strain component (Fig. 22c). In some heavily deformed metamorphic rocks (mylonites), 001 pole figures are often asymmetric, contrary to March predictions (Fig. 22d). This has been attributed to a simple shear component in the deformation history [O'BRIEN &AL. 1987].

1.6 Ice (H_2O)

Deformation of hexagonal ice–I has been the subject of numerous investigations and it has long been known that preferred orientation develops during the flow of glaciers and deformation of the large polar ice sheets, with c-axes oriented perpendicular to the surface [e.g. DUVAL 1979, KAMB 1959, PERUTZ & SELIGMON 1939]. These textures are largely attributed during flattening in a glide and climb regime. Locally components of simple shear have been documented and in polar ice sheets there are layers with pervasive recrystallization (see Chapter 14). In experimental studies it was observed that during deformation and dynamic recrystallization c-axes align parallel to the compression direction [DUVAL &AL. 1983, KAMB 1972, WILSON & RUSSELL-HEAD 1982]. The development of anisotropy was considered in flow laws [GOODMAN &AL. 1981, LILE 1978]. Most of these preferred orientation studies were performed close to the melting point in the polymorph ice–I. Only recently could ice be deformed at low temperature and high pressure [DURHAM &AL. 1988] in the stability field of other polymorphs which are particularly relevant for planetology since they supposedly compose satellites of outer planets. BENNETT &AL. [1997] have analyzed ice textures by neutron diffraction at 77 K and observed that ice–I deformed at low temperatures also has c-axes parallel to the compression direction, whereas the high pressure polymorph ice–II has a-axes parallel to the compression direction (see Chap. 4 Fig. 26b).

1.7 Polymineralic rocks

Most texture research in geology was done on monomineralic rocks such as quartzites and marbles. By contrast most of the naturally occurring deformed rocks are polymineralic and this adds great complexities to characterization and interpretation. For example granite, consisting of feldspars, quartz and mica, shows initially an almost random orientation distribution [WENK & PANNETIER 1989, Fig. 23a]. With increasing deformation at metamorphic conditions mica and quartz attain a strong texture in mylonite (Fig. 23b). During progressive deformation to phyllonite, which

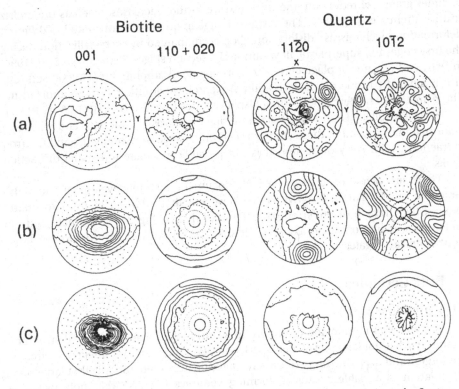

Fig. 23. Pole figures of biotite (a mica mineral) and quartz in granite from the Santa Rosa mylonite zone in Southern California (a) which was progressively deformed to mylonite (b) and phyllonite (c). Determination by neutron diffraction. Equal-area projection.   Contours for biotite are 0.5,1,1.5,2,2.5,3,4,...14 m.r.d., for quartz 0.5,0.75,1,1.25...m.r.d., dot pattern below 1 m.r.d. [WENK & PANNETIER 1989].

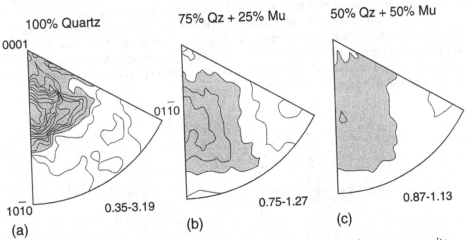

Fig. 24. Inverse pole figures for quartz in experimentally deformed quartz-muscovite (a platy mica mineral) aggregates measured by TOF neutron diffraction. Deformation conditions are: 800°C, 1200 MPa and 10^{-6} s^{-1}. Equal-area projection, contour interval is 0.2 m.r.d., shaded above 1 m.r.d. Minimum and maximum pole densities are indicated below each diagram [TULLIS & WENK 1994].

includes grain size reduction, the mica texture further increases, whereas the quartz texture attenuates (Fig. 23c). The texture change in quartz was attributed to different deformation mechanisms: dislocation glide accompanied by recrystallization during the first stage, and superplastic flow during the second stage. SIEGESMUND &AL. [1994] investigated textures of plagioclase and hornblende in amphibolite, DORNBUSCH &AL. [1994] studied triclinic plagioclase and pyroxene in mylonitic pyroxene anorthosite. These are two of the few reports on feldspar textures [see also ULLEMEYER &AL. 1994, JI & MAINPRICE 1988]. LEISS &AL. [1994] described textures of dolomite and calcite in deformed marbles. All of these studies relied on U-stage measurements and neutron diffraction. With x-ray diffraction it is not possible to isolate individual diffraction peaks.

There are a few experiments which document texture development in poly-mineralic rocks. TULLIS & WENK [1994] observed that addition of mica to quartz reduces the strength of preferred orientation of quartz (Fig. 24), presumably due to preferential sliding on sheet silicates. JORDAN [1987] studied deformation of lime-stone-halite aggregates.

2. Bulk Ceramics

2.1 α-Alumina (Al$_2$O$_3$)

Trigonal aluminum oxide (point group $\bar{3}2/m$) is the most widely used structural ceramic, such as for lamp envelopes, spark plugs and substrates for integrated circuits. Hot forging and deep drawing experiments have documented not only that large scale deformation is a viable means of forming ceramics but also that polycrystalline alumina developed a pronounced crystallographic texture [HEUER &AL. 1969]. In hot forging various mechanisms are active, including intracrystalline slip [HEUER 1970, HEUER &AL. 1980] and superplasticity [CHEN & XUE 1990]. MA & BOWMAN [1991] observed a prevalent basal texture with c-axes aligned parallel to the compression direction (Fig. 25). They conclude that basal slip may be responsible for the texture development in alumina that does not display a grain shape anisotropy. In alumina powders with plate-like shape, textures can be produced through green state process-ing. Various techniques have been used such as die pressing of dry powders, slip casting , tape casting [BÖCKER &AL. 1991], pressure filtration [SANDLIN & BOWMAN 1992], centrifugal casting, or extrusion [SALEM &AL. 1989] all followed by sintering to produce cohesion by grain growth which often enhances the texture.

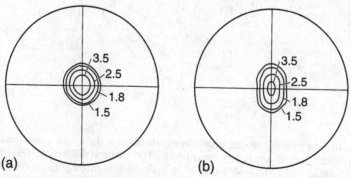

Fig. 25. (0001) pole figures of alumina, recalculated from the ODF for (a) hot-forging and (b) plane-strain compression. Contours are in m.r.d. [MA & BOWMAN 1991].

Since thermal expansion is anisotropic, small residual stresses are introduced during thermal cycling in highly textured alumina which can lead to microfractures and eventual brittle failure [LEE &AL. 1993].

2.2 Zirconia (ZrO₂) transformation textures

Zirconia is also used for structural applications. Zirconia alloys occur in a cubic (when doped with Y), a tetragonal (high-temperature) and a monoclinic (lower-temperature) structure (baddeleyite). The tetragonal to monoclinic phase transformation can be stress-induced (martensitic) and is greatly influenced by crystallite orientation [MUDDLE & HANNINK 1986]. REYES-MOREL & CHEN [1988] demonstrated texture in deformed zirconia with powder diffraction. While the texture of the monoclinic phase is most relevant to generating strains and plastic work, the texture of the tetragonal phase can be advantageously tailored to enhance transformability and toughness [BOWMAN & CHEN 1993, LI &AL. 1988, REIDINGER & WHALEN 1987]. In martensitic transformations there is a rigorous crystallographic relationship between the two phases. The parent phase transforms into the monoclinic phase by simple shear on the plane (100) in the [001] direction. In contrast to other deformation textures where the texture is inherited from a prior texture of the parent phase, in zirconia it is produced even in a polycrystal without initial preferred orientation directly by the selective stress-induced transformation (Fig. 26a). This is illustrated by differences in intensities in a powder pattern for a sample of random texture where the phase transformation was thermally induced with large (111)mS peaks (Fig. 26b) and a sample in which the transformation was mechanically induced with reduced (111)mS peaks (Fig. 26c) [BOWMAN &AL. 1988]. Fig. 26d illustrates apparent volume fractions of the tetragonal and monoclinic phase as derived from (111) diffraction intensities. The

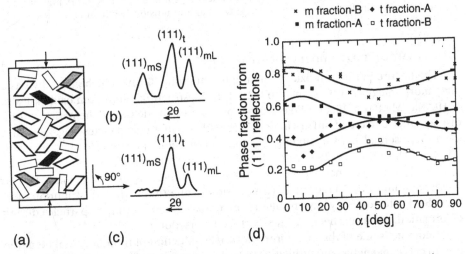

Fig. 26. Deformation of zirconia in compression. (a) Schematic view of the sample. (b) Diffractometer scan of a powder with no preferred orientation. The tetragonal (111)T peak splits into two monoclinic peaks (111)MS and (11-1)ML. (c) Diffractometer scan of a stressed sample with a change in relative peak intensities due to texture. (d) Volume fractions of tetragonal and monoclinic phases, derived from (111) diffraction intensities for a moderately (A) and a highly strained sample (B). The variation of specimen orientation relative to the stress axis is due to texture [BOWMAN &AL. 1988].

variation with the angle α from the stress axis is due to texture. The inverse pole figures in Fig. 27 demonstrate a model for the transformation sequence in tetragonal zirconia. The first orientations which are expected to transform in tension or compression are those with (101) parallel to compression and tension axis (corresponding to $\alpha=45°$ in Fig. 26d). These orientations should produce the largest strains along tensile or compressive stress axes. Since the toughness enhancement is directly related to the texture, it would be advantageous to produce a strong (100) texture to optimize the texture in tensile deformation [BOWMAN 1991]. The zirconia system would lend itself well for a quantitative texture study which has not been done yet.

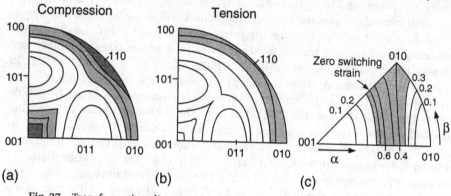

(a)   (b)   (c)

Fig. 27. Transformation diagrams for zirconia for (a) compression and (b) tension. Contoured is the normalized stress assistance W/Wmax. The dark shaded area in (a) represents orientations which receive no positive stress assistance and should not transform. Contours are 0, 0.2, 0.4, 0.6, 0.8, 0.9, light shading below 0.4 [BOWMAN & CHEN 1993]. (c) Stereographic projection indicating volume fractions of orientations which may transform in compression (gray area) and tension (white area) [BOWMAN 1991].

2.3 Hematite (Fe_2O_3) and magnetite (Fe_3O_4)

The texture in the iron oxide hematite (point group $\bar{3}2/m$), which generally has a platy morphology, is very similar to that of the isostructural aluminum oxide [SIEMES & HENNIG-MICHAELI 1985]. Pole figures of a naturally deformed hematite texture display great circle and small-circle girdles with varying occupancies (Fig. 28). In a COD, represented here in oblique sections, all sections are more or less identical (Fig. 29). This illustrates that **a**-axes have a rotational degree of freedom about the **c**-axis which one would not expect from looking at the pole figures, particularly (11$\bar{2}$0). The maximum in the (11$\bar{2}$0) pole figure is simply due to the girdle distribution of **c**-axes. Such a texture may have resulted from passive rotations of rigid particles during deformation, just as in sheet silicates discussed in Section 1.5.

Hematite is one of the major iron ores. The reduction of rhombohedral hematite ores to cubic magnetite and ultimately to Fe is commercially important. The chemical reactions are influenced by structural relationships and therefore by texture. Several crystallographic relationships have been established between hematite (H) and magnetite (M). At high temperature (800°C) $(111)_M \parallel (0001)_H$ and $[110]_M \parallel [10\bar{1}0]_H$, at lower temperature (650°C) a second orientation relationship with $(112)M \parallel (0001)_H$ and $[110]_M \parallel [10\bar{1}0]_H$ is also observed [e.g. BECKER &AL. 1977, WITHERS & BURSILL 1980, MODARESSI &AL. 1991]. The topotactic reduction reaction produces a system of

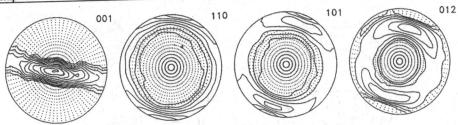

Fig. 28. Naturally deformed hematite. Pole figures measured by neutron diffraction are represented in equal-area projection. Logarithmic contour intervals, 0.5, 0.71, 1 m.r.d., etc., dotted below 1 m.r.d. (Data courtesy of W. Schaefer, Jülich).

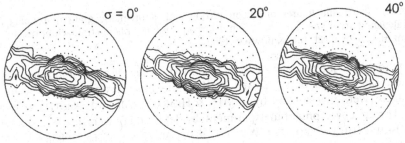

Fig. 29. Naturally deformed hematite. Oblique σ-sections of the COD calculated from the pole figures in Fig. 26 with the WIMV method. The lack of pattern change with σ indicates that **a**-axes spin freely about the **c**-axis. Logarithmic contour intervals, 0.5, 0.71, 1 m.r.d., etc., dotted below 1 m.r.d.

crystallographically aligned pores which let the reducing gas reach the interface [HEIZMANN &AL. 1988a].

2.4 Silicon nitride (Si_3N_4)

Silicon nitride is a ceramic material which has found numerous new applications in automotive components and machining tools due to its high stiffness, strength and hardness, coupled with good fracture toughness and thermal shock and corrosion resistance. It occurs in two hexagonal forms, α and β. The high-temperature β-phase forms rod-shaped needles, elongated along the **c**-axis, which can be oriented to some degree by hot-pressing or forging [e.g. LANGE 1973, LEE & BOWMAN 1992, WALKER &AL. 1995]. The textured materials have very anisotropic fracture properties [WESTON 1980, WILLKENS &AL. 1988]. In accordance with the grain shape, the preferred orientation is 'inverse' to that of alumina, i.e. (0001) poles are at high angles to the direction of principal compression, in axial and plane-strain compression (Fig. 30). The fracture toughness generally increases and becomes more anisotropic with sharpening texture [WALKER &AL. 1995].

2.5 Alkali halides and related structures

The deformation behavior of halite ('rock salt', NaCl) and related minerals has been extensively investigated, mainly because salt mines are being considered as nuclear waste repositories and their long term stability needed to be evaluated. Most of these studies emphasize mechanical properties but texture plays obviously an important role. The review of KERN & RICHTER [1985] still gives a fairly up-to-date

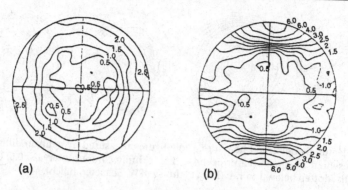

Fig. 30. 0001 Pole figures of silicon nitride with a prismatic grain morphology, recalculated from the ODF for (a) hot-forging and (b) plane-strain compression. Contours are in m.r.d. [LEE & BOWMAN 1992].

overview to which interested readers are referred. A list of some important publications on salt deformation follows: ductility of single crystals and deformation mechanisms were investigated by CARTER & HEARD [1970] and SKROTZKI & HAASEN [1981]. Texture development in experimentally deformed polycrystals was studied by KERN & BRAUN [1973], SKROTZKI & WELCH [1983], CARTER & HANSEN [1983], FRANSSEN & SPIERS [1990] and SKROTZKI &AL. [1995]. In compression experiments {110} fiber textures are generally produced and in extrusion experiments a {100} fiber dominates. Salt textures in natural settings described by GOEMAN & SCHUMANN [1977], KÄMPF &AL. [1986] and SCHEFFZÜK [1996] document a {100} maximum perpendicular to the foliation plane. In Chapter 14 halite will be used to explain some complications of polycrystal plasticity theory for highly anisotropic systems.

Cubic silver chloride is isostructural with halite and, like all cubic crystals, optically isotropic. However, when stress is applied, silver chloride displays birefringence. DIETZ & GIELESSEN [1995] relate textures in rolled polycrystalline AgCl and anisotropic photo-elastic effects due to internal stress.

2.6 Ceramic matrix composites

During many investigations of composite ceramics, it became apparent that reinforcement textures are extremely important for mechanical properties. Yet a quantitative characterization of textures is generally lacking. Experimental work on alumina-SiC whisker composites has shown that little texture development is observed in the alumina matrix under conditions that have been shown to produce strong preferred orientation in pure alumina [SANDLIN & BOWMAN 1992], rather similar to the case of quartz-mica mixtures described above. By contrast, whiskers attain strong preferred orientation due to their elongated grain shape. Surprisingly the orientation distribution is more complex than predicted by the MARCH [1932] model (Fig. 31a, SANDLIN &AL. 1992). In alumina, reinforced with Si_3N_4 whiskers, the preferred orientation of Si_3N_4 increases with compression as predicted by the JEFFERY [1923] – MARCH [1932] relationship (Fig. 31b). In platelet and whisker reinforced ceramics the preferred orientation of the second phase is often assessed by image processing based on stereology [SANDLIN &AL. 1994]. Particularly zirconia ceramics have been reinforced with alumina platelets to increase fracture toughness by crack deflection and crack arresting at platelets [ROEDER &AL. 1995, LI & SORENSEN 1995].

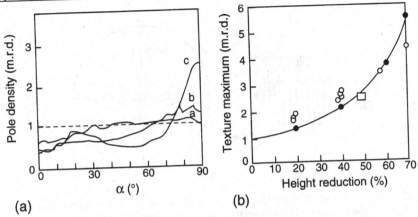

(a)   (b)

Fig. 31. (a) Preferred orientation of SiC whiskers in alumina. Shown are (111) pole density profiles (2.51Å) from parallel ($\alpha=0°$) to perpendicular ($\alpha=90°$) to the compression axis. a: green presses, b: hot pressed, c: hot forged. Units are m.r.d. [SANDLIN &AL. 1992]. (b) Texture maxima for compression of ceramics containing rod-like grains as a function of height reduction. Open circles are Si_3N_4, box is SiC whisker reinforced Al_2O_3. Solid circles are predictions based on March theory [MA &AL. 1994].

These mechanisms depend on textures. In textured composites mechanical properties such as fracture toughness can be very anisotropic as illustrated in Table I for alumina, reinforced with SiC whiskers.

Table I. Fracture toughness of alumina, reinforced with SiC whiskers (w) and platelets (pl). Values are in MPa$\sqrt{}$m [LEE &AL. 1993].

Material	Maximum	Minimum
Al_2O_3 (Coors Co. Golden Co.)	3.9	2.9
Al_2O_3 with 30 vol% SiC w, hot-pressed, 1850°C	6.2	5.2
Al_2O_3 with 30 vol% SiC w, hot-forged 40%, 1850°C	6.8	4.1
Al_2O_3 with 30 vol% SiC w, hot-forged 55%, 1850°C	6.8	2.4
Al_2O_3-SiC, w, Commercial tool	6.5	5.6
Al_2O_3-SiC, w, Commercial tool	7.8	4.6
Al_2O_3 with 30 vol% SiC pl	7.0	3.3

Ceramic composites consisting of a SiC matrix, reinforced by SiC fibers have recently been developed for thermostructural applications (Fig. 32a). Their advantage lies in the low density, the high mechanical strength and rigidity and their chemical inertness. DIOT & ARNAULT [1991] documented that the SiC matrix has ⟨111⟩ directions preferentially aligned parallel to the fiber axis of the composite (Fig. 32b). Such a texture has the lowest surface energy between matrix and fibers and is preferred.

2.7 Bulk high-temperature superconductors

The conductivities of the high-temperature cuprate superconductors are highly anisotropic and largely confined to the (001) Cu-O plane [e.g. TUOMINEN &AL. 1990]. Techniques for developing strong preferred orientations in macroscopic samples have been highly successful in improving critical current densities in many of these

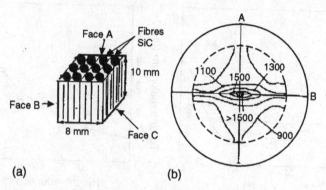

Fig. 32. Composite specimen with SiC fibers in a SiC matrix (a). (b) Unnormalized 111 pole figure of the matrix material illustrating an alignment of 111 in the fiber direction [DIOT & ARNAULT 1991].

materials [e.g. KUMAKURA 1991], yet also in this field quantitative texture descriptions are scarce.

Among the cuprate oxides, the production and description of texture is most highly developed for the Y-123 compounds ($YBa_2Cu_3O_{7-x}$). Significant degrees of texture can be produced by high-temperature plastic deformation of polycrystalline pellets [WENK &AL. 1989c] and by extrusion and hot-pressing of powder compacts [CHEN &AL. 1987, 1993]. [001] axes of Y-123 align themselves with the compression direction, due to intracrystalline slip (Fig. 33a). Other techniques to produce texturing in Y-123 are melt textured growth [SELVAMANICKAM & SALAMA 1990, PELLERIN &AL. 1994] and alignment in a magnetic field [CHOI &AL. 1989, KALLEND &AL. 1991b, DERANGO &AL. 1991] even though the desired result to have [001] of Y-123 aligned parallel to the magnetic field is not always achieved [CHATEIGNER &AL. 1997a, Fig. 33b]. Yet these experiments to produce texture in high-temperature superconductors so far do not lend themselves to large industrial applications, partially due to problems with weak links.

Other compounds of interest have been Bi-2223 ($Bi_2Sr_2Ca_2Cu_3O_x$) and Bi-2212 ($Bi_2Sr_2CaCu_2O_x$). Also in this system melt-textured growth in a magnetic field has produced preferred orientation [NOUDEM &AL. 1995]. The platy crystal morphology makes these materials suitable for green-body processing. ASARO &AL.[1992] demonstrate a large anisotropy on the punch faces of cylindrical and channel die compacts, STEINLAGE &AL. [1994] use centrifugal slip casting to produce texturing and WENK & PHILLIPS [1992] observe strong textures in cold pressed powders of Bi-2223. Plate-like crystals are preferentially oriented with their **c**-axes (normal to the plane of the plate) parallel to the compression direction (Fig. 34a). The microstructure of these Bi-2223 compounds is very reminiscent of sheet-silicates in geological materials such as shales, and preferred orientation develops largely by rigid body rotations. At low compaction strains texture development shows a behavior as predicted by the MARCH [1932] theory, modified for compaction; at higher strains texture evolution becomes less efficient because of grain interaction and the platelet pole density distribution flattens out at maxima of 6–8 m.r.d. (Fig. 35a from SINTUBIN &AL. 1995). This is very similar to experiments with phyllosilicates [TULLIS 1976] where maxima are slightly higher (Fig. 35b) because in Bi-2223 inclusions of second-phase particles further impede texture development. Processing of fairly pure powders at higher

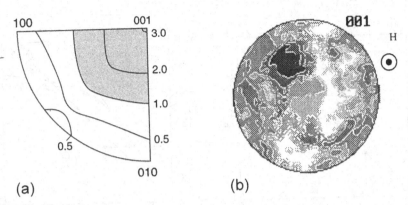

Fig. 33. Texture development in Y-123 high-temperature superconductors. (a) Inverse pole figure for hot compression, assuming tetragonal crystal symmetry; equal-area projection, shaded above 1 m.r.d.[WENK &AL. 1989c], (b) (001) pole figure illustrating texture produced during growth in a magnetic field. Contrary to expectations the c-axis is not aligned with the axis of the magnetic field (center of pole figure). The pole figure was recalculated from the OD, determined by very incomplete neutron diffraction data [CHATEIGNER &AL. 1997a].

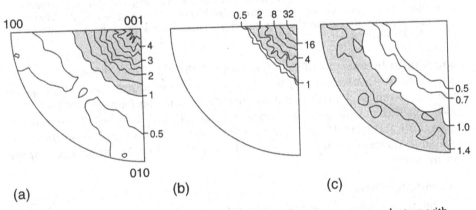

Fig. 34. Textures in bulk Bi-2223 and 2212 high-temperature superconductors with platy morphology, represented in inverse pole figures, assuming tetragonal crystal symmetry. (a) Compaction of Bi-2223 powders [WENK & PHILLIPS 1992], (b) Bi-2212 multifilamentary tape encased in silver, (c) Bi-2212 wire encased in silver [WENK &AL. 1996b]. Equal-area projection; contour spacing linear in (a), logarithmic in (b).

temperature, where diffusion and recrystallization alleviate mechanical grain interaction, may develop stronger textures with more favorable electrical properties.

Unfortunately techniques using axial stress produce a relatively low degree of crystallite orientation. More successful has been a method to sheath Bi-super-conductors in silver tubes and fabricate a tape or wire using wire manufacturing techniques, then thermally treat it [SANDHAGE &AL. 1991, HALDAR &AL. 1992]. Silver is used as the sheath material due to its chemical compatibility with Bi-2212 and 2223. It also supplies ductility and protection for the tape or wire for use in potential applic-ations such as long power transmission lines and superconducting coils. Textures, particularly in monocore and multifilamentary tapes (Fig. 34b) are very strong with

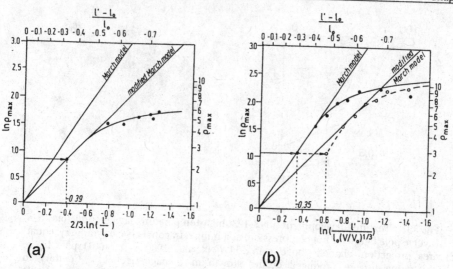

Fig. 35. Maximal (001) pole densities (in m.r.d.) versus strain for (a) cold pressed Bi-2223 aggregates [SINTUBIN &AL. 1995] and (b) fluorophlogopite (a platy mica mineral) aggregates [TULLIS 1976]. The strain scale is adjusted so that the first data point falls on the line predicted by the March model. In the case of fluorophlogopite adjustment is made with regard to both the original March model (full circles) and the March model modified for compaction (open circles). Straight lines give behavior predicted by the March theory.

maxima up to 46 times that of a random sample [WENK &AL. 1996c] and high Jc's have been obtained [GRASSO &AL. 1993]. Multifilamentary conductors are advantageous since the current is shared by multiple parallel conducting paths [MOTOWIDLO &AL. 1994]. Textures in wires are considerably weaker (Fig. 34c) and this may account for the low critical current densities. In monocore and multifilamentary tapes the **c**-axes are oriented perpendicular to the rolling plane, in wires they are perpendicular to the wire direction.

2.8 Cement minerals

Concrete consists of particles of gravel and sand ('aggregate') which are connected by a variety of hydrated minerals occurring in the cement paste. Of particular importance for the strength of concrete is a thin layer of calcium hydroxide directly adjacent to aggregate particles. DETWILER &AL. [1988] have documented that this layer shows a strong growth texture with **c**-axes aligned perpendicular to the aggregate surface (Fig. 36a). The strength of the texture increases with aging (Fig. 36b) which appears to be preferable. Also, TEM studies of corrosion products around steel reinforcements in concrete suggest oriented growth of iron hydroxide [GLASSER & SAGOE-CRENTSIL 1989]. The orientation of these microcrystalline phases is not only related to the strength of the concrete but also the permeability by solutions that influence the durability.

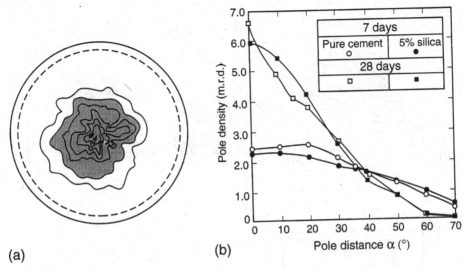

Fig. 36. Texture of calcium hydroxide on cement paste- aggregate interface. (a) (0001) pole figure for a 28 day old specimen. Contour interval is 1 m.r.d., shaded above 2 m.r.d., equal-area projection. The projection plane is the cement paste-substrate interface, (b) diagram of 0001 pole density versus angle to interface surface, to illustrate the effect of adding 5% silica fume to the paste (none) and of aging (increase in texture strength) [DETWILER &AL. 1988].

3. Thin Films and Coatings

In thin films the importance of texture is obvious (see e.g. SZPUNAR [1996]). Texture influences elastic properties and thermal expansion which are essential for the mechanical stability of films. Also, in the case of superconducting films, electrical properties are intimately linked to crystal orientation. Qualitative texture information on thin films has been extracted from powder diffractometry. Only recently have pole figure measurements and orientation distribution analysis been used to quantify crystallite orientation. There are two types of film textures. The first type consists of films and coatings on polycrystalline or amorphous substrates in which the influence of the substrate is minor and the texture is formed mainly by anisotropic growth. These are generally axially symmetric fiber textures. Of a second type are epitaxial films deposited on a structurally related single crystal. In those the texture is usually extremely strong, approaching a single crystal and is controlled by the match between the two crystal structures.

3.1 Silicon and diamond

The broad range of applications of polycrystalline silicon films in microelectronics [e.g. KAMINS 1988] and in the fabrication of micromechanical structures for use in actuators and sensors [e.g. HOWE 1988], demands that the material be well characterized. From the mechanical point of view, the elastic properties and residual (intrinsic) stress of a film govern the behavior of a thin film structure. In these materials, texture is correlated with surface roughness [HARBEKE &AL. 1983], rate of impurity diffusion [KAMINS &AL. 1972], and the residual stress [FAN & MULLER 1988], and therefore is of great importance.

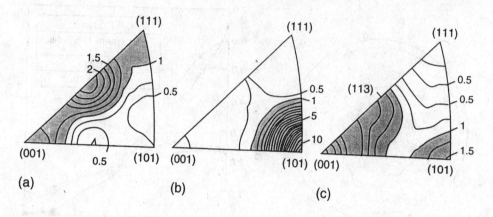

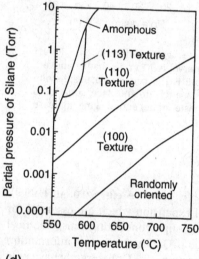

Fig. 37. Si films deposited on a single crystal Si wafer covered with amorphous silicon dioxide. (a-c) Typical textures represented in inverse pole figures. Stereographic projection, contour interval for (a,c) 0.25 m.r.d., for (b) 0.5 m.r.d., shaded above 1 m.r.d. (d) Temperature/silane-pressure diagram, illustrating the variation in texture [WENK &AL. 1990].

   Silicon films deposited from vapor on (111) single crystal substrates on an amorphous silicon dioxide layer show a large variation in texture which is mainly controlled by temperature and silane pressure [e.g. HUANG &AL. 1990, JOUBERT &AL. 1987]. At high silane pressure and low temperature, Si-films are amorphous (Fig. 37d). With increasing temperature and lower pressure textures change from a strong {113/112} fiber (Fig. 37a), to {110} (Fig. 37b), {100} texture components (Fig. 37c) and finally random. Low-temperature films show tensile stresses (up to 700 MPa), high-temperature films display compressive stresses (up to 600 MPa). Texture types in polycrystalline Si-films can be correlated with microstructures (KRULEVITCH &AL. 1991). The {110} texture in the 620 to 650°C range is due to the columnar grains, which share ⟨110⟩ as the growth direction. DROSD & WASHBURN [1982] showed that ⟨110⟩ and ⟨112⟩ are fast growth directions for silicon that crystallizes out of an amorphous state. Therefore, when the rate of crystallization exceeds the deposition rate, twinned grains oriented with the ⟨110⟩ direction close to the film normal survive and become columnar in shape. The tendency for growth normal to the film surface explains the observed {112} and {110} texture components in the low-temperature films. When

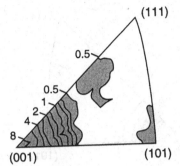

Fig. 38. Diamond film on MgO. The inverse pole figure displays a strong (100) fiber texture. Logarithmic contours 0.5, 0.7, 1, 1.5 m.r.d. etc., maximum is 12.9 m.r.d., shaded above 0.5 m.r.d., equal-area projection [BRUNET &AL. 1996].

twinning does not accompany crystallization, ⟨100⟩ is the direction of fastest crystal growth. For low stress 700 °C films deposited at 300 mTorr, there is a lower driving force for twinning, resulting in columnar grains elongated along ⟨100⟩ and consequently a predominantly {100} texture.

Recently diamond coatings have received a lot of interest. Also here the texture varies with fabrication conditions and can be very strong. Fig. 38 illustrates an axial {100} texture with a very minor {113} and {110} component [BRUNET &AL. 1996]. With bias-voltage techniques epitaxial diamond textures, e.g. on silicon single crystal substrates have been produced [HELMING &AL. 1995, SCHRECK &AL. 1994].

3.2 Nitride, carbide and oxide coatings

Coatings are often applied to metal and ceramic substrates to improve properties for various applications. They may improve heat and wear resistance, lower frictional forces, prevent chemical corrosion resistance or simply add to the decorative appearance. Carbides and nitrides of transition metals are particularly useful [TOTH 1971] and coatings have led to significant improvements of cutting tools for turning and milling operations. Coatings are often prepared by arc evaporation and deposited from a gaseous phase on to a substrate, e.g. tungsten carbide. LEONHARDT &AL. [1982] describe quantitatively textures in TiC. Figure 39a illustrates in an inverse pole figure a strong {111} fiber texture which is typical of high-temperature deposition. At lower deposition temperature, a {100} fiber is also observed. They document that the texture

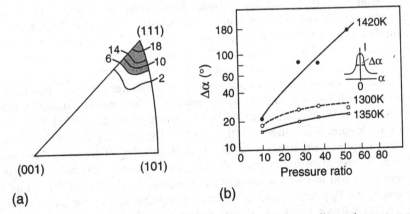

Fig. 39. TiC coating on tungsten carbide. (a) inverse pole figure illustrating a strong (111) fiber texture. Stereographic projection, contours 2, 6, 10, 14, 18 m.r.d., shaded above 6 m.r.d. (b) Variation of the width of the (111) texture peak Δα with processing conditions (temperature and vapor pressure) [LEONHARDT &AL. 1982].

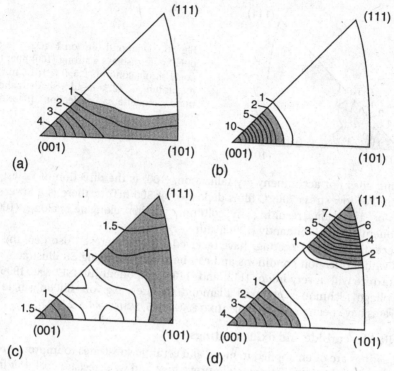

Fig. 40. Texture variations in HfN (a,b) and TiN tool coatings (c,d). Inverse pole figures, stereographic projection. (a) Contour interval 0.5 m.r.d. shaded above 1 m.r.d., (b) contour interval 1 m.r.d., shaded above 2 m.r.d., (c) contour interval 0.25 m.r.d., shaded above 1 m.r.d., (d) contour interval 0.5 m.r.d., shaded above 1 m.r.d. (Samples courtesy of J. Haygarth, Wah Chang.)

depends greatly on conditions such as temperature and vapor pressure, as expressed, for example, in the sharpness of the {111} texture peak (Fig. 39b). Similar textures are observed in TiN and HfN coatings on WC and Fig. 40 displays some inverse pole figures. SURI &AL. [1980] and SUE & TROUE [1987] have documented the strong influence of texture on erosion rates which are reduced for {111} textures (Fig. 41).

Alumina coatings have been applied to metals in orthopedic applications to reduce reactivity and improve fixation. The plasma sprayed coatings consist of γ-Al_2O_3 and a strong {100} fiber texture is observed [CARREROT &AL. 1991]

Textures of oxide coatings on polycrystalline substrates appear to be mainly controlled by growth kinetics and not by the texture of the substrate material, although there are reports of Nb_2O_5 films on a rolled Nb sheet where the oxide film inherits the orthorhombic sample symmetry of the substrate [HEIZMANN &AL. 1988b]. Zirconium oxide has a thermal expansion coefficient similar to that of stainless steel and thus only a small thermal mismatch when deposited on metallic substrates. Because of its high hardness and toughness, cubic yttrium stabilized zirconia (YSZ) is one of the most widely used protective coatings in the automotive, aeronautical and cutting tool industry (RHYS-JONES 1990). At room temperature films grow preferentially in the [111] direction, above 500°C the preferred growth direction is [100] [LUTTEROTTI &AL. 1994].

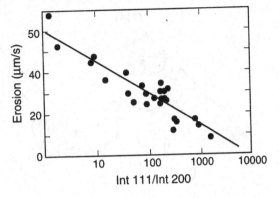

Fig. 41. Dependence of the alumina particle jet erosion rate on a TiN coating as a function of texture, evaluated as (111)/(200) diffraction peak intensity ratio [SUE & TROUE 1987].

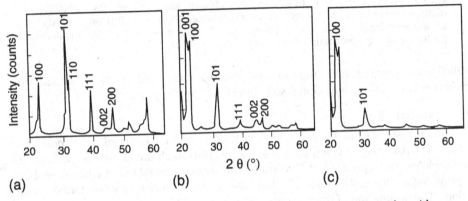

(a)   (b)   (c)

Fig. 42. X-ray diffractometer patterns of PbTiO$_3$. (a) Gel-derived powder with no preferred orientation as reference, (b) 500 nm thick film on (100) MgO, (c) 200 nm thick film on MgO. Changes in peak intensities are due to texture [CHEN &AL. 1989].

Films of ferroelectric PbTiO$_3$ on a (100) MgO substrate have a preferential alignment of [100] of perovskite normal to the film as documented by powder diffractometry (Fig. 42) [CHEN &AL. 1989]. The different relative intensities in 2θ-scans for different processing conditions are a qualitative expression of preferred orientation. Such textures can be much more quantitatively characterized by pole figure measurements and OD calculations. A survey of Pb(Zr,Ti)O$_3$ (PZT) and Pb$_2$ScTaO$_6$ (PST) films on various substrates revealed a relatively weak 001 and 111 fiber texture for PZT on a (100) silicon crystal coated with platinum (Fig. 43a), a very strong 111 fiber texture for PST on the same substrate (Fig. 43b), and a strong 001 fiber texture for PST on (110) Al$_2$O$_3$ (Fig. 43c), the latter with a weak in-plane alignment [CHATEIGNER &AL. 1997b]. MANSOUR & VEST [1992] and KIM &AL. [1994] found that the ferroelectric and fatigue behaviors of PZT films depends on crystallite orientation. There is no doubt that quantitative texture analysis will become an important method in the characterization of electronic devices. It has been mentioned in Chapter 4 that anomalous scattering produced by synchrotron x-rays may make it possible to determine the absolute orientation of these non-centric crystals.

Zinc oxide is another non-centric material (point group 6mm) of considerable technological interest because of its relatively strong electro-mechanical and thermo-mechanical interactions. Thin films of ZnO are used in the manufacture of the ultrasonic transducers. Texture and polarity are important and JOHNSON & FERRARI

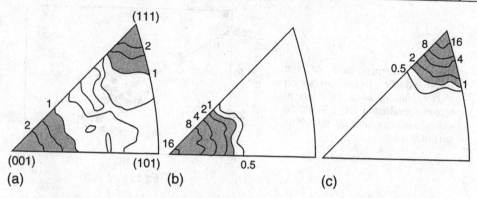

Fig. 43. Inverse pole figures of PZT and PST films on various substrates. (a) PZT on platinum coated (100) Si, (b) PST on platinum coated (100) Si, (c) PST on (110) Al$_2$O$_3$. Logarithmic contours 0.5, 0.7, 1, 1.5 m.r.d. etc., shaded above 1 m.r.d., equal-area projection [CHATEIGNER &AL. 1997b].

[1988] used piezoelectric properties to determine the odd coefficients (and thus directionality) of the orientation distribution function.

3.3 Epitaxial films

Much research has been invested in producing thin epitaxial *superconducting films* on various single crystal substrates, in an effort to obtain electronic devices with favorable electrical properties. TEM studies have documented that the films consist of a polycrystal with small crystallites but with a very high degree of preferred orientation. The orientation of the crystallites is controlled by epitaxy and growth velocity. The balance between the two factors is primarily dependent on the distance from the substrate surface (and thus the foil thickness) and the temperature and partial pressures at which the crystals grow. Because superconductivity in HTS ceramics is restricted to the Cu-plane, it is important to have the corresponding lattice plane (001) parallel to the film. The goal is to find conditions at which relatively thick films have a favorable crystal alignment with the substrate.

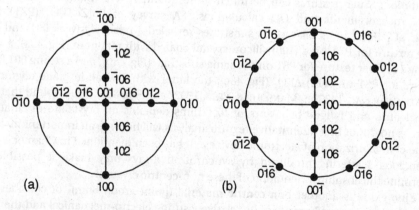

Fig. 44. Stereographic projection illustrating disposition of some lattice planes of YBCO (Y-123) for the C orientation (a) and A orientation (b), projected on the film plane.

Methods which have become established to measure texture in deformed metals and ceramics have been modified and adapted to the special requirements of thin epitaxial films prepared, for example, by laser ablation (see Chap. 4 Sec. 2.3). Most work so far has been done on Y-123 films on various substrates. In this system it has been useful to measure the combined 102 and 012 reflection of Y-123 because this reflection does not coincide with the substrate. Two orientations are of primary interest: the C-orientation has c-axes of Y-123 parallel to the foil normal and the A-orientation has c-axes in the foil plane and a-axes of Y-123 in the foil normal (Fig. 44) [HEIDELBACH &AL. 1992]. The two orientations are expressed by 102 peaks at pole distances of 56° (C-orientation) and 34° (A-orientation) respectively. The a-axes of Y-123 crystallites in the C-orientation are aligned either parallel or at 45° to those of the substrate depending on the best match in the oxygen sublattice of film and substrate in the perovskite, rock salt and fluorite structures [Fig. 45 a-c, TIETZ &AL. 1989]. The 102 pole figures document these orientation relationships and show that under some deposition conditions there may be secondary C components rotated 45° about the c-axis (C_2 in films on MgO and ZrO_2 substrates) [Fig. 46, WENK &AL. 1996]. In these very strong epitaxial textures where widths of texture peaks are less than 5°, pole figures measured with 5°×5° angular increments give only qualitative information about orientation relationships. To get quantitative values texture peaks need to be scanned on a much finer grid. Figure 47 shows for a $LaAlO_3$ substrate such scans of the vicinity of the Y-123 102 peaks for different deposition temperatures. Counts for the A-orientation first increase and then decrease with deposition temperature. The A-peak almost disappears at 850°C. C-intensities increase with temperature with only a small decrease at the highest temperature. Integrating over these texture peaks one can assess orientation volume fractions. The increase in total counts for the C-orientation changes by a factor of ten between 710 and 850°C. This increase is attributed not only to texture but also to poor crystallinity at low temperatures which affects the electrical properties (Fig. 48a). The ratio of the volume of crystals with c-axes in the foil plane to that of crystals with a-axes parallel to the foil plane shows a steady increase with deposition temperature (Fig. 48b).

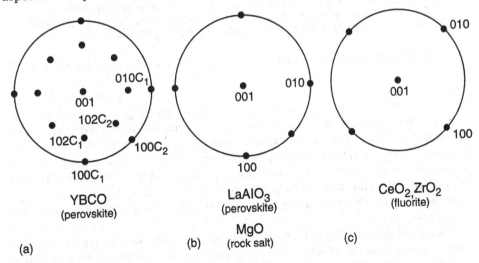

Fig. 45. Stereographic projection showing major orientation relationships between epitaxial YBCO films (a) and various substrate crystals (b,c).

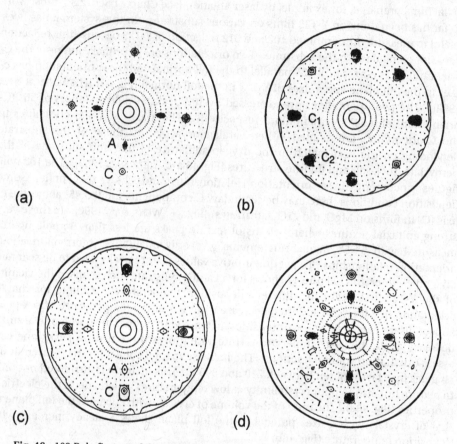

Fig. 46. 102 Pole figures of thin films of Y-123, laser ablated on various substrates. A, C, C_1 and C_2 orientations are indicated. (a) (001) $LaAlO_3$ 710°C, (b) (001) MgO 850°C, (c) (001) CeO_2 coated (01-12) Al_2O_3 745°C, (d) (001) ZrO_2 [WENK &AL. 1996b].

Texture measurements on epitaxial films provide useful data for quantitative characterization. They are easier to perform than transmission electron microscopy and are non-destructive. Also, unlike electron microscopy which provides local qualitative information, texture analysis averages over large areas and is more representative. We have emphasized high-temperature superconductors where quantitative texture analysis is becoming increasingly appreciated. There are other epitaxial systems with large anisotropy of properties due to preferred orientation. An example is nonlinear optical materials such as potassium titanyl phosphate (KTP) with an extreme variation of the second-harmonic signal intensity as a function of the rotation about the film normal [Fig. 49, XIONG &AL. 1994]. So far no texture information is available.

A related system is liquid crystals. If they are deposited on polished surfaces they can enhance texture effects. Their orientation is controlled by orientation, composition and microstructure of the substrate [TOMILIN 1990]. Few applications have been made in texture research.

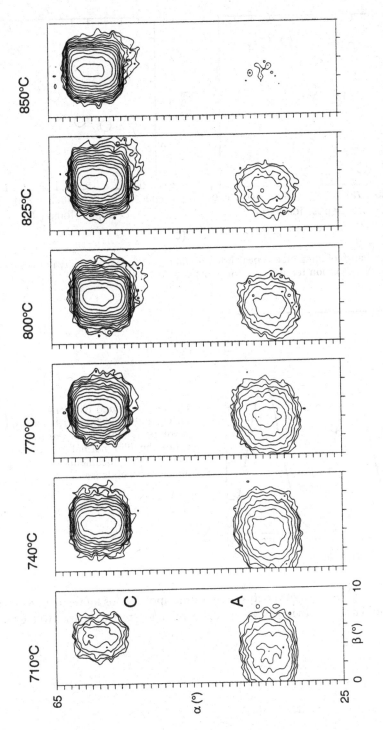

Fig. 47. Y-123 thin films on LaAlO$_3$. Detailed α–β (102) scans of the C- and A-orientation texture peaks, as a function of deposition temperature. Increments for α are 0.5°, for β 0.2° [WENK &AL. 1996b].

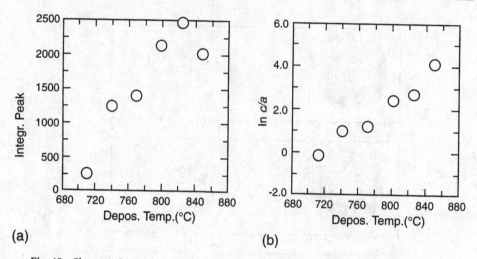

(a)   (b)

Fig. 48. Change of integrated C peak intensity (a) and C/A peak intensity ratio (b) as function of deposition temperature for Y-123 films deposited on LaAlO$_3$ [WENK &AL. 1996b].

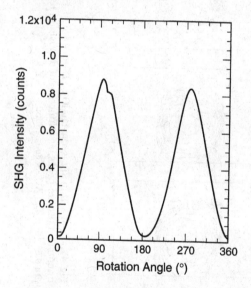

Fig. 49. Second-harmonic signal (SHG) intensity for KTP film on sapphire as function of the rotation about the film normal [XIONG &AL. 1994]. It documents a strong direction dependence.

The survey reveals a large variety of interesting applications for texture analysis in ceramic materials. It obviously is an open field for research and will gain in importance.

Chapter 7

TENSOR PROPERTIES OF TEXTURED POLYCRYSTALS

TENSOR PROPERTIES OF TEXTURED POLYCRYSTALS

Carlos Tomé

In a broad sense, a material property links an action exerted upon the material with the associated reaction. When action and/or reaction are tensorial magnitudes, and are linearly related (such as stress and elastic strain), the property also has a tensorial character. Linear relations between conjugate magnitudes usually arise when a quadratic expansion suffices to describe the energy of the system over the range of interest. Beyond such range, either higher-order terms may be required (e.g., third-order elastic constants) or the property may not be tensorial in character (e.g.: yield stress or magnetic saturation). The physical evidence of anisotropy is given by the difference in the reactions to the same action, when the latter is rotated with respect to the sample axes and exerted along a different direction in the material. In the tensorial representation of the property, the evidence of anisotropy is given by the difference between some of the tensor components which would, in the case of an isotropic material, be equal or exhibit a definite relation.

In practice, elastic, thermal, plastic, electric, magnetic and transport properties of single crystals are always anisotropic, although sometimes only slightly. In addition, while the tensors representing these properties are the same for every grain when expressed in crystal axes, they differ for grains of different orientations when referred to the same set of axes. As a consequence, aggregates of crystalline grains exhibit local variations of the properties which arise from the distribution of orientations (texture) at the polycrystalline level. In addition, the overall properties of the polycrystal depend on the particular texture, and have to be determined by an adequate volume average of the constituent grains' properties.

In general, a straightforward average of the single-crystal properties is not correct or consistent. A well known example in this respect is the elastic properties: the elastic characteristics of the aggregate that result from averaging the single-crystal stiffness tensor are different from the ones given by the average of their inverses, the compliances. The former average relies on the assumption that the strain field is uniform, while the latter average requires the stress field to be uniform. Both averaging procedures ignore the local perturbations to those fields induced by the grain interactions. Alternative averaging schemes for calculating overall elastic, thermal and viscous properties of aggregates will be discussed in this chapter.

1. Grain Averages

We present in this Section the notation and definitions used here to describe tensor averages over an aggregate of grains (scalar quantities are a particular case of zero order tensors). The averaging of material properties is treated in Sections 4 and 5. The volume average of a tensorial quantity t(x), a local function of the position, is defined as the volume average of each tensor component and will be denoted by $\langle t \rangle$:

$$\langle t \rangle = \frac{1}{V} \int_V t(x)\, dV \tag{1}$$

For polycrystal calculations, we are usually interested in tensors *that are constant over the grain domain*, in which case the above integrals adopt the form of an algebraic mean, expressed as a summation over the crystals c:

$$\langle t \rangle = \sum_c t^c w^c \tag{2}$$

where $w^c = V^c/V$ is the volume fraction or 'weight' associated with each grain, and g represents the grain's orientation, for example by the set of angles $\{\Psi,\Theta,\phi\}$. It is important to remark that the inverse of the arithmetic average defined by Eq. 2 differs, in general, from the average of the individual inverse tensors (harmonic average) defined by

$$\left\langle t^{-1} \right\rangle = \sum_c \left(t^c \right)^{-1} w^c \neq \langle t \rangle^{-1} \tag{3}$$

As a consequence, when the average is meant to represent a physical property of the aggregate, it is crucial to decide which is the correct averaging procedure. The issue will be addressed in the following Sections of this Chapter.

In the continuum limit, the weight associated with each orientation becomes infinitesimal and is described by the Orientation Distribution Function (ODF) f(g) defined in Chapter 2 as

$$\frac{dV}{V} = f(g)\, dg \tag{4}$$

For tensors that remain constant within the domain of the grain, the volume integral (Eq. 1) can be transformed into an integral defined over the domain of the orientation angles, and the average becomes

$$\langle t \rangle = \int t(g)\, f(g)\, dg \tag{5}$$

For actual calculations one must have a way of representing the ODF. One possibility is by means of a finite series expansion in terms of the generalized spherical harmonics defined in Eq. 5 of Chapter 3, as

$$f(\Psi,\Theta,\phi) = \sum_{l=0}^{L_o} \sum_{m=-l}^{l} \sum_{n=-l}^{l} W_{lmn} Z_{lmn}(\cos\Theta)\, e^{im\Psi}\, e^{in\phi} \tag{6}$$

where the coefficients W_{lmn} depend upon the texture of the polycrystal, and the index L_0 is in general large but finite. It is convenient to expand the tensor t in the basis of the spherical harmonics as well:

$$t(g) = \sum_{l=0}^{L'_0} \sum_{m=-l}^{+l} \sum_{n=-l}^{+l} t_{lmn} Z_{lmn}(\cos\Theta) e^{im\Psi} e^{in\phi} \tag{7}$$

where t_{lmn} are the tensorial coefficients of the expansion. The number of terms L'_0 required in the expansion is determined by the symmetry of the property. When Eqs. 6 and 7 are replaced in Eq. 5, and the orthonormality relation for the generalized spherical harmonics is accounted for, the average (5) is given by the sum of the expansion coefficients products

$$\langle t \rangle = 4\pi^2 \sum_{l=0}^{L''_0} \sum_{m=-l}^{+l} \sum_{n=-l}^{+l} t_{lmn} W_{lmn} \tag{8}$$

where $L''_0 = \min\{L_0, L'_0\}$. In some cases, such as for the elastic tensor, L''_0 is small and, as a consequence, Eq. 8 contains only a few terms. It is possible, in such cases, to determine the texture coefficients W_{lmn} up to an order L''_0, by measuring the physical property (the elastic constants of the aggregate in this case), using Eq. 7 to calculate the coefficients t_{lmn}, and solving the linear system (8). The reader is referred to the work of Bunge (see Chap. 25 of WENK [1985a]) for a detailed discussion of the harmonics formulation as it applies to the description of average elastic and plastic properties of polycrystals. In particular, Bunge derives relatively simple expressions for cubic crystals and orthotropic polycrystal symmetry, and shows how the main texture coefficients can be obtained from elastic measurements done along different directions in rolled polycrystals. However, it has to be borne in mind that the method is based on the assumption of the validity of a straightforward arithmetic average of the single-crystal properties to represent the aggregate properties, which is appropriate when what matters is to characterize the upper and lower bounds to the polycrystal property. More complex relations result when this assumption is abandoned, and the analytic advantage of using the Spherical Harmonics Method is partially lost. A good example in that respect is the work done by MORRIS [1970] on cubics, where the texture coefficients are correlated with the elastic constants through a self-consistent elastic formulation. We will see in Section 3.4 that describing the ODF by means of a discrete collection of orientations provides more numerical flexibility than the Spherical Harmonics Method for implementing polycrystal averaging procedures.

2. Crystal Elastic Properties[1]

Elastic properties provide a good case for analyzing aggregate averages, because stress and strain are linearly related and an analytic treatment of the problem is possible. Hooke's law states that

$$\sigma_{ij} = \mathbb{C}_{ijkl} \, \varepsilon_{kl} \tag{9}$$

---

[1] Since stress, strain and elastic tensors used in this Section are always referred to single crystals, we will omit the superscript 'c' throughout Section 2; all $\mathbb{C}$ and $\mathbb{S}$ components are presumed written in crystallographic coordinates. See also footnote 3.

where $\mathbb{C}$ is the single-crystal stiffness tensor. Equation 9 can be alternatively expressed in terms of the elastic compliance $\mathbb{S} = \mathbb{C}^{-1}$

$$\varepsilon_{ij} = \mathbb{S}_{ijkl}\,\sigma_{kl}$$
(10)

For what follows, it is useful to introduce the representation of second- and fourth-order tensors in a basis of second-order orthonormal symmetric tensors $\mathbf{b}^{(\lambda)}$ with the following properties [LEIBFRIED & BREUER 1978]:

$$b_{ij}^{(\lambda)} = b_{ji}^{(\lambda)}$$

and

$$b_{ij}^{(\lambda)}\,b_{ji}^{(\lambda')} = \delta_{\lambda\lambda'} \qquad (\lambda,\lambda'=1,6)$$
(11)

The symmetry under index permutation ($\sigma_{ij} = \sigma_{ji}$, and $\mathbb{C}_{ijkl} = \mathbb{C}_{jikl} = \mathbb{C}_{ijlk} = \mathbb{C}_{klij}$) permits one to represent the second- and fourth-order tensors in terms of $\mathbf{b}^{(\lambda)}$, as

$$\sigma_{ij} = \sum_{\lambda=1}^{6} \sigma_\lambda b_{ij}^{(\lambda)}$$

and

$$\mathbb{C}_{ijkl} = \sum_{\lambda=1}^{6} C_{\lambda\lambda'}\,b_{ij}^{(\lambda)}\,b_{kl}^{(\lambda')}$$
(12)

where the coefficients σ_λ and $C_{\lambda\lambda'}$ are given by the internal products

$$\sigma_\lambda = \sigma : \mathbf{b}^{(\lambda)} = \sigma_{ij}\,b_{ij}^{(\lambda)}$$

and

$$C_{\lambda\lambda'} = \mathbf{b}^{(\lambda)} : \mathbb{C} : \mathbf{b}^{(\lambda')} = b_{ij}^{(\lambda)}\,\mathbb{C}_{ijkl}\,b_{kl}^{(\lambda')}$$
(13)

These coefficients constitute the so called *matrix representation of the tensor*. The result of tensor addition, contraction and inversion is given by the corresponding operations of matrix addition, matrix product and matrix inversion, respectively. In particular, Eq. 9 becomes

$$\sigma_\lambda = C_{\lambda\lambda'}\,\varepsilon_{\lambda'} \qquad (\lambda,\lambda'=1,6)$$
(14)

A particular basis is the one formed by the eigentensors $\mathbf{b}^{(\lambda)}$ that diagonalize the matrix $C_{\lambda\lambda'}$. The eigentensors are defined by the equation

$$\mathbb{C}_{ijkl}\,b_{kl}^{(\lambda)} = C^{(\lambda)}\,b_{ij}^{(\lambda)}$$
(15)

Expressed in this basis the stiffness tensor adopts the form

$$\mathbb{C}_{ijkl} = \sum_{\lambda=1}^{6} C^{(\lambda)}\,b_{ij}^{(\lambda)} b_{kl}^{(\lambda)}$$
(16)

The eigenvalues $C^{(\lambda)}$ are invariants of the stiffness tensor, since a rotation will only affect the tensors $\mathbf{b}^{(\lambda)}$ in the linear combination Eq. 16. It is evident, comparing Eqs. 9 and 15, that a strain state ε proportional to any of the eigentensors $\mathbf{b}^{(\lambda)}$ will induce a stress state σ which is also proportional to it. An advantage of using the eigentensors of $\mathbb{C}$ as a basis is that the eigenvalues of the inverse tensor are simply the reciprocal of the eigenvalues of the direct tensor. It is easy to prove, using the property Eq. 11, that the elastic compliance tensor given by

$$S_{ijkl} = \sum_{\lambda=1}^{6} \left[C^{(\lambda)} \right]^{-1} b_{ij}^{(\lambda)} b_{kl}^{(\lambda)} \tag{17}$$

is the inverse of $\mathbb{C}$ given by Eq. 16.[2]

For illustration and for later applications, we present in what follows the explicit expressions of the eigenvalues and the eigentensors for cubic and hexagonal symmetry [LEIBFRIED & BREUER 1978; here we use a different notation]. It is easy to verify, by replacing in Eq. 15, that for crystals with cubic symmetry ($\mathbb{C}_{11}=\mathbb{C}_{22}=\mathbb{C}_{33}$; $\mathbb{C}_{12}=\mathbb{C}_{13}=\mathbb{C}_{23}$; $\mathbb{C}_{44}=\mathbb{C}_{55}=\mathbb{C}_{66}$) the eigenvalues of the stiffness tensor are

$$C^{(1)} = C^{(2)} = \mathbb{C}_{11} - \mathbb{C}_{12}$$
$$C^{(3)} = C^{(4)} = C^{(5)} = 2\,\mathbb{C}_{44} \tag{18}$$
$$C^{(6)} = \mathbb{C}_{11} + 2\mathbb{C}_{12}$$

while the associated eigentensors are

$$b^{(1)} = \frac{1}{\sqrt{6}} \begin{pmatrix} -1 & 0 & 0 \\ 0 & -1 & 0 \\ 0 & 0 & 2 \end{pmatrix}; \; b^{(2)} = \frac{1}{\sqrt{2}} \begin{pmatrix} -1 & 0 & 0 \\ 0 & 1 & 0 \\ 0 & 0 & 0 \end{pmatrix}; \; b^{(3)} = \frac{1}{\sqrt{2}} \begin{pmatrix} 0 & 0 & 0 \\ 0 & 0 & 1 \\ 0 & 1 & 0 \end{pmatrix}$$

$$b^{(4)} = \frac{1}{\sqrt{2}} \begin{pmatrix} 0 & 0 & 1 \\ 0 & 0 & 0 \\ 1 & 0 & 0 \end{pmatrix}; \quad b^{(5)} = \frac{1}{\sqrt{2}} \begin{pmatrix} 0 & 1 & 0 \\ 1 & 0 & 0 \\ 0 & 0 & 0 \end{pmatrix}; \quad b^{(6)} = \frac{1}{\sqrt{3}} \begin{pmatrix} 1 & 0 & 0 \\ 0 & 1 & 0 \\ 0 & 0 & 1 \end{pmatrix} \tag{19}$$

The first five eigentensors are deviatoric and, since the associated stress tensor is proportional to them, the eigenvalues give the shear modulus associated with each of them. Equations 18 indicate that they are degenerate, with only two being independent. The eigentensors $b^{(1)}$ and $b^{(2)}$ involve shears in the {110} planes: the first one represents the superposition of (101) [$\bar{1}$01] and (011) [0$\bar{1}$1] shears (which gives axisymmetric deformation along the axis 3), while the second one represents simple shear in the (110) [$\bar{1}$10] system. The eigentensors $b^{(3)}$, $b^{(4)}$ and $b^{(5)}$ are associated with {001} ⟨010⟩ shear modes. As for $b^{(6)}$, it represents a dilatational strain and, as a consequence $C^{(6)}$ is proportional to the bulk modulus of the cubic crystal. When $C^{(2)}=C^{(3)}$ the eigenvalues describe an isotropic elastic tensor, which suggests that the departure of the ratio $C^{(2)}/C^{(3)}$ from one is a measure of the elastic anisotropy of the cubic crystal. The anisotropy ratio of several cubic crystals, calculated with the room temperature elastic constants, is listed in Table III of Chapter 1. The results go against the intuitive belief that cubic properties are not strongly anisotropic: with the exception of aluminum and tungsten, the rest of the materials show a marked departure from the isotropy condition.

---

[2] All fourth-order tensors discussed in this Chapter exhibit the symmetry $A_{ijkl} = A_{jikl} = A_{ijlk}$. As a consequence we will often denote their components according to the established contracted index convention: 11→1, 22→2, 33→3, 23→4, 13→5, 12→6 (e.g. $A_{1122}\equiv A_{12}$). Note that we use this convention for both the stiffness and the compliance components, without multiplying the latter by factors 2 or 4, as is common in the Voigt representation.

The eigentensors given by Eq. 19 represent a general basis of symmetric second-order tensors. Besides diagonalizing elastic tensors with cubic symmetry, it is shown below that they diagonalize fourth-order tensors associated with incompressible hexagonal crystals. In addition, since they uncouple deviatoric and hydrostatic components, they represent a natural basis to formulate plasticity in Chapters 8 to 11. They do not, however, diagonalize elastic tensors with hexagonal symmetry.

For crystals with hexagonal symmetry ($\mathbb{C}_{11}=\mathbb{C}_{22}$; $\mathbb{C}_{13}=\mathbb{C}_{23}$; $\mathbb{C}_{44}=\mathbb{C}_{55}$; $\mathbb{C}_{11}-\mathbb{C}_{12}=2\mathbb{C}_{66}$) the solutions to Eq. 15 are the eigenvalues

$$C^{(1)} = \frac{\mathbb{C}_{33}+\mathbb{C}_{11}+\mathbb{C}_{12}}{2} - \frac{\mathbb{C}_{13}}{2}\sqrt{\alpha^2+8}$$

$$C^{(2)} = C^{(5)} = \mathbb{C}_{11} - \mathbb{C}_{12} = 2\mathbb{C}_{66}$$

$$C^{(3)} = C^{(4)} = 2\mathbb{C}_{44} \tag{20}$$

$$C^{(6)} = \frac{\mathbb{C}_{33}+\mathbb{C}_{11}+\mathbb{C}_{12}}{2} + \frac{\mathbb{C}_{13}}{2}\sqrt{\alpha^2+8}$$

and the associated eigentensors are

$$\mathbf{b}^{(1)} = \frac{1}{\sqrt{2+\beta_1^2}}\begin{pmatrix} -1 & 0 & 0 \\ 0 & -1 & 0 \\ 1 & 0 & \beta_1 \end{pmatrix}; \mathbf{b}^{(2)} = \frac{1}{\sqrt{2}}\begin{pmatrix} -1 & 0 & 0 \\ 0 & 1 & 0 \\ 0 & 0 & 0 \end{pmatrix}; \mathbf{b}^{(3)} = \frac{1}{\sqrt{2}}\begin{pmatrix} 0 & 0 & 0 \\ 0 & 0 & 1 \\ 0 & 1 & 0 \end{pmatrix}$$

$$\tag{21}$$

$$\mathbf{b}^{(4)} = \frac{1}{\sqrt{2}}\begin{pmatrix} 0 & 0 & 1 \\ 0 & 0 & 0 \\ 1 & 0 & 0 \end{pmatrix}; \qquad \mathbf{b}^{(5)} = \frac{1}{\sqrt{2}}\begin{pmatrix} 0 & 1 & 0 \\ 1 & 0 & 0 \\ 0 & 0 & 0 \end{pmatrix}; \quad \mathbf{b}^{(6)} = \frac{1}{\sqrt{2+\beta_2^2}}\begin{pmatrix} 1 & 0 & 0 \\ 0 & 1 & 0 \\ 0 & 0 & \beta_2 \end{pmatrix}$$

where

$$\alpha = \frac{\mathbb{C}_{11}+\mathbb{C}_{12}-\mathbb{C}_{33}}{\mathbb{C}_{13}}$$

$$\beta_1 = \frac{\alpha+\sqrt{\alpha^2+8}}{2} \tag{22}$$

$$\beta_2 = \frac{-\alpha+\sqrt{\alpha^2+8}}{2}$$

As expected from the hexagonal symmetry of the crystal, the eigentensors representing shears of the prismatic planes ($\mathbf{b}^{(2)}$ and $\mathbf{b}^{(5)}$) are associated with the same shear modulus. The same is true for the shears in the basal plane ($\mathbf{b}^{(3)}$ and $\mathbf{b}^{(4)}$).

Observe that when $\alpha=1$ the cubic and hexagonal eigentensors given by Eqs. 19 and 21 coincide, although the eigenvalues are different. In particular, only under such condition, a dilatational strain will induce a purely hydrostatic stress in the hexagonal crystal. For any other value of α, a dilatational strain will also induce deviatoric stress components and vice versa. For the eigenvalues to describe an elastically isotropic crystal, the further conditions $C^{(1)}=C^{(3)}$ and $C^{(2)}=C^{(3)}$ have to be realized. As a consequence, the departure of the parameters α, $C^{(1)}/C^{(3)}$ and $C^{(2)}/C^{(3)}$ from unity can be used as a measure of the elastic anisotropy of hexagonal crystals. These parameters are listed in Table I for several hexagonal elements, together with the independent elastic constants at room temperature. It can be seen that in Co, Hf, Mg, Re, and Ti a dilatation strain will induce a nearly hydrostatic stress state. In addition, judging from the values of the anisotropy parameters, Hf and Mg can be regarded as nearly elastically isotropic, while Be and Zn are highly anisotropic.

Table I. Single crystal elastic constants at room temperature (units of GPa; expressed in contracted index notation) and anisotropy parameters for several hexagonal materials. (Extracted from "Single Crystal Elastic Constants and Calculated Aggregate Properties", G. SIMMONS & H. WANG (Cambridge, MA: M.I.T. Press, 1971).

	C_{11}	C_{33}	C_{12}	C_{13}	C_{44}	C_{66}	α	$C^{(2)}/C^{(3)}$	$C^{(1)}/C^{(3)}$
Be	292.3	336.4	26.7	14.0	162.5	132.8	-1.24	0.82	0.94
Cd	114.5	50.8	39.5	39.9	19.8	37.5	2.58	1.89	0.66
Co	306.3	357.4	165.1	101.9	75.3	70.6	1.12	0.94	1.72
Hf	181.1	196.9	77.2	66.1	55.7	51.9	0.93	0.93	1.16
Mg	59.4	61.6	25.6	21.4	16.4	16.9	1.09	1.03	1.24
Re	618.2	683.5	275.3	207.8	160.6	171.4	1.01	1.07	1.48
Ru	562.6	624.2	187.8	168.2	180.6	187.4	0.75	1.04	1.22
Tl	40.8	52.8	35.4	29.0	7.3	2.7	0.81	0.37	1.50
Ti	162.4	180.7	92.0	69.0	46.7	35.2	1.07	0.75	1.07
Y	77.9	76.9	28.5	21.0	24.3	24.7	1.41	1.02	1.20
Zn	163.7	63.5	36.4	53.0	38.8	63.6	2.58	1.64	0.39
Zr	143.5	164.9	72.5	65.4	32.1	35.5	0.787	1.103	1.479

In the case of an incompressible material, the tensorial properties must be such that they conserve the volume during deformation. This case may not be of interest for the elastic regime, but is relevant to the viscous response to be described in Section 4 of this Chapter and to the plastic response to be introduced in Chapter 8. Since the anisotropy depends only on the crystal symmetry and not on the particular property being described, we illustrate the case for the elastic properties. The condition of incompressibility is better expressed in terms of compliances using Eq. 10:

$$\text{tr} \, \varepsilon \equiv \sum_{i=1}^{3} \varepsilon_{ii} = \sum_{i=1}^{3} \mathbb{S}_{iikl} \sigma_{kl} = 0 \tag{23}$$

and has to be valid for any arbitrary stress state. In particular, the three tensile stresses along each crystal axis ($\sigma_{kl} = \sigma \, \delta_{kl}$) give the system of equations:

$$\sum_{i=1}^{3} \mathbb{S}_{iikk} = \sum_{i=1}^{3} \mathbb{S}_{ik} = 0 \quad (k=1,2,3) \tag{24}$$

where we have used the contracted index convention (see footnote 2) to simplify the notation. For crystals with cubic, hexagonal or orthogonal symmetry, Eq. 24 imposes one, two, or three further conditions, respectively, on the three, five and nine independent components. As a consequence, the tensors representing incompressible properties of cubic, hexagonal and orthotropic crystals have only two, three and six independent components, respectively.

For crystals with cubic symmetry ($\mathbb{S}_{11}=\mathbb{S}_{22}=\mathbb{S}_{33}$; $\mathbb{S}_{12}=\mathbb{S}_{13}=\mathbb{S}_{23}$; $\mathbb{S}_{44}=\mathbb{S}_{55}=\mathbb{S}_{66}$) Eq. 24 imposes the further conditions over the elastic compliances:

$$\mathbb{S}_{33} = -2\mathbb{S}_{13} \tag{25}$$

It is easy to verify that the eigentensors given by Eq. 19 are a solution to the eigenvalue Eq. 15 for the cubic incompressible tensors defined by the previous relations. The corresponding eigenvalues are

$$S^{(1)} = S^{(2)} = \tfrac{3}{2}\mathbb{S}_{11}$$

$$S^{(3)} = S^{(4)} = S^{(5)} = 2\mathbb{S}_{44}$$

$$S^{(6)} = 0 \tag{26}$$

For crystals with hexagonal symmetry ($\mathbb{S}_{22}=\mathbb{S}_{11}$; $\mathbb{S}_{23}=\mathbb{S}_{13}$; $\mathbb{S}_{55}=\mathbb{S}_{44}$; $2\mathbb{S}_{66}=\mathbb{S}_{11}-\mathbb{S}_{12}$) Eq. 24 imposes two further conditions over the elastic compliances:

$$\mathbb{S}_{33} = -2\mathbb{S}_{13} \quad \text{and} \quad \mathbb{S}_{12} = -\mathbb{S}_{11}-\mathbb{S}_{13} \tag{27}$$

It is easy to verify that the same set of tensors given by Eq. 19 are also eigentensors of the hexagonal incompressible compliances defined by the previous relations. The corresponding eigenvalues are

$$S^{(1)} = \tfrac{3}{2}\mathbb{S}_{33}$$

$$S^{(2)} = S^{(5)} = \mathbb{S}_{11} - \mathbb{S}_{12} = 2\mathbb{S}_{66}$$

$$S^{(3)} = S^{(4)} = 2\mathbb{S}_{44} \tag{28}$$

$$S^{(6)} = 0$$

It is apparent why it is preferable to work in terms of compliances, since the inverse of the eigenvalue $S^{(6)}$ is not defined. If σ_λ are the components of the stress tensor in the basis of symmetric tensors $b^{(\lambda)}$ defined by Eq. 19, the components of the strain tensor in the same basis are given by the product

$$\varepsilon_\lambda = S^{(\lambda)} \sigma_\lambda \qquad (\lambda=1,6) \tag{29}$$

In particular, the components ε_6 and σ_6 associated with the tensor $b^{(6)}$ are the dilatation and the hydrostatic pressure, respectively. It is evident that $S^{(6)} = 0$ will give always zero dilatation, independently of the value of the pressure σ_6.

3. Elastic and Thermal Properties of Polycrystals[3]

HILL [1967] demonstrates that, under very general conditions, the volume averages of the local stress and strain tensors have to coincide with the overall strain ε and the overall stress σ:

$$\varepsilon = \langle \varepsilon(\mathbf{x}) \rangle \equiv \frac{1}{V}\int_V \varepsilon(\mathbf{x})\, dV \tag{30a}$$

$$\sigma = \langle \sigma(\mathbf{x}) \rangle \equiv \frac{1}{V}\int_V \sigma(\mathbf{x})\, dV \tag{30b}$$

Since the elastic and thermal properties are constant within the domain of each grain, and since polycrystal models assume that the stress and the strain are uniform within the grain, the expressions derived in Section 1 of this Chapter are suitable for calculating the average elastic and thermal moduli.

The problem is formulated as follows. The elastic stiffness and the thermal expansion tensors of the single crystal, $\mathbb{C}^c$ and α^c respectively, are known. The texture is also known, either as an ODF or as a discrete distribution of orientations with

---

[3] From here onward, all $\mathbb{C}$ and $\mathbb{S}$ components are presumed written in a common coordinate system. For each grain (or 'crystal', superscript c), the components of $\mathbb{C}^c$, for example, must be obtained in sample coordinates from the crystallographic components (primed below) by tensor transformation, using the transpose of the orientation matrix g defined in Chap. 2 Eq. 6: $\mathbb{C}^c_{ijkl} = g^T_{ii'} g^T_{jj'} g^T_{kk'} g^T_{ll'} \mathbb{C}^c_{i'j'k'l'}$.

weights. $\mathbb{C}$ and α are the overall stiffness and thermal tensors of the polycrystal, which are to be determined. Assume that the stress σ and the total strain ε (elastic plus thermal) are uniform in each grain, that an initially stress free polycrystal is subjected to a uniform temperature increment δT and to an overall stress σ, and that ε is the overall strain in the polycrystal. The total strain in each grain is given by the superposition of the elastic and the thermal contributions as

$$\varepsilon^c = \mathbb{C}^{c^{-1}} : \sigma^c + \alpha^c \, \delta T \tag{31a}$$

and for the aggregate it is defined by

$$\varepsilon = \mathbb{C}^{-1} : \sigma + \alpha \, \delta T \tag{31b}$$

When combined with the condition Eq. 30 over the average strain, these equations lead to an expression that couples the crystal moduli with the overall moduli

$$\left\langle \mathbb{C}^{c^{-1}} : \sigma^c \right\rangle + \left\langle \alpha^c \right\rangle \delta T = \mathbb{C}^{-1} : \sigma + \alpha \, \delta T \tag{32}$$

The inverse of Eqs. 31 provides the alternative set of equations

$$\sigma^c = \mathbb{C}^c : (\varepsilon^c - \alpha^c \, \delta T) \tag{33a}$$

$$\sigma = \mathbb{C} : (\varepsilon - \alpha \, \delta T) \tag{33b}$$

which, combined with the condition Eq. 30b over the volume average of the local stress, lead to

$$\langle \mathbb{C}^c : \varepsilon^c \rangle - \langle \mathbb{C}^c : \alpha^c \rangle \delta T = \mathbb{C} : \varepsilon - \mathbb{C} : \alpha \, \delta T \tag{34}$$

The conditions Eqs. 32 and 34 are completely general, and the only assumption invoked in their derivation is that the strain and the stress are uniform inside the domain of each grain. In the following we describe different approaches for solving those equations for the unknowns $\mathbb{C}$ and α.

3.1 Upper and lower bounds

The easiest (but not the only) solutions to Eqs. 32 and 34 follow from assuming uniform strain or uniform stress. Fulfilling compatibility (but not necessarily equilibrium) gives an upper bound on $\mathbb{C}$, while fulfilling equilibrium (but not necessarily compatibility) provides an upper bound on $\mathbb{S}$. The upper and lower bounds are important because they define the interval where the overall elastic constants of the aggregate must be. Sometimes, the upper or lower bounds are used as estimates for the overall mechanical properties of polycrystals. The upper bound follows from the Voigt assumption that the total strain is uniform within the aggregate: $\varepsilon^c = \varepsilon$, which in a sense amounts to prioritizing strain compatibility over stress equilibrium. If Eq. 34 has to be valid for an arbitrary strain ε and an arbitrary temperature increment δT, the overall moduli are given by:

$$\mathbb{C}^V \equiv \langle \mathbb{C}^c \rangle \tag{35}$$

and

$$\alpha^V = \mathbb{C}^{V^{-1}} : \left\langle \mathbb{C}^c : \alpha^c \right\rangle$$

where the averages are calculated using the equations of the previous section. Observe that the overall thermal moduli are coupled to the elastic moduli in the Voigt approach.

The lower bound, on the other hand, emphasizes stress continuity across the aggregate and follows from the original Reuss assumption that the stress is uniform within the polycrystal: $\sigma^c = \sigma$. If such a condition has to be fulfilled and if Eq. 32 has to hold for an arbitrary stress and temperature increment δT, the overall moduli are given by:

$$\mathbb{C}^R \equiv \left\langle \mathbb{C}^{c^{-1}} \right\rangle^{-1}$$

and

$$\alpha^R \equiv \langle \alpha^c \rangle \tag{36}$$

In this extreme case the overall thermal properties are independent of the elastic properties.

The elastic constants and thermal expansion coefficients that result from the Voigt and the Reuss assumptions (Eqs. 35 and 36) are different because the average of the inverses does not, in general, coincide with the inverse of the average. Only when the grain elastic properties are isotropic, then the tensors are independent of grain orientation and both averaging procedures give the same stiffness and thermal tensors. Moreover, MURA [1991] shows that the Voigt and Reuss approximations provide upper bounds for *each* component of the 'true' average elastic stiffness $\mathbb{C}$ and its inverse $\mathbb{S}$, respectively, as follows:

$$\mathbb{C}_{ijij} \leq \mathbb{C}_{ijij}^V$$
$$\mathbb{S}_{ijij} \leq \mathbb{S}_{ijij}^R \tag{37}$$

The upper bound for the compliances defined by the second equation is usually interpreted in terms of the inverses and regarded as a lower bound for the elastic stiffness. Care must be exercised in such a case, though, because the term 'lower bound' does not necessarily imply that all the components of $(\mathbb{S}^R)^{-1}$ are lower than the components of the true stiffness tensor $\mathbb{C}$. Examples in such respect are given in the numerical Tables of Section 4.

Better upper and lower bounds for the elastic properties have been derived by HASHIN & SHTRIKMAN [1962]. As said above, only when the single-crystal elastic properties are isotropic, the Voigt and Reuss average coincide. Since they are usually not far apart for most metallic polycrystals, HILL [1952] has suggested the use of the arithmetic mean of the Voigt and Reuss elastic constants as a reasonable empirical estimate of the overall moduli, namely

$$\mathbb{C}^H \equiv \frac{1}{2}\left[\langle \mathbb{C}^c \rangle + \left\langle \mathbb{C}^{c^{-1}} \right\rangle^{-1} \right]$$
$$\mathbb{S}^H \equiv \frac{1}{2}\left[\langle \mathbb{S}^c \rangle + \left\langle \mathbb{S}^{c^{-1}} \right\rangle^{-1} \right] \tag{38}$$

From a formal point of view, however, Hill's empirical estimate fulfills neither the general condition Eq. 30 over the average strain and the average stress, nor the condition that the polycrystal stiffness $\mathbb{C}^H$ is the inverse of the polycrystal compliance $\mathbb{S}^H$

defined by Eq. 38. A geometric average that gives an average stiffness which is consistent with the inverse of the average compliance is reviewed briefly in the following Section. In addition, the self-consistent approach described in Section 3.4 also fulfills this condition, and does not rely on the assumption that either the stress or the strain are uniform throughout the polycrystal.

3.2 The geometric mean

In 1966, Aleksandrov and Aisenberg proposed an averaging procedure constructed around the condition of the commutation of inverse and averaging operations. Their formulation was later improved by MORAWIEC [1989] and MATTHIES & HUMBERT [1993], and is reviewed in what follows in a form which is consistent with the notation used in this book.

In the orthonormal basis of eigentensors, the natural logarithm of the elastic stiffness tensor Eq. 16 is defined as the linear combination of the logarithm of its eigenvalues as

$$\left(\ln \mathbb{C}^c\right)_{ijkl} = \sum_{\lambda=1}^{6} \ln\left[C^{(\lambda)}\right] b_{ij}^{(\lambda)}\, b_{kl}^{(\lambda)} \tag{39}$$

And the polycrystal average over orientation space of the previous expression can be performed using Eq. 5.

$$\left\langle \ln \mathbb{C}^c\right\rangle_{ijkl} = \int f(g)\, R_{ii'}^c\, R_{jj'}^c\, R_{kk'}^c\, R_{ll'}^c \left(\ln \mathbb{C}^c\right)_{i'j'k'l'} dg = \sum_{\lambda=1}^{6} \ln\left[C^{(\lambda)}\right] \mathbb{F}_{ijkl}^{(\lambda)} \tag{40a}$$

where

$$\mathbb{F}_{ijkl}^{(\lambda)} = \int f(g)\, R_{ii'}^c\, R_{jj'}^c\, R_{kk'}^c\, R_{ll'}^c\, b_{i'j'}^{(\lambda)}\, b_{k'l'}^{(\lambda)}\, dg \tag{40b}$$

is a tensor that depends on the texture and the single-crystal symmetry, through f(g) and $b^{(\lambda)}$, respectively. The logarithm of the polycrystal stiffness, expressed in terms of its (yet unknown) eigenvectors $B^{(\lambda)}$ and eigenvalues $L^{(\lambda)}$ adopts the form

$$\left(\ln \mathbb{C}\right)_{ijkl} = \sum_{\lambda=1}^{6} L^{(\lambda)}\, B_{ij}^{(\lambda)}\, B_{kl}^{(\lambda)} \tag{41}$$

and the polycrystal stiffness is given by the exponential of the previous expression as

$$\mathbb{C}_{ijkl} = \sum_{\lambda=1}^{6} \exp\!\left(L^{(\lambda)}\right) B_{ij}^{(\lambda)}\, B_{kl}^{(\lambda)} \tag{42a}$$

while its inverse, the polycrystal compliance, is given by the reciprocal of the stiffness eigenvalues and the same eigentensors as

$$\mathbb{C}_{ijkl}^{-1} = \sum_{\lambda=1}^{6} \exp\!\left(-L^{(\lambda)}\right) B_{ij}^{(\lambda)}\, B_{kl}^{(\lambda)} \tag{42b}$$

So far, no assumption has been made to relate the overall stiffness to the single-crystal moduli. What ALEKSANDROV & AISENBERG [1966] propose is that the logarithm of the overall stiffness (Eq. 41) is equal to the average of the logarithms of the single-crystal stiffness (Eq. 40). As a consequence, the eigenvalues can be identified with the contracted product

$$L^{(\lambda)} = B_{ij}^{(\lambda)} \left\langle \ln \mathbb{C}^c \right\rangle_{ijkl} B_{kl}^{(\lambda)}$$

$$= \sum_{\lambda'=1}^{6} \ln\left[C^{(\lambda')}\right] B_{ij}^{(\lambda)} \mathbb{F}_{ijkl}^{(\lambda')} B_{kl}^{(\lambda)} \tag{43a}$$

$$= \sum_{\lambda'=1}^{6} \ln\left\{ \left[C^{(\lambda')}\right]^{F^{(\lambda\lambda')}} \right\}$$

where

$$F^{(\lambda\lambda')} = B_{ij}^{(\lambda)} \mathbb{F}_{ijkl}^{(\lambda')} B_{kl}^{(\lambda)} \tag{43b}$$

Replacing Eq. 43 in Eq. 41 permits one to express the eigenvalues of $\mathbb{C}$ and $\mathbb{C}^{-1}$ in terms of the single-crystal eigenvalues as

$$\exp\left(L^{(\lambda)}\right) = \prod_{\lambda'=1}^{6} \left[C^{(\lambda')}\right]^{F^{(\lambda\lambda')}} \tag{44a}$$

$$\exp\left(-L^{(\lambda)}\right) = \prod_{\lambda'=1}^{6} \left[\frac{1}{C^{(\lambda')}}\right]^{F^{(\lambda\lambda')}} = \prod_{\lambda'=1}^{6} \left[S^{(\lambda')}\right]^{F^{(\lambda\lambda')}} \tag{44b}$$

where $S^{(\lambda')}$, the reciprocals of $C^{(\lambda')}$, are the eigenvalues of the single-crystal compliance tensor. The factorial form of these expressions explain the denomination of 'geometric mean' assigned to this averaging procedure. Equations 42 and 44 indicate that the stiffness tensor $\mathbb{C}$ is given by the geometric average of the single-crystal stiffness eigenvalues, while its inverse is given by the reciprocals of those eigenvalues, which are the single-crystal compliance eigenvalues. The latter result indicates that the geometric mean leads to the same polycrystal elastic properties, independently of whether the single-crystal stiffness or the single-crystal compliance is used to calculate the average. There may not be any physical basis for this method but it exhibits the attractive feature that the averaging of the inverse is automatically consistent with the inverse of the average property.

3.3 Average bulk modulus

The average bulk modulus of polycrystals is sometimes erroneously assumed to be the inverse of the compressibility. A closer look at how these magnitudes are defined indicates that the apparently simple concept of average bulk modulus involves assumptions which are sometimes overlooked. A discussion of such assumptions serves to clarify some of the concepts exposed in the previous sections.

According to the thermodynamic definition, the bulk modulus is the ratio between the increment of the hydrostatic stress component and the imposed dilatation increment, while keeping constant the deviatoric strain components and the temperature. That is to say

$$B = \frac{\delta p}{\delta V / V}\bigg|_{\varepsilon',T} = \frac{\operatorname{tr}(\delta\sigma)/3}{\operatorname{tr}(\delta\varepsilon)}\bigg|_{\varepsilon',T} = \frac{\left\langle \delta\sigma_{11}^c + \delta\sigma_{22}^c + \delta\sigma_{33}^c \right\rangle}{3\left\langle \delta\varepsilon_{11}^c + \delta\varepsilon_{22}^c + \delta\varepsilon_{33}^c \right\rangle}\bigg|_{\varepsilon',T} \tag{45}$$

where $\delta\sigma$ and $\delta\varepsilon$ are the overall stress and strain increments. If one did invoke the Voigt assumption that the dilatational strain increment is the same in every crystal ($\delta\varepsilon^c = \delta\varepsilon = \delta\varepsilon\, \mathbf{I}$), one would get

$$\left\langle \sum_{i=1}^{3} \delta\sigma_{ii}^{c} \right\rangle = \left\langle \sum_{i,k=1}^{3} \mathbb{C}_{iikk}^{c} \right\rangle \delta\varepsilon \tag{46}$$

$$\left\langle \sum_{i=1}^{3} \delta\varepsilon_{ii}^{c} \right\rangle = 3\,\delta\varepsilon$$

Replacing Eq. 46 in 45 gives,

$$B^{V} = \frac{1}{9} \left\langle \sum_{i,k=1}^{3} \mathbb{C}_{iikk}^{c} \right\rangle \tag{47}$$

which provides an upper bound for the bulk modulus, according to Eq. 37. We will show that the particular combination of stiffness moduli that appears in Eq. 47 is an invariant of the stiffness tensor and, as a consequence, the upper-bound bulk modulus is independent of the texture. Using the expansion of $\mathbb{C}^{c}$ in the basis of eigentensors Eq. 16 we obtain that the contribution of each grain to the average is

$$\sum_{i,k=1}^{3} \mathbb{C}_{iikk}^{c} = \sum_{\lambda=1}^{6} C^{(\lambda)} \sum_{i=1}^{3} b_{ii}^{(\lambda)} \sum_{k=1}^{3} b_{kk}^{(\lambda)} = \sum_{\lambda=1}^{6} C^{(\lambda)} \left[\mathrm{tr}\!\left(\mathbf{b}^{(\lambda)} \right) \right]^{2} \tag{48}$$

Although the eigentensors $\mathbf{b}^{(\lambda)}$ are different for every crystal when expressed in the sample reference system, both $C^{(\lambda)}$ and $\mathrm{tr}\,\mathbf{b}^{(\lambda)}$ are invariants and, as a consequence, the average Eq. 47 is invariant.

As for the compressibility modulus, it is defined (see Chap. 1, Eq. 21) as the ratio between an increment in the volumetric dilatation associated with an increment in the hydrostatic pressure, while keeping constant the deviatoric stress components and the temperature. That is to say

$$\kappa = \left. \frac{\delta V / V}{\delta p} \right|_{\sigma',T} = \left. \frac{\mathrm{tr}(\delta\varepsilon)}{\mathrm{tr}(\delta\sigma)/3} \right|_{\sigma',T} = \left. \frac{3\left\langle \delta\varepsilon_{11}^{c} + \delta\varepsilon_{22}^{c} + \delta\varepsilon_{33}^{c} \right\rangle}{\left\langle \delta\sigma_{11}^{c} + \delta\sigma_{22}^{c} + \delta\sigma_{33}^{c} \right\rangle} \right|_{\sigma',T} \tag{49}$$

If one invokes now the Reuss assumption of uniform stress in the aggregate, and uses the fact that the external stress increment is purely hydrostatic ($\delta\sigma^{c} = \delta\sigma = \delta p \times \delta_{ij}$), the compressibility modulus adopts the form

$$\kappa^{R} = \left\langle \sum_{i,k=1}^{3} \mathbb{S}_{iikk}^{c} \right\rangle \tag{50}$$

where $\mathbb{S}^{c}$ is the elastic compliance tensor of the crystal. According to Eq. 37 this latter expression constitutes an upper bound for the compressibility modulus of the aggregate. Using the property of the eigenvalues of the inverse tensor given by Eq. 17 we can write

$$\sum_{i,k=1}^{3} \mathbb{S}_{iikk}^{c} = \sum_{\lambda=1}^{6} \left[C^{(\lambda)} \right]^{-1} \left[\mathrm{tr}\!\left(\mathbf{b}^{(\lambda)} \right) \right]^{2} \tag{51}$$

which shows that the individual contribution of each grain to the average Eq. 50 is an invariant. As a consequence, the compressibility modulus given by the Reuss assumption is independent of the texture. The inverse of Eq. 50 gives the bulk modulus associated with the Reuss assumption as

$$B^{R} = \left\langle \sum_{i,k=1}^{3} \mathbb{S}_{iikk}^{c} \right\rangle^{-1} = \left\langle \sum_{\lambda=1}^{6} \left[C^{(\lambda)} \right]^{-1} \left[\mathrm{tr} \left(\mathbf{b}^{(\lambda)} \right) \right]^{2} \right\rangle^{-1} \tag{52}$$

Observe that, although B^{R} is also an invariant independent of the texture, it is in general different from B^{V} given by Eqs. 47 and 48. *Only for cubic crystals* both expressions give the same value, because only the eigentensor $\mathbf{b}^{(6)}$ in Eq. 19 has non-zero trace. As a consequence, Eq. 47 becomes

$$B^{V} = \frac{1}{9} C^{(6)} \left[\mathrm{tr} \left(\mathbf{b}^{(6)} \right) \right]^{2} = \frac{1}{3} \left(\mathbb{C}_{11} + 2 \mathbb{C}_{12} \right) \tag{53}$$

while Eq. 52 becomes

$$B^{R} = \left(\left[C^{(6)} \right]^{-1} \left[\mathrm{tr} \left(\mathbf{b}^{(6)} \right) \right]^{2} \right)^{-1} = \left(3 \left[\mathbb{C}_{11} + 2 \mathbb{C}_{12} \right]^{-1} \right)^{-1} = B^{V} \tag{54}$$

For hexagonal (or lesser) symmetry, on the other hand, Eq. 21 indicates that both $\mathbf{b}^{(1)}$ and $\mathbf{b}^{(6)}$ have non-zero trace and, as a consequence, the Voigt and Reuss bulk moduli given by Eqs. 47 and 52, respectively, are in general different. Two conclusions that can be extracted from the foregoing exercise are:

•If non-cubic single crystals are subjected to a hydrostatic stress (dilatational strain), they will develop deviatoric strain (deviatoric stress) components. This can also be expressed by the observation that the identity is not an eigentensor of the stiffness or the compliance for symmetry less than cubic.

•For aggregates of cubic crystals, the Voigt and the Reuss compressibilities coincide, which suggests that they are independent of the polycrystal model used for their calculation. For aggregates of non-cubic crystals, the Voigt and the Reuss compressibilities are different. They are, however, independent of the texture because they only depend on invariants of the single-crystal elastic tensor.

3.4 Self-consistent estimates

In the Appendix we discuss the problem of elastic ellipsoidal inclusions and inhomogeneities embedded in an infinite elastic medium. In particular, the stress (and strain) induced in the inclusion by a stress applied far from it is deduced, assuming that the elastic moduli of the inclusion and the medium are known (the Eshelby problem). A self-consistent model that makes use of this result was first proposed by KRÖNER [1961], who assumed that the grains can be regarded as inclusions embedded in a Homogeneous Effective Medium having the average elastic properties of the aggregate.

The difference between the elastic inclusion problem and the self-consistent approach is that in the former case the matrix and inclusion properties are defined beforehand, and the problem admits an exact solution. In the self-consistent case, on the other hand, the overall properties follow from imposing Eqs. 30, which state that the average response of the grains has to be the same as the overall response of the material. The method is called 'self-consistent' because the assumed HEM is adjusted to coincide with the average of the 'inclusions'. The embedding assumption in the self-consistent procedure is that the response in the vicinity of the grain (inclusion) is adequately described by the average moduli of the HEM, independently of the actual neighborhood of the grain. As a consequence, the self-consistent framework is statistical, in the sense that a given orientation represents all the grains with the same orientation, and the HEM represents the average neighborhood of such set of grains. When neighbor orientations are correlated, or when the discontinuity in physical properties between neighboring grains is large, the HEM assumption may not provide a realistic representation of the intergranular interactions.

The previous concepts can be extended to include the effective thermal properties of an aggregate of grains: the aggregate is replaced by a HEM exhibiting the same thermal response, and each grain is treated as an inhomogeneity with different elastic and thermal properties. Within this approach the thermal dilatation induced by a temperature increment δT is regarded as a transformation strain. If Ω is the domain of a given grain and $(V-\Omega)$ is the domain of the surrounding HEM, the stresses in grain, matrix and at the external boundary Γ_V are given by Eqs. 33:

$$\sigma^c_{(x)} = \mathbb{C}^c : \left(\varepsilon^c_{(x)} - \alpha^c \, \delta T \right) \qquad \text{in} \, \Omega \tag{55a}$$

$$\sigma_{(x)} = \mathbb{C} \; : \left(\varepsilon_{(x)} - \alpha \, \delta T \right) \qquad \text{in} \left(V - \Omega \right) \tag{55b}$$

$$\sigma \;\; = \mathbb{C} \; : \left(\varepsilon - \alpha \, \delta T \right) \qquad \text{in} \, \Gamma_V \tag{55c}$$

where ε is *the total strain* formed by the elastic and the thermal contribution, $\mathbb{C}^c$ and α^c are the elastic stiffness and the thermal expansion tensors of the grain, $\mathbb{C}$ and α are the corresponding (unknown) properties of the HEM. The distance from the boundary to the inclusion is assumed to be much larger than the dimensions of the inclusion, in order to neglect image stresses effects. The problem defined by Eqs. 55 and the condition of local stress equilibrium ($\sigma_{ij,j} = 0$), is the one of the inhomogeneous elastic inclusion solved in Section 4 of the Appendix. Equation A32 condenses the most relevant result: for ellipsoidal inclusions, stress and strain are uniform within the domain Ω of the inclusion, and they are linearly related to the stress and strain at the boundary through the so called *interaction equation*:

$$\tilde{\sigma} = - \, \mathbb{C} : \mathbb{R} : \tilde{\varepsilon} \tag{56a}$$

where

$$\tilde{\varepsilon} \equiv \varepsilon^c - \varepsilon$$
$$\tilde{\sigma} \equiv \sigma^c - \sigma \tag{56b}$$

define the deviations in strain and stress with respect to the average magnitudes. The tensor

$$\mathbb{R} = (\mathbb{I} - \mathbb{E}) : \mathbb{E}^{-1} \tag{57}$$

is called in what follows the *reaction tensor*, because it 'modulates' the stress reaction induced by a strain inhomogeneity. The Eshelby tensor $\mathbb{E}$ is a function of the overall elastic constants $\mathbb{C}$ and of the ellipsoid aspect ratios, and is derived in Section 2 of the Appendix. The importance of this result is that it does not require solving the local stress and strain variations around the inclusion. Instead, replacing Eqs. 55 in Eq. 56 gives the *total strain* in the grain as:

$$\varepsilon^c = (\mathbb{C}^c + \mathbb{C} : \mathbb{R})^{-1} : \left[(\mathbb{C} + \mathbb{C} : \mathbb{R}) : \varepsilon + (\mathbb{C}^c : \alpha^c - \mathbb{C} : \alpha) \, \delta T \right] \tag{58}$$

Averaging both members over all the grains and imposing the average condition Eq. 30a over the strain leads to:

$$\varepsilon = \langle (\mathbb{C}^c + \mathbb{C} : \mathbb{R})^{-1} : (\mathbb{C} + \mathbb{C} : \mathbb{R}) \rangle : \varepsilon + \langle (\mathbb{C}^c + \mathbb{C} : \mathbb{R})^{-1} : (\mathbb{C}^c : \alpha^c - \mathbb{C} : \alpha) \rangle \, \delta T \tag{59}$$

Since this equation has to hold for arbitrary ε and δT, the first coefficient in the second member has to be the identity and the second has to be zero. These conditions give, after some algebraic manipulation, the following equations from which the overall elastic and thermal moduli can be calculated

$$\mathbb{C} = \langle (\mathbb{C}^c + \mathbb{C} : \mathbb{R})^{-1} : (\mathbb{C} + \mathbb{C} : \mathbb{R}) : \mathbb{C} \rangle \tag{60a}$$

and

$$\alpha = \langle (\mathbb{C}^c + \mathbb{C} : \mathbb{R})^{-1} : \mathbb{C} \rangle^{-1} : \langle (\mathbb{C}^c + \mathbb{C} : \mathbb{R})^{-1} : \mathbb{C}^c : \alpha^c \rangle \tag{60b}$$

Equation 60a gives the overall elastic constants as a weighted average of the single-crystal elastic constants. The weighting factor is not only the grain's volume fraction but it also contains a term that depends upon the relative stiffness of grain and matrix, and upon the grain shape through $\mathbb{R}$. As a consequence, Eq. 60a is a non-linear implicit equation, because $\mathbb{R}$ depends upon the unknown $\mathbb{C}$ through the Eshelby tensor $\mathbb{E}$. Equation 60a has to be solved iteratively, starting with a guess value of $\mathbb{C}$ to evaluate the second member and reinserting the result of this evaluation as a new guess until the guess and the second member coincide to a specified accuracy. This procedure is called *self-consistent* because, eventually, the overall stiffness used to evaluate the second member yields the same stiffness in the first member.

Once the overall $\mathbb{C}$ has been obtained, the effective thermal properties of the aggregate are calculated in a straightforward manner using Eq. 60b. Observe that the effective thermal properties are coupled to the elastic inhomogeneity of the medium, except when there is no elastic inhomogeneity (either because all the grains have the same orientation or because they are elastically isotropic). In such a case $\mathbb{C}^c$ is the same for all grains and equal to $\mathbb{C}$, and then $\alpha = \langle \alpha^c \rangle$, which is the same result given by the upper and lower bound estimates (Eqs. 35 and 36).

Equation 56 can also be expressed in terms of the stresses using Eq. 31, as:

$$\sigma = \mathbb{C}^c : (\mathbb{C}^c + \mathbb{C} : \mathbb{R})^{-1} : [(\mathbb{C} + \mathbb{C} : \mathbb{R}) : (\mathbb{C}^{-1} : \sigma + \alpha \, \delta T) + (\mathbb{C}^c : \alpha^c - \mathbb{C}:\alpha) \, \delta T] - \mathbb{C}^c : \alpha^c \, \delta T \tag{61}$$

Performing the grain average and imposing the condition Eq. 30b that $\langle \sigma^c \rangle = \sigma$, leads to a set of equations from which $\mathbb{C}$ and α can be obtained:

$$\left\langle \left(\mathbb{C}^{c^{-1}} + [\mathbb{C}:\mathbb{R}]^{-1} \right)^{-1} : \left(\mathbb{C}^{-1} + [\mathbb{C}:\mathbb{R}]^{-1} \right) \right\rangle = \mathbb{I} \tag{62a}$$

$$\left\langle \left(\mathbb{C}^{c^{-1}} + [\mathbb{C}:\mathbb{R}]^{-1} \right)^{-1} : \left(\alpha - \alpha^c \right) \right\rangle = 0 \tag{62b}$$

It can be shown that these equations are equivalent to Eqs. 60 only if $\mathbb{R}$ can be factorized from the averages. This requires the Eshelby tensor to be the same for all the grains, which is the case if all the ellipsoids representing the grains have the same shape and orientation. If not, Eqs. 60a and 62a have to be solved simultaneously [HILL 1965]. Only when all the grains have the same shape is the condition that the average grain strain is equal to the overall strain (Eq. 30a) equivalent to the condition that the average grain stress is equal to the overall stress (Eq. 30b).

3.5 Numerical applications

We present in this Section several examples of the relation between texture, single-crystal anisotropy, and the overall elastic and thermal properties of poly-crystalline aggregates. We use for this purpose the upper bound, lower bound and self-consistent estimates described in the previous Sections. In particular, we calculate the elastic and thermal moduli of textured and non-textured copper, zirconium, calcite and uranium polycrystals. Next, we study how the texture affects the directional Young's modulus in the same aggregates. Finally, we present an application of how the texture affects the directional wave propagation in textured olivine.

Average elastic and thermal properties

We have chosen four materials having different crystallographic symmetry for illustrating the relation between texture and anisotropy. These materials are: copper (cubic), α-zirconium (hexagonal), calcite (trigonal) and α-uranium (orthorhombic). For a detailed analysis of the relation between the crystal symmetry and the symmetry of the tensors representing the mechanical properties of the material, the reader is referred to the book of NYE [1957] and to Chapter 1 of this book. Here it suffices to say that Cu, α-Zr, calcite and α-U have, respectively, three, five, six, and nine independent elastic moduli; and, respectively, one, two, two, and three independent thermal expansion moduli. The corresponding numerical values are listed in Table II. Zirconium is the least anisotropic of the four materials as far as the crystal elastic properties are concerned. In the thermal properties, all the crystals but copper are strongly anisotropic.

For comparative purposes, and in order to evidence the relationship between texture and anisotropy, we are going to consider two distributions of crystal orient-ations associated with each of these materials. The first is a random distribution of orientations corresponding to a non-textured polycrystal. Since by definition such aggregate exhibits isotropic properties, the overall elastic tensor and the thermal expansion tensor are fully characterized by two and one independent moduli, respectively. The second distribution corresponds to a polycrystal with a rolling texture. The properties of this aggregate exhibit orthotropic symmetry: the elastic tensor has nine independent moduli and the thermal expansion tensor has three. Observe that the symmetry of the overall properties is determined by the texture, and not by the single-crystal symmetry. However, since the moduli of the random and the textured aggregates can be written in terms of the same single-crystal moduli using any of the averaging procedures discussed in the previous Section, it is possible, in principle, to infer the latter from a knowledge of the texture, the single-crystal symmetry, and the measured overall properties.

The rolling textures that we use for our calculations are not experimental, but have been produced using the large strain simulation techniques that are explained later in this book. The non-textured polycrystal is represented using a distribution of 1000 randomly generated orientations. The textured polycrystals are represented by means of 681 (Cu), 631 (Zr), 500 (calcite) and 2916 (U) orientations. One pole figure for each of the materials is plotted in Fig. 1, just for the sake of illustrating the ortho-rhombic symmetry and the resemblance with experimental pole figures.

Table II. Single crystal elastic constants (units of GPa; expressed in contracted index notation) and thermal expansion coefficients (units of 10^{-6} K^{-1}).

	C_{11}	C_{22}	C_{33}	C_{12}	C_{13}	C_{23}	C_{44}	C_{55}	C_{66}	C_{14}	α_{11}	α_{22}	α_{33}	Ref.
Copper	168.0	168.0	168.0	121.4	121.4	121.4	75.4	75.4	75.4	0.0	16.00	16.00	16.00	[a,b]
Zirconium	143.5	143.5	164.9	72.5	65.4	65.4	32.1	32.1	35.5	0.0	5.70	5.70	11.40	[a,c]
Calcite	146.3	146.3	85.3	59.7	50.8	50.8	34.1	34.1	43.3	-20.8	-5.60	-5.60	25.00	[a,d]
Uranium	214.8	198.6	267.1	46.5	21.8	107.6	124.4	73.4	44.3	0.0	25.41	0.65	20.65	[e,f]

[a] SIMMONS, G. & WANG, H. (1971) *Single Crystal Elastic Constants and Calculated Aggregate Properties* (Cambridge MA: M.I.T. Press).
[b] NYE, J.F. (1957). *Physical Properties of Crystals* (Oxford: Clarendon Press).
[c] MACEWEN, S.R., TOMÉ, C.N. & FABER JR, J. (1989), *Acta Metall.* **37**, 979-989.
[d] *International Critical Tables* (New York: McGraw Hill, 1929).
[e] FISHER E.S. (1966), *J. Nucl. Mater.* **18**, 39-54.
[f] LLOYD, L.T. & BARRETT, C.S. (1966), *J. Nucl. Mater.*, **18**, 55-59.

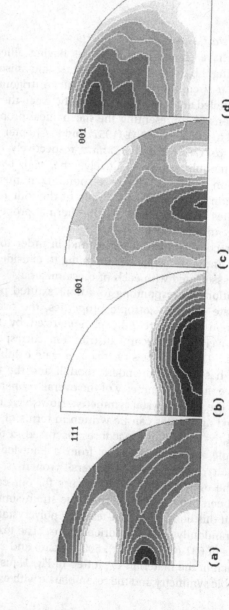

Fig. 1. Pole figures corresponding to the discrete sets of orientations used for the calculation of thermal and elastic properties of aggregates in Section 3.5: (a) copper; (b) zirconium; (c) calcite; (d) uranium. Equal-area projection; RD (=1) up, TD (=2) right.

For illustrating the coupling between the texture and the associated macroscopic properties, the calculated textures serve the same purpose as the experimental ones.

Besides considering different crystal and polycrystal symmetries, we also investigate here the influence of grain shape on the macroscopic properties of the polycrystal. For this purpose we consider two different grain geometries: equiaxed and flat grains. In the first case we use a sphere for representing the grain shape, while in the latter case we use an ellipsoid with a ratio 5:1:0.2 between the rolling, transverse and normal directions. Although these aspect ratios are not typical for heavily rolled materials, they are large enough to highlight the effect of grain shape upon polycrystal properties. Notice that in elasticity grain shape effects are incorporated into the self-consistent scheme through the inclusion formalism, but are not a feature of the upper and lower bound calculations.

The results of this Section have been condensed in Tables III to VI, and are discussed in what follows. Table III reproduces the elastic and thermal moduli of the rolled (orthotropic) polycrystal, as predicted using the upper bound (UB), the lower bound (LB) and the self-consistent (SC) averaging schemes, and is meant to highlight the dependence of the calculated moduli upon the averaging scheme being used. An equiaxed grain was assumed for the SC calculation, in order to separate grain shape effects from the effects of the approximation. For zirconium, the three schemes give very similar results, due to the rather weak elastic anisotropy of the single crystal. For uranium, the differences between the UB and the LB predictions are of the order of 10%, while for calcite and copper they show a spread of about 20% for some constants. It is clear that more reliable estimates of the elastic constants, such as SC or Hill estimates, are required for calcite, copper and uranium. The Hill estimate gives values that are in between the upper and lower bound predictions, and which do not differ much from the more rigorous SC estimates. It is interesting to remark, in connection with Eq. 37, that the off-diagonal components $\mathbb{C}_{12}$, $\mathbb{C}_{13}$ and $\mathbb{C}_{23}$ given by the 'upper-bound' estimate are smaller than the ones predicted with the 'lower-bound' estimate. As for the average thermal moduli, the difference in the UB and LB predictions is rather large for calcite and uranium, and the more rigorous estimate of the SC approach seems to be necessary for these materials.

Next, in Table IV, we investigate the effect of texture upon the associated polycrystal moduli. For this purpose, we compare the prediction of the SC estimate for non-textured and textured polycrystals, assuming equiaxed grain shape. The weak anisotropy of zirconium and the high symmetry of copper explain why similar predictions are obtained for the random and the textured polycrystal. The elastic constant of textured calcite differ by about 10% from the non-textured polycrystal, and for uranium a rather large difference of about 20% is predicted. Except for copper, which is thermally isotropic, very large differences are obtained for the thermal coefficients of rolled zirconium, calcite and uranium, with respect to the non-textured material. This latter result indicates that the texture has to be accounted for if reliable estimates of the thermal moduli are required.

Finally, the effect of grain morphology upon the polycrystal moduli is assessed in Table V, where the SC estimates for the rolled polycrystal are presented, assuming either equiaxed or flat grains. The differences in elastic constants are small, not exceeding 3%, and can probably be neglected. A slightly larger effect of grain shape upon the average thermal moduli is predicted, especially in the case of the calcite polycrystal.

Table III. Effect of calculation scheme on polycrystal moduli. Polycrystal elastic constants (units of GPa; expressed in contracted index notation) and thermal expansion coefficients (units of 10^{-6} K^{-1}), corresponding to rolling texture and equiaxed grain shape, calculated with the upper bound (UB), the lower bound (LB) and the self-consistent (SC) schemes.

		C_{11}	C_{22}	C_{33}	C_{12}	C_{13}	C_{23}	C_{44}	C_{55}	C_{66}	α_{11}	α_{22}	α_{33}
Copper	UB	209.2	211.6	206.8	101.4	106.3	103.8	54.8	57.3	52.4	16.00	16.00	16.00
	SC	199.8	202.3	197.4	106.1	111.0	108.5	47.8	50.4	45.3	16.00	16.00	16.00
	LB	188.3	190.3	186.3	112.2	116.3	114.2	38.7	41.1	36.7	16.00	16.00	16.00
Zirconium	UB	142.8	142.9	147.9	70.2	68.8	73.5	40.1	34.1	34.6	5.43	7.83	9.69
	SC	142.7	142.4	146.8	70.3	69.1	73.6	39.5	34.0	34.5	5.67	7.79	9.40
	LB	142.6	141.9	145.8	70.4	69.4	73.8	38.8	33.9	34.4	5.85	7.75	9.19
Calcite	UB	128.1	132.6	120.5	53.5	53.5	52.6	36.0	36.9	37.0	1.95	0.81	5.35
	SC	118.1	122.6	110.9	54.8	53.9	53.3	30.8	31.8	31.5	2.72	1.41	6.27
	LB	108.3	112.2	102.4	54.8	53.4	53.1	26.5	27.5	26.8	3.85	2.33	7.62
Uranium	UB	214.4	249.4	216.4	67.2	58.3	50.5	89.8	87.8	94.6	11.00	16.57	18.47
	SC	208.8	240.9	212.7	67.6	58.7	51.3	86.3	84.0	90.1	10.45	17.22	18.49
	LB	204.1	233.1	209.6	68.2	59.3	52.2	83.2	80.7	85.8	9.76	18.40	18.55

Table IV. Effect of texture on polycrystal moduli. Polycrystal elastic constants (units of GPa; expressed in contracted index notation) and thermal expansion coefficients (units of 10^{-6} K^{-1}), corresponding to a random and a textured (rolled) polycrystal, calculated with the self-consistent (SC) scheme and assuming equiaxed grain shape.

		C_{11}	C_{22}	C_{33}	C_{12}	C_{13}	C_{23}	C_{44}	C_{55}	C_{66}	α_{11}	α_{22}	α_{33}
Copper	rand.	202.1	202.1	202.1	107.3	107.3	107.3	46.4	46.4	46.4	16.00	16.00	16.00
	text.	199.8	202.3	197.4	106.1	111.0	108.5	47.8	50.4	45.3	16.00	16.00	16.00
Zirconium	rand.	143.5	143.5	143.5	71.2	71.2	71.2	36.1	36.1	36.1	7.63	7.63	7.63
	text.	142.7	142.4	146.8	70.3	69.1	73.6	39.5	34.0	34.5	5.67	7.79	9.40
Calcite	rand.	117.3	117.3	117.3	53.9	53.9	53.9	31.7	31.7	31.7	3.44	3.44	3.44
	text.	118.1	122.6	110.9	54.8	53.9	53.3	30.8	31.8	31.5	2.72	1.41	6.27
Uranium	rand.	229.4	229.4	229.4	54.6	54.6	54.6	82.1	82.1	82.1	15.39	15.39	15.39
	text.	208.8	240.9	212.7	67.6	58.7	51.3	86.3	84.0	90.1	10.45	17.22	18.49

Table V. Effect of grain shape on polycrystal moduli. Polycrystal elastic constants (units of GPa; expressed in contracted index notation) and thermal expansion coefficients (units of 10^{-6} K^{-1}), calculated for a textured (rolled) polycrystal using the self-consistent approach, and assuming flat (a_1=5, a_2=1, a_3=0.2) and equiaxed (a_1=a_2=a_3=1) grains.

		C_{11}	C_{22}	C_{33}	C_{12}	C_{13}	C_{23}	C_{44}	C_{55}	C_{66}	α_{11}	α_{22}	α_{33}
Copper	flat	205.0	202.5	194.9	102.1	109.7	112.2	45.7	48.2	48.0	16.00	16.00	16.00
	equi.	199.8	202.3	197.4	106.1	111.0	108.5	47.8	50.4	45.3	16.00	16.00	16.00
Zirconium	flat	142.6	142.3	146.4	70.3	69.2	73.7	39.4	34.0	34.5	5.75	7.78	9.31
	equi.	142.7	142.4	146.8	70.3	69.1	73.6	39.5	34.0	34.5	5.67	7.79	9.40
Calcite	flat	121.3	123.6	107.7	53.5	53.0	53.5	29.9	31.2	33.0	2.08	1.10	7.67
	equi.	118.1	122.6	110.9	54.8	53.9	53.3	30.8	31.8	31.5	2.72	1.41	6.27
Uranium	flat	209.6	241.2	212.3	66.6	58.3	51.2	86.2	84.3	90.9	10.83	17.17	18.38
	equi.	208.8	240.9	212.7	67.6	58.7	51.3	86.3	84.0	90.1	10.45	17.22	18.49

Table VI. Effect of grain shape on the shape factor. Components of the reaction tensor $\mathbb{R}$ in Eq. 57 (non-dimensional; expressed in contracted index notation), calculated for a random polycrystal using the elastic constants listed in Table V, and assuming flat (a_1=5, a_2=1, a_3=0.2) and equiaxed (a_1=a_2=a_3=1) grains. The calculated Young's modulus (E, units of GPa) and Poisson's ratio (v) of the random polycrystal are also listed.

		$\mathbb{R}_{11}$	$\mathbb{R}_{22}$	$\mathbb{R}_{33}$	$\mathbb{R}_{12}$	$\mathbb{R}_{13}$	$\mathbb{R}_{23}$	$\mathbb{R}_{44}$	$\mathbb{R}_{55}$	$\mathbb{R}_{66}$	E	v
Copper	flat	29.036	2.660	0.014	-0.204	0.084	0.037	0.143	0.110	2.527	127.5	0.346
	equi.	0.921	0.921	0.921	-0.237	-0.237	-0.237	0.586	0.586	0.586		
Zirconium	flat	29.787	2.719	0.016	-0.188	0.087	0.051	0.136	0.110	2.581	96.25	0.332
	equi.	0.929	0.929	0.929	-0.212	-0.212	-0.212	0.570	0.570	0.570		
Calcite	flat	24.015	2.381	0.026	-0.108	0.077	0.066	0.157	0.115	2.615	83.54	0.310
	equi.	0.930	0.930	0.930	-0.182	-0.182	-0.182	0.561	0.561	0.561		
Uranium	flat	30.728	2.826	0.036	0.167	0.173	0.150	0.115	0.109	2.531	208.5	0.192
	equi.	0.978	0.978	0.978	0.008	0.008	0.008	0.511	0.511	0.511		

The grain morphology does, however, make a difference in the non-dimensional reaction tensor $\mathbb{R}$ that couples strain and stress deviations in the inclusion (Eq. 56). For isotropic materials $\mathbb{R}$ exhibits a weak dependence upon Poisson's ratio and, although it does depend strongly upon the grain shape, as the values quoted in Table VI demonstrate, such dependence is somewhat smoothed by Eqs. 60 giving the average moduli. The dependence of $\mathbb{R}$ on the grain shape has more important consequences for plasticity, where it provides the formal basis of the Relaxed Constraints approach discussed in Chapter 8. As TIEM &AL. [1986] demonstrate, Eq. 56 indicates that those strain components which are coupled with the stress through small components of the reaction tensor $\mathbb{R}$ do not lead to large stress deviations with respect to the average. As a consequence, those strain components can be relaxed without causing much stress disturbance. For the particular case of rolling (see Table VI), $\mathbb{R}_{44}$ and $\mathbb{R}_{55}$ are reduced substantially when passing from an equiaxed to a flat grain morphology. As a consequence, plastic deviations in the strain components $\varepsilon_4 \equiv \varepsilon_{23}$ and $\varepsilon_5 \equiv \varepsilon_{13}$ can be accommodated elastically without inducing large reaction stresses.

Directional Young's modulus

Although Tables II to VI contain all the relevant information about the elastic anisotropy of the rolled aggregate, a frequently used indicator of such anisotropy is the directional Young's modulus in the rolling plane. The directional Young's modulus $E(\alpha)$ measures the elastic stiffness of the material when subjected to a tensile stress (all other stress components being kept zero) along a direction which is in the rolling plane, at an angle α with respect to the RD. From an operational point of view, E is calculated as the ratio between the stress and the deformation measured along the sample axis (superscript α)

$$E(\alpha) = \frac{\sigma_{11}^{\alpha}}{\varepsilon_{11}^{\alpha}} = \frac{1}{\mathbb{S}_{1111}^{\alpha}}$$

where

$$\mathbb{S}_{1111}^{\alpha} = R_{1i}^{\alpha}\, R_{1j}^{\alpha}\, R_{1k}^{\alpha}\, R_{1l}^{\alpha}\, \mathbb{S}_{ijkl}$$

(63)

Here $\mathbb{S}$ is the overall elastic compliance expressed in the reference system where axis 1 is along the RD and axis 2 along the TD, and R^{α} is the matrix that rotates around the ND by an angle α.

Two factors affect the variation of E with α: the single-crystal elastic anisotropy and the texture. Two extreme cases that can be envisaged are: a random aggregate of highly anisotropic grains, and a highly textured polycrystal of elastically isotropic grains. In both cases the aggregate is elastically isotropic and there is no dependence of the directional Young's modulus with the angle α: however, a large discrepancy between the UB and the LB is to be expected in the former case, while in the latter case the UB and the LB coincide. In a broad sense, it can be stated that the difference between the UB and the LB estimate of $E(\alpha)$ in a given direction is controlled by the single-crystal anisotropy.

We have calculated $E(\alpha)$ for the four materials considered here, using the aggregate elastic constants listed in Table III, which correspond to the UB, LB, and SC schemes. The results are plotted in Fig. 2. Zirconium is the most isotropic at the single crystal level and, as a consequence, the UB estimate is close to the LB estimate, while the SC one is in-between (see Fig. 2b). The texture is sharp enough, though, to produce a non-negligible dependence of Y on α (notice however that the scale used in Fig. 2b is not the same as for the other materials). The curves for rolled copper (Fig. 2a) exhibit a large discrepancy between the UB and the LB, and a non-negligible directional dependence. The differences with similar calculations reported by other authors are not significant, and can be attributed to differences in the texture or the elastic constants used for doing the calculation. A detailed analysis of the dispersion that the latter factors introduce in the calculated Young's modulus has been performed by WRIGHT &AL. [1993b]. Calcite does not show much variation with α but, on the other hand, it exhibits a large difference between the UB and the LB, as a consequence of the marked anisotropy of the single crystal (Fig. 2c). Finally, uranium shows both a strong directional dependence and a marked effect of the single-crystal anisotropy, which reflects in the discrepancy between the UB and the LB (Fig. 2d).

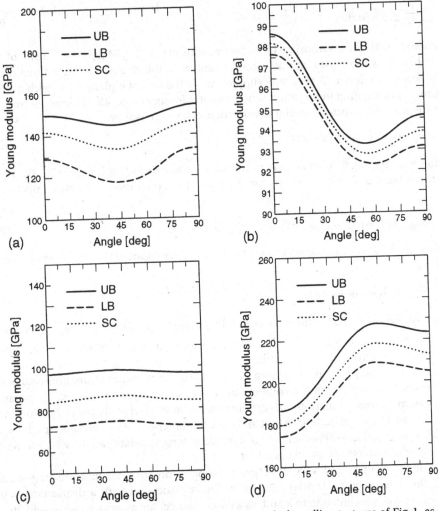

Fig. 2. Directional Young's modulus associated with the rolling textures of Fig. 1, as predicted using each of the three elastic estimates listed in Table III (UB: upper bound; LB: lower bound; SC: self-consistent). (a) copper; (b) zirconium; (c) calcite; (d) uranium.

Elastic waves propagation

   We resume in this Section some basic concepts concerning wave propagation in elastically anisotropic media (see LANDAU &LIFSHITZ [1959], Chapter 3), and present a brief application for the case of orthorhombic olivine. The problematic of elastic waves is of particular interest in the case of geological materials, because it describes the propagation of seismic waves. A comprehensive discussion of the relation between texture and wave propagation in geological media is presented in Chapter 14.

   Geological aggregates in general, and those composed of olivine in particular, are formed by very anisotropic single crystals and are heavily textured. As a consequence, they exhibit marked elastic anisotropy, and a strong directional dependence of the wave velocity. Elastic wave propagation is described by the dynamic equilibrium equation

$$\rho \frac{\partial^2 u_i}{\partial t^2} = \mathbb{C}_{ijkl}\, u_{k,lj}$$

(64)

combined with the appropriate boundary conditions. In Eq. 64, ρ is the density of the medium, $\mathbb{C}$ is the elastic stiffness tensor, and $\mathbf{u}$ is the displacement, a function of position $\mathbf{x}$ and time t. A relatively simple case is the one of a plane wave with a wave vector $\mathbf{k}$ propagating in an infinite medium at a frequency ω. The functional form of the displacement vector associated with such a wave is given by

$$u_i(\mathbf{x}) = u_i^o \exp\left[i\left(\mathbf{k}\cdot\mathbf{x} - \omega t\right)\right]$$

(65)

Differentiating Eq. 65 with respect to time and position, and replacing in Eq. 64, leads to three homogeneous equations for the pre-exponential displacement components u_i^o

$$\left(\mathbb{C}_{ijlm}\, k_1\, k_j - \rho\, \omega^2\, \delta_{im}\right) u_i^o = 0$$

(66)

This system has non-trivial solutions only if the determinant of the coefficients is zero, that is:

$$\left|\mathbb{C}_{ijlm}\, k_1\, k_j - \rho\, \omega^2\, \delta_{im}\right| = 0$$

(67)

The eigenvalues $\lambda = \rho\omega^2$ of this equation define three frequencies $\omega = \sqrt{\lambda/\rho}$ associated with a given wave vector $\mathbf{k}$. Replacing these eigenvalues in Eq. 66 gives three orthogonal eigenvectors $\mathbf{u}^o$, describing the direction of the displacements associated with each frequency. It is clear from Eq. 67 that the eigenvalues λ are homogeneous functions of degree two of the components k_i. As a consequence, the eigenfrequency ω is a homogeneous function of degree one, and the wave velocity $\mathbf{v} = \partial\omega/\partial\mathbf{k}$ does not depend on the magnitude of $\mathbf{k}$ but only on its direction. For isotropic media $\mathbf{v}$ is parallel to $\mathbf{k}$, and the displacements $\mathbf{u}^o$ are either longitudinal (parallel with the wave vector) or transverse (perpendicular to it).

   For anisotropic media the direction of propagation does not coincide, in general, with the wave vector. The largest eigenvalue is associated with a displacement $\mathbf{u}^o$ which is closest in direction to $\mathbf{k}$ and, as a consequence, the associated wave velocity is called 'longitudinal'. In general, the velocity of a plane wave along a direction $\mathbf{x}$ parallel to $\mathbf{k}$ can be inferred from Eq. 65 to be:

$$v = \frac{\omega}{k} = \sqrt{\frac{\lambda}{\rho}\frac{1}{k}} \tag{68}$$

As an application, we compare the longitudinal wave velocities in a single crystal of olivine and in a textured olivine aggregate. Olivine has orthorhombic crystallographic symmetry, a density $\rho = 3.324$ g/cm^3, and the single-crystal elastic constants are given in Table VII.

Table VII. Single crystal elastic constants of olivine (units of GPa; expressed in contracted index notation) [DANDEKAR 1968].

	$\mathbb{C}_{11}^c$	$\mathbb{C}_{22}^c$	$\mathbb{C}_{33}^c$	$\mathbb{C}_{12}^c$	$\mathbb{C}_{13}^c$	$\mathbb{C}_{23}^c$	$\mathbb{C}_{44}^c$	$\mathbb{C}_{55}^c$	$\mathbb{C}_{66}^c$
Olivine	324.0	198.0	249.0	59.0	79.0	78.0	66.7	81.0	79.3

Predicted 100, 010 and 001 pole figures of an aggregate deformed by simple shear at high temperature to an equivalent strain of 0.5 (γ=0.866) represented by means of 500 discrete orientations are shown in Fig. 3 (see also Fig. 29 in Chap. 4). The elastic constants of the aggregate, calculated using the discrete orientations file and the self-consistent procedure described in Section 3.4 are, in contracted index (see footnote 2) notation:

$$\mathbb{C}\,[\text{GPa}] = \begin{bmatrix} 245.7 & 78.7 & 79.2 & -0.2 & 1.3 & 11.0 \\ 78.7 & 232.9 & 77.0 & -0.7 & -0.4 & 8.6 \\ 79.2 & 77.0 & 234.9 & -0.3 & 1.1 & 1.6 \\ -0.2 & -0.7 & -0.3 & 79.2 & 4.3 & 0.2 \\ 1.3 & -0.4 & 1.1 & 4.3 & 82.6 & -0.2 \\ 11.0 & 8.6 & 1.6 & 0.2 & -0.2 & 83.1 \end{bmatrix}$$

These values reflect the fact that the main texture component is rotated with respect to the principal axes and that the 500 discrete orientations do not have perfect orthotropic symmetry.

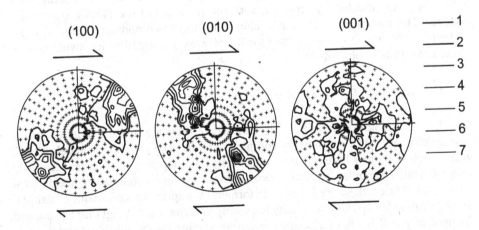

Fig. 3. Predicted 100, 010, and 001 pole figures of an olivine aggregate deformed in simple shear up to γ=0.866. (Equal-area projection; levels correspond to multiple of random orientations).

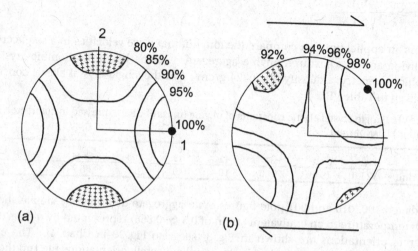

Fig. 4. Longitudinal elastic wave velocities as a function of the propagation direction, corresponding to: (a) single-crystal olivine (maximum = 9857 m/s); (b) the textured aggregate of Fig. 3 (maximum = 8900 m/s). (Equal-area projection; level values correspond to percent of maximum wave velocity.)

The wave velocities for the single crystal and the polycrystal are depicted in Fig. 4, using the direction of the wave vector **k** for representing them as contour lines in a pole figure. For the single crystal the maximum velocity is 9857 m/s along the stiffer direction x_1. The minimum velocity, along the more compliant direction x_2, is 7708 m/s (about 22% slower) and shows the effect of elastic anisotropy upon the propagation of elastic waves. These two values represent upper and lower bounds for the wave velocity attainable in any olivine aggregate in any direction. In a random aggregate the longitudinal wave velocity is the same in any direction and equal to 8499 m/s. For the particular case of the polycrystal represented in Fig. 3 the maximum is 8900 m/s and coincides, as expected, with the direction of the (100) texture component. The minimum is 8170 m/s, along the texture component (010). A comprehensive discussion of the characteristics of wave propagation in olivine can be found in Section 3.1 of Chapter 14.

4. Visco-elastic Properties of Polycrystals

There are some time effects in elasticity: they are usually referred to as 'anelasticity' and are expressed by complex moduli. The topic of 'visco-elasticity', which we will address here, refers to a different time-dependent situation: elasticity coupled with Newtonian viscous creep (linear). The one dimensional mechanical equivalent of visco-elasticity is a system formed by a spring in series with a dashpot. The equivalent system for anelasticity is a spring in series with a Voigt unit (spring and dashpot in parallel). Both systems present an instantaneous elastic response upon application of a force, but visco-elasticity differs from anelasticity in two respects: the final strains can be much greater than the elastic ones, and the changes are irreversible after removing the force.

Newtonian creep may take place under conditions where mass flow can occur by the diffusion of vacancies to grain boundaries through the volume or along the boundaries. It can also occur during neutron irradiation, as a result of stress induced changes in the diffusivity of vacancies and self interstitials. Newtonian creep is also different from the 'creep' associated with plasticity (especially at high temperatures), where the relation between stress and strain-rate is usually grossly non-linear, even when steady-state has been reached.

This Section deals with the superposition, at the single crystal level, of linear, time-independent elasticity, and linear creep (Maxwellian crystals). We will frequently refer to this mechanism as 'linear visco-elasticity', in order to differentiate it from the 'creep' associated with plasticity. There are non-linear transients that occur in the polycrystal response, which are due to the coupling between the two deformation mechanisms. As a consequence, aggregates of Maxwellian crystals do not exhibit, in general, a Maxwellian response.

The linear visco-elastic regime exhibits several features which differ markedly from the purely elastic response of aggregates, and which justify a separate treatment. In the first place, it describes a more general regime, of which the elastic response is only a limit case. Secondly, the anisotropy in the polycrystal response is generally much larger than in the case of elasticity, and the difference between an upper and a lower bound estimate can be quite marked. Finally, the problem admits an exact self-consistent solution when the HEM embedding assumption introduced in Section 3.4 is assumed. This solution gives the purely elastic and the purely viscous formulations of polycrystals as limit cases, together with the evolution with time of the internal stress in the grains due to creep. It also gives the evolution with time of the overall visco-elastic properties, which in this case depend not only upon intrinsic properties, such as the grain anisotropy and the polycrystal texture, but are also a function of the stress imposed to the aggregate. The formulation that we present below applies to a general linear visco-elastic material, and accounts for the anisotropy of elastic and creep properties at the grain and at the polycrystal level.

There are several works addressing the modeling of linear and non-linear creep of polycrystals. BROWN [1970] proposes one of the first self-consistent models for treating transient thermal creep in an elastically isotropic random polycrystal, but his formulation leads to an upper bound approach for the steady state, indicating that the grain-matrix interaction is not properly accounted for. HUTCHINSON [1976] develops a non-linear viscous model for analyzing the steady-state thermal creep of polycrystals. His model gives the linear creep response as a particular case but, since the formulation does not include elastic effects, it is not applicable to the transient state. LAWS & McLAUGHLIN [1978] model the transient creep response of a system composed of ellipsoidal particles embedded in a uniform visco-elastic matrix, from initial loading to steady state, assuming that the visco-elastic properties of the matrix are known in advance and that the particles only deform elastically. WOO [1985] implements a Kröner–Budiansky–Wu type formulation (see BUDIANSKY & WU 1962) which accounts for polycrystal texture, but which converges to an upper limit in the steady state (as BROWN'S [1970] formulation) because no plastic accommodation is allowed in the matrix. Later, WOO [1987] proposes a fully visco-elastic formulation based on an empirical constitutive law, and derives approximate analytic results for the steady-state creep and growth of the polycrystal; Woo's steady state formulation was implemented numerically by TOMÉ &AL. [1993]. A model of visco-elastic aggregates

which allows for viscous accommodation in the matrix has been proposed by
KOUDDANE &AL. [1993,1994]. Their formulation is based on an approximate constitutive law for the overall response, and only applies to an aggregate of visco-elastic
isotropic inclusions, under the assumption that elasticity is incompressible and
homogeneous. TURNER & TOMÉ [1993] and ROUGIER &AL. [1994] develop a fully
anisotropic formulation for visco-elastic polycrystals based on the Carson–Laplace
transform. In a later paper TURNER &AL. [1994] show that the empirical expressions
used as starting point by WOO [1987] and by KOUDDANE &AL. [1993] can be obtained as
approximations of the exact formulation by TURNER & TOMÉ [1993].

In Sections 4.1 to 4.4 that follow we present a completely general self-consistent
formulation of visco-elasticity in polycrystals. Such a formulation accounts for
anisotropic elasticity, anisotropic linear creep, and their coupling. As a technologically meaningful application, in Section 4.5 we calculate the response to hydrostatic pressure of nuclear reactor tubes made of a zirconium alloy, from the initial
loading until they reach steady-state conditions.

4.1 Single crystal and polycrystal constitutive equations

The formulation that follows applies to creep strains that can eventually be much
larger than the elastic component (typically of the order of 2%), but not so large as to
be concerned with plastic rotations. As a consequence of this requirement the stress
rate and the strain rate will be referred to a fixed frame, and partial time derivatives
coincide with total time derivatives.

At the crystal level it is possible to decompose the strain rate into an elastic and an
inelastic contribution, where the latter accounts for the creep contribution. In the
approach of small strain rates, these terms are uncoupled and the total strain rate is
given by the following differential equation, which describes the strain and stress
evolution in the domain of each grain:

$$\mathbf{D}^c(t) = \mathbb{S}^c : \frac{d\boldsymbol{\sigma}^c(t)}{dt} + \mathbb{K}^c : \boldsymbol{\sigma}^c(t) \tag{69}$$

In writing this equation we explicitly state that strain rates within each grain vary
with time. The elastic component relates the strain rate to the stress rate through the
elastic compliance tensor $\mathbb{S}^c$. The creep rate is assumed to exhibit a linear dependence with the stress acting in the crystal, through the single-crystal creep compliance
tensor $\mathbb{K}^c$. Under certain conditions of stress, temperature and neutron flux, Eq. 69
provides a good description of creep in irradiated materials. In such a case, $\mathbb{K}^c$ is
mainly determined by the dislocation and grain boundary structure, and by the
vacancy and interstitial dynamics. In what follows $\mathbb{S}^c$ and $\mathbb{K}^c$ are assumed to be anisotropic but constant in time.

In visco-elasticity, the more general constitutive law for the single crystal is
expressed as an integral, known as the *Stieltjes convolution or hereditary integral*:

$$\varepsilon_{ij}^c(t) = \left(\mathbb{M}^c * \boldsymbol{\sigma}^c\right)_{ij} \equiv \int_{-\infty}^{t} \mathbb{M}_{ijkl}^c(t-t') \frac{d\sigma_{kl}^c(t')}{dt'} dt' \tag{70}$$

where $\mathbb{M}^c(t)$ is defined as the visco-elastic tensor for the single crystal. When the
elastic compliance $\mathbb{S}^c$ and the creep compliance $\mathbb{K}^c$ are constant in time, the visco-
elastic tensor adopts the simple form:

$$M^c(t) = S^c + K^c t \tag{71}$$

It is easy to prove that Eq. 69 follows from replacing Eq. 71 in Eq. 70. In principle, the problem is well defined from the point of view of the grain, because the elastic and the creep compliances of the single crystal are known or can be inferred from experimental measurements.

The formulation can be extended to deal with more general conditions than the ones presented above. For example, the elastic and viscous properties of the grains do not need to be constant provided that the evolution with time of each tensor is known 'a priori'. In addition, a term representing a transformation rate (such as a thermal or growth rate) may be added to the constitutive law Eq. 69. Such a term has to be independent of stress but may vary with time in a known way. Finally, the constitutive law Eq. 69 may include a 'history' term proportional to the strain. See TURNER & TOMÉ [1993] for a discussion of these features.

Although each single crystal that composes the aggregate obeys a Maxwell differential law Eq. 69, the constitutive response of the polycrystal is not of the Maxwell type because the overall elastic and inelastic rate components are coupled and cannot be separated. The coupling arises from the grain interactions, which are responsible for the generation of intercrystalline stresses. Without loosing generality, one can postulate that the overall response of the polycrystal is also described by a visco-elastic constitutive equation having the form of a *hereditary integral*:

$$\varepsilon_{ij}(t) = \int_{-\infty}^{t} M_{ijkl}(t-t') \frac{d\sigma_{kl}(t')}{dt'} dt' \tag{72}$$

where σ and ε are the overall stress and strain tensors. $M(t)$ is the visco-elastic tensor *of the aggregate*, and cannot be decomposed into an elastic and a Newtonian component, as is the case for the single crystal (Eq. 71). The strain rate tensor can be derived from Eq. 72 using the Leibniz Theorem:

$$D(t) = \frac{d\varepsilon(t)}{dt} = M(t=0) : \frac{d\sigma(t)}{dt} + \int_{-\infty}^{t} \frac{dM(t-t')}{dt'} : \frac{d\sigma(t')}{dt'} dt' \tag{73}$$

A special case of the constitutive laws Eqs. 72 and 73, relevant to creep, is when the load is applied instantaneously and remains constant throughout the duration of the experiment. In such a case the time dependence of the imposed stress is of the form

$$\sigma(t) = \sigma^0 \, h(t) \tag{74}$$

where $h(t)$ is the Heaviside step function. Replacing $\sigma(t)$ in Eq. 72 and using the result that the time derivative of the Heaviside function is Dirac's $\delta(t)$ function, gives a straightforward linear relation between overall stress and strain, and a simple interpretation for the visco-elastic tensor, namely:

$$\varepsilon(t) = M(t) : \sigma^0 \tag{75}$$

The associated strain rate for this particular case follows from Eqs. 73 and 74 as

$$D(t) = \frac{dM(t)}{dt} : \sigma^0 + M(t=0) : \sigma^0 \, \delta(t) \tag{76}$$

The last term is a fictitious infinite strain rate associated with the instantaneous stress loading at $t=0$, and can be ignored for practical purposes. TURNER &AL. [1994] show that when the restrictions required by the *quasi-elastic* method apply (evolution of functions with log t is quasi-linear), the integral Eq. 73 giving the overall strain rate can be written in an approximate form

$$D(t) \cong \frac{d\mathbb{M}(t)}{dt} : \sigma(t) + \mathbb{M}(t=0) : \dot{\sigma}(t)$$

(77)

It is clear that neither the exact creep rate Eq. 76 associated with a constant stress, nor the approximate creep rate Eq. 77 associated with a time varying stress, represent the constitutive response of a Maxwell solid, because the viscosity is not constant with time. Only in the limit when $t \to \infty$ and under the assumption of steady-state conditions the time derivative of the visco-elastic tensor becomes constant and can be identified with an intrinsic viscosity.

4.2 The Carson transform: definition and properties

The Carson transform is particularly useful for solving the Stieltjes convolution integrals. For this reason, here we state briefly its essential properties, and the relations to be used in the Sections that follow. The Carson transform of a function (or a tensor) f(t) is defined by:

$$\hat{f}(s) = s \int_0^\infty e^{-st} f(t)\, dt$$

(78)

and differs from the Laplace transform by the multiplicative factor s in front of the integral. It is easy to demonstrate, based on the definition Eq. 78, that the transform of the derivative is:

$$\widehat{\frac{df}{dt}}(s) = s\left[\hat{f}(s) - f(t=0)\right]$$

(79)

The latter property is what permits a reduction of the linear differential equations into linear algebraic equations in the transformed space. The Carson transform is better suited than the Laplace transform for dealing with integrals with the Stieltjes convolution form which arise in visco-elastic theory:

$$f(t) = \left(f_1(t) * f_2(t)\right) = \int_0^t f_1(t-t') \frac{df_2(t')}{dt'}\, dt'$$

(80)

The Carson transform of f(t) is simply the product of the transform of each function in the integrand, without any extra factor s:

$$\hat{f}(s) = \hat{f}_1(s) \times \hat{f}_2(s)$$

(81)

The previous equations also apply to the components of tensors (of any rank n). By definition, the tensors $f_1(t)$ and $f_2(t)$ are Stieltjes inverses if their Stieltjes convolution obeys:

$$\left(f_1(t) * f_2(t)\right) = h(t)\,\mathbb{I}$$

(82)

where h(t) is the Unit Step (Heaviside) function and $\mathbb{I}$ is the n-th rank identity tensor. Since the Carson transform of h(t) is simply one, Eqs.(80) and (81) lead to the relation:

$$\hat{f}_1(s) : \hat{f}_2(s) = \mathbb{I}$$

(83)

indicating that when the tensors f_1 and f_2 are Stieltjes inverses in the direct space, their Carson transforms are tensorial inverses, which is an useful property for their numerical manipulation in transformed space.

Other useful properties of the Carson transform [SCHAPERY 1974] refer to some limit forms, such as the initial value theorem

$$\lim_{s \to \infty} \hat{f}(s) = \lim_{t \to 0} f(t) \tag{84}$$

the final value theorem

$$\lim_{s \to 0} \hat{f}(s) = \lim_{t \to \infty} f(t) \tag{85}$$

and the limit relation

$$\lim_{s \to 0} s\hat{f}(s) = \lim_{t \to \infty} f(t)/t \tag{86}$$

All three limits are consistent with the fact that the Carson transform of t is the reciprocal of s.

4.3 The visco-elastic inhomogeneity

The creep tensor of the aggregate $M(t)$ that appears in the constitutive equations of Section 4.1 is not known a priori. It will be calculated here using the assumption of the HEM previously employed for the elastic problem. Within the HEM approach, each grain is regarded as an inclusion embedded in and interacting with the average medium represented by the other grains. As an intermediate step, in this section we present the solution to the problem of a visco-elastic inhomogeneity embedded in a visco-elastic matrix [LAWS &McLAUGHLIN 1978; MURA 1987]. A standard technique in linear visco-elasticity consists in transforming the integral equations associated with the problem into algebraic equations by means of the Carson transform, described in Section 4.2 above. In the transformed space the formulation adopts the same form as the one of the elastic inhomogeneity, and the same formal solution applies, something which is known in continuum mechanics as *the correspondence principle*.

The inclusion (domain Ω) has a governing equation of the form Eq. 70, and outside the domain of the inclusion (V-Ω), the visco-elastic compliance of the HEM is assumed to be uniform and equal to the overall compliance defined by Eq. 72. As a consequence, the local and the overall response are described by the equations:

$$\varepsilon^c(\mathbf{x},t) = \int_{-\infty}^{t} M^c(t-t') : \frac{d\sigma^c(\mathbf{x},t')}{dt'} dt' \qquad \mathbf{x} \in \Omega \tag{87a}$$

$$\varepsilon(\mathbf{x},t) = \int_{-\infty}^{t} M(t-t') : \frac{d\sigma(\mathbf{x},t')}{dt'} dt' \qquad \mathbf{x} \in (V-\Omega) \tag{87b}$$

$$\varepsilon(t) = \int_{-\infty}^{t} M(t-t') : \frac{d\sigma(t')}{dt'} dt' \qquad \text{on } \Gamma_V \tag{87c}$$

The local dependence of the stress and the strain has been explicitly indicated, and the last equation describes the overall response of the medium. Since the three integrals have the form of a Stieltjes convolution, their Carson transform gives a linear relation between the stress $\hat{\sigma}$ and the strain $\hat{\varepsilon}$ in Carson space, of the form:

$$\hat{\sigma}^c(\mathbf{x},s) = [\hat{M}^c(s)]^{-1} : \hat{\varepsilon}^c(\mathbf{x},s) \qquad \mathbf{x} \in \Omega \tag{88a}$$

$$\hat{\sigma}(\mathbf{x},s) = [\hat{M}(s)]^{-1} : \hat{\varepsilon}(\mathbf{x},s) \qquad\qquad \mathbf{x} \in (V-\Omega) \qquad (88b)$$

$$\hat{\sigma}(s) = [\hat{M}(s)]^{-1} : \hat{\varepsilon}(s) \qquad\qquad \text{on } \Gamma_V \qquad (88c)$$

where we have inverted the equations in order to give the stress as a linear function of the strain, as in the elastic problem. The Carson transform of the single-crystal compliance Eq. 71 is simply

$$\hat{M}^c(s) = \mathbb{S}^c + \mathbb{K}^c s^{-1} \qquad (89)$$

In writing Eqs. 88 we are assuming that the stress and the initial strain in the grains is zero at $t=0$. An extra term has to be added when an instantaneous transformation takes place at $t=0$ or when residual stresses are present in the aggregate at $t=0$ [TOMÉ &AL. 1996]. Since the Carson transform defined by Eq. 78 operates on the variable time and not on the position, the stress equilibrium equation also holds for the transformed stress components

$$\frac{\partial \hat{\sigma}_{ij}(\mathbf{x},s)}{\partial x_j} = 0 \qquad (90)$$

The problem described by Eqs. 88 and 90 is formally identical to the one of the elastic inhomogeneity analyzed in Section 4 of the Appendix, when the stress and the total strain are replaced by the transformed of those magnitudes, and when the elastic stiffness tensor is replaced by the inverse of the transformed creep tensor. This result, known as the *correspondence principle*, states that the tensorial relationships that hold for the elastic inhomogeneity will be the same for the equivalent tensors of the visco-elastic problem for a parametric value of s. In particular, the stress and the strain are uniform inside the inclusion, and their deviations from the overall stress and strain are described by the equivalent, in Carson space, of the *interaction equation* Eq. 56:

$$(\hat{\varepsilon}^c(s) - \hat{\varepsilon}(s)) = -\mathbb{A}(s) : \hat{M}(s) : (\hat{\sigma}^c(s) - \hat{\sigma}(s)) \qquad (91a)$$

where

$$\mathbb{A}(s) = (\mathbb{I} - \mathbb{E}(s))^{-1} : \mathbb{E}(s) \qquad (91b)$$

is the *accommodation tensor*, which couples the strain deviations with the stress deviations. $\mathbb{E}(s)$ is the Eshelby tensor and can be calculated for a general ellipsoidal inclusion embedded in an anisotropic medium using the integrals derived in Section 2 of the Appendix, replacing the elastic stiffness by the creep stiffness $[\hat{M}(s)]^{-1}$. As a consequence, it is possible to exploit the inclusion formalism for developing a fully anisotropic numerical treatment of the problem. Replacing the constitutive laws Eqs. 88a and 88c into the interaction law Eq. 91 leads to a linear relation between the stress in the inhomogeneity and the stress applied at the boundary of the material:

$$\hat{\sigma}^c(s) = \mathbb{B}^c(s) : \hat{\sigma}(s) \qquad (92a)$$

where

$$\mathbb{B}^c(s) = (\hat{M}^c(s) + \mathbb{A}(s):\hat{M}(s))^{-1} : (\hat{M}(s) + \mathbb{A}(s) : \hat{M}(s)) \qquad (92b)$$

is called the *localization tensor*. The above equations provide a complete solution (in Carson space) of the linear visco-elastic inhomogeneity problem, provided that the inclusion properties $\mathbb{S}^c$, $\mathbb{K}^c$, the matrix visco-elastic properties $\mathbb{M}(t)$, and the external loading history $\sigma(t)$ are known. The Carson transform of the overall strain and the strain in the inhomogeneity is obtained by inverting Eqs. 88a and 88c.

4.4 Self-consistent polycrystal model

While an aggregate of visco-elastic grains also exhibits a visco-elastic response, its overall properties are a function of the grain's properties and the grain's response, and are not known a priori. A solution to this problem can be derived using the approach of the Homogeneous Effective Medium derived in Section 3.4 for the elastic poly-crystal. Within this approach we assume that each grain is an inhomogeneity embedded in a HEM exhibiting a response equal to the average response of all grains. Applying the Carson transform Eq. 78 to Eqs. 30, which express the overall strain (stress) as the volume average of the grain strains (stresses), gives:

$$\hat{\varepsilon}(s) = \langle \hat{\varepsilon}^c(s) \rangle \tag{93a}$$

$$\hat{\sigma}(s) = \langle \hat{\sigma}^c(s) \rangle \tag{93b}$$

After replacing Eqs. 88a and 92a in Eq. 93a, the transformed of the average strain adopts the form:

$$\hat{\varepsilon}(s) = \langle \hat{\mathbb{M}}^c(s) : \mathbb{B}^c(s) \rangle : \hat{\sigma}(s) \tag{94}$$

A comparison of this expression with the polycrystal constitutive law Eq. 88c leads to an implicit equation that the overall visco-elastic tensor has to fulfill:

$$\hat{\mathbb{M}}(s) = \langle \hat{\mathbb{M}}^c(s) : \mathbb{B}^c(s) \rangle \tag{95}$$

This equation is formally identical to the one derived for the purely elastic polycrystal in Section 3.4, except that: (i) the single-crystal compliance is now $\hat{\mathbb{M}}^c(s) = (\mathbb{S}^c + \mathbb{K}^c s^{-1})$ instead of $\mathbb{S}^c$ alone and, (ii) the formulation is stated in the Carson space. The overall creep compliance in Carson space $\hat{\mathbb{M}}(s)$ may be derived solving Eq. 95 iteratively for fixed values of the parameter s. Our final goal, however, is to calculate the time evolution of stress and strain in grains and polycrystal, and also of the polycrystal properties. When all the ellipsoids representing the grains have the same *shape and orientation* the Eshelby tensor is the same for all the grains and the accommodation tensor $\mathbb{A}(s)$ can be factorized from the averages, in which case starting from the average stress condition Eq. 93b also leads to Eq. 95.

Before transforming the tensors defined in the Carson space back into the direct space, it is instructive to investigate the polycrystal response for two limit cases: initial loading and steady state. Applying the initial value theorem Eq. 84 to the Eq. 95 and defining the initial compliance $\mathbb{S} = \lim\limits_{t\to 0} \mathbb{M}(t)$, gives:

$$\mathbb{S} = \langle \mathbb{S}^c : (\mathbb{S}^c + \mathbb{A}:\mathbb{S})^{-1} : (\mathbb{S} + \mathbb{A}:\mathbb{S}) \rangle \tag{96}$$

where the accommodation tensor has the same functional form: $\mathbb{A} = (\mathbb{I}-\mathbb{E})^{-1}\mathbb{E}$, but the Eshelby tensor $\mathbb{E}$ is only a function of the elastic stiffness $\mathbb{C} = \mathbb{S}^{-1}$. Equation 96 is the self-consistent equation that defines the polycrystal elastic compliance, and it can be shown to be the inverse of Eq. 60a, giving the overall elastic stiffness. Such a result was to be expected, since the initial response of the aggregate to instantaneous loading is elastic, while the process of relaxation by creep has associated a characteristic time delay. The initial stress in each grain can also be derived by applying the initial value theorem Eq. 84 to Eq. 92, which gives

$$\sigma^c(t=0) = \lim_{t \to 0} \sigma^c(t) = \left(\mathbb{S}^c + \mathbb{A}:\mathbb{S}\right)^{-1} : (\mathbb{S} + \mathbb{A}:\mathbb{S}) : \sigma(t=0) \tag{97}$$

This expression depends only upon the elastic properties of the material and $\sigma^c(t=0)$ is the instantaneous load applied to each grain at $t=0$.

For the limit case of steady-state conditions, the applied stress remains constant in time and the internal stresses achieve an asymptotic equilibrium value. The steady-state form of the visco-elastic compliance, defined by the creep compliance $\mathbb{K} = \lim_{t \to \infty} \mathbb{M}(t)/t$, can be calculated by applying the final value theorem Eq. 85 to Eq. 95:

$$\mathbb{K} = \langle \mathbb{K}^c : (\mathbb{K}^c + \mathbb{A}:\mathbb{K})^{-1} : (\mathbb{K} + \mathbb{A}:\mathbb{K}) \rangle \tag{98}$$

where $\mathbb{A} = (\mathbb{I}-\mathbb{E})^{-1}:\mathbb{E}$ is the viscous accommodation tensor, and the Eshelby tensor $\mathbb{E}$ is a function of the creep stiffness $\mathbb{K}^{-1}$. Equation 98 is a particular case of the creep compliance derived by HUTCHINSON [1976] for steady-state power law creep, and coincides with the one found by WOO [1987] for the case of steady-state irradiation creep. The steady-state stress in each grain is obtained by applying the final value theorem Eq. 85 to Eq. 92:

$$\sigma^c(t=\infty) = \lim_{t \to \infty} \sigma^c(t) = \left(\mathbb{K}^c + \mathbb{A}:\mathbb{K}\right)^{-1} : (\mathbb{K} + \mathbb{A}:\mathbb{K}) : \sigma(t=\infty) \tag{99}$$

Equations 98 and 99 indicate that the steady-state creep compliance and the grain stress are functions of only the single-crystal creep tensors, and are independent of the single-crystal elastic properties. As a consequence, only the deviatoric stress components can be calculated from Eq. 99, and the information about the hydrostatic stress component in the grain is lost in the limit process. This does not imply that the final pressure in each grain is zero, but that it can only be calculated following the stress evolution with time, as will be shown below.

The limit cases emphasize the fact that the overall response of an aggregate of Maxwell type single crystals is, initially, purely elastic and, once the steady state has been reached, purely viscous. In addition, the previous limits provide the corresponding tensorial magnitudes in direct space. However, to investigate the evolution of the polycrystal from the elastic loading up to the steady state, one has to be able to transform the solution obtained in the Carson space back into direct space. The analytic inverse of the Carson transform adopts the form of integrals in complex space and is very difficult to treat numerically, even when elastic and viscous isotropy are assumed at the grain level [ROUGIER &AL. 1993]. There is, though, an approximate inversion procedure which is applicable to the case of stress relaxation processes, and which only requires to evaluate the Carson transform at a discrete set of points. Such a procedure, called the *collocation method*, is described in detail by SCHAPERY [1962, 1974] and is used by LAWS & MCLAUGHLIN [1978].

The collocation method was originally proposed by BIOT [1954] and consists in expressing the creep function as a finite Dirichlet series of the form:

$$\mathbb{M}(t) = \mathbb{S} + \mathbb{K}\, t + \sum_{n=1}^{N} \mathbb{M}^{(n)}\left(1 - e^{-t/\tau_n}\right) \tag{100}$$

where the range of summation N depends on the particular system, the τ_n are character-istic retardation times (positive and constant) and the expansion coefficients $\mathbb{M}^{(n)}$ are constant fourth order tensors. It is evident, from the limits Eqs. (96) and (98) that this expression gives the correct initial and final values of $\mathbb{M}$, since:

$$\lim_{t \to 0} \mathbb{M}(t) = \mathbb{S}$$

$$ \tag{101}$$

and

$$\lim_{t \to \infty} \left(\mathbb{M}(t)/t\right) = \mathbb{K}$$

The unknown tensorial coefficients $\mathbb{M}^{(n)}$ of the Dirichlet series can be determined as follows. The Carson transform of Eq. 100 is:

$$\hat{\mathbb{M}}(s) - \mathbb{S} - \mathbb{K}\frac{1}{s} = \sum_{n=1}^{N} \mathbb{M}^{(n)}\left(1 + s\,\tau_n\right)^{-1} \tag{102}$$

If $\hat{\mathbb{M}}(s)$ is known at N points s_n, chosen as the reciprocals of the characteristic times ($s_n = 1/\tau_n$), the expression Eq. 102 leads to a linear system of N equations of the form

$$\hat{\mathbb{M}}(\tau_i^{-1}) - \mathbb{S} - \mathbb{K}\,\tau_i = \sum_{n=1}^{N} \mathbb{M}^{(n)}\left(1 + \frac{\tau_i}{\tau_n}\right)^{-1} \tag{103}$$

from where the N unknown tensorial coefficients $\mathbb{M}^{(n)}$ can be calculated performing a straightforward matrix inversion. The relaxation times τ_n (and as a consequence the points s_n) have to be chosen such as to give a good representation of the tensor $\mathbb{M}(t)$ in the transition interval. They are usually taken as equispaced in a logarithmic scale for describing stress relaxation processes.

The complete calculation procedure is as follows: first evaluate the elastic compliance $\mathbb{S}$ and the creep compliance $\mathbb{K}$ for the pure elastic and the pure viscous regimes, respectively. Then, a discrete set of values $\{s_n, n=1,N\}$ is selected and the single-crystal creep tensor $\hat{\mathbb{M}}^c(s_n) = (\mathbb{S}^c + \mathbb{K}^c s_n^{-1})$ is calculated. For each value, the polycrystal visco-elastic compliance $\hat{\mathbb{M}}(s_n)$ is derived by solving iteratively the self-consistent Eq. 95. Afterwards, the collocation method described above permits one to calculate the coefficients $\mathbb{M}^{(n)}$ and with them it is possible to calculated the creep function at an arbitrary time t using the Dirichlet series Eq. 100.

An expansion of the same form as Eq. 100 can also be used to represent the stress in the grains $\sigma(t)$. The initial and the steady-state values are given by Eqs. 97 and 99, and provide the first two terms of the expansion, as in the case of the visco-elastic tensor. The co-efficients of the series are calculated evaluating $\hat{\sigma}^c(s)$ at N points using Eq. 92 and the collocation method explained above. Once $\mathbb{M}(t)$ and $\hat{\sigma}^c(t)$ are obtained under the form of Dirichlet series, the grain strain $\varepsilon^c(t)$ and the overall strain $\varepsilon(t)$ are calculated, integrating Eqs. 70 and 72, respectively. Those integrals may admit an analytic solution, depending on the functional form of the loading history $\sigma(t)$.

4.5 Numerical applications

Arbitrary creep and elastic tensors can be chosen for the purpose of performing a numerical application of the visco-elastic theory described above. In what follows, however, we have selected an example that corresponds to an actual technological application for which the creep response of the grains is proportional to the applied stress. The formulation described above is used to analyze irradiation induced creep in a cylindrical tube subjected to internal pressure. We assume that the material is a zirconium alloy of hexagonal structure, of the type used in nuclear reactors, with a texture typical of extruded Zr alloy tubes. The polycrystalline sample is modeled as a discrete collection of 147 orientations with weights chosen to reproduce the experimental texture. The 0002 basal pole figure, presented in Fig. 5, shows an important basal component close to the transverse direction. For comparison purposes we also consider a non-textured sample represented by means of 227 randomly generated orientations of equal weights. In what concerns the grain shape, and in order to simplify the discussion, here we consider equiaxed grains with a ratio of ellipsoid axes 1:1:1. The effect of grain shape upon the visco-elastic response of pressure tubes is discussed in detail by TOMÉ &AL. [1993] and by TURNER and TOMÉ [1993].

The pressurization of the tube induces an instantaneous distribution of stress in the grains, dictated by the elastic characteristics of the aggregate. These stresses are in equilibrium with the external stress but are not uniform because of the elastic anisotropy at the grain level. These stresses are relaxed through creep and evolve until a steady-state configuration is achieved which is controlled by the creep properties of the aggregate. At steady-state conditions the stress and the creep rate in the grains remain invariant.

The process described above is relevant to the subject of "Texture and Anisotropy" because it highlights the relation between single-crystal anisotropy and the anisotropy of the mechanical response of the aggregate for two completely different regimes: the purely elastic and the purely viscous, and for intermediate states as well. For this particular aggregate texture, the anisotropy of the elastic properties is not marked but, as the transition to steady state takes place, the increasing contribution of creep to the

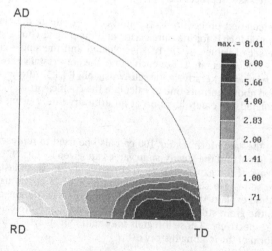

Fig. 5. Basal pole figure for pressure tube. (Equal-area projection; levels are in multiples of a random distribution).

response makes the latter increasingly anisotropic. For the purpose of comparison, the response of a non-textured aggregate is also analyzed. As stated at the beginning of Section 4.1, grain reorientation during deformation is not accounted for in the calculation, since the strains considered are not so large.

The hexagonal elastic constants used are the ones reported in Table II for zirconium single crystals at room temperature, and are expected to be similar to the elastic constants of zirconium alloys. They are reported again in Table VIII below.

Table VIII. Zirconium single-crystal elastic stiffness (units of GPa; in contracted index notation).

$\mathbb{C}^c_{11}$	$\mathbb{C}^c_{22}$	$\mathbb{C}^c_{33}$	$\mathbb{C}^c_{44}$	$\mathbb{C}^c_{55}$	$\mathbb{C}^c_{66}$	$\mathbb{C}^c_{12}$	$\mathbb{C}^c_{13}$	$\mathbb{C}^c_{23}$
143.5	143.5	164.9	32.1	32.1	35.5	72.5	65.4	65.4

In the case of irradiation creep, vacancies and interstitials produced through neutron collisions with lattice atoms are annihilated at dislocations and grain boundaries. Intrinsic anisotropic diffusion and stress induced anisotropic diffusion produce a bias in the rate of interstitials or vacancies that are annihilated at a given sink. This bias depends on the crystallographic orientation of the sink, such as: basal or prismatic dislocation loops, grain boundaries parallel or perpendicular to the basal planes, etc. The net effect is a dimensional change in the crystal, with volume conservation. It has been shown [WOO, 1985] that the associated deformation rate is proportional to the applied stress at reactor operation temperatures ($T < 0.3\ T_m$) and for a wide range of neutron fluxes. The single-crystal creep compliances that we use are the ones reported by CHRISTODOULOU &AL. [1993], from fitting steady-state irradiation creep measurements done in Zr-Nb pressure tubes. The creep tensor $\mathbb{K}^c$ has the hexagonal symmetry of the lattice, but the additional condition of non-dilatational deformation reduces the number of independent components from five to three, as we discussed at the end of Section 2. For this particular case, the eigenvalues of the single-crystal compliance are given by Eq. 28 as:

$$K^{(1)} = \frac{3}{2}\mathbb{K}^c_{33} = 0.2\ 10^{-6}\ \text{GPa}^{-1}\ \text{h}^{-1}$$

$$K^{(2)} = K^{(5)} = \mathbb{K}^c_{11} - \mathbb{K}^c_{12} = 2\ \mathbb{K}^c_{66} = 5.0\ 10^{-6}\ \text{GPa}^{-1}\ \text{h}^{-1} \qquad (104)$$

$$K^{(3)} = K^{(4)} = 2\ \mathbb{K}^c_{44} = 2.0\ 10^{-6}\ \text{GPa}^{-1}\ \text{h}^{-1}$$

$$K^{(6)} = 0$$

The creep compliances are customarily expressed in units of neutrons per hour per square meter, because irradiation creep is proportional to the neutron flux. The eigenvalues given above have been expressed in the more usual units of hours used in visco-elasticity, by assuming a reference flux of $\Phi=1.22$ neutrons per hour per square meter. The associated (non-zero) cartesian components, given by Eq. 17, are listed in Table IX and show a marked anisotropy, related to the lack of efficient creep mechanisms for accommodating deformation along the c-axis of the crystal. In particular, the creep produced by the deformation mode $\mathbf{b}^{(1)}$, associated with the eigenvalue $K^{(1)}$, comes from the sole contribution of pyramidal dislocations. The modes $\mathbf{b}^{(2)}$ and $\mathbf{b}^{(5)}$ correspond to shear in the prismatic planes, and are mainly associated with the contribution of prismatic dislocations. The modes $\mathbf{b}^{(3)}$ and $\mathbf{b}^{(4)}$ correspond to shear in the basal plane, and require activity of basal dislocations. The creep properties of the crystals and their anisotropy depend on the relative density of these dislocations, generated during the fabrication of the tube. The eigenvalues listed in Eq. 104 are consistent with the typical dislocation densities observed in Zr-Nb pressure tubes [TOMÉ &AL. 1996].

Table IX. Zr-2.5%Nb single-crystal creep compliance (units of 10^{-6} [GPa h]$^{-1}$, corresponding to a reference flux: $\Phi=1.22$ neutrons/hour/m^2; in contracted index notation. (See also footnote 2.)

K_{11}^c	K_{22}^c	K_{33}^c	K_{44}^c	K_{55}^c	K_{66}^c	K_{12}^c	K_{13}^c	K_{23}^c
2.533	2.533	0.133	1.0	1.0	2.5	-2.467	-0.067	-0.067

We first use the limits of the visco-elastic formulation to calculate the overall elastic and viscous polycrystal properties given by the self-consistent Eqs. 96 and 98. The independent components of the elastic stiffness tensor associated with the orthotropic symmetry of the sample are listed in Table X, while the steady-state creep compliances are listed in Table XI. For the latter case, only six components are independent because of the incompressibility condition. Differences between the elastic constants of the extruded tube and of the rolled zircaloy sheet (Table III) are due to differences in the associated textures. For comparison purposes, the upper and lower bound estimates which result from assuming homogeneous stress or homo-geneous strain rate throughout the aggregate are also shown in Tables X and XI. As a consequence of the relatively low elastic anisotropy of the single crystal, the three estimates give similar values for the overall elastic stiffness (Table X).

Table X. Calculated elastic stiffness for pressure tubes, in the Upper-Bound, Self-Consistent and Lower-Bound estimates (units of GPa; expressed in contracted index notation).

	$\mathbb{C}_{11}$	$\mathbb{C}_{22}$	$\mathbb{C}_{33}$	$\mathbb{C}_{44}$	$\mathbb{C}_{55}$	$\mathbb{C}_{66}$	$\mathbb{C}_{12}$	$\mathbb{C}_{13}$	$\mathbb{C}_{23}$
U B	144.4	150.5	143.0	33.8	34.9	37.8	71.3	70.5	68.4
S C	143.8	149.4	142.8	33.7	34.8	37.2	71.6	70.6	68.7
L B	143.2	148.2	142.7	33.6	34.7	36.7	71.8	70.7	69.0

In the case of the overall creep moduli listed in Table XI, on the other hand, the discrepancy between the three methods is substantial, and the averages depend strongly on the single-crystal creep anisotropy. Contrary to the elastic case, where the upper and lower bounds gives estimates which are close to the prediction of the self-consistent scheme, using these bounds is not advisable for describing the viscous response of the aggregate [TOMÉ &AL. 1993].

Table XI. Calculated pressure tube creep compliance for the Upper Bound, Self-Consistent and Lower Bound estimates (units of [10^6 GPa h]$^{-1}$; in contracted index notation. See footnote 2).

	$\mathbb{K}_{11}$	$\mathbb{K}_{22}$	$\mathbb{K}_{33}$	$\mathbb{K}_{44}$	$\mathbb{K}_{55}$	$\mathbb{K}_{66}$	$\mathbb{K}_{12}$	$\mathbb{K}_{13}$	$\mathbb{K}_{23}$
U B	0.589	0.280	0.609	0.879	0.988	0.249	-0.130	-0.459	-0.150
S C	1.325	0.649	1.535	1.219	1.545	0.565	-0.219	-1.105	-0.429
L B	1.877	1.193	2.472	1.492	1.916	0.921	-0.298	-1.578	-0.893

Next, we calculate the time evolution of the visco-elastic tensor, and analyze the influence that texture has upon its behavior. The loading conditions considered here correspond to a thin-walled tube with no initial internal stresses, pressurized at $t=0$ with a pressure of 9.3 MPa. This load induces a biaxial stress state which depends on the stiffness of the aggregate and the thickness of the tube. For the present case the diagonal stress components are:

$$\sigma_1^0 = 0.0\,\mathrm{MPa} \qquad \sigma_2^0 = 115.0\,\mathrm{MPa} \qquad \sigma_3^0 = 57.5\,\mathrm{MPa}$$

The shear components are zero, and the main axes are taken in the radial, transverse and axial directions of the tube, respectively. We have shown (Eq. 76) that for the case of instantaneous uniform loading the overall creep rate is simply proportional to the applied stress:

$$\mathbf{D}(t) = \frac{d\mathbf{M}(t)}{dt} : \sigma^0 \qquad (t > 0) \tag{105}$$

The evolution with time of the creep rate $\mathbf{D}(t)$ and of the visco-elastic tensor $d\mathbf{M}(t)/dt$ is the subject of the analysis that follows. The expression (83) provides a clue to the expected behavior: initially, the elastic compliance dominates, while at large times the linear term proportional to the creep compliance controls deformation. The form and the duration of the transition depends upon the exponential terms in the Dirichlet expansion Eq. 100. In Fig. 6 we plot the predicted evolution of the non-zero components of $d\mathbf{M}/dt$ for the tube texture (Fig. 6a) and for the random aggregate (Fig. 6b). In both cases the aggregate is loaded with the same overall stress. Only one component is independent for the isotropic aggregate because $\mathbf{M}_{11} = \mathbf{M}_{12} = \mathbf{M}_{33}$, $\mathbf{M}_{44} = \mathbf{M}_{55} = \mathbf{M}_{66}$, $\mathbf{M}_{12} = \mathbf{M}_{13} = \mathbf{M}_{23}$, $\mathbf{M}_{11} + \mathbf{M}_{12} + \mathbf{M}_{13} = 0$ and $2\mathbf{M}_{44} = (\mathbf{M}_{11} - \mathbf{M}_{12})$, as opposed to six for the orthotropic one. In addition, the difference in the response of the textured and the random aggregate is evident, and is due to the presence of a strong basal component in the texture, combined with the marked creep anisotropy at the crystal level. Similar differences will show up again in relation to the associated creep rates.

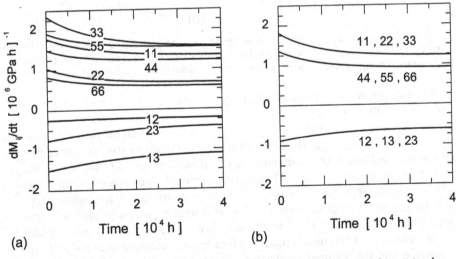

Fig. 6. Time derivative of the overall visco-elastic tensor predicted for: (a) tube texture; (b) random aggregate. The components are identified using the contracted index notation. The overall creep compliance is defined by the limit $\mathbb{K} = \lim_{t \to \infty} d\mathbf{M}/dt$. Equiaxed grain shape was used in the calculations.

The evolution with time of the visco-elastic tensor components M_{11} and M_{12} is plotted in Fig. 7 for the textured and the random aggregates. For comparison purposes, in addition to the self-consistent (SC) predictions, the ones of the lower (LB) and the upper (UB) bound models are also depicted. The overall compliance associated with the LB estimate is given by the average of the single-crystal compliances (Eq. 71): it exhibits a strictly linear time dependence and tends to underestimate the intergranular constraints. The UB amounts to averaging the inverse of the single-crystals compliances given by Eq. 71: it tends asymptotically to a linear dependence on time when the creep term dominates, and overestimates the grain-matrix interaction. These bounds provide a continuous transition between the purely elastic and the purely viscous values reported in Tables X and XI. As discussed above, the discrepancy between the UB and the LB estimates increases as deformation increases and the creep properties dominate.

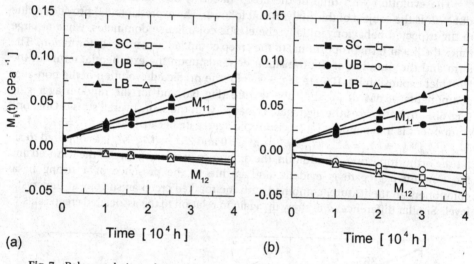

Fig. 7. Polycrystal visco-elastic components M_{11} and M_{12} predicted by the Upper Bound (UB), the Lower Bound (LB) and the Self-Consistent (SC) schemes for: (a) tube texture and (b) a random aggregate. Equiaxed grain shape was used in the calculations.

Although the visco-elastic moduli of Fig. 6 contain all the necessary information about the evolution with time of the system, the effect of texture over the response of the polycrystal is better evidenced by the evolution of the strain rate components. They are plotted in Fig. 8a for the tube texture and in Fig. 8b for the random aggregate. Only the diagonal components are non-zero for these textures and loading conditions, and they add up to zero because of volume conservation. It can be seen that the faster stress relaxation takes place during the first 5000 hours (0.6 years), while past that point the transition rates are much slower and tend to an asymptotic steady-state value. This transition is related to the evolution of the initial stresses induced by loading. The effect of texture is made evident by the comparison of the response of the random and the textured aggregates. The random polycrystal does not deform along the axial direction, and shows opposite rates in the transverse and the radial directions (Fig. 8b). This is consistent with the fact that the overall deviatoric stress is zero along the axial direction and has opposite signs along the other two directions.

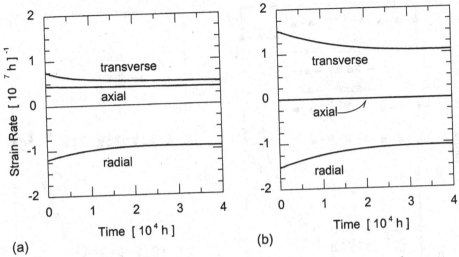

Fig. 8. Overall strain rate components (radial, transversal and axial) calculated for: (a) tube texture and (b) a random aggregate. Equiaxed grain shape was used in the calculations.

The textured tube, on the other hand, exhibits a positive axial strain rate and a lower transverse rate (Fig. 8a). The relationship between texture and creep response is of particular interest in nuclear applications, because it permits one to control – to a certain extent – the dimensional changes of reactor pressure tubes and, as a consequence, to optimize their performance under operating conditions.

Although the stress and strain evolution in each individual grain may be difficult to interpret, grain statistics are simple to derive and provide valuable information concerning the characteristics of the system. The evolution from a low anisotropy visco-elastic compliance (where the elastic properties dominate) to a more anisotropic creep-dominated behavior at the steady state is reflected in the departure of the individual grains from the overall polycrystal response. In Fig. 9 we plot the deviation of the stress and the strain rate components in the grains normalized by the norm of the average tensor, for the case analyzed in Fig. 8. The stress deviations are defined as:

$$\Delta\sigma_i = \sqrt{\frac{\left\langle\left(\sigma_i^c - \sigma_i\right)^2\right\rangle}{\sum_j \sigma_j\sigma_j}} \tag{106}$$

and similarly for the strain rate components. Both the textured and the random aggregates show similar deviations in stress as a function of time. At the beginning of deformation, when the tube is pressurized and elasticity dictates the response of the system, the diagonal stresses exhibit a low dispersion of about 3%, which is consistent with the low elastic anisotropy. After about 500 hours the effect of internal creep relaxation starts showing up and, as a result, the stress deviations increase to about 15%. This behavior reflects the transition towards the state where deformation is controlled by the creep properties of the system. The diagonal strain rate components, on the other hand, show a larger relative deviation which remains essentially constant throughout deformation. The creep anisotropy can be seen in that axial, transverse and radial rate deviations are very different for the textured sample, while

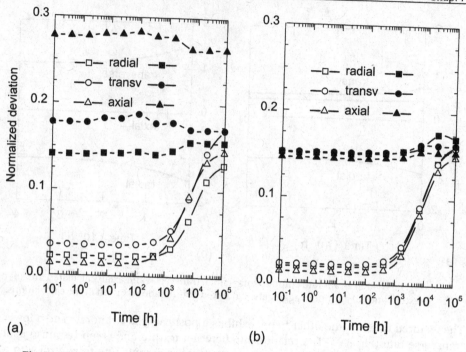

Fig. 9. Normalized stress and strain rate deviations in the grains, predicted by the SC model for the cases of Fig. 8: (a) tube texture and (b) a random aggregate. Open symbols denote diagonal components of stress; solid symbols denote strain rate components. Equiaxed grain shape was used in the calculations.

they are practically the same for the random aggregate. The non-diagonal stress and strain rate components exhibit a similar behavior and for that reason they are not plotted here.

A more complete application of the linear visco-elastic approach for textured aggregates which includes the relaxation of thermal stresses, the presence of a constant rate term in the single-crystal equation (growth rate) and the effect of grain shape upon the response of the material can be found in TURNER and TOMÉ [1993]. A further application, where the model is used for analyzing growth under irradiation of specimens with different initial configurations of internal stresses, can be found in TOMÉ &AL. [1996]. The extension of the self-consistent polycrystal model for non-linear visco-elasticity is discussed by TURNER &AL. [1994] and ROUGIER &AL. [1994].

Chapter 8

KINEMATICS AND KINETICS OF PLASTICITY

1. **Slip and Twinning in Crystals**
    1.1 Kinematics of crystallographic glide and deformation twinning
    1.2 The yield surface and plastic potential of single crystals
        Yield surface versus plastic potential
    1.3 Closing the single-crystal yield surface
        Independent deformation mode sets. Climb. Scaling
    1.4 Symmetry properties of the yield surface
        Vector notation. Sign dependence. Isotropy. The irreducible region of a yield surface.
        Effect of symmetry operators. Irreducible set of vertices of the SCYS.
        Topological domains and deformation mode maps
    1.5 Strain hardening
        Slip system interaction
    1.6 Kinetics of slip. Flow potential for rate-sensitive materials
        Thermal activation. Many slip systems: flow potential and explicit tensor relation.
        Flow stress scaling
    1.7 Constitutive relations for crystal plasticity
        Grains in a polycrystal. Stress and strain-rate scaling. Operational procedure.
        Linearization of the constitutive law. Elasticity. Restrictions

2. **Grain Interaction and Polycrystal Plasticity**
    2.1 Modes of polycrystal yielding: physics
        Band propagation versus bulk plasticity. Elasto-plastic transition.
        Grain size effects. Very heterogeneous materials
    2.2 Grain interaction: mechanics
        The flat-grain limit. Upper and lower bounds. Models of grain interaction
    2.3 Modeling bulk plasticity
        Microstructure. Grain-scale variations of stress and strain.
        Energy and work principles. Simulation versus modeling

3. **Kinematics in Polycrystalline Bodies**
    3.1 Frames of reference, 'material frames'
    3.2 Updating the grain shape
    3.3 Updating the texture
    3.4 Rotation fields, orientation jumping, orientation streaming
    3.5 Summary

KINEMATICS AND KINETICS OF PLASTICITY

U. F. Kocks

Plasticity is one of the most anisotropic of material behaviors. To some extent, this is true for the yield strengths in different directions under prescribed boundary displacements. Much more important is the converse case: when the displacements on some boundaries are free (and the boundary tractions prescribed, for example, zero) – the shape change may be very anisotropic.

Plastic anisotropy of polycrystals arises from two causes: anisotropy of the single-crystals properties, and texture. In this Chapter, we address the single-crystal properties and the way in which grains interact in a polycrystal. This leads to a formulation of the constitutive relations to be used in simulations of polycrystal plasticity.

Plastic deformation is also one of the main causes of texture formation and change. This is due to the kinematics of slip and twinning: the modes by which grains deform plastically. The rotations connected with the deformations are to be incorporated into schemes to update the state of the material during simulations.

This Chapter, then, provides the background for treatments of texture evolution and plastic properties of polycrystals in Chaps. 9 through 12. Plasticity in polycrystals is such a complex problem that it can only be solved by modeling and simulation. The models to be used depend on the degree of heterogeneity of the local material properties. Chapters 9 and 10 address cases where an assumption of quasi-homogeneity yields results that are internally consistent and, in many respects, in reasonable agreement with experiment. Chapters 11 and 12 deal with cases where heterogeneity is of the essence. The general features of polycrystal plasticity modeling are summarized in the present Chapter.

It should be noted at the outset that *all* the descriptions of plastic properties, from the constitutive relations for kinetics and evolution to the models for grain interaction, are extremely simple in comparison with the plethora of microstructural mechanisms that underlie these phenomena. It is one of the major, often re-experienced surprises of materials science, how few equations and parameters do in fact suffice to provide a meaningful representation of macroscopic material behavior.

1. Slip and Twinning in Single Crystals

1.1 Kinematics of crystallographic glide and deformation twinning

Dislocation glide on a plane (thought of as horizontal, Fig. 1a) causes a displacement of the upper half of the body with respect to the lower half by the Burgers vector **b**, in proportion to the increment of area δA swept out by the dislocation (lightly shaded). The average effect at the surfaces can be described by a simple shear in the amount of

$$\delta\gamma = b\,\delta A/V \tag{1}$$

where b is the magnitude of the Burgers vector and V is the total volume of the specimen. Only when a dislocation arrives at the surface does a slip step occur (with the step displacement vector equal to the Burgers vector, Fig. 1b); the macroscopic straining, however, is a continuous function of the area swept. When many dislocations move on many planes (as is necessary for significant macroscopic straining to occur) one describes this process as *slip*, on slip planes in slip directions, rather than as the movement of individual dislocations. The slip planes and slip directions are crystallographic in nature.

Deformation *twinning* is similar in its kinematic aspects. Figure 2a shows a thin lenticular twin (intersecting the front surface). It is of finite extent in the twinning plane, and its propagation can be described by the movement of twinning dislocations. In Fig. 2b, the twin has already propagated to the surface (or to a grain boundary). It is limited by plane interfaces – which can, by some mechanism, move perpendicular to themselves, thus increasing the twinned volume. The inside of the twin has suffered, with respect to the matrix grain, a homogeneous simple shear γ_t. Either form of the twin leads to a macroscopic simple shear in the amount of

$$\delta\gamma = \gamma_t\,\delta V/V \tag{2}$$

where $\delta V/V$ is the increase in the volume fraction of twinned material. Note that twinning leaves new boundaries behind, whereas dislocation glide leaves no evidence of its occurrence inside the matrix (except through stored debris, see Sec. 1.5).

It is essential that both slip and twinning are 'simple shears' and not 'pure shears': they correspond to a displacement in the direction $\hat{b}$ on one side of a plane perpendicular to $\hat{n}$, not also including an equal displacement by $\hat{n}$ on the plane perpendicular to $\hat{b}$. Simple shear induces *rotations*, and this is the entire basis for the formation of textures during plastic deformation.

A simple tensile deformation of a symmetric bicrystal serve as an instructive example. Figure 3a shows the sample, along with some slip plane traces and one schematic edge dislocation in each crystal. Dislocation movement in the direction indicated will lead to identical strains in both crystals, but opposite rotations. After tensile deformation, these two dislocations, as well as all similar ones in parallel slip planes, will have moved to the boundary, where they partially annihilate. The remnant describes exactly the change in boundary misorientation needed to accommodate the difference in rotation between the two grains. In this sense, one may ascribe all texture formation to dislocation accumulation within the boundaries [AERNOUDT &AL. 1993]; the point, however, is that this dislocation accumulation is an *alternative* description, not an additional effect.

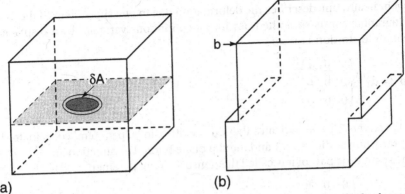

(a)   (b)

Fig. 1. (a) Dislocation glide: an incremental extension of slipped area, dA, leads to an incremental displacement of the far surfaces (not shown). (b) Slip: after completion of dislocation glide to the boundary (free surface or grain boundary), a step appears of displacement vector **b**.

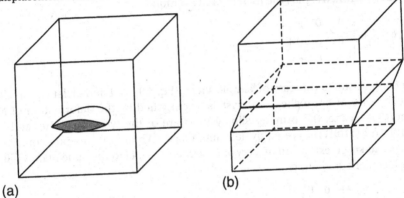

(a)   (b)

Fig. 2. (a) A twin lamella. An increase in the twinned volume by extension of the lamella leads to a displacement of the far surfaces (not shown). (b) When a twin extends throughout the cross section (of a crystal or a grain), a surface step appears; lateral displacement of the coherent interfaces also increases the twinned volume and therefore the surface displacements.

Fig. 3. Schematic representation of a longitudinal bicrystal under tension. Edge dislocations move in the direction of the arrows. After entering the grain boundary, they are partially annihilated; the remainder precisely accounts for the change in misorientation of the boundary necessary as a consequence of the (equal and opposite) rotation of the two crystals.

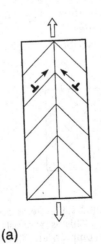

(a)   (b)

Algebraically, one describes the deformation due to one slip system (if the discrete crystallographic slip is smoothed out into a continuous variable) by a displacement gradient (or 'distortion') tensor

$$d\beta^{(s)} = d\gamma^{(s)} \begin{pmatrix} 0 & 0 & 1 \\ 0 & 0 & 0 \\ 0 & 0 & 0 \end{pmatrix} \tag{3}$$

where the superscript (s) indicates that $d\beta$ is written in 'slip system' coordinates (with the slip direction as direction 1 and the slip plane normal as direction 3).

The symmetric part of it is called the (tensor) strain increment

$$d\varepsilon^{(s)} = d\gamma^{(s)} \begin{pmatrix} 0 & 0 & \frac{1}{2} \\ 0 & 0 & 0 \\ \frac{1}{2} & 0 & 0 \end{pmatrix} \tag{4}$$

and the skew symmetric part the incremental rotation

$$d\omega^{(s)} = d\gamma^{(s)} \begin{pmatrix} 0 & 0 & \frac{1}{2} \\ 0 & 0 & 0 \\ -\frac{1}{2} & 0 & 0 \end{pmatrix} \tag{5}$$

For comparison to this case of 'simple shear' (Eq. 3, Figs. 1 and 2), let us consider a very special case of 'pure shear' in crystals, exemplified by the deformation of NaCl single crystals. Here, the primary slip systems are of the type $\{110\}\langle\bar{1}10\rangle$, and both $(110)[\bar{1}10]$ and $(\bar{1}10)[110]$ are equivalent members. When both operate equally, the result is plane-strain extension of a crystal oriented parallel to the cube planes (Fig. 4). The displacement gradient is

$$d\beta^{(s)} = d\gamma^{(s)} \begin{pmatrix} -1 & 0 & 0 \\ 0 & 0 & 0 \\ 0 & 0 & 1 \end{pmatrix} \tag{6}$$

The strain increment $d\varepsilon$ is identical to $d\beta$; the rotation is nil. Note that the lines drawn are like scratches inscribed on the surface; the *lattice planes* remain at ±45° – but each slip plane at +45° is cut by all those at -45° and vice versa; this is schematically illustrated by the jagged central lines.

Usually, one wants to describe straining in a coordinate system that is not necessarily bound to the slip elements; for example, in cubic axes for all cubic crystals (not

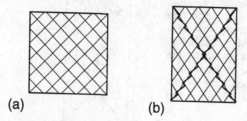

(a)   (b)

Fig. 4. When two orthogonal slip systems operate simultaneously, a shape change ensues, but no rotation. This is variously called 'pure shear' or 'plane-strain extension' (or plane-strain compression). The zig-zag lines represent the final appearance of initial diagonal surface scratches.

just NaCl); or, in general, in an arbitrary sample coordinate system, in which one may want to sum the effects of simultaneous slip on many systems. Tensor transformation rules provide the scheme:

$$d\beta_{ij}^{(c)} = R_{ik}\, R_{jn}\, d\beta_{kn}^{(s)} \tag{7}$$

where the rotation matrices R_{ij} are, simply, the direction cosines for transformation from the system (s) to the 'crystal' coordinates (c). A different way of writing this equation, which is more compact, assumes that the unit vectors $\hat{b}^{(s)}$ in the slip direction and $\hat{n}^{(s)}$ normal to the slip plane are written in (c) coordinates. Then,

$$d\beta_{ij}^{(c)} = d\gamma^{(s)}\, \hat{b}_i^{(s)}\, \hat{n}_j^{(s)} \qquad \text{(no sum on s)} \tag{8}$$

(This can also be written $d\beta = d\gamma(s)\, \hat{b}^{(s)} \otimes \hat{n}^{(s)}$, see Eq. 13.) For the strain and rotation increments, respectively, one defines the symmetric and skew parts of the transformation operator:

$$m_{ij}^{(s)} = \frac{1}{2}\left(\hat{b}_i^{(s)}\, \hat{n}_j^{(s)} + \hat{b}_j^{(s)}\, \hat{n}_i^{(s)}\right) \qquad \text{(no sum on s)} \tag{9}$$

$$q_{ij}^{(s)} = \frac{1}{2}\left(\hat{b}_i^{(s)}\, \hat{n}_j^{(s)} - \hat{b}_j^{(s)}\, \hat{n}_i^{(s)}\right) \qquad \text{(no sum on s)} \tag{9'}$$

In this way, the incremental strain and incremental rotation due to a single slip system become (in crystal coordinates)

$$d\varepsilon^c = \sum_s d\gamma^{(s)}\, m^{(s)} \tag{10}$$

$$d\omega^c = \sum_s d\gamma^{(s)}\, q^{(s)} \tag{10'}$$

Note that we have assumed a differential (or a virtual) increment in displacement and thus in the ensuing tensor variables; this is appropriate for plasticity where the integrated variables (i.e., 'finite strain measures' such as 'deformation gradient') are not state variables. When one divides each of the preceding equations by the time increment dt, one obtains 'rates' which, similarly, are not rates of change of a state variable; for this reason, the Mechanics community has defined non-dotted symbols for these variables. These will now be summarized and, at the same time, expressed as sums over all active slip systems (s) in a crystal (c). The strain rate is

$$D^c = \sum_s \dot{\gamma}^{(s)}\, m^{(s)} \tag{11}$$

(We will always write the sum over *slip systems* explicitly, even when the summation convention would make this unnecessary.) The scalar $\dot{\gamma}^{(s)}$ is truly a time rate (which enters into the kinetic equations) whereas the $m^{(s)}$ characterize the 'direction', i.e. the tensor character, of straining in a single slip system.

Analogous to Eq. 11, one can write for the rotation rate ($d\omega/dt$ from Eq. 5), with Eq. 10,

$$W^c = \sum_s \dot{\gamma}^{(s)}\, q^{(s)} \tag{12}$$

and for the sum of D and W, the rate of distortion ($d\beta/dt$ from Eq. 8):

$$L^c = \sum_s \dot{\gamma}^{(s)}\, \hat{b}^{(s)} \otimes \hat{n}^{(s)} \tag{13}$$

This is a particular form of the velocity gradient **L** introduced in Chap. 1 Eq. 11, with its skew part, the spin **W**. (In particular, $\mathbf{W}^c$ is usually called the 'plastic spin'.) For the symmetric part, **D**, we will continue to use the term '(tensor) strain rate' or 'straining rate'.

In preparation for the forthcoming treatment of plastic deformation in polycrystals, we have attached a superscript c to the quantities describing single-crystal behavior (and super-s implies a particular slip system in a particular crystal). It is easy to ascertain [HILL 1967] that, for a macroscopically homogeneous body like a polycrystal (no superscript), **D** is the volume average (marked by ⟨ ⟩) of the local values of **D**; if the latter are characterized by $\mathbf{D}^c$, for all or part of a crystal, but in any case linked to its crystal orientation, then

$$\mathbf{D} = \langle \mathbf{D}^c \rangle \tag{14}$$

What is difficult, and will occupy much of the rest of this book, is the inverse problem: what is the partitioning of a given macroscopic **D** into the grain-level quantities $\mathbf{D}^c$; this is sometimes termed 'localization'.

Just as we saw in Eq. 11 that $\mathbf{D}^c$ for each slip system is composed of a (scalar) rate and a (tensor) direction, it is useful in general to write

$$\mathbf{D} = \dot{\varepsilon}\,\hat{\mathbf{D}} \tag{15}$$

where $\dot{\varepsilon}$ is a true scalar rate, whereas $\hat{\mathbf{D}}$ represents the tensor character of straining, and will be called the 'straining direction'. There is a certain arbitrariness, by a *constant*, in the separation of a quantity into a scalar and a dimensionless part; this will be further discussed in connection with the overall constitutive relations (Sec. 1.7).

Writing the last few expressions as 'rates' may appear to imply a rate sensitivity of plasticity. While this is indeed usually present, one can use the same separation of scaling factor and direction in rate-independent plasticity; then a strain *increment* dε replaces the scalar rate [Hill 1950]:

$$d\varepsilon = d\varepsilon\,\hat{\mathbf{D}} \tag{15'}$$

where $d\varepsilon \equiv \dot{\varepsilon}\,dt$.

Plastic straining by slip or twinning is isochoric; thus, the quantities **D** and $\hat{\mathbf{D}}$ are identical to their deviators, and the most general component space in which they can be described has five, not six, dimensions.

1.2 The yield surface and plastic potential of single crystals

The Cauchy stress **σ** in a macroscopically homogeneous polycrystal (no superscript) is the volume average of the local stresses, again characterized by those that would result from the crystal properties, $\boldsymbol{\sigma}^c$:

$$\boldsymbol{\sigma} = \langle \boldsymbol{\sigma}^c \rangle \tag{16}$$

The local stress $\boldsymbol{\sigma}^c$ (in all or part of a crystal c) does work through the strain increment due to one system (Eq. 10) according to

$$dw = \boldsymbol{\sigma}^c : d\varepsilon^s = d\gamma^s\,\mathbf{m}^s : \boldsymbol{\sigma}^c \tag{17}[1]$$

---

[1] For tensor notation conventions used in this book, see the appendix on Notation.

The inner product $\mathbf{m}^s : \sigma^c$ is the projection of the applied stress onto the straining direction of the particular slip system; it is called the 'resolved shear stress' on that slip system and must reach a certain value for it to become active; this value is called the 'critical resolved shear stress (CRSS)' τ^s:

$$\mathbf{m}^s : \sigma^c = \tau^s \qquad \text{(plasticity, one system s)} \qquad (18)$$

This equation is called the (generalized) 'Schmid Law'; its essence is that the component of the applied stress that does the plastic work determines the kinematic behavior.

A geometric description of these facts is shown in Fig. 5a, in a two-dimensional projection of a five-dimensional space in which stress and strain-increment components are superposed. The solid line is meant to be the trace of a (hyper)plane perpendicular to the strain increment due to one slip system, $d\varepsilon$, or just its 'direction', $\mathbf{m}^s$. Its distance from the origin is proportional to the CRSS on this system, τ^s. Then, all stresses whose representative 'vectors' lie on this plane, such as σ_1, will have the appropriate resolved stress to potentially activate system s. Stresses 'inside' the (hyper)plane, such as σ_2, do not activate this system.[2]

In general, there is more than one slip system available: each is represented by a plane perpendicular to its $\mathbf{m}^s$. Equation 18 must then be generalized to read

$$\mathbf{m}^s : \sigma \leqslant \tau^s \qquad \text{(for \textit{all} systems s)} \qquad (19)$$

The *inequality* expressed in Eq. 19 is necessary in order for all *inactive* system to have a resolved stress less than their CRSS. This is a peculiarity of the strict rate independence postulated above.

The criterion 19 gives rise to the concept of a *yield surface* as an *inner envelope* of all the facets corresponding to particular deformation modes. (More properly phrased, this is a description of the 'plastic potential' rather than the yield surface; the difference will be discussed below.) Figure 5b shows an example of three (symmetrically disposed) slip directions in one slip plane. (The three stress components shown are not an independent set.) Figure 5c shows an expansion into a third dimension: for example, allowing a tensile (or compressive) stress in the [111] direction of an fcc metal: it gives a corner where the facets of six slip systems meet. A general polyhedron in five-dimensional space is bounded by hyperplanes, hyperedges, and hypercorners of various dimensions. We will use the special designations *facet* for a plane corresponding to a single slip system and *vertex* for a true corner in five-dimensional deviatoric stress space, at which five independent systems intersect. In highly symmetric situations, such as Fig. 5c, more than five facets can meet, but they are not independent. (Such highly symmetric situations require that the τ^s values of the various slip systems are exactly equal, or in a precise ratio – a situation that is not likely to be fulfilled in real crystals.)

An important consequence of the 'yield condition' (Eq. 19) is that not all arbitrary combinations of slip modes can be simultaneously activated (with arbitrary signs).

---

[2] The stress could have arbitrary additional components perpendicular to the plane show in *projection*. In a *section* through the 5-D or 6-D stress space, conversely, the indicated trace would be the *intersection* of the plane perpendicular to $\mathbf{m}^s$ with the 2-D space shown, and there could be arbitrary additional strain components activated, e.g., by the stress σ_i; τ^s would still be defined as the shortest perpendicular distance of the $\mathbf{m}^s$-plane to the origin and would thus, in general, not be seen in the plane.

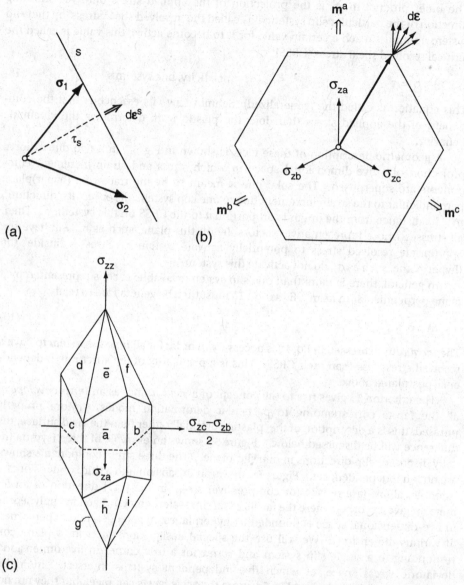

Fig. 5. Illustration of the Schmid law. (a) The projection of the stress vector on the strain-increment vector $d\varepsilon^s$ for a single slip system (s) must reach a critical value τ^s for this system to operate; all stresses σ_1 that lie on a facet perpendicular to $d\varepsilon$ fulfill this condition. (b) In a 2-D projection of stress space, in which the shear stresses in a plane are displayed, all stresses on the inner envelope of the facets belonging to potential slip systems lead to yield; in this case, three coplanar slip systems (a, b, c) are assumed, each operable in the forward and reverse direction. Stress states in a corner where facets meet can activate any combination of those slip systems (in a strictly rate-independent theory), giving rise to any strain increment vector $d\varepsilon$ in the cone of normals of the yield surface corner. (c) Extension of (b) into a third dimension. In this particular case, six slip systems meet in a vertex (corresponding, for example, to tension of a ⟨111⟩ single crystal of copper). (The facets carry slip system designations.)

For example, in Fig. 5c, where we have labeled the facets with arbitrary letters, it would not be possible to activate systems d, e, and c at the same time: they do not join at a corner. Similarly, while the combination d, e, ā (meaning system a operating in the negative direction) can evidently be activated, it turns out that d, e, and (+)a cannot.

For a single system, a facet was defined to be perpendicular to its straining direction (Eq. 18). For many facets, or any closed surface, the generalization is that the overall straining direction is perpendicular to the yield surface. If the yield surface has sharp corners (as all the ones shown so far), the generalization is that the straining direction must be within the cone of normals of the corner. This is indicated by the bundle of arrows in Fig. 5b at each corner. The fact that $\hat{D}$ may be *anywhere* within this cone of normals is somewhat counterintuitive, since one would expect all systems that have reached their critical resolved shear stress to behave the same way and produce the same shear increment. This intuition is correct in any material with a finite *rate sensitivity*, which will be discussed in detail in Section 1.6. Within the present framework of strict rate independence, however, *any* strain rate within the cone of normals can be obtained under the corner stress state.

The yield surface in stress space may be looked upon as indicating what straining directions ($\hat{D}$) may occur when a stress is prescribed. Equally, however, it can be used to derive what stress is necessary when a straining direction is prescribed. In this light, the single-crystal yield surfaces discussed so far exhibit an interesting duality. If one successively picked stressing directions from a continuous spectrum in stress space, only a discrete set of straining directions would ensue (corresponding to the positive and negative slip systems); only an infinitesimally small fraction of stresses would be precisely in a corner and thus allow other values of $\hat{D}$. Conversely, if one successively picked *straining* directions from a continuous spectrum in $\hat{D}$-space, only a discrete set of stresses would be demanded (those in the vertices); only an infinitesimally small fraction of arbitrary selected values of $\hat{D}$ would happen to exactly correspond to a slip system and thus allow a non-discrete spectrum of stresses. BISHOP & HILL [1951a, BISHOP 1953] were the first to point out that the Taylor model of uniform straining requires vertex stresses in the yield surface of each grain. Thus, the Taylor model in its full form is also sometimes called the Bishop–Hill model.

The problem of finding the *correct* stress σ for a given straining direction $\hat{D}$ has been made algorithmically easy by the 'maximum work principle' proposed by BISHOP & HILL [1951a]. If one has a list of all the vertices σ* that follow from a kinematic analysis of all possible slip systems, the correct one maximizes the work:

$$(\sigma - \sigma^*) : \hat{D} \geq 0 \qquad (20)$$

Figure 6 illustrates how a plane perpendicular to an arbitrary $\hat{D}$, when brought in from

Fig. 6. The maximum-work principle [BISHOP & HILL 1951a]. If a straining direction dε is prescribed, it will in general require a vertex stress state; among vertex stress states, the one that maximizes the work through dε is the one that is kinematically correct: dε lies within its cone of normals.

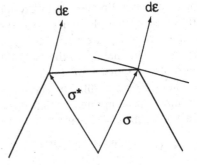

infinity, touches the yield surface first at the correct vertex; *viz.* the one that contains $\hat{D}$ in its cone of normals. (TAYLOR [1938b] achieved the same result, without consideration of the single-crystal yield criterion, by minimizing the algebraic sum of shears needed to attain a given D [CHIN & MAMMEL 1969, KOCKS 1970].)

Equation 20, when applied to *all* stresses σ* also expresses the fact that the yield surface is convex – which was already evident from its definition of the inner envelope of all facets.

The normality of the straining direction on the yield surface can be formally expressed by

$$\hat{D}_{ij} = \frac{\partial \Phi(\sigma)}{\partial \sigma_{ij}} \tag{21}$$

where Φ is called the 'plastic potential'. It is usually written with D instead of $\hat{D}$ on the left side, and a scalar multiplier (our $\dot{\varepsilon}$) on the right; then, ($\Phi \dot{\varepsilon}$) is the work rate, $\dot{w}$: a potential in the sense usually used in physics.

In Eq. 21, the normalized work rate $\Phi = \dot{w}/\dot{\varepsilon}$ is expressed as a function of σ; it could be similarly expressed as a function of $\hat{D}$ and used to derive the stress:

$$S_{ij} = \frac{\partial \Phi(\hat{D})}{\partial \hat{D}_{ij}} \tag{22}$$

However, this equation is fundamentally different from Eq. 21: only those stress components (here called S_{ij}) can be derived for which the dependence of Φ on $\hat{D}$ is defined. For example, when the strain (rate) is isochoric, as is usual in plasticity, the hydrostatic pressure has no influence on Eq. 21, and it cannot be derived from Eq. 22. In this case, S is the deviatoric stress tensor, and in this sense it will be used henceforth.

The plastic potential is frequently used for visual demonstrations. An alternative potential, in strain-rate space, has been elaborated by VAN HOUTTE [1994]. These potentials are not usually known as such: the physical mechanisms (to be discussed in detail in Sec. 1.6) typically give the D versus S relation directly.

In single slip, the stress (deviator) can be described as proportional to the CRSS, τ^s. When there are many systems operative, we can use a scaling factor τ^c for the whole crystal; it is equal to τ^s when all systems have the same value, or a reference value when they are not. Then,

$$S^c = M^c(\hat{D}^c)\,\tau^c \qquad \text{(no sum)} \tag{23}$$

The symbol $M^c(\hat{D}^c)$ signifies a dimensionless stress tensor in the direction demanded for a straining direction $\hat{D}$ in crystal c – in general, a vertex vector when the yield surface is faceted. Equation 23 is the converse of the Schmid law: it assumes a prescribed straining direction, whereas the Schmid law assumed a prescribed stressing direction. In analogy to the 'Schmid factor' – or rather the Schmid *tensor*, m – one defines a 'Taylor factor' (or Taylor tensor) M, as the directional part of the stress (deviator)[Kocks 1970]. The classical *scalar* Taylor factor M for the crystal c is defined by only that component which does the work (through $\dot{\varepsilon}$ in Eq. 15 or dε in Eq. 15′) [Taylor 1938b, 1956, TOMÉ &AL. 1984]:

$$S^c : \hat{D}^c = M^c : \hat{D}^c \tau^c \equiv M^c\,\tau^c \qquad \text{(no sum)} \tag{24}$$

Note that these definitions of 'Taylor factors' do not imply the Taylor [1938a] model of polycrystals deformation; they apply to individual crystals. For example, HOSFORD [1965] derived and measured them for wire-drawing of single crystals (approximating a tensile test with prescribed *strains*).

Equation 23 achieves a certain separation of the flow stress for a general test into a scalar factor τ^c and a directional quantity $\mathbf{M}^c$. The term τ^c specifies the *size* of the yield surface, for a given state of the material; the set of all $\mathbf{M}^c(\hat{\mathbf{D}})$ for various tests specifies its *shape*.

Equation 18, which we called the 'Schmid Law', is a generalization of the originally proposed scalar relation

$$m\ \sigma = \tau$$

for tensile deformation of single crystals in single slip only, in which the inverse of the 'Schmid factor' m was supposed to describe the entire orientation dependence of the tensile stress σ for a given τ. It was, in fact, first assumed that the entire stress/strain curves of single crystals of all orientations could be made into a single universal $\tau(\gamma)$ relation by multiplying σ by m and dividing ε by m. This would demand an orientation independence of *strain hardening*, which was soon proved to be an untenable assumption.

Yield surface versus plastic potential

'Non-Schmid' effects were, in a similar vein, originally associated with departures from the expected orientation dependence of single-crystal tension (and compression). In the modern interpretation of the Schmid Law, Eq. 18, they are associated with a dependence of the *critical* stress value τ on the current state of stress in the material [ARGON 1973, KOCKS 1987]. Much as the CRSS obviously depends on temperature T:

$$\tau^s = \tau^s(T)$$

we could have a dependence of τ^s on *pressure* – or indeed on *any* stress:

$$\tau^s = \tau^s(\sigma) \tag{25}$$

An example would be an effect of a normal stress on the core glide resistance of dislocations. In fact, all very *localized* glide resistances are of this type; one may also ascribe the general phenomenon to any significant *non-linearity* of dislocation/obstacle interactions.

In such cases, the stress enters into Eq. 18 (or 19) in two ways: through the work it does (on the left-hand side), and through its effect on the properties of the material (on the right-hand side). Note that a yield function $f(\sigma)$ is sometimes written such that Eq. 19 would be replaced by

$$f(\sigma) \equiv \mathbf{m}^s : \sigma - \tau^s \leqslant 0$$

Then, both effects would be combined in $f(\sigma)$. We shall see later (Sec. 1.6) that a more physically meaningful combination is actually the *ratio* of $\mathbf{m}^s{:}\sigma$ to τ^s, not the difference between them: it determines the *rate* of shearing on a slip system,

$$\dot{\gamma}^{(s)} = \mathcal{K}\left(\frac{\mathbf{m}^{(s)}{:}\sigma}{\tau^{(s)}(\sigma)}\right)$$

The normality of strain rates to the yield surface was based on τ^s being a constant; if τ^s is stress dependent, the plastic potential (on which the strain rates are perpendicular *by definition*) is no longer 'homothetic' to the yield surface [KOCKS 1987]. Another way of saying this is that the function Φ, as written in Eq. 22, is proportional to τ, but the differentiation in Eqs. 20–21 is meant only with respect to the *directional* part of σ, or 'at constant state'.

For *deformation twinning*, the work rate criterion, Eq. 18, should apply just as well as in glide; however, one would expect [LEBENSOHN & TOMÉ 1993b], and has occasionally observed [CHRISTIAN & MAHAJAN 1995] that the CRSS is influenced by other stresses, so that the 'Schmid Law' may not be obeyed. In order to assess valid deformation mode combinations, only the plastic potential is needed, and it is not influenced by the stress effects on the CRSS.

Throughout this book, we often use the term 'yield surface' as if it were a plastic potential.

1.3 Closing the single-crystal yield surface

One problem for which the concept of a yield surface (or actually of a plastic potential) is useful is to ascertain whether indeed any arbitrary strain(-rate) can be supplied by a crystal (or imposed on it). This amounts to the question of whether the single-crystal yield surface (SCYS) is 'closed' or not. A better formulation is actually: in what subspace is it closed? For example, if plastic deformation is such as to never cause volume changes, the yield surface is always open in the direction of hydrostatic stresses. Then, one usually addresses *deviatoric* stress space only; it is five-dimensional. To close it, one needs five independent deformation modes – the so-called von Mises criterion [1928]. (Note, however, that not all plastic strains are isochoric; e.g., when one treats martensitic transformations: they are similar to twinning, but are generally associated with volume change [ONO & SATO 1988].)

Figure 7 shows a case very similar to Fig. 5c – but the difference is crucial. Here, we imagine that Fig. 5b represented basal glide in a *hexagonal* material and that Fig. 7 adds the effect of one twinning system. Twinning is a unidirectional deformation mode; for example, the particular twinning mode incorporated in Fig. 7 gives an

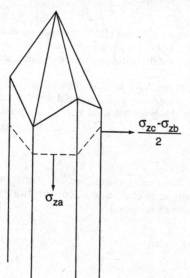

$$\frac{\sigma_{zc} - \sigma_{zb}}{2}$$

σ_{za}

Fig. 7. Extension of Fig. 5b into a third dimension, with the activation of twinning systems. (Example: zinc with basal glide and pyramidal twinning.) Since twinning can operate in only one direction, the single-crystal yield surface in this case remains open in one direction ('one-half dimension').

extension in the **c**-direction; but a compression in the c-direction cannot be achieved by either basal glide or this twinning mode. Thus, this three-dimensional yield surface is open in one direction of one dimension – 'open in 1/2 dimension', as one may say colloquially. (If only basal glide were considered, this yield surface would also be open in the two remaining deviatoric stress components; the twinning mode considered here would close these dimensions again unidirectionally, so that the total, five-dimensional yield surface for this case is 'open in 3/2 dimensions'.)

Independent deformation mode sets

An important consideration in assessing whether a yield surface is closed (without actually deriving it) is to establish how many *independent* deformation modes there are. For example, the three modes shown in Fig. 5b have one dependency relation between them: they are co-planar. The six facets meeting in the apex of Fig. 5c have one dependency relation between them (due to the same basic reason). Another reason for systems not to be 'independent' in this sense is if they provide the same strain, but different rotations; for example, (100)[010] and (010)[100].

A general algebraic method to assess whether a set of equations is overdetermined is to investigate the rank of the matrix [GROVES & KELLY 1963]. This method, however, does not take into account the *sign* of straining in each mode. When the sign is not arbitrary, such as when there are some twinning modes, a set of five 'independent' modes does not necessarily close the yield surface. Deriving and plotting the actual yield surface is general [KOCKS 1964a].

Sometimes, a yield surface is not 'open' in any direction, but requires very high stresses to activate far-out vertices; then, it is called a very 'anisotropic' SCYS. Similarly, when a yield surface is expected to be 'open' in one direction on physical grounds, one may close it off by an artificial high-stress mode, so as to have a fully determined problem.

Climb

Climb of dislocations can, in principle, contribute additional deformation modes [GROVES & KELLY 1969]; they would respond to certain deviatoric stresses, but would not lead to orientation changes. Since climb modes are not volume conserving, a total of six independent modes is needed. Climb modes are sometimes invoked only as (infinitely easy) relaxation modes for the dislocations that would be stored when only three independent slip systems are available (e.g., in NiAl [MARGEVICIUS & COTTON 1995]). This requires the three Burgers vectors to be distinct and non-coplanar [GROVES & KELLY 1969].

Scaling

All the above discussion has been concerned with the *shape* of the yield surface. For single crystals, the shape is governed by the ratios of the CRSSs of all the deformation modes. If one characterized its *size* by a scalar strength parameter τ^c (for the given crystal), then one may define all the CRSSs by

$$\tau^s = \alpha^s \tau^c \tag{26}$$

It is common to use the CRSS of one particular mode (such as the easiest one) as the measure of the scale (so that for this system $\alpha^{(1)} \equiv 1$).

1.4 Symmetry properties of the yield surface

Vector notation

Stress, strain, and strain rate are symmetric second-rank tensors, with six independent components. For some applications, it is easiest to keep them in 'tensor' notation; but for others, it is convenient to define 1×6 arrays (sometimes called 'vectors', though not in the sense of a first-rank tensor). This was done first by Voigt, for applications in elasticity. (See Chap. 1 Sec. 4.2). The elastic constants then become 6×6 matrices. For this reason, this is also called the 'matrix notation'.

A general way (somewhat different from Voigt's) to set up such a scheme is to first define 'basis tensors' for symmetric second-rank tensors. For deviatoric tensors, one needs five, and they may be defined as follows (Chap. 7 Eqs.19):

$$\mathbf{b}^{(1)} \equiv \frac{1}{\sqrt{6}} \begin{pmatrix} \bar{1} & 0 & 0 \\ 0 & \bar{1} & 0 \\ 0 & 0 & 2 \end{pmatrix} \qquad \mathbf{b}^{(2)} \equiv \frac{1}{\sqrt{2}} \begin{pmatrix} \bar{1} & 0 & 0 \\ 0 & 1 & 0 \\ 0 & 0 & 0 \end{pmatrix}$$

$$\mathbf{b}^{(3)} \equiv \frac{1}{\sqrt{2}} \begin{pmatrix} 0 & 0 & 0 \\ 0 & 0 & 1 \\ 0 & 1 & 0 \end{pmatrix} \quad \mathbf{b}^{(4)} \equiv \frac{1}{\sqrt{2}} \begin{pmatrix} 0 & 0 & 1 \\ 0 & 0 & 0 \\ 1 & 0 & 0 \end{pmatrix} \quad \mathbf{b}^{(5)} \equiv \frac{1}{\sqrt{2}} \begin{pmatrix} 0 & 1 & 0 \\ 1 & 0 & 0 \\ 0 & 0 & 0 \end{pmatrix} \tag{27}$$

The *second* row are the three pure unit shears in the {XYZ} system. The base tensor $\mathbf{b}^{(2)}$ is also a pure shear, but it is turned 45° around Z from $\mathbf{b}^{(5)}$; it is expressed as plane-strain extension/compression in the Z-plane. Finally, tensor $\mathbf{b}^{(1)}$ represents unit deviatoric tension in the Z-direction. The components $\lambda=1$ and $\lambda=2$ make up the so-called 'π-plane': a projection parallel to the 'hydrostatic' direction of the space made up of the three diagonal components. If 'isotropic' parts such as hydrostatic pressure and volume change were considered, one would need one more base tensor: it is added as $\mathbf{b}^{(6)} \equiv 1/\sqrt{3}\,\mathbf{I}$ (the unit tensor) in Chaps. 7 and 11.[3]

Using these basis tensors, the vector components of an arbitrary stress σ and strain (increment) ε are then defined by

$$\sigma_\lambda \equiv \sigma : \mathbf{b}^{(\lambda)} \qquad\qquad \varepsilon_\lambda = \varepsilon : \mathbf{b}^{(\lambda)} \tag{28a}$$

and the total quantities can be expressed in terms of their components as

$$\sigma = \sigma_\lambda \mathbf{b}^{(\lambda)} \quad \text{and} \quad \varepsilon = \varepsilon_\lambda \mathbf{b}^{(\lambda)} \quad \text{(sum on } \lambda\text{)} \tag{28b}$$

Explicitly, the 'vector' components of the deviatoric stress tensor **S** and of the plastic strain-*rate* tensor **D** (*defined* as deviatoric) become[4]:

$$\{S_\lambda\} = \left\{ \frac{(\sigma_{33}-\sigma_{11})+(\sigma_{33}-\sigma_{22})}{\sqrt{6}},\ \frac{\sigma_{22}-\sigma_{11}}{\sqrt{2}},\ \sqrt{2}\,\sigma_{23},\ \sqrt{2}\,\sigma_{31},\ \sqrt{2}\,\sigma_{12} \right\}$$

$$\{D_\lambda\} = \left\{ \frac{(D_{33}-D_{11})+(D_{33}-D_{22})}{\sqrt{6}},\ \frac{D_{22}-D_{11}}{\sqrt{2}},\ \sqrt{2}\,D_{23},\ \sqrt{2}\,D_{31},\ \sqrt{2}\,D_{12} \right\} \tag{29}$$

---

[3] Note that in Chaps. 7 and 11, the same tensors, with the same symbols, are used as *eigen*tensors for *fourth*-rank tensors, in which case they hold only under cubic symmetry conditions; in Eq. 27, they are general *base* tensors for *second*-rank tensors.

[4] LEQUEU &AL. [1987a] used a similar vectorization scheme (with the first two components interchanged).

Note that, in this definition, the forms for σ and D are exactly the same and, further, their product is the work rate. In *general*, the ortho-normal definition of the bases in Eqs. 27 ensures for the work rate

$$\sigma_{ij} D_{ij} = \sigma_\lambda D_\lambda \tag{30}$$

Sign dependence

We saw in Fig. 7 that the yield surface does not necessarily have a center of symmetry in stress space; for example, when twinning modes are involved. Another example is the so-called 'Bauschinger effect': when a sample has been deformed along a particular path (say, in tension) and has strain hardened and then been unloaded – the flow stress upon reloading in the reverse direction (compression) is lower than it would be upon reloading in the same direction (tension). This is often a transient effect (and thus depends sensitively on the definition of the flow stress – see Sec. 1.5); but it can be of persistent significance, for example in two-phase materials.[5]

Such cases are referred to as a 'sign dependence' of yield. We re-emphasize (Chap. 1) that the stress itself, and also the resulting strain rate, are 'centrosymmetric' variables, in the sense that a change in sign of all coordinate axes does not affect them. The terms centrosymmetry and inversion symmetry refer only to position space, not to stress space.

Isotropy

By way of contrast to the various examples of single-crystal yield surfaces already shown, let us now look first at an *isotropic* material, where the symmetry properties can be derived in a qualitative way. For example: since any 'turning' of the sample cannot matter, the yield surface in the shears space $\{\sigma_{23}, \sigma_{31}, \sigma_{12}\}$ (λ = 4, 5, and 6) must be a sphere. Second, permutation of the sample axes cannot matter – including their negatives, because of the centrosymmetry of the stress tensor itself. Therefore, the π-plane sub-surface needs to be specified only in one 60°-sector: say, between $+S_{33}$ and $-S_{11}$ (Fig. 8). If, in addition, yield is sign independent, only a 30°-sector is needed: say, between $+S_{33}$ and $(S_{33}-S_{11})/2$. Finally, the latter quantity is a shear (turned 45° from S_{31}) and thus has to have the same yield stress as σ_{31}; in other words, corresponding to the radius of the sphere in the shears subspace.

It is worth emphasizing that the π-plane for isotropic materials can be plotted for any arbitrary sample axes: they need not be the principal axes of stress, as is usually done. For *anisotropic* materials, in which the principal axes of stress and strain-rate do not generally coincide, it is in fact necessary to choose normal axes that correspond to the special properties of the anisotropy.

The 'von Mises' circle (Fig. 8), which is often used for isotropic materials in the π-plane, demands that there is a certain relation between uniaxial deviatoric paths and shear paths. There is no reason for such a presumption; the von Mises yield criterion is merely one of convenience. (To ensure convexity, the true yield surface must lie in the shaded region of Fig. 8 for sign independence. Its lower bound is the 'Tresca yield surface' [TRESCA 1864].)

---

[5] The Bauschinger effect is sometimes described (without experimental justification) by a *translation* of the yield surface ('kinematic hardening' [PRAGER 1959]); in that case, it would not affect the symmetry of the (translated) body of the yield surface itself.

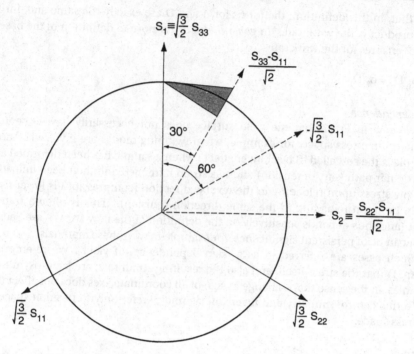

Fig. 8. The 'π-plane', which contains the diagonal components of the deviatoric stress tensor S. For isotropic materials, only one 60° sector is needed; when yield is sign independent, only one 30° sector. The von Mises circle is the result of an assumption.

The irreducible region of a yield surface

We have encountered some examples in which the specification of a *part* of the yield surface was sufficient to define the whole. In all, the application of symmetry rules can have three effects, all of which reduce the region of stress space that need be covered, to the 'irreducible region':

- a decrease in the number of dimensions of stress space that need be plotted (two for isotropy: one of the values there holds also for the other three dimensions);
- a decrease in the extent within a certain (sub-)space that need be specified (only one sector of the π-plane for isotropy); and finally
- a rule concerning which (if any) subspaces may be 'contained'.

The last effect has yet to be defined and discussed. The concept was introduced by CANOVA &AL.[1985] (but the special subspaces were called 'closed' there). A subspace is called 'contained' if an arbitrary σ (or S) within it cannot cause a D that has a component outside it. In a special sense, it is 'orthogonal' to all other dimensions. In the isotropic isochoric material, the π-plane is contained: the diagonal elements of a stress tensor cannot cause any off-diagonal shears. In addition, every one of the shear dimensions is contained: no σ_{12} can cause any D_{23}, for example. A contained subspace is identical whether it is obtained by a *section* or by a *projection* of the full yield surface (i.e., by setting certain *stress* or certain *strain-rate* components to zero).

Effect of symmetry operators

For general anisotropies, it is best to derive the symmetry properties of the yield surface by formal methods: apply all applicable symmetry operators; and demand that a symmetrically equivalent stimulus (say, stress) lead to a symmetrically equivalent response (strain rate) [CANOVA &AL. 1985]. The formal prescription reads as follows. If the nonlinear relation between **D** and **S** is expressed as

$$\mathbf{D} = \mathcal{F}(\mathbf{S}) \tag{31}$$

then we require that

$$\mathbf{H}\,\mathbf{D}\,\mathbf{H}^{\mathsf{T}} = \mathcal{F}(\mathbf{H}\,\mathbf{S}\,\mathbf{H}^{\mathsf{T}}) \tag{32}$$

for any symmetry operator **H** applicable to the system (Chap. 1 Sec. 1.2). (Only if the relation were linear could the symmetry operators be directly applied to the fourth-rank tensor that connects **D** and **S**.)

We illustrate the principle in some detail for one mirror plane (perpendicular to the sample axis 2, as was done for a second-rank tensor *property* in Chap. 1 Sec. 1.1). The application of this symmetry element to an arbitrary stress gives

$$\begin{pmatrix} 1 & 0 & 0 \\ 0 & -1 & 0 \\ 0 & 0 & 1 \end{pmatrix}\begin{pmatrix} \sigma_{11} & \sigma_{12} & \sigma_{31} \\ \sigma_{12} & \sigma_{22} & \sigma_{23} \\ \sigma_{31} & \sigma_{23} & \sigma_{33} \end{pmatrix}\begin{pmatrix} 1 & 0 & 0 \\ 0 & -1 & 0 \\ 0 & 0 & 1 \end{pmatrix} = \begin{pmatrix} \sigma_{11} & -\sigma_{12} & \sigma_{31} \\ -\sigma_{12} & \sigma_{22} & -\sigma_{23} \\ \sigma_{31} & -\sigma_{23} & \sigma_{33} \end{pmatrix} \tag{33}$$

and to an arbitrary strain rate

$$\begin{pmatrix} 1 & 0 & 0 \\ 0 & -1 & 0 \\ 0 & 0 & 1 \end{pmatrix}\begin{pmatrix} D_{11} & D_{12} & D_{31} \\ D_{12} & D_{22} & D_{23} \\ D_{31} & D_{23} & D_{33} \end{pmatrix}\begin{pmatrix} 1 & 0 & 0 \\ 0 & -1 & 0 \\ 0 & 0 & 1 \end{pmatrix} = \begin{pmatrix} D_{11} & -D_{12} & D_{31} \\ -D_{12} & D_{22} & -D_{23} \\ D_{31} & -D_{23} & D_{33} \end{pmatrix} \tag{33'}$$

Observe that, if σ_{12} and σ_{23} were zero, but all other stress components arbitrary, the two stress states (before and after the application of the symmetry operator) would be identical, and they should induce identical strain rates. This requires that D_{12} and D_{23} be zero in Eq. 33'. Consequently, the subspace $\{\pi, \sigma_{31}\}$ is contained. Similarly, if σ_{12} and σ_{23} were the *only* non-zero components, σ and **D** would *both* be reversed in sign through the operation; thus, the subspace $\{\sigma_{12}, \sigma_{23}\}$ is contained. In both cases, sign independence would allow one to reduce the extent of the subspace needed by a factor of 2.

It is worth emphasizing, however, that an *arbitrary* stress (even in the presence of this mirror plane) requires a full 5-D description: one cannot decompose it into two parts, one for each subspace, and then superpose the result.

Application of other mirror planes and rotation axes may generally increase the number of contained subspaces and thus reduce the dimensions of stress space needed for special (though common) stress states; this is illustrated in Table I. In addition, higher symmetries may reduce the extent of any subspace that must be covered and, in extreme cases, even reduce the number of dimensions required for a *general* stress state (four for fiber symmetry, two for isotropy). Table II shows results for sign-independent yield surfaces. The extent necessary in any contained space is cut in half by a center of symmetry if sign independence applies. (For more detail, see CANOVA &AL. [1985].)

Table I. Contained yield surface subspaces. (All the subscripted stress components are meant to be shear stresses in the sample symmetry axes; i.e., i≠j, k≠l; but k=i or l=j.)

Sample symmetry	Contained subspaces
1 mirror plane (M_2)	$\{\pi, \sigma_{31}\}$; $\{\sigma_{12}, \sigma_{23}\}$
orthotropic	$\{\pi\}$, $\{\pi, \sigma_{ij}\}$, $\{\pi, \sigma_{ij}, \sigma_{kl}\}$; $\{\sigma_{ij}\}$, $\{\sigma_{ij}, \sigma_{kl}\}$, $\{\sigma_{12}, \sigma_{31}, \sigma_{12}\}$

Table II. Information required for a sign-independent yield surface.

Sample symmetry	Irreducible space
orthotropic	$\frac{1}{2}\{\pi\} + \frac{1}{8}\{\sigma_{23}, \sigma_{31}, \sigma_{12}\}$
cubic	$\frac{1}{12}\{\pi\} + \frac{1}{48}\{\sigma_{23}, \sigma_{31}, \sigma_{12}\}$
hexagonal ($C_6 \| X_3$)	$\frac{1}{6}\{\pi, \sigma_{12}\} + \frac{1}{12}\{\sigma_{23}, \sigma_{31}\}$
fiber ($C_\infty \| X_3$)	$\frac{1}{2}\{\pi\} + \{\sigma_{ij}, \sigma_{kl}\}$
isotropic	$\frac{1}{12}\{\pi\}$

A particular case of wide interest is that of *orthotropic* symmetry. Here, the application of three mutually perpendicular mirror planes leads, first, to the definition of a sensible *coordinate system*; namely, all axes that are in the intersection of the mirror planes. In this coordinate system, the π-plane is contained, and so is any space consisting of the π-plane and one or two shear components. Also, the space of all shears is contained. These statements hold for all higher symmetries as well (i.e. those that contain orthotropic symmetry). The irreducible region consists of the π-plane plus one quadrant of each of the shears spaces. This region is further decreased when higher symmetries apply. Figure 9 shows one octant of the shears space (which is needed for sign independent yield in an orthotropic material) and (shaded) the part of this octant needed when the material has cubic symmetry.

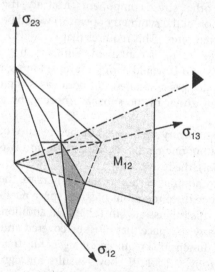

Fig. 9. The subspace of the three shears in a coordinate systems aligned with orthotropic axes. The quadrant shown is all that is needed in this space when yield is sign-independent. The region shaded in dark suffices when the material has cubic symmetry.

Irreducible set of vertices of the SCYS

A useful application of the concept of the irreducible stress space, under certain symmetries, is the derivation of a single-crystal yield surface (SCYS). In the rate-independent limit, this amounts to an enumeration of all vertices within (and on the borders of) the irreducible region, their connectivity, and perhaps the specific deformation modes that contribute to each vertex [KOCKS &AL. 1983]. An additional gain can be realized by expressing these vertices in some algebraic form that may depend on parameters such as the c/a ratio in hexagonal materials and on the specific ratios of the critical stresses to activate various systems [TOMÉ & KOCKS 1985]. Such an algebraic form would hold for a continuous set of cases so long as the *yield surface topology* does not change.

We shall illustrate the principle on the case of γ-TiAl, an intermetallic alloy of tetragonal lattice structure [MECKING &AL.1996a]. While the tetragonality of the lattice itself is minimal (c/a=1.02), the ordered nature of the structure leads to a clear distinction, for example, between the slip systems (111)[1$\bar{1}$0] and (111)[10$\bar{1}$] or, generically, between {111}⟨110] and {111}⟨101]. These two slip systems are the major deformation modes. One of them requires superdislocation motion, the other ordinary dislocation motion. In a cubic material, they would be the same. There is also a twinning mode, which we shall ignore for this demonstration. Further, we assume sign independence for the slip.

Figure 10a shows the subspace of the three shears, Fig. 10b the π-plane plus the shear $S_3 = \sqrt{2}\,\sigma_{23}$. The 'super'-mode is shown solid, the 'ordinary'-mode dashed. We use the 'super' mode as the reference mode and have only one parameter, $\alpha = \tau^o/\tau^s$ for the ratio of the critical shear stresses. The situation shown is for α=1. As α increases, the dashed planes move out. It is immediately evident that the 'ordinary' mode would not

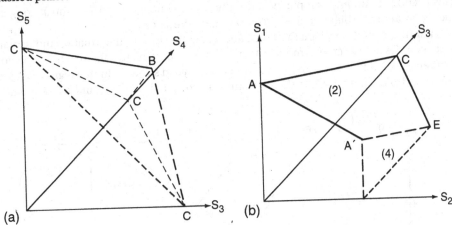

(a)   (b)

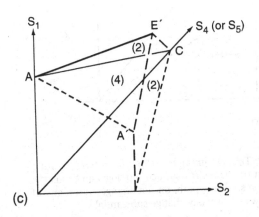

(c)

Fig. 10. Yield surface subsections for single crystals of γ-TiAl, a tetragonal material. The slip systems are as in fcc metals, but separated into two groups: eight systems requiring superdislocation glide (solid lines) and four ordinary dislocation glide (dashed; hidden lines have *thin* dashes). The situation shown is where the two systems have equal flow stresses. It is easy to visualize the qualitative changes in the yield surface when the dashed planes move in or out.

be part of the yield surface if its CRSS were in excess of twice that for the 'super' mode. There appear to be three topological domains: $\alpha > 2$, $1 < \alpha < 2$, and $\alpha < 1$; the equalities are special cases. All the vertices seen are in fact 5-D vertices (although this is not obvious). They are vertices of different types, labeled with different capital letters.

An interesting observation can be made when a further subspace is plotted: Fig. 10c, again for $\alpha=1$. The topology has little resemblance to that of Fig. 10b, although in both cases the π-plane is plotted plus one shear. Note that the vertex E' (and also B in Fig. 10a) would not appear in any two-dimensional section along the S_λ coordinates. This illustrates the unfortunate truth that one cannot in general derive a 5-D yield surface by making lower-dimensional sections only. A general computer algorithm is required [TOMÉ & KOCKS 1985].

Topological domains and deformation mode maps

When one moves the dashed planes in or out in Fig. 10a, the topology of the yield surface changes, since the point C is a vertex only for $\alpha=1$. Similarly, in Fig. 10b, the point E is a special vertex, and in Fig. 10c, there are two. A thorough analysis shows, however, that the topology then remains the same for a large range of α. While the location of the new vertices changes, their number and character does not; for example, each one of the vertices maintains the same combination of slip systems. One can therefore construct a 'deformation mode map', in which regions are delineated in which the slip system combination does not change [CHIN &AL. 1969]. This is convenient, for example, because the texture changes are the same so long as the slip system combination for a given $\hat{D}$ remain the same.

For the example of γ-TiAl discussed above, Fig. 11 shows a deformation mode map that includes the twinning mode. In the bottom part of the figure, the CRSS for twinning is so high that it never contributes to the yield surface. In this regime, we see confirmed that there are but three regimes: for $\alpha < 1$, $1 < \alpha < 2$, and $\alpha > 2$. When

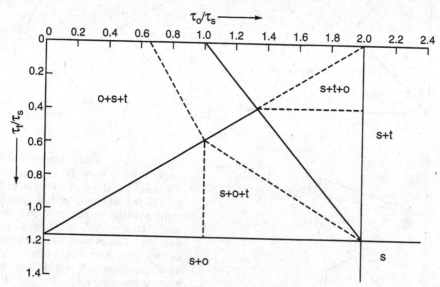

Fig. 11. A deformation mode map for γ-TiAl, assuming two slip system types, s and o (as in Fig. 10) and one twinning system (t). The yield surface topology, and thus the deformation modes activated for a given straining path, do not change within any of the regions, even though the ratio of flow stresses may change substantially.

twinning does contribute, one can derive the boundaries between topologically identical regions by a systematic search of possible intersections of the three separate yield surfaces for each mode [MECKING &AL. 1996a].

A particular aspect worth noting in Fig. 11 is that there is no region in which superdislocations do not contribute. The yield surface that describes ordinary slip and twinning only is not closed – even though these are five 'independent modes': some of them are 'half-modes' and leave one side of a particular direction in stress space open.

The deformation mode maps can be used to ascertain which deformation modes actually contribute in a particular case. Given a polycrystal model, one predicts the texture changes that should be observed if a particular region were activated. If textures are predicted that are never observed, this cannot be the actual region. In this way, the choices can be narrowed substantially and then used for other predictions. This 'inverse method' obviates the need for a determination of all CRSS values by way of single-crystal experiments.

Two examples in which this technique has led to clear results are that of ⟨c+a⟩-slip in hexagonal metals, which cannot be a significant deformation mode [TOMÉ &AL. 1991a]; and an explanation of a change in texture type with temperature in calcite [TAKESHITA &AL. 1987].

1.5 Strain hardening

Figure 12 displays typical 'stress/strain curves' for single crystals at two different scales. In the first (Fig. 12a), the elastic loading part is clearly seen in the initial

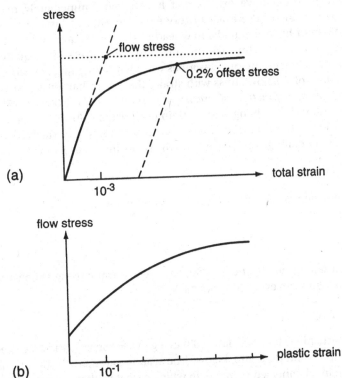

Fig. 12. (a) The relation between stress and strain near yield. (b) The relation between flow stress and plastic (pre-)strain for large strains. Note the different scales on both abscissa and ordinate.

straight line. Then, there is a transition to a part that appears, for all practical purposes, to be a constant stress asymptote. If there were no smooth transition, but an abrupt change from the elastic line to a constant stress, this would be called an 'elastic/ideally-plastic' material, with a 'flow stress' (or 'yield strength') at the asymptote. The experimentally observed transition ends around about the 0.2% 'offset' (i.e., total strain minus elastic unloading strain). For this reason, a characterization of the 'flow stress' comes out about the same whether it is quoted as the 0.2% offset yield or as the intersection of the asymptotic plastic line with the elastic one (for cubic materials). A definition of yield as a first departure from linearity ('proportional limit'), or at a very small offset, does not give results of the same reproducibility, and is questionable in terms of a physical mechanism. The asymptotic stress, on the other hand, may be identified with long-range dislocation motion, or the 'percolation threshold' [KOCKS 1985].

The asymptote in Fig. 12a is, in reality, not exactly at constant stress but has, in most cases, a small positive slope – of the order of 1/100 of the elastic slope. This is due to the phenomenon of 'work hardening' or 'strain hardening' (these two terms being used interchangeably in the materials science literature). When this slope is made visible, by changing the scales (Fig. 12b), the elastic line appears vertical [KOCKS 1975].

Work hardening is characterized by a 'memory': a piece of material that was previously strained into the fully plastic regime 'remembers' the last flow stress reached (more exactly, the flow stress defined by back-extrapolation of the stress/strain curve it would have exhibited if it had been immediately reloaded). This 'memory' is not one of the exact history experienced, but instead of the 'state' last reached; different historical paths may lead to the same state. While the strain is, in general, a history parameter (like the time), the flow stress often qualifies as a *state parameter*. For this reason, one describes work hardening most appropriately only as a *rate of change* of the flow stress with strain: the 'strain-hardening curve' is the *locus of flow stresses as a function of prestrain* [COTTRELL 1953, KOCKS 1975]. And if this strain-hardening rate θ changes with strain (as it usually does), one describes this fact as a change of θ with the current *flow* stress. Figure 13a shows such a diagram, for two cases. Strain hardening depends only on the stress increase *since yield*,

$$\tau_\varepsilon \equiv \tau^s - \tau_0^s \tag{34}$$

where τ_0^s may depend on many structural variables, but τ_ε is related primarily to the dislocation density ρ:

$$\tau_\varepsilon \propto \mu b \sqrt{\rho} \tag{35}$$

The hardening slope generally decreases with strain; using state variables, this is better expressed as an evolution with τ_ε:

$$\theta = \theta_0 \cdot \mathcal{E}\{\tau_\varepsilon / \tau_v\} \tag{36}$$

Here, the hardening slope was labeled θ in a generic sense; it will be defined properly, for single slip and multiple slip, below. The scaling parameter θ_0 may be chosen as the initial value in an annealed material, in which case it is often of the order of μ/100 and makes the function $\mathcal{E}$ monotonically decreasing (Fig. 13a). There is no generally valid form, but it can often be a unique function for a given material at all temperatures and

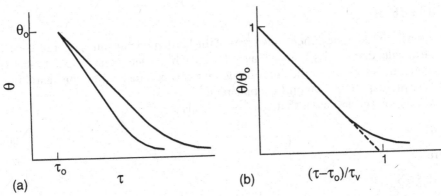

Fig. 13. (a) The strain-hardening rate, θ, as a function of the current flow stress, τ (beyond the initial yield stress, τ_0), for two different temperatures. (b) The same curves, normalized with the material parameters θ_0 and τ_v. Note that θ is the slope in Fig. 12b; if the slope were taken from Fig. 12a, it would rise sharply near the yield point. The values of τ_v and θ_0 must be chosen in conjunction to lead to a good 'universal' curve in Fig. 13b.

strain rates, if the stresses are scaled by $\tau_v = \tau_v(T,\dot{\varepsilon})$. This is illustrated schematically in Fig. 13b.

A common case is that the initial decrease of θ is linear:

$$\theta/\theta_0 = 1 - \tau_\varepsilon/\tau_v \tag{37}$$

This is equivalent, in the state parameter description [KOCKS 1976] to the 'Voce law' of exponential saturation with strain. The 'law' rarely holds at very large strains; but the extrapolation of the initial decay to $\theta=0$ is sometimes used to characterize the scaling parameter (called τ_v for this reason above). The heuristic equation

$$\theta/\theta_0 = (1 - \tau_\varepsilon/\kappa\tau_v)^\kappa \tag{37'}$$

is sometimes useful (see also BRONKHORST &AL.[1991]). When steady state is asymptotically approached, at τ_v or some higher stress τ_s (=$\kappa\tau_v$ above), a relation between the strength parameter and the strain rate follows from Eqs. 36 for $\theta=0$. Scaling for different temperatures follows from an Arrhenius equation, as discussed in detail, for the flow stress itself, in Sec. 1.6. It also involves the temperature dependent shear modulus.[6]

Slip system interaction

For a single slip system (number 1) define

$$\theta \equiv d\tau^1/d\gamma^1 \tag{38}$$

In this book, we will be primarily concerned with polycrystal deformation, which always demands the operation of many slip systems. In general, they influence each other, so one may write a generic matrix relation

---

[6] For an overview of the physical aspects of strain hardening, including dislocation storage ('stage II'), dynamic recovery ('stage III'), see MECKING[1977], MECKING & KOCKS [1981], KOCKS [1985], and GIL SEVILLANO [1993]; for a discussion including 'stage IV' (for which a physical law has been proposed instead of the empirical Eq. 37'), see the last of the above, and ROLLETT &AL. [1987b].

$$d\tau^s = \theta\, h^{st}\, d\gamma^t \tag{39}$$

where we have taken the dimensioned part of the hardening rate into a reference value θ (for example, that for slip system 1, as in Eq. 38, if h^{11} is defined to have the value 1). In general, the elements of the cross-hardening matrix $\mathbf{h}$ depend on the values of τ^s and τ^t, so that Eq. 39 is linear only in appearance.

A common special case is that *all* $h^{st}=1$. It leads to

$$d\tau^s = \theta\, d\Gamma^c \tag{40}$$

where

$$d\Gamma^c \equiv \sum_s d\gamma^s \tag{41}$$

is the 'algebraic sum of shears' [TAYLOR 1938b] in the crystal c. Equation 40 is the same for all s; thus, the excess of the current flow stress over that at zero strain, τ_e, remains the same for all systems. Equation 40 is usually used for the definition of the hardening rate, θ, which, itself, is generically expressed by Eq. 35.

The simplest case is that all *initial* values of τ^s are the same; for example, if there is but one type of slip system. In this common case, an equal hardening rate leads to a simple increases in scale of the yield surface: *proportional* hardening (Fig. 14a). The case of 'kinematic hardening' [PRAGER 1959], a *translation* of the yield surface (Fig. 14b) is rare in practice and occurs only when there are ordered internal stresses, typically only in two-phase materials.

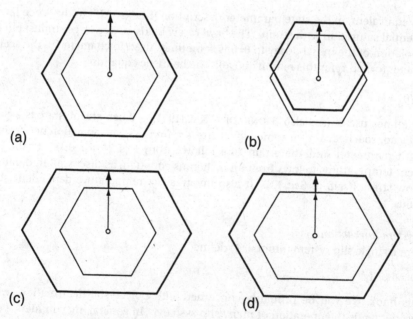

(a)   (b)

(c)   (d)

Fig. 14. Various 'hardening rules' for single crystals, describing the change in the yield surface – (a) proportional hardening: the yield surface only changes size; (b) 'kinematic' hardening: the yield surface changes only location; (c) anisotropic hardening: in the case shown, the 'latent' systems harden more than the active one (which is typical); (c) 'isotropic hardening': the rate of hardening is the same in all slip systems, regardless of their initial yield stress.

A change in *shape* of the yield surface is, however, not uncommon. There are two special cases. Figure 14c exemplifies 'latent hardening': the effect of an active system on an inactive, 'latent' one. It has been studied in some detail in fcc materials. The overall conclusion is that inactive systems harden the same or more than active ones, and that the ratio of CRSSs is usually no larger than 1.4 [KOCKS 1964b, KOCKS &AL. 1991a]. (Evaluations that give ratios less than 1.0 (which would tend to eliminate high-symmetry vertices) are based on a definition of the CRSS as a proportional limit [WU &AL.1991].)

Figure 14d demonstrates the case that the initial values of τ^s were different (such as they typically would be in non-cubic materials), but that all harden at the same rate, according to Eq. 40. Then, the *ratios* of the various τ^s decrease and the SCYS will become *more* isotropic. (Note that use of the term 'isotropic hardening' for the situation when the shape of the yield surface does not change, is misleading because the *rate* of hardening is then not isotropic but *proportional*: $d\tau^s \propto \tau^s$. This is unrealistic from the perspective of dislocation theory; if anything, the rate of hardening is *inversely* proportional to the current value of the flow stress, because the dislocation density is proportional to $\tau^s d\tau^s$ [MECKING & LÜCKE 1969].)

It is noteworthy that the algebraic sum of shears, $d\Gamma^c$, is the appropriate scalar measure of strain in crystals whenever the rate of hardening is isotropic (Eq. 40) – regardless of whether the initial CRSSs were the same [MECKING &AL. 1996b]. When they were *not* the same, $d\Gamma^c$ is no longer the work conjugate to the scale parameter of the yield surface, τ^c, and must therefore be determined separately from the simulation. Formally, one can define a second 'Taylor factor' M_ε by

$$d\Gamma^c = M_\varepsilon^c : d\varepsilon^c \tag{42}$$

(When all $\tau^s = \tau^c$, then $M_\varepsilon^c = M^c$.) In its scalar form

$$d\Gamma = M \, d\varepsilon \tag{42'}$$

this was actually the way in which TAYLOR [1938b, 1956] first introduced what has come to be known as the 'Taylor factor'. For the single grain c, M_ε^c has the fortunate property that it remains the same (for a given orientation and prescribed strain increment) so long as the SCYS *topology* remains the same, even when the actual ratios of the τ_s change and, thus, the standard (stress related) Taylor factor changes continually. This leads to an easy identification of the increase in *stress* (deviator), in macroscopic coordinates. From Eq. 23, it would be

$$dS^c = d(M^c \, \tau^c) \tag{43}$$

with some reference value τ^c. But it can also be written in terms of M_ε^c, which does not change during the step (only 'after' the step, through updating the orientation) as

$$dS^c = M_\varepsilon^c \, d\bar{\tau} \tag{44}$$

if one defines the work conjugate of $d\Gamma^c$ for the general case (i.e. when the τ^s are not all the same) as

$$\bar{\tau} \equiv \frac{\sum \tau^s \, d\gamma^s}{\sum d\gamma^s} \tag{45}$$

$\bar{\tau}$ weights the CRSS values by the extent to which the system contributes to straining. Since this depends on the direction of straining, $\bar{\tau}$ is *not* a good reference value for the scale of the yield surface, τ^c. However, for the *increases*

$$d\bar{\tau} = d\tau^s \tag{46}$$

holds, and therefore

$$\bar{\tau} - \bar{\tau}_0 = \tau^s - \tau_0^s = \tau_\varepsilon \tag{47}$$

Then, one can write for the local strain-hardening law, instead of Eqs. 40 and 35:

$$d\bar{\tau} = \theta(\bar{\tau})\, d\Gamma^c \tag{48}$$

where $\theta(\bar{\tau})$ is the same function as $\theta(\tau_\varepsilon)$ and can thus be integrated. It must be assumed to be a universal grain-level hardening law.

Equations 42, 44, and 48 together give:

$$dS^c = M_\varepsilon^c\, \theta\, M_\varepsilon^c : d\varepsilon^c \tag{49}$$

On the other hand, the initial yield stress at zero strain remains

$$\sigma_0 = M^c\, \tau_0^c \tag{50}$$

with the 'standard' Taylor factor; M^c and τ are jointly defined so that τ is an arbitrary scalar measure of the yield surface (not necessarily $\bar{\tau}$).

The function $\theta(\bar{\tau})$ cannot usually be obtained directly (say, on single crystals), but may be determined by an 'inverse method', consisting of trial and error such that Eq. 44 predicts at least one of the measured stress/strain curves from a simulation in which the evolution of the Taylor factors has been taken account of. All other stress/strain curves (for different paths, for example) must then follow from the same grain-level hardening law.

1.6 Kinetics of slip. Flow potential for rate-sensitive materials

Thermal activation

The temperature and the time-rate of straining affect the flow stress of a material. The influence is sometimes great, sometimes small, but never absent. Its very existence changes the phenomenology of yield and especially of slip system selection [CANOVA &AL. 1988]. In this section, the fundamentals of the physical mechanisms underlying this dependence will be reviewed, as well as the way in which they are translated into phenomenological relations. The concern is with the influence of temperature and strain rate on the flow stress *in a given state*; steady-state flow will be addressed only parenthetically. Further, only 'normal' behavior will be discussed, in which the flow stress decreases with temperature and increases with strain rate.

The physics of dislocation glide suggest that this dependence should follow from an Arrhenius law for the (scalar) rate of straining, $\dot{\varepsilon}$:

$$\dot{\varepsilon} = \dot{\varepsilon}_0 \exp\left(-\frac{\Delta G(\tau^s/\hat{\tau}^s)}{kT}\right) \tag{51}$$

where $\dot{\varepsilon}_0$ is a material parameter (of order $10^7 s^{-1}$), τ^s is the 'flow stress' of system s under current conditions, and $\hat{\tau}^s$ may be considered a parameter of the current state. (It is the flow stress at $T = 0$ K and has been called the 'mechanical threshold' [KOCKS &AL. 1975].) This law expresses the fact that the overcoming of obstacles by dislocations, while primarily governed by stress, can be aided by *thermal activation* at the temperature T, with an 'activation energy' (exactly: an activation free enthalpy) ΔG. (k is the Boltzmann constant.) Thermal activation is a local process; Eq. 51 is meant to already express the result of an appropriate averaging procedure over all events occurring during the percolation of dislocations through the slip plane at the stress τ^s. Although the activation process refers to the behavior on a particular slip system (super-s), Eq. 51 is meant to hold in the presence of slip on many systems.

Equation 51 was written such as to imply that τ^s is given, and the strain rate to be derived. Actual boundary conditions are often the reverse (both in test and in calculations, say, of the finite-element kind): the strain rate is prescribed, and the 'flow stress' (i.e., the percolation stress) under these conditions is asked for. Then, one may wish to write Eq. 51 in the reverse form:

$$\tau^s = \hat{\tau}^s \; \mathcal{K}\{kT \ln(\dot{\varepsilon}_0/\dot{\varepsilon})\} \tag{51'}$$

where $\mathcal{K}$ is a function which is generally *not* known a priori, but to be determined from experiments; it is the inverse of the function $\Delta G(\tau^s/\hat{\tau}^s)$ in Eq. 51. These equations link the effects of temperature and strain rate on the flow stress. We will later apply a small correction which, however, is of great practical importance: the flow stress is almost always proportional to the shear modulus, and this causes an additional dependence on T (though not on rate) [KOCKS &AL. 1975].

> The strain rate was written as the scalar measure of the macroscale strain rate $\dot{\varepsilon}$ (Eq. 15); its value can be externally prescribed to some extent. The quantity one would have expected on the left side of Eq. 51, on the other hand, is the rate of shearing on slip system s – which, however, cannot be externally prescribed under any circumstances. To within the accuracy needed within the logarithmic term in Eq. 51, it is therefore prudent to insert the operational quantity $\dot{\varepsilon}$. (The exact definition of $\dot{\varepsilon}$ will be discussed again in Sec. 1.7.)

> Equation 51 is a 'kinetic law' for slip. The term 'dynamics' is, in the context of plastic deformation, more commonly used when dislocation inertia is involved, and perhaps a general viscous drag; in other words, in the context of an equation of motion for individual (straight) dislocations [KOCKS &AL. 1975]. In the thermal-activation regime, dislocations are at rest most of the time; they proceed to the next rest position, in an overdamped fashion, after each local thermal-activation event. The thermal-activation regime is dominant in most cases – except at very high rates ($>10^5 s^{-1}$ at least [REGAZZONI &AL. 1987]). In fact, the demarcation line is the 'mechanical threshold' (which can be measured as the back-extra-polated flow stress at zero absolute temperature): below this stress, thermal-activation kinetics controls rate effects; above it, dislocation drag and inertia do.

'Rate independence' can physically occur only in a limiting sense at absolute zero temperature: here the strain rate is indeterminate. In the thermal-activation regime, the operating stress is always *below* the one that could be associated with rate-*independent* 'yield'. Figure 15a shows this behavior schematically. (It is very similar to one published by OROWAN [1934].) The actual curve in most realistic cases is even much steeper than the one shown; this means that (whenever the strain rate is substantial) yield is rate-*insensitive*, though not rate-*independent*. The rate sensitivity is often quantified by the parameter m which, for a single system s is defined as

$$m^s \equiv \left(\frac{\partial \ln \tau^s}{\partial \ln \dot{\varepsilon}}\right)_{T,\hat{\tau}^s} \tag{52}[7]$$

> Note that the differentiation is taken at constant T and in a particular *state of the material*; thus, m^s is called the *instantaneous rate sensitivity* of the flow stress (in system s). In general, m^s is not a constant: it depends on the values of T and $\hat{\tau}^s$ and, in addition, on the variable τ^s unless ΔG is logarithmic in $\tau^s/\hat{\tau}^s$. The rate sensitivity may, in general, be different for different types of system.

---

[7] This is a scalar m^s, not to be confused with the Schmid tensor $\mathbf{m}^s$. This physical scalar rate sensitivity will be discussed *only* in the present Section.

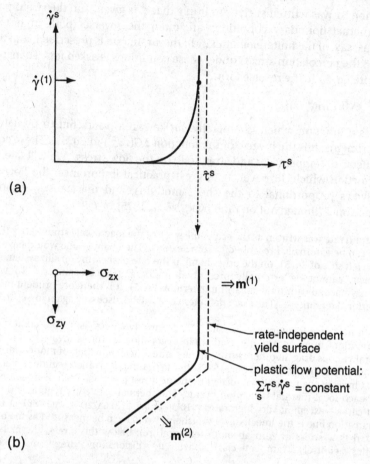

Fig. 15. Rate-dependent plasticity. (a) A typical relation between the resolved stress on a slip system (τ^s) and the ensuing strain rate $\dot{\gamma}^s$. When a strain rate $\dot{\gamma}_1$ is prescribed, the 'flow stress' for that rate is lowered by thermal activation from the rate-independent value $\hat{\tau}^s$. (b) A portion of a single-crystal yield surface under rate independence (dashed) and under rate dependence according to (a) (solid). In the region where more than one slip system is significantly active, the rate for each is lowered and this its flow stress. The normality of the straining direction on the 'yield surface' (or better 'flow surface' or 'plastic potential') is retained when the locus is chosen at constant work rate.

Typical values of the rate sensitivity in metals are between 0.03 and 0.003. This is normally true when dislocation glide controls the kinetics. There is a special case where $m^s=1/3$ and a power-law description is exact [KOCKS & CHEN 1993], at elevated temperatures in certain alloys. This refers to *steady-state* deformation, not to a constant state. When values are quoted for the rate sensitivity between 1/4 and 1/9, they usually relate to cases where the deformation is neither at steady state nor at constant state, and the rate sensitivities both of flow and of strain hardening are involved [KOCKS 1976].

Many slip systems: flow potential and explicit tensor relation

When the rate dependence is low, it is frequently neglected; however, its very existence, small or large, has an essential influence on the yield surface [KOCKS 1975]. This is demonstrated in Fig. 15b. The rate-independent yield surface, at the mechanical threshold $\hat{\tau}$, is shown dashed, for two slip systems (each in one particular direction only). The solid line that lies parallel to this yield surface (*inside* it) along its straight portions represents the flow stress at a given strain rate $\dot{\gamma}^{(s)}$ on this slip system. The interesting novelty appears where the rate-independent yield surface has a vertex: here, the 'flow surface' at a finite ($<\infty$) strain rate has a more-or-less sharp *curvature* – but the slope is *continuous* so long as $m>0$ (no matter how small). The difference between continuous and discontinuous behavior is that between a rate-dependent and a rate-independent material. We use the term 'rate-insensitive' for materials in which the rate sensitivity may be 'negligible' in magnitude, but is nevertheless greater than zero.

It is obvious that, when both systems operate equally, each may slip at a lower rate than when only one is operative: that's why the solid curve dips, near the vertex, even farther inside than it does near the flat portions. When the stress points directly toward the corner, both systems will in fact operate at the same rate, as one would expect intuitively; but if the stress favors one system ever so slightly, this system will operate at a much faster rate. The value of m (which we have assumed the same for all active systems) is an approximate measure of radius of curvature at what used to be a previous vertex, divided by the stress.

This behavior can be expressed by a *flow surface* for a rate-sensitive material: a plastic potential much like the one introduced in Eq. 21. Now, however, the scalar rate of straining, $\dot{\varepsilon}$, does matter [TÓTH &AL. 1988]. We must leave the plastic potential equal to the work rate $\dot{w}$ (not normalized by $\dot{\varepsilon}$) and obtain the whole straining rate **D**:

$$D_{ij}^c = \frac{\partial \dot{w}}{\partial \sigma_{ij}^c} \tag{53}$$

In words: in order for $\mathbf{D}^c$ to be perpendicular to a surface (the 'plastic potential'), this surface must be a contour at constant plastic work rate (not plastic strain-rate).

A quantitative description of the directional relation between the total stress and strain rate in any crystal (labeled c, assumed quasi-homogeneous) also follows directly from Eq. 11 (without explicit use of a flow potential):

$$\mathbf{D}^c = \dot{\varepsilon} \sum_s \left(\frac{\dot{\gamma}^s}{\dot{\varepsilon}} \right) \mathbf{m}^s \tag{54}$$

This superposition law is here re-written in the spirit of separating scalar from directional effects according to Eq. 15. Forward and backward operation of a slip system are best treated using the same $\mathbf{m}^s$ but different signs of $\dot{\gamma}^s$; $\dot{\varepsilon}$ is always positive, by definition.

For the dependence of the actual shear rate $\dot{\gamma}^s$ in system s on the resolved stress acting in it, $\mathbf{m}^s{:}\sigma^c$, one typically employs a *power law* [CANOVA & KOCKS 1984, ASARO & NEEDLEMAN 1985]:

$$\frac{\dot{\gamma}^s}{\dot{\varepsilon}} = \left| \frac{\mathbf{m}^s{:}\sigma^c}{\tau^s} \right|^{n^s} \mathrm{sgn}\left(\mathbf{m}^s{:}\sigma^c\right) \tag{55}$$

τ^s is the 'flow stress' at the reference strain rate $\dot{\varepsilon}$. The exponent is usually assumed independent of s, but this need not be so. Equation 55 may be looked upon as a

τ^s is the 'flow stress' at the reference strain rate $\dot{\varepsilon}$. The exponent is usually assumed independent of s, but this need not be so. Equation 55 may be looked upon as a development of the actual curve in Fig. 15a around the point $\{\dot{\varepsilon}, \tau^s\}$. Inserting Eq. 55 into Eq. 54 gives

$$\mathbf{D}^c = \dot{\varepsilon} \sum_s \left| \frac{\mathbf{m}^s : \boldsymbol{\sigma}^c}{\tau^s} \right|^{n^s} \mathbf{m}^s \, \mathrm{sgn}\!\left(\mathbf{m}^s : \boldsymbol{\sigma}^c\right) \tag{56}$$

This is an explicit equation for $\mathbf{D}^c$ in terms of $\boldsymbol{\sigma}^c$ or (to the extent that it is invertible) an *implicit* equation for $\boldsymbol{\sigma}^c$ in terms of $\mathbf{D}^c$. The deviatoric stress $\mathbf{S}^c$ may, in analogy to Eq. 23, be expressed as scaled by a crystal reference flow stress, τ^c (say, one of the τ^s):

$$\mathbf{S}^c = \mathbf{M}^c(\hat{\mathbf{D}}) \, \tau^c \tag{57}$$

The exponent in the phenomenological kinetic relation, Eq. 55, was labeled n; it is not meant to be identical to the inverse rate sensitivity, $1/m^s$ of any system. To illustrate the potentially important difference, the scalar relation of Fig. 15a is plotted again in Fig. 16a, with a number of differences: first, the axes are interchanged and plotted logarithmically; second, the temperature dependence is added, so that the abscissa in Fig. 16a is proportional to the effective slip-plane activation energy ΔG; and third; both the stress and the activation energy are normalized with $\mu(T)$ [KOCKS &AL.1975]. It is plotted for two crystallographically different slip systems, s=1 and s=2. The curves in Fig. 16a relate different values of τ^s and $\dot{\varepsilon}$ to each other; for example, in different tests, different values of T and $\dot{\varepsilon}$ may be prescribed and, as a result, the τ^s are different.

Flow stress scaling

The physical relation involves, as the most important parameter, the mechanical threshold, $\hat{\tau}$. However, it is an experimentally awkward quantity: it can be measured only by extrapolation to absolute zero temperature (from a temperature regime that belongs to the same mechanism as the actual one). It is more appropriate, operationally, to define a regime of interest and use, as the state parameter, the flow stress under some standard conditions within this regime (usually chosen at the lowest temperature of interest, and a strain rate that is easily measured). If we use the index 1 for the standard conditions, we may rewrite Eq. 51' as follows (setting $\tau^s \equiv \tau$, for simplicity):

$$\frac{\tau}{\mu} = \frac{\tau_1}{\mu_1} \frac{\mathcal{K}\!\left(\dfrac{kT}{\mu\,b^3} \ln \dfrac{\dot{\varepsilon}_0}{\dot{\varepsilon}} \right)}{\mathcal{K}\!\left(\dfrac{kT_1}{\mu_1\,b^3} \ln \dfrac{\dot{\varepsilon}_0}{\dot{\varepsilon}} \right)} \tag{58}$$

Again, the corrections due to the temperature-dependent shear modulus μ has been incorporated.

If the lines in Fig. 16a were straight, one could express this relation as a power law – albeit with a T-dependent value of the stress exponent m^s from Eq. 52. But such a power law is rarely a good enough approximation. On the other hand, we have made use of a power law in Eq. 55 and, as a consequence, in Eq. 56. The rationale for this difference is as follows (Fig. 16b). We only wish, in this instance, to describe the different strain rates in the contributing systems – which are of interest only when they are within one order of magnitude. Furthermore, the temperature and state are constant. For this limited application, a power law is a useful approximation even when the exponent is very large. Finally, n takes on the role of a computational variable when Eq. 56 is solved as an implicit equation. The higher n, the more difficult it is to obtain a solution by iteration. It is often

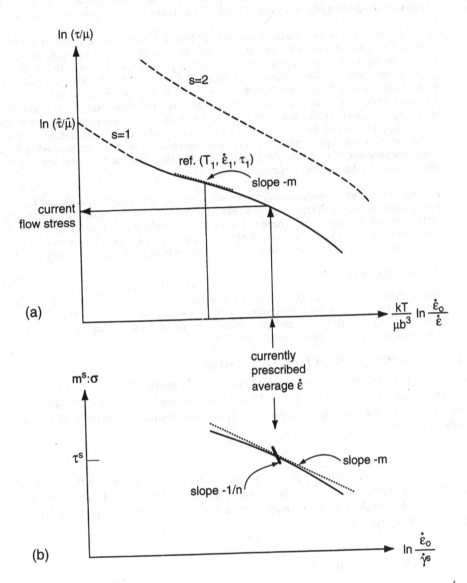

Fig. 16. It is useful to separate the effects of different straining rates on different slip systems (at the same macroscopic rate) from those of differing macroscopically prescribed rates. Figure (a) describes the latter, in conjunction with the temperature dependence of yield. For applications at different temperatures and (macroscopic) strain rates, it is best to choose a set of reference conditions, $\{T_1, \dot{\varepsilon}_1\}$ leading to a reference value τ_1 of the material strength. For a particular test, the value of the flow stress τ^s on a single slip system (if it were operating alone) follows from the material relation depicted in Fig.(a). The rate sensitivity m is typically quite small. Figure (b) depicts the power law (power $1/n$) used to describe the distribution of straining rates among the slip systems; in other words, the curvature of the flow potential near a vertex. The exponent n can be chosen much smaller than $1/m$ to facilitate numerical solutions – if these do not in fact depend on such a change (as is typical).

found that the results do not depend on n when it is greater than, say, $n=20$. Then, it is opportune to replace a physical value of $1/m^s=200$ (not unusual at all) with a value $n=20$.

It is worth noting that the rate-*independent* limit ($n\rightarrow\infty$) is contained in Eq. 56: it requires $m^s:\sigma^c = \tau^s$ (Eq. 18) and would seem, at first sight, to lead to multiple solutions of Eq. 56 when there are more systems than necessary. It has been shown, however, that a proper limiting procedure for $1/n\rightarrow0$ does give a unique solution just like for any $n<\infty$ [NEALE & ZHOU 1991; ZHANG & JENKINS 1993; ANAND & KOTHARI 1996]. It deserves emphasis that the so-called 'ambiguity' of solutions occurs only when strict rate-independence is assumed – which is physically unreasonable. On the other hand, for some applications (such as those in Secs. 1.3 through 1.5), the rate-independent, faceted single-crystal yield surface is useful. In fact, even for an approximate, but very close, determination of the stress needed for a given deformation, the discrete yield surface is adequate – just not for a derivation of the distribution of deformation rates among the systems.

Finally, a word about steady-state deformation. One can treat the kinematics and kinetics by the same formalism of Eqs. 54 and 55, respectively, so long as one treats only the steady-state limit, not the transients. There are, however, two slight differences in interpretation. First, the exponent n is now the actual physical value. (It is typically 3.) Second, τ^s is no longer a state parameter, but merely a combination of physical constants; for example,

$$\tau^s = \mu \exp(Q/nkT)$$

(59)

with a *constant* Q. Similarly, the scalar $\dot{\varepsilon}$ should be replaced by a physical constant A involving, for example, the pre-exponential D_o of the diffusion coefficient [KOCKS & CHEN 1993]. In order to keep $\dot{\varepsilon}$ as a standard scalar measure of the strain rate (say, $10^{-2}s^{-1}$), one may multiply the right-hand side of Eq. 59 by $(\dot{\varepsilon}/A)^{1-n}$.

1.7 Constitutive relations for crystal plasticity

The physical basis of constitutive relations for plasticity lies in the behavior of single crystals. In this Section, the results of the previous Sections will be summarized and applied to the ensuing behavior of polycrystals – without assuming any specific polycrystal model.

Grains in a polycrystal

The kinematics and kinetics of single-crystal plasticity are condensed in the relation (Eq. 56) between the crystal straining rate $\mathbf{D}^c$ and the crystal stress σ^c, given its s glide (and twinning) systems, each characterized by its Schmid factor $\mathbf{m}^s$ and its reference flow stress τ^s:

$$\mathbf{D}^c = \dot{\varepsilon} \sum_s \left| \frac{\mathbf{m}^s : \sigma^c}{\tau^s} \right|^{n^s} \mathbf{m}^s \, \mathrm{sgn}\left(\mathbf{m}^s : \sigma^c\right) = \dot{\varepsilon} \, \hat{\mathbf{D}}^c$$

(60)

$\dot{\varepsilon}$ is a scaling factor which will be defined precisely below; the essence of Eq. 60 is the relation between the straining direction $\hat{\mathbf{D}}^c$ (Eq. 15) and the crystal-level resolved stress. The latter is also scaled, by the strength level on the slip systems; when it is different for different slip systems, one may define a reference crystal strength τ^c, and each $\tau^s \equiv \alpha^s \tau^c$.

Equation 60 seems to emphasize *rate* effects in plasticity. This is particularly appropriate in *steady-state creep* when the stress exponent is typically $n^s=3$ for all systems; then, the scaling factors $\dot{\varepsilon}$ and τ^s are well defined material constants (see the end of Sec. 1.6).

However, Eq. 60 also applies to rate-insensitive plasticity. Here, the τ^s are typically state parameters, which evolve with strain (Eqs. 36, 39):

$$\left.\frac{d\tau^s}{d\Gamma^c}\right|_{T,\dot{\varepsilon}} = \theta_0 \, \mathcal{E}\left\{\tau^s_\varepsilon / \tau_v\right\} \tag{61}$$

where $d\Gamma^c$ is a measure of the strain increment in the grain (Eq. 41). $\mathcal{E}$ is a generic 'evolution' function of the strain-dependent part of the flow stress, τ_ε (which is due to the increasing dislocation density, see Sec. 1.5). It scales, for different temperatures and strain rates, with a parameter $\tau_v(T,\dot{\varepsilon})$. The initial hardening rate (when $\tau_\varepsilon=0$) is θ_0, and the function $\mathcal{E}$ typically decreases monotonically from 1 to 0. The limit of no hardening ($\mathcal{E}{\rightarrow}0$) corresponds to steady state; then $\tau_v(T,\dot{\varepsilon})$ determines the kinetics.

In classical plasticity theory, it was typically assumed that one had either steady state (and thus rate dependence) or strain hardening (but then no rate dependence). The physical basis condensed into Eqs. 60 and 61 provides for hardening *and* rate dependence. In fact, there are two kinds of rate dependence: the one of hardening – and, consequently, of steady state – (Eq. 61), and the one at constant state (Eq. 60). The latter is expressed as a power law in Eq. 60. Such a power law is not usually a good description of material behavior over a wide range of the variables. Then, Eq. 60 may be looked upon as an expansion around a set of *reference values* $\{\tau^s,\dot{\varepsilon}\}$; they themselves may be related to each other by an appropriate physical law, such as Eq. 51' (modified to include effects of the temperature dependence of the shear modulus μ, see Sec. 1.6):

$$\tau^s = \hat{\tau}^s \, \mathcal{K}\{kT/\mu b^3 \cdot \ln(\dot{\varepsilon}_0/\dot{\varepsilon})\} \tag{62}$$

with the 'mechanical threshold' $\hat{\tau}$. Equivalently Eq. 58 may be used, with the flow stress under a standard set of conditions as a state parameter. Such a physical description of the rate sensitivity of the reference flow stress frees the exponent n^s in Eq. 60 from having to be physically correct over a large range; it can be chosen in part for numerical convenience. (Sec. 1.6)

Equation 60 is an explicit expression for the local strain rates, when the local stresses are known. It is often used, instead, as an implicit equation for the stresses, when the strain rates are known. This is done numerically by an iteration scheme such as the Newton–Raphson method. Only those components of the stress can be derived that actually influence $\mathbf{D}^c$; for example, for isochoric plasticity, one can derive from an inversion of Eq. 60 only the *deviatoric* stress $\mathbf{S}^c$. The result scales with the crystal reference strength, τ^c (Eq. 23):

$$\mathbf{S}^c = \mathbf{M}^c \, \tau^c \tag{63}$$

The directional term $\mathbf{M}^c$ depends on the straining direction $\hat{\mathbf{D}}$ and the texture; the scalar term τ^c depends on the temperature, the strain rate $\dot{\varepsilon}$, perhaps the stress state σ^c in the grain (including the pressure, Eq. 25) – and on the state of the material, such as it follows from structure parameters like the dislocation density, the volume fraction of inclusions, the grain size, etc.

Stress and strain-rate scaling

The stress and the straining rate in a polycrystalline aggregate follow from a volume average of the crystal quantities (Eqs. 14, 16):

$$S = \langle S^c \rangle \qquad\qquad D = \langle D^c \rangle \tag{64}[8]$$

The scaling properties of the single-crystal behavior carry through to the polycrystal. We define

$$S = \sigma_f \hat{S} \qquad\qquad D = \dot{\varepsilon} \hat{D} \tag{65}$$

Note that we have used the same symbol $\dot{\varepsilon}$ in relation 65 as in Eq. 60 – where we had postponed a precise definition. This scalar strain rate will now be specified through work conjugacy to the strength parameter σ_f. The work rate is

$$\dot{w} = S : D \equiv \hat{S} : \hat{D} \, \sigma_f \dot{\varepsilon} = \sigma_f \, \dot{\varepsilon} \tag{66}$$

In other words, the stress direction and the straining direction are scaled jointly such that

$$\hat{S} : \hat{D} \equiv 1 \tag{67}$$

This leads to a connection between $\dot{\varepsilon}$ and σ_f:

$$\dot{\varepsilon} = \dot{w}/\sigma_f = \hat{S} : D \tag{67'}$$

Typically, it is $\dot{\varepsilon}$ that is macroscopically prescribed in a given test, but $\dot{w}$ that must be used to compare different tests.

It remains to specify the precise meaning of the polycrystal strength parameter σ_f. It is conventionally derived from the result of a tensile test (on a long thin wire, in the Z-direction). This is operationally meaningful, since here all stress components are prescribed zero except σ_{ZZ}. The strain rate components depend on the anisotropy of the material – but the only one through which work is done is D_{ZZ}. Thus, one jointly measures σ_f as σ_{ZZ} and $\dot{\varepsilon}$ as D_{ZZ}. In this way, the *functional relation* $\sigma_f(\dot{\varepsilon})$ for a rate-sensitive material may be established. [σ_f may be called the 'flow stress'. It is equivalent to k_f in German usage, R (for *résistance)* in DIN norm, Y of HILL [1950], and τ in KOCKS &AL. [1975] – where the difference between polycrystals and single crystals was glossed over, which is of the essence here.]

It needs to be emphasized that the scaling parameter of the stress, *viz.* the strength parameter σ_f, is not the norm of the stress, i.e., the 'length' of the stress vector [HART 1976]. The latter is often used in isotropic materials, in its deviatoric form:

$$\sigma_{vM} = \sqrt{\tfrac{3}{2} \, S{:}S} \tag{68}$$

and then the 'von Mises yield criterion' is $\sigma_{vM} = \sigma_f$: all stresses of the same norm are 'equivalent'. In an anisotropic material, one may identify σ_f with the length of the stress deviator vector *for tension in the Z-direction*; but note that different values would hold for tension in the X- and Y-directions; by convention, only the former specifies σ_f. And in any other test, the norm of S may have any other value.[9]

---

[8] Inasmuch as the scalar properties are characteristic of the whole material, one may rewrite Eqs. 64 in terms of the directional parts only:

$$\hat{S} = \langle \hat{S}^c \rangle; \qquad \hat{D} = \langle \hat{D}^c \rangle \tag{64'}$$

[9] The norm of D has no meaning for an anisotropic material: the relation between $\dot{\varepsilon} = D_{ZZ}$ and D:D depends on the anisotropy.

Physically speaking, the strength parameter σ_f is linked to an average of the crystal strengths τ^c. To derive this relation, we again consider the work rate, this time for a single crystal:

$$\dot{w}^c = S^c : D^c = M^c : \hat{D}^c \, \tau^c \, \dot{\varepsilon} \equiv M^c \, \tau^c \, \dot{\varepsilon}$$

It defines the standard, scalar Taylor factor for a crystal, M^c [TAYLOR 1956]. Using the above definition of σ_f as the yield strength in tension (in the Z-direction), we get

$$\sigma_f \equiv \langle\, M^c_{ten} \, \tau^c \,\rangle \tag{69}$$

This relation is useful for the scaling of model simulations. The strength parameter σ_f depends on texture, but it also depends on the kinetic variables and the state of the material (due to microstructural features other than texture, and the state of stress):

$$\sigma_f = \sigma_f \{\dot{\varepsilon}, T; \text{state}\} \tag{70}$$

As a consequence of the universal scaling with σ_f, the directional part of the crystal deviatoric stress tensor can now be defined as

$$\hat{S}^c = \frac{M^c(\hat{D}^c)}{M^c_{ten}} \tag{71}$$

It characterizes the shape of the flow potential and must be computed based on the current orientation distribution. It must be updated according to changes in the texture (see Sec. 3.3).

The scalar strength parameter σ_f changes with straining both for geometrical reasons (due to changes in texture, affecting M^c_{ten}) and because of physical strain hardening (affecting all the τ^s). The latter is usually not isotropic even if the texture were not to change. It depends on the slip distribution and the interaction between slip systems. In general, the slip distribution must be determined for each grain (by inserting the solution stress into the power part under the sum of Eq. 60) and a hardening law used (Eq. 39). For the simple but plausible case that the *increase* in flow stress in all systems is the same (even when their initial flow stresses, typically dominated by the Peierls stress, are not), this $d\tau^s = d\tau^c$ becomes proportional to the algebraic sum of shears, $d\Gamma^c$: this is the strain pacing parameter used in Eq. 61. We will now link it to the macroscopic strain increment defined in Eq. 67' as

$$d\varepsilon \equiv \dot{\varepsilon}\, dt \tag{72}$$

by setting

$$d\Gamma^c \equiv M^c_\varepsilon \, d\varepsilon \tag{73}$$

Only when all τ^s are the same is the term M^c_ε the same as the Taylor factor defined in Eq. 69; otherwise, it depends on the distribution of deformation mode contributions and is *defined* by Eq. 73 [MECKING &AL. 1996b]. It is a function only of the yield surface topology, not of its actual shape, and therefore does not, in most cases, change during a straining step. This gives rise to the possibility of writing, for the *increase* $d\sigma_f = d\langle M^c \tau^c\rangle = M_\varepsilon \langle d\tau^c\rangle$ (where $M_\varepsilon \equiv \langle M^c_\varepsilon \rangle$), and thus

$$\Theta \equiv \left.\frac{d\sigma_f}{d\varepsilon}\right|_{T,\dot{\varepsilon}} = M_\varepsilon^2\,\theta_0\,\Big\langle \mathcal{E}\big\{\tau_\varepsilon^c / \tau_v\big\}\Big\rangle \equiv \Theta_0\,\mathcal{H}\{\sigma_f / \sigma_v\} \tag{74}$$

where $\mathcal{H}$ again goes from 1 toward 0 (but is otherwise unknown). Equation 74 is handy for estimating the single-crystal hardening law from the polycrystal hardening law; especially θ_0, which is often a fixed fraction of the shear modulus μ (e.g., $\mu/200$).

An important aspect of constitutive relations is that dimensionless quantities are defined, which combine variables and material properties on the basis of fundamental mechanisms. We have encountered the combination $\Phi \equiv kT/\mu b^3\cdot\ln\dot{\varepsilon}_0/\dot{\varepsilon}$, which is a physically scaled activation energy for plastic flow and links the temperature and rate dependence in the thermally activated regime. Another combination, which has not yet been spelled out in this summary, is that all parameters of stress dimensions scale with the shear modulus μ (Eq. 58). When its temperature dependence matters (which is the normal case), one must use $\mu(T)$ as the scaling parameter, and this provides for an additional temperature dependence (which is *not* linked to a rate dependence). Under these circumstances, it is also important to normalize the activation energy parameter by the temperature dependent μb^3. (The T-dependence of b^3 is usually negligible by comparison.)

These scaled variables serve as the coordinates for appropriate plots; however, the functional relation between them is often not known with as much certainty and may have to be determined by experiment. In the case of strain hardening, a unique functional form must be assumed at the crystal level and then used to predict macroscopic stress/strain curves that incorporate the effects of texture change. One such prediction can be matched to experiments, thus fixing the microscopic form; others should then follow if the basic premise of a unique crystal hardening law is correct.

Operational procedure

At this point, it may be helpful to summarize the operational procedure for the determination and prediction of anisotropic plasticity properties. For a particular material in some well defined initial state, one needs to

- measure the texture;
- decide on a convenient standard sample axis **Z**;
- cut out a 'wire' sample in this **Z**-direction (making sure that the cross section contains enough grains so that the fraction of surface grains is negligible – and also keeping the ratio of gauge length to diameter sufficiently high);
- measure the stress/strain curve on this wire at a value of the true strain rate $\dot{\varepsilon}$ that can be easily prescribed (for example, $10^{-2}\mathrm{s}^{-1}$), until just before necking;
- repeat this procedure (if needed) for a range of $\dot{\varepsilon}$ and T;
- find $\mu(T)$;
- jointly determine the initial yield stress σ_f and the initial hardening rate Θ_0 (perhaps by smoothing and differentiating the stress/strain curve) – preferably so as to make Θ_0/μ a constant for all T and $\dot{\varepsilon}$;
- plot σ_f/μ versus the scaled activation energy $kT/\mu b^3\cdot\ln\dot{\varepsilon}_0/\dot{\varepsilon}$ (picking $\dot{\varepsilon}_0$ such as to unify the T- and $\dot{\varepsilon}$-dependencies) and describe this empirical function.

So far, the measurement of properties was model independent. To predict the behavior under other conditions, one must have a model for polycrystal deformation. There are two types of prediction one may wish to make: concerning the scalar behavior (flow stress and strain hardening as a function of all variables); and the behavior under different tensorial straining paths, starting with the current state. For the former, one uses Eqs. 60 through 70, for the latter Eq. 71. The texture needs to be known at every state; its evolution can be predicted as part of the polycrystal model (see Sec. 3).

Linearization of the constitutive law

Finally, we wish to state, in a summary fashion, a way in which the tensorial relation between strain rates and stresses has been abstracted for use, for example, in finite-element codes or 'self-consistent' calculations (which will be exposed in Chaps. 12 and 11, respectively). In Eq. 60, one may factor out one σ^c and write for the straining direction, in components,

$$\hat{D}_\mu^c = \left(\sum_s \left| \frac{m_\lambda^s \sigma_\lambda^c}{\tau^s} \right|^{n^s-1} \frac{m_\mu^s m_\nu^s}{\tau^s} \right) \sigma_\nu^c \equiv \mathbb{P}_{\mu\nu}^c \sigma_\nu^c \tag{75}$$

where the second part is a definition of a 'plasticity modulus' $\mathbb{P}$ (as it was first defined by HART [1976]). It depends strongly on σ^c. If it did not, this last part of Eq. 75 would describe an ellipsoid in stress space. (Multiplication of both sides with σ_μ gives the work rate on the left and a square form on the right.) The fact that single-crystal yield surfaces have little resemblance to an ellipsoid is just an expression of the fact that $\mathbb{P}^c$ is a very strong function of σ^c. For polycrystalline aggregates, the angularity of the yield surface is not as pronounced (unless the texture is *very* sharp). Still, a description is terms of

$$\hat{D} = \mathbb{P} : S \tag{76}$$

(where we have also replaced the Cauchy stress σ by its deviatoric part S) with some *constant* $\mathbb{P}$ would amount to an ellipsoid that would fail to describe some important features of the plastic behavior. Leaving $\mathbb{P} = \mathbb{P}(S)$ in Eq. 76 is merely a formal equation that seems linear, but is not.

What an equation like 76 can be used for, though, is to describe the situation in the neighborhood of one particular point $\{S_1, \hat{D}_1\}$; in other words, the *trend* of the changes resulting in one variable upon an imposed change in the other. One would typically express this trend in form of a Taylor expansion:

$$\hat{D} - \hat{D}_1 = \mathbb{P}^{tg}(S_1) : (S - S_1) + \dots \tag{77}$$

where a 'tangent plasticity modulus' $\mathbb{P}^{tg}$ has been defined at the position S_1. Since we are discussing a particular grain with constant properties, including its current 'state', the only variations Eq. 77 describes are those along a given yield surface or flow potential. Inasmuch as $\hat{D}$ is the direction of the *normal* at S, $\mathbb{P}^{tg}$ characterizes the *curvature* of the flow potential [KOCKS 1994]. It may be visualized as the parameter for the 'tangent ellipsoid', which is centered at S_0 and has the equation

$$\hat{D} = \mathbb{P}^{tg}(S_1) : (S - S_0) \tag{76'}$$

This ellipsoid has its principal axes in the direction of $\hat{D}$ and along the eigenvectors of the curvature tensor of the yield surface.

In analogy to the 'tangent' modulus in Eqs. 77, one may call $\mathbb{P}$ in Eq. 76 a 'secant' modulus. It, too, is useful for the description of the trend in the neighborhood of a point S_1. This is at first counter-intuitive, but may be understood by looking at the two-dimensional case (Fig. 17). Here, the 'tangent ellipsoid' is a circle centered at S_0 with the radius of curvature at the point S_1. The 'secant ellipsoid'

$$\hat{D} = \mathbb{P}(S_1) : S \tag{76''}$$

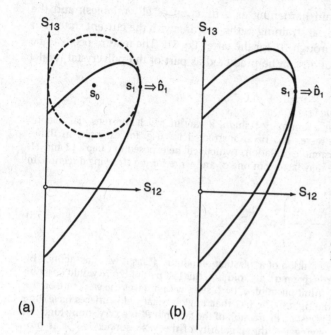

(a)   (b)

Fig. 17. A 'linearization' of the constitutive law leads to ellipsoidal yield surfaces. This may be an adequate description under some circumstances for small deviations from a known plastic state $\{S_1, \hat{D}_1\}$. The 'tangent' formulation corresponds to a curvature ellipsoid (in two dimensions, a circle) centered at $S_o \neq 0$; the 'secant' formulation to an ellipsoid centered at the origin. Both have the same local curvature.

is centered at the origin; its three parameters can be fitted to fulfilling $\hat{D}_1 = \mathbb{P}(S_1):S_1$ *and the curvature* at $\{S_1, \hat{D}_1\}$. Figure 17b demonstrates this for two different local curvatures. It can be shown that the number of parameters available in n dimensions is precisely the right one ($n^2/2 + n/2$) to match the value and all slopes and curvatures of the yield surface at one point.

Figure 17a allows a comparison of Eqs. 77 and 76′. In the close neighborhood of the point, they give identical results. Not so far away, neither one of them gives anywhere near a correct description. The difference is that the 'secant' ellipsoid provides *some* answer for stresses in all directions, whereas the 'tangent' description allows for stresses only in the vicinity of S_1. Conversely, the tangent form can give arbitrary $\hat{D}$ for such stresses near (though not at) S_1, whereas in the secant form the $\hat{D}$-response is bounded for that range of stresses. Both effects would seem to make the *approximation* in Eq. 76′ safer than the *truncated* series expansion in Eq. 77. However, the tangent method may converge faster.

It is evident that the more the curvature changes along the yield surface the narrower is the range of usefulness of any 'linearized' description like Eq. 77 or 76′. Yield surfaces that are very 'angular' (i.e., consist of quasi-vertices on the one hand and large flat spots on the other) are equivalent to very *nonlinear* visco-plasticity in the sense that $\mathbb{P}$ is not at all constant, and $\mathbb{P}^{tg}$ even less so: both depend strongly on S_1. In general, flow surfaces become less angular the higher the rate sensitivity and the larger the number of deformation modes; this makes 'linear' approximations more likely to be appropriate. At the other extreme, for a very angular yield surface, the tangent ellipsoid becomes very small near the vertices and thus the tiniest variations in stress would yield a *sign reversal* of $\hat{D}$. The secant ellipsoid, on the other hand, will make the stress and strain in the average *coaxial* around a vertex; while this is not correct, it is better than to have an indeterminacy.

All the above moduli are proportional to $1/\sigma_f$ (and a proportionality to the reference strain rate $\dot{\varepsilon}$ has already been left out by discussing only the strain-rate *direction* $\hat{D}$). The essence of Eqs. 77 and 76′ is in the *directional* relation and a term such as 'inverse viscosity' is misleading in that it implies *time* effects.

Elasticity

Elasticity may be needed under some circumstances; e.g.,
(a) when treating a big body (rather than a representative material element) and finding some regions with no plastic deformation (or, in rate-sensitive materials, very little);
(b) in the elasto-plastic transition region (and thereby for cyclic hardening, fatigue, and the transient aspects of the so-called Bauschinger effect; and
(c) if *elastic anisotropy* matters.

Whenever plasticity is the dominant mode of deformation, it is appropriate (or at least possible) to determine the stress from the plasticity relation at a given instant and then use the stress to derive the elastic strains:

$$\varepsilon^{el} = \mathbb{C}^{-1}\sigma \tag{78}$$

with $\mathbb{C}$ the correct elastic constant tensor at the level one wishes to describe (polycrystal or single crystal). The elastic contribution can be evaluated at every step and the increase used to relate the plastic strain increment $\mathbf{D}\,dt$ to the total strain increment. Note that Eq. 78 contains the 'true' or 'Cauchy' stress

$$\sigma \equiv \mathbf{S} - p\,\mathbf{I} \tag{79}$$

To obtain it, one needs to know the pressure p (which is not necessarily constant throughout the medium).

It is worth emphasizing that in the description given here, the stress is determined from $\mathbf{D}$ through the plasticity relation (assuming the plastic strain rate $\mathbf{D}$ to be equal to the total strain rate), and *then* the elastic strains calculated (which may subsequently be used in an iteration to correct the total strain rates to the plastic ones[10]). In many conventional descriptions, the stress is determined from the elastic strains; but since these must be superposed on the plastic strain *rates*, this leads to the need for carefully defining stress *rates* (which are not needed in our description at all) [PEIRCE &AL 1982, HAVNER 1992].

Restrictions

These constitutive relations were meant for a wide, but not an infinite range of materials, material states, temperatures, strain rates, etc. – and not for all aspects of behavior. For example, the constitutive relations were developed for metals and alloys, and do hold for some ductile ceramics; but their applicability for each type of material must be ascertained experimentally. The range of variables covers most conventional testing; but not, for example, temperatures near absolute zero (where dislocation inertial effects become important), or strain rates above about 10^3sec^{-1} (where additional hardening mechanisms occur). The grain interactions treated above do not include grain boundary sliding, such as it occurs in superplastic materials. The temperature and strain-rate regime in which diffusion actually controls plastic flow is excluded. And, as a final example, cyclic deformation in the range where elastic and plastic deformation are commensurate, have not been addressed.

2. Grain Interaction and Polycrystal Plasticity

2.1 Modes of polycrystal yielding: physics

Band propagation versus bulk plasticity.

The very first plastic event as the stress on a polycrystal is increased from zero is likely to be slip on a single slip plane in one grain. This produces the equivalent of a shear crack in the grain or, in dislocation language, a pile-up (Fig. 18a). There is a strong tendency for this 'crack' to propagate; in other words, for the pile-up stresses in

---

[10] For finite step sizes, one must also properly account for elastic rotations [LEE 1969].

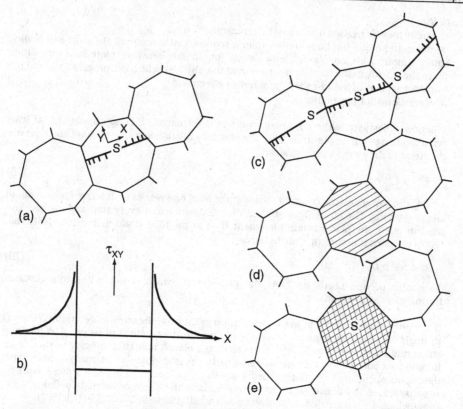

Fig. 18. Schematic of polycrystal yielding: (a) pre-yielding, one slip active in one grain; (b) ensuing distribution of internal stress; (c) one mode leading to yielding: propagation of slip into the next grain; (d) another consequence of internal stresses due to process (a): spread of slip to parallel planes in same grain; (e) another mode leading to yielding: polyslip in each grain.

the next grain to activate a nearby source and produce slip on a single plane (or a few neighboring ones) which is oriented favorably (Fig. 18c). Once the sheared region has increased in size, the tendency for propagation increases even more. The distribution of slip stresses on the slip plane (and its extension into the neighboring grains) is schematically shown in Fig. 18b: there is a uniform backstress inside the slipped region, and an outside forward stress that decays as $r^{1/2}$. As a consequence, the whole cross section will yield.

This situation has been observed experimentally in a mild steel [SUITS & CHALMERS 1961], with a slight variation: the first grain to yield activates slip on a whole set of parallel slip planes, before it propagates into the next grain (Fig. 18d). This can also be explained on the basis of pile-up stresses *inside* the grain after only a single plane has operated.

The next step in this process is lateral propagation of the deformed region in the form of a 'Lüders band'. The deformation of the polycrystal occurs under a constant stress until the band has reached the ends of the gauge length. This stress plateau is usually preceded by an initial 'upper yield stress', especially when there are no stress concentrations in the specimen and the machine is hard.

An alternative progression toward macroscale deformation is illustrated in Fig. 18e. In the absence, for some reason, of forward relief of the pile-up stresses, the interior of the grain deforms first (again) on many parallel slip planes and then on another set of slip planes. This leads to *bulk plasticity* – mechanistically, the 'polyslip' [KOCKS 1970] mode of polycrystal deformation – and to the Taylor model in first instance. The initial transition to this mode will be described below.

Which of these two modes of macroscale yield, Lüders band propagation or bulk plasticity, is operative depends on many details of the material and conditions [MECKING 1980]; two effects can be highlighted. First, the polyslip mode depends on the potential availability of a sufficient number of slip systems. The propagation mode is also somewhat easier when many slip systems are available, because then a well aligned plane in the next grain may be found. The second and major criterion is that the propagation mode is expected when the material is 'notch sensitive'; i.e., when a local stress concentration and the ensuing local dislocation generation near a grain boundary is severe enough to make slip go through the whole grain. If, on the other hand, the 'friction' stress controls deformation, local effects may not lead to macro-scale yield.

Elasto-plastic transition

The elasto-plastic transition to the 'polyslip' mode of polycrystal plasticity was first treated by KRÖNER [1961] on the basis of the ESHELBY [1957] model of 'stress-free strain' in an ellipsoidal inclusion embedded in a homogeneous matrix. Kröner identified the matrix with a 'homogeneous effective medium' (HEM) that has the properties of the average polycrystal, which are yet to be determined in a 'self-consistent' way. The stress-free strain is set to the difference in plastic strain between the inclusion and the HEM. The inclusion is meant to be representative of many grains of essentially the same orientation, so that even the immediate neighbors of the individual grains form an ensemble that in the average may be representative of the polycrystal as a whole.

If the initial strain in a grain is as depicted in Fig. 18d, the back-stress in the grain leads to the activation of a second system (Fig. 18e). The schematic stress-space, Fig. 19, shows how the back stresses inside the grain (represented by the dashed lines) lead to movement of the local stress to a corner, as the applied stress is increased from σ_1 to σ_2. The 'corner' shown actually corresponds to a two-system 'edge' but the physical process will, by the same token, lead to a polyslip vertex. As HUTCHINSON [1964a] and BUDIANSKY & WU [1962] have shown, virtually all grains have reached a vertex after $\approx 5\, \varepsilon^{el}$. To the extent that this model is applicable, it would justify use of the Taylor model for 'large' plastic strains: then, large deviations in strain from grain to grain are not allowed, because they would lead to forbiddingly high back stresses.

The model has, however, serious shortcomings; particularly that the reaction of the matrix is treated as elastic. Figure 18b illustrated the high stress concentration that would be generated in this elastic model just outside the grain. It would exceed the gross yield stress in many cases over a sufficiently large volume to render the assumption of an elastic response of the HEM inappropriate. Attempts have been made to decrease the effective elastic stiffness of the HEM [BERVEILLER & ZAOUI 1978] and to actually make it elasto-plastic [HILL 1965, HUTCHINSON 1970, TURNER & TOMÉ 1994]. In the extreme, the elastic response is *neglected* with respect to the plastic one, and the latter is described as *visco*-plastic [MOLINARI &AL.1987, LEBENSOHN & TOMÉ 1993b]. Then, considerable variations in strain rate are allowed from grain to grain, and the number of significantly active slip systems may be less than five. This model will be dealt with in detail in Chap. 11.

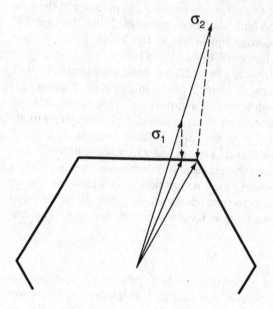

Fig. 19. Schematic illustration of the movement of a local stress vector (bold) along the yield surface to a vertex, under the action of the back stresses (dashed) generated by slip on an insufficient number of systems.

Grain size effects on flow stress, hardening, and yielding mechanism

In the Lüders band deformation mode, the (lower) yield strength depends on the grain size d according to the 'Hall–Petch' relation

$$\sigma_f = \sigma_0 + k_y\, d^{-1/2} \tag{80}$$

where σ_0 and k_y are constants (see LESLIE [1981]). This relation comes directly from the square-root decay of the forward stresses at the head of a (shear) crack (which can also be described by 'geometrically necessary dislocations'). The mode of stress relief in the bulk plasticity mode is different (Fig. 19), and does *not* give rise to a dependence of yield on grain size. It is indeed observed experimentally that, when the yield stress is plotted according to a Hall–Petch law (Eq. 80), the constant k_y comes out much smaller when no yield drop is observed. Such a mild dependence can be described in more appropriate ways. For example, it can be due to a mild dependence of *strain hardening* on grain size [MECKING 1981b, NARUTANI & TAKAMURA 1991]; some further discussion of this effect will be given in Sec. 2.2.

The most important effect of grain size is the likelihood of a qualitative change in mechanism when the grain size gets too small. For example, when it is less than the typical spacing of discrete slip planes ($\approx 1\ \mu m$), which is averaged over in the standard slip models, easy sources of dislocations within grains may not be available.

A special case of plastic deformation, when the grain size is small, is connected with grain boundary sliding at high temperatures ('superplasticity'). It requires accommodation by bulk mechanisms such as slip or diffusion. In the absence of a simple-shear mechanism, one would not expect any grain rotations and thus no texture change.

The entire book from here on out assumes <u>bulk plasticity</u>, *and a grain size that is not* <u>too</u> *small to cause a change in mechanism.* (The grains also cannot be so large as to make any realistic volume element consist of just a few grains.) Any grain size effect (whether on yield or strain hardening) has to be carried inside the stress/strain law for the representative grain.

Very heterogeneous materials

Polycrystals are heterogeneous materials, in the sense that the properties (if measured in a single macroscopic frame) change from place to place because of the change in orientation. A more stringent case of heterogeneity is when the material itself is heterogeneous, such as when it is a two-phase mixture (e.g., Ni and Al), in which the strength parameter of the SCYS can change drastically from grain to grain [KOCKS 1994]. An intermediate case occurs when the material is homogeneous, but the SCYS is very anisotropic: then, its extremities essentially serve as sources of internal stress which vary widely in 'direction' (in stress space) [KOCKS & WESTLAKE 1967].

In the milder cases (such as in cubic metals) the heterogeneity may not be of much concern, and one can treat the material as 'quasi-homogeneous', in the same way that individual dislocations are ignored when treating 'slip', and individual slip planes are ignored when considering the deformation modes in a polycrystal. In the more severe cases, however, heterogeneity is of the essence. These will be dealt with in detail in Chaps. 11 and 12.

A qualitative model for a very heterogeneous material, in an extreme situation, is that of non-deformable inclusions in a fluid. Here, the primary mechanism to be considered is the 'flow' of the matrix around the inclusions. Indeed, a composite of a copper polycrystal with embedded tungsten wires behaves just about in this manner. Fig. 20a shows a finite element calculation of the expected deformation pattern [BOLMARO &AL. 1993], Fig. 20b an experimentally observed one [POOLE &AL. 1994]. It is clear that, in such situations, a 'homogeneous effective medium' (HEM) model cannot be effective.

An important general concept in this context is whether the 'hard' phase 'percolates', i.e. is continuous throughout the sample, or consists of insular inclusions as above. In three dimensions, it is possible for both the hard and the soft phase to be continuous. The material behavior should be quite sensitive to whether one or the other or both phases percolate. All treatments so far have assumed the hard phase to be insular. A case of special interest is that of a very anisotropic SCYS, where an extension of the notion of 'percolation' has been proposed: a percolation of the elastic dipoles that signify the internal stresses generated at the extremities of the SCYS [KOCKS & WESTLAKE 1967].

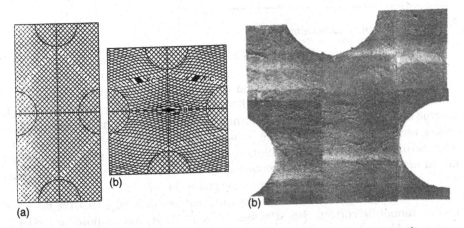

Fig. 20. Altered flow patterns in a metal-matrix composite. (a) Finite-element simulation [BOLMARO &AL. 1993]; (b) one observation [POOLE &AL. 1994].

2.2 Grain interaction: mechanics

The central problem of polycrystal averaging is the interaction of grains across their interfaces: interactions that require both equilibrium and compatibility conditions to be met.

The flat-grain limit

There is one case of grain interaction for which an *exact* solution exists: a plane interface. If the interface is perpendicular to the 3-direction, and the two grains are labeled A and B (Fig. 21), the local equilibrium and compatibility requirements (regardless of the far boundary conditions) are [KOCKS 1964a]

$$
\begin{aligned}
\sigma_{33}^A &= \sigma_{33}^B & D_{11}^A &= D_{11}^B \\
\sigma_{31}^A &= \sigma_{31}^B & D_{22}^A &= D_{22}^B \\
\sigma_{23}^A &= \sigma_{23}^B & D_{12}^A &= D_{12}^B
\end{aligned}
\tag{81}
$$

The components of stress σ and of strain-rate D on which they act are complementary: this is what makes an exact solution obtainable. The former three make up the surface traction *on* the interface; the latter three the distortion *in* the plane. Note that there is *no* condition on the continuity of D_{31}, for example; the shape of the two grains shown in Fig. 21 could have arisen from independent deformations in them, without constraints at the boundary. This amounts to a *relaxation of constraints*, as compared to the *fully constrained* Taylor limit.

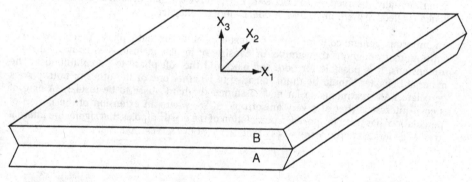

Fig. 21. A flat bicrystal, schematic.

Now it is true that no interface in a real material is an infinite plane. However, there are certain forms of deformation in which all grains get flatter and flatter with strain. One may assume that a larger and larger fraction of each grain will be subject to plane-interface boundary conditions [KOCKS & CANOVA 1981, MECKING 1981a,b]. It is also true that, in the limit of infinite deformation, real grains may have broken up into smaller units that no longer behave as one. Nevertheless, the temptation of the exactness of the limiting flat-grain solution, according to Eqs. 81, is to set as a standard that all *models* for more-or-less homogeneous deformation must tend to the exact plane-interface solution in the limit when the grains get very flat.

The same argument also demands the corollary: for equiaxed grains, the flat-grain solution cannot be correct. Nevertheless, HERSHEY [1954] has proposed a model in which mixed boundary conditions of the style of Eq. 81 act on every grain. We will find (in Chap. 9) that Taylor-type models are more appropriate in the limit of equiaxed grains (provided the grain-to-grain heterogeneity is not too great). Since the grains do change shape during deformation, a transition has to be found to the flat-grain limit whenever the grains get flatter. This is the essence of the 'relaxed constraints (RC' model of polycrystal plasticity [TOMÉ &AL. 1984].

An interesting result in this context was found from an elasto-plastic self-consistent calculation [TIEM &AL. 1986]: the reaction stresses gradually decrease to zero for those components that the flat-grain limit demands; for others, they change at most by a factor of two or so. Figure 22 shows a new calculation in a visco-plastic framework [CANOVA, priv. com.] for the reaction to shear differences in compression and tension (one component each). They are typical for other cases of flat and non-flat grains. This topic will be taken up again in connection with self-consistent and finite-element modeling (Chaps. 11 and 12).

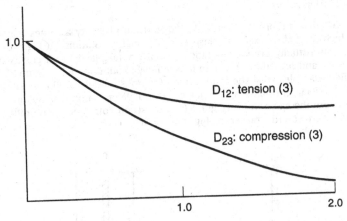

Fig. 22. Reaction stresses in an ellipsoidal inclusion in an elastic medium for D_{23} in 3-compression (solid) and for D_{12} in 3-tension. The aspect ratio is assumed 1.0 at strain zero and then changes with strain (horizontal axis) [Canova, priv. com.1994].

Upper and lower bounds

A powerful tool of mechanics is the derivation of upper and lower bounds to a solution. They have the advantage of being exact, whereas *models* provide approximations. When the upper and lower bound to a particular property are close together, as they often are in elastic problems, their average may be sufficient and a model may not be necessary (Chap. 7).

The upper bound for the stresses (or actually the work) necessary to achieve a certain macroscopic strain or strain rate is obtained by assuring compatibility throughout the body and ignoring local equilibrium conditions. A trivial way to obtain compatibility is to assume the strain uniform. The Taylor model for plasticity makes this assumption and is thus tantamount to assigning a higher priority to compatibility than equilibrium consideration. This can be justified in a qualitative way for large plastic strains [KOCKS 1970]: any disequilibrium generated that corresponds to stresses of the order of the overall flow stress could be fixed by the superposition of an elastic strain field – and the resulting changes in elastic strains will always be small compared to the plastic ones, thus not disturbing compatibility unduly.

A special property of an upper-bound solution is that it requires the yield surface to be closed in every element of the material. When the flow stress in some grains is, at least in some directions, very much higher than the average, or even 'infinite' (i.e., the grains are non-deformable, at least in some modes), uniform strain cannot be demanded and upper bounds in general are not likely to be reasonable.

A lower bound to the work necessary to achieve a given macroscopic average strain or strain rate is obtained when equilibrium is assured throughout the body and local compatibility conditions are ignored. A trivial way to assure equilibrium everywhere is to make all stresses uniform. In a strict sense, this can never be reasonable for large-strain plasticity: at any instant, when the orientation distribution is random, only one grain would deform on only one slip system. Only when a lower-bound model is coupled with strong strain hardening or 'strain-rate hardening' (i.e., a high rate dependence of the flow stress) can it potentially provide large macroscopic strains [HOSFORD 1993, CHASTEL & DAWSON 1994]: while only one grain deforms at every instant, all grains contribute in turn.

Models of grain interaction

We have seen that the primary model for large-strain plasticity assumes that the upper bound (which as such is exact) is close to a reasonable solution. In light of this, it is surprising that virtually every new paper on polycrystal plasticity discusses the 'Sachs model' as well, and sometimes refers to it as a 'lower-bound model'. Figure 23 illustrates on simplified examples what the implication of various models is. For tension of a thin, long wire, Fig. 23a represents a true lower bound: in a 'bamboo structure', all stress components are the same in every grain (and all strain components are different). The Sachs model corresponds, rather, to Fig. 23b: all stress components are zero except the

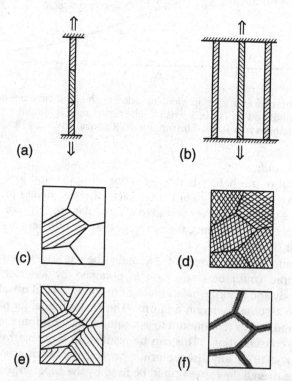

Fig. 23. Schematic description of various polycrystal plasticity models: (a) a true lower bound for a linear serial polycrystal; (b) the Sachs model (independent parallel grains); (c) a true lower bound for a 3-D polycrystal (only one grain deforms at any instant); (d) a true upper bound (also the Taylor model); (e) the Kochendörfer model (single slip plus bending); (f) the Ashby model (polyslip plus 'geometrically necessary dislocations').

normal stress in the wire axis, but the latter varies from grain to grain and is averaged for the polycrystal yield stress (assumed to be at the same longitudinal strain offset). Thus, equilibrium is not guaranteed (except through the grips) and, since single slip will generally ensue, compatibility is severely violated. This qualitative failing is true also for various modifications of the Sachs model [LEFFERS 1979].

The three-dimensional situation is idealized in Figs. 23c through 23f (using just a few grains in 2-D). The first represents the true lower-bound model used by CHASTEL &DAWSON [1994] and discussed above; the second (Fig. 23d) illustrates the uniform-strain upper bound used by Taylor. Finally, two 'mixed' models have been proposed: KOCHENDÖRFER [1941] added a constant stress to the Sachs average to account for 'bending' stresses near the grain boundaries (Fig. 23e); and ASHBY [1970] added, in a similar way, a contribution from 'geometrically necessary dislocations' to account for incompatibilities. Such corrections should not be used in the framework of a Taylor model (Fig. 23f) [MECKING 1980, 1981b]: it is already an upper bound.

Some of these various models may have merits for a description of the elasto-plastic transition region, and thus perhaps for cyclic deformation; for large-strain plasticity, the Taylor model has proven itself very resilient indeed.

The Taylor model (including its various modifications) is expected to fail, when the plastic heterogeneity is too great. We have encountered two such cases above: the altered flow patterns (Sec. 2.1); and the lower bound with strong strain hardening or rate sensitivity. Both would typically involve finite element simulations.

A model that deals with an insufficient number of slip systems is the 'projection method' proposed by PARKS & AHZI [1990]. The externally imposed strain is projected (in strain space) into the subspace in which the yield surface is closed. The correct average strain is obtained by scaling all strains – provided that the texture is broad enough so that the yield surface of the *aggregate* is closed. For a detailed assessment of this method, see PRANTIL &AL. [1995b].

An algebraic model that has been used with great success in elastic and visco-elastic problems (Chap. 7) is the 'self-consistent' solution of an interaction problem between an inclusion in a matrix consisting of a homogeneous effective medium (HEM). An application to problems in visco-*plasticity* requires that the interaction equation be linearized. With an 'effective plasticity modulus' $\mathbb{P}^{\text{eff}}$, it can be written as

$$\tilde{D} = -A : \mathbb{P}^{\text{eff}} : \tilde{\sigma} \tag{82}$$

where $\tilde{D}$ and $\tilde{\sigma}$ are the local deviations (of the 'inclusion' from the 'matrix') in strain rate D and Cauchy stress σ, respectively. The interaction parameter A is a function of the 'Eshelby tensor', which still depends on the *anisotropy* of $\mathbb{P}^{\text{eff}}$ as well as on the inclusion shape. Phenomenologically, Eq. 82 incorporates the upper and lower bounds as limiting cases for $A \rightarrow 0$ and $A \rightarrow \infty$, respectively. However, when 'self-consistency' is fully implemented, A is not an adjustable parameter.

Equation 82 supplements the material constitutive law, which may be written (see Eq. 76) as

$$\hat{D} = \mathbb{P}(\sigma) : \sigma \tag{83}$$

Note that the sign in this equation is positive, whereas in Eq. 82 it was negative: an increased local strain rate, for example, will tend to give a *back*-stress. Equation 83 holds at the crystal level, where it is known, and presumably at the aggregate level, where it is to be determined.

The plasticity modulus $\mathbb{P}$ usually is a grossly non-linear function of σ. While the real behavior of the grains can be inserted in Eq. 83, in Eq. 82 it is necessary to represent $\mathbb{P}(\sigma)$ by some constant $\mathbb{P}^{eff}$. The most obvious way to do this is *for small deviations only*: say, from a value S_1 of the deviatoric stress. It was shown in Sec. 1.7 that this can be done either by setting $\mathbb{P}^{eff}=\mathbb{P}^{tg}(S_1)$, the 'tangent modulus', or by simply using $\mathbb{P}^{eff}=\mathbb{P}(S_1)$, the 'secant modulus'. For the tensorial part of the deviations, these were illustrated in Fig. 17. In addition, the scalar strength parameter (at a given reference strain rate $\dot{\varepsilon}$) may be different in inclusion and matrix. Incidentally, for the *scalar* relations, we have used an approximation in the neighborhood of a reference point that amounts to a tangent modulus on a *double logarithmic* plot.

The scalar effect is orthogonal to the tensorial ones. A tangent modulus has conventionally been used for both parts in 'self-consistent' calculations, the secant modulus in finite element calculations. The solution is truly self-consistent only if indeed the predicted variations in stress and strain-rate are small enough to fall within the range of applicability of the Taylor expansion, Eq. 75. Chapter 11 will deal in detail with this model.

2.3 Modeling bulk plasticity

A viable model of polycrystal plasticity must be based on physical insight – and must be tractable. Plasticity is too complex a phenomenon that all its aspects could be captured in a single model. Furthermore, this is not necessary, since qualitatively different length scales are involved [KOCKS 1991]. The task then is to judge which aspects to emphasize and which to downplay in a particular effort.

In the context of "Texture and Anisotropy", it is the geometric relations that are to be addressed. They are but a small part of the whole problem – but they are a part that can be done with rigor, once the basic elements are presumed given (such as the available deformation modes and the appropriate grain interaction model). It is therefore sensible to treat this part first, before addressing aspects that require more speculation. A case in point is polycrystal strain hardening. One factor in it is texture evolution, and this factor can be treated with confidence. The remaining part requires subtle dislocation modeling. If one does not know this 'physical' part of the strain-hardening law, one can derive it from observations by deconvoluting the effects of texture change.

A part of the geometry of the problem is, in principle, the *topology* of the grain arrangement. This aspect is implicitly ignored by treating the polycrystal as a quasi-homogeneous medium, as it is done, for example, in the Taylor model. In some cases of very heterogeneous materials, this may not be appropriate; FEM models for specific cases are then the only recourse (so far).

In the process of establishing a model that makes physical sense *and* fits a wide variety of experiments, one often needs some combination of insight and empiricism: this alternation between model and verification will pervade all topics throughout the rest of this book. It is, however, important to remember that those aspects that were deliberately not included in the model cannot be used to disprove it. For example, the mere existence of intragranular variations of structure, strain, and stress [BOAS & HARGREAVES 1948, BOAS & OGILVIE 1954] does not invalidate the Taylor model (or any other model that treats only grain-average quantities); the question is rather, whether such variations do or do not have a significant influence on the range of macroscopic phenomena to be treated by the model.

Note that polycrystal deformation models must assume the operative deformation modes. Sometimes, a discrepancy between predictions and experiments can be resolved by assuming a different set, i.e. a different single-crystal yield surface. The addition of a twinning mode for copper-zinc alloys to the slip modes of copper is an example. Another example is the observed variation of textures from one pure fcc metal to another – even though the purported deformation modes are identical. This is just not allowed by the geometry of the problem. Mentioning the difference in stacking-fault energies (SFE) does not solve the dilemma; it is only through a specific influence of the SFE on deformation modes (or on grain interaction) that different textures can be achieved. Such a case will be treated in Chap. 9.

In cases where little is known from metallographic information, the observed textures can serve as an indicator of the operative deformation modes: this is the 'inverse method' to be used later in detail for materials of hexagonal lattice structure. An extreme case of diagnostic value is the occurrence of microcracks at grain boundaries in a case where only three independent systems are expected in first order (as in NiAl at room temperature) – and the disappearance of such microcracking at higher temperatures when other deformation mechanisms may come in [MARGEVICIUS & COTTON 1995].

Microstructure

The above macroscopic indications of operative deformation modes in polycrystals are often more reliable than either single-crystal experiments or microscopic observations. The latter fall into two classes: surface observations, and transmission electron microscopy (TEM). Neither is a 'bulk' experiment. In addition, TEM observations yield information on *dislocations* that were stored during previous deformation, not on *slip*, which depends on how far they moved. By way of a gross summary of both the surface and the TEM observations, it may be stated that more typically they indicate the presence of many slip systems and many dislocation types over the volume of one grain, rather than of just one system (unless the propagation mode of polycrystal plasticity, Sec. 2.1, is operative; then, large-scale pile-ups are observed). For example, a careful investigation of brass polycrystals by etching [FLEISCHER 1987] showed that, in the interior, the average number of slip *planes* is about 2.5. Since this is likely to mean 'between four and five slip systems', it is more a confirmation than a denial of the Taylor hypothesis (particularly if the activity on any one system is small).

Another kind of microscopic information is important in the context of the grain interaction model to choose; for example, 'flat' grains would suggest the use of a relaxed Taylor model; 'small' grains would suggest caution about the mechanisms; 'large' grains the need for very large specimens – or else finite element calculations for the specific sample.

Microstructures are interesting for reasons that are, in a sense, 'parallel' to texture development. As an example, consider strain-hardening mechanisms: dislocation accumulation in the grains, and orientation changes of the grains, do not interact. They also occur on different scales and in different strain regimes. Nevertheless, dislocation structures have sometimes been used in an attempt to ameliorate a disagreement between a macroscopic-geometrical prediction (such as using the Taylor model) with experiment [JUUL JENSEN & HANSEN 1990, ROLLETT &AL. 1992]. It would seem appropriate that such techniques be used only when all possibilities of altering the geometric model have been exhausted. It is these geometric and kinematic aspects only that are addressed in this book.

A different question is what (if any) microstructural features might be *predicted* from the polycrystal plasticity models. In general, this is not an aspect to be expected of these macroscopic models. However, some insight has been gained by a finite element simulation of a few grains (with many elements per grain, in 3-D) [BEAUDOIN &AL.1996]. Figure 24 shows that different regions of a grain may deform differently and, as a consequence, form new interior boundaries – either a single boundary (Fig. 24a) or a double-boundary or 'band'(Fig. 24b).

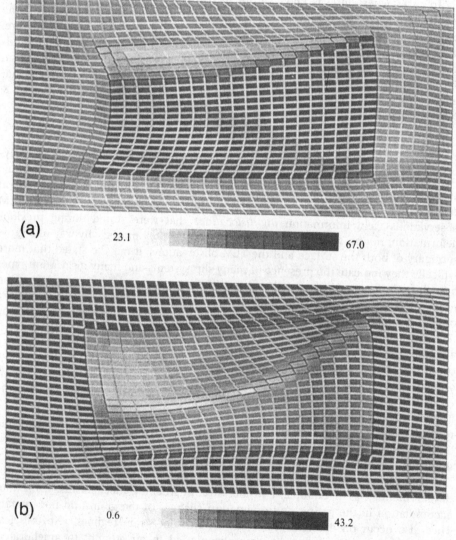

Fig. 24. The break-up of grains due to specific neighbor interactions, as modeled by a finite element simulation [BEAUDOIN &AL. 1996]: (a) one boundary between misoriented domains; (b) a misoriented band within one grain. The shade of each element is an indicator of its orientation; the shade of the element boundary an indicator of the mis-orientation.

Grain-scale variations of stress and strain

The overwhelming question in assessing a polycrystal plasticity model is: what is the distribution of stresses and strains? We have seen that a *uniform* distribution of either one or the other gives a *bound*, not a solution, and that a uniform distribution of strains (or strain rates) – the Taylor model – is a better estimate for quasi-homogeneous materials than a uniform distribution of stresses. Thus, the *dispersion* of strain rates, for example, gives a good measure of the departure from the Taylor model. Conversely, a *low* dispersion of stresses resulting from a Taylor calculation would tend to indicate that the violation of equilibrium conditions is not very serious.

Figure 25 illustrates the results of a finite-element simulation of rolling in an fcc polycrystal, using a weak form of both the equilibrium and the constitutive relations (SARMA & DAWSON [1996], Chap. 12). One can see that the deformation is homogeneous on a gross scale, but nevertheless varies significantly from grain to grain. Figure 26 shows a schematic representation of the kind of statistics expected in various cases (in a way similar to that used by ARMINJON & IMBAULT [1996]). The upper row displays the relative variation strain rate, the lower that in stress – both in the compressive component of a plane-strain compression test. Fig. 26a represents an upper-bound solution; the stresses were derived from a LApp simulation and fulfill the criterion that the internal stresses be no more than about half the flow stress (see Chap. 9). Fig. 26c, on the other hand, shows the results of the lower-bound simulation described above. The strain rate was that during one step after a strain of 0.2, in a strongly hardening and mildly rate-sensitive material. Figure 26b is a schematic of a case where strain rates and stresses vary in about the same way from place to place.

Grain-to-grain variations must, in reality, be due to the specific interaction of a particular grain with its particular surroundings. This introduces a stochastic effect into any model that assumes quasi-homogeneity. It tends to broaden textures from those calculated simply (a welcome, realistic effect). It appears from experience, however, that predictions of the overall stress/strain response are *not* sensitive to these variations.

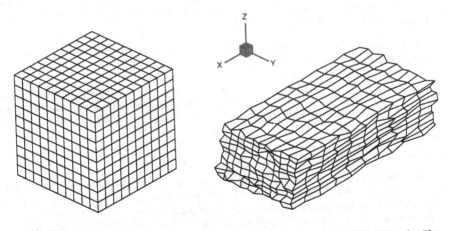

Fig. 25. Finite-element polycrystal (one element per grain); rolled fcc metal. The different element shapes indicate the dispersion of strains; the shading the orientation [SARMA & DAWSON 1996].

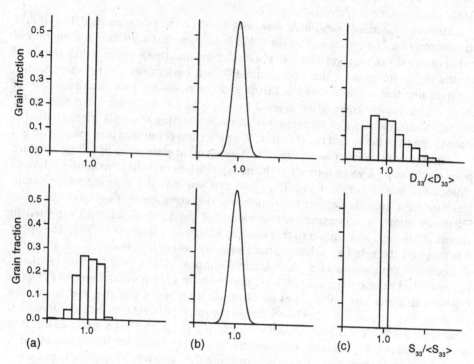

Fig. 26. Distributions of strain rate (upper row) and stress (lower row) for (a) an upper-bound model (actually from a LApp simulation) and (c) a lower-bound model [SARMA & DAWSON 1996]; (b) represents a schematic of equal variability. The open bars symbolize a delta-function; the simulations were for plane-strain compression of an fcc metal (3: compression direction).

There may, however, be a certain *correlation* between neighboring orientations, for example due to twinning or just because the texture is strong [AERNOUDT & STÜWE 1970, MECKING 1981b]. This could lead to specific grain interaction models which have, so far, only been formulated in an abstract way [ADAMS & FIELD 1991] or to 'pattern formation' [AERNOUDT &AL. 1993, HUGHES & KUMAR 1996].

Also note that a certain stochasticity has often been introduced in an artificial way: by first assuming rate independence, and then resolving the resulting indeterminacy of slip system selection in highly symmetric cases by random choices of various kinds [DEZILLIE &AL.1988]. A rate dependent model leads to strict determinacy. While rate dependency is realistic, strict determinacy is not: it must be eliminated by some physically stochastic mechanism. In Chap. 12, some simulation will be reviewed that address the 'ensemble' nature of polycrystals.

Energy and work principles

It may be helpful to review very briefly a number of different methods that employ extremum principles on energy or work. First, there is the thermodynamic principle that a system will tend to minimize its free energy under the given constraints (and will then be in thermodynamic – perhaps metastable – equilibrium). Since many of the actual constraints are local, this principle is most usefully applied at a local scale. But it holds true, primarily, for entire systems under certain macroscopic boundary conditions (such as constant temperature, constant stress). Different forms of the potential to be miminized are derived in classical thermodynamics.

A common extension of this principle does not refer to equilibrium situations, but to kinetic behavior (sometimes expressed as 'irreversible thermodynamics') [KOCKS &AL. 1975]. If two 'flow' mechanisms are considered competitively, the one that requires the lower driving force will actually occur. Conversely, at a given driving force, the mechanism that gives the highest flow rate dominates. (Note that, in the former case, the power dissipation is minimized, in the latter it is maximized.) Note that there is *no* principle that restricts possible avenues of *evolution* (as proposed by RENOUARD & WINTENBERGER [1981]): this would amount to teleology.

An entirely different set of criteria surrounds the 'principle of virtual work' in mechanics. It states that the average of the internal work equals the external work – but not because of any 'energy conservation' law. It is a geometric (or algebraic) consequence of averaging under two conditions only: equilibrium and compatibility are fulfilled everywhere (including at the boundaries) – but the stress and strain fields used need not be linked by any constitutive law. This principle is used in a variational form; for example, if a continuous displacement field is postulated, and a stress field is derived from the constitutive behavior of the material, minimizing the residual of the virtual work inequality amounts to minimizing departures from equilibrium.

Recently, a number of investigators have proposed a derivation of actual strain distributions in polycrystals on the basis of minimizing the 'internal energy' or the 'accommodation work' [VAN HOUTTE 1995, WAGNER & LÜCKE, UNPUBLISHED]. It is not clear which (if any) of the above principles are used for justification.

Finally, we have used the 'maximum work principle' proposed by BISHOP & HILL [1951a]. It is not based on either of the above principles, but expresses the geometric properties of normality and convexity on the yield surface. Taylor's 'minimum work principle' to derive active slip system combinations is justified only by being equivalent to Bishop and Hill's maximum work principle.

Simulation versus modeling

Explicitly and implicitly, we have mentioned *simulations* – by computer, that is – of complex polycrystal problems throughout this Section 2. It deserves emphasis that *modeling* is distinct from simulation. Modeling enters the process before simulation in setting up a solvable problem. Its essence is judgment: which experimental observations and theoretical principles to include, and which to leave out; which aspects to simplify and which to treat with special care; etc.

Modeling enters again *after* the simulation: in abstracting the results into a set of rules that can be applied to general cases, rather than the specific one that was simulated. In a successful effort, many iterations between modeling, simulation, and experimentation take place. The aim is not so much to solve one distinct problem exactly as to describe a large set of phenomena adequately and without contradictions.

For the remainder of this book, we will address problems of texture development and its influence on properties on the basis of the geometry of slip (and twinning), in a polycrystalline aggregate that is being treated as having homogenous average properties.

3. Kinematics in Polycrystalline Bodies

3.1 Frames of reference, 'material frames'

Throughout this book, we have encountered two main frames of reference for the description of material behavior: the 'crystal' frame and the 'sample' frame; the relation between them is what we call 'orientation' (for one crystal) or 'texture' (for many). By implication, the frames have always been taken to be cartesian. In *crystals* that do not have a cubic unit cell, this requires some conventions, which have long been worked out in crystallography (see Chap. 2). The *sample* frame requires some additional thought. It is not innate to the material, but is typically fixed by the user for convenience in a particular application – and often different for different applications of the same piece of material. In addition, any set of material-connected axes, chosen initially to be cartesian, do not in general remain orthogonal when the material is deformed. This requires some judicious choices for a material reference frame.

For a metallographer, the obvious frame of reference for orientations is the normal to the material plane lying on the stage and the material line in this plane that is butted against a holder. The 'stage' may be on a microscope, when the orientation of the grain shapes is being determined; or it may be on a texture goniometer, when the orientations of crystals are to be measured. It is useful to define sample axes that directly connect to these normal presumptions; for distinction against other sample frames to be discussed, let us call this the *metallographic sample frame* for now. A convenient coordinate system would then consist of (1) the normal to the 'plane of view', (2) the marked edge (which is necessarily perpendicular to direction 1), and (3) a direction that is perpendicular to both.

To illustrate the change in sample shape of an anisotropic material, Fig. 27 shows a cube (light lines) that has been deformed, say, by compression between two friction-less parallel platens perpendicular to the Z-direction; the final shape is indicated by heavy lines. Not only has the cube exhibited its anisotropy by expanding more in the X-direction than in the Y-direction; the angles have not remained orthogonal. In the particular example shown, this is true for two of the angles, labeled β and γ. In this example, the orientation and spacing of the planes originally perpendicular to **Y** have not changed. If the sample was indeed deformed in *free* compression, as described above, this behavior would be due to a particular symmetry inherent in the structure of the sample. Alternatively, the final shape could have been achieved by 'channel-die compression', in which it is the external boundary conditions that restrict the shape change in the Y-direction, and only the remaining free shape changes are dictated by the material anisotropy.

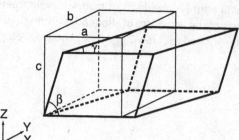

Fig. 27. Shape change of a cube (thin lines) under compression. The shape after the test (solid lines) is not completely general: there have been no displacements in the Y-direction (either because of a special sample anisotropy, or because of lateral constraints).

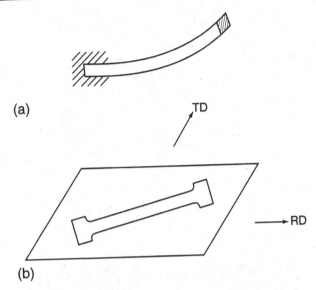

Fig. 28. (a) A 'sample' (shaded) at the end of a plastically bent bar. (b) A 'sample' cut out of a rolled sheet.

The point of the illustration is that an initial cartesian frame does not remain so and, in fact, *no* cartesian frame may be easily identifiable in the final sample shape. Then, a marked plane and a marked line in it still identify an orthogonal 'metallographic sample frame'.

Two examples shall illustrate the relation of this sample coordinate system to some background frame. Figure 28a shows a bent strip. Its end has rotated with respect to the vise, but it can be cut off and mounted on a stage. This 'sample' may not have all right angles; but a plane parallel to the bend plane, and a direction parallel to the (local) long dimension of the strip can be identified and used to define the metallographic sample frame; its relation to the vise can only be known from a macroscopic record.

For a second example, Fig. 28b depicts a rolled sheet and the outline of a tensile sample that may be cut out of it. Its behavior is best characterized in the frame of the rolling and transverse directions, in which the texture is most easily specified. When a sample is known to have orthotropic symmetry, the frame of choice is clearly the set of axes of orthotropy. Note, however, that extensive straining in a tensile test would destroy the initial orthotropic symmetry; only the one mirror plane in the rolling plane would be maintained. Thus, a sample symmetry frame is not suitable for describing processes in which this symmetry changes. Only tests conducted on a material that was sufficiently isotropic to start with and is then subjected to a 'radial' test (in which the ratios of straining components remain constant) will retain the symmetry of the test throughout.

It is therefore important to ascertain what the *actual* symmetry of a given sample may be. Fig. 29a displays the result of a *measurement* undertaken on a sample from a rolled sheet, presumably with the rolling direction aligned with the sample holder. But it is evident by inspection of the pole figure that the sample was not actually well aligned: whereas orthotropic symmetry can be gleaned from the figure, it is along axes that are somewhat tilted and rotated from the axes of symmetry. This is a frequent

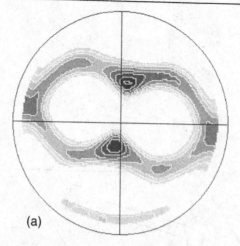

(a)

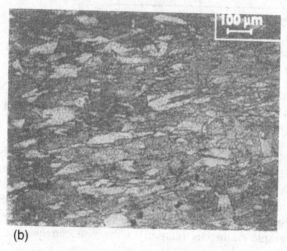

(b)

Fig. 29. (a) Presumed texture measurement from sample cut out of a rolled sheet of copper. Axes of symmetry are evident, and they are best chosen as properties reference axes. (b) Presumed metallographic section from a deformed piece: the grains appear elongated. Sufficient sections can define a aspect ratio and an 'orientation' of an ellipsoid that characterizes the grain shape.

occurrence in practice. It is appropriate to 'correct' this pole figure before symmetrizing it for further analysis.

The 'axes of maximum sample symmetry' (much like in crystals, though statistical here) are useful because any property described in these axes can be specified by a minimum number of parameters. In this sense, the term 'sample axes' was used in Chap. 1, for example. One may call this the 'sample symmetry frame' or the 'properties reference frame'. (It has been called the 'principal axes of anisotropy' in the literature [HILL 1950], even though they may not be 'principal axes' in the sense used in the description of tensors or ellipsoids sense, and furthermore, the term 'anisotropy' is meaningful only with respect to a stated property.)

Finally, yet another set of axes is sometimes of interest: the *grain axes*. Figure 29b displays a micrograph that suggest some preference for the alignment of grain shapes.

The periphery of the grain consists of elements of material planes which, in general, rotate independently. To the extent, however, that the grain shape can be approximated as an ellipsoid, there will always be a set of (orthogonal) principal axes. The grain shape frame has the advantage that it can sometimes be directly related to a finite strain measure (Sec. 3.2). Also, the grain shape can be relatively easily measured (although perhaps not with great accuracy). Its main drawback as a sample frame is that it may be influenced by recrystallization, grain fragmentation, and the like.

Any of the above three frames could be used as a 'sample frame' or 'material reference frame'. In general, they will rotate with respect to each other during plastic deformation, and they will also rotate with respect to some macroscopic 'background' or 'laboratory' frame. Only the latter can, in the most general case, be chosen to be *fixed*. A judicious choice is, however, often appropriate even among material frames.

In general, then, one must know how these material-bound sample axes relate to each other and to a macroscopic coordinate system. Material *testing* is a special case: it is usually done in an easily identifiable, constant frame in which the boundary conditions are prescribed. If these are not parallel to symmetry axes, anisotropy becomes obvious.

All of the above frames (except the 'test frame') are bound to the material and seem therefore to be particularly suitable for the specification of material properties. This is easy to see for polycrystals with observable grain boundaries, or any material with observable anisotropies; it may seem artificial for isotropic continua. It is for the latter that so-called 'material frames' have been commonly used that are not bound to the material in the above, observable sense; instead, they trace back all properties to the original, *undeformed* (orthogonal) configuration. The preference for this *Lagrange frame* is due to, and most appropriate for, the treatment of primarily *elastic* deformations, where the *undeformed* configuration (not the same as the *unloaded* one when plasticity is present) has *state* properties. But note that we have had no need to even introduce the Lagrange frame in the current context. Ensuring 'frame indifference' needs attention when frames of reference are used that are not bound to the material. When material-bound frames are used, the material constitutive behavior is always independent of relative motions (rotations or translations) between material and observer.

One short remark on nomenclature: the 'orientation' of one frame with respect to another (whether of a crystal, a sample, or a grain shape, with respect to each other or to the laboratory frame) is often treated as a 'rotation' between frames, and labeled R (as any orthonormal transformation matrix). We shall use the term *orientation* g when no motion is implied (though a *reference* frame is required), the term *rotation* R for an integrated *motion* from a given *initial* configuration. Note also that the *rate* of rotation,

$$\dot{\mathbf{R}} = \mathbf{W} \cdot \mathbf{R} \tag{84}$$

is usually specified by the 'spin' W (Chap. 1 Eq. 12″).

3.2 Updating the grain shape

The grain shape can often be approximated as an ellipsoid, and in this form its evolution with strain can be described by standard methods of solid mechanics. It is in this context that the mechanics concept of 'finite strain measures' (see, e.g., ASARO & NEEDLEMAN [1985], HAVNER [1992]) becomes useful in polycrystal plasticity (for which, as we have seen, the constitutive relations are always described in terms of *differential* strains).

The deformation gradient matrix F may, for our purposes, be defined as the integral, from F = I (the identity matrix), over the actual path, of the velocity gradient L introduced in Chap. 1 (Eq. 11), so that

$$\dot{F} = L \cdot F$$

(85)

F thus contains information on both strains (integrated strain rates D) and rotations (integrated spins W). It may be used to describe the shape change of the parallel-epiped in Fig. 27: an initial *line* of a length and direction described by p^o becomes

$$p = F \cdot p^o$$

(86)

and an initial plane described by its normal q^o (in both length and direction) becomes

$$q = q^o \cdot F^{-1}$$

(86')

(See, for example, CHIN &AL.[1966].)

An initial sphere changes into an ellipsoid (the 'stretch ellipsoid') which may be described by a matrix V defined by

$$V^2 \equiv F \cdot F^T$$

(87)

If grains were initially equiaxed (i.e., quasi-spherical), their shape will evolve like the stretch ellipsoid. If the shape was initially not quasi-spherical but quasi-ellips-oidal, the *evolution* of this shape would still follow the same evolution as the stretch ellipsoid [KOCKS &AL.1994b].

An ellipsoid can be described by a matrix G (Fig. 30). Its properties can be specified in terms of its principal axes λ_i and its orientation (with respect to the laboratory frame, for example), which we will call g^E. The λ_i are the eigen*values* of G, the rows of g^E are its eigen*vectors*. Both may change with straining.

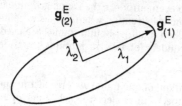

Fig. 30. Description of the grain shape by an ellipsoid with principal axes λ_i and orientation g^E. It is updated just like a 'stretch ellipsoid': according to the Euler spins.

In the simplest case, the principal axes of the strain *rate* D are aligned with those of G; then g^E does not change, and we need only the equation

$$\dot{\lambda}_i = D_{ii} \lambda_i \qquad \text{(no sum)}$$

(88)

which describes a change in the aspect ratios. When the principal axes of D are *not* aligned with those of G, Eq. 88 still holds for the eigenvalues when D is written in grain axes .

In addition, the grain shape will change its orientation. In solid mechanics, the rate of reorientation of the ('left') stretch ellipsoid defined by Eq. 87 is called the Euler spin W^E; then [HILL 1978]:

$$\dot{g}^E = W^E \cdot g^E$$

(89)

Note that the grain shape is a state parameter, whereas the stretch ellipsoid is a history variable; the *rates of change* are the same. The components of the Euler spin *in grain axes* are given by [KOCKS &AL.1994b]

$$W_{ij}^E = W_{ij} - \frac{\lambda_i^2 + \lambda_j^2}{\lambda_i^2 - \lambda_j^2} D_{ij} \equiv W_{ij} - \alpha_{ij}' \, D_{ij} \quad \text{(no sum, i} \neq \text{j)} \tag{90}$$

The interesting part of Eq. 90 is the second term: a spin that is proportional to a shear strain rate. It may be made transparent by noting that W^E and D are both time rates; thus the proportionality constant may be called a 'shear spin': a change of orientation (say, $d\alpha = dg \cdot g^T = W^E \, dt$) *with strain* $d\varepsilon = D \, dt$. The shear spin has been labeled α' in Eq. 90.

The rotation of the axes of the grain shape ellipsoid becomes particularly relevant when some shears are not the same inside the grain and in its surroundings, such as in the 'relaxed constraints' model. Since the principal axes of the grain and of the 'hole' in which it sat must remain aligned, the Euler spins must be equal whether derived for the outside or the inside [KOCKS &AL.1994b]. One may take this to be a part of the assumption (or necessity) of displacement continuity; but its failure would not, in contrast to strain incompatibility, affect the *stresses* necessary for the deformation.

If we now interpret Eq. 90 as describing the Euler spin of the 'hole', i.e. of the *surroundings* of the grain, and we write it again for the grain itself (using a superscript g for *its* spin W^g and strain rate D^g), the postulate of coincident Euler spins W^E yields, by subtraction of the right-hand sides:

$$W_{ij}^g - W_{ij} = \alpha_{(ij)}' \left(D_{ij}^g - D_{ij} \right) \qquad \text{(grain axes)} \tag{91}$$

Equation 91 means that the assumed coincidence of the Euler spins of grain and hole forces a spin on the grain, with respect to whatever coordinates L is written in, which is different from W whenever there are differences in shear strain. Note that, for a given shear difference, the spin difference *decreases* as the grain eccentricity increases, according to Eq. 90.

In the 'fully constrained (FC)' model [TAYLOR 1938a, 1956], all strain rates are assumed equal; as a *consequence* of the postulate of locally equal Euler spins (Eq. 22), all spins are equal, too, regardless of the grain shape.

In Fig. 31, the shear spin is plotted for the specific case of $\alpha'_{(12)}$, and assuming $\lambda_1 \lambda_2 = 1$, as in plane strain at constant volume. The abscissa is then equal to the true strain if the straining direction is never changed, and the function is actually a hyperbolic cotangent. The pole at the origin reflects the fact that the eigenvectors themselves are here undefined.

Figure 31 also shows that the aspect ratio need not be very large for the grain to be effectively 'flat'.

Finally, we must discuss the case $\lambda_1 = \lambda_2$ (and cyclic equivalents). While α' is here singular in terms of the instantaneous grain shape, this degeneracy will immediately resolve itself, since the principal strain *rates* (times the time step) will become the principal *strains* (the principal axes of G) – and in these axes, the shear and thus the shear spins are zero.

The eigenvalues may nevertheless remain very similar, so that the shear spins vary rapidly. Under such circumstances, the integration according to Rodrigues (equivalent to Eq. 89 and discussed in more detail in the next Section) becomes problematical: one would require smaller and smaller time steps the closer one is to the degeneracy. In practice, one will need to use linear updating, according to Eq. 85.

3.3 Updating the texture

To update the texture, one needs to calculate the reorientation of every grain. The constitutive relations call for a *differential* formulation of such orientation changes: the 'lattice spins'. Simulations occur, on the other hand, in finite steps. This section

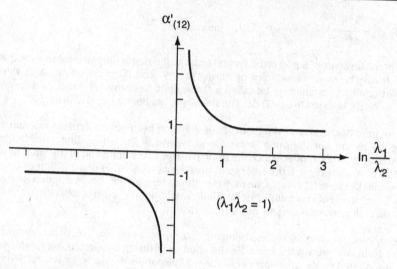

Fig. 31. Dependence of the shear spin on the aspect ratio of a flat grain [KOCKS &al. 1994b].

will address both the definition of appropriate spins needed, and the algorithms used for finite increments.

We have encountered a particular spin when discussing slip kinematics (Eq. 12):

$$W^c = \Sigma_s \dot{\gamma}^s q^s \tag{92}$$

This 'plastic spin' describes the differential rotation of the sample axes of a grain in terms of its crystallographic axes. It is different from grain to grain. If one wanted a single sample frame for all grains, the spin of their cubic axes or the like would be the negative of Eq. 92. Finally, if one allowed, in general, the grain shape itself to rotate with respect to some external frame (and called this spin W^g, Sec. 3.2), the rate of reorientation of the crystal axes (the 'lattice spin' W^*) becomes

$$W^* = W^g - W^c \tag{93}$$

When the grain (i.e. the grain shape ellipsoid) rotates the same way as the whole body, which is often the case (deviations were discussed in Section 3.2) then one gets the more common equation (see, e.g. AERNOUDT [1978])

$$W^* = W - W^c \tag{94}$$

where W is the 'rigid-body spin' (a concept that must be carefully defined in a body that isn't rigid).

The current crystal orientation, g (often called R^*, in the 'rotation' view of orientations) follows, in principle, from integration of

$$\dot{g} = W^* \cdot g \tag{95}$$

Integration is not trivial, because small finite increments may accumulate into a break of orthonormality. So-called 'exact updating' can be obtained by the Rodrigues scheme:

$$g = g_0 \exp(W^* \Delta t) \tag{96}$$

where $\mathbf{g}_0$ is the initial value and Δt the time increment since then. This works only, however, when the spin can be regarded as constant during the time increment Δt. (We encountered a case in which this not appropriate in Sec. 3.2.) Where this is in doubt, a safer procedure is to update the deformation gradient linearly (from Eq. 85):

$$\mathbf{F} = (\mathbf{I} + \mathbf{L}\,\Delta t)\,\mathbf{F}_0 \tag{97}$$

(where $\mathbf{I}$ is the identity matrix and $\mathbf{F}_0$ the initial value) and then diagonalize $\mathbf{F} \cdot \mathbf{F}^T$ to assure orthogonality of $\mathbf{g}$. While this procedure would accumulate non-orthogonalities when done for too many steps, it has the advantage of giving smooth variations. The scheme according to Eq. 96 is faster and is preferable when appropriate.

A useful alternative set of reference coordinates exists when the sample frame is a metallographic or test frame (see last Section). Here, only two elements are specified: a material plane plus a material line in it; they follow Eqs. 86 and 86' above; then, the third axis is defined *ex post facto* to be perpendicular. If the material plane normal is called 3 and the line in it 1, this means [KOCKS & CHANDRA 1982] that the velocity gradient must have the form

$$\mathbf{L} = \begin{pmatrix} L_{11} & L_{12} & L_{13} \\ 0 & L_{22} & L_{23} \\ 0 & 0 & L_{33} \end{pmatrix} = \begin{pmatrix} D_{11} & 2D_{12} & 2D_{13} \\ 0 & D_{22} & 2D_{23} \\ 0 & 0 & D_{33} \end{pmatrix} \qquad \textit{(sample axes)} \tag{98}$$

and its spin the form

$$\mathbf{W} = \frac{1}{2}\begin{pmatrix} 0 & L_{12} & L_{13} \\ -L_{12} & 0 & L_{23} \\ -L_{13} & -L_{23} & 0 \end{pmatrix} = \begin{pmatrix} 0 & D_{12} & D_{13} \\ -D_{12} & 0 & D_{23} \\ -D_{13} & -D_{23} & 0 \end{pmatrix} \qquad \textit{(sample axes)} \tag{99}$$

For the solution of complex boundary value problems (such as Fig. 28a), it is usually convenient to use a 'general', 'external', 'laboratory', or 'background' frame: one that remains fixed throughout the process and for all elements of the body. In this frame, an arbitrary velocity gradient $\mathbf{L}$ may, of course, have *any* component non-zero. This can be achieved, without any loss of generality, by adding to Eq. 99 a rigid-body spin

$$\mathbf{W}^s = \begin{pmatrix} 0 & -L_{21} & -L_{31} \\ L_{21} & 0 & -L_{32} \\ L_{31} & L_{32} & 0 \end{pmatrix} \tag{100}$$

This is the spin $\mathbf{W}^s$ of the 'sample axes' with respect to the chosen 'background' frame; together, Eq. 99 (written in sample axes) and Eq. 100 make up the standard 'rigid-body spin' of a volume element with respect to the laboratory frame. As emphasized in Section 3.1, the term 'rigid-body spin' may give the mistaken impression that the rest is pure straining in some 'material frame'. The trouble is that in a non-rigid body, $\mathbf{W}$ can be defined only when a frame has been defined that remains orthonormal.

It is perhaps worth noting that, under some boundary conditions, not all components of the velocity gradient tensor, $\mathbf{L}$, can be externally prescribed, but they follow from the material response. For example, a sample under free compression between two platens (with or without friction) may suffer simple shear in the face that is being compressed, as a consequence of sample anisotropy. The extent to which this is to be described by strain and rotation (even for a differential increment) is arbitrary in macroscopic coordinates and only determined by the sample axes of anisotropy.

3.4 Rotation fields, orientation jumping, orientation streaming

The discussion so far has assumed that orientation changes are infinitesimal – or, in practice, of the same order as the strain step in the simulation. There are some cases where this is not so: deformation twinning and phase transformations are two examples. In these cases, the jump in orientation is finite. The new orientation follows, in principle, from the application of the transformation operator to the old orientation. In the limit of small strain steps, it is the volume transferred that may be small, rather than the change in orientation. This leads to practical problems with maintaining a tractable number of grains, and schemes have been devised to enable simulation of such cases. They will be discussed in connection with twinning in hcp metals in Chap. 11.

When the orientation changes are *smooth*, one may describe them as vector fields in orientation space [KLEIN &AL.1988, MORAWIEC 1990, BACZMANSKI &AL. 1993]. This is a useful concept in principle, but it does not obviate the need for simulation, since a different spin occurs for each different incremental straining path. The concept also relies on the orientation of a grain being its *only* state parameter. For example, the current flow stress (the size parameter of the yield surface) is likely to evolve different-ly in each grain; in polycrystal models in which the interaction between grains is locally specific (even if the grains deformed uniformly), this may cause the orientation change to depend on the current flow stress. Then, different grains with the same orientation but different histories might not rotate in the same way.

For these reasons, rotation fields have served not so much for quantitative pre-dictions as for a qualitative, visual description of the dynamics of texture changes: the sweeping-out of large areas of orientation space and, conversely, the formation of stable end orientations are especially obvious. Of particular interest, however, have been *deviations* from the presumed existence of stable textures. The reason is that the underlying equation used for rotation fields,

$$\dot{g} = W^c(g,D) \cdot g \tag{101}$$

is actually correct only when the 'rigid-body spin' W is zero. (Compare Eqs. 94 or 93 with Eq. 95 [BOLMARO & KOCKS 1992].) Equation 101 is a differential equation for the orientation **g**, given the strain rate **D**, which leads to the formation of 'hills' and 'valleys' in orientation space. When a homogeneous term W is added to the crystal spin (or 'plastic spin') W^c, to make the lattice spin $W^* = \dot{g} \cdot g_T$ (Eq. 95), this corresponds to a continual 'streaming' of orientations through orientation space. Thus, if one has a rotation field for *pure* shear (and it shows stable orientations or fibers) and now adds a 'rigid-body spin', all orientations have a tendency to spin around – and only reside longer, but not *permanently*, in the 'stable' regions.

The case of torsion has been analyzed in detail by TÓTH &AL. [1989]. Figure 32 shows a particularly interesting case: the neighborhood of the 'C' orientation (center) on the 'B' fiber (vertical in center). (For the nomenclature, see Chap. 5, Fig. 6.) There is a continuous streaming of orientations (from right to left in this figure). In the absence of rate sensitivity (not shown), the rate of streaming would be zero along the entire B fiber, so that it becomes stable. With a rate sensitivity of $m=0.05$, however, the rate of streaming never vanishes in the lower half of the figure. Taking other sections of orientation space into account, it turns out that the B fiber becomes completely unstable, and the C orientation is a singular stable point (which will therefore never be reached).

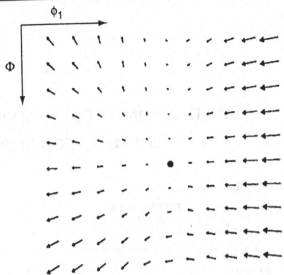

Fig. 32. A rotation field for simple shear in a rate-sensitive fcc polycrystal [TÓTH &AL. 1989].

Finally, all descriptions so far have presumed that the straining path **D** was prescribed constant. In practical cases, the principal axes of **D** may change during the straining process. If this change occurs smoothly, one may define a 'spin' of the principal axes of strain-rate, $\mathbf{W}^d$, which has a similar influence as the 'rigid-body spin' in fixed axes. BOLMARO & KOCKS [1992] have combined the two effects by defining a 'relative spin'

$$\mathbf{W}^{rel} \equiv \mathbf{W} - \mathbf{W}^d \tag{102}$$

which, when inserted into Eq. 94 in place of merely **W**, has led to effective descriptions of the qualitative change in behavior upon path changes.

A modern use of rotations fields employs finite-element modeling in orientation space, using Rodrigues space as a basis (see Chap. 2 Sec. 2.6) [KUMAR & DAWSON 1995]. All rotation effects can be taken account of, and interpolations are easy to perform.

3.5 Summary

Grains may change their orientation because of straining and also because of 'rigid-body rotations'. The concept of a rigid-body rotation is a difficult one in bodies that aren't rigid. When a material-bound coordinate system does not remain orthogonal during deformation, it is appropriate to define a set of axes that does (and then account for *its* rotations with respect to a laboratory frame). For constitutive descriptions, this set of axes should still be bound to the material in its evolving state (and not to a historic configuration); it is best chosen to reflect the maximum symmetry in the sample: then, the information needed to describe anisotropic properties is minimal.

When the straining of a polycrystal does not occur in a strictly uniform manner, and the grains are not equi-axed, individual grains experience an additional spin with respect to their surroundings; this needs to be accounted for in the derivation of texture changes. The rotation of the grain *shape* can then also be important; it can be derived on the basis of the evolution of the 'stretch' associated with the deformation.

When rotations are important, texture changes may not converge to a fixed end state, but rather achieve a steady state that includes continual streaming.

Chapter 9

SIMULATION OF DEFORMATION TEXTURE
DEVELOPMENT FOR CUBIC METALS

SIMULATION OF DEFORMATION TEXTURE
DEVELOPMENT FOR CUBIC METALS

U. F. Kocks

In Chapter 8, we outlined the general problem of polycrystal plasticity and the various models that have been proposed to deal with it. In this and the following Chapters, we select one particular model for the simulation of material properties and comparison of the results with experiments. In each case, an attempt is made initially to choose a basic model that appears appropriate for a certain set of properties in a certain class of materials; the interaction with experimental experience then invariably leads to modifications and additions to the basic model. This evolution leads to a 'matured' set of model assumptions which, it is hoped, will have predictive value for the chosen set of problems. The beginning of this era of a *quantitative* understanding of deformation texture development, and of some specific shortcomings of the prime models, is marked by two superb reviews: MECKING [1985] and VAN HOUTTE [1986].

The present Chapter uses a 'modified Taylor model' which was presented in Chap. 8 (Sections 2.2 and 2.3) as a reasonable model for bulk plasticity in cubic metals. Moreover, the Taylor model has been used as a starting point for virtually all treatments of polycrystal plasticity – even in cases where its assumptions are doubtful. It has often been found to be 'surprisingly' successful, so long as only macroscopic properties such as stress/strain behavior and texture development are addressed (and not microscopic ones such as the occurrence of local heterogeneities).

The two modifications to the original Taylor model that are deemed essential in any case, inasmuch as their physical basis is well grounded and their effects significant, are (a) the incorporation of material rate sensitivity; and (b) the incorporation of grain shape effects. The former mechanism avoids the problem of non-uniqueness that occurs in classical Taylor codes. The latter modifies the Taylor prescription of equi-partitioning of the strain increment to a case of mixed boundary conditions on the grain. In the limit of infinitely flat grains, an exact solution is known and should be approached in the simulation. Since the grain shape changes during deformation, a transition algorithm must be found that retains the basic assumption of the Taylor model: 'compatibility first'. The ensuing strain dependence of texture development is an important element of the study of deformation textures; this will be addressed in Section 2.

All implementations of modern polycrystal models are done by computer *simulation*, and this requires a whole set of additional assumptions, techniques, and algorithms. While this book is not the right place to address these in detail, it seems advisable to at least enumerate some problems that have arisen and been solved. We demonstrate the principles of what any code must do, in Section 1, on the example of the LApp (Los Alamos polycrystal plasticity) code developed by KOCKS, CANOVA, TOMÉ, ROLLETT & WRIGHT [1988a].

In the further sections, we summarize various specific results of LApp simulations that relate to *texture development* and compare them with experiment. In Chap. 10, the predictions of macroscopic, anisotropic mechanical *properties* will be addressed and compared with experiment.

1. Simulation Procedures

1.1 The two principal algorithms

The original, rate-independent Taylor model was expressed by BISHOP & HILL [1951a] in terms of a 'maximum work principle', which ensures that the yield condition is met everywhere (Chap. 8 Fig. 6 and Eq. 20). From among all the vertex stress states σ^* of a single-crystal yield surface (SCYS), one can select the correct one, σ, for a prescribed straining direction, $\hat{D}$, by using the condition

$$(\sigma - \sigma^*) : \hat{D} \geqslant 0$$

(1)

This remains a powerful, fast method for the determination of the stress necessary in each grain to produce a given straining direction – and thus, in general, to derive the polycrystal yield surface from the single-crystal yield surface. This is so even when the material is mildly rate sensitive ($m < 0.05$, say), since the rounding of the vertices will change the stress only by the order of m: less than the experimental accuracy of most measurements. In cases where only some components of $\hat{D}$ are prescribed (and in the other directions the stress components are prescribed), Eq. 1 holds for the prescribed components of $\hat{D}$.

What is needed to apply this method is information on the SCYS. A complete set of relations for fcc metals was published by KOCKS &AL. [1983]. It includes topological information such as which facets (slip systems) contribute to which vertex, and which vertices are connected by (hyper-)edges that activate a subset of systems. This becomes important for the case of mixed boundary conditions (Sec. 1.2). The same SCYS can be used for bcc materials in restricted glide (Chap. 5 Sec.2.1).

The aspects of the problem that cannot be achieved with the Bishop–Hill ('BH') method are related to the deformation rates on the various system and, as a consequence, to the evolution of hardness and texture. For this, we use the kinetic relation for the strain rate tensor D in a crystal c (Chap. 8 Eq. 56):

$$D^c = \dot{\varepsilon} \sum_s \left| \frac{m^s : \sigma^c}{\tau^s} \right|^{n^s} m^s \, \mathrm{sgn}(m^s : \sigma^c)$$

(2)

It is explicit in D, implicit in σ. Since, in all problems in this Chapter (and also in FEM codes, Chap. 12), it is D that is prescribed, σ (or rather S, its deviatoric part; see Chap. 8 Sec 1.7) must be obtained by an iterative scheme such as the Newton–Raphson ('NR') method. It requires an initial guess for the stress: the vertex value derived from the BH-method is well suited. Thus, a combination of the two principal algorithms, Eqs. 1 and 2, serves the problem perfectly, for rate insensitive (though not rate independent) materials – if the SCYS is known.

An alternative initial guess – necessary when the SCYS is not known – is to assume the stress deviator to be parallel to the straining direction and lying on the rate-independent facet (i.e. the CRSS for the single slip system that will usually be activated in this way). This amounts to the 'linear-programming' algorithm, which actually has some similarity to the physical sequence of events (Chap. 8 Fig. 19). One must consider, however, whether the *converged* result (after whatever criterion has been met) is unique: the 'result' may depend on the direction from which it was approached. Reproducibility may be assured by first proceeding to a vertex and only *then* applying a Newton–Raphson technique [VAN HOUTTE 1996b]; this is equivalent to the combined BH–NR method.

The SCYS can always be calculated so long as all deformation modes, and their respective CRSSs, are known. This may not be worth doing when one is exploring a variety of cases, such as different slip or twinning system types of varying CRSS ratios. For the alternative method outlined above, it suffices to specify the crystallographic elements of each deformation mode, along with the CRSS in the forward and backward direction. If the backward direction is 'impossible', such as in twinning, one merely specifies a high CRSS ratio, chosen such that other systems will actually close the SCYS.

1.2 Boundary conditions

In a general mechanics problem, some tractions and some displacements may be prescribed on various parts of the surface of a body. Since we deal here with 'samples' or test specimens, which are assumed quasi-homogeneous in material and without significant gradients in any field variable, it is permissible to describe the boundary conditions in terms of constant stress or strain components [HILL 1967, RICE 1970, KOCKS &AL.1975]. Of each component, in a sample coordinate system, either the stress or the strain (increment or rate) is prescribed, the other is derived from the material response.

In the Taylor model, as well as in each step of an FEM code, one usually assumes all strains prescribed, all stresses to be derived. However, for practical test specimens, the actual boundary conditions may correspond more nearly to prescribed *stresses*. For example, in a tension test on a long, thin wire, all stress components except one are prescribed zero; or in a 'free-ends' torsion test, the axial stress is assumed zero. When the simulation algorithm is based on fully prescribed straining components, one may loop through adjustments of these strains until the prescribed stress components are matched, to a prescribed accuracy (which needs to be about 2% of the stress norm); this procedure is adopted in LApp [KOCKS &AL.1988a]. It is also possible to incorporate mixed boundary conditions directly into the NR-algorithm [MATHUR &AL. 1990].

These boundary conditions apply to the specimen as a whole, at the *macroscopic* level. At the *microscopic* level, the effective boundary conditions on each grain are certainly in fact mixed [HONNEFF & MECKING 1981, KOCKS & CANOVA 1981]. In the classical Taylor model, all local boundary conditions are prescribed in terms of strain (-rate) components. However, in the 'relaxed constraints' (RC) model (Chap. 8 Sec. 2.2), certain strain-rate components are assumed open and the corresponding stress components prescribed. One usually assumes these stress components to be zero throughout the body and considers the corresponding *section* through the SCYS [KOCKS & CHANDRA 1982]. When the macroscopic average value of this stress component is not prescribed, it is still assumed uniform throughout the body, but its magnitude is varied until the average *straining* component matches the macroscopically prescribed one. This works when the prescribed straining component is zero ('orthogonal relaxation'), but is difficult – and usually not done – when it is the primary component, as in torsion ('parallel relaxation' [VAN HOUTTE 1996a]).

Finally, in test samples, it is usually prescribed that one particular plane, or one line, or both, remain parallel to itself. This fixes the skew-symmetric components of the velocity gradient, so that only the strains need be prescribed. (See 'sample axes' in Chap. 8 Sec. 3.3)

1.3 Initial and final conditions

The presumed or known state of the material must be specified as an input. The same state parameters can be listed as an output file and then be re-used as input for a further test: this is the essence of a state parameter description. Part of the simulation is to update the state (see Chap. 8 Secs. 3.2 and 3.3).

Among the necessary input parameters are the orientations of a set of grains. The conventional, and most compact way of doing this is by a set of three Euler angles for each grain. It is appropriate to also assign a weight to each grain. In principle, this could reflect the volume fraction occupied by a particular grain. In practice, one can use this 'weight' to specify, in a textured sample, the volume fraction of a whole assembly of grains (at different locations) that fall within a certain range of the specified orientation. This algorithm has been used to input a measured initial texture (see Chap. 2). When such a procedure is used, it is important that the initial grains file (to be weighted by the measured texture) cover the whole of orientation space as uniformly as possible. This demands a representation in an 'equal-area' (or actually, equal-volume, or 'isochoric'), and not in a cartesian space (Chap. 2); and it requires care in including various special orientations, which are in fact likely to occur in practice. A random file may not be the best choice for this procedure. Moreover, its fluctuations in density are of the order of a factor of 2 for 1000 grains.

Other parameters of importance are those that specify the grain shape (when it is not equiaxed). We have summarized in Chap. 8 that the deformation gradient matrix (F), derived from a presumed initial equiaxed state, serves this purpose well. It can be updated easily (Chap. 8 Sec. 3.2).

Finally, the current flow stress of each grain is a state parameter. When it is initially the same in each grain, as is to be expected in a well annealed (single-phase) material, it will in general evolve differently with deformation, and the end state can be retained for a further test. Initialization of the flow stress distribution in an unknown material is perhaps possible in an approximate form through neutron line-broadening analysis; this technique has not been used widely, and is unlikely to ever be practicable on a routine basis. Within the framework of a Taylor theory, flow stress differences from grain to grain do not affect texture development.

Apart from parameters of state, there are, of course, general properties, such as the rate sensitivity, the strain-hardening law, etc., as well as computational parameters, that need to be specified for every simulation.

1.4 Choice of computational parameters

A number of computational parameters always need to be specified, such as convergence criteria and the maximum number of iterations. We highlight two of a more physical nature, which are subject to re-adjustment according to feedback from the results.

The first is the rate sensitivity. It is in fact often very low, corresponding to a very high stress exponent in Eq. 2. This makes for very 'stiff' behavior and potential numerical problems. However, it was pointed out in Chap. 8 that, so long as only the distribution between different deformation modes is of interest, the constitutive behavior need not be described with the same accuracy as would be needed for variations in strain rate by many orders of magnitude. Experience has shown that an exponent of $n=20$ gives the same results as all larger ones – but an exponent of 10 does not.

The second judgment of considerable influence on the time required for solutions is the step size. When the Bishop–Hill method is used, at least for an initial guess of the stress state, a convenient criterion is that there should be no more than a single vertex change for any grain during one strain step. It is therefore opportune to output information on the active vertex for at least a few crystals. Experience with fcc metals has shown that a strain increment of 2.5% is not too large.

1.5 Updating the state

During deformation, the state of the material changes continually with straining. The simulations, on the other hand, proceed in steps of finite strain increment. All the changes we discuss are so gradual that it is sufficient to execute a step under the assumption of constant state, and then update the state before the next step.

The CRSS changes on each slip system, due to strain hardening. Since the shear increment during one step is known for all systems, the increase in flow stress can be calculated, if an appropriate hardening law is known (Chap. 8 Sec. 1.5). In the cubic materials discussed in the present Chapter, the grains slip on many systems simultaneously, and under such conditions, it is common to merely update the *size* parameter of the entire SCYS, in proportion to the total sum of shears on all systems ('Taylor hardening rule' [TAYLOR 1938b]). This is a good first approximation. If further details are to be modeled, they are taken as departures from this average behavior, and are labeled 'latent hardening'.

A similar homogenization takes place in connection with the grain *shape*. Firstly, this shape is approximated by an ellipsoid (i.e., a sphere when it is 'equiaxed'). Secondly, the ellipsoid is assumed to be the same in orientation and aspect ratios for all grains (even when the grains are allowed to deform somewhat differently). The updating of this shape follows the routine known for the 'Euler stretch ellipsoid' (Chap. 8 Sec. 3.2); the deformation gradient matrix (**F**) is updated according to the average velocity gradient for the polycrystal.

Finally, the crystallographic orientations of the grains are updated according to the finite (though small) rotation increments during a strain step. This requires care, since departures from the case of infinitesimal increments can accumulate and lead to loss of orthogonality of the axes (Chap. 8 Sec. 3.3). For quasi-homogeneous test samples – which are the principal object of our simulations – it is easiest to use a sample reference system in which one plane, and one line in this plane, remain parallel to each other; this is equivalent to specifying certain components of the velocity gradient to be zero, and thus to specifying the 'rigid-body spin' inherent in the particular mode of deformation (Chap. 8 Sec. 3.3). (The degenerate case of *uni*axial deformation must be dealt with separately.) This scheme is embodied in LApp.

1.6 Other output

The applicability of the Taylor model is predicated upon a sufficiently small variation in stress from grain to grain. The standard deviation of all stress components that emerges from the simulation provides a measure of this criterion. Its maximum is typically between 0.2 and 0.4 of the stress norm, in cases where the results seem reasonable and self-consistent. Standard deviation in excess of 0.5 should be taken as a warning sign that the method may not be appropriate to this case. A more detailed question concerns the shape of the distribution. In most cases, it is more or less Gaussian; however, sometimes one expects a bimodal result, or at least a very wide spread. Calculation of the fourth moment (and normalization by the square of the second moment) provides a check on this parameter. (An example will be given in Chap. 10.)

A quantity that has drawn much interest is the actual repartition of the strain in each grain over the various deformation modes: the number of significantly active slip systems or slip planes, or the twinning activity. These parameters are straightforward to calculate as part of the simulation.

2. Effects of Relaxing Constraints due to Grain Shape. Strain Dependence

Most of the practical interest is in textures at *large* strains (certainly greater than one), and most of the data are in this regime. From the modeling perspective, it is, on the other hand, interesting to see the *development with strain*, which will be emphasized in this Section. In addition to a general sharpening of the texture, changes in the *location* of peaks and fibers may occur, owing to the continual change in grain shape and, thus, in the boundary conditions on each grain.

As was derived in Chap. 8, there is no question about the limit of infinitely flat grains: an exact solution exists, which requires continuity of three strain components across the flat face. For each grain, then, three strain components are prescribed, and this requires, in general, the activation of three independent deformation modes [RENOUARD & WINTENBERGER 1976]. For equiaxed grains, on the other hand, five independent modes are generally needed – not only in the Taylor model, but whenever compatibility with the neighbors is considered the prime condition to be fulfilled [VON MISES 1928, KOCKS & CANOVA 1981].

The change, with strain, from five to three independent modes has profound effects on texture changes. It has to occur in all models. The fact that it is due to changes in the local boundary conditions on a grain, as given by its interaction with its surroundings, was modeled using elastic interactions by HONNEFF & MECKING [1981] and was later accounted for in visco-plastic self-consistent models (Chap. 11) and in finite-element simulations (Chap. 12). In the present Chapter and Chap. 10, a Taylor model (with rate sensitivity) is used for equiaxed grains, and is labeled 'full constraints (FC)'. The transition to fully 'relaxed constraints (RC)' according to the flat-grain limit is the topic of the next Section.

Some investigators have employed 'relaxed constraints models' that are of a heuristic nature: it has long been known that certain shear relaxations would produce certain texture components [LEFFERS 1979, HONNEFF & MECKING 1978, HIRSCH 1990]. MECKING [1981a] and VAN HOUTTE [1981, 1986, 1996a] have given extensive overviews.

It must be noted that *grain-shape* arguments always demand the relaxation of *both* the shears on the flat plane; the 'lath model' has no such basis. In the present treatment, only grain-shape arguments are used to relax constraints.

2.1 The transition from full constraints to relaxed constraints

In the flat-grain limit, only two components of strain are 'relaxed', i.e. not pre-scribed by the surroundings: the two shears *on* (not the one *in*) the plane of the large grain boundary. In the current model, it is only these two components for which a transition from FC to RC is modeled – and it is done in the same way for both (even when the grain shape changes anisotropically).

A flat grain is thought of as consisting of an interior that obeys the flat-boundary conditions, and a rim that doesn't (Fig. 1a). The proportion of the two regions changes as the aspect ratios change. While the evolving aspect ratios of the grain shape are derived algorithmically by assuming them to be *ellipsoids* (spheres initially; see Chap. 8 Sec. 3.5), the model for partitioning the grain into a rim and an interior is based on a 'brick' shape. In this picture, a flat interior appears only after a finite amount of strain. To quantify the proportions, we use one adjustable parameter: the width of the rim (w); a width equal to the thickness (t) is a reasonable starting point that has often served well [TOMÉ &AL. 1984]; this choice leads to the evolution depicted in Fig. 1b. Observed quantitative changes as a function of strain can serve to determine this parameter.

While the boundary conditions for the interior are clear, those for the rim should change from place to place, and serve to effect a transition to the different deform-ations at the adjoining 'interiors' [VAN HOUTTE 1984]. (The notion that the rims must 'accommodate' the difference, with minimum work [VAN HOUTTE 1995, ARMINJON & IMBAULT 1996], is counter to the basic view of the flat-grain interiors as 'relaxed'.) In any case, the rims should require five independent deformation modes. (The proposal in TOMÉ &AL. [1984] to require four at the edges, five at the corners, has been aban-doned.) In LApp, the rim regions are treated as if they were subject to the *macroscopic*

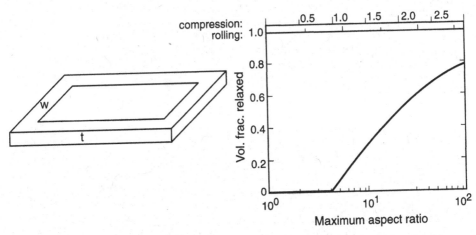

Fig. 1. (a) Schematic of a 'flat' grain (thickness t) with a 'rim' (width w) around its periphery. (b) Volume fraction of the non-rim region (which is assumed to be under 'relaxed constraints'), as a function of the largest aspect ratio. The upper scale indicates the von Mises strain corresponding to this aspect ration in compression and in rolling.

boundary conditions. The algorithm assumes the polycrystal to be made up of two grain populations: a certain volume fraction under fully relaxed conditions (according to the flat-grain limit), the remainder fully constrained.

We translate the volume fraction relaxed in every grain into a fraction of *grains* that is treated fully relaxed. The algorithm picks them anew every time from the whole set. This causes statistical fluctuations, e.g. in the Taylor factors and thus the stress/strain curve – because each grain switches back and forth between being FC and RC. The alternative of keeping the same grains relaxed and merely adding to their number produces two populations, leading to an uncharacteristic splitting of the texture into two parts.

2.2 Rolling in fcc metals

The term 'rolling' is meant as short for idealized sheet rolling: plane-strain compression, in the normal direction N, with the rolling direction R free, and no displacements in the transverse direction T. Thus, two macroscopic boundary conditions are $D_{TT}=D_{TN}=0$. (This would not have to be true for a plate in which the width is not large compared to the thickness.) Friction on the rolls will, in real practice, lead to a non-uniform distribution of D_{RN}; however, the average must be zero if the sheet is to come out without curling around one of the rolls. Furthermore, in the midplane, the macroscopic boundary conditions demand $D_{RN}=0$. The third shear, D_{RT}, is also zero macroscopically because of the conditions in the roll gap.

On the microscopic scale, flat grains develop with strain and would, in the limit, leave D_{RN} and D_{TN} entirely free. We combine macroscopic and microscopic conditions into demanding the *average* of these two shear components over all grains to be zero at all times, but letting the *local* value depend on the FC–RC transition scheme.

In Fig. 2, we use the representation introduced in Chap. 2: a polar crystal orientation distribution (COD) arranged so that the 'beta-' and 'alpha-'fibers are easily seen in the sections $\phi=45°$ to 90°; the ϕ-*projection* is the 001 pole figure (with the rolling

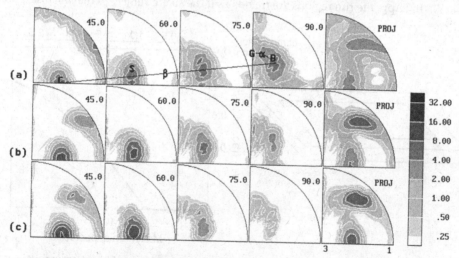

Fig. 2. Simulated rolling textures after a von Mises equivalent strain of (a) 1.0, (b) 2.0, and (c) 3.0. Displayed are some sections of the COD (ϕ), and its projection, the 001 pole figure. The β- and α-fibers are indicated in (a), along with the 'texture components' Copper, S, Brass, and Goss. Simulations based on a random initial texture, and gradually increasing relaxed constraints (1000 grains, smoothed 10°). RD on right.

direction pointing to the right). 1000 random grains were 'rolled' by LApp to von Mises equivalent strains of 1.0, 2.0, and 3.0, using the standard FC-to-RC transition scheme, which gives a volume fraction of relaxed grains of 0.07, 0.57, and 0.82, respectively, for these three strains. It is seen that some regions of orientation space empty out rapidly; in fact, this occurs during the first 50% strain [HIRSCH & LÜCKE 1988b]. At 100%, both the β- and α-fibers are clearly seen, and their occupation is about even along the fibers. Thus, the Taylor model gives rise to an initial 'brass' component along with all other classical rolling components (see Chaps. 2 and 5). The main effect of increasing strain is a decrease in the brass component in favor of the other orientations along the β-fiber. This shift may be attributed to the increasing flatness of the grains shape which, in fact, *reinforces* the constraint against the 'in-plane shear' D_{RT}, which would result from rolling of a 'brass orientation' single crystal [MECKING 1981a].

All these are qualitatively the observed trends in aluminum (see Chap. 5 Fig. 5), nickel, and copper. It is true that the observed textures are all generally less sharp than simulated ones; the data shown were smoothed by a 10° Gaussian filter [KALLEND &AL. 1991a]. Yet the transition scheme seems to give at least qualitatively the correct answers.

Literature data are usually available only at strains in excess of two, which is too late to see the transition; furthermore, initial textures are often not specified. Notable exceptions are the investigations of HIRSCH &AL. [1984], HIRSCH & LÜCKE [1988a] and DEZILLIE &AL. [1988]. The data set that was available to us in sufficient detail to make a quantitative comparison was on copper [NECKER 1997]. It is presented in Fig. 3, in a form quite similar to Fig. 2. While the gross features are similar, there are clear differ-

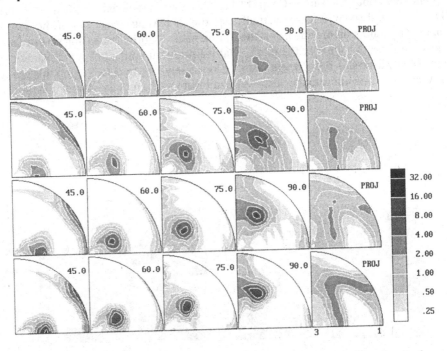

Fig. 3. Experimental textures in the same display as Fig. 2. The first row refers to the initial texture, the subsequent ones to similar strains as Fig. 2 (reductions of 58%, 82% and 90%, respectively).

ences in detail between the predictions in Fig. 2 and the copper experiments in Fig. 3. First, there is a weak cube component, which may be associated with the fact that the specimens had become quite warm during 'cold' rolling, and thus may be partially recrystallized. The remaining discrepancy is not likely to be due to the mild initial texture (top row), but is an expression of the well known fact (Chap. 5) that deformation textures depend on material, even within the same lattice structure. (Data on aluminum or nickel should show better agreement.) This material dependence has been a challenge to interpret for a long time and cannot be assessed as solved; we will discuss it further in Section 4.5.

Returning to the strain dependence, as modeled by simulations, Fig. 4 shows a straight Taylor (FC) calculation for the maximum strain (top row). It is seen that, even here, there is an initial 'brass component'. The second row displays a difference COD between the FC–RC scheme (bottom row of Fig. 2) and the FC case. Finally, the bottom row of Fig. 4 presents results of a simulation in which fully relaxed constraints were imposed from the beginning (which would be physically reasonable only if the grains were flat from the beginning). The difference COD (against the FC–RC case, bottom row of Fig. 2) shows that the end texture, at the same strain and the same grain shape, depends on the *history* of the grain shape development in a non-negligible way: *it is path dependent*.

We end this Section by noting that detailed FEM simulations [BEAUDOIN &AL. 1995a] recover the general trends embodied in the FC–RC scheme: they tend (as they must) toward the flat-grain limit; but they also give similar results for equiaxed grains, despite the fact that all shears are allowed from the beginning. The interesting result is that the deviations of the *in-plane* shear *decreases* as the grain gets flatter, while the others stay about the same (Chap. 12 Fig. 5), as was predicted by MECKING [1981a]. Thus, it would seem that it is actually not the increasing *relaxation* of the compatibility constraints on the shears *on* the rolling plane, but the increasing severity of the constraints on the shear *in* the plane, that is important (and causes, for example, the move away from the 'brass orientation'). Conversely, one may look upon this effect as an increase of the importance of stress equilibrium as the face gets larger (Chap. 11).

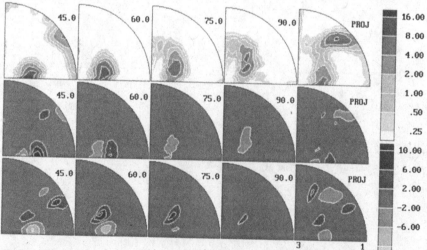

Fig. 4. Simulations as in Fig. 2c, but under full constraints (top row). Difference CODs, against this FC base, under standard RC, as in Fig. 2c (middle row), and under fully relaxed constraints from the beginning (bottom row).

2.3 Torsion in fcc metals

The term 'torsion' is meant to apply to a (thin-walled) tube or, in some cases, including the surface layer of a solid round bar; as such it is equivalent to 'simple shear'. It was shown in Chap. 2 (Fig. 31) that the developed textures have some similarity to the 'pure shear' textures, which correspond to plane strain viewed from the T-direction (and turned by 45°). There are also differences, notably the absence, in general, of orthotropic symmetry. In addition, as was shown in Chap. 8, the 'rigid-body rotation' inherent in simple shear leads to a continual motion of orientations – with a steady-state, rather than a static end texture. Chapter 5 (Secs. 1.5 and 1.6) highlighted some experimental observations for various fcc metals and discussed effects of strain reversal. In the present Section, we will show simulated effects of the boundary conditions in torsion tests ('free ends' versus 'fixed ends') and then concentrate on simulated and observed strain dependencies.

It is difficult to set the experimental boundary conditions on torsion tests precisely (as will be discussed in some detail in Chap. 10). By simulation, one can easily assess what effect they might have. In Chap. 10, we will discuss the influence on anisotropic macroscopic behavior; here we show the effect on texture development. In a typical Taylor code, one prescribes all *straining* components; this is also what we normally do unless otherwise stated. LApp allows you also to iterate the strain prescriptions until a stress criterion is met (Sec. 1.2). This needs to be done to treat 'free-ends' torsion in a thin-walled tube. Figure 5 shows a *difference pole figure* for the effect of free ends as opposed to fixed ends. The effect is mild, but not negligible; it can be approximately described by a rotation of the whole texture around the radial direction (2), in the direction of the sense of shear [LOWE &AL. 1987].

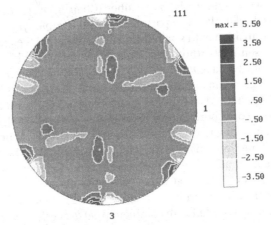

Fig. 5. Difference pole figure for torsion (to ε_{vM}=2.0) under free-end and fixed-end boundary conditions. Axial direction up, shear to left on top; equal-area mapping.

In torsion, there could be two reasons for a dependence of texture on strain: the continual 'rigid-body rotation' with respect to the 'sample' coordinates (tangential, radial, and axial) [GIL SEVILLANO &AL. 1975]; and the developing grain shape. The grains become lath-shaped, as in other plane-strain tests (though the orientation of the grain-shape ellipsoid also changes; i.e., its principal axes rotate). However, we

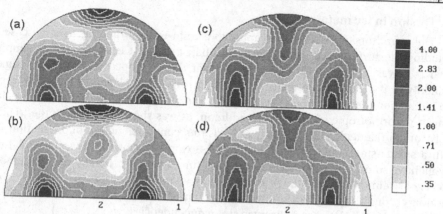

Fig. 6. Torsion textures displayed as 001 pole figures. (a) and (b): simulated to ε_{vM}=4.0 and 5.0, respectively, assuming fixed ends, random initial texture, RC (69% and 75%, respectively); 512 grains, smoothed by 10°. (c) and (d): experimental results for nickel (c) and Ni-30%Co (d), at similar strains. (The Ni and Ni-30%Co series for various strains had given very similar results [HUGHES & WENK 1988].) Axial direction up. shear to left on top. Equal-area mapping.

relax only one of the two shear components, since the other is the one that is *prescribed* (in the average)[CANOVA &AL. 1984a].[1]

Figure 6 shows torsion textures for two strains (top and bottom), both simulated (left) and measured (right). For representation, we use 001 pole figures with the radial direction in the center; they discriminate well between standard components and fibers. Of particular interest is the 'A' fiber (Chap. 5 Fig. 6): the only one among the common types that displays a departure from orthotropic symmetry.

There is a transition with strain from an 'asymmetric' (i.e., monoclinic) pole figure to one that is essentially orthotropic. This transition occurs between (von Mises equivalent) strains of 4.0 and 5.0 (left vs. right half of Fig. 6). The simulations were undertaken under relaxed constraints, and only these are responsible for the change in character of the texture. The experimentally measured textures are from a nickel alloy at approximately corresponding strains [HUGHES & WENK 1988]; the initial texture was so weak that a meaningful orientation distribution could not be derived from it. The agreement with the simulations is satisfactory; in particular, the departure from orthotropy becomes minimal at about the same strain. Thus, the relaxed constraints are apparently being brought in at about the correct rate. (CANOVA &AL. [1984a] had already shown that RC improves the agreement between theory and experiment at large strains, but they started with completely flat grains from the beginning.)

2.4 Compression in fcc metals

Simulations of compression to a strain of 1.0 (Fig. 7a) show qualitative agreement with experiments; they are about the same whether RC conditions are phased in or not. (At a strain of 1.0, 31% of the grains are relaxed.)

The situation becomes interesting at a strain of 2.0. Relaxing constraints according to Sec.2.1 makes a major difference as compared to FC (Fig. 7c versus Fig.7b): the

---

[1] In the work of HARREN &AL. [1989], a set of boundary conditions was used that do not fulfill the symmetry conditions derived by CANOVA &AL. [1984a] and implicitly amount to RC, but not only due to grain shape.

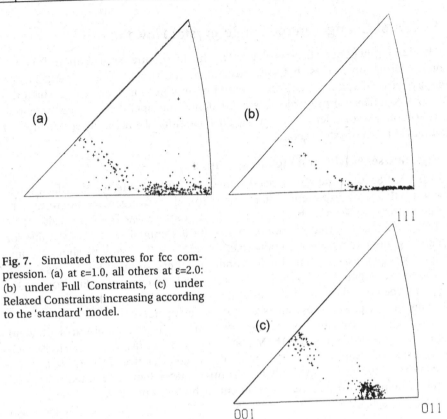

Fig. 7. Simulated textures for fcc compression. (a) at ε=1.0, all others at ε=2.0: (b) under Full Constraints, (c) under Relaxed Constraints increasing according to the 'standard' model.

{110}-component disappears – and this is *not* in agreement with any experiments (see Chap. 5 Fig. 1).[2]

There is a basic explanation as to why compression of a material that has already developed a {110} texture should be a special case. In free compression of a single crystal of this orientation, the actual deformation would be in plane strain (using four slip systems). To enforce axisymmetric straining, one has to supply a compression stress that is higher by a factor of 1.5. Furthermore, and perhaps even more important, severe lateral stresses would develop by such axisymmetric deformation. Therefore, this may well be a case where the Taylor model should not apply (see Chap. 8). Indeed, the 'self-consistent' scheme produces the right texture (Chap. 11).

In *bcc tension* which, under {110}⟨111⟩ slip, should be equivalent to fcc FC–compression, the local deformation is known to occur in plane strain, with the result that the grains 'curl' around each other [Hosford 1964]; but while this preference for plane strain is evident through the formation of ribbons in tension, it is less likely to be observable for plates in compression. There has been an *ad hoc* solution by Van Houtte [1984], with some 'relaxation of constraints' other than those caused by grain shape; it is essentially heuristic. *Compression in bcc* (which is like tension in fcc, but with RC) will be briefly discussed in connection with deformation-mode dependence.

---

[2] One could retard this effect, and thus keep reasonable agreement at a strain of 2.0, by bringing in RC conditions more slowly than in the standard LApp scheme; actually, a comparison with stress/strain behavior (Chap. 10) also works better with this retardation.

3. Effects of Changes in the Single-crystal Flow Potential

The development of textures is dictated by the quantitative contributions from the various deformation modes that are activated by the stress. They depend on the single-crystal yield surface (SCYS) or – more properly put for materials in which the flow stress is, at least in principle, rate dependent – the single-crystal flow potential. In this Section, we consider effects on this flow potential of a high rate sensitivity and of changes in the slip system geometry.

3.1 High rate sensitivity and high temperatures

In the last Section, the rate dependence was assumed quite small; in fact, it has been shown that stress exponents n in excess of 20 yield convergent texture results. In the other extreme of Newtonian viscous flow ($n=1$), CANOVA &AL.[1988], and also TÓTH &AL. [1988] have shown that texture development is minimal. The lowest value for dislocation glide (and climb) at high temperatures is $n = 3$. Figure 8a shows a compression texture for fcc metals under this condition, after a strain of 1.0. In this limit, a lower-bound calculation may also be of interest; it is shown in Fig. 8b [KOCKS &AL. 1994a]. It is seen that there is not a strong effect – perhaps because there are sufficient slip modes and the plastic stress states are not widely separated.

Such high rate sensitivities are expected only at high temperatures (if then) [BACROIX 1986, CHEN & KOCKS 1991b, MAURICE &AL. 1992, ENGLER &AL. 1993]. At such temperatures, new slip system types may also become active [LEHAZIF &AL. 1973, MARTIN 1993, MAURICE & DRIVER 1993]. Various workers have attempted to explain changes in deformation texture development at high temperatures on the basis of such mechanisms [BACROIX & JONAS 1988, KOCKS &AL. 1994]; Figs. 8c and d show two

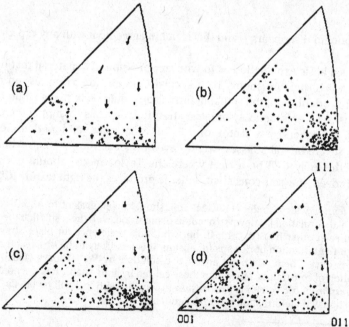

Fig. 8. Fcc compression textures for $n=3$ and $\varepsilon=1.0$ simulated under various special assumptions: (a) full constraints; (b) lower-bound model; (c) slip on {111} and {100} planes; (d) slip on {100} and {110}.

examples. But the predicted effect of different slip modes is not great, unless one assumes *disappearance* of conventional {111} slip [KOCKS &AL. 1994a].

One of the most puzzling experimental observations in the temperature range where $n=3$ is appropriate, is the rapid development of a {001} fiber texture in both compression and tension [CHEN & KOCKS 1991b]. The conclusion was that this can only be connected with dynamic recrystallization [KOCKS &AL. 1994a].

3.2 Possible effects due to latent hardening

The SCYS can also change in the course of deformation as a consequence of 'latent hardening' – meaning different rates of hardening on different slip systems [KOCKS &AL. 1991a]. This is not very likely when there are many systems operating. In any case, even in free single crystals, it has been shown that the ratio of CRSSs on different slip systems is typically between 1.0 and 1.4 [KOCKS 1964b] or at most 2.0 [JACKSON & BASINSKI 1967], beyond the easy glide regime . It might, however, be expected to depend on material, through the stacking-fault energy for example.

In order to assess the maximum effect such latent hardening assumptions could have on deformation texture development, we have simulated a rolling and a compression texture on the basis of assuming inter-*plane* latent hardening only, of a ratio of 2.0, and found that they do not exhibit sufficient differences to warrant further investigation. As an example, Fig. 9 shows the difference between the resulting rolling texture, at $\varepsilon_{vM}=3.0$ and the equivalent Fig. 2c: a small amount, of about 1 m.r.d., is shifted in one spot. We conclude that realistic levels of latent hardening have minimal effects on texture development.

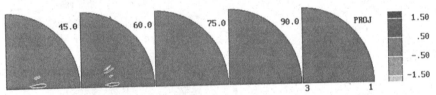

Fig. 9. Difference COD sections for rolling to $\varepsilon_{vM}=3.0$: assuming extreme latent hardening, versus the standard in Fig. 2c.

3.3 Pencil glide. Bcc rolling textures

Bcc metals often deform by slip on the {110}⟨111⟩ systems. A general observation is that such 'restricted glide' may occur in bcc *alloys* and at low temperatures, but that other slip planes become increasingly important at higher temperatures and for purer materials [RAABE & LÜCKE 1994]. In the limit, this is often described as 'pencil glide': no crystallographic preference for any plane, so long as it contains the ⟨111⟩ slip direction [TAYLOR 1956; KOCKS 1970; PIEHLER & BACKOFEN 1971; DILLAMORE & KATOH 1971; ROLLETT & KOCKS 1988; GILORMINI &AL. 1988a,b; MAURICE & DRIVER 1993; BECKER 1995]. We will discuss only the two limits of restricted glide and pencil glide. A postulate of '12 slip planes' (of the {110},{211}, and {321} types, and of equal CRSS) for each slip direction is equivalent to pencil glide (but sometimes less expensive in simulations). Slip on six slip planes for each direction (namely those of {110} and {112} character) has been claimed to be physically meaningful; however, the nature of the pencil glide yield surface would also lead to an apparent preference for {211} planes [KOCKS 1967]. In any case, the behavior could be gleaned from interpolation between the extremes.

As was derived in Chap. 5 and discussed further in Chap. 8, there should be a certain correspondence between fcc and bcc textures – if the bcc material deforms in restricted glide [TAYLOR 1956]. In rolling, the correspondence is achieved by interchanging the rolling direction (RD) and the rolling plane normal (ND) [DILLAMORE & KATOH 1971]. The exact correspondence breaks down when the grain *shape* orientation also matters; for example, in rolling, the grains always become flat in the *actual* rolling plane. It is instructive to represent bcc textures by making use of this correspondence, and then analyze deviations in terms of grain shape or pencil glide effects.

Figure 10 shows experimental textures *before* (top row) and *after* (bottom row) cold rolling, to 80% reduction, of a low-carbon steel [RAABE 1992]. The initial texture exhibits departures from randomness that turn out to be significant. All the rows in between show simulation results to the same strain, based on the actual initial texture. The second row is under full constraints (FC); it should look like the fcc textures in Fig. 2, were it not for the initial texture in the present case. Indeed, one sees the long ridge in the 90° section, where the initial texture was strong. (The comparison with Fig. 2 should be made with the first row, sharpened up from a strain of 1.0 to 2.0; or to the second row, by ignoring the effects of RC.)

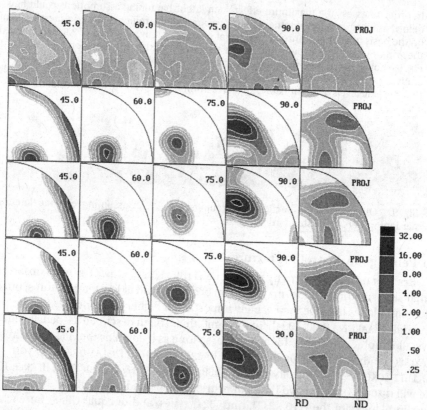

Fig. 10. COD sections and the 001 pole figure for iron (plotted such that in restricted glide they would look like fcc metals, with the RD to the right). Top and bottom rows: experiments [RAABE 1992], for the initial texture, and for that after a rolling reduction of 80%. Intermediate rows display simulation results, based on the initial texture, assuming, respectively, restricted glide under FC and RC, and pencil glide under RC. The increase in relaxed constraints was twice as fast as in the standard model. Both the experimental and the theoretical textures were smoothed by 10°. The 'initial texture' in the top row is from a representation as a file of weighted grains, derived from the measured ODF [RAABE 1992].

Next, we take account of grain shape effects. A scheme where the RC conditions were brought in twice as fast as 'standard' was found to be somewhat more successful; it gives the middle row. Finally, the fourth row adds pencil glide to the 'fast RC' conditions. At this point, the agreement with the experiments is very satisfactory. RAABE [1992] also found that both pencil glide (which he represented by 12 slip planes) and 'pancake' RC (which, however, he started from the beginning, and named 'CS') were necessary to achieve agreement. Our accelerated FC–RC transition gave slightly better results than starting RC from the beginning.

Experiments that include the strain dependence are even sparser in bcc metals than in fcc. They have been found to fall into two categories, probably correlated with whether the slip mode is restricted glide or pencil glide [RAABE & LÜCKE 1994].

4. General Discussion and Assessment

The examples presented above have shown many agreements and some disagreements between Taylor-type simulations and experimental textures in cubic metals. Among the successes are many qualitative features of deformation textures; e.g., the relatively rapid formation of fibers in orientation space, at more-or-less the right locations and, of course, with the right symmetry properties. It is also well-known (and will be dealt with in detail in Chap. 10) that the 'Taylor factors', i.e. the effective orientation factors for polycrystals under different straining paths, give reasonable agreement between theory and experiment [KOCKS 1958]. One might conclude that "texture theories are very forgiving" [DILLAMORE & KATOH 1971]. At present, however, more rigorous assessments are possible.

For a quantitative comparison between prediction and experiment, it seems mandatory that *initial* textures be measured as a matter of course, and that textures be measured at a few distinct strains (including low ones). It is the lack of such experimental data (at least in a useable format) that has restricted our examples to just a few. We will expand on the importance of initial textures in Section 4.1; it is by no means true that they tend to become unimportant as the strains get larger.

On the theoretical side, we shall re-examine the model assumptions in terms of the results (Sec. 4.2), and assess the importance of stochastic variations and heterogeneities (Secs. 4.3 and 4.4). The most troubling aspect of current simulations is the frequently observed material dependence, for the same lattice structure (such as the difference between the behavior of the fcc metals Al, Cu, Ag, for example). This will be discussed in Sec. 4.5.

4.1 The influence of initial texture

In the comparison of experimental rolling textures with simulated ones (Sec. 3.3, Fig. 10), it became clear that the initial texture of the samples remained important to large strains. This is a general phenomenon which will be illustrated by other examples and general considerations in the present Section. (See also FREDA & CULLITY [1959], LEFFERS & JUUL JENSEN [1986], DEZILLIE &AL.[1988], HIRSCH [1990].)

Uniaxial straining

Figure 11a shows a simulated 'rotation field' (Chap. 8 Sec. 3.4) for fcc crystals under tension: a number of special orientations (large + signs) were subjected to four increments of 0.025 strain (decreasing symbol size). It is obvious that they will collect

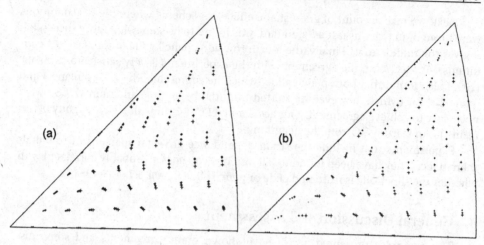

Fig. 11. Rotation fields (from larger to smaller symbols, for a total of 10% strain) for compression in bcc metals under (a) restricted glide, (b) pencil glide.

in two end-components: $\langle 111 \rangle$ and $\langle 001 \rangle$. Figure 11b displays results for bcc in compression, under full constraints; they should be identical to Fig. 11a if restricted glide had been assumed, but Fig. 11b represents a *pencil glide* simulation. There are very small differences between the two.

Drainage volumes

These results are most easily discussed in terms of a picture of 'potential' contours, in which orientations fall 'downhill' (rather than of intensity contours, in which orientations collect on peaks). The orientations in Fig. 11, then, drain toward two sinks. Each sink has a certain 'drainage volume' associated with it. In this case, it is greater for $\langle 111 \rangle$ than for $\langle 001 \rangle$. Therefore, if the initial distribution were uniform (for example, random), then the final intensity in $\langle 111 \rangle$ will be greater than in $\langle 001 \rangle$ (as is indeed typically observed in wire-drawn samples, of aluminum for example). However, if the initial density in the $\langle 001 \rangle$ drainage volume should have been greater, the situation could be reversed.

In particular, two extreme cases are worth exploring. First, if there is a peak in the initial texture, this peak may well move under the new deformation conditions, but it will almost certainly *intensify*. Second, if there is a hole in the initial texture, a peak that would have drawn on the density in that area cannot develop. Both cases indicate that the influence of an initial texture is *permanent*, not transient [SREE HARSHA & CULLITY 1962]. This is the reason why a comparison between theory and experiment is meaningless when the initial texture is not known: it must be measured, and it must be incorporated as an initialization of the state parameters in any simulation.

The limits of the 'drainage volumes' depend, to some extent, on the polycrystal model; primarily they depend on the single-crystal yield surface (SCYS). Figures 11a and 11b referred to slightly different SCYSs: in pencil glide, the hexagonal cylinders (corresponding to the three slip systems on {110} planes that have a particular $\langle 111 \rangle$ slip direction) are replaced by circular cylinders [KOCKS 1970]. In the particular case of these figures, the movement in the boundary between the two drainage volumes is almost imperceptible. One is tempted to generalize that changes in the single-crystal behavior (unless they are too drastic) have less effect on the final texture than changes in the initial texture.

Figure 11 also illustrates that the drainage of the large volumes occurs rather rapidly. (The total strain here is 10%.) This has an interesting consequence on the development of *fibers* in orientation space, for example after rolling. The fibers themselves form rather rapidly [HIRSCH & LÜCKE 1988b]; only afterwards do the intensities shift within them. Again, the influence of initial textures can be substantial. Even subtle differences in the distribution in a particular drainage volume can populate the different parts of a fiber differently at first – even if they all collect to the same end components. Thus, it is important to study the development of deformation textures *with strain*.

There is only one case in which the initial texture does not influence the final texture, and that is if all orientations congregate in an 'end texture' that corresponds to a *single component*: then, only the *rate* at which this end texture is approached would depend on the initial texture. This is a rare case.

Finally, it is worth emphasizing that the initial texture matters in one more respect: if the initial *symmetry* of the texture was lower than that of the test, the lower symmetry may be retained for ever. A particular case of such behavior was reported in STOUT &AL. [1988].

4.2 The model assumptions re-checked

Stress variations

The fundamental assumption of all Taylor-like models is that it is most important, when the plastic strains are large compared to the elastic ones, that compatibility between the grains be fulfilled. Regardless of whether the strain is actually *uniform*, as in the original Taylor model, or merely locally *prescribed*, the local stresses that follow from the single-crystal response are not likely to be in equilibrium from grain to grain. The implied assertion is, therefore, that the necessary equilibrium can be achieved by elastic adjustments. These consideration were set out in detail in Chap. 8 Sec. 1.2.

It is interesting, then, to check on the internal stresses that are actually set up within a Taylor-type simulation: are they small or large? If they came out to be of the same order as the yield stress, the model would not be internally consistent. In materials with very anisotropic single-crystal yield surfaces, one would expect the grain-to-grain stress differences to be too large; that is why hexagonal metals will be treated with the 'self-consistent' or 'finite-element' methods in Chaps. 11 and 12.

A coarse measure of the internal stresses following from the simulation is the standard deviation of the grain stresses. In the LApp code, this is output for each stress component separately. For the cubic metals that were discussed in the present Chapter, they are typically between 20% and 40% of the average polycrystal stress. This is a general result for more-or-less random orientation distributions under any straining path. It would seem to verify the internal consistency of the Taylor model for these materials.

Some simulation results have, however, been found to be in unsatisfactory agreement with experiment. These are generally on materials with a fairly strong texture, and only under particular straining conditions. An example is uniaxial straining of a ⟨110⟩ texture, discussed above. As another example, in Chap. 10, we will encounter difficulties with rolling of a 'Goss' texture, and with tension in rolled sheets with a strong 'S' component. In these cases, a case-by-case analysis of the single-crystal yield surface shows why there should be problems with the Taylor model: in all these cases, the maximum standard deviation of any component was 40% or more. This is probably too coarse a marker for a general test of the applicability of the Taylor model, but it may serve as an easy warning flag. (Some simulations in which this marker was exceeded, nevertheless gave reasonable results.)

Number of significantly active systems

Another consequence of locally prescribed strains is the necessary activation of five slip systems, in the general case. This has often been stated to be in disagreement with experiments. Unfortunately, most of these experiments relate to *surface* grains, which are less constrained, and to slip *planes*, not slip *systems*. Furthermore, deformation in any grain is not homogeneous, and not expected to be so. Our model treats only the *average* behavior of the grain.

Simulations can easily provide the number of slip systems or slip planes that were in fact 'significantly' active in any grain; for example those that provided at least 5% of the strain. Figure 12 shows the number of such planes for a variety of straining paths, starting with a random sample, under full constraints. It is seen that, in most cases, only three sets of slip traces should be observed over the whole grain, and in torsion, after significant strain, only two. We conclude that the simulations are *not* in disagreement with experiment on this front.

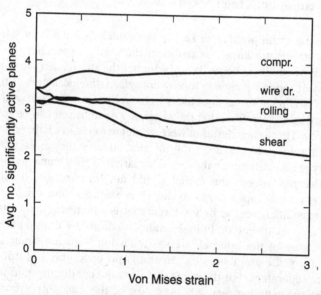

Fig. 12. The number of significantly active slip planes (i.e., which contribute more than 10% of the strain in the step), as a function of von Mises equivalent strain.

4.3 The sharpness of textures. Stochastic effects

It has been a long-standing observation that predicted deformation textures tend to be sharper than experimental ones. Put more succinctly: the amount of smoothing one needs to introduce to make theoretical textures agree with experimental ones is typically between 5° and 10°. Note, on the other hand that one would expect a smoothing of at least 5° to be inherent in the experiments and their evaluation (if undertaken on a 5° grid). This discrepancy is, generally speaking, a consequence of the fact that calculations are deterministic, whereas nature has many stochastic influences. It is perhaps possible to introduce some stochastic elements into simulations; so far, this has not proven very successful [DEZILLIE &AL. 1988]. It is also questionable whether the results would be worth the effort; we shall see in Chap. 10 that the influence of the texture sharpness on the macroscopic plastic *properties* is not great. (However, the influence on recrystallization could be substantial.)

There are two distinct reasons why the orientation change of a particular grain may be subject to some variability. One is that a *given* tensorial strain rate may often be achieved by a variety of slip combinations; this effect runs under the heading of 'ambiguity'. The other effect is that the strain rate itself may be influenced, in each particular grain, by its particular surroundings; we will review this as 'neighbor interactions'.

Ambiguities

Ambiguities may exist whenever there are more than five slip systems available (in a fully constrained case). They are heightened by crystal symmetry which leads, in the case of fcc crystals for example, to many stress states that have the same resolved stress on six or eight slip systems. Of course, the value of the *critical* resolved shear stress (CRSS) could in fact vary stochastically; then, the SCYS would not obey the crystal symmetry [LEFFERS &AL. 1988], and the effect could be different from grain to grain.

Once symmetric states are presumed to exist, they lead to *ambiguities* only under an assumption of strict rate independence. The use of the physically established rate dependence makes the problem completely deterministic. Even if ambiguity were assumed, taking the *average* of the presumed range of indeterminacy is just as deterministic and amounts to assuming that the straining vector is always in the middle of the cone of normals (which is especially unnatural when this cone itself is not centered around the stress vector). Making a *random* choice at every step does not provide the necessary spread either [DELZILLIE &AL. 1988]. However, making an *initial* random choice and then staying with it may correspond to the actual developing asymmetry of CRSSs in a particular grain and, in this way, lead to a reasonable dispersion between grains.

Neighbor interactions

The other, and more likely, reason for stochastic effects is based on specific *neighbor interactions*. If one started from a Taylor model, it is the stress in a particular grain that would be modified by the neighbors. In a rate dependent model, this would lead to a change in straining direction and, for *this* reason, to a change in slip distribution. The stress may well stay in the vicinity of the same vertex, and thus not disturb the basic model too much.[3]

In Chapter 12, we will discuss finite-element solutions of the grain interactions. (A preview was given in Chap. 8 Sec. 2.3.) In these, compatibility is also exactly fulfilled, but an attempt is made to minimize the violations of local equilibrium. The concomitant local strain variations do lead to a spread in textures. One hopes that the

---

[3] Adding random *stresses* [LEFFERS 1979], presumably outside the small range allowed by the vertex curvature, *would* completely destroy the Taylor basis (and doesn't help much if you start from Sachs: you'd still be on the same facet, unless you made the stress variations *very* large).

results from such detailed calculations can eventually be abstracted into a meaningful algorithm that incorporates stochastic effects into Taylor-type calculations for quasi-homogeneous deformation.

We note in passing that another method that has been used to resolve the ambiguities caused by strict rate independence, proposed by RENOUARD & WINTENBERGER [1981], is not tenable on physical grounds: a grain is supposed to behave according to what will be more energy-saving for it in the *future*. In addition, this postulate is contradicted by many practical examples; e.g., a $\langle 111 \rangle$ single crystal in tension does not deform in single slip, as this 'principle' would demand.

4.4 Deformation heterogeneities and instabilities

Deformation in polycrystals is never uniform (as assumed by the rigid form of the Taylor model); the question is whether any of the heterogeneities that do arise cause any substantial changes in the developing deformation textures and the resulting properties. There is *no* question but that subsequent recrystallization can be severely impacted by any local *deviations* from the average behavior [ENGLER 1996]. On the other hand, *bulk* properties are often not sensitive to such details – and that is why the Taylor model has been so successful (and Taylor's abstraction so far-sighted).

We will mention some of the widely observed deformation heterogeneities, in the order from macroscopic to microscopic. (A more detailed overview can be found in AERNOUDT &AL. [1993].) At the sample level, *shear bands* develop in orientations that are not correlated with the crystallography of the grains – although they may well *start* from a plastic instability at the grain level in highly textured or in 'layered' materials [MORII &AL. 1985, HIRSCH &AL. 1988]. They occur only at high strains, and predominantly in plane-strain deformation [VAN HOUTTE &AL. 1979, WAGNER & AL. 1995]. The volume fraction of material inside shear bands is usually much too small to affect macroscopic averages significantly. In addition, an interesting observation concerning the macroscopic shear bands observed in rolling was made by GIL SEVILLANO &AL. [1977]: a simple-shear texture in macroscopic bands at the observed angle of $\pm 35°$ to the rolling direction would produce a texture very similar to the rolling texture.

At the intermediate size scale, deformation heterogeneities originate from the different interactions of one grain with its various neighbors, which lead to different slip system combinations – and thus to different orientation changes – in various parts of each grain. Such domains are sometimes separated by single boundaries, and sometimes by 'transition bands'. (See, e.g., the simulation by BEAUDOIN &AL.[1996].) Both kinds tend to start at grain edges [LARSON & KOCKS 1963]. Similarly, at the grain scale, it has long been realized that certain favored orientations tend to decompose into two or more *variants* [AERNOUDT & STÜWE 1970; KOCKS & CHANDRA 1982; HIRSCH 1990].

Finally, a plethora of substructural features is observed in heavily deformed materials, many of them connected with the discreteness of slip and, indeed, of dislocations; often, they can be described as band-like in nature. (See, e.g., HUGHES & HANSEN [1991], BAY &AL. [1992].) They are not primary carriers of orientation-change mechanisms; however, ROLLETT &AL. [1992] have made use of such structures as obstacles that influence local slip system selection – and therefore texture changes.

4.5 Material dependence

If deformation textures were merely a consequence of slip geometry, then all fcc metals should develop the same textures with strain. This is often reported to be untrue in practice (see Chap. 5); the differences become accentuated with strain and also depend on temperature. Many explanations have been proposed, usually connected with the fact that the stacking-fault energy (SFE) varies with material [see, e.g. KALLEND & DAVIES [1971a]). Before we address such basic physical mechanisms, we summarize our concerns that the quoted observations may have causes other than an altered deformation mechanism.

In first instance, we have emphasized (Sec. 4.1) that different *initial* textures may have a severe effect on end textures – and they have usually not been reported. They may be expected to be different for different materials, for example, because of different recrystallization behavior during the sample preparation (which can, incidentally, also lead to different densities of annealing twins). Different initial textures are expected to have an especially strong influence on uniaxial textures (Sec. 4.1, [WASSERMANN & GREWEN 1962]); in Chap. 5 Fig. 1, compression experiments were shown that dispel all but the most extreme material differences once the initial textures have been taken account of [STOUT &AL. 1988]. Thus, the most-quoted results, the wire-drawing experiments by ENGLISH & CHIN [1965], may well have been due to the same cause, and not to a genuine material dependence. The same effect was pointed out already by AHLBORN & WASSERMANN [1962], in a critique of the early experiments by SCHMID & WASSERMANN [1927].

As a second effect, partial recrystallization may well occur during deformation; for example, wire-drawing is usually undertaken at very large strain rates, which would tend to heat up the specimens [DILLAMORE & ROBERTS 1965]; and this effect is the more pronounced the stronger the material (and thus, at a given strain, the lower the SFE).

As a third example, the deformation heterogeneities that develop could (and in fact do) depend on material; but we have argued (Sec. 4.4) that they, by themselves, are not likely to have a strong effect on textures. Indirectly, they may influence the deformation pattern in a wider neighborhood.

Despite all these *caveats*, it is likely that extraneous effects do not account for *all* of the observed material differences. Thus, it is opportune to review the general phenomenology of the observations and the interpretations. It would be very desirable to have information for all straining paths; but since most of the work has been devoted to *rolling*, we imply this deformation mode when a specification is necessary.

A comparison of textures in different materials *at the same strain* may lead to misinterpretations. In *rolling* of fcc materials, the development of fibers (primarily the 'β-fiber', but also the 'α-fiber') is common to all materials; it is the distribution *along* the fibers that is observed to vary with material [MECKING 1985] – but this distribution varies with strain (Sec. 2.2). Figure 13 displays a small excerpt from an exhaustive experimental study by HIRSCH & LÜCKE [1988a]. Both the strain and the alloying dependence of the (generalized) β-fiber in Cu-Zn alloys is displayed. It demonstrates that the textures are quite similar up to a reduction of about 50%, and include a 'brass' component (on the right side of the charts). The latter in fact increases with strain in all cases. At large strains, it is overtaken by orientations around the middle of the fiber in Cu and Cu-5%Zn. At the highest strains, the 'brass' component is fully developed

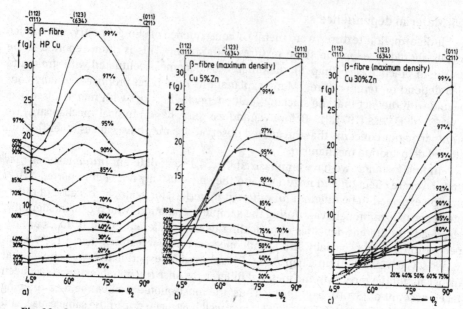

Fig. 13. Strain and alloy dependence of the β-fiber in Cu-Zn alloys [HIRSCH & LÜCKE 1988a]. The fiber location was defined by the maximum intensities.

already in the 5% alloy. Deformation twinning is observed in the high alloy, but not in the low.

The orientations of neighboring areas have to be well correlated. (After AERNOUDT & STÜWE [1970].)With these important introductory points in mind, and abstracting from many experimental and theoretical observations, we offer the following brief assessment of the nature of any genuine material dependence.

(1) The sequence of texture differences often follows the sequence of stacking-fault energies; in addition, lower-temperature behavior is often equivalent to lower-SFE behavior [AHLBORN & WASSERMANN 1963, HU & GOODMAN 1963]. The transitions are not abrupt, but *gradual*: from aluminum to nickel to copper to silver; from pure copper through bronzes and brasses of increasing solute content to the fully develop-ed (70:30) 'brass' (Chap. 5) [MECKING 1981a]. Sometimes, the extremes are called 'brass' or 'alloy' textures, versus 'copper' or 'pure metal' textures [WASSERMANN & GREWEN 1962]; but note that pure copper is actually not one extreme: pure aluminum is. Substantial deformation twinning is only observed in the extreme cases. Finally, the SFE-dependence is not unique or universal: it may be different for alloys and pure materials [ENGLISH & CHIN 1965]; and substantial texture changes may occur when there is *no* change in SFE (Cu-Mn [ENGLER &AL. 1994c]).

(2) The Taylor model (in the grain-shape-sensitive form) is quite successful in modeling the 'aluminum' extreme. In fact, the occasionally observed mismatch of, say, one observation out of 20 [JUUL JENSEN & HANSEN 1990] serves more to emphasize the incredible goodness of fit than any failing. More experiments on aluminum and nickel, which include a measurement of the initial texture and texture measurements as a function of strain, would certainly need to be published for a final conclusion. There is no model in which the agreement with the 'brass' extreme is anywhere nearly

as complete. We are therefore looking for reasons for a departure from the Taylor model in the 'lower-SFE' cases [HIRSCH &AL. 1984].

The older interpretation was, conversely, that the 'brass' texture is 'normal', and one must look for *high* SFE (and high *T*) to explain the behavior of copper and aluminum. (See the exhaustive review by DILLAMORE & ROBERTS [1965].) It was claimed that cross-slip from one plane to another, which is favored by high SFE and high temperature, makes the Taylor model more applicable; but in fact it is only the *number* of slip systems operating that is needed for the Taylor model to apply, not the intimacy of their connection by a thermally activated process. Another feature that has drawn much attention is the *planarity* of slip in the case of 'brass'-type materials; but again, the planarity would be of no concern, if there were enough slip systems operating (of planar or wavy nature). It seems that the real distinction that was made was between *single slip* and polyslip – or, more accurately, between slip on one *plane*, or a *few* slip systems, on the one hand and *many* (three to five) on the other. For single slip, the Sachs model was used as a basis, for polyslip the Taylor model was conceded. It may then be a matter of taste whether the gradual transitions are treated as modifications (toward Taylor) of the Sachs model [LEFFERS 1979], or modific-ations of Taylor (toward Sachs) [MECKING 1981b]. Inasmuch as Taylor has a more solid basis (see Chap. 8) and is demonstrably successful at one extreme, we favor the latter view. The question then is, why (and indeed whether) there is an increasing tendency toward single slip as one proceeds from aluminum toward brass. It is not a foregone conclusion, though, that a tendency toward single slip will always give the 'brass' component.[4]

The simulations reported in Sec. 2.2 showed that a movement of intensities *along* the β-fiber should occur with strain, especially under increasing relaxed constraints. It appears that this movement toward the 'S' orientation does not take place in the lower-SFE materials (see Fig. 13): the 'brass' component remains stable or it may, in fact, move along the α-fiber toward the Goss component. (The common statement that the 'brass' component only develops in the low-SFE materials is not accurate: as we have seen, at low strains, with random initial texture and under full constraints, it develops in both high-SFE materials and in the simulations.)

The stability of (or the movement along) the α-fiber in low-SFE materials cannot be blamed on initial textures. A different behavior at the grain level has been used, and abandoned, on a number of fronts. The most common one is the postulate of substantial deformation twinning [WASSERMANN 1963]. The data in Fig. 13 and in Chap. 5 Fig. 3, as examples, argue against this: the change in Cu-Zn alloys is gradual, with increasing Zn content; we judge the preponderance of the evidence to imply that material differences do exist in the complete absence of deformation twinning (al-though opinions differ on this [HIRSCH &AL. 1988]). If small amounts of deformation twinning should take place, the volume fraction cannot be large enough to affect tex-tures substantially [LEFFERS 1996]. For this case, a model of 'bundling' of slip parallel to twins [LEFFERS & JUUL JENSEN 1991, AERNOUDT &AL. 1993] has been postulated.

Another possibility of long standing is an influence of latent hardening [BISHOP 1954, BACROIX 1986]. It would have the advantage that an increased level of latent hardening might well correlate with a lower SFE (because the forest dislocations are more extended, and their being cut by the mobile dislocations would give the appropriate temperature dependence). However, in Sec. 3.2, we showed in one

---

[4] For an understanding of the older literature on deformation textures, it is worth noting that the tensorial nature of stress and strain was not properly recognized in many instances: rolling was treated as a superposition of tension and compression (at right angles). This gives the correct result for resolving the stresses, but not for determining the rotations.

example that the most extreme assumptions on latent hardening cannot explain any significant texture change. Figure 9 demonstrated this on one example of many we have tried.

Finally, an attempt was made [KOCKS & NECKER 1994] to explain the results on the basis of a change in the SCYS that corresponds to an influence of the stress resolved onto ⟨211⟩ (partial) dislocations. It showed that the texture evolution was very sensitive to this parameter, and a reasonable value could be found that would 'explain' the results for rolling in a copper for which the initial texture was known (and in which no heterogeneities developed). However, the model was not successful for torsion or wire-drawing (which, as seen above, would probably be explained by initial texture differences). Also, a more accurate analysis has shown that a movement along the α-fiber is observed, but not explained by this model.

All the proposals discussed so far have attempted – without success – to associate the material dependence of textures on changes in the single-crystal yield surface (SCYS). The remaining alternative is to assume that the best-suited *polycrystal model* depends on details of the material. The grain-to-grain interactions could in fact depend on the planarity of slip: if the stress concentrations at the head of individual slip planes, at grain boundaries, were strong enough, they could favor the 'propagation' mode of polycrystal deformation (Chap. 8 Sec. 2.1). Then, the Taylor model, or most variations on it, would be inapplicable. However, it might prove difficult to build a model that allows for a smooth transition between these two mechanisms.

Finally, we need to discuss a different model of polycrystal deformation that may have a chance for further development and generalization; we will call it the 'pattern formation' model. Let us introduce it by quoting a special case: flat grains of brass that have a {110} rolling plane. Under a rolling *stress*, they would produce large shears of the kind forbidden in flat grains (*in* the plane); if they were allowed this shear, they would be stable [HONNEFF & MECKING 1978]. AERNOUDT & STÜWE [1970] proposed that overall compatibility could be maintained by alternating twinned an untwinned lamellae in this plane: the twinning shear precisely compensates for the 'forbidden' shear due to glide. They generalized this concept in the form of Fig. 14, taken from their work.

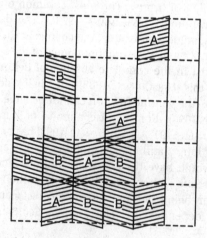

Fig. 14. Possible pattern for single-slip deformation.

Such solutions require a special relation between neighboring regions of single slip (or twinning), which is conceivable *within* grains and perhaps also *between* grains in a highly textured material. The necessary topology could develop as a consequence of such a mechanism [HIRSCH &AL. 1988, HIRSCH 1990]. The special neighborhood relations might be more easily accomplished in an alternating-*plate-like* morphology, as it has been observed by HUGHES & KUMAR[1996].

The mechanism to be investigated for its viability might be described algorithmically as follows. Uniform stress is prescribed initially. The most favored single system (or single slip plane) in any one region begins to operate. So far this is the same as in the transition to the Taylor model at small strains. However, this system will remain the only one operating in that region. Neighboring regions now feel a combination of external and internal stresses and again operate only the single most favored system; but it may be a different one (even if it's within the same grain). The question is: can a spatial pattern of such regions be set up where equilibrium is enforced strictly, and compatibility is satisfied only in an average sense over ordered 'clusters'? The larger the cluster, the more stringent becomes the need to satisfy compatibility [KOCKS &AL. 1986]. This is a case where 'geometrically necessary' dislocations may play a role [ASHBY 1970]. In the end, the point to be investigated is: why would pattern formation be favored over a more homogeneous deformation as the SFE or the temperature decreases?

A particular case of an implied formation of organized structures is that in which a single twinning system operates and subsequently favors slip in 'bundles' on the parallel plane only [LEFFERS & BILDE-SØRENSEN 1990; AERNOUDT &AL. 1993; LEBENSOHN & TOMÉ 1994]. As these authors showed, this mechanism would tend to give a 'brass' texture – but only for materials in which twinning occurs on a significant scale.

4.6 Recommendations

The era of quantitative simulations of deformation texture development, even in cubic materials, has not drawn to a close. To foster further progress, simulations as well as experiments should be undertaken at the level of sophistication that has already been achieved, and they should cover a wide range of conditions.

We present below a series of recommendations, which address experimental problems as well those of simulation: to verify (or disprove) a model of texture development, the demands on experimental expertise are as great as those on modeling.

Simulations
- For input, a description of the initial texture must be incorporated that is sufficiently accurate and does not leave out orientations of special symmetry.
- If the flatness of the grains' morphology changes during deformation, some algorithm must be used to incorporate its indubitable effects.
- A sensitivity analysis of the predicted results should be undertaken with respect to all assumptions and approximations used in the model that are not unambiguously grounded in the basic physics.
- As part of the output, ancillary results should be recorded that may aid in an assessment of the self-consistency of the method used; e.g., standard deviations of the grain stresses (and strains).

Mechanical tests and texture measurement
- It is imperative that initial textures be measured before all mechanical tests, and that the relation between the specimen faces in the test and in the goniometer be meticulously recorded.

- The range of deformation paths should be as diverse as possible, and the strain dependence of texture development should be recorded. Note that the actual deformation path may depart from ideal assumptions, and the developing texture may be quite sensitive to such deviations [HONNEFF & MECKING 1981].
- The deformation geometry should be well controlled. (See, e.g., BACKOFEN 1972].) Ideally, it should cause homogeneous deformation; when it does not, the subsequent texture specimens must be taken from well-characterized sections.

Texture analysis and comparison
- The first mandate on any experiment is that its results be recorded in a form that can be shared with other investigators. Today, this means ASCII files, which incorporate corrections inherent in the experimental setup, but no further analysis; e.g. pole figures with background and defocusing correction (not coefficients from a harmonic analysis).
- Although a three-dimensional orientation-space analysis is necessary to assess details of a texture, a comparison of simulation and experiment is best undertaken at the level closest to the experiments; that is, e.g., on pole figures. The predictions should be smoothed by at least the experimental mesh size. An elimination of experimental spread and lack of precise symmetry is also appropriate; thus *recalculated* pole figures should be compared with *smoothed* predictions.
- The sharpness of a texture is indicated not only by the maximum intensity in the peaks, but also by the range of *low* intensities. Inasmuch as the maximum intensities are subject to many smoothing effects, differences should not be taken too seriously. On the other hand, the distribution of intensities below 1 m.r.d. is of great interest.

Chapter.10

EFFECTS OF TEXTURE ON PLASTICITY

1. **Experimental Techniques and Phenomenology**
   1.1 Tension and wire-drawing
   1.2 Compression and plane-strain compression
   1.3 Torsion and simple shear
   1.4 Strain anisotropy determination
   1.5 Yield-surface measurement
        Torsion of long tubes. Biaxial stretching of sheet. 'Forging' of cubes
   1.6 Continuously changing paths

2. **Yield Surface Shapes: Predictions and Experimental Results**
   2.1 Yield-surface symmetry
   2.2 Isotropy
   2.3 Simulation of plane-strain compression
        Strain dependence. Yield surface shapes
   2.4 Rolled cubic metals: yield-surface measurements and interpretation
   2.5 Rolled cubic metals: strain anisotropy
        Simulations. Observations. Latent hardening in polycrystals. Microstructural obstacles
   2.6 Sharp textures and plastic instabilities
   2.7 Torsion: free versus fixed ends
   2.8 Dependence on deformation modes
        Pencil glide. Twinning

3. **Stress/strain Curves: Experimental Results and Analysis**
   3.1 Dependence on straining path
   3.2 Determination of the grain-level hardening law

4. **Summary and Assessment**

EFFECTS OF TEXTURE ON PLASTICITY

M. G. Stout and U. F. Kocks

All plasticity has two separable aspects: evolution of the state, and properties at a given state. The most well-known aspect of plasticity is perhaps strain hardening: it is due to an evolution of the state of the material; for example, the density of dislocations increases. After deformation, the material is in a new state; for example, it has a new yield strength. Other properties at a given state are the dependencies of the current yield strength on strain rate and temperature. The most important one in the context of anisotropy is the dependence of the current yield strength on the direction in which the specimen is being strained: not only forward or backward, but in the X- or the Y-direction of the given sample, for example. In addition, tension and shear paths behave differently: it is the entire form of the stress or the strain-rate tensor that matters. This general anisotropy of the yield strength is expressed by a 'yield surface'. We have encountered yield surfaces for *single crystals* in Chap. 8; this Chapter will deal extensively with yield surfaces in *polycrystals*. An excellent text that introduces the concepts and provides many applications was written by HOSFORD & CADDELL [1983]. A lucid early review highlighting the importance of these anisotropies for industrial forming processes was written by WILSON [1969].

Texture is relevant for both the evolution properties and the state properties. Texture itself evolves with straining, and this was the topic of Chap. 9. This evolution of texture causes an additional contribution to strain hardening, which is not based on changes in the dislocation structure, but on the change in the orientation factors needed to convert macroscopic stress and strain rate to crystal-level stresses and strain rates. This fact contributes, for example, to differences in stress/strain curves between tension and compression, since texture evolves differently in the two cases. It also leads to a difference between tension and torsion stress/strain curves, merely on the basis of the differently evolving textures – which forces a re-examination of the concepts of 'equivalent strain'. These topics are not properly labeled as consequences of 'anisotropy'; but they *are* macroscopic effects that can be explained on the basis of texture and the single-crystal yield surface. They will be dealt with in Section 3 of this Chapter.

Section 2 deals with yield surfaces and straining anisotropies – presenting theoretical concepts and simulations first, and then comparing them to the scant experimental data available. We will be using Taylor-type simulations in this Chapter and point out applications in which this assumption is not expected, or not observed, to be adequate; such cases will be dealt with in Chaps. 11 and 12. The particular, modified Taylor simulation used ('LApp') was described in detail in Chap. 8 Sec. 1.

The experimental concepts and techniques are of fundamental importance for the whole Chapter; they will therefore be reviewed in Section 1. The authors have repeatedly found that first-cut disagreements between theory and experiment called for improvements in both the modeling and the experimental technique.

1. Experimental Techniques and Phenomenology

The two aspects of plasticity – evolution, and properties at constant state – put different demands on the experimental technique: one, the ability to impart large strains under 'good' boundary conditions; the other, to measure flow stresses 'instantaneously' in a meaningful way. By 'good' boundary conditions it is meant that the straining occur as uniformly as possible, so the properties of the whole sample represent the behavior of the material itself. For an investigation of the monotonic stress/strain response and the accompanying texture development, four types of experiment are used: uniaxial extension; uniaxial compression; channel-die compression; and simple shear. Two of these deformation paths produce uniaxial stresses; two, plane-strain deformation. Since uniaxial and plane-strain deformation patterns may involve physically distinct mechanisms (e.g. in the minimum number of slip systems required), it is generally advisable to have experimental information from at least one in each category.

'Instantaneous' measurements must assure that the state of the material is not significantly disturbed by the 'probe'; on the other hand, they must also allow for the elasto-plastic transient to be considered completed during the probe, so that the flow stress is well defined. We highlight some of these difficult techniques and merely mention others in passing.

1.1 Tension and wire-drawing

The most reliable test of plastic properties is the most common one: *uniaxial tension*. So long as the gauge length is long compared to the sample diameter (or the largest of its lateral dimensions), the boundary conditions for the test cause a uniaxial stress. For the determination of the yield strength and initial strain-hardening parameters, this remains the preferred technique. Its major drawback is the formation of a necking instability when strain hardening does not persist to large strains, or other limits to ductility, such as cleavage fracture.

For such cases, *wire-drawing* can provide supplementary data. In this case, lateral normal stresses as well as shear stresses are allowed, and the tensile stress in the axial direction cannot be adequately measured by the drawing force. However, it is possible to alternate between imparting strain to the specimen by wire-drawing and then testing its flow stress in tension. This does not involve a significant change in deformation path. (The cylindrical symmetry of the wire-drawing operation is much more favorable for testing than the industrially common swaging process, which should be avoided even as pre-processing.) The difficulties in wire-drawing, which will be described in some detail in the following, lie in minimizing the heterogeneity of deformation that results from friction, and in keeping the strain *rates* comparable in wire-drawing and tensile testing.

In order to minimize the effects of redundant shear deformation in wire-drawing, one should use a special set of dies. The dies should have a very small included angle, around 8°, and a large reduction in area per die, say 25%. This results in a logarithmic axial extension of 0.30 per die. A typical sequence would be to begin with a 15 mm diameter rod, and draw this rod through eight increments ending with a diameter of 5 mm. The total axial extension realized in this process is 240%. We have found that the best lubricant for the purposes of this laboratory drawing is a molybdenum disulfide lubricant in a lacquer carrier, Molykote[1]. This particular form of the lubricant is not extruded from the die/work-piece interface by the die pressure.

The drawing need not be done with a standard draw bench; there are many advantages to using a conventional screw-driven testing machine. The latter technique allows one to accurately control the drawing rates to produce the same average true strain rate for each die. In addition, one can choose this rate to be the same as is subsequently used to evaluate the drawn material in uniaxial tension in order to avoid the convolution of the two effects [STOUT &AL. 1987] . Note that to achieve this match, very long times are required for the drawing operation.

The tensile specimens should be machined from the center of the drawn wire to evaluate its mechanical properties. In this way, the surface layer of the wire is removed in preparing the specimens, eliminating material that had potentially experienced shear deformation in addition to the uniaxial extension. Care must be taken that the diameter of the gauge section remain large compared to the through-thickness grain size. Finally, the gauge sections of the wire need to be given a light electro-polish to remove any scratches, notches, or other defects that might induce an early necking instability during testing.

1.2 Compression and plane-strain compression

Uniaxial compression has two advantages over tension in the context of textures: the specimen shape provides more surface for texture measurements (without having to construct composite samples); and one can reach much higher strains. A potential difficulty is the maintenance of uniform temperature in the specimen, especially at high strain rates when adiabatic heating occurs during deformation.

The chief difficulty with compression experiments is the procedure to minimize friction, which typically causes 'barreling' of the specimen, non-uniform stress distribution, and increased apparent strain hardening. With modern methods, all of these effects can be avoided. Specimens are best chosen squat; in practice, it is best to stay in the neighborhood of a 1:1 aspect ratio and re-machine when necessary (typically every 10–30%).

Compression experiments do not, however, produce a reliable measure of the elastic modulus and details of a yield point may be masked. This is so because in squat specimens, direct strain measurement is impossible and displacement measurements are inaccurate in view of the large machine and fixture compliance (even in a 'stiff' frame); also, the specimen surface next to the platen is often not perfectly smooth and will adjust during the initial stages of the test. For these reasons, an 'offset' yield stress would have to be defined at a strain that is typically much larger than the common value 0.2%. The *back-extrapolated* flow stress (Chap. 8 Fig. 12) can nevertheless be reliably determined when the specimen end faces are well plane and parallel. (Note that polishing methods that produce a smooth, but not a plane surface, are inadequate.)

One can begin these experiments with reasonably large specimens, say 10 mm in diameter which, for example in aluminum alloys, will not exceed the load capacity of most testing machines. The initial height is then well chosen at 14 mm, and the height at interruption at about 9 mm (which corresponds to an aspect ratio of 1/1.4 and a true strain of about 25%).

[1]Molykote is a trademark of Dow Corning Corporation, Midland, MI.

Testing can be done at a constant displacement rate in a conventional screw-driven loading frame, or with a varying displacement rate, such as to keep the true strain rate constant, in a programmable servo-hydraulic loading system.

When tests are performed near room temperature, it is easiest to insert a ('cast') Teflon[2] film, about 0.1 mm thick, between specimen and platen, which has first been sprayed with Molykote[1] [LOVATO & STOUT 1992]. At other temperatures, one can use a mixture of vacuum grease and colloidal graphite for lubrication; at high temperatures various glassy lubricants of matched softening temperature can be found [CHEN & KOCKS 1991a]. For fluid lubricants, it is best to engrave concentric grooves in the specimen end surfaces, for example by electro-discharge machining (EDM). This prevents the lubricant from being squeezed out during the test. A single depression in the faces, with the specimen dimensions given above, is also effective and has been widely used (Fig. 1a) [RASTEGAEV 1940].

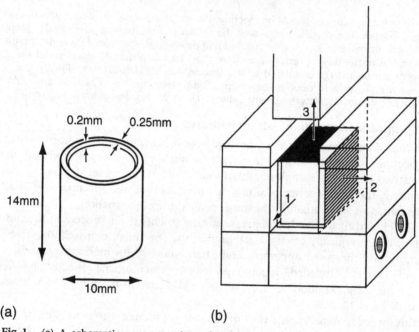

Fig. 1. (a) A schematic representation of a compression specimen with a single 'pocket' on the ends. This specimen geometry originated with RASTEGAEV [1940] and is a DIN norm. (b) An example of a channel-die fixture used to perform plane-strain compression experiments. The specimen is depicted wrapped in a Teflon[2] film.

Compression in a channel die is a means of prescribing *plane strain* (although two shears may occur if the specimen does not possess orthotropic symmetry aligned with the axes of the channel die). The channel-die fixture is shown in Fig. 1b. This test is best suited for measuring material response in monotonic deformation, although the technique can be used for path-change experiments as well [FRANCIOSI &AL. 1987]. Lubrication is even more critical in the channel-die experiment. It is possible to use Teflon[2] film for this purpose. CHEN & KOCKS [1991b] have also machined specimens by EDM to contain grooves on all sides. Finally, one must take care to machine the specimen to precise tolerances for fitting in the die. If the specimen does not fill the channel die it will initially deform in free, or uniaxial, compression rather than plane

---

[2]Teflon is a trademark of E.I. Du Pont de Nemours & Co., Inc., Wilmington, DE.

strain. It is because of this situation that the channel-die test is better suited for obtaining plastic-flow information rather than yield data. The only stress component that can be easily measured is the one that does the work; this is sufficient to obtain strain-hardening information. In any case, the textures produced should be reliable and better than ill-controlled rolling experiments can provide.

Another test that has sometimes been called 'plane-strain compression', or better 'knife-edge compression' [WATTS & FORD 1955], consists of loading a small length, but the entire width, of a sheet with a punch. This test produces very non-uniform deformation.

1.3 Torsion and simple shear

A different kind of test in which large strains can be obtained is torsion. It is a special case of simple shear (for which other, small-strain tests have also been designed [IOSIPESCU 1967, PIERRON &AL. 1995]). Simple shear is a plane-strain test with prescribed macroscopic rotations. There are essentially three kinds of torsion experiments: solid cylinders, short tubes, and long tubes.

Torsion of a solid cylinder leads to non-uniform stresses and strains. In fact, while the outer regions of the cylinder undergo plastic deformation, the center may remain elastic. When a specimen of this configuration is unloaded, a state of residual stress will be present, making any use of the specimen for subsequent evaluation difficult. Nevertheless, it is possible to derive the relation between shear stress and shear strain at the periphery by testing a series of rods with different diameters [CANOVA &AL. 1982]. The torque, M, at each degree of twist, θ, depends on the specimen radius, a. When this relation is fitted with a polynomial, it can be differentiated to yield the shear stress

$$\tau(a) = \frac{1}{2\pi a^2} \frac{dM}{da} \tag{1}$$

which can be evaluated, say, for the midpoints between tested radii. The shear strain at this value of a and θ is simply

$$\gamma = \frac{a\theta}{L} \tag{2}$$

where L is the gauge length of the specimen. A major advantage of the method is that one can obtain continuous finite deformations with such a geometry, without limitations of necking, buckling or large geometric shape changes.

> This method presumes that the plastic properties are a function of the present, local state only; it does not require that γ be a state function [NADAI 1931] or that τ be related to γ and $\dot{\gamma}$ by power laws [FIELDS & BACKOFEN 1957]. CANOVA &AL. [1982] applied their technique to ETP copper, using ten specimens for verification. They found that, by comparison with their stress/strain curves, data evaluated using only two specimens deviates at low strain, whereas those data evaluated according to FIELDS & BACKOFEN [1957] deviated at large strains. No systematic investigation appears to have been made of the number of samples needed for *sufficient* accuracy.

Torsion of a *thin-walled tube* provides another means of obtaining stress/strain data. One can use Eq. 2 for the shear strain at the average radius and, assuming the shear stress, to sufficient accuracy, to be constant across the thickness,

$$\tau = \frac{3M}{2\pi\left(a_o^3 - a_i^3\right)} \qquad\qquad (3)$$

where a_o and a_i are the outer and inner radius, respectively. WOOLLEY [1953] has proposed a formulation to account for the effects produced by a thicker tube wall.

Equation 2 indicates that a *long* tube would provide high resolution in shear strain. In addition, this geometry is insensitive to gauge-length changes, and affords one a reasonably large volume of material for subsequent analysis, e.g. texture measurements. In addition, one can carry out subsequent deformation experiments, e.g. combinations of tension and internal pressure, on a tube preloaded by torsion. Experiments of this type will be summarized in Sec. 1.5. For attaining large strains, unfortunately, this type of experiment is limited by instabilities, primarily buckling.

When the tube is *short*, large strains can be easily obtained. The massive ends of the short thin-walled tube, outside the gauge area, prevent buckling instabilities. It is noteworthy that even in a *short* tube, the boundary conditions lead to uniform stress [READ 1950; LIPKIN &AL. 1988]. It is still possible for deformation to concentrate in bands perpendicular to the torsion axis; a scribe line on the specimen surface (initially inclined in the direction opposite to the shear angle) allows one to assess whether such an instability has occurred. Unfortunately, it is very difficult to strain-gauge such a specimen. As a consequence, yield and other small-strain behavior cannot be reliably investigated. Texture measurements must be made on composite specimens.

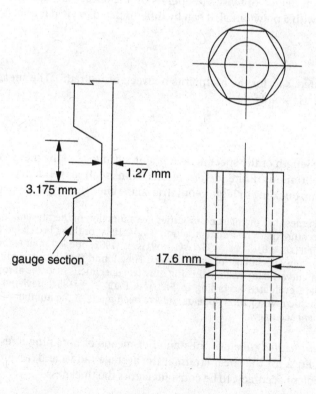

Fig. 2. The short thin-walled tube geometry used by HUGHES [1986].

One such short thin-walled geometry is shown in Fig. 2. This particular specimen is that of HUGHES [1986]. One should note that there are many variations to this design. DUFFY &AL. [1971] and his students were some of the first to employ this type of specimen with their split torsional pressure bar. The use of the short thin-walled tube for this purpose was popularized by LINDHOLM &AL.[1981]. For some materials it is necessary to increase the thickness of the test section, and the correction according to WOOLLEY [1953] should be used. For the particular specimen geometry shown in Fig. 2 the correction is small.

The short gauge length produces additional restrictions. It has been proposed to use this geometry to investigate and quantify the length change, first noted by SWIFT [1947], that occurs in torsion. However, an increase in length during torsion is accompanied by a decrease in tube diameter – which is restricted by the massive end sections. Finally, the initially short gauge length makes any absolute length changes very small and extremely difficult to measure. LIPKIN & LOWE [1989] have had the greatest success of using this specimen to evaluate the Swift effect. They were able to obtain the necessary resolution in length by using an LVDT mounted internally to the short thin-walled tube.

1.4 Strain anisotropy determination

It is in practice impossible to prescribe all straining components on a test sample; i.e., to enforce displacement boundary conditions on the entire surface. 'Free' surfaces, on which the tractions are prescribed zero, will in general be displaced in an anisotropic fashion. For example, an initially circular cross-section of a tension or compression sample will become elliptical, unless the texture possesses certain symmetry elements. Similarly, an initially cubic specimen in a channel die will in general develop certain angles between the surfaces (Fig. 1). In the rolling of a sheet sample that does not have orthotropic symmetry aligned with the rolling axes from the beginning, this manifests itself by the sample emerging at an angle to the rolling plane or to the rolling direction. As a final example, torsion of a thin-walled tube with 'free ends' will in general lead to an extension (or shrinking) of the tube: the so-called 'Swift effect' is expected at least in part as a consequence of plastic *anisotropy*. (Other contributions to the 'Swift effect' that have been frequently discussed are second-order; the present effect is first-order when the tube is initially textured. For a detailed discussion, see HARREN &AL. [1989].)

From an experimental point of view, it is an important part of test evaluation to record such behavior. This is easy to do at the end of an extended test; it would be valuable information as a function of strain. For modeling efforts, this provides a rather sensitive tool for verification. It will be discussed in connection with yield-surface predictions in Sec. 2. It should be noted that, in a similar vein, reaction stresses at boundaries where the *displacement* is prescribed zero are similarly affected by anisotropy. And the tensile stress in a twisted tube need not be zero, when the ends are fixed. These reaction stresses are linked to the equivalent free strains by the yield surface (not by elasticity). Thus, their independent measurement, although difficult, may on occasion be quite useful.

The plastic anisotropy of rolled sheet has, for many years, been characterized by the contraction ratios observed in a tension test on strips cut, at various angles, in the plane of the sheet [KEELER & BACKOFEN 1963]. A typical test arrangement is shown in Fig. 3. The conventional parameter is called the 'R-value' or 'Lankford Coefficient' [LANKFORD &AL. 1950], which is defined as the ratio of the width-strain ε_w (in the sheet plane at 90° to the tensile axis) to the through-thickness strain, ε_t:

$$R = \varepsilon_w^p / \varepsilon_t^p \qquad (4)$$

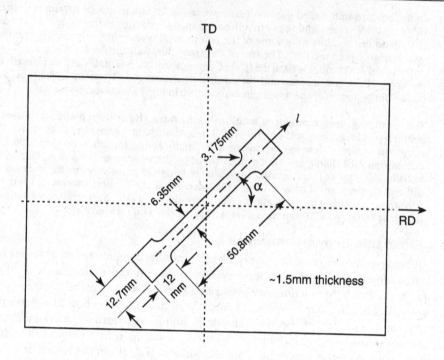

Fig. 3. A tensile coupon geometry appropriate for measuring R-values from rolled sheet. The orientation of the tensile bar with respect to the sheet rolling direction is specified by the angle α.

(where the superscript p emphasizes that only the *plastic* strains are to be used). It is commonly measured after the specimen has failed (or at *half* the fracture strain, or after a certain strain, such as 20% [ASTM E517]). It is therefore instructive first to inquire into the strain dependence of R [WELCH &AL. 1983]. Figure 4a shows three such tests, in a mildly rolled copper (texture: Fig. 4b); the values at apparently zero strain come from tensile tests. It is seen that, at least in the 45° test, there is a significant strain dependence, even at very small strains. Thus, even when failure occurs at strains considerably less than 1%, as it may well in cold-rolled sheet, the conventionally measured value of R could depend on exactly when the failure occurred.

In such situations, R is not a well-chosen parameter. A quantity of more physical meaning, in any case, is the ratio of the respective components of the strain *rate*, **D**, in any given state:

$$r = D_w/D_t \tag{5}$$

(The symbol r is commonly used for this quantity, but also sometimes in place of R.) This parameter still inherits another unfortunate property from R: its range of values goes from 0 (no widening) to infinity (no thinning). For this reason, the 'contraction ratio'

$$q = -D_w/D_l \tag{6}$$

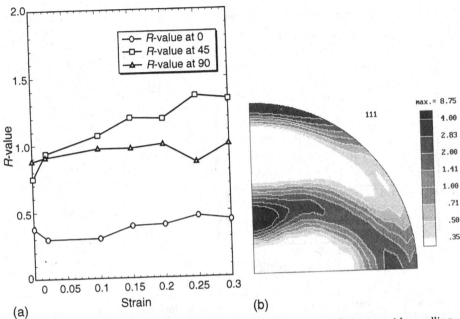

Fig. 4. (a) Results of compression experiments on a cube of copper with a rolling texture. The value of the Lankford coefficient, R, particularly for the 45° test, is a function of the prestrain. (b) The 111 pole figure (recalculated) for the copper specimen used to perform the R-value compression experiments shown in (a). This is a classic rolling texture of copper. The rolling direction is up; equal-area mapping.

has been introduced (with D_l the rate of length change) [HOSFORD & BACKOFEN 1964]; it is more like a Poisson's ratio in elasticity (though anisotropic). Its range is between 0 and 1. It is easily seen that $q=r/(1+r)$, and the *other* 'Poisson's ratio' (negative thickness strain rate over length strain rate) is $1-q$, at constant volume. Both parameters are 0.5 for isotropy.

These parameters are used in applications in two ways: to characterize the resistance against thinning (and thus perhaps obtain an indication of formability); and to provide a measure of in-plane anisotropy (leading to 'earing' in a cup drawing operation) after various thermo-mechanical rolling process histories. The former is often quantified by an 'average R', $\bar{R} = (R_0+R_{90}+2R_{45})/4$, the latter by a parameter $\Delta R = (R_0+R_{90}-2R_{45})/2$. These conventional definitions, apart from being based on the integral R-values, have no basis in any sensible averaging technique. It is more appropriate to calculate a *differential* $\bar{r}$ from the average of q over all measured (or calculated) angles, according to

$$\bar{r}_{\text{eff}} = \bar{q}/(1-\bar{q}) \tag{7}$$

The average of q is taken, for all angles, at the same tensile strain.

An $\bar{r}$ greater than one, or a $\bar{q}$ greater than 1/2, means that the sheet material has undergone less through-thickness than in-plane deformation. This is often taken to mean that a sheet of material will have good formability. High-quality steels manufactured for sheet forming applications have an R-value of about 2.0. Such high R-

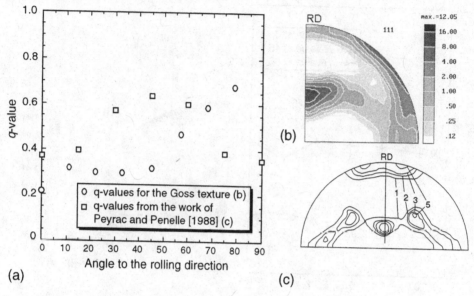

(c)

Fig. 5. (a) Measured values of the contraction ratio q for two aluminum sheets with very different textures:; (b) a texture with a strong Goss component, obtained by the authors; and (c) a near-classical rolling texture from work by PEYRAC & PENELLE [1988].

values are also not atypical of sheet materials with a hexagonal crystal structure. One finds that the average R-values characterizing face-centered cubic metals, e.g. aluminum sheet, are rarely greater than one and often substantially less.

However, in-plane anisotropy can also be important for formability. Figure 5 shows results from two investigations on aluminum sheet that had different textures. One has a typical cold-rolling texture [PEYRAC & PENELLE 1988], the other has a high 'Goss' component[3]. It is evident that the in-plane anisotropy is completely different – while the *average q* is near 0.5 (the average R near 1) in both cases. Textures with a strong Goss component have been found to be especially detrimental to formability, presumably because of the large in-plane anisotropy [BRYANT &AL. 1994]. It is also evident from this graph that the conventional way of characterizing in-plane anisotropy by ΔR would not capture the essence in this case. A simple measure would be

$$\Delta q = q_{max} - q_{min} \tag{8}$$

We have used R (or q) as a parameter to validate the predictions of different crystal-plasticity theories and as a measure of the in-plane anisotropy present in sheet material processed by different thermo-mechanical histories. We found that the requirements on data acquisition for these purposes are more sophisticated than those typically required for sheet forming applications. In the latter case the strain ratios are often taken from micro-meter measurements and scribe lines, after the tensile test is over and the specimen has failed, or at best by extensometer measurements. Only one value for R is quoted for each angle: there is often not a continuous record of deformation throughout the experiment.

---

[3] The 'aluminum' used for these experiments was an ALCOA 99.99% base alloy with 0.17% Fe and 0.8% Si (weight fractions); this is equivalent to a controlled form of 1100 Al. It was rolled to 73% reduction and annealed at 300°C for 1 hour, which produced a 20% recovery in yield stress but no texture change.

In our experiments we have used strain gauges to measure both the axial and the transverse strain throughout the tensile test. This form of instrumentation gives a continuous record of deformation and sufficient resolution to determine elastic as well as plastic strain as a function of tensile load. Strain gauges are limited to small deformations, about 5% plastic strain, because of failure of the adhesive bonding of the gauge to the specimen, which in some instances might be a problem. However, for experiments on rolled sheet the specimens invariably necked and failed before the gauges separated from the tensile sample. One must be aware that when using strain gauges in biaxial strain states the transverse sensitivity of the gauge must be accounted for.[4] For a two-gauge 90° rosette, not necessarily aligned with the axes of principal strain, the corrected strains are obtained from a simple relation:

$$\varepsilon_{11} = \frac{(1 - \nu K_t)(\tilde{\varepsilon}_{11} - K_t \tilde{\varepsilon}_{22})}{1 - K_t^2} \tag{9}$$

and

$$\varepsilon_{22} = \frac{(1 - \nu K_t)(\tilde{\varepsilon}_{22} - K_t \tilde{\varepsilon}_{11})}{1 - K_t^2} \tag{9'}$$

where, $\tilde{\varepsilon}_{11}$ and $\tilde{\varepsilon}_{22}$ are the uncorrected strains in the 1 and 2 directions, ν is Poisson's ratio (assuming elasticity to be isotropic), and K_t is the transverse sensitivity for the rosette. For the case of a 'delta' rosette, the equations are much more complicated.

The measurements of these strain ratios are surface measurements that represent a material response averaged across the entire specimen thickness. One must take care that the properties of this section are uniform. Rolling will introduce redundant shear deformations on the surfaces of the rolled sheet or plate that alter both the deformation texture and mechanical response. One can remove the surface layers of rolled material to achieve uniformity. It is also important to conduct the R-value measurement with a tensile specimen that has a sufficiently high length-to-width ratio to produce uniform tensile stress and strain fields. We have determined, based on independent strain-gauge measurements in the tensile mid-section and at the end of the gauge section, that an aspect ratio of 8:1 is adequate for this purpose. If the specimen aspect ratio is less than 8:1 there are also strain gradients across the specimen width that should be accounted for.

1.5 Yield-surface measurement

The most complete information about the plastic anisotropy of a material in a given state, under arbitrary loading or straining, is contained in the yield surface and plastic potential. As discussed in Chap. 8, these two coincide under most conditions for metals. The yield surface (or 'yield criterion') describes the dependence of the current yield stress on the stressing 'direction' (in 5-D stress deviator space); the plastic potential describes the straining direction (in 5-D plastic-strain-rate space) when the yield surface is reached at any particular point. This straining direction is represented by the normal to the yield surface (or, more generally, the plastic potential) at that point. For the purposes of this Section, we will assume the yield surface and plastic potential to be coincident.

The measurement of strain-rate ratios observed during reloading of a pre-processed material, described in the last Section, is thus tantamount to a partial description of the plastic potential. The present Section is devoted to a review of other effective methods of determining parts of the yield-surface information. It appears

---

[4] "Errors Due to Transverse Sensitivity in Strain Gauges". *Measurements Group Tech. Note, TN-509*, Measurements Group Inc., Raleigh, NC (1982).

that a complete determination of the yield surface is virtually impossible; partial information of any reliability is hard enough to obtain. It is for this reason that physically and phenomenologically meaningful polycrystal-plasticity *models* are so important. The primary use of the measurements is then to verify – or improve the reliability of – theoretical predictions of the yield surface, and the material's actual current texture.

Of particular interest is the potential presence of vertices on the yield surface – as they exist in single crystals (so long as they are considered rate independent) – or at least of regions of sharp curvature, as they are present in rate-dependent single crystals and, as we shall see, in some strongly textured polycrystals. Here, there exists the potential for large variations in straining direction under virtually constant stress, which may lead to a constitutive plastic instability. Similarly, the presence of facets or regions of very low curvature – a necessary adjunct of (quasi-)vertices – is of interest: the stress response to a prescribed strain rate may be subject to large uncertainties.

A variety of experimental techniques will be briefly reviewed below: testing of cubes in compression and plane-strain compression; testing of strips of various widths in tension or punching; hydraulic bulge testing; and torsion of long tubes. First, we will use the last of these techniques to illustrate the importance of an agreed-upon definition of 'yield'. In Chap. 8, it was pointed out that the elasto-plastic transition region is subject to many variations of material, state, testing technique, and so forth, and that the large-scale definition of yield (corresponding to dislocation percolation) is the most appropriate for our purposes [KOCKS 1987]. (See also SHIRATORI &AL.[1979], HECKER [1976], and MICHNO & FINDLEY [1976].) In uniaxial tests, it can often be easily obtained by back-extrapolation of the stress/strain curve or by an offset of 0.2% plastic strain (Chap. 8 Fig.12a). Yield-surface determination, however, amounts to a reloading test *with a change in path*. This may invoke special problems [LI & BATE 1991, BATE 1993]. For example, it is known that dislocation structures are sometimes unstable under such conditions, which gives rise to an initial low (or even negative) strain-hardening rate [KOCKS 1964b, JACKSON & BASINSKI 1967]. On the other hand, *reversal* of the loading direction tends to give an initial *hardening* transient, from a low starting stress.

Figure 6a shows two 'yield surfaces' of a pre-twisted specimen, at two different offset strains [HELLING &AL. 1986, STOUT &AL. 1985]. In Fig. 6b, it is compared to stress/strain curves in a Bauschinger test on two different materials, both Al-Cu alloys. The solution-treated material has a short transient upon reloading in compression; the one that contains second-phase particles exhibits a 'permanent offset'. It is concluded that the 'translated' nature of the low-offset yield surface in Fig. 6a is a transient phenomenon only, related to transient nature of the Bauschinger effect. A large-offset or back-extrapolated definition of yield is preferable also for a discussion of yield surfaces, as it was for flow stresses (Chap. 8 Fig. 12). Only in materials with macroscopic residual stresses or other internal stresses due to second-phase particles, in which the elasto-plastic transition is much longer or even 'permanent', is there any merit to the notion of 'kinematic hardening' and to an offset definition of yield [HELLING &AL. 1986].

The above experiments were undertaken on combined loading of long tubes. Other techniques are based on testing of sheet or 'forging' of cubes. All of these will now be reviewed in turn.

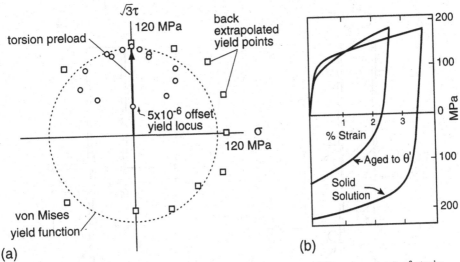

Fig. 6. (a) Two experimental yield surfaces, defined by a different amount of strain offset, from the same aluminum tube that was pre-twisted. The surfaces have vastly different shapes depending on the yield definition. (b) Stress/strain curves measured in a Bauschinger test on an Al-Cu alloy (after MOAN & EMBURY [1979]).

Torsion of long tubes

A long thin-walled tube geometry is ideally suited for conducting stress-path-change experiments. This geometry of tube can be loaded using combinations of tension, internal pressure, and torsion. The disadvantage of the tube geometry is that it is most difficult to modify the initial material state. One can alter rolling and heat treatment schedules for example when working with sheet. However, with tubes one is limited to what is delivered or can be achieved by annealing; the possibilities of pre-deforming tubes to large strains are severely limited, since a tube is much less geometrically stable than sheet. For example, in balanced-biaxial tension the tube will experience a necking instability at about a third of the strain ($\varepsilon_1 = \varepsilon_2 = 0.2$ to 0.3) that would be expected from sheet [STOUT & HECKER 1983]. We have tried wrapping sheet into tubes for testing, but with limited success.

Torsion of a thin-walled cylinder does not result in a necking instability but rather a buckling instability. One can use either a solid or undercut lubricated internal mandrel to prevent the buckling instability. Examples of a thin-walled tube geometry with both types of mandrels are illustrated in Figs. 7a and 7b respectively. In both cases the interior wall of the tube must be true and honed to a precise tolerance with the mandrel. In experiments with a solid mandrel, the friction between the specimen and mandrel is unknown, making the determination of shear flow stress an upper bound. However, this experiment can be used for preparing tubes to be subsequently tested using combinations of tension and internal pressure, i.e., in plastic-strain path-change experiments. The advantage of the undercut mandrel is that friction will not significantly influence the determination of the shear stress. In addition, the undercut mandrel acts as a subpress to maintain alignment during torsion. This postpones the initiation of buckling but will not stop its spread once it has begun. The undercut mandrel has been employed for the determination of yield loci in combinations of tension and torsion. We have been able to reproducibly obtain shear strains as high as 0.55 with this pure torsion technique. After attaining these levels of shear, we were subsequently able to perform the full tension/torsion yield locus evaluation.

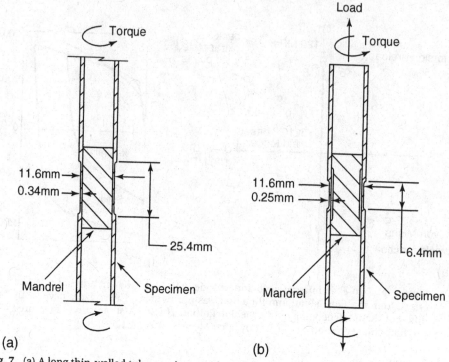

Fig. 7. (a) A long thin-walled tube specimen with an internal mandrel used to postpone a buckling instability. (b) A long thin-walled tube geometry appropriate for yield-surface measurements. The internal mandrel is undercut to eliminate effects from friction of the test section.

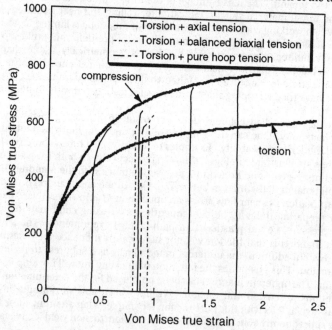

Fig. 8. Results of stress path-change experiments on brass tubes. The tubes were first deformed in pure torsion to different shear strains; subsequent deformation was in uniaxial and balanced-biaxial tension.

Figure 8 shows the results of uniaxial and balanced-biaxial tension experiments on brass tubes after they had been deformed in torsion to von Mises equivalent strains of $\varepsilon_{vM} = 0.5, 0.9, 1.3,$ and 1.75. These tubes had been annealed prior to the torsion prestrain and were very long (100mm). Thus, the principal stress experiments after shear were truly uniaxial and biaxial, respectively. The path-change experiments are compared with the stress/strain curves from continuous torsion and compression test, using a von Mises definition of effective stress and strain. Clearly, as the ultimate (uniaxial or biaxial) strength of the pre-deformed tubes exceeds the torsional flow stress, the von Mises criterion is not an appropriate definition of material state.

Tests using tubes have also been done at different strain rates and temperatures. An extensive series of creep experiments on tubes of 304 stainless steel [OHASHI &AL. 1982] and 316 stainless steel [OHASHI &AL. 1986] are among the most notable. These experiments examined the effects of multiple cycles in tension/torsion stress space on material behavior, and were conducted at a temperature of 650°C.

Biaxial stretching of sheet
Strip specimens. Tensile specimens, cut from sheet, with a gauge width greater than the gauge length will develop a region of plane strain in the center of the gauge section. WAGONER & WANG [1979] have investigated a number of such geometries with the intent of establishing a standard that gives as large a region as possible of uniform strain state. One such design that they investigated is shown in Fig. 9a.

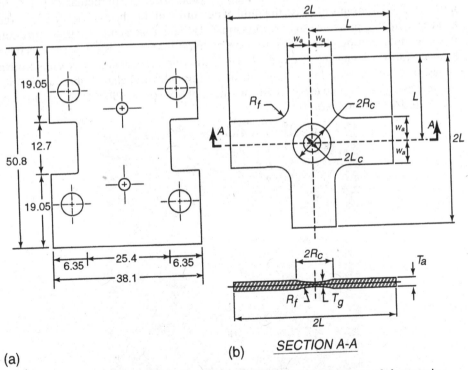

(a)   (b)   **SECTION A-A**

Fig. 9. (a) A schematic of a tension specimen designed to obtain a state of plane strain. The holes in the grip section are for locator pins and clamping bolts. The design is one originally proposed by WAGONER & WANG [1979]. (b) A cruciform specimen geometry with which one can obtain biaxial stress states. This particular specimen geometry has been proposed by MAKINDE &AL. [1992].

Cruciform specimens. Another approach is to use a specimen in the shape of a cross and apply biaxial tension; a recent, comprehensive design of experiments of this nature is due to MAKINDE &AL.[1992]. A typical specimen proposed by these authors is shown in Fig. 9b. A variety of stress states is possible with such a specimen, by changing the ratio of applied tensile stresses. As in the case of the plane-strain tensile specimen, there are appreciable strain gradients that must be minimized and accounted for. In addition, a very specialized loading device is necessary. To properly design these specimens and to know how to convert applied load to stress in the gauge section, an elastic-plastic finite-element analysis is required. This analysis pre-supposes a given yield function, the very property one is attempting to evaluate. Another limitation of these experiments is that it is difficult to achieve large deformations during the experiment. Strains tend to concentrate at the grips or in specimen corners as a result of designing for a uniform strain in the gauge section. These concentrations in turn lead to a premature specimen failure, particularly for cold-rolled material.

Hydraulic bulge tests are another means of measuring the properties of sheet in biaxial tension. Figure 10 shows a photograph of such a specimen from the 'inside;' this particular sheet was square gridded and subsequently bulged to failure. Most of these testing machines are designed for circular bulging producing an equi-biaxial stress state. However, one can use elliptic dies rather than the circular die to produce stress states ranging from near plane strain to equi-biaxial tension. An advantage of bulge tests is that one can deform sheet to relatively large strains without experiencing a necking instability, particularly in biaxial tension. On the other hand, it is very hard to make accurate measurements of material yield. There is particular difficulty in measuring small strains because the radii of curvature of the sheet at yield are so large and the hydraulic pressures so low. One must also note that a constant pressurization rate does not correspond to a constant true strain rate. Many classic experiments have been performed using the hydraulic-bulge technique, both to measure constitutive behavior and to investigate the limits of sheet formability in balanced-biaxial tension. Notable examples of such experiments are those of MELLOR [1956], DUNCAN & JOHNSON [1968], and WOODTHORPE & PEARCE [1970].

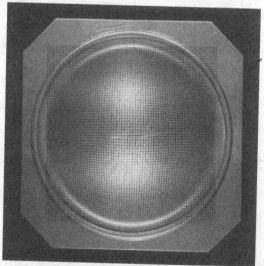

Fig. 10. The results of a balanced-biaxial tension experiment on a gridded sheet using the hydroforming technique. This particular sheet was tested to failure.

A variation of the hydraulic bulge technique has also been used to study plane-strain constitutive behavior of sheet [AZRIN & BACKOFEN 1970; GHOSH & BACKOFEN 1973]. They used a solid hemispherical punch to deform a sheet of material. The center of the tested sheet had a groove (section of reduced thickness) that was machined with an end mill. A polyethylene spacer prevented the punch from making contact with the groove as the sheet was being stretched. Thus this section of the sheet with the groove remains planar and deforms in plane strain. This technique was used for both monotonic stressing and also as a preloading technique for stress path-change experiments. In the latter case, tensile specimens were cut from the sheet, such that the gauge section of the tensile specimens contained the pre-stressed groove.

'Forging' of cubes.
Figure 11 shows how the plane-strain compression test seems ideally suited for making many measurements by turning the sample [FRANCIOSI &AL. 1987]. In addition, free compression gives the three yield stresses along the negative axes. In practice, these are very difficult tests to perform and to evaluate. FRANCIOSI & STOUT [1988] reported on some experiments in aluminum and copper, in which the strain dependence was emphasized. CHAKI & LI [1986] did one of these tests (and a reversal) on copper. The only general conclusion we would report is that the flow stress in the new direction (if not the *reverse* of the old) was always larger, by an amount of the order of 10%. CHAKI & LI [1986] interpreted this as a 'latent hardening' effect, similar to that in single crystals. This would be a possibility especially if their specimen had indeed been 'isotropic' as they interpreted from the equality of three stress/strain curves. However, even weak textures, such as those measured by FRANCIOSI &AL.[1987], could give the effect, as it is indeed interpreted by these authors. SCHMITT [1986] has generalized the concept of straining path changes, again without consideration of texture. (See AERNOUDT &AL.[1987].)

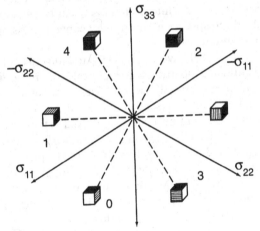

Fig. 11. A schematic showing in the π–plane how different stress states (with respect to the material symmetry axes) can be obtained using a channel die. The test labeled '0' shows the original compression face black, the original transverse face containing lines in the free direction. These shadings are maintained on the specimen to show its subsequent rotations in the channel die [FRANCIOSI &AL. 1987].

1.6 Continuously changing paths

The measurement of a yield surface in a given state consists of (at least partial) unloading and reloading; it is a series of *probes* using different straining paths. A different kind of test is one in which the prescribed straining direction, or the prescribed stress path, is changed *continuously*. This ascertains the constitutive

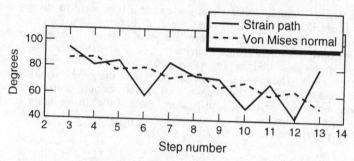

Fig. 12. Experimental results of HECKER [1972] from tubes loaded along prescribed stress paths. The strains that resulted from stress path changes were measured 'on-the-fly'. The fact that the measured paths have a much greater variation in straining direction than what would be expected under a von Mises assumption indicates the presence of a corner in the yield surface.

relations 'on the fly', and especially the degree of *stability* of a plastic state. This is well illustrated by a classic result of HECKER [1972]. He imposed biaxial-stress loading conditions, in different stressing directions, on a tube and measured the resulting hoop and axial strains. One of Hecker's results is plotted in Fig. 12. It shows that the variation in loading direction can produce a change in straining direction that is much larger than what one would expect based on the assumption of a von Mises yield surface shape: a clear indication of a 'corner' in the yield surface; or, more exactly, a region of high curvature. Hecker interpreted his result to be associated with the fact that he did not unload the specimen, but rather changed stressing direction with the specimen under a continuous load.

ASARO & NEEDLEMAN [1985] have also investigated the effect of corners in the yield surface on the constitutive properties of polycrystals (their Fig. 26). They simulated deformation of a random polycrystal to a von Mises equivalent strain of 0.23 in plane-strain tension, using their crystal-plasticity model, and then changed the deformation-gradient rate to combinations of pure shear and axial extension, without unloading. This amounts to a 'tangential' change along the yield surface. When the yield surface is smooth, as in strain-rate dependent solids, the initial response must be elastic. This was found to be the case at $\sigma_{12} = 0$. However, at very small additional shear stress the predicted slope of the stress/strain curve drops dramatically. This effect is the strongest for conditions near a proportional path; i.e., loading paths closest to the original plane-strain tension. Thus, even for smoothed-out yield surfaces, the behavior is the same as the characteristic corner behavior for rate-independent solids [STÖREN & RICE 1975].

If there were an actual discontinuity in the slope of the plastic potential, a given stress state could elicit a spectrum of different straining responses, and more than one of these may be compatible with the boundary conditions [STÖREN & RICE 1975]. Essentially the same is true when the slope is continuous – there is merely a sharp curvature. Spontaneous instabilities are of great importance for formability. In sheets, localized necking is always associated with a change in path to plane strain. In some theoretical interpretations [MARCINIAK & KUCZYŃSKI 1967], the path change is assumed to be caused by a convenient pre-existing groove (of a size rare in practice); in the above model, it may be *spontaneous*: due to the 'constitutive instability' at a texture-caused yield corner [STÖREN & RICE 1975]. (See Sec. 2.6.)

2. Yield Surface Shapes: Predictions and Experimental Results

In this Section, it will be assumed that the 'state' of the material is known; the problem at hand is the prediction and verification of its plastic properties in the given state. The state is characterized by the texture and the grain shape (described by the grain aspect ratios and 'principal axes', both assumed uniform throughout the sample). Strain-hardening properties do not enter at this stage; they will be discussed in Section 3. They will generally lead to a development of differences in flow stress from grain to grain which, if they are correlated with the orientations, could affect the anisotropies to be discussed; such effects will be ignored. One reason for this choice is that, although the individual grain flow stresses could be predicted as further state parameters from a simulation beginning at the annealed state, they cannot be measured in a deformed material in any reliable way.

The plastic properties at a given state are characterized by the yield surface and plastic potential. The concept of a yield surface was introduced in Chap. 8. It was pointed out that the phenomenon of 'yield' must be quantified in two ways: the 'strength' level – a scalar that increases as the material hardens, through dislocation accumulation, for example; and a 'surface' whose shape in stress space specifies the dependence of yield on the tensorial nature of the stress tensor, i.e. the 'direction' of the five-dimensional deviatoric stress vector. The yield surface (or an equivalent algebraic criterion) may always be scaled by the 'strength'; typically, the yield stress in tension in a well-defined '3'-direction, σ_f. Thus, we will use the term 'the yield surface' for the *scaled* version: to characterize its *shape*. It depends both on anisotropy (such as a difference in yield stress for tension in the 1- and 2-directions) and on the tensor character (such as the distinction between uniaxial and multiaxial stresses, or just between tension and shear).

As was also discussed in Chap. 8 (Sec. 1.2), the 'plastic potential' usually coincides with the yield surface (as scaled by the strength parameter). It is defined such that the *normal* at any point is coincident with the 'straining direction' $\hat{D}$. (This requires that the surface is a locus of constant work rate, Chap. 8). For the purposes of this Chapter, we treat 'yield surface' and 'plastic potential' as synonymous; equivalently, the 'normality rule' is presumed to hold.

Yield surfaces are displayed as two-dimensional *sections* or *projections*. The latter is easier to obtain and preserves *normality* of the straining direction in the 2-D sub-space: there is no component 'out-of-plane' (though there may be a *stress* component out-of-plane). A *section* requires iterations to eliminate these stress components (Chap. 9 Sec. 1.2, VAN HOUTTE [1987]); but there may be out-of-plane *strain*-rate components. Most plots that will be shown are 'contained': the sections and projections coincide [CANOVA &AL.1985].

2.1 Yield-surface symmetry

The yield surface is subject to symmetry constraints [CANOVA &AL.1985, VAN HOUTTE 1987]. First, there is the symmetry of the stress and strain-rate tensors, which reduces the total space for second-rank tensor properties from nine to six. Second, we assume constant volume throughout, as well as independence from hydrostatic stress, so that only five dimensions are necessary. Third, we will assume in this Chapter, unless otherwise stated, that straining or stressing in the 'forward' or 'backward' directions causes a change in sign of the response and nothing else; then, the yield surface has a center of symmetry in stress space. Finally, the presence of symmetry elements in the sample restricts the regions of the space that need be mapped and, in some cases, causes some subspaces to be 'contained' (so that a stress component in one does not cause a strain-rate component in the other). For example, Fig. 6 of Chap. 8 showed that, for an orthotropic material, half the π-plane and the positive octant of the 3-D shear space suffice, and that they are uncoupled.

2.2 Isotropy

In an isotropic material, tensile yield must be the same for all material directions. In addition, the yield stress in compression equals that in tension, under the assumption of sign independence. However, yield in *shear* is not simply related to yield under uniaxial conditions. This is shown in Figure 13a, where the circle is merely the result of an *assumption* (that of VON MISES [1913]). The data points show a simulation of a random polycrystal (first undertaken by BISHOP AND HILL [1951b]). Finally, the inscribed hexagon (the assumption of TRESCA [1864]) provides a lower bound (for convex yield surfaces with given tensile yield stress).

The particular way of plotting the yield surface used in Fig. 13a is called the 'π-plane'; we will use this terminology *without* implying that it is the space of deviatoric principal axes – merely that of diagonal components. Then, Fig. 13a must be supplemented by other yield-surface sections (or projections), for the off-diagonal components. Here, the situation is different: isotropy demands, without further assumptions, that all the shear stresses – of the form σ_{12} – must be the same, and that they also be the same as the shear stresses of the form $(\sigma_{22}-\sigma_{11})/2$, which occur in Fig. 13a. Figure 13b shows this on a scale linked to that of Fig. 13a: the yield surfaces are all circles; but their radius is different depending on the assumption made for the π-plane. (The scale factor is σ_f, the tensile yield stress, in Fig. 13a; it is $\sigma_f/\sqrt{3}$ in Fig. 13b.)

The straining direction is, then, even in isotropic materials, generally parallel to the stressing direction only in the shear subspaces; under *arbitrary* stress states, including uniaxial ones, it holds *only* when the von Mises assumption is applicable in addition to the assumption of isotropy.[5] One of the primary consequences of anisotropy is that the principal axes of stress and of strain rate do not, in general, coincide. It is therefore necessary to choose as a coordinate system for plotting the yield surface and plastic potential one that is attached to the material; e.g. the axes of orthotropy in a rolled sheet. The 'π-plane' is now best chosen in coordinates that coincide with axes of symmetry of the texture; this will be done in Fig. 15. (See Chap. 8 Sec. 1.4.)

Another plot frequently used is shown in Fig. 13c: here, only *two* normal stresses are used, in a cartesian coordinate system; this is useful when the third normal stress (Cauchy stress, not deviatoric stress) is zero. Again, the 'von Mises circle' (an ellipse in this space), the Tresca hexagon, and the contour derived for an isotropic polycrystal deforming on {111}⟨110⟩ slip systems are shown. However note that this figure represents a yield-surface *section* only, and not a *projection*: the normal at one of the pure tensile stresses, for example, does not indicate that there is also a straining

---

[5] When the 'normality rule' (i.e., the coincidence of yield surface and plastic potential) is not taken for granted, the combination of a circular yield surface with a parallelism of stress and strain-rate is called the 'Lévy–Mises' assumption [HILL 1950]. In this Section, we use the term 'yield surface' to encompass that of 'plastic potential'; thus, in the figures, it is arbitrary whether a series of *stress* points is plotted or a series of *facets* that envelop the yield surface. (See Chap. 8 Sec. 1.2.)

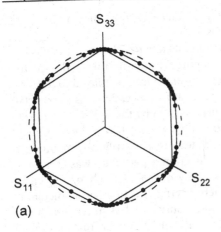

(a)

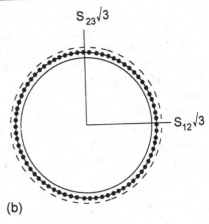

(b)

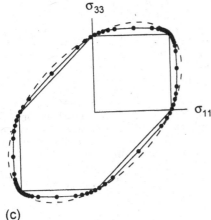

(c)

Fig. 13. Yield-surface sections for an isotropic material: (a) in the 'π-plane' consisting of the diagonal deviatoric stresses (whether principal or not); (b) in a subspace spanned by any two shears; and (c) in a space of two normal stresses, assuming the third one to be zero.

component perpendicular to the plane of the plot. (In the π-plane, which is both a section and a projection, in orthotropic materials, one can derive *all* the straining components by *projecting* the normal on the three axes, using a technique similar to that derived for hexagonal crystals in Chap. 2 Fig. 2.)

2.3 Simulation of plane-strain compression

Strain dependence

Since we have seen, in Sec. 1, that experimental determinations of yield surfaces are very difficult, and cover at best small regions of stress space, we will now first discuss the results of *simulations* for a variety of cases. For this we choose plane-strain compression (i.e., idealized rolling) in face-centered cubic (fcc) metals, deforming by {111}⟨110⟩ slip only. The simulations are undertaken with the Los Alamos polycrystal plasticity (LApp) code, on the basis of the Taylor model with two essential modifications (Chap. 9): a relaxation of constraints for flat grains (in some monotonic function of their aspect ratios); and a finite rate dependence of flow. For 'rate-insensitive' materials (the norm), we use a creep stress exponent of $n=33$ and plot the plastic potential as the inner envelope of facets; as an extreme for rate sensitive materials, we use a stress exponent of $n=5$ and plot the stresses (as little arrowheads). We quote the 'von Mises strain' (see Chap. 5 Sec. 1.2) of 1.0, 2.0, and 3.0, instead of the rolling reductions of 58, 83 and 93%, respectively. In all cases to be shown, the 'yield surfaces' are in fact constant-work-rate potentials.

Even if a material were initially isotropic (a difficult state to achieve in practice!) it would, in general, become anisotropic with straining. Primarily this is due to the development of a texture; but even a change in grain shape from equiaxed to flat, say, can cause macroscopic anisotropies. A specific mechanism by which this is expected to occur is the change in boundary conditions experienced by the individual grain to 'relaxed constraints', which requires fewer deformation modes (Chap. 8). Figure 14 shows the result of a simulation in which a random texture was artificially maintained and only the grain shape effect accounted for. The grain aspect ratio was chosen as 8:1:1/8, as it would be after a reduction in rolling by 87.5%. The change in yield-surface shape (Fig. 14) is perceptible near S_{33} (where 3 is the direction perpendicular to the flat face); it appears as a reduction in the radius of curvature.[6] A similar result was obtained by VAN HOUTTE [1987]. We will see immediately, however, that the major effect of grain shape comes through its influence on texture development.

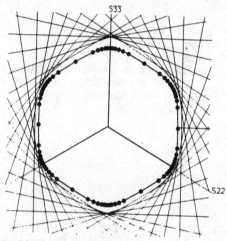

Fig. 14. The effect of grain shape on the yield surface of a polycrystal with random orientations. The Bishop-Hill result under full constraints is drawn in as a solid line.

Figure 15 demonstrates the effect of strain on yield-surface shape in a simulation of plane-strain compression (on the 3-plane in the 1-direction). The yield surface of a rate-insensitive material is shown as the inner envelope of tangent planes. The part-icular model chosen to phase in the effects of grain shape is ineffective at a von Mises strain of 1.0, 60% effective at a strain of 2.0, and 85% effective at a strain of 3.0 (Chap. 9 Sec. 2.1). The qualitative change in texture (top row), apart from its general sharpen-ing, is due to this effect. The second row exhibits π-plane yield surfaces for the tex-tures in the top row, but without accounting for grain-shape effects in the calculation of the current yield surface itself; this would be appropriate, for example, if the grains had 'broken up' during deformation or if partial recrystallization had led to a return to equiaxed grain shape without change in texture. The third row in Fig. 15, on the other hand, assumes that the grain shape achieved for geometric reasons during the rolling history actually influences deformation in the current yield-surface probes also. Note that the shapes are quite different depending on the detailed assumptions made. This is particularly evident at the intersection of the yield surface with the axis S_{22}.

---

[6] Note that, although the grain shape was presumably obtained by compression, once it is established, it will affect tension and compression in the 3-direction equally.

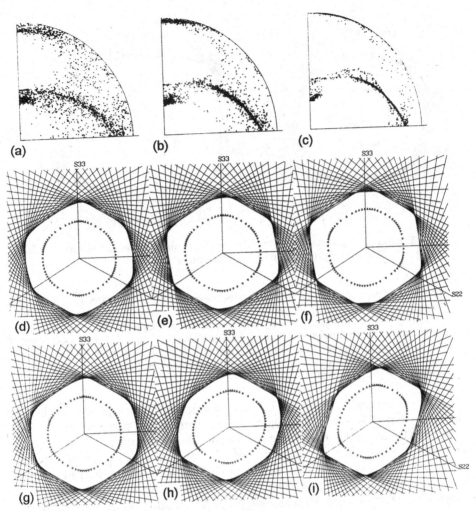

Fig. 15. Simulated rolling textures (using the RC-model) after a von Mises strain of (a) 1.0, (b) 2.0, and (c) 3.0. The corresponding predicted yield surfaces are shown underneath. In (d) through (f), the FC model is used for the property simulation; in (g) through (i), the effect of RC is also accounted for in the yield-surface calculation. The yield surface envelopes refer to a rate sensitivity of 1/33, the inner points to 1/3.

Yield surface shapes

As another example, Fig. 16 shows yield surfaces equivalent to that in Fig. 15f, at a strain of 3.0, but now in different projections of the yield surface – all shears. Remember that all shear-only yield-surface projections would have to be circular for isotropy (Fig. 13b): the deviations are substantial.

The principal point to be made from this exposition of simulated yield surface shapes is that these shapes, for rate-insensitive materials, are quite varied: they depend on the strain (through the texture and grain shape evolutions) and on the operative mechanisms. Therefore, it does not seem likely that one will ever find an algebraic form of some generality to describe them all: simulation is of the essence.

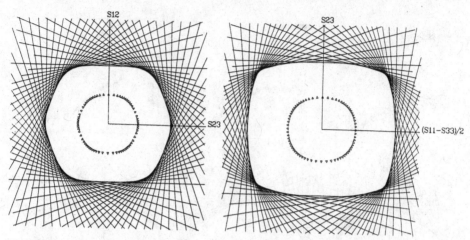

Fig. 16. Yield-surface sections in two different shear subspaces for the texture shown in Fig. 15c. The property simulation includes RC conditions.

The phenomenological forms that have been proposed [HILL 1948,1979; HOSFORD 1979b, 1993] can at best capture the cases of very high rate sensitivity: the smooth yield surfaces shown by triangles in Fig. 15. A semi-phenomenological approach using texture input for fcc metals has been proposed by BARLAT & RICHMOND [1987], and by LEQUEU &AL. [1987b]: the so-called Continuum Mechanics of Textured Polycrystals (CMTP). Again, it may capture the anisotropy of *stresses* well enough in many cases, but the straining directions remain poorly represented.

A very important characteristic of the yield surfaces for stronger textures in rate-insensitive materials is that they appear to have *vertices*. To some extent, these are artificial. The finite rate dependence in even the case $n=33$ implies that the apparent vertices are actually not *discontinuities* of slope, but merely regions of very high curvature.[7] This still means that very small changes in stress can lead to very substantial changes in straining direction. Conversely, there are other regions of very *low* curvature ('flat spots'); here, the common problem of finding a stress for a given strain rate becomes almost indeterminate. In reality, of course, various causes such as statistical variations will lead to a unique solution in any case; but such regions do provide a warning flag – not only for problems to be expected in simulations, but also for a sensitivity to details in practical application.

It is worth emphasizing that all the yield surfaces shown in this Section represent yield stress probes in stress space at a certain *state* of the material, as characterized by the current texture and grain shape. This state was produced by monotonic deformation in plane strain but it could, in principle, have been obtained in other ways. On the other hand, there can be no expectation that arbitrary histories would produce yield surface shapes that are in any way similar. We will see some examples for torsion in Sec. 2.7, where even the symmetry is different. Even deformation paths that retain the orthotropic symmetry of rolling, such as cross-rolling in an arbitrary sequence, will develop textures that are different from case to case, and thus have distinct yield surfaces.

---

[7] HOSFORD [1979a] has pointed out that differential work hardening in different grains also leads to a rounding of the polycrystal yield surface.

HOSFORD [1979a] and HILL & HUTCHINSON [1992] have proposed a phenomenological model in which yield under different paths in stress space is compared at constant total work done. Inasmuch as the plastic work is not a state parameter, such an evaluation does not correspond to what is called a 'yield surface' or 'flow potential' here. It was used in HILL &AL.[1994]. (Note that a constant current work *rate* (or increment) for all the probes along different paths is used by all for the definition of the plastic potential.)

2.4 Rolled cubic metals: yield-surface measurements and interpretation

In this Section, we will present some efforts aimed at determining a yield surface experimentally, by measurement of both stresses and straining directions for different prescribed straining paths; the next Section will concentrate on strain anisotropies in cases where some surface displacements are free. We will take the position that a given texture has been *measured* (and perhaps the grain shape, too) and we now attempt to predict – and verify – some aspects of the current yield surface and plastic potential.

The very essence of the definition of the yield surface as a locus of probes in stress space *at a given state* presages the problem of measurement: all probes but one correspond to a *change in path*. One must distinguish between changes in path that occur under continuous plastic deformation, and those after at least partial unloading. Here, we will deal with the latter. (The former have been addressed in Sec. 1.6 and will again be mentioned in Sec. 2.6). Changes in path are problematical since the rate of hardening, even in uniaxial monotonic tests, tends to show an instability upon a change in path; the dislocation structure, in metastable equilibrium under a given stress (or zero stress), is known to be unstable under certain other stresses (or strain rates) [JACKSON & BASINSKI 1967].

In the current context, we are interested only in the effect of *texture* (and perhaps grain shape) on the yield surface. One should imagine a test in which the *strength* level in all grains is returned to the 'annealed' value – without a change in texture. The study of a texture developed through deformation followed by extensive recovery provides an example of such a test.[8]

In an early test, KALLEND & DAVIES [1971b] determined the 0.2%-offset tensile stress for rolled copper and brass as a function of the angle to the RD. They found the behaviors to be quite different – and well in accord with a simple Bishop–Hill prediction based on the respective measured textures.

FRANCIOSI & STOUT [1988] have examined the yield loci of 1100 aluminum and 270 nickel sheet after 83% and 60% reduction, respectively. They used a combination of experimental techniques: uniaxial tension, plane-strain tension, and wrapping the sheet into tubes for biaxial tension and near-biaxial tension. In all cases, the specimens were instrumented with strain gauges to measure the direction of plastic strain increment in addition to the yield stress. An offset definition of yield was used for these experiments, particular offsets are indicated at each particular yield point. Their results, plotted in the π-plane, are shown in Fig. 17. Both the aluminum and nickel surfaces show many of the same features. They are flattened in the direction of prior rolling and there is a region of sharpened curvature in in-plane balanced-biaxial tension (equivalent to through-thickness compression). For uniaxial tension, the trans-

---

[8] There is an additional reason for this: the strength level is not expected to evolve identically in all the grains; this can be simulated, but it is very difficult to measure.

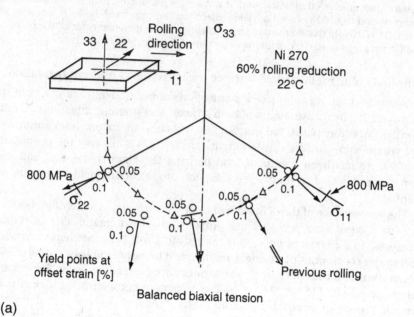

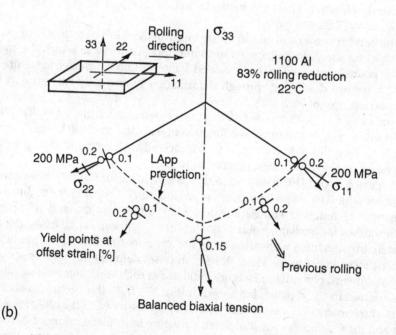

Fig. 17. Experimental results for the yield surface (determined at a variety of offset strains, listed in units of 10^{-6}) and the plastic potential (indicated by arrows for the measured straining direction: (a) Ni 270 rolled to 60% reduction; (b) 1100 Al rolled to 83% reduction [ROLLETT &AL. 1992]. The dashed lines refer to a LApp simulation.

verse direction is stronger than the rolling direction; this effect is particularly notice-able for the nickel. The overall impression, though, is the most important: there clear-ly are 'flat spots' and regions of high curvature. As one can see from the figures, the direction of the plastic strain increment vectors is crucial for an interpretation of the data.

Figure 17 also shows the results of LApp simulations based on the same texture [ROLLETT &AL. 1992]. They are in qualitative agreement everywhere; quantitatively, the 'vertex' that could be inferred from the experimental data is shifted from the predicted one, in the direction away from the stress state of the previous rolling. This observation is not unusual and will be discussed below, under *latent hardening*.

FRANCIOSI & STOUT [1988] also reported experiments on copper that had been prestrained to 7% and 50% using a channel die. After the prestrain, specimens were removed from the channel die, re-machined, and then re-tested with different orient-ations. Experiments done in this manner, together with the standard tension and compression tests, give a measurement of the yield locus following the prestrain. One disadvantage of these tests is that the direction of plastic strain increment was not measured, and neither was the lateral pressure in the channel die. However, the copper results also showed the features of a flattened region in the direction of pre-strain and indications of sharpened curvatures.

Finally, KANETAKE &AL. [1981] and TOZAWA [1978] have observed similar features on a number of materials: including 1100Al, 70:30 brass, and steel, prestrained by rolling. Their results were presented without plastic strain increment vectors and they do not give a detailed description of their experimental technique.

2.5 Rolled cubic metals: strain anisotropy

Some early comparisons of predicted and measured R-values showed good agreement [VIANA &AL. 1979, SOWERBY &AL. 1980]; since then, difficulties have been encountered in some cases. We will highlight a few examples for fcc metals using current simulations and measurements. (For a cogent discussion of bcc metals, see DANIEL & JONAS [1990].)

Simulations

Figure 18 shows two π-plane yield-surface predictions; both were obtained using the same procedures, for the same piece of material, but from slightly different tex-tures: one measured by x-ray diffraction, the other by microtexture measurements [KOCKS &AL. 1996]. (The difference was in the intensity distribution along the 'α-fiber'.) The yield surfaces look similar – until one concentrates on the slope of the normal at particular points. At σ_{33}, Fig. 18a would predict equal lateral straining in a tensile test, Fig. 18b would give more thinning in the 1-direction. Such 'ovaling' (of an initially circular cross-section) has been observed, but it has not been studied with great care. Inasmuch as it provides valuable information on the material's plastic anisotropy without an additional test, it should be routinely measured in both tension and compression tests.

Two other straining directions are highlighted in Fig. 18: for tensile tests in the 1- and 2-directions (corresponding to the previous RD and TD, respectively). Such tests are routinely used for determination of the 'R-value' (Sec. 1.4). It is seen to be quite sensitive to details of the texture (actual or perceived) that is used for the yield locus calculation. Conversely, this result demonstrates that actual R-value measurements, using a tensile test, can be quite sensitive to slight changes in texture.

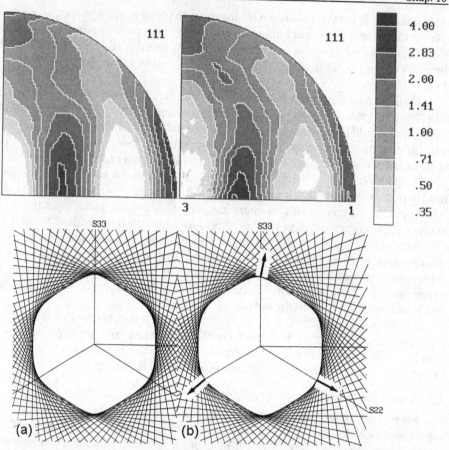

Fig. 18. Two texture measurements on the same material: (a) by x-ray diffraction; (b) by Orientation Image Microscopy™. The ensuing predicted yield surfaces differ in the sharpness of the quasi-vertex [KOCKS &AL. 1996]. The rolling direction is to the right; equal-area mapping.

Observations

In Section 1.4, it was pointed out that strain anisotropy measurements are especially difficult on cold-rolled material, in which the ductility is often very small. We will therefore start our comparison of predictions and observations on recryst-allized material. In the two cases to be discussed, the recrystallization texture is different from the deformation texture.

A texture of commercial importance is the 'cube' component, where all cubic axes are essentially aligned with the rolling coordinates. Figure 19 shows a prediction using a 10° spread around a pure cube orientation [ROLLETT &AL. 1987a]. Superposed are measurements on Cu by MOLS &AL.[1984] who obtained the anisotropy values from the specimen dimensions after 15% elongation. Predictions and observations agree in that the R-values are a little less than 1.0 in the rolling and transverse directions, and that the minimum is at 45°. The actual value at this point is a function of the severity of the cube texture. MOLS &AL.[1984], who had a severe texture, reported a value of $R=0$ at 45°; DAVIES &AL.[1975] measured a minimum R-value of 0.2, again at 45°.

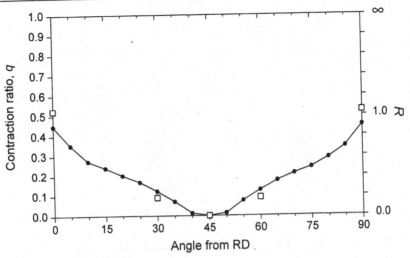

Fig. 19. *R*-values for a cube texture, predicted and measured [MOLS 1984].

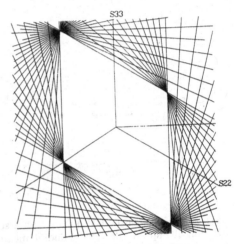

Fig. 20. Predicted π-plane yield surface for an artificial texture made up from a Goss orientation and a 10° spread. It is indistinguishable from that for a Goss single crystal. Loading in the S_{22}-direction leads to no thinning in the 33-direction: $R_{90} = \infty$ [KOCKS &AL. 1996].

Another important texture component is a cube orientation rotated 45° around the rolling direction (RD), called the 'Goss texture'. It is frequently observed as a component in aluminum after hot rolling. Figure 20 represents the predicted π-plane yield surface for a 10° spread around the pure Goss orientation. It is noteworthy that this figure looks exactly like that of a single crystal: the significant orientation spread does not influence this yield-surface section at all. It is also identical for all the symmetrically equivalent variants of this orientation. The crucial result is that tension in the 2-direction leads to no thinning at all: the normal to the yield surface at S_{22} has no component in the 33-direction. Thus, a calculation of *R* for any texture with a Goss component would involve many crystals with values of infinity. It is for these cases that *R* is an especially inappropriate parameter, and we shall use *q* (Sec. 1.4), which has the same trend as *R*, but goes to a maximum value of 1.0. Figure 21a shows an actual texture along with the ideal Goss component as a 111 pole figure. Figure 21b shows simulations based on the measured texture, along with the relevant data set from Fig. 5 (the one with the strong Goss component).

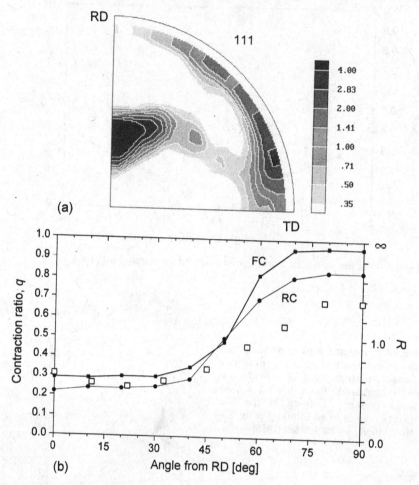

Fig. 21. (a) A measured Al texture with a strong Goss component (equal-area projection; RD up); (b) the corresponding measurements of the contraction ratio q and predictions according to the FC and RC models.

Finally, we will discuss 'classical' fcc rolling textures, such as those shown in Figs. 22a and 22b: one is for a 99.9% copper (with 0.01% P), rolled 93% and annealed at 600°C, which exhibits a strong 'copper component' [PEROVIĆ & KARASTOJKOVIĆ 1980]), the other for a 99.99% pure cathode copper cold rolled to a reduction of 96% [STEPHENS 1968]. Figure 22c presents measured q-values for these materials. The trend is similar to that shown for aluminum in Fig. 5 (the one without the strong Goss component).

The typical strain anisotropy response in these cases has high values of q (or R) at 45° – but different for the different experiments. Many comparisons between experiments and simulations have had difficulties in this regime. Some simulations for a typical fcc rolling texture are also shown in Fig. 22c. It can be seen that the addition of relaxed constraints improves the match – but not enough.

The reason for these difficulties can be understood in a similar way as for the Goss component. One of the most characteristic attributes of the standard fcc rolling

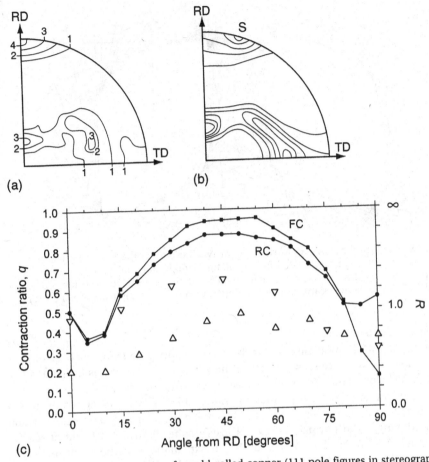

Fig. 22. Two measurements for cold-rolled copper (111 pole figures in stereographic projection): (a) according to PEROVIĆ & KARASTOJKOVIĆ [1980]; (b) according to STEPHENS [1968]. (c) The reported R-value measurements (∇ and Δ, respectively), along with FC and RC predictions.

texture is the 'S-component', marked in Fig. 22b. It has four symmetrically equivalent representations. If we take one at a time and turn it 45° around the sheet normal, we get two types of response to tensile stress in the 1-direction (Fig. 23a and b): half the grains allow no thinning, half do. (For tension in the 2-direction, the halves reverse.) One would assume that a Taylor-like model cannot describe this situation well and that, similarly, the results of experiments may be very sensitive to details of the grain arrangement.

In summary, all the cases discussed showed significant in-plane anisotropy, and its degree is very sensitive to details, both in experiment and in modeling. Since this in-plane anisotropy is the root cause of the commercially significant problem of 'earing' in beverage can production [TUCKER 1961], it deserves further quantitative study.

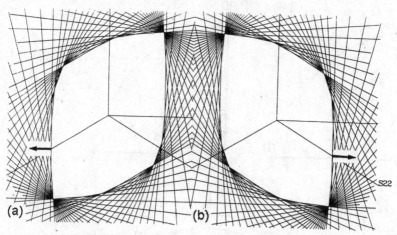

Fig. 23. Simulated π-plane yield surfaces for two variants of the *S*-orientation, in a coordinate system that is rotated in the rolling plane by (a) +45° and (b) -45° with respect to the rolling direction. Each predicts infinite *R*-value for one test. It is concluded that these variants are too incompatible to warrant application of a Taylor model.

Latent hardening in polycrystals

One aspect of the modeling to which the results may be somewhat sensitive is latent hardening. We have shown before (Chap. 9 Sec. 3.2) that it has little influence on texture development; however, at a given texture, on the yield surface, latent hardening might change local predictions noticeably. The *trend* is demonstrated schematically in Fig. 24a (which was actually derived by using excessive single-crystal latent hardening): the normals at special axes might change and could affect the lateral contraction ratio; this could then be an explanation for the slight mismatch between simulation and experiment that was observed in Fig. 16.

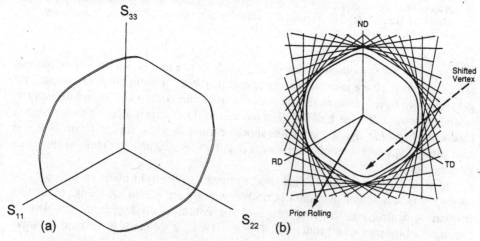

Fig. 24. (a) The outer yield surface assumes excessive latent hardening (LHR=9), the inner isotropic hardening: the difference is negligible. LApp simulated rolling to ε_{vM}=2.0. (b) Modification of yield surface for randomly oriented, flat grains (facets) by a simulated effect of bands in the microstructure [ROLLETT &AL. 1992].

Microstructural obstacles

A similar modification of the simulations was proposed by ROLLETT &AL. [1992]: it was assumed that microstructural features such as (non-crystallographic) microbands could serve as additional obstacles that modify the CRSS of each slip system according to the angle the slip *plane* makes with the band. The result of a particular simulation is shown in Fig. 24b; again, the modification (inner line) effected a subtle movement of the quasi-vertex, which could have an important influence on some particular strain anisotropy values.

2.6 Sharp textures and plastic instabilities

We have found (Fig. 15) that sharp textures lead to predictions of sharp vertices (in rate-insensitive materials). Experimental determinations of the 'reloading' yield surface also indicate the existence of 'flat spots' and regions of high curvature, and those investigated during continuous changes tend to confirm the existence of vertices.

Vertices and regions of high curvature have one important characteristic: small fluctuations in stress ('internal stresses') or in material state (and thus the location of one facet) can lead to large variations in straining direction [STÖREN & RICE 1975; CANOVA &AL. 1994b]. Figure 25 shows three π-plane yield-surface sections, all derived for a strong wire-drawing texture. Figure 25a used a simulated texture. At σ_{33}, there is a sharp vertex; the two arrows indicate the limits of the 'cone of normals' on the vertex. If the vertex were a true discontinuity in slope, then an arbitrarily small change in stress state could lead to a change in straining direction from one to the other of these arrows. Figure 25b comes from an experimental texture. At S_{33}, there is a sharp curvature, but clearly not a discontinuity in slope. The two arrows are now placed at the stress point that activates the respective straining direction; they differ by a 'sideways' component of stress that is only a few percent of the main component. A small internal stress could thus change the straining direction very significantly even here. Figure 25c is based on the same experimental specimen, but the texture data were evaluated with a harmonic method truncated at order four; this represents the 'texture' as it would have been measured by elastic anisotropy. The (quasi-)vertex is gone; it would take a variation in stress of about 10% to effect a substantial change in straining direction.

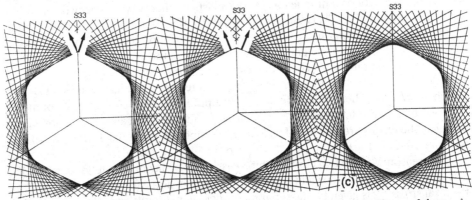

Fig. 25. At a sharp vertex (a), many straining directions are possible; if one of them leads to softening, an instability will develop. When there is no vertex but a sharp curvature (b), a small variation in stress may lead to a large variation in straining direction and therefore potentially to an instability. Complete rounding-off (c) makes this process unlikely.

Two points will be made from these observations. The first relates to the sharpness of textures. It is well known that most predictions give too sharp a texture. Thus, they would tend to exaggerate 'vertex' effects. However, the experimental texture could have been reproduced by a little smoothing of the predicted one [KOCKS &AL. 1996]; on the other hand, too much smoothing can hide such effects even if they should be there. We judge the situation in Fig. 25b to be typical: there is no true vertex (because the texture is not infinitely sharp, and because the rate sensitivity is finite [PAN & RICE 1983]); but nevertheless there may be sharp curvatures. In any case, the difference between an internal stress amplitude of zero and of a few percent is of no practical consequence. Thus both cases, in practice, show 'vertex effects'.

The second point relates to plastic instabilities. An instability may be characterized by the situation where a small, 'allowable' fluctuation in one variable leads to a large, permanent change in the response. The 'vertex effects' described above fulfill one of the three criteria: if one lets the 'small stress change' be a *fluctuation* (in time or space) of the local stress, the strain-rate response is a *large* change. It can become *permanent*, if it accelerates (or, equivalently, if it leads to a *softening* of the material). Finally, the fluctuation in both variables must be *allowable*. This amounts to making the fluctuation *compatible* with its surroundings. For example, if an entire cross-section of the wire used in Fig. 25 were to deform more rapidly and in a simple-shear mode at a certain angle to the axis, this would not disturb the other elements of the wire. The situation just described is in fact a mechanism for 'localized necking' in tension. It is always connected with a change in path of the straining direction. Its most important application is to the 'forming limit diagram' of rolled sheets – primarily under biaxial, rather than uniaxial, tension. Note that in the present 'instability' treatment, a pre-existing groove [MARCINIAK & KUCZYŃSKI 1967] is not required [STÖREN & RICE 1975]. Even when one is postulated, the sensitivity of predictions to details of the yield surface shape is great [NEALE & CHATER 1980; BARLAT & RICHMOND 1987].

2.7 Torsion: free versus fixed ends

In Fig. 6, we reported on some torsion measurements in connection with defining 'yield'. Long thin-walled tubes of 1100 aluminum had been pre-twisted over an undercut mandrel to $\gamma=0.5$ and then reloaded in different combinations of torsion and tension (or low values of compression) [STOUT &AL. 1985]. The large-offset yield stress, which was compared in Fig. 6 with a von Mises yield surface, is shown in Fig. 26 together with a LApp prediction. The agreement between the yield locus measured by the back extrapolation technique and crystal-plasticity theory is quite satisfactory.

Represented in Fig. 27a is the predicted yield surface associated with a fixed-ends torsion test carried out up to $\varepsilon_{13} = 0.5$ ($\gamma \approx 1$). This $\{\sigma_{33}, \sigma_{13}\}$ projection displays the axial effects. First, if the condition $d\varepsilon_{33} = 0$ has to be obeyed, the corresponding stress state must include a compressive component $\sigma_{33} < 0$.[9] This compressive axial stress was evaluated as a function of the deformation and is reported in Fig. 27b normalized by the corresponding shear component σ_{13}. Second, in a free-ends test, the condition $\sigma_{33}=0$ would give rise to a positive increment $d\varepsilon_{33}$ and thus to an axial elongation of the sample. The results are shown in Fig. 27b, superimposed on the axial stress associated with the fixed-ends test.

The reason why the two curves exhibit dissimilar behavior is as follows: while $d\varepsilon_{33}$ is related to the slope of the yield surface at the intersection with the σ_{13} axis (Fig. 27a), the axial stress σ_{33} is given by the separation of the left summit from the σ_{13} axis. One could say that one effect is related to a 'rotation' of the yield surface, the other to a 'shear'. Thus it is possible for ε_{33} to exhibit a monotonic variation, whereas this is not

---

[9] Under other circumstances, it may be a *tensile* stress [MONTHEILLET &AL. 1984].

required for σ_{33}. Incidentally, the *textures* developed under fixed and free ends conditions are quite similar but not identical: the components of the texture are the same, the only difference is a greater tilting of the whole texture about the radial axis in the free-ends case (Chap. 9 Sec. 2.3). As GIL SEVILLANO &AL. [1975] first observed, it is the 'asymmetry' (i.e., lack of *orthotropic* symmetry) that causes the length effects, and this asymmetry peaks at intermediate deformations (Chap. 9 Sec. 2.3).

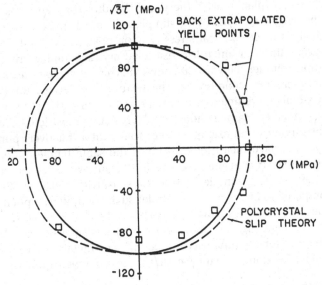

Fig. 26. Comparison of yield-surface measurements (squares) in combined torsion and tension on a pre-twisted Al specimen with a LApp prediction (dashed line). The departures from the von Mises circle (solid line) are well described [HELLING &AL. 1986].

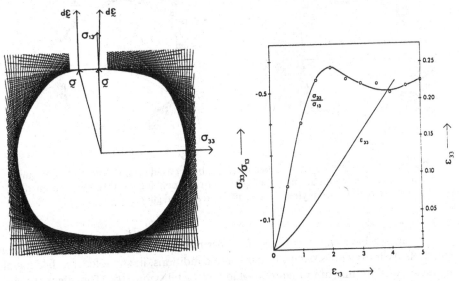

Fig. 27 (a) Simulated yield surface of a specimen twisted in the 13-direction. A a subsequent pure σ_{13} stress induces a strain component of 33-elongation. Conversely, a pure $d\varepsilon_{13}$ strain increment would induce a 33-compression stress component. The two effects are quite different, as shown in (b) as a function of prestrain.

2.8 Dependence on deformation modes

Pencil glide

The preceding sections have dealt with fcc materials for which the slip mode is well established. As has been pointed out before (Chap. 5), bcc materials would behave exactly the same, only with the sign of straining reversed, if they deformed in restricted glide on {110}⟨111⟩ (and if the grain shape played no role). In fact they often deform on more slip planes, with the same slip direction, and this is sometimes idealized by 'pencil glide', in which the slip plane is merely the one (containing ⟨111⟩) that has the highest resolved shear stress [TAYLOR 1956, KOCKS 1970, PIEHLER & BACK-OFEN 1971]. Since this is a much less restricted crystallographic criterion, the yield surfaces become rounder. Figure 28 shows the results for *one texture* under two different subsequent test conditions. The texture was derived under pencil-glide conditions by simulated rolling to a strain of $\varepsilon_{vM}=1.0$. To derive the subsequent yield surfaces, we assumed the same conditions in Fig. 28a, but restricted glide in Fig. 28b. This could correspond to an example of hot rolling of a (relatively pure) bcc metal, followed by yield-surface measurements (a) at the same temperature, (b) at a low temperature (where restricted glide occurs). Both figures were derived under identical conditions otherwise (for example, with a rate sensitivity of 0.03). The rounding of Fig. 28a as compared to 28b is obvious. Under such conditions, a phenomenological description becomes feasible [LOGAN & HOSFORD 1980]. For restricted glide, it is instructive to compare Fig. 28b and Fig. 15a: an exchange of axes 33 and 11, as suggested (Chap. 5) for a comparison of fcc and bcc metals makes the two almost identical – which goes along with the fact that the textures were quite similar.

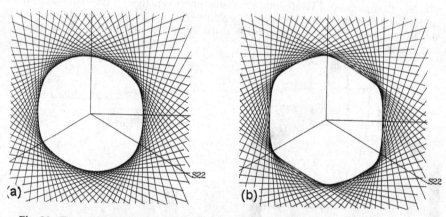

Fig. 28. Two yield surfaces for a bcc material, both based on the same rolling texture (which was derived assuming pencil glide). In (a), pencil glide was assumed for the calculation of the current yield surface, in (b), restricted glide.

Twinning

So far, we have not accounted for possible twinning modes in deriving yield surfaces. Twinning does occur, under some conditions, in some fcc and bcc metals. However, it is prominent in materials of lower crystal symmetry. The effect of twinning on the polycrystal yield surface has been investigated in some detail for zirconium alloys [TOMÉ & KOCKS 1985; LEBENSOHN & TOMÉ 1993a,1996]. Here, we will demonstrate the effect by a simulation for the case of Ti, in which basal and prism glide are treated

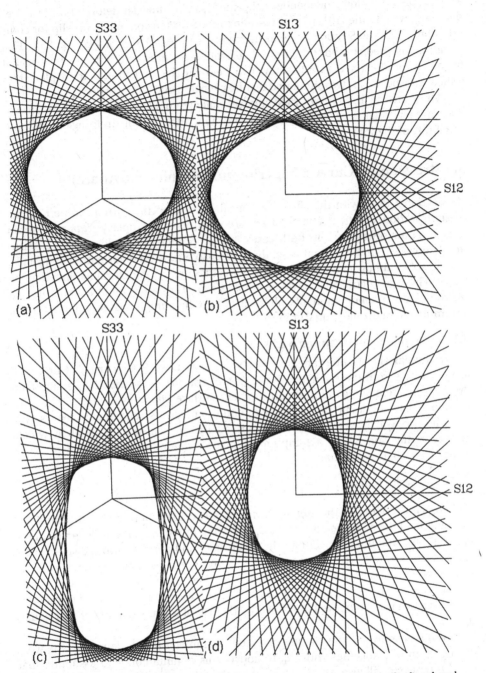

Fig. 29. Simulated yield surfaces, in two subspaces, for Ti deforming by basal and prism glide. Uniaxial deformation in the 33-direction (the ⟨c⟩-direction) is accommodated by easy twinning in tension and by hard ⟨c+a⟩-glide in compression. The presence of twinning causes the lack of centrosymmetry of these yield surfaces. The texture was derived by simulated extension to 20% in (a) and (b), by simulated compression to 20% in (c) and (d).

as easy, providing four independent slip modes. For uniaxial deformation in the c-direction, we add the $\{10\bar{1}2\}\langle10\bar{1}1\rangle$ twinning mode for tension and $\langle c+a\rangle$-slip for compression. (The CRSS ratios were, in the sequence presented, 0.5, 1.0, 3.0, 4.5). This combination predicts about the right deformation textures (and it can still be treated by the Taylor model). Figure 29 shows two yield-surface projections – the π-plane and a shear subspace – for two simulated textures: tension and compression.

These are the first cases of yield surfaces that are not centrosymmetric (because of the unidirectionality of twinning). In addition, they demonstrate again that the variability of shapes is great.

3. Stress/strain Curves: Experimental Results and Analysis

Section 2 dealt with the effects of a given texture on the current properties; the present Section will treat effects of texture *evolution* on property *changes* – strain hardening (or softening). In both cases, the properties to be addressed are anisotropic; however, in the present case, there are also effects on deformation in a single, monotonic straining path – though, in general, they are different for different straining paths. These geometric effects will be emphasized. They are geometric in nature and convolute the phenomenon of 'physical' strain hardening, based on dislocation accumulation in the individual grains.

3.1 Dependence on straining path

In multiaxial tests, one can measure a variety of 'stress/strain curves'. One would like to extract from these a single curve that characterizes (scalar) 'hardening' (and describe the remaining information as yield surface *shape* evolution). The conventional way of doing this is to use 'equivalent stress' and 'equivalent strain' according to von Mises. They were defined for an isotropic material in which, moreover, the von Mises assumption is made, which sets the equivalent stress (Chap. 8 Eq. 68) to a *constant*:

$$\sigma_{vM} = \sqrt{\tfrac{3}{2}\,\mathbf{S}\!:\!\mathbf{S}} = \sigma_f$$

(10)

Here, σ_{vM} is a function of the *applied* deviatoric stress $\mathbf{S}$ (proportional to the norm, i.e. the 'length' of the stress vector), while σ_f is a strength parameter that depends on the material, its state, the temperature, the strain rate (and perhaps also the stress state, see Chap. 8 Sec. 1.2). The whole point of anisotropy, and even of deviations from the von Mises law for isotropic materials, is that σ_{vM} does *not* describe 'equivalent' states. This was demonstrated convincingly in Fig. 8. Thus, one would not expect the general hardening behavior to be well described by the use of a stress parameter that equals σ_{vM}.

Even less appropriate is the use of a strain parameter that is defined as the work conjugate to σ_{vM}: together with Eq. 10 above, this is tantamount to assuming that equal total work will give equal states – even though most of the plastic work is dissipated and can thus not be 'remembered' by the material.

Finally, assuming a plot of σ_{vM} versus ε_{vM} to provide a universal hardening curve presumes that the rate of hardening itself is independent of the path in stress (or strain-rate) space. All of these assumptions are inappropriate. Nevertheless, it is common to use such a plot to describe the actual behavior, by one plot for each path;

they generally turn out to be *different*. TOMÉ &AL.[1984] were some of the first to observe and quantify the inadequacies of this approach and propose the use of crystal-plasticity theory to define an equivalent-stress criterion. Their early data were taken using OFE copper and most clearly showed the von Mises criterion could not normalize the stress/strain behavior between uniaxial compression and torsion.

From the point of view of crystal plasticity, one might expect the hardening law in the individual crystals, or at least on the individual slip systems, to be universal (Chap. 8). It is expressed differentially: as a rate of change of flow stress τ with some strain measure. Then, the macroscopic stress/strain curve follows from an integration over the path: using the current texture – the current (average) Taylor factor M – at every point [TOMÉ &AL. 1984, KOCKS &AL. 1988b, KALIDINDI &AL. 1992]. MECIF &AL. [1997] have recently investigated the principles of this superposition in single crystals.

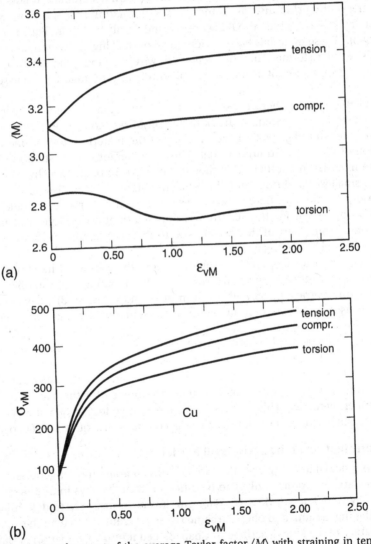

Fig. 30. (a) Development of the average Taylor factor $\langle M \rangle$ with straining in tension, compression, and torsion. (b) Resulting stress/strain curves based on an assumed unique grain-level hardening law (the one shown for Cu in Fig. 32) [KOCKS &AL. 1988b].

Figure 30a shows the simulated development of M for an fcc metal with an initially random orientation distribution. It is seen, for example, that tension and compression are the same at the beginning (as they must, under these assumptions) but evolve differently: in tension, a sharp $\langle 111 \rangle$ texture develops (which has the highest Taylor factor); in compression, the spread of orientations is much wider (around $\langle 110 \rangle$). The initial value for torsion reflects the departure of the random polycrystal yield surface from the von Mises assumption. (Note that the Taylor factor is defined in terms of the work, not the shear stress; see Chap. 8 Eq. 24, TOMÉ &AL. [1984].) Figure 30b shows the effect of this on the stress/strain curve, assuming a unique hardening law at the crystal level (which will be discussed below).

The texture development along any particular straining path also depends on material (Chap. 9). This produces an effect on the stress/strain curves of different materials that is superimposed on different dislocation accumulation laws. The true plastic stress/strain response, as a function of deformation mode, is plotted for three different fcc metals (1100 Al, OFE copper, and 70:30 brass) in Figs. 31a-c [KOCKS &AL. 1988b]. These materials have a wide range of stacking fault energies (SFEs). The materials were deformed monotonically in wire-drawing plus tension, in compression, and in torsion of a short thin-walled tube, in the manner discussed in Sec. 1.3.

In all these cases, there is but one stress component to plot. The curves for different tests are compared on the basis of von Mises equivalent stress and strain. One first notes that these data never, for any of the materials, fall on one line. The separation of the stress/strain data between the different deformation modes varies from one material to another. It is the largest for 70:30 brass (which has the lowest SFE) and smallest for aluminum (which has the highest). One also observes that, in some cases, the different hardening curves for a given material are parallel (1100 Al and OFE copper) but for the 70:30 brass they continue to diverge with increasing deformation. Some part of this can be due to different texture evolutions, some to different physical hardening laws.

One approach to separating the physical from the geometric hardening is to try and define a new set of 'equivalent' stress and strain, based on crystal plasticity; for example, some average CRSS and the sum of shears Γ (Chap. 8). One would then convert measured stress/strain curves into this tentative set and see whether it is universal [JONAS &AL. 1982].

The approach we used, instead, postulates a unique physical hardening law at the grain level, in differential form (Chap. 8). If this were known for the given material, one could integrate the stress/strain curve for an arbitrary path, using the measured initial texture and the texture evolution that is simulated for this path (and perhaps verified experimentally). This procedure was in fact used for Fig. 30b: it was derived to simulate the behavior of copper, Fig. 31b. The comparison is quite satisfactory.

3.2 Determination of the grain-level hardening law

The physical hardening laws that apply within a grain are not generally known a priori; they are, for example, likely to be different from the laws in free single crystals (even if they were known) [KOCKS 1958]. An 'inverse' procedure to determine grain-level hardening (assumed to obey the same law in all grains) is to hypothesize a differential relation of the kind described in Chap. 8, feed it into a simulation of, say, a tension test on a specimen with known initial texture and then compare the prediction

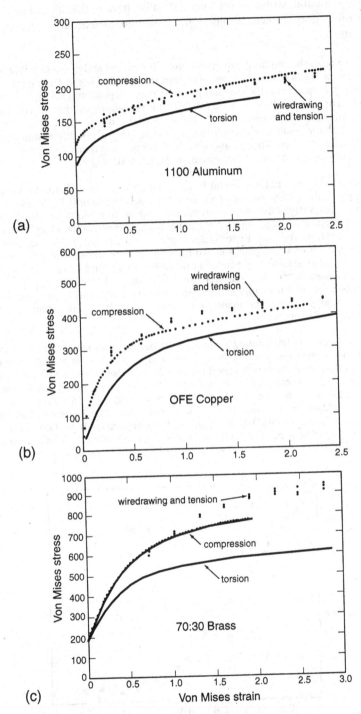

Fig. 31. Measured stress/strain curves for (a) 1100 Al, (b) OFE Cu, and (c) 70:30 brass, in tension, compression, and torsion. The tension experiments were supplemented by wire-drawing experiments with subsequent determination of the tensile flow stress after various prestrains.

with the experimental stress/strain curve; finally, iterate the local law until the macroscale prediction is accurate enough. It should then hold for all other tests [KOCKS &AL. 1988b].

Figure 32 displays the resulting hardening curves for copper and aluminum that led to the 'predictions' in Fig. 30b, and to agreement with the experiments in Fig. 31a and b (using appropriate parameters for the simulation – see Chap. 8 Eqs. 34 and 36; e.g., $\tau_0 = 10$ MPa, $\theta_0 = 210$ MPa, and $\tau_v = 88$ MPa for the Cu at R.T.) This procedure can then be combined with laws concerning the temperature and strain-rate dependence of τ_0, θ_0, and τ_v for the material – with the result that the behavior for a wide range of conditions and textures can be described by a single scheme; and since it is based on physical laws, it can be extrapolated beyond the range of the measurements [CHEN & KOCKS 1991a].

If this procedure were not successful in describing the stress/strain behavior along all straining paths, then the postulate of a unique grain-level hardening law would have to be abandoned. Suggestions that have been made, which require additional assumptions, are that local hardening is greater when the number of significantly active slip planes is greater [TOMÉ &AL. 1984] or when the Taylor factor is greater [MILLER &AL. 1995]. Inasmuch as the minimum number of active planes is two for uniaxial tests, but only one for plane-strain tests, one could postulate *two* fundamental (differential) hardening laws for the each material. All these procedures should be used only when it has been clearly established that a single hardening law, applied to appropriate simulations with the measured initial texture as an input, cannot be made to agree with experiments that fulfill the 'quality' criteria developed in Section 1.

A detailed recipe for determining grain-level hardening is to get scaling parameters in tension (since it the most accurate test at low strains, and there is no influence of grain shape). For the rest of the curve, a good compression test is easiest (but may not be free of assumptions for fcc, because of 'grain curling'). The results must match tensile tests at low strains. For larger strains, one must know how to phase in grain shape effects in order to be able to use the data. Torsion (on a short tube, for both fcc and bcc) would be ideal, because these tests can be run to virtually the complete flat grain-limit, where there is no question as to the model to be used; then, one can match the experimental texture transition with predictions according to various assumed FC-to-RC transition schemes [KOCKS & NECKER 1994].

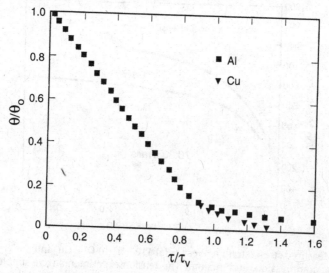

Fig. 32. The differential hardening law that could, when folded into the texture development simulation, explain all observed stress/strain curves for Al and Cu.

4. Summary and Assessment

Two general conclusions can be drawn from this Chapter: (1) simulations predict anisotropic behavior of great variety and often great severity, such as it is not captured by any extant phenomenological description (except in the special cases of high rate sensitivity and of pencil glide); and (2) experiments of a sufficiently wide range to characterize this anisotropy quantitatively are very difficult. Thus, if the problems are indeed as portentous as the simulations indicate, then simulations are the only practical means to assess them for engineering applications. This puts high demands on the reliability of the simulations.

The root cause of the macroscopic plastic anisotropy lies in the polycrystalline nature of the materials. The plastic properties of the individual crystals are severely anisotropic, even when the lattice structure is cubic – and they can be experimentally assessed. The anisotropy is carried forward to the polycrystal when the texture is sufficiently strong – and it can be measured. Neither of these effects is captured in phenomenological laws of plasticity, but both are inherent in any polycrystal-plasticity model that may be used. Thus, the physically based nature of such simulations, coupled with measured textures and verifiable constitutive laws, does lend polycrystal simulations substantial credence.

The question to be addressed is really how sensitive the predictions are to the details of the assumptions and procedures used to implement the model. We can group these sensitivities into three classes.

- The prediction is certain. This holds, for example, for symmetry properties: once they are established for a given sample, the axes for an appropriate description of the yield surface are given (and they are normally *not* the principal axes of stress or strain-rate), and the range over which the behavior must be characterized is also given (and often much reduced from the most general, though always greater than that under isotropy).

- The prediction is insensitive to details. This holds, in many cases, for the description of the anisotropy of the yield stress (which is typically subject to an experimental uncertainty of about 5% anyway). In such cases, the yield surface can often be adequately represented by a smoothed-out form that is simple. Similarly, the effect of texture evolution on the path dependence of stress/strain curves is well predicted by polycrystal models (and significantly distinct from the phenomenological equivalent-stress/equivalent-strain assumptions).

- The prediction is sensitive to many details. This holds for many predictions of the anisotropy of strain-rate, in cases where some surface displacements are free. This is the primary area in which anisotropy is of industrial interest: for forming operations. In particular, plastic *instabilities* may arise from such freedom of variation for strain rates, when the texture is severe.

The sensitivity can be to experiment as well as to simulation. As an example, the in-plane anisotropy of the 'R-value' appears to be very sensitive to certain components of a texture. In this case, small variations in measurement can lead to wide variations in predicted behavior. It is for these cases that further work in this field is indicated. A combined physical/phenomenological approach may be the most appropriate. As a general principle, it seems advisable to predict a general yield surface shape (or a series of them for various detailed assumptions) on the basis of a polycrystal model, and then to 'anchor' this yield surface by one or two experiments at those places that are of particular interest. For example, we find the existence of vertices and flat spots on the yield surface of strongly textured materials almost certain from our simulations and the scant experiments; but the precise location of the vertices seems more subject to question and may, for the time being at least, have to be settled heuristically.

The emphasis in this Chapter has been on cubic metals, and all simulations were based on the Taylor model with its various modifications. The success, where it could be experimentally verified, followed the scheme above: it was least convincing in the cases that are most sensitive. In a number of cases, we have identified a specific reason for this lack of success: the Taylor model itself should be inapplicable when it leads to local disequilibria of the same order as the flow stress. This can occur even in cubic metals when, for example, the texture contains essentially only one component, but in symmetrically equivalent variants that are incompatible with one another for certain straining paths.

We end this Chapter with two recommendations that are obvious from this exposition:

• We would urge that initial textures be measured as part of any mechanical test, and that resulting anisotropic shape changes in the specimen be recorded routinely.

• To assess the applicability of any Taylor-type simulation, it is of value to record the standard deviation of the predicted grain stresses; it should not exceed about 0.5 of the flow stress.

Chapter 11

SELF-CONSISTENT MODELING
OF HETEROGENEOUS PLASTICITY

SELF-CONSISTENT MODELING
OF HETEROGENEOUS PLASTICITY

C. N. Tomé and G. R. Canova[†]

In this Chapter we will be mainly concerned with the heterogeneity of the plastic response of quasi-homogeneous bodies, and with the local variations in stress and strain that such heterogeneity induces during plastic deformation. The term 'quasi-homogeneous' is used in the sense that the displacements and tractions at the surface of the body can be represented by the volumetric average of stress and strain inside the body. The term 'heterogeneity' is used in the sense of a space variation of a property in the material. The difference between plastic heterogeneity and the heterogeneity of elastic, thermal and viscous properties treated in Chapter 7, is that in plasticity the constitutive relations are usually grossly non-linear.

We are going to distinguish between two types of heterogeneous materials: single-phase and polyphase. In single-phase materials, the properties of the constituent domains (grains) are identical with respect to the crystallographic reference system. Heterogeneity arises from the differences in the orientation of the grains, since their properties are different when referred to an 'absolute' reference system. Within these materials, polycrystalline aggregates represent the best example of **Texture** and **Anisotropy** being intertwined, precisely because heterogeneity is a consequence of the orientation distribution and the single-crystal anisotropy. The combination of both determines the anisotropy of the overall response. Polyphase materials are treated in some detail only for the case of polycrystalline aggregates, in which the anisotropy of the constituent crystals is of principal concern.

For composites (metal or ceramic matrix composites) and reinforced materials anisotropy is usually high, but is mainly controlled by the morphology and distribution of the hard phase, and not so much by the texture of the matrix. In this case, the main source of heterogeneity is not the texture, but the difference in the intrinsic properties of the constituent phases. For this reason, composites and reinforced materials are beyond the scope of this book and will not be treated here.

1. Background

The usual requirement of polycrystal models is that the loading and displacement conditions at the boundary of the polycrystal are uniform, and that the volume average of stress, strain and strain rate over all grains must coincide with the overall stress, strain and strain rate at the boundary [HILL 1967]. There are, however, several possible configurations of stress and strain in the grains that fulfill such overall requirements: the essential characteristic of a model is the assumption that it uses to choose between them.

The Taylor assumption of uniform strain throughout the polycrystal ensures compatibility and provides an upper bound to the stress. The so called Sachs model, on the other hand, assumes proportional (not uniform) loading in the grains, and satisfies neither compatibility nor stress equilibrium. As a matter of fact, the Taylor model has been favored by materials scientists because it leads to multiple slip in the grains and to reasonable texture predictions. Because of its characteristics, it has been mainly used for modeling cubic materials, which are in general mildly anisotropic plastically. Enforcing local compatibility over local equilibrium, though, may be a severe assumption in some situations. The Relaxed Constraints (RC) approach imposes a mixture of both conditions based on grain shape considerations, and has proven a better alternative for the prediction of rolling and torsion textures at large strains. Both the Full Constraints (FC) and the RC approaches have been thoroughly described in Chapters 8 and 9.

Polycrystal plasticity and texture development of low-symmetry materials are characterized by a variety of active deformation modes present in each grain, non-negligible twinning activity, and markedly anisotropic single crystals with 'soft' and 'hard' orientations which induce highly directional grain interactions. These crystals cannot accommodate certain deformation components because they either lack the necessary deformation systems or because those systems require high activation stresses. In such a case multislip is restricted to fewer than five systems and the stress build-up tends to induce intergranular fracture, or to be relaxed by inducing extra plastic accommodation in the neighboring grains. As a consequence of such aniso-tropy, a more complex approach, other than the standard polycrystal models, is required in order to predict the plastic response and the deformation textures of such materials.

An empirical way of tackling the scarcity of deformation systems without sacrificing the numerical and conceptual simplicity of the Taylor assumption is the Constrained Hybrid (CH) model of PARKS & AHZI [1990], which was formulated to model semi-crystalline polymers. The CH model consists in projecting the externally imposed strain along the directions (in strain space) where the grain can yield. The average deformation clearly falls short of the externally imposed strain because only part of the strain is accommodated by every grain. The authors circumvent this difficulty by scaling the strain tensor that is applied to the grains, in order for the average to give the correct overall strain. The CH model reduces to the FC approach when the single-crystal yield surface is closed in the 5-D deviatoric stress space.

A more rigorous approach to the modeling of heterogeneous deformation of polycrystals is given by self-consistent and FEM methods. Self-consistent polycrystal models aim at deducing the overall response of the aggregate from the known properties of the constituent grains and an assumption concerning the interaction of each grain with its environment. Because of their statistical nature they generally do not account for the precise topology of the aggregate: as all models, they fulfill com-patibility and stress equilibrium in an average sense. FEM calculations, on the other hand, fulfill local compatibility and are better suited for describing the local response. However, local equilibrium is imposed only in a weak sense and the results usually depend on the topology used.

Self-consistent models have already been discussed in Chapter 7 in connection with elastic and Newtonian media, in which the response is linear. The main characteristic of self-consistent models applied to polycrystal *plasticity* are:

(i) The constitutive equation that describes the local response is non-linear and is known explicitly. The non-linear overall response of the aggregate is approximated using a linearization procedure, which is assumed to represent the material behavior within a certain range of stresses and strain rates.

(ii) Heterogeneity at the grain level leads to localized intergranular interactions, which are nearly impossible to describe realistically. A simplifying hypothesis is to assume a Homogeneous Equivalent Medium (HEM) with uniform properties, whose overall response to external loading conditions is the same as the aggregate's. Each grain (or cluster of grains) is assumed to be embedded in and interacting with such a homogeneous matrix. Each grain is assumed to be representative of all the grains with the same orientation, and the HEM represents the average neighborhood of all these grains. A disadvantage of this approach is that it overlooks topology, correlation and localization effects (such as percolation). This disadvantage is partially addressed by the 'cluster' method to be discussed in Section 6 of this Chapter.

(iii) The deformation of the inclusion (or cluster) embedded in an effective medium has to fulfill local equilibrium and compatibility. In practice, except for the exact Eshelby solution for ellipsoidal inclusions in linear media (see the Appendix), these conditions cannot be fulfilled exactly for non-ellipsoidal inclusions or non-linear media. It is usual, however, to linearize the response of the medium, and to assume that the grain can be characterized as essentially ellipsoidal, which guarantees that the stress and the strain are uniform within the domain of the grain. By extending the Eshelby result it is possible to use the interaction equation, which relates the stress and the strain in the grain to the overall stress and strain in the aggregate. This is an important characteristic of self-consistent models, since it permits one to account for the relative anisotropy of both grain and matrix.

(iv) The previous relation is assumed to apply to every grain (or cluster) considered individually. When combined with the condition that the volume averages of stress and strain must be equal to the macroscopic stress and strain, it leads to a non-linear equation that permits a calculation of the unknown overall moduli that describe the HEM response.

By accounting explicitly for the relative anisotropies of the grains and the matrix, self-consistent models give a better description of material response than classical upper-bound models. It seems evident that the strength of the interaction of the grain with its surroundings dictates how much of the plastic deformation will be accommodated by the grain and how much by the surroundings. The relative anisotropy of grain and matrix will be directly responsible for such response and, as a consequence, both have to enter explicitly in the formulation.

The approach that we present here for addressing the problem of large plastic deformation of polycrystals neglects the elastic effects. In Section 2 we introduce the visco-plastic constitutive response of the grains and the overall material response. In Section 3 we solve the interaction of the grain with the rest of the polycrystal by treating the former as an inhomogeneity embedded in a homogeneous effective medium that represents the polycrystal. Such a formulation leads to an interaction equation that couples the stress and strain rate in the grain with the overall stress and strain rate of the effective medium. The results of Section 3 are used in Section 4 to develop a self-consistent polycrystal model based on this interaction equation. The formulation explicitly accounts for the plastic anisotropy of grain and polycrystal, grain shape effects, and their evolution with deformation. A critical discussion of the formulation and its consequences upon the predicted material behavior is made in Section 4.

Section 5 focuses on the modeling of single-phase aggregates: we analyze the visco-plastic response of single-phase polycrystals of cubic, hexagonal, trigonal and orthorhombic structure, emphasizing the correct treatment of crystal anisotropy and analyzing its influence upon the response of the material. Quasi-homogeneous materials (fcc) are treated with the techniques used for modeling heterogeneous materials, in order to assess the departure from the predictions made using simpler theories (FC and RC).

In Section 6 we analyze the visco-plastic response of a heterogeneous poly-crystalline aggregate (granite), which exhibits severe discontinuities in properties across the interfaces, using a generalized self-consistent formulation adapted to account for non-uniform intragranular strain.

2. Constitutive Equations for Grain and Polycrystal[1]

The kinematic equation that relates stress and strain rate for a single crystal in which plastic shear is controlled by a rate sensitive mechanism, discussed in Chap. 8 Sec. 1.6, is:

$$D_\lambda^c = \dot{\gamma}_o \sum_s m_\lambda^s \left(\frac{\mathbf{m}^s : \mathbf{\sigma}}{\tau^s} \right)^n \tag{1}$$

where the sum is carried over all the systems in the grain. Since the dilatational component m_6^s of the Schmid tensor is zero, the component D_6^c is identically zero and the strain rate $\mathbf{D}^c$ is deviatoric. For the same reason, Eq. 1 and, in general, all the equations in this Chapter, are independent of the hydrostatic stress component σ_6. Equation 1 represents a non-linear system of five equations with ten unknowns: the five deviatoric strain-rate components and the five deviatoric stress components. Within the FC or the RC approach, this system is solved for each grain independently, imposing either the five strain-rate components or a combination of five stress and strain-rate components, respectively. The self-consistent approach that is described below, instead, relates the strain rate and the stress in each grain to the average strain rate and the average stress in the polycrystal. This condition provides another set of five equations for each grain. In addition, the requirement that the overall stress is given by the average over the grains provides another five equations which couple the response of all grains. This system of $10N+5$ coupled equations permits the calculation of the five deviatoric stress and strain-rate components in each grain, and any

---

[1] In this Chapter, we will denote contraction of second- and fourth-rank symmetric tensors, in equations like $A_{ij} = \mathbb{B}_{ijkl} C_{kl}$ or $\mathbb{D}_{ijmn} = \mathbb{E}_{ijkl} \mathbb{F}_{klmn}$ either by using the basis of symmetric tensors $\mathbf{b}^{(\lambda)}$ and the associated matrix representation, as described in Chap. 8 Sec. 1.4 (i.e. $A_\lambda = B_{\lambda\mu} C_\mu$ or $D_{\lambda\lambda'} = E_{\lambda\mu} F_{\mu\lambda'}$), or by the symbolic tensor notation outlined in the Notation Chapter at the end of the book (i.e. $\mathbf{A} = \mathbb{B} : \mathbf{C}$ or $\mathbb{D} = \mathbb{E} : \mathbb{F}$). The indices in cartesian tensor components run from 1 to 3 when they are Latin, from 1 to 5 (or 6) when they are Greek.

combination of five overall stress and overall strain-rate components (the other five are given by the boundary conditions).

Equation 1 can be expressed in a pseudo-linear form:

$$D_\lambda^c(\sigma) = \left\{ \dot\gamma_o \sum_s \frac{m_\lambda^s m_\mu^s}{\tau^s} \left(\frac{m^s : \sigma}{\tau^s} \right)^{n-1} \right\} \sigma_\mu = \mathbb{P}_{\lambda\mu}^{c(sec)}(\sigma)\, \sigma_\mu \tag{2}$$

where $\mathbb{P}^{c(sec)}$, *the secant visco-plastic compliance tensor* of the grain, is not an intrinsic property of the crystal since it depends on the stress state (except when $n=1$, in which case $\mathbb{P}^{c(sec)}$ stands for a material property). Note that $\mathbb{P}^{c(sec)}$ is identical to the tensor $\mathbb{P}^c$ defined in Chap. 8 Eq. 75. However, throughout this Chapter we will denote explicitly either the secant or the tangent characteristic of the visco-plastic tensor. When the stress is uniform in the grain, Eq. 2 is exact and does not represent any approximation, but just an alternative way of writing Eq. 1. A linear relation valid in *the vicinity of the point* σ^0 is obtained by doing a first order Taylor expansion of Eq. 1 around σ^0:

$$D_\lambda^c(\sigma) = D_\lambda^c(\sigma^o) + \left.\frac{\partial D_\lambda^c}{\partial \sigma_\mu}\right|_{\sigma^o} (\sigma - \sigma^o)_\mu$$

$$= \mathbb{P}_{\lambda\mu}^{c(sec)}(\sigma^o)\, \sigma_\mu^0 + n\, \mathbb{P}_{\lambda\mu}^{c(sec)}(\sigma^o)(\sigma - \sigma^o)_\mu \tag{3}$$

$$= \mathbb{P}_{\lambda\mu}^{c(tg)}(\sigma^o)\, \sigma_\mu + D_\lambda^{c^o}(\sigma^o)$$

where the *tangent modulus* $\mathbb{P}^{c(tg)}$ and the back extrapolated term D^{c^o} are defined by:

$$\mathbb{P}^{c(tg)} = \left.\frac{\partial D^c}{\partial \sigma}\right|_{\sigma^o} = n\, \mathbb{P}^{c(sec)}$$

$$D^{c^o} = (1-n)\, \mathbb{P}^{c(sec)} : \sigma^o = (1-n)\, D^c(\sigma^o) \tag{4}$$

At the macroscale level the constitutive law relating the overall stress σ to the overall strain rate D is going to be non-linear, but it may be assumed to be described by a secant or a tangent expression similar to Eq. 2 and Eq. 3, valid for particular values of overall stress and strain rate.

$$D = \mathbb{P}^{(sec)}(\sigma) : \sigma \tag{5}$$

and

$$D = \mathbb{P}^{(tg)}(\sigma) : \sigma + D^0 \tag{6}$$

with:

$$\mathbb{P}_{\lambda\mu}^{(tg)} = \left.\frac{\partial D_\lambda}{\partial \sigma_\mu}\right|_\sigma \tag{7}$$

HUTCHINSON [1976] demonstrates, using the principle of virtual work and a visco-plastic potential function, that the polycrystal tangent tensor defined by Eq. 7 fulfills the same relation with the secant tensor as the corresponding single-crystal moduli:

$$\mathbb{P}^{(tg)} = n\, \mathbb{P}^{(sec)} \tag{8}$$

The overall secant and tangent tensors have a mathematically precise meaning. The tangent tensor, in particular, does not necessarily coincide with the 'effective' visco-plastic tensor discussed in Chap. 8 Sec. 1.7. One has to bear in mind that Eqs. 5 and 6 represent two equivalent forms of describing the same strain rate induced in the polycrystal by a given stress imposed at the boundary. Whenever the stress σ acting on the aggregate changes, the secant and tangent moduli have to be modified, because they do not represent intrinsic material properties and depend on the stress.

In connection with the 'linearization' discussed in Chap. 8 Sec. 1.7, the argument exposed there is that a first order Taylor expansion is not necessarily more accurate than other linear approximations that tend to capture the contribution to the strain rate of higher order terms in the expansion. As a consequence, the 'tangent' compliance defined by Chap. 8 Eq. 77 has to be interpreted as an 'effective tangent compliance', to which the relation Eq. 8 does not apply. In this Chapter we will implement the polycrystal formalism assuming that the first order Taylor expansion describes the constitutive response when the local stresses are in the vicinity of an imposed stress σ. We will also explore the consequences of assuming a stiffer constitutive response, described by the secant modulus.

3. The Visco-plastic Inclusion Formalism

Equation 6 describes the overall response of a homogeneous medium loaded at the boundary Γ_V with a stress σ. Assume an inhomogeneous domain Ω inside such medium, where the constitutive response is given by a law of the form (3), with a visco-plastic compliance $\mathbb{P}^{c(tg)}$. The presence of the inhomogeneity will disturb the otherwise uniform stress field and induce a local field $\sigma(x)$ where x is a general co-ordinate. The linearization assumption requires one to assume that the local response of the medium in the vicinity of the inclusion is linear and is described by the same visco-plastic compliance $\mathbb{P}^{(tg)}(\sigma)$ and the same reference strain rate $D^0(\sigma)$ that describe the overall response at the boundary. The local stresses in such a medium are given by (3) and (6) as:

$$\sigma(x) = \left[\mathbb{P}^{c(tg)}\right]^{-1} : \left[D(x) - D^{c^0}\right] \qquad \text{in } \Omega \tag{9a}$$

$$\sigma(x) = \left[\mathbb{P}^{(tg)}\right]^{-1} : \left[D(x) - D^0\right] \qquad \text{in } V - \Omega \tag{9b}$$

while the condition at the boundary is

$$\sigma = \left[\mathbb{P}^{(tg)}\right]^{-1} : \left[D - D^0\right] \qquad \text{in } \Gamma_V \tag{9c}$$

The local stress $\sigma(x)$ induced by a σ imposed at the boundary has to satisfy the equilibrium equation

$$\sigma_{ij,j} = 0 \tag{9d}$$

The problem stated by (9) leads to a system of second-order differential equations on the displacement rates. This problem is completely equivalent to the one of the elastic inhomogeneous inclusion solved in Section 4 of the Appendix, with the inverse of the visco-plastic compliance playing the role of the elastic constants and the strain rate replacing the total strain (see Eq. A26). As a consequence, the solution is formally identical: the stress and strain rate inside the domain Ω are uniform, and their deviations from *the overall* stress and strain rate are related through the interaction equation

$$\tilde{D} = -A : \mathbb{P}^{(tg)} : \tilde{\sigma} \tag{10a}$$

where

$$\tilde{D} \equiv D^c - D$$
$$\tilde{\sigma} \equiv \sigma^c - \sigma \tag{10b}$$

and A is the *accommodation tensor* defined as:

$$A = (I - E)^{-1} : E \tag{10c}$$

Here E is the *visco-plastic* Eshelby tensor. According to Eq. 10a, the accommodation tensor determines the extent to which stress deviations are amplified into rate deviations, and plays a central role in the embedding procedure discussed in the following Section. By similitude with the elastic case, the local deviation in the rotation rate $\tilde{W}$ of the ellipsoid is given by Eq. A32 as:

$$\tilde{W} = W : E^{-1} : \tilde{D} \tag{11}$$

The functional dependence of E and W upon the ellipsoid axes and the visco-plastic stiffness $Q^{tg} = [\mathbb{P}^{(tg)}]^{-1}$ is the same as for the elastic Eshelby tensors described by Eqs. A13 to A19 of the Appendix.

There is a fundamental difference between the elastic and the 'linearized' visco-plastic inclusion problems. In the latter case the medium is incompressible and, as a consequence, the visco-plastic compliance $\mathbb{P}^{(tg)}$ is singular and cannot be inverted. Such a characteristic is evident when the tensor $\mathbb{P}^{(tg)}$ is represented in the basis $b^{(\lambda)}$ (see Chaps. 7 and 8), which gives a matrix with zero elements in the sixth row and the sixth column. Formally, the condition of incompressibility can be incorporated as another equation to be solved simultaneously with Eq. 9, as proposed by MOLINARI &AL. [1987]. In this Section, however, we opt for using the approximate *penalty method*, consisting in assigning a small fictitious compressibility to the medium (element $\mathbb{P}_{66}^{(tg)}$ of the compliance matrix). This procedure permits an inversion of the visco-plastic compliance and the use of the formalism of the elastic anisotropic inclusion for calculating the Eshelby tensor that enters into the definition of the accommodation tensor (Eq. 10c).

An important remark has to be made at this point regarding the extension of the elastic inclusion formulation to the visco-plastic problem. While the elastic moduli are uniform throughout the medium, the visco-plastic compliances (and so their inverses), can be regarded as uniform only when the stresses induced by the inclusion in its neighborhood do not exceed the range where the tangent constitutive Eq. 9b is valid. As a consequence, the visco-plastic formulation based on the tangent modulus is formally valid only when such condition is met. The implications associated with using either the tangent or the secant moduli to describe the local response of the matrix are discussed in Chap. 8 Sec. 1.7. The practical consequences on the deformation textures of fcc materials are presented below (Sec. 5.1).

From a physical point of view, it is meaningful that the tangent (instead of the secant) plastic modulus enters into the interaction equation, because it describes the *tendency* in the response of the HEM at points where the stress differs from the average value. The relation Eq. 8 between the macroscopic secant and tangent moduli, though, permits one *to express the equations* in terms of the secant moduli. Even when the Eshelby tensor is formally a function of the *tangent moduli*, since it is a homogeneous function of degree zero (see Appendix B), and since $\mathbb{P}^{(tg)} = n\mathbb{P}^{(sec)}$, the same result is obtained when the *secant moduli* are used to calculate it. As a consequence, the interaction tensor can *be expressed* in terms of the secant moduli as:

$$\mathbb{A} : \mathbb{P}^{(tg)} = n\,(\mathbb{I}-\mathbb{E})^{-1} : \mathbb{E} : \mathbb{P}^{(sec)}$$

$$(12)$$

A formalism which takes into account the interaction of each grain with the nearest neighbors, and assumes a HEM beyond such a cluster has been proposed by MOLINARI &AL. [1987] and is described in Section 6 of this Chapter. Such a formalism leads to a more general form of the interaction equation, which reduces to Eq. 10 in what is called the *one-site* approximation, in which the neighboring grains are ignored and only the HEM is considered.

4. Self-consistent Polycrystal Formulation

In the previous Section we relate the stress and the strain rate induced in a visco-plastic inclusion embedded in a visco-plastic homogeneous medium with the stress applied at the boundary of the medium. This relation is a function of the visco-plastic moduli of the inclusion and of the HEM. The inclusion compliance, though, is a function of the stress in the inclusion, which in turn depends on the macroscopic moduli through the interaction Eq. 10. Such relation is evident when Eqs. 2 and 5 are combined with Eq. 10 to express the stress as:

$$\sigma^c = \mathbb{B}^c : \sigma$$

Here:

$$(13a)$$

$$\mathbb{B}^c = (\mathbb{P}^{c(sec)} + \mathbb{A} : \mathbb{P}^{(tg)})^{-1} : (\mathbb{P}^{(sec)} + \mathbb{A} : \mathbb{P}^{(tg)})$$

$$(13b)$$

describes the partitioning of stress and is called the *localization tensor*. Its departure from the identity measures the degree of heterogeneity between inclusion and HEM.

This result can be used to formulate a polycrystal scheme, by assuming that each grain of the aggregate is an inclusion embedded in the HEM that describes the average behavior of the aggregate. The difference with the previous problem is that the overall visco-plastic compliance of the HEM is not known a priori. However, the condition that the weighted average of stress (and strain rate) over the grains has to coincide with the corresponding macroscopic magnitudes provides an expression from which the overall tensor $\mathbb{P}^{(sec)}$ can be calculated in a self-consistent iterative way. The condition that $\langle \sigma^c \rangle = \sigma$ leads to:

$$\langle \mathbb{B}^c \rangle = \langle (\mathbb{P}^{c(sec)} + \mathbb{A} : \mathbb{P}^{(tg)})^{-1} : (\mathbb{P}^{(sec)} + \mathbb{A} : \mathbb{P}^{(tg)}) \rangle = \mathbb{I}$$

$$(14)$$

which can also be written in the alternative form:

$$\mathbb{P}^{(sec)} = \langle \mathbb{P}^{c(sec)} : \mathbb{B}^c \rangle$$

$$(15)$$

Equation 15 indicates that the polycrystal compliance is given by an average of the single-crystal compliances, weighted by the associated volume fraction and the localization tensor $\mathbb{B}^c$, defined by Eq. 13. Since $\mathbb{B}^c$ is a function of $\mathbb{P}^{(sec)}$, both explicitly and through the dependence of the accommodation tensor $\mathbb{A}$ upon $\mathbb{P}^{(sec)}$, Eq. 15 represents an implicit equation from which $\mathbb{P}^{(sec)}$ has to be obtained self-consistently.

4.1 Comparison of the tangent and secant modulus assumptions

The self-consistent polycrystal formulation presented above is formally equivalent to the elastic one of Chapter 7 (except that it is stated in terms of compliances instead of stiffnesses) because a linear response is forced upon the medium. This approximation limits the applicability of the formulation to those systems for which the linear expansion around the average stress is a valid representation of the local response. This assumption is exact for the case $n=1$, which represents the limit of Newtonian viscosity analyzed in Chap. 7 Sec. 5. It is to be expected that the linearization assumption becomes more critical as the power n and/or the degree of inhomogeneity between the grains increase. LEBENSOHN & TOMÉ [1993a] show that the previous formulation approaches asymptotically a lower bound as n increases (rate insensitive limit). Such a response can be qualitatively deduced by inspecting the interaction equation (10) associated with the tangent approximation: as n increases, the matrix becomes more compliant because $\mathbb{P}^{(tg)}=n\mathbb{P}^{(sec)}$ increases. As a consequence, stress deviations are strongly amplified and lead to increasingly large strain-rate deviations. The tendency of the system is to correct for the latter by decreasing the stress discontinuities between grains. In the limit of an infinitely compliant medium the stress tends to be uniform and the lower bound is achieved. The upper bound, on the other hand, would correspond to an infinitely stiff medium (zero compliance). In this case Eq. 10 shows that the rate deviations will be zero independently of the stress deviation in the inclusion. Since it is apparent that neither limit describes the actual response of a real aggregate, in what follows we study what is the consequence of using different effective compliances for a practical case.

We apply a tensile deformation to an fcc aggregate represented by 1000 random orientations. Deformation is assumed to be accommodated by $\langle 110 \rangle (111)$ slip with a CRSS τ_c, and calculations are done for parametric values of the rate sensitivity n applying the formulation described above. The stress and strain rate are calculated in each grain for an imposed strain increment using two forms of the visco-plastic SC formulation: the first one is based on the tangent assumption described in the previous Section and will be denoted SC-tan. The second formulation consists in assuming that the interaction between the matrix and the inclusion is described by the secant modulus $\mathbb{P}^{(sec)}$ instead of the tangent $\mathbb{P}^{(tg)}$. The only modification introduced by the second approach is in the interaction Eq. 10, which now adopts the form

$$\tilde{D} = -A : \mathbb{P}^{(sec)} : \tilde{\sigma} \qquad (16a)$$

where the accommodation tensor

$$A = (I - E)^{-1} : E \qquad (16b)$$

is the same as before because $\mathbb{E}(\mathbb{P}^{(tg)}) = \mathbb{E}(n\mathbb{P}^{(sec)}) = \mathbb{E}(\mathbb{P}^{(sec)})$. We denote the latter approximation as SC-sec.

Several statistics are performed that help to analyze the results. An obvious one gives the overall stress σ and strain rate D as the weighted average of the corresponding grain's tensors. In addition, we calculate the deviation of each stress and strain-rate component with respect to the average, normalized by the norm of the average tensor:

$$\Delta\sigma_\lambda = \sqrt{\frac{\left\langle \left(\sigma_\lambda^c - \sigma_\lambda\right)^2 \right\rangle}{\sum_\mu \sigma_\mu \, \sigma_\mu}} \qquad (17a)$$

$$\Delta D_\lambda = \sqrt{\frac{\left\langle \left(D_\lambda^c - D_\lambda \right)^2 \right\rangle}{\sum_\mu D_\mu D_\mu}} \tag{17b}$$

The stress deviations are zero in the lower-bound approximation, while the strain-rate deviations are zero in the upper-bound (FC) approximation. Clearly, their departure from those limits is an indication of the stiffness of the assumption used to represent the polycrystal. Another statistic refers to the AVerage number of significantly ACtive Systems per grain (AVACS), defined as the *weighted* average of the active systems in each grain $\langle n^c \rangle$, where n^c is computed as follows: after identifying the system where the shear increment $\Delta\gamma$ is maximum, all the systems in the grain where the shear is at least 5% of that maximum are counted as active.

The Taylor factor, given by the ratio (σ/τ_c) between the overall tensile stress and the CRSS in the slip systems, measures the 'stiffness' of the polycrystal under an imposed deformation. The Taylor factor calculated with the SC-tan and the SC-sec models is plotted in Fig. 1a as a function of the rate sensitivity. For comparison purposes, the result associated with an upper-bound calculation, where the same strain increment is imposed on every grain, is also plotted in Fig. 1a. The standard deviations for the tensile stress component and the tensile strain-rate component are plotted in Fig. 1b. It has to be borne in mind, when analyzing these statistics, that they refer to the vector components σ_1 and D_1, defined by Chap. 8 Eq. 29. As a consequence, the tensor components are affected by a factor ($\sigma_1 = \sqrt{\frac{3}{2}}\, S_{33}$ and $D_1 = \sqrt{\frac{3}{2}}\, D_{33}$). As expected, the self-consistent results fall systematically below the upper-bound, but they exhibit a completely different trend as n increases: within the SC-tan scheme the tensile stress achieves a maximum at $n \simeq 10$, decreases slowly for n between 10 and 20, and shows an accelerating decrease for values of n larger than about 20 ($1/n = 0.05$). The predictions of the SC-sec model, on the other hand, increases monotonically with n and tends asymptotically to the upper bound. Coincidentally, within the SC-tan model the standard deviation of the tensile stress component and the AVACS decrease, while the corresponding strain-rate component increases as n increases. The difference between the two models becomes extreme when $n > 20$ (see Fig. 1b). These results indicate that fewer grains that are favorably oriented tend to accommodate the applied strain increment, while more grains become progressively inactive as n increases. The prediction of the SC-sec calculation, on the other hand, is consistent with an upper-bound result for increasing values of n: the deviation in strain rate decreases, the AVACS stabilizes at about 5.5 and the deviation in stress remains fairly uniform at about 15%. Although the previous results depend on the single-crystal anisotropy, the texture of the aggregate, and the deformation mode imposed, we expect that the conclusions will remain qualitative valid in general. As a consequence, the plots in Fig. 1 seem to indicate that when using the tangent self-consistent scheme the value of n should be lower than at most $n = 20$.

It is appropriate, at this point, to discuss the *incremental* self-consistent visco-plastic formulation proposed by HUTCHINSON [1976]. Although Hutchinson claims that his formulation is based on a tangent approach, it is evident from the evolution of the Taylor factor in Fig. 1a that his result is consistent with using the secant and not the tangent compliance in the interaction Eq. 10. The difference between Hutchinson's and the SC-tan formulation presented above is that Hutchinson starts from the definition (7) of the

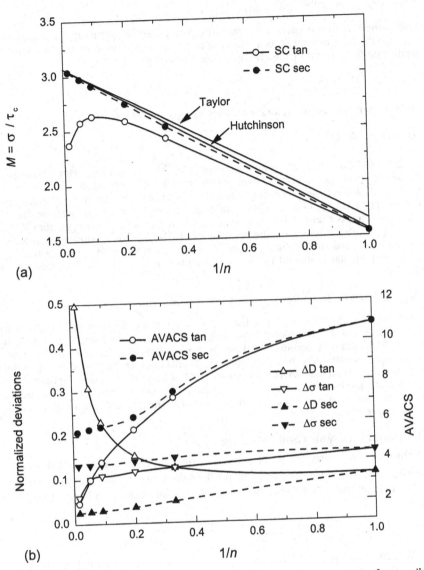

Fig. 1. (a) Taylor factor as a function of the rate sensitivity $m=1/n$ for tensile deformation of a random cubic aggregate, predicted by the Self Consistent scheme using the tangent and the secant modulus. Comparison with predictions of the Full Constraint approach and HUTCHINSON'S [1976] tangent approach. (b) Normalized deviation of the strain-rate and stress components along the tensile direction ($D_1 = \sqrt{3/2}\, D_{33}$ and $\sigma_1 = \sqrt{3/2}\, S_{33}$), and Average Number of Active Systems per grain (AVACS), predicted by the SC tangent and the SC secant schemes.

tangent modulus, which linearly relates *overall* stress increments with *overall* strain-rate increments

$$(\mathbf{D}_{(\sigma+\delta\sigma)} - \mathbf{D}_{(\sigma)}) = \mathbb{P}^{(tg)} : \delta\sigma \tag{18}$$

and uses the interaction Eq. 10a deduced for the inclusion, to relate deviation between increments as:

$$(\delta \mathbf{D}^c - \delta \mathbf{D}) = -\mathbb{A} : \mathbb{P}^{(tg)} : (\delta \sigma^c - \delta \sigma) \tag{19}$$

Within the SC-tan formulation, on the other hand, the tangent modulus is assumed to describe the *local* response. As a matter of fact, the relation between the *local* stress and strain-rate deviations from the average values is given by Eq. 9 as

$$(\mathbf{D}_{(x)} - \mathbf{D}) = \mathbb{P}^{(tg)} : (\sigma_{(x)} - \sigma) \tag{20}$$

and the interaction equation for the inclusion adopts the form Eq. 10

$$(\mathbf{D}^c - \mathbf{D}) = -\mathbb{A} : \mathbb{P}^{(tg)} : (\sigma^c - \sigma) \tag{21}$$

Equation 19 follows from 21 only if the compliance $\mathbb{P}^{(tg)}$ does not vary. Hutchinson applies the result of the inclusion formalism not to an inclusion embedded in a homogeneous medium loaded with a stress σ at the boundary but, instead, to the change induced in the inclusion by a change in the loading condition. However, one has to be aware that when the loading stress changes from σ to $(\sigma + \delta \sigma)$, the *overall modulus* changes from $\mathbb{P}^{(tg)}{}_{(\sigma)}$ to $\mathbb{P}^{(tg)}{}_{(\sigma + \delta \sigma)}$. As a consequence, the compliances to be used in each case for solving the equilibrium Eq. 9d are not the same and the result (19) is not valid. The relation induced by an incremental change of the loading condition may be obtained by differentiating Eq. 21 as:

$$(\delta \mathbf{D}^c - \delta \mathbf{D}) = -\mathbb{A} : \mathbb{P}^{(tg)} : (\delta \sigma^c - \delta \sigma) - \mathbb{A} : \delta \mathbb{P}^{(tg)} : (\sigma^c - \sigma) \tag{22}$$

where we have assumed that the Eshelby tensor (which enters in $\mathbb{A}$) exhibits a weak dependence with $\mathbb{P}$. Equation 22 contains an extra term with respect to Hutchinson's Eq. 19. This extra term accounts for the dependence of the compliance on the stress and is of the order $(n-1)\mathbb{P}^{(tg)}:\delta \sigma$. As a consequence, it turns out that the interaction equation for the increments assumed by Hutchinson (Eq. 19) is $(n-1)$ times less compliant than the relation (22) predicted by the SC-tan formulation. This latter result explains the difference in the response depicted in Fig. 1a, and the fact that Hutchinson's formulation converges to the upper bound as n increases.

4.2 Reorientation by slip and twinning

Within the self-consistent scheme the deformation of each grain is different from the matrix and, as a consequence, the reorientation due to slip has to include the effect of the local rotation. The lattice rotation rate for each grain is given by Chap. 8 Eq. 93 as:

$$W_{ij}^{*} = W_{ij} + \tilde{W}_{ij} - \sum_s \frac{1}{2} \left(b_i^s n_j^s - b_j^s n_i^s \right) \dot{\gamma}^s \tag{23}$$

where W is the anti-symmetric component of the macroscopic distortion rate, $\tilde{W}$ is the local deviation with respect to the average, and the last term is the anti-symmetric component of the plastic distortion rate W^c [KOCKS &AL. 1994b]. The form of the local deviation $\tilde{W}$ is given specifically by Eq. 11 and was derived in Eq. A19 of the Appendix within the inclusion formalism. This rigid rotation of the ellipsoid that represents the grain is due to the fact that the deformation of the inclusion differs from the one of the surrounding medium. This component is null for spherical inclusions, and increases as the ellipsoid becomes more distorted, as TIEM &AL. [1986] show for the elastic inclusion case. For the particular cases of rolling and axisymmetric tension of zirconium alloys presented in the following Section, this term is negligible at the beginning of deformation, but represents about 40% and 20%, respectively, of the total lattice rotation rate given by Eq. 23 at a true strain of 1.0.

In what concerns twinning, the plastic shear associated with it is not homogeneous but localizes in either parallel lamellae or in a portion of the grain. The twinned region represents a fraction $\Delta\gamma/\gamma_t$ of the grain (here $\Delta\gamma$ is the shear in the twinning system and γ_t, typically lower than one, is the characteristic shear of the twin), and an overall volume fraction $f^g = w^g \Delta\gamma/\gamma_t$ (where w^g is the weight of the particular orientation). The twinned fraction adopts a characteristic orientation with respect to the matrix crystal, which amounts to splitting the original grain in two different orientations. Since the number of orientations represented by the twinned fractions increases geometrically during deformation, the simulation of this process becomes quite rapidly intractable from a numerical point of view.

As a consequence, for the calculations that follow we use the Volume Fraction Transfer (VFT) scheme to describe reorientation due to slip and twinning [TOMÉ &AL. 1991a]. While local effects are accounted for in the calculation of the crystallographic reorientation through the term $\widetilde{W}$ in Eq. 23, the orientation and shape of the ellipsoid that represents each grain is only updated in an average sense. Ellipsoid shape and orientation are assumed to be the same for all the grains and are determined by the overall rotation rate W. The reason for this has to do with the assumptions of the VFT scheme and is discussed below.

The Volume Fraction Transfer scheme

While the purpose of this method is to account for all twinning reorientations and their associated volume fractions, it is also applicable to reorientation by slip. The polycrystal is represented as a set of orientations with weights but, contrary to the classical polycrystal models, the orientations are kept fixed while the associated volume fractions are allowed to evolve during deformation. Within the Volume Fraction Transfer (VFT) scheme the Euler space is divided into identical and approximately equiaxed cells of volume $\Delta\phi, \Delta\sin\theta, \Delta\omega$. Each cell g contains an initial volume fraction u^g and represents an orientation defined by the coordinates of its center, the Euler angles $(\phi^g, \theta^g, \omega^g)$. When only the slip systems are active, the reorientation that this grain will experience as a consequence of imposing a strain increment can be represented as a vectorial displacement $(\delta\phi, \delta\theta, \delta\omega)$ in Euler space. If the dimensions of the cell are small, one may assume that every point within the cell will shift by about the same amount, which results in a rigid displacement of the cell as a whole. (Alternatively, one may also consider the distorted cell that results from the non-equal displacement of each corner, instead of assuming a rigid displacement of the cell.) In this process, a portion of the shifted (or, alternatively, distorted) cell g overlaps with the neighboring cells g'. The basis of the VFT scheme is to assume that the volume fraction $\Delta f^{gg'}$ contained in the overlapping portions is transferred from the cell g into the cell g' as a consequence of the strain increment. If the material is assumed to be uniformly distributed within the cell, then $\Delta f^{gg'}$ will be proportional to the overlapping volume. (If the piecewise uniform distribution is replaced by an interpolation algorithm, the transferred fraction has to be calculated as an integral over the overlapping volume). This process of transference, repeated after every strain increment, leads to a gradual variation of the volume fraction associated with each cell. The situation is sketched in Fig. 2a for a two-dimensional Euler space, although for numerical purposes the method is implemented in a three-dimensional space.

When twinning systems are also active inside the grains, one has to account for the twinned fraction $f^g = u^g \Delta\gamma/\gamma_t$. Since the crystallographic orientation of the twinned fraction is completely different from the grain orientation, its representative point in Euler space represents a finite and not an incremental displacement, and the volume fraction f^g has to be transferred to a non-contiguous cell g' of the grid. The situation is sketched in Fig. 2b.

A disadvantage of the VFT scheme is that when volume fractions associated with different orientations are accumulated into a given element of the grid, the deformation history of the reoriented fractions is lost. As a consequence, the grid orientation has to be regarded as an average representative of all grains having similar orientations though generally

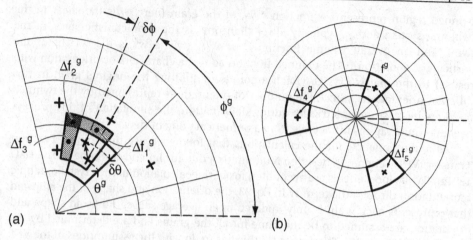

Fig. 2. Schematic representation of two-dimensional cells in Euler space showing the volume fractions transferred from cell n: (a) to the neighboring cells due to reorientation by slip; (b) to non-neighboring cells due to twinning reorientation.

different histories. The grain shape is one of the parameters that has to be regarded in an average sense. That is why grain shape effects enter in the formulation through the overall rotation rate W. The advantage of the VFT method is that it permits one to account exactly for the twinned volume fraction without having to increase the initial number of orientations. In addition, it is also applicable to reorientation by slip alone.

5. Applications and Discussion

The SC model has been applied to materials of various symmetries, and some of these applications are reviewed in this Section and the predicted textures are compared with the experimental ones. For cubic metals, textures are rather similar to those obtained with the FC and the RC models [MOLINARI &AL. 1987], and will be discussed in Section 5.1. Significant differences are observed in plastically more anisotropic hexagonal metals [TOMÉ &AL. 1991a, LEBENSOHN & TOMÉ 1991, 1993a, 1994, LEBENSOHN &AL. 1994, 1996], although sometimes the reorientation due to twinning obscures the comparison between the FC and the SC methods. Examples from hexagonal metals will be discussed in some detail in Section 5.2.

The SC model has been particularly useful for anisotropic minerals, and examples will be illustrated in Chap. 14. For halite, texture predictions based on the Taylor theory are entirely different from those based on the SC model [WENK &AL. 1989b]. In the case of calcite, differences between models are minor due to the large number of available slip and twinning systems [TOMÉ &AL. 1991b]. Calcite is a good example to illustrate the topology analysis of the single-crystal yield surface and its application to texture variations [TAKESHITA &AL. 1987], as will be shown in Section 5.3. Also for dolomite, FC and SC texture predictions are similar and agree with experiments, but the TEM evidence about slip systems agrees better with the number of active systems predicted by the SC model (see BARBER &AL. [1994] and Chap. 6 Fig. 9.). For quartz, the SC model was used to predict the relation between grain shape and grain orientation [WENK &AL. 1989a], and is potentially useful for an interpretation of recrystallization textures [WENK &AL. 1997a]. An extremely anisotropic case, discussed in Section 5.4, is orthorhombic olivine, with very few available slip systems and an open single-crystal yield surface [PARKS & AHZI 1990; TAKESHITA &AL. 1990; WENK &AL. 1991b].

The SC model has also been used to describe plastic deformation of highly anisotropic orthorhombic and monoclinic polymers [PARKS & AHZI 1990]. A generalized SC formulation has been applied to modeling polyphase materials such as $(\alpha+\beta)$ titanium alloys [DUNST & MECKING 1996], olivine-pyroxene [WENK &AL. 1991a] and quartz-mica [CANOVA &AL. 1992]. The extended model and the latter application will be discussed in detail in Section 6.

The SC formulation permits one to account for deviations in the grain response from the average behavior of the polycrystal. Such deviations are significant for materials with marked plastic anisotropy; as a consequence, important differences with respect to the predictions of classical formulations should be expected for low-symmetry materials. However, one has to bear in mind that if the heterogeneity between the grain and the HEM properties is large, the stress deviation with respect to the average in the vicinity of the grain can be expected to be large too, and the linear expansion represented by the tangent assumption will provide a qualitatively correct but quantitatively approximate description of the matrix-grain interaction.

In this Section, we present deformation textures predicted by the SC formulation, and compare them with the corresponding experimental textures. An exponent of $n=19$ is used to describe the rate dependence of the slip and twinning systems in all the calculations. The choice of this value represents a compromise between describing a relatively rate insensitive material, and staying within the range of validity of the tangent approximation discussed in Section 4.1. In Section 5.1 we simulate rolling, tension and compression of the more symmetric material, a generic fcc aggregate. Next, in Section 5.2, we simulate rolling, axisymmetric tension and axisymmetric compression of the less symmetric hexagonal zirconium. Deformation textures associated with pure shear of trigonal calcite and orthorhombic olivine are calculated in Sections 5.3 and 5.4, respectively.

In order to assess the influence of the interaction assumption on the results, calculations are made using the SC-tan and the SC-sec formulations. The latter, described in Section 4.1, uses the overall secant modulus (instead of the tangent) in the interaction equation (16). As we have shown, such an assumption leads to a stiff matrix-grain interaction and gives results which are close to the ones of the FC model, although it has the advantage of accounting (in a weak sense) for the relative anisotropy of grain and matrix. In addition, the relaxation of some of the strain components is built into the model through the dependence of the grain-matrix interaction with the evolving grain shape. As a consequence, it is not necessary either to introduce an 'ad hoc' transition from the FC to the RC regime, or to force some of the stress components to be strictly zero in every grain.

The ratio of CRSSs is kept constant throughout the deformation process: i.e., the change in yield surface is homothetic. At the beginning of deformation the grains are assumed to be equiaxed (the representative ellipsoid is a sphere) and, as deformation proceeds, the grain shape is updated *in average* but not individually. The grain shape enters into the calculation of the Eshelby tensors $\mathbb{E}$ and $\mathbb{W}$, which are a function of the ellipsoid axes.

We use the grain statistics presented in Section 4.1 to complement the analysis, namely: the normalized deviation of each stress component and each strain-rate component defined by Eq. 17, and the AVerage number of ACtive Systems per grain

(AVACS). It has to be borne in mind, when analyzing these statistics, that they refer to the vector components of the tensor, defined by Chap. 8 Eq. 29. As a consequence, the shear components σ_{23}, σ_{13} and σ_{12} are affected by a factor $\sqrt{2}$, and $\sigma_1 = \sqrt{\frac{3}{2}} S_{33}$ (same for the strain rate). Since in this Section we study materials that present more than one crystallographic deformation system, we also analyze the relative activity of each deformation mode. The latter is computed as the ratio of two *weighted* averages over the grains: the sum of the plastic shears contributed by a given mode (i.e. prismatic systems) divided by the sum of the shears contributed by all the deformation modes (i.e. prismatic+pyramidal+basal systems).

5.1 Cubic textures

Tension, compression and rolling of fcc crystals up to a true strain of 100% along the main deformation direction are analyzed in this Section. An aggregate of 500 initially random orientations is used and deformation is assumed to be accommodated by $\langle 110 \rangle \{111\}$ slip. Calculations are performed using the SC-tan and SC-sec formulations discussed in the previous Section.

Rolling

Rolling was simulated imposing plane-strain conditions and using a reference frame where $x_1 \equiv$ RD, $x_2 \equiv$ TD and $x_3 \equiv$ ND. Figure 3a,b displays the distribution of (111) poles after 100% true strain in the normal direction (64% thickness reduction) as predicted by the SC-sec and the SC-tan schemes respectively. While both approximations correctly reproduce the Cu component along the RD associated with the relaxation of D_{13} and D_{23}, they differ in the position and relative intensity of the texture component that develops between the ND and the RD. An important difference appears in the average number of active systems per grain required in each case to achieve deformation (Fig. 3c). They are about three within the SC-tan scheme, indicating that the tendency is for the grains to accommodate deformation by deforming in plane strain *with respect to the crystal axes*. The stress in most grains is along edges and not vertices of the five-dimensional SCYS. The AVACS starts at around five for the SC-sec case and decreases as the grains become flatter and some of the shear strains can deviate from zero without inducing large stress deviations. The evolution of the deviation of the shear stress components (Fig. 3d,e) is consistent with the RC picture: $\Delta\sigma_{13} \sim \Delta\sigma_4$ and $\Delta\sigma_{23} \sim \Delta\sigma_3$ decrease with accumulated strain. The evolution of the corresponding strain-rate deviations is more complex: while ΔD_{13} increases, ΔD_{23} decreases. The latter deviation is a consequence of the grain-matrix interaction becoming stiffer for the shear ε_{23}, and shows the effect that an increasing directional anisotropy in the matrix properties (tensors $\mathbb{P}$ and $\mathbb{A}$ in Eqs. 10 and 16) has over the grain's response. Such an effect cannot be captured by a model without interaction.

Although the latter result may be perceived to be in contradiction with the RC model, it is not so: the basis of the RC assumption is that grain shape evolution permits some of the strain components in the grains to deviate from the average because in doing so they do not induce high interaction stresses. However, it is not a requirement of the RC model that they *have* to deviate, only that they *can* deviate. As a matter of fact, the condition imposed by the RC model is that the shear stresses σ_{13} and σ_{23} have to be zero in the rolled grain, and the deviation exhibited by the corresponding shear strain components is only a consequence of imposing such a condition.

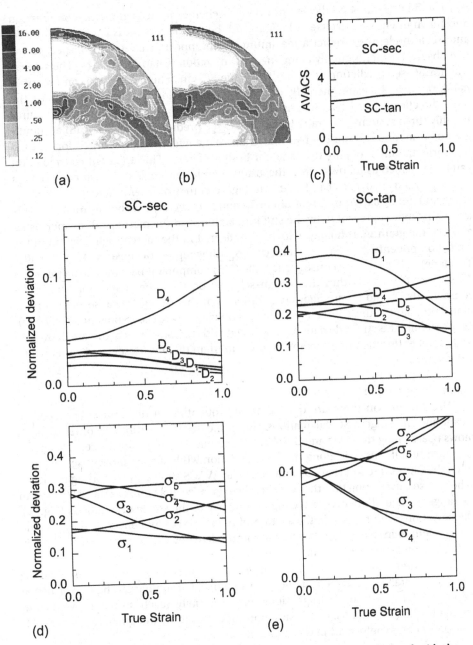

Fig. 3. 111 pole figures after 100% true strain ($\varepsilon_{33} = -1$) in rolling calculated with the Self Consistent scheme using $n=19$ and: (a) the secant modulus, (b) the tangent modulus, (c) Average Number of Active Systems per grain (AVACS). Normalized deviations of the vectorial stress and strain-rate components Eq. 17 predicted by: (d) the SC-secant, and (e) the SC-tangent schemes.

Axisymmetric compression

The final textures predicted with the SC schemes for axisymmetric compression are shown in Fig. 4 as inverse pole figures of the tensile axis. It is known that the FC and RC formulations predict a maximum which appears shifted with respect to the one observed experimentally along the $\langle 101 \rangle$ direction (Chap. 9 Sec. 2.4). The same is true for SC-sec predictions (Fig. 4a) while the SC-tan scheme (Fig. 4b), on the other hand, predicts the experimentally observed concentration of inverse poles along the $\langle 101 \rangle$ direction. In what concerns the AVACS (Fig. 4c), the SC-tan scheme requires roughly three systems per grain, while the SC-sec predicts an increase with accumulated deformation despite the fact that grain shape effects are accounted for and the final orientation is roughly the same for both schemes. This apparent contradiction can be explained as follows: while the grains tend to minimize the shear components $\sigma_3 \sim \sigma_{23}$ and $\sigma_4 \sim \sigma_{13}$, as can be seen in Fig. 4d,e, they also tend to deform by plane strain and not by axisymmetric strain (ΔD_{13} and ΔD_{23} increase with deformation). This latter tendency is prevented by the stiff interaction with the matrix and the increasing effect of the grain morphology, which are reflected in the increasing deviation of the stress components $\sigma_2 \sim (\sigma_{22} - \sigma_{11})$ and $\sigma_5 \sim \sigma_{12}$ (they represent shears at 45° in the RD-TD plane) in Fig. 4d. As a consequence, the (101) component becomes unstable in the SC-sec simulation. Within the SC-tan scheme, on the other hand, the strain-rate components D_2 and D_5 are allowed to differ from the average because the matrix is more compliant in these directions, which stabilizes the (101) component (see Fig. 4e). As for rolling, the stress deviations associated with the SC-tan scheme are systematically smaller (for the same accumulated strain) than the ones associated with the SC-sec scheme.

Axisymmetric tension

Also for tension there are qualitative and quantitative differences between the secant and the tangent SC calculations. In both cases the distribution of tensile axes shows peaks along the $\langle 111 \rangle$ and the $\langle 001 \rangle$ directions. However, for the secant scheme the maximum intensity is along the $\langle 111 \rangle$ direction while for the tangent scheme it is along the $\langle 001 \rangle$ direction (Figs. 5a and 5b). The AVACS shows an increasing tendency in both cases, indicating that the stress tends to localize at a vertex of the SCYS even in the tangent calculation, due to the high symmetry of the stable orientations (Fig. 5c). The deviations with respect to the average of the stress components $\sigma_2 \sim (\sigma_{22} - \sigma_{11})$ and $\sigma_5 \sim \sigma_{12}$ seem to indicate a tendency to accommodate deformation by grain curling, in both the tangent and secant schemes (Fig. 5d,e).

It is of interest to note that softer $\langle 001 \rangle$ grains will accumulate more strain than the harder $\langle 111 \rangle$ grains. As a consequence, if a significant strain hardening is assumed at the crystal level, the $\langle 001 \rangle$ orientations will eventually reach strength levels comparable to the $\langle 111 \rangle$ grains. Such situation may lead to a strengthening of the $\langle 111 \rangle$ fiber after enough deformation is accumulated.

5.2 Zirconium alloys textures

In this Section we present textures calculated with the visco-plastic SC tangent formulation described above for zirconium alloys deforming in plane strain, axisymmetric tension and axisymmetric compression. Experimental textures and textures calculated with an FC scheme are also presented for comparison purposes.

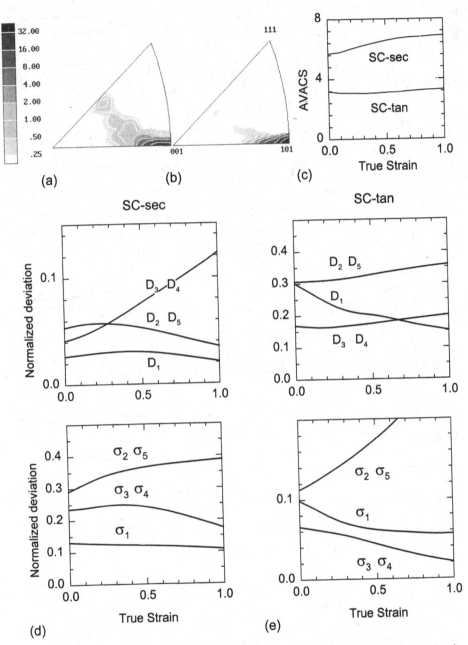

Fig. 4. Inverse pole figure of the compressive axis calculated after 100% true strain ($\varepsilon_{33}=-1$) in axisymmetric compression calculated with the Self Consistent scheme using $n=19$ and: (a) the secant modulus, (b) the tangent modulus, (c) AVACS. Normalized deviations of the vectorial stress and strain-rate components Eq. 17 predicted by: (d) the SC-secant, and (e) the SC-tangent schemes.

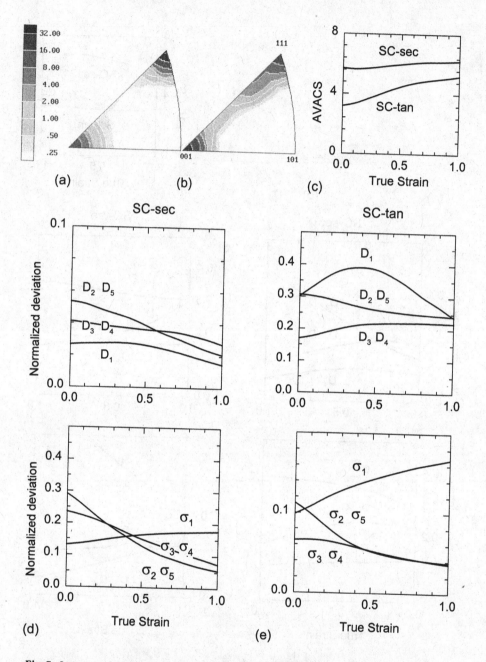

Fig. 5. Inverse pole figure of the tensile axis calculated after 100% true strain ($\varepsilon_{33}=1$) in axisymmetric tension calculated with the Self Consistent scheme using $n=19$ and: (a) the secant modulus; (b) the tangent modulus. (c) AVACS. Normalized deviations of the vectorial stress and strain-rate components Eq. 17 predicted by: (d) the SC-secant, and (e) the SC-tangent schemes.

According to the experimental evidence [TENCKHOFF 1974, 1978; AKHTAR 1973a, 1973b, 1975; POCHETTINO &AL. 1992], the following deformation systems are found to be active at low temperatures in zirconium alloys: (a) $\{10\bar{1}0\}\langle1\bar{2}10\rangle$ prism slip (pr$\langle$a$\rangle$); (b) $\{10\bar{1}2\}\langle10\bar{1}1\rangle$ tensile twinning (ttw) plays an important role in deformation; and (c) $\{2\bar{1}\bar{1}2\}\langle2\bar{1}\bar{1}3\rangle$ compressive twins (ctw) and/or $\{10\bar{1}1\}\langle11\bar{2}3\rangle$ pyramidal slip (pyr$\langle$c+a$\rangle$) play a secondary role at low temperatures. While pr$\langle$a$\rangle$ is by far the easiest and most active system, some activity is required on the other modes in order to provide deformation along the c-axis of the hexagonal crystals. It must be noted, however, that the closure of the SCYS is not a necessary condition within SC models, and that deformation can be accommodated all the same with fewer than five active systems per grain.

The relative values of CRSSs assigned to the deformation modes determines which systems are active and to what extent they contribute to the grain deformation and, as a result, to its reorientation and to the final texture. As a consequence, the comparison of experimental and predicted textures provides a way of inferring the active systems and their associated CRSSs if the polycrystal model gives a reliable description of plastic deformation. A detailed discussion of the relation between the active systems and the resultant textures in these materials can be found in the works of TOMÉ &AL. [1991a], LEBENSOHN & TOMÉ [1991, 1993a], LEBENSOHN &AL. [1993, 1994, 1996], POCHETTINO &AL. [1994].

Since in the case of hexagonal crystals the diversity of deformation systems makes it unfeasible to try every combination of CRSSs, it helps to identify the *topological domains* of the SCYS introduced in Chap. 8. Within those domains the CRSSs adopt values that lead to no change at all in the calculated textures when an FC model is applied. When an SC model is used, the sharp texture transition that the FC model predicts when crossing the domain boundaries is replaced by a gradual transition. A detailed analysis of the SCYS topology of Zr alloys can be found in the works of TOMÉ & KOCKS [1985], TOMÉ &AL. [1991a] and LEBENSOHN & TOMÉ [1996].

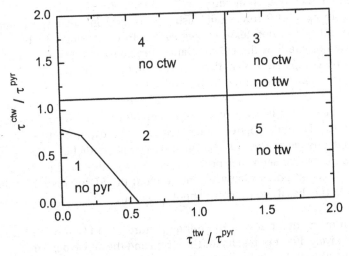

Fig. 6. Topological domains of the SCYS for the systems observed to be active in Zircaloy. In all the cases τ^{pr} is taken as half the minimum of $\{\tau^{pyr}, \tau^{ttw}, \tau^{ctw}\}$ in order to guarantee that it is always an active deformation mode.

For the particular deformation systems considered here, the topological map that combines the CRSS of pyramidal slip, tensile twins, and compressive twins, contains five domains of CRSSs (see Fig. 6). For calculating the topological domains, the CRSSs of the twinning systems, τ^{ttw} and τ^{ctw}, have been normalized to τ^{pyr}, while τ^{pr} has been taken as half the minimum CRSS of the other three systems, in order to guarantee that it is always present in the SCYS. In addition to prism slip, only tensile and compressive twins are active in domain 1, while all three modes add to the prismatic contribution within domain 2. When $\tau^{ttw} > 1.23\tau^{pyr}$ and $\tau^{ctw} > 1.11\tau^{pyr}$, then only pyr⟨c+a⟩ is active (domain 3); and when either the former or the latter condition is fulfilled, pyr⟨c+a⟩ is present in combination with compressive twins (domain 5) or tensile twins (domain 4), respectively. In what follows, we will see that observed Zircaloy textures are compatible with CRSSs within domain 1 and, occasionally, within domain 4.

For the calculations that follow, the ratio of CRSSs is kept constant throughout the deformation process. This is equivalent to assuming homothetic hardening, and implies that the representative point in the topologic domain does not change with deformation. A rather rate-insensitive exponent $n=19$ is used to describe the rate dependence of the slip and twinning systems in the calculations, and reorientation by twinning is treated using the Volume Fraction Transfer (VFT) scheme described in Section 4.2. All the SC results are derived using the tangent modulus.

Rolling

Within domain 1 the following CRSS ratios, $\tau^{pr} : \tau^{ttw} : \tau^{ctw} = 1.0 : 1.25 : 2.5$ give the best agreement with experimental textures, and are used to simulate rolling up to a true strain of 1.0 (64% thickness reduction). Figures 7a and 7b show typical experimental 0002 and 10$\bar{1}$0 pole figures of rolled Zircaloy-4. The basal poles exhibit a maximum at an angle between 35° and 40° measured from the ND towards the TD, while the prism poles exhibit a weaker maximum at the RD. Figures 7c and 7d are the basal and prism pole figures, respectively, predicted by the FC approach in combination with the VFT scheme. Figures 7e and 7f correspond to the SC-tan calculation, also implemented in combination with a VFT scheme. Both simulations, FC and SC, are carried up to $\varepsilon_{33} = -1.0$ true strain, using a 10° grid in Euler space and initial weights consistent with a non-textured aggregate. In what concerns the basal pole distribution, we observe that the FC simulation predicts a maximum in the ND–TD plane, shifted by comparison with the experimentally observed one, and two unobserved texture components along the RD and the TD. The latter are a consequence of a relatively high activity of the compressive twins forced by the FC formulation, as will be discussed later. For the case of the SC calculation, the basal poles show a larger tilting towards the TD and a better agreement with the experiment, while the unrealistic components along the RD and the TD are not predicted. Concerning the prismatic poles, it can be seen that both FC and SC textures predict the maximum along the RD. The component between the ND and the RD predicted by the FC is associated with the basal poles component along the TD and is not observed experimentally.

In addition to the difference in the rolling textures, the FC and the SC schemes also give qualitatively different results for the strain and the active deformation modes in each grain. Within the FC scheme, the requirement of homogeneous deformation forces high stress discontinuities between grains (the standard deviation of the stress

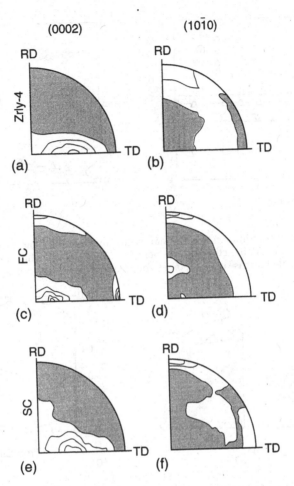

Fig. 7. (a) 0002 and (b) 10$\bar{1}$0 measured pole figures of rolled Zircaloy-4 [TOMÉ &AL. 1988]; (c) 0002 and (d) 10$\bar{1}$0 pole figures calculated with the FC formulation, after deforming up to ε_{33}=−1.0 true strain in rolling using τ^{pr}=1.0, τ^{ttw}=1.25, τ^{ctw}=2.5 ; (e) 0002 and (f) 10$\bar{1}$0 pole figures calculated with the SC-tan formulation for the same true strain and CRSSs. (Lines correspond to multiples of random orientations. Shaded area indicates intensity values lower than 1. Equal-area projection).

components with respect to the average is typically of the order of 30%), an average of about five active systems per grain and a non-negligible activity of the deformation modes with high CRSS. Within the SC scheme, on the other side, each grain deforms differently, depending on its orientation and its anisotropy relative to that of the matrix. As a consequence, the stress distribution is smoother (the standard deviation is about 15% for the stress components, and 25% for the strain-rate components), deformation is accommodated with fewer active systems, and the 'soft' deformation modes are favored with respect to the 'hard' ones. This situation is evident in Figs. 8a and 8b, where the relative contribution of each deformation mode and the average number of active systems per grain are plotted against accumulated deformation, for the FC and the SC cases, respectively. It can be seen in Fig. 8a that within the FC case the activities of the tensile and compressive twinning systems are comparable and

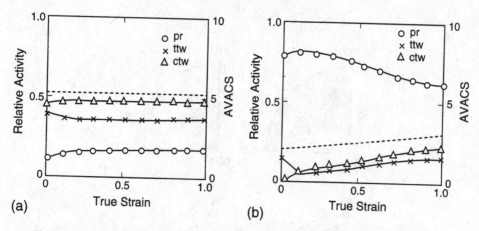

Fig. 8. Relative contribution of each deformation mode to the total shear (solid lines) and Average Number of Active Systems per grain (AVACS, dashed line) as a function of deformation. Rolling case of Fig. 7: (a) FC formulation; (b) SC-tan formulation.

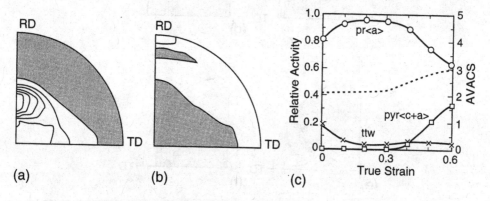

Fig. 9. (a) 0002 and (b) $10\bar{1}0$ pole figures calculated with the SC-tan formulation after rolling to $\varepsilon_{33}=-0.6$ true strain using $\tau^{pr}=1.0$, $\tau^{ttw}=1.5$, $\tau^{pyr\langle c+a\rangle}=4.0$ (Multiples of random orientations. Equal-area projection). (c) Relative contribution of each mode to the total deformation and AVACS as a function of the strain.

much higher than the prismatic activity, and the number of significantly active systems is around five. The implication of this result is that tensile and compressive twins should be simultaneously present in a given grain, a microstructural feature which is rarely observed experimentally. For the SC calculation, on the other hand, most of the deformation is accommodated through prismatic slip, and three active systems per grain suffice for that matter (Fig. 8b), which is in better agreement with the experimental evidence.

Next, we consider a combination of CRSSs belonging to the topological domain 4 of Fig. 6, with the ratios $\tau^{pr} : \tau^{ttw} : \tau^{pyr\langle c+a\rangle} = 1.0 : 1.5 : 4.0$. The basal and prism pole figures after 45% thickness reduction in rolling ($\varepsilon_{33} = 0.6$), are plotted in Figs. 9a and 9b, while the statistic on system activity is reported in Fig. 9c. This combination of active systems illustrates the complexity of plastic deformation in this material. Observe that deformation is overwhelmingly accommodated through prismatic slip (Fig. 9c), while tensile twinning is operative at the beginning of deformation but is gradually replaced by pyramidal slip activity, as the grains reorient. All this, together with the absence of compressive twins, leads to a completely different distribution of basal poles, with a maximum in the ND–RD plane, although the distribution of prism poles is similar as before. A texture like the one in Fig. 9 has been reported by CHARQUET &AL. [1987] in samples of Zircaloy-4 cold rolled under different processing conditions from the ones leading to the experimental texture shown in Fig. 7.

Axisymmetric tension

We have shown above that the SC model and a combination of CRSSs within domain 1 give rolling textures which are consistent with the experimental evidence. For the model to be reliable, it must also reproduce the experimental observations associated with other macroscopic deformation modes. It is for this reason that in what follows we analyze the predictions of the SC approach for the cases of axisymmetric tensile and compressive deformation of Zr alloys, and compare them with experimental rod textures.

Experimental Zircaloy-2 rod textures have been reported by MACEWEN &AL.[1988,1989]. Although these textures are obtained after a cold swaging process, followed by a 4 hour anneal at 925 K, no relevant differences between them and the texture associated with an axisymmetric strain path are expected. Textures of swaged Zr-2.5%Nb are reported by SALINAS & JONAS [1992], SALINAS & ROOT [1991] and SALINAS [1995]. However, Zr-2.5%Nb is a two-phase aggregate with about 15% volume fraction of β-phase (bcc) at temperatures lower than 610°C. Above this temperature the β-phase (hcp) transforms into β, until at about 850°C all the α-phase has transformed. The presence of the β-phase and the fact that part of the swaging is performed at temperatures above the transus, makes it difficult to analyze the measurements on Zr-2.5%Nb as coming from the deformation of a single phase aggregate. As a consequence, only the textures of Zircaloy-2 are going to be compared with the calculated textures.

In what follows, the FC and the SC-tan formulations are used to simulate axisymmetric deformation in tension, up to 100% true deformation and starting from a non-textured polycrystal. Axisymmetry makes it unnecessary to plot complete pole figures because the iso-intensity lines consist of circles, concentric with the axial direction. As a consequence, the basal and prismatic pole intensities are plotted in Fig. 10 as a function of the tilt angle of the poles with respect to the tensile axis.

The basal pole intensities measured in Zircaloy-2 are shown in Fig. 10a, super-imposed to the basal pole intensities calculated with the FC and SC schemes. Both simulations correctly predict the basal component perpendicular to the axial direction, but a relative maximum at 20^0 that appears when using FC is not predicted by the SC simulation. As for the case of rolling, this maximum is a consequence of the high activity of compressive twins required by the FC calculation, as will be discussed later. Figures 10b and 10c depict the experimental FC and SC intensities vs. tilt angle for the $(10\bar{1}0)$ and the $(1\bar{2}10)$ poles, respectively. The comparison of the predicted and the experimental prismatic pole intensities shows that the SC calculation reproduces the experimental features much better than the FC predictions.

The deformation mode activity associated with axisymmetric tension complements the previous results and is reported in Fig. 11. The activities corresponding to the FC calculation (Fig. 11a) and the SC calculation (Fig. 11b) are completely different. It can be seen that while the FC requires a substantial amount of tensile and compressive twinning, the SC accommodates most of the deformation through the

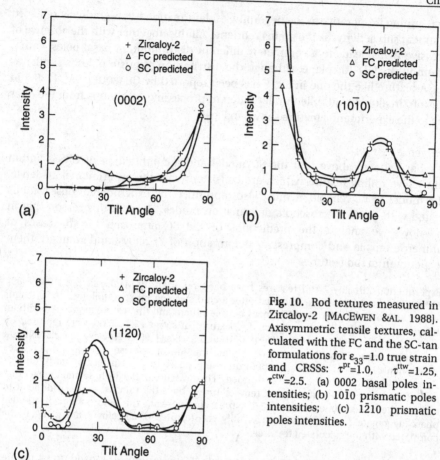

Fig. 10. Rod textures measured in Zircaloy-2 [MACEWEN &AL. 1988]. Axisymmetric tensile textures, calculated with the FC and the SC-tan formulations for $\varepsilon_{33}=1.0$ true strain and CRSSs: $\tau^{pr}=1.0$, $\tau^{ttw}=1.25$, $\tau^{ctw}=2.5$. (a) 0002 basal poles intensities; (b) $10\bar{1}0$ prismatic poles intensities; (c) $1\bar{2}10$ prismatic poles intensities.

easier prismatic slip, with a minor contribution of tensile twinning and a negligible amount of compressive twinning. The high reorientation rate associated with twinning in the FC scheme prevents the prism poles from achieving a stable orientation and is responsible for the flat appearance of the FC intensity curve. The high prism activity predicted by the SC is, on the other hand, a known experimental feature in Zirconium alloys. The average active systems are around five in the FC case, while only two systems per grain suffice to accommodate the deformation in the SC case (Fig. 11). These results emphasize another qualitative difference between both schemes, namely, that within the SC calculation the SCYS does not have to be closed, because the matrix accommodates deformation when the grain lacks five independent systems or when some of them are too 'hard' for being activated.

Axisymmetric compression

The following case corresponds to an axisymmetric compression test performed on the swaged Zircaloy-2 bar of the previous case [MACEWEN &AL. 1988]. Compression experiments performed in Zr-2.5%Nb by SALINAS & JONAS [1992], SALINAS & ROOT [1991] and SALINAS [1995] are bound to be affected by the presence of the β-phase and are not used in the comparison that follows. The initial rod texture used for the

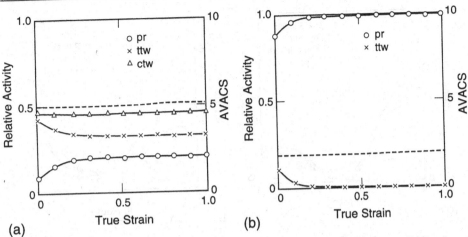

Fig. 11. Relative contribution of each deformation mode to the total shear (solid lines) and AVACS (dashed line) as a function of deformation. Axisymmetric tension case of Fig. 10: (a) FC formulation; (b) SC-tan formulation.

simulation is the one calculated for 100% axisymmetric tension and is reported in Fig. 12. The experimental (0002) pole intensities after 20% compressive deformation, and the corresponding predictions of the FC and the SC-tan calculations are reported in Fig. 12a, while Figs. 12b and 12c show the $(10\bar{1}0)$ prism and the $(11\bar{2}0)$ prism intensities, respectively. The strong reorientation of basal poles along the axial direction and the reorientation of the prismatic poles are accurately reproduced both by the FC and the SC calculations. However, such a reorientation is achieved via completely different deformation system activities in each calculation scheme, as can be seen in Figs. 13a and 13b. Once again, the FC scheme requires high tensile and compressive twinning activity, while the experimental evidence reported by BALLINGER [1979] indicates that compressive twins are not usually observed in Zircaloy-2. The SC calculation, on the other hand, involves mainly prismatic slip, very low tensile and even lower compressive twin activity. The average active systems are five for FC and around three for SC (Fig. 13). The standard deviation in the stress components is of about 30% for FC and of about 15% for SC. In the latter case the relative deviation of the strain-rate components with respect to the average is around 20%.

5.3 Calcite textures

The deformation textures of calcite rocks (like limestone and marble), and their relevance to geological applications have been discussed in some detail in Chap. 6 Sec. 1.1. The purpose of the present Section is to complement the aforementioned discussion with the modeling of the plastic response of calcite.

Calcite constitutes an interesting and challenging material for simulating the plastic behavior, not only because it has a less symmetric trigonal crystal structure, but also because it exhibits high plastic anisotropy, presents a variety of deformation modes with CRSSs that depend strongly on temperature, and twins profusely at room temperature. As a result, the observed calcite textures are a consequence not only of the deformation mode but also of the temperature, and can be used as an indicator of the geological conditions under which deformation took place.

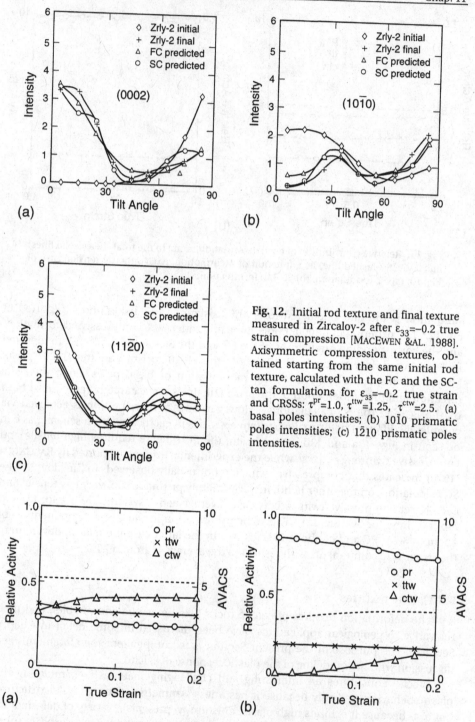

Fig. 12. Initial rod texture and final texture measured in Zircaloy-2 after $\varepsilon_{33}=-0.2$ true strain compression [MACEWEN &AL. 1988]. Axisymmetric compression textures, obtained starting from the same initial rod texture, calculated with the FC and the SC-tan formulations for $\varepsilon_{33}=-0.2$ true strain and CRSSs: $\tau^{pr}=1.0$, $\tau^{ttw}=1.25$, $\tau^{ctw}=2.5$. (a) basal poles intensities; (b) $10\bar{1}0$ prismatic poles intensities; (c) $1\bar{2}10$ prismatic poles intensities.

Fig. 13. Relative contribution of each deformation mode to the total shear (solid lines) and Average Number of Active Systems per grain (AVACS, dashed line) as a function of deformation. Axisymmetric compression case of Fig. 12: (a) FC formulation; (b) SC-tan formulation.

The following deformation modes have been observed in calcite [TAKESHITA &AL. 1987]: three twinning systems of type $\{01\bar{1}8\}<0\bar{4}41>$ (to be denoted e^+), three slip systems $\{10\bar{1}4\}<20\bar{2}1>$ (r^-) plus the opposite ones (r^+), and six slip systems $\{01\bar{1}2\}\langle 20\bar{2}\bar{1}\rangle$ (f^-). Slip of the type r^- is usually the easiest to activate, while the systems with the opposite Burger vector (r^+) exhibit a systematically higher CRSS. Depending on the deformation temperature, slip f^- or twinning e^+ may be the next easiest systems, and slip f^+ is usually not observed, indicating that the associated CRSS is high compared to the other modes. As a consequence of these characteristics, the SCYS of the calcite crystal is not only very anisotropic but, in addition, it is non centrosymmetric. Just as for hexagonal materials, a topological analysis of the SCYS is of much help for analyzing the combinations of active modes and associated CRSSs before attempting the simulation of plastic deformation and texture development. A complete analysis of the SCYS for this material can be found in the work of TAKESHITA &AL. [1987].

The main topologic domains for the active systems mentioned above are shown in Fig. 14, where the CRSSs τ^{f^-} and τ^{e^+} are referred to the CRSS τ^{r^-}, while a constant ratio $\tau^{r^+} = 1.5\ \tau^{r^-}$ is assumed for the opposite systems. Within each of the domains indicated in the figure a certain combination of deformation modes is active. In the extreme cases of the topological domain K the SCYS excludes e^+ twinning and r^- slip, while in the domain B deformation occurs mainly by r^- slip and abundant e^+ twinning. Domains F and H are characterized by contributions from all four modes in varying amounts. In the applications that follows, CRSS within domains B and H are chosen to calculate texture development.

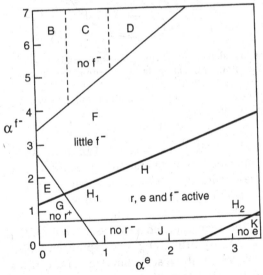

Fig. 14. Simplified SCYS topology diagram for calcite represented as an $\alpha^{f^-} = \tau^{f^-}/\tau^{r^-}$ versus $\alpha^{e^+} = \tau^{e^+}/\tau^{r^-}$ section at constant $\tau^{r^+} = 1.5$ and $\tau^{r^-} = 1.0$. Main domains are indicated, together with associated active systems.

Figures 15a and 15b show two experimental textures of calcite deformed in plane-strain compression up to 0.5 true strain (40% finite shortening), measured at 100°C and 400°C, respectively [TAKESHITA &AL. 1987]. Calcite presents a non-negligible strain-rate sensitivity and for the purpose of the calculation an exponent $n=9$ is assumed, which is the experimentally observed value for marble [HEARD & RALEIGH 1972]. The sets of CRSSs chosen are : (1) $\tau^{r^-} = 1.0$, $\tau^{r^+} = 1.5$, $\tau^{e^+} = 0.2$ within domain B, and (2) $\tau^{r^-} = 1.0$, $\tau^{r^+} = 1.5$, $\tau^{f^-} = 1.0$, $\tau^{e^+} = 3.0$ within domain H. The latter is close to the topological transition to domain K. The predicted textures are plotted in Figs. 15c and

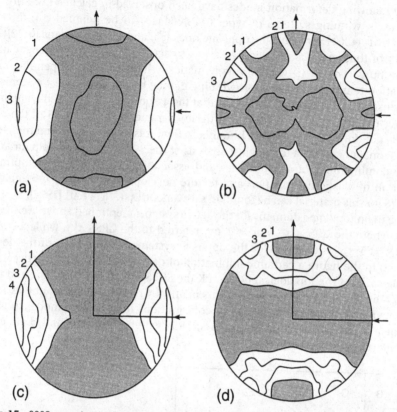

Fig. 15. 0002 experimental pole figures for calcite deformed in pure shear (rolling) at (a) 100°C and (b) 400°C [TAKESHITA &AL. 1987]; (c) 0002 pole figures predicted after $\varepsilon_{33} = -0.5$ true strain using $\tau^{r^-} = 1.0$, $\tau^{r^+} = 1.5$, $\tau^{e^+} = 0.2$; (d) Same using $\tau^{r^-} = 1.0$, $\tau^{r^+} = 1.5$, $\tau^{r^+} = 1.0$ and $\tau^{e^+} = 3.0$. (Equal-area projection. Multiple of random orientations. Shaded area indicates intensity values lower than one.)

15d, while the relative system activity and the AVACS are depicted in Fig. 16. It can be seen that the differences in the experimental textures at different temperatures can be explained in terms of a deactivation of the e^+ twinning and its replacement by f^- slip at high temperatures. The replacement of twinning by slip at high temperatures is a feature also observed in hexagonal materials. The predicted textures and the relative system activity are not too different from the ones predicted by TAKESHITA &AL. [1987] using a Taylor approach, and by TOMÉ &AL. [1991b] using a SC approach which assumes isotropic response in the matrix. We attribute such similarity between the predictions of different models to the scarcity of deformation modes present in each grain which, by reducing the choice of active systems, leads to similar reorientations in all the cases. However, as in previous applications discussed in this Chapter, the SC differs from the FC approach in the average number of active systems required to accommodate the imposed deformation. While within the former about three systems are active per grain, the FC approach requires more than five systems [TOMÉ &AL. 1991b].

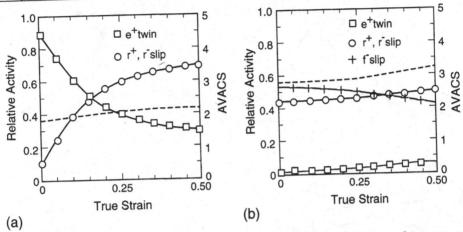

Fig. 16. Relative contribution of each mode to the total shear (solid lines) and Average Number of Active Systems per grain (AVACS, dashed line) as a function of strain for the calcite case of Fig. 15. (a) Case τ^{r^+} =1.5, τ^{r^-} =1.0, τ^{e^+} =0.2 ; (b) Case τ^{r^+} =1.5 , τ^{r^-} =1.0, τ^{f^+} =1.0 and τ^{e^+} =0.2.

5.4 Olivine textures

Olivine is a mineral of orthorhombic structure which enters in the composition of the upper mantle of the Earth. Experimental deformation textures are presented in Chap. 6 Sec. 1.4, and their relevance to seismic waves propagation in the upper mantle is analyzed and discussed in Chap. 7 Sec. 3.5 and Chap. 14 Sec. 3. In the present Section we model axisymmetric compression of olivine as an example that highlights the complexity of this material.

Olivine single crystals lack enough deformation systems to accommodate an arbitrary deformation and, as a consequence, the associated SCYS is open. This latter feature complicates the modeling of plastic deformation of olivine polycrystals and excludes the possibility of using a direct Taylor approach. Previously, deformation of pure olivine has been modeled with the relaxed Taylor theory [TAKESHITA &AL. 1990] and the Constrained Hybrid theory of PARKS & AHZI [1990]. WENK &AL. [1991a] have modeled olivine and the two-phase olivine-pyroxene system peridotite using a cluster self-consistent model which is described in Section 6 of this Chapter. Some geological applications of self-consistent simple shear simulations of olivine aggregates will be discussed in Chapter 14.

In what follows we present a calculation done with the tangent SC model introduced in Section 4, in order to demonstrate how the scheme handles a material with such extreme plastic characteristics. The lattice parameters of the olivine crystal are a=0.4799 nm, b=1.0393 nm and c=0.6063 nm. The active slip systems are (010)[100], (001)[100] and (010)[001], and can only accommodate shear in the facets of the orthorhombic cell, but not in diagonal planes. The associated CRSSs are $\tau^{(010)[100]}$ = $\tau^{(001)[100]}$ = 0.4 $\tau^{(010)[001]}$ and the rate sensitivity parameter is n=3 [WENK &AL. 1991a]. Plastic deformation was simulated up to 50% axisymmetric compression, starting from a random distribution of 1000 orientations. The predicted 100, 010 and 001 pole figures are presented in Figure 17, and compare favorably with textures of olivine deformed experimentally in axial compression (Chap. 6 Fig. 21). The activity of the three slip modes, shown in Fig. 17d, is very similar despite the difference in CRSS, and

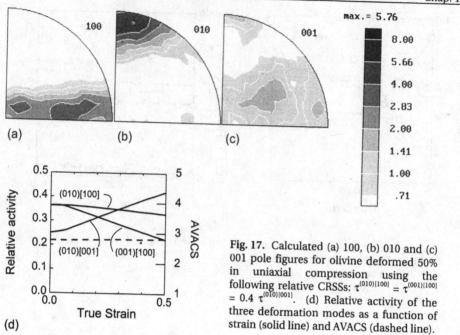

Fig. 17. Calculated (a) 100, (b) 010 and (c) 001 pole figures for olivine deformed 50% in uniaxial compression using the following relative CRSSs: $\tau^{(010)[100]} = \tau^{(001)[100]} = 0.4\ \tau^{(010)[001]}$. (d) Relative activity of the three deformation modes as a function of strain (solid line) and AVACS (dashed line).

an average of 2.7 systems is active in each grain throughout deformation. The reason for this behavior is to be found in the scarcity of deformation modes available. The result highlights the flexibility of the SC scheme in handling plastic deformation of aggregates presenting a severe limitation in deformation modes.

6. Self-consistent Modeling of Two-phase Aggregates

Many materials are composed of different phases with relatively strong and weak plastic properties. Multiphase materials are becoming more utilized in modern technologies because they offer the possibility of better tailoring their mechanical properties. In addition, they are commonly found in geological formations, since most rocks are a combination of two or more mineral phases. The complexity of these materials requires more refined modeling tools, in order to understand their mechanical response and predict the outcome of forming operations or plate tectonics. As we discuss above, single-phase materials exhibit an intrinsic anisotropy at their crystal level, and a texture at the aggregate level. As a consequence, they are heterogeneous and some crystal orientations are harder to deform than others, which can induce highly non-uniform strain fields. Depending on its orientation, a grain may be hard with respect to certain strain conditions, and soft for other strain conditions. A difference between single-phase and multi-phase materials is that in the former case the crystallographic orientation of the grain surroundings are in general not correlated with the grain orientation, in which case it may be appropriate to assume that the grain is surrounded by a Homogeneous Effective Medium. For multi-phase materials, on the other hand, heterogeneities are enhanced by the fact that crystals with different symmetry and hardness coexist (each with its own anisotropy, its own set of slip systems and associated critical stresses) and, most important, that the phases exhibit spatial correlation (usually, the grains that constitute one of the phases are surrounded by grains of a different phase). Capturing these features in a model requires a different approach from the one described in Section 4 for modeling the plastic deformation of single-phase polycrystals. In particular, the SC model has been expanded to model deformation of polyphase materials such as $(\alpha+\beta)$ Ti–Al–V [DUNST &AL. 1994, DUNST & MECKING 1996], olivine-pyroxene [WENK &AL. 1991a] and quartz-mica [CANOVA &AL. 1992]. The latter application will be described at the end of this Chapter in some detail.

By far the most investigations of deformation textures have been done on single phase material, and there is relative little information in the literature about the effect of a given phase on the texture development of the other phases, or on the anisotropy of multiphase materials. WASSERMANN &AL. [1978] and BERGMANN &AL. [1978] document wire drawing textures for metal–metal, metal–ceramic and metal–glass composites and, in addition, they review previous works on texture formation in polyphase material. MORII &AL. [1988] study α and β-phase texture of Ti-6Al-4V, SALINAS & JONAS [1992] and GANGLI &AL. [1995] report α and β textures of Zr-Nb systems, BOLMARO &AL. [1994] measure textures of individual phases in a Ni-Ag composite for different amounts of accumulated deformation and for different phase fractions, POUDENS &AL. [1995] characterize wire-drawing textures of an Al matrix composite with different fractions of SiC inclusions. An important result which is common to all of the aforementioned works is that, when compared to the deformation texture of the single phase metal, the texture of the matrix sharpens in the

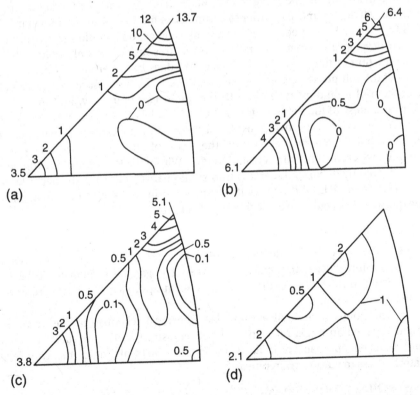

Fig. 18. Inverse pole figure of the drawing axis for Al/Al$_2$O$_3$ composites with: (a) 0%, (b) 1%, (c) 5%, and (d) 15% volume fractions of alumina particles, deformed by wire drawing. [Reproduced from WASSERMANN &AL. 1978].

presence of a small fraction of particles, while it becomes less pronounced when the matrix contains a large fraction of particles. Such a situation is illustrated in Fig. 18, reproduced from the work of WASSERMANN &AL. [1978]. Figure 18 depicts four reciprocal pole figures of the tensile axis, measured after wire-drawing an Al matrix composite with 0%, 1%, 5% and 10% volume concentration of alumina particles (Al_2O_3). Compared with the alumina-free specimen, a 1% concentration enhances the $\langle 100 \rangle$ fiber, while a 5% concentration tends to weaken both the $\langle 100 \rangle$ and the $\langle 111 \rangle$ fibers, and a 10% concentration of particles leads to a nearly random distribution.

A similar behavior is found for the case of a quartz-mica aggregate discussed in Chapter 6 (see Chap. 6 Fig. 24). While both phases exhibit hexagonal crystal symmetry, the mica phase represents a non-negligible volume fraction and is much harder to deform than the quartz phase (SiO_2). When an aggregate of single-phase quartz is deformed in compression to 35% strain at 800°C a fiber texture develops. However, when this quartz phase coexists with a 25% fraction of hard mica flakes, and the composite is deformed to the same overall strain of 35%, the quartz develops a weaker texture.

This effect has also been documented at a macroscopic level by BOLMARO &AL. [1993] and POOLE &AL. [1994a; 1994b]. The previous authors analyze the local deformation in a composite formed by parallel tungsten wires embedded in a copper matrix, being deformed in plane strain.

This behavior is somewhat counterintuitive, because one would expect that, for a given total deformation, the more ductile phase (i.e.: Al, quartz or Cu in the previous examples) will deform more in order to compensate for the non-deforming hard phase (i.e.: SiC, mica or W, respectively). As a consequence, one would expect the ductile phase to develop a sharper texture. However, this argument only holds when the volume fraction of the hard phase is small and does not alter substantially the flow pattern of the soft phase. Otherwise the presence of the hard phase induces localized deformation of a random type close to the interphase boundary, and increases the background noise of the deformation texture. A model of this system should account for the connectivity of the two phases and the strain localization. In Section 6.1 we describe a more general version of the visco-plastic self-consistent treatment presented in Section 4, called a 'cluster' model. Within the latter model vicinity effects are considered by dividing the grains into subdomains, and explicitly accounting for the interaction of each subdomain with the neighboring ones (cluster). An application to the quartz-mica aggregate is presented in Section 6.2.

6.1 Cluster scheme

The full visco-plastic SC theory will not be re-exposed here, but only the elements necessary for introducing the 'cluster' scheme. The general formalism of the Green function is given in the Appendix and the tangent approach has been presented in Section 3 of this Chapter.

For what follows we shall assume that the local response of the aggregate is described by the non-linear relation Eq. 1 between stress and strain rate. We will also assume a tangent approach, as in Eq. 3, and express the local tangent behavior relating deviatoric stress to plastic strain rate as:

$$S(x) = Q(D(x)) : D(x) + S^0(D(x))$$

(24)

This is the inverse of Eq. 3, where $\mathbb{Q}=(\mathbb{P}^{tg})^{-1}$ is the local tangent stiffness, S^0 is the stress intercept at zero strain rate, and both depend on D. A HEM is defined so as to have homogeneous properties and provide the same response as the real aggregate. Its behavior is assumed to be described by a tangent law like Eq. 6

$$S = \mathbb{Q}(D) : D + S^0(D) \tag{25}$$

where $\mathbb{Q}$ is the overall tangent stiffness of the medium, S and D are the overall stress and strain rate. Defining the local deviations with respect to the overall magnitudes as

$$\delta S(x) = S(x) - S$$
$$\delta D(x) = D(x) - D \tag{26}$$
$$\delta \mathbb{Q}(x) = \mathbb{Q}(x) - \mathbb{Q}$$
$$\delta S^0(x) = S^0(x) - S^0$$

permits one to combine the previous two equations into one as

$$\delta S(x) = \mathbb{Q} : \delta D(x) + (\delta \mathbb{Q}(x):D + \delta S^0) \tag{27}$$

(second-order terms have been eliminated from Eq. 27). Imposing the equilibrium equation for the Cauchy stresses ($\sigma = S + p I$), together with the condition of incompressibility, leads to a set of differential equations in the velocity $v(x)$ of the form

$$\mathbb{Q}_{ijkl} v_{k,lj} - p_{,i} = -f_i \tag{28}$$

$$v_{k,k} = 0$$

Here $p(x)$ is the local pressure and

$$f_i = (\delta \mathbb{Q}_{ijkl} v_{k,l} + \delta S^0_{ij})_{,j} \tag{29}$$

is a function of the local deviation with respect to the overall properties. This latter term accounts for local inhomogeneity effects via a fictitious force field in an otherwise homogeneous medium. The system of partial differential Eqs. 28 can be integrated by means of the Green function method [MOLINARI &AL. 1987]. The strain rate at any location of the heterogeneous material is given by an interaction equation of the same form as Eq. A25 derived in the Appendix:

$$D(x) = D + \int_{V'} G(x - x') : [\delta \mathbb{Q}(x') : D(x') + \delta S^0(x')] dx' \tag{30a}$$

which can equivalently be expressed in terms of the local stress and strain-rate deviations using (26) as:

$$D(x) = D + \int_{V'} G(x - x') : [\delta S(x') - \mathbb{Q} : \delta D(x')] dx' \tag{30b}$$

Here G is a fully symmetric tensor and equivalent for plasticity to the tensor defined in Eq. A25 for elasticity, function of the derivatives of the Green tensor of the system,

$$G_{klmj} = \frac{1}{4}\left[G_{km,lj} + G_{lm,kj} + G_{kj,lm} + G_{lj,km}\right] \tag{31}$$

The expression Eq. 30b represents an implicit non-linear integral equation. It may be solved approximately, using a discretization technique, which consists in defining domains 'V' within which the material properties are assumed uniform. While in the self-consistent models of the previous Section the basic volume element where the properties are assumed uniform is an ellipsoidal domain (grain), the present approach to polyphase materials requires one to consider basic elements smaller than the grains in order to capture the intragranular strain-rate heterogeneity. The volume shape of the elements is arbitrary, provided that they completely fill the space. Within each domain, the material properties are assumed to be homogeneous, and the material response is assumed to be uniform. The latter condition is enforced by defining an average stress and strain-rate for the volume element V as

$$\mathbf{D}^{v} = \frac{1}{V}\int_{V}\mathbf{D}(\mathbf{x})\,d\mathbf{x} \quad \text{and} \quad \mathbf{S}^{v} = \frac{1}{V}\int_{V}\mathbf{S}(\mathbf{x})\,d\mathbf{x}$$

(32)

Notice that the hypothesis of discretization introduces discontinuities of stress and displacement across the element boundaries. As a consequence, compatibility and equilibrium are only fulfilled approximately. Along the same line, it is possible to define a tensor $\Gamma^{vv'}$ that mechanically couples the volume elements V and V', as:

$$\Gamma^{vv'} = \frac{1}{V}\int_{V}\int_{V'}\mathbf{G}(\mathbf{x}-\mathbf{x}')\,d\mathbf{x}'\,d\mathbf{x}$$

(33)

where $\mathbf{G}(\mathbf{x}-\mathbf{x}')$ is given by Eq. 31. This coupling tensor depends on $\mathbf{Q}$, on the shape of the volume elements V and V', and on their relative separation and size, but not upon the absolute dimensions and position of the elements. In the application that follows the aggregate is partitioned into a set of N^{3} identical cubic domains. The numerical procedure for solving the integrals Eq. 33 within cubic domains is described by CANOVA &AL. [1992]. The discretization procedure imposed through Eqs. 32,33 reduces the integral Eq. 30b to a sum over the material domains where the material properties are assumed to be uniform, as follows:

$$\mathbf{D}^{v} = \mathbf{D} + \sum_{v'}\Gamma^{vv'}:\left[\left(\mathbf{S}^{v'}-\mathbf{S}\right)-\mathbf{Q}:\left(\mathbf{D}^{v'}-\mathbf{D}\right)\right]$$

(34)

Equation 34 is an implicit non-linear equation function of the stress in each domain, of the form

$$F^{v}(\mathbf{S}^{1},...,\mathbf{S}^{v},...)=0$$

(35)

from which the local stress (and as a consequence the local strain-rate) may be derived as a function of the overall properties of the medium and the boundary conditions. Solving this non-linear system of equations requires us to know explicitly the topology of the domains, their spatial correlation, and the orientation of the crystallographic structure associated with each domain. This means that a topology has to be defined that represents the actual distribution of the phases in the multiphase aggregate. This will be done in the next section for the particular case of the quartz-mica aggregate.

For solving Eq. 34 a tentative value of the aggregate stiffness $\mathbf{Q}$ is assumed and, once it has been solved, it must be checked whether the average conditions:

$$\langle \mathbf{D}^{v}\rangle = \mathbf{D} \quad \text{and} \quad \langle \mathbf{S}^{v}\rangle = \mathbf{S}$$

(36)

are fulfilled. If not, then $\mathbf{Q}$ is adjusted accordingly and the procedure is repeated until convergence is achieved. The numerical procedure is as follows:

(1) The macroscopic velocity gradient **L** is prescribed, and therefore the symmetric and skew- symmetric components **D** and **W**.

(2) Assuming **Q** known, the overall stress is calculated in terms of the secant stiffness [HUTCHINSON 1976] as $S = 1/n \; Q : D$.

(3) Instead of solving simultaneously the set of non-linear Eqs. 34, an iterative procedure can be used as follows: call k the index of the iteration, and $S^{v,k}$, the stress calculated in each volume v at the end of this iteration. Those solutions are used as a new guess for the iteration k+1, entering them into the equation $F^1(S^{1,k}, S^{2,k}, ..., S^{v,k}, ...)=0$, which provides a corrected solution $S^{1,k+1}$ for the element 1. The non-linear equations $F^{v+1}(S^{1,k+1}, ..., S^{v,k+1}, S^{v+1,k}, S^{v+2,k}, ..., S^{N,k})=0$ are solved sequentially for the other elements, making use of the corrected solutions. At the end of the sequence a residue Res= $\Sigma_v(S^{v,k+1}-S^{v,k})^2$ is evaluated and convergence is considered to be achieved when the residue becomes smaller than a certain allowed tolerance.

(4) A scalar error function is defined associated with the boundary conditions Eq. 36 as

$$Err(Q) = \left\{ \frac{\left\| \langle D^v \rangle - D \right\|}{\left\| D \right\|} + \frac{\left\| \langle S^v \rangle - S \right\|}{\left\| S \right\|} \right\} \tag{37}$$

where $\| \; \|$ indicates the norm of the bracketed tensor. The minimum of this function with respect to the components of **Q** provides the best approximation to the fulfilling of Eq. 36. The stiffness **Q** is updated using the conjugate gradient method and the procedure restarts at step (1) until the minimum of Eq. 37 is achieved.

Equation 34 allows the definition of an interaction range between volume elements by considering a 'cluster' formed by the N closest neighbors around each element V. Only the neighboring elements contained inside the 'cluster' interact with the element indexed as v and, beyond such cluster, the expansion in Eq. 34 is truncated and the HEM is assumed. This formulation reduces to the one of the single inclusion embedded in the HEM, described in Section 4 of this Chapter, if only the central domain is included in the cluster and the interaction with the neighbors is neglected. The latter amounts to assume $\Gamma^{vv} = \delta_{vv} \Gamma^{vv}$ and leads to an interaction equation of the same form as the one for the elastic inhomogeneity (Eq. A33):

$$\left(S^v - S \right) = \left(\left[\Gamma^{vv} \right]^{-1} + Q \right) : \left(D^v - D \right) \tag{38}$$

In what follows we will refer to the latter as the 'one-site approach', and will compare its predictions with the ones of the 'cluster approach' defined by the more general interaction Eq. 34.

After having found the solution of Eqs. 34 and 36, the state of the material (crystallographic rotations and hardening) can be updated. In particular, the skew-symmetric equivalent of Eq. 34 is used to calculate the local spins in the discretized medium as:

$$W^v = W + \sum_{v'} B^{vv'} : \left[\left(S^{v'} - S \right) - Q : \left(D^{v'} - D \right) \right] \tag{39}$$

where the tensor $B^{vv'}$ is the skew symmetric equivalent of $\Gamma^{vv'}$. The lattice reorientation inside the volume element V is given by Eq. 23, as the difference between the local spin W^v and the plastic spin:

$$W_{ij}^{v*} = W_{ij}^{v} + \sum_{s} \frac{1}{2} \left(b_i^s n_j^s - b_j^s n_i^s \right) \dot{\gamma}^s \tag{40}$$

The shear rates are given by the rate sensitive Eq. 1 once the stresses have been solved. In the implementation of the calculation that follows, only the crystal orientations but not the shape of the domains V is updated during deformation.

6.2 Quartz–mica textures

The system that we analyze here is a two-phase aggregate of quartz and mica, where the quartz grains exhibit a rather equiaxed morphology [CANOVA &AL. 1992]. The mica grains, on the other hand, represent a volume fraction of 25%, are randomly distributed, and have a flake-type morphology. The partition of the space is done in three steps: first, a compact arrangement of 100 irregular polyhedra (Voronoi cells) which represent the quartz grains is generated with a 3-D periodicity; second, mica flakes are randomly assigned to the interfaces between quartz grains and, third, a 3-D arrangement of 16^3=4096 elementary cubes is superimposed to the Voronoi cells.

The Voronoi cells are constructed as follows: a set of 100 random triplets (x,y,z) are generated within the interval [−1,+1] in order to obtain a rather uniform distribution. Each triplet represents the center of a Voronoi quartz grain. The segments that join each center with all other centers (including the periodic ones) are bisected with planes. The inner envelop of those planes defines a convex polyhedron (Voronoi cell). The complete set of polyhedra so generated produces a compact arrangement of irregular cells, and any 2-D section of such a 3-D structure will also be a compact arrangement of convex polygons. Figure 19a shows a 2-D section of four periods of such a structure. It turns out that each cell has an average of about 13 neighbors in 3-D space.

For representing the second phase, some planes separating the quartz grains are chosen at random, and a parallel plane is built at a distance somewhat smaller with respect to the center of the cell. The flat domain that results is assumed to represent a

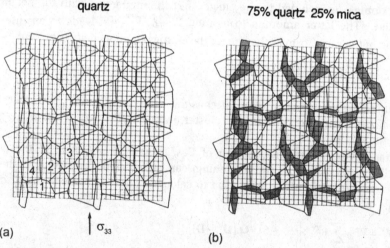

Fig. 19. (a) YZ sections at X=0 through the microstructure illustrating Voronoi cells representing quartz grains (heavy lines) and cube cells. The compression direction is vertical (arrow). (b) In the quartz-mica composite the mica platelets are shaded. Four periodic cells are depicted that result from repeating the microstructure by translation along Y and Z.

mica flake. A total of 92 such regions are generated, and the interplanar distance is set such as to obtain 25% volume fraction of mica. Figure 19b shows a 2-D section of such a structure.

Another problem consists in determining the volume elements v to be used in Eq. 34. In fact the grains themselves could be used as volume elements, but the volume integrals to be performed in those regions can be very difficult and, in addition, the modeling is expected to reproduce the intragranular strain heterogeneity. As a consequence, a finer mesh composed by 16^3 small cubes is superimposed to the Voronoi structure. This means that, in average, each quartz and mica grain is composed of about 30 and 10 cubes, respectively, which are going to describe the local strain variations inside the grain.

In the present case, only the effect of the first neighbors on the element response is investigated. This means that a cube is in direct interaction with the cluster formed by its 26 neighboring cubes and, past those, it is assumed that the HEM is found. It is, of course, possible to neglect the interaction with the first neighbors, in order to investigate the predictions of a one-site model. It is also possible to account for longer range interactions.

Here we present two cases for the purpose of comparison: the single-phase quartz aggregate, and the two-phase quartz–mica composite. In turn, each of these cases is analyzed in two different ways: using a one-site approach and using the 'cluster' approach.

The slip systems assumed for each phase, and their relative CRSSs are given in Table I. It can be seen on this Table that both phases have a hexagonal symmetry, and

Table I. Slip systems (characterized by slip plane normal **n** and slip direction **b** – here not normalized), number of crystallographically equivalent systems, and relative CRSSs assumed in the deformation modeling.

phase	**n**	**b**	n	τ_c
quartz	[0001]	$\langle \bar{1}2\bar{1}0 \rangle$	3	1
	[10$\bar{1}$0]	$\langle \bar{1}2\bar{1}0 \rangle$	3	4
	[10$\bar{1}$1]	$\langle \bar{1}\bar{1}23 \rangle$	3	12
mica	[0001]	$\langle \bar{1}2\bar{1}0 \rangle$	3	10

that the mica is assumed to be harder than quartz. Texture results after 50% compression are presented in Fig. 20 as 0001 pole figures (c-axes) of quartz with the compression axis at the center. In all cases the tendency is for the c-axes to rotate towards the compression axis. The texture for mica is not given because in the present calculation the harder mica flakes do not undergo significant deformation by slip and, as a consequence, do not rotate significantly. The pole figures on the left-hand side of Fig. 20 are obtained with the one-site approximation, in which case all the cubes contained in a given Voronoi cell (grain domain) see only the HEM. And since they experience the same reorientation, their representative points all superimpose in the pole figure. The one-site results should show a concentration of the texture when the mica phase is added, because the two phases do not interact, and the volume fraction effect makes the softer quartz phase deform more when the mica is present than in the single-phase case. For this particular case, 75% of soft phase undergoes 50/0.75=67% strain, as compared to 50% for the single phase case. However, the small number of

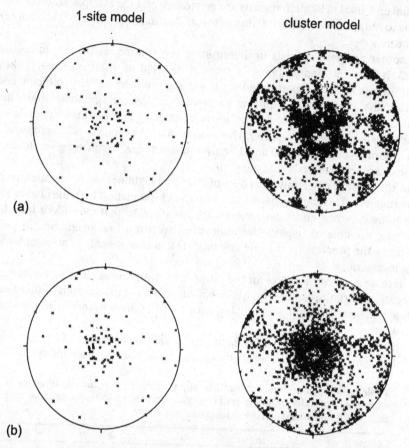

Fig. 20. 0001 pole figures illustrating texture development of quartz after 50% deformation in axial compression, as predicted by the one-site scheme and the cluster scheme. (a) Pure quartz aggregate. (b) 75% quartz-25% mica aggregate. (Equal-area projection. Compression axis is at the center of the pole figure.)

points that enter in the pole figures corresponding to the one-site model make it difficult to derive a meaningful conclusion.

The pole figures on the right-hand side of Fig. 20 are obtained with the cluster scheme, and the 4096 individual cube orientations that enter in the plot exhibit a systematic spread around the 100 orientations displayed on the left-hand side of Fig. 20. This is due to the fact that, when accounting for the intragranular strain, each cube in the Voronoi cell deforms differently, which leads to different reorientation and, therefore, more spread in texture. Observe that the 'cluster' prediction for the two-phase aggregate (bottom right) shows that the addition of mica results in a more diffuse texture than the single phase material (top right). This is a consequence of the increased strain heterogeneity due to the presence of the hard phase.

In order to investigate the texture heterogeneity associated with the cluster approach, the reorientation of the cubes contained in each of four particular Voronoi cells (labeled 1 to 4 in Fig. 19) are analyzed in Fig. 21. Grain 1 is favorably oriented for accommodating compression but is surrounded on three sides by mica, grain 4 is hard

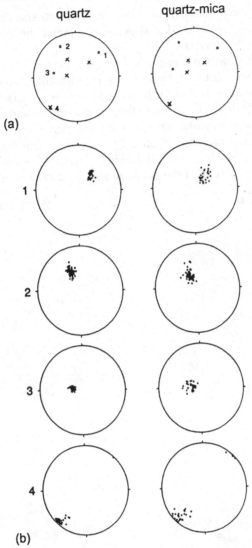

Fig. 21. Reorientation of the (0001) poles for the four quartz grains indicated in Fig. 19 after 50% compression. Results for the pure quartz aggregate (left) and the 75% quartz-25% mica aggregate (right). (a) Predictions of the one-site model. The starting orientation is indicated by a square symbol; the crosses show the end orientations of the quartz grains. (b) Predictions of the cluster model. The crosses show the final orientations of the cube cells associated with each of the same four grains. (Equal-area projection).

for deforming under compression, grains 2 and 3 are soft and are predominantly surrounded by soft quartz phase. The pole figures show the final orientations of the cubes associated with each Voronoi cell, for the single phase case (left) and for the two-phase material (right). The upper graph (Fig. 21a), shows the end orientations predicted by the one-site scheme for the same grains: it is evident that the two-phase one-site texture develops faster than the single phase one-site case. As for the cluster calculations (Fig. 21b), they show a scatter of the c-axes of the individual cubes when compared to the one-site prediction (Fig. 21a). Those results also show a larger spread in texture in the two-phase case, that appears as a larger dispersion of the points associated with a given Voronoi cell.

Finally, the heterogeneity of the strain field can be directly visualized in the 2-D sections of Fig. 22. One period of the structure is represented here with its tiling of cubes. The shades are made darker the larger the compressive strain-rate component D_{33}. The graphs on the left correspond to the one-site model and show, as expected, that strain-rates are uniform within Voronoi cells, and that quartz grains tend to have higher strain-rates in the presence of mica (bottom left). The results of the cluster approach, shown on the right, exhibit smoother strain-rate profiles, as is to be expected from the partitioning of the grains in smaller domains. Observe, however,

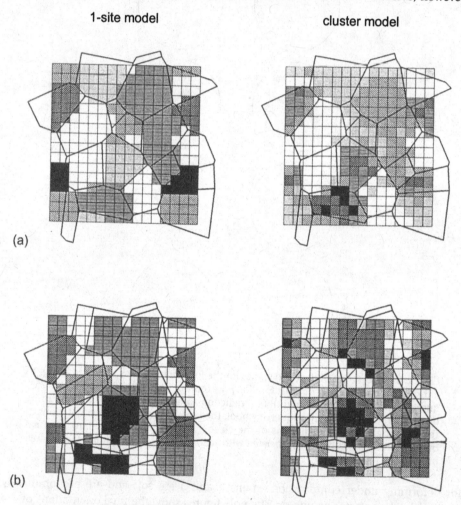

Fig. 22. Same microstructure as in Fig. 19 showing only one period for the two different models. (a) Pure quartz aggregate. (b) 75% quartz-25% mica aggregate. The shading indicates the magnitude of the compressive strain-rate component D_{33} in each cell. Darker shades indicate higher rates. Shading levels are the same in the four figures.

grain one (top left). For the case of the two-phase material, on the other hand, the strain-rate gradients increase in the presence of mica, the heterogeneity is more marked, and the cluster approach makes a difference with respect to the one-site approach (compare bottom figures). A similar analysis done for the compressive stress component S_{33} is consistent with larger variations in the presence of the second phase [CANOVA &AL. 1992].

These results indicate that the 'cluster' version of the SC model can, to some extent, account for the influence that a hard phase exerts on the texture development of a soft phase. The result is, as suspected, a combination of two mechanisms: one induces sharpening of the texture due to enhanced plasticity, since the same overall deformation is accommodated by a smaller volume; the other tends to 'spread' the texture and increase the background component, through the heterogeneous intragranular strain field that the hard phase induces in the soft one.

7. Conclusions

We discuss in this Chapter two self-consistent approaches for modeling the plastic deformation of polycrystals. The first one consists in treating each grain as an ellipsoidal inclusion embedded in a Homogeneous Effective Medium. The second, more general but also numerically more involved, is a cluster scheme that accounts for the specific neighborhood of the grains. Self-consistent approaches are more complex than the classical FC or RC models: they explicitly account for the interaction of the grain (or subdomains in the grain) with its surroundings, for their relative anisotropies, and for grain shape effects. As a consequence, they permit one to address problems that cannot be tackled with the simpler FC or RC models, such as the plastic deformation of highly anisotropic materials with a limited number of slip systems (like the case of orthorhombic olivine treated in Section 5.4), or the heterogeneous deformation of polyphase materials (like the case of quartz-mica developed in Section 6.4).

The conditions under which deformation takes place are different in the SC scheme than they are in the FC or RC schemes. The closure of the SCYS is not a necessary condition within SC models because the matrix can accommodate deformation when the grain lacks five independent systems or when some of them are too 'hard' for being activated. As a matter of fact, the SC scheme accommodates deformation mostly by means of the soft systems and, in general, fewer than four active systems per grain are required.

The consequences of the embedded ellipsoid model are studied in detail in Sections 4 and 5. In particular, in Section 5.1 we analyze the linear assumption used to describe the intergranular interaction: we find that even for the not so anisotropic cubic materials, results obtained using the tangent or the secant plastic compliance are different. The discrepancy is expected to increase when more anisotropic and less symmetric materials are considered, although we do not carry on a similar comparison for the hexagonal, trigonal and orthorhombic crystals analyzed in Sections 5.2 to 5.4. We do observe, however, that deformation is accomplished activating the softer deformation modes, and that the predicted textures are in reasonable agreement with the experiments.

The second SC approach, the cluster method, is also based on the linearization assumption and uses the tangent modulus. However, by breaking the grain into smaller domains, it is more apt for addressing intragranular heterogeneous deformation and localization in the vicinity of hard phases. As a matter of fact, heterogeneity can be severe in the latter case, and the assumption of uniform deformation within the grain may not be realistic, as Fig. 22 illustrates. The cluster method has the advantage over the one-site approach of better accounting for interactions and heterogeneity. Still, it is difficult within the cluster method to account for the rotation of rigid platelets, the localized plastic flow around them, or to update the grain shape.

All SC approaches have the limitation of relying on a linearization assumption. Here we favor the tangent, but the formulations can be presented in terms of the secant moduli as well, and it is likely that neither of them describes completely the true response of the material: the tangent approach seems to be too compliant and the secant seems to be too stiff. As a consequence, since there is no exact solution for the inclusion embedded in a non-linear material, it would be important to develop an empirical criterion for deciding which is the effective stiffness that best describes the non-linear response of the HEM. Clearly, such an approach requires first a knowledge of the solution for the non-linear problem. A procedure suggested by MOLINARI and TÓTH [1994a, 1994b] is to use FEM calculations to tune the stiffness of the interaction equation. Their FEM results show that, for low values of n, the response of the inclusion is in between the ones derived using the secant and the tangent schemes.

Some limitations of the polycrystal SC models, such as the assumption of an effective linear constitutive response or the treatment of localized plastic flow, can be overcome, in principle, by decomposing the grains in subdomains and using FEM methods to solve the intergranular and intragranular equilibrium. The use of FEM methods to study heterogeneous deformation of polycrystals is the subject of the next Chapter.

Chapter 12

FINITE ELEMENT MODELING
OF HETEROGENEOUS PLASTICITY

FINITE ELEMENT MODELING
OF HETEROGENEOUS PLASTICITY

P. R. Dawson and A. J. Beaudoin

The finite element method is a powerful complement to polycrystal plasticity theory for modeling the non-uniform deformation of crystalline solids. Polycrystal plasticity provides a micromechanical model for slip dominated plastic flow and serves as a constitutive theory for engineering analyses. The finite element method offers a numerical means to solve partial differential equations, such as the field equations of elasticity or plasticity. The two can be combined in different ways depending on the goals of a modeling effort.

1. Heterogeneous systems

1.1 Length scales

One possible combination is to embed polycrystal theory within a finite element formulation for macroscopic bodies. Polycrystal plasticity serves as a constitutive theory in essentially the same way as continuum elasto-plasticity models. Examples of this combination are the topic of Chapter 13. A second combination of finite elements and polycrystal plasticity is associated with more detailed analyses of the crystal ensembles themselves. In this case, finite elements discretize the crystals, and balance laws are applied at the level of individual crystals (Fig. 1). For this case, a crystal ensemble is called an aggregate. We refer to these as large-scale and small-scale applications, respectively. With respect to characterizing these applications, it is useful to define a geometric parameter, ζ, as the relative sizes of a finite element and a crystal. Allowing h to be a characteristic dimension of an element (taken from 'h-type' elements) and d to be the representative grain size (determined, for example, from the mean linear intercept), the parameter ζ simply is h/d. Here, large ζ implies large numbers of crystals in each element; small ζ implies many elements within each crystal. Chapter 13 deals with the use of finite element formulations for large-scale applications. The focus of the present chapter is on the latter, small-scale applications, and the use of a finite element formulation to better understand materials with considerable heterogeneity of deformation over the dimension of an aggregate. At the same time, the characteristic dimension of an element must be much larger than the distance between slip planes for the process of slip to be homogenized in a meaningful way.

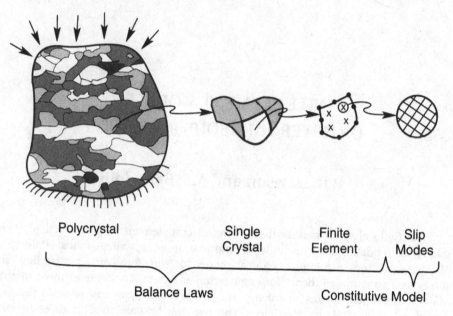

Fig. 1. Relation of finite elements to crystals in small-scale simulations.

1.2 Background

The manner in which deformations are partitioned among the crystals of an ensemble affects the average stress exhibited by the ensemble and the subsequent evolution of its crystallographic texture. For materials comprised of crystals that exhibit low anisotropy in the single-crystal yield surface, the assumption of uniform straining put forward by TAYLOR [1938a] is adequate for predicting the stress response and gross features of the texture, but is deficient in describing various texture details. For materials having high anisotropy in the single-crystal yield surface, the assumption of equal straining is less satisfactory, and in many instances is not plausible. For example, for crystals with fewer than five independent slip systems, equal straining is not possible because each grain lacks sufficient deformation modes to accommodate a general deformation as discussed in Chapter 11. The Taylor hypothesis is an example of a partitioning assumption that admits analysis without requiring detailed definition of the aggregate topology: partitioning of the macroscopic deformation among crystals of an ensemble is accomplished without regard to which crystals are neighbors. The inclusion of topology, however, may be important in obtaining more accurate predictions due to the influence of each crystal's specific neighborhood on its response. Ideally, the abstraction of specific results then leads to the introduction of topology into portioning rules – in a 'phenomenological' sense.

Numerical methods for discretizing the deformation of aggregates prove valuable in either of these cases to better understand the influence of the aggregate topology on the material behavior under loading. From numerical simulations, it is possible to quantify the mechanical responses of stress and texture evolution for aggregates that are permitted to exhibit complex internal deformation patterns. Using finite element methodologies, the crystals of an aggregate may be discretized with elements, using one or more elements for every crystal. In small-scale applications finite elements typically have volumes similar to that of a single grain, and in many cases are one-to-one. If an assumption of nearly uniform straining over a crystal is adequate then a single element may be designated for each crystal. However, if details of the inhomogeneity of straining within an element are sought, then the discretization can be made sufficiently fine to capture sharp gradients in the deformation. The crystal properties may be prescribed to allow for different crystal geometries, active slip systems, and slip system kinetics and hardening. In defining the mesh, one also is prescribing the topology of the aggregate. The balance laws are applied directly to an aggregate via the weighted residuals of the finite element formulation. The solution to the field equations renders the partitioning of deformation among the crystals. Examples of finite element analyses of aggregates of crystals have been reported in the literature [ASARO & NEEDLEMAN 1985; BECKER 1991a, 1991b; BRONKHORST &AL. 1991, 1992; BEAUDOIN &AL. 1993].

The crystal anisotropy is the source both of interesting behaviors (and thus the focus of many studies) and of numerical difficulties in the solution of the model equations. For example, the non-uniformity of straining is of particular importance. The tendency of deformations to localize into shear bands as the preferred mode of deformation has received considerable attention. Also, an understanding of crystal-to-crystal variations in the deformation on texture development is necessary for improvements to models which employ simpler mean field assumptions such as the Taylor or Sachs hypotheses. However, the discontinuity of properties across the elemental boundaries that is the source of interesting material behavior also may be a source of difficulty for simulation. Conventional kinematically based finite element formulations rely on increased resolution of the discretization to reduce traction discontinuities at elemental interfaces, and thereby to approach convergence. As property discontinuities are an inherent feature of interfaces within an aggregate, special attention to the element boundaries is necessary to assure converged solutions.

In the following, we concentrate on problems in which an element represents a single grain ($\zeta = 1$). A three-dimensional aggregate is constructed using eight-node isoparametric elements. Making use of large-scale parallel computer architectures, aggregates of 10^3 to 10^4 elements may be analyzed routinely. While the finite element model has considerable generality and flexibility, we do not attempt to model a number of possibly important physical phenomena, as described more fully in the next section. Rather, we use this tool to explore the implications of aggregate topology for materials undergoing intra-crystal slip. The results are intended to help develop a more comprehensive understanding of polycrystal behavior and motivate better models for plastic response. The applications we examine include both fixed-state stress response and texture evolution under macroscopic conditions of plane-strain compression for fcc and hcp crystals. The effects of the variability of straining from crystal to crystal are evident. The stress computed using the finite element model is compared to that given by several models employing various mean field assumptions.

1.3 Model equations and solution procedure

The equations that describe the slip of single crystals have been reviewed in Chapter 8. Here, the particular equations used in the finite element formulation are summarized for clarity and completeness in understanding the finite element results. The plastic flow is assumed to be isochoric; volumetric responses such as void growth, are neglected. The elastic response is not included in the results shown here, although the hybrid formulation presented in the following section provides a direct route for incorporating elasticity. The plastic flow is assumed to be a consequence of rate dependent slip, occurring on a restricted number of slip systems and diffusely within the crystals. Other deformation modes, such as grain boundary sliding, diffusion, twinning, or micro shear bands, are not addressed. The crystal boundaries are cohesive, so that compatibility is always maintained through continuity of the displacements at the boundaries. As the aggregate will be discretized into elements, each representing an individual crystal, there is no requirement for a mean field assumption, such as the Taylor assumption. The balance laws, as applied in the next section, accomplish the task of connecting crystals into an aggregate.

Briefly, as applied to the analysis of aggregates, the finite element formulation has several features. Each crystal is discretized by one or more finite elements. Standard finite element methodology is utilized to impose compatibility both within crystals and across their boundaries. Equilibrium is enforced by requiring that, in a weighted residual sense, the interface tractions vanish. This is accomplished using a hybrid finite element formulation in which the full body is divided into physically identifiable domains and the traction constraint is applied between domains. Within the context of a hybrid formulation, domains are defined so that finite elements correspond to individual crystals or parts of those crystals [BEAUDOIN &AL. 1995a]. The single-crystal slip relation is satisfied approximately over each crystal domain via a weighted residual using trial functions for the stress that a priori satisfy equilibrium within an element. In this way the aggregate is a polycrystalline body to which the balance laws and constitutive relations are applied. The instantaneous velocity field can be computed for specific boundary conditions; based on the velocity field, the geometry can be updated and the crystal lattice within each element re-oriented. The spatially varying velocity field quantifies the non-uniformity of straining among the crystals. Some additional details of the formulation are provided in the following section, which is not necessary reading for understanding the applications that appear later.

Hybrid element formulation

The relations within a crystal that quantify intra-crystal slip are used to define the material properties needed in the finite element formulation. As described in Chapter 8, the 'flow stress', τ^s, on any slip system is related to the shear rate knowing the kinetics of plastic flow. For plastic deformation, the resolved shear stress must equal the flow stress on each system, giving a nonlinear relationship between the crystal stress and the deformation rate (Eq. 14 of Chap. 8). This relation is written in linearized form as:

$$\mathbb{P}^c : \mathbf{S}^c - \mathbf{D}^c = 0 \ , \tag{1}$$

where $\mathbb{P}^c$ serves as a crystal visco-plastic matrix, $\mathbf{S}^c$ is the crystal deviatoric stress, and $\mathbf{D}^c$ is the crystal deformation rate. In the following, the volumes are taken over crystals or parts of crystals, so that crystal quantities are identical to the corresponding macroscopic quantities. The superscript, c, to designate crystal quantities is unnecessary, and will be omitted. A residual is constructed over the domain:

$$\int_{\Omega_e} (\mathbb{P} : \mathbf{S} - \mathbf{D}) : \Psi \, d\Omega = 0 \tag{2}$$

using weighting functions, Ψ. Consistent with the order of Eq. 1, Ψ is a second-rank quantity. The traction equilibrium residual, after utilizing the divergence theorem, is given as:

$$\sum_e \left[\int_{\Omega_e} (S - pI) : \nabla(\Phi) \, d\Omega - \int_{\Gamma_e} \Phi \cdot t \, d\Gamma \right] = 0 \tag{3}$$

where t is the surface traction, I is the identity tensor, and p is the pressure. Φ are vector weights and Γ_e is the surface bounding Ω_e. Finally, a weighted residual on the incompressibility condition provides a constraint equation using scalar weights, Υ:

$$\int_{\Omega} \Upsilon \, \mathrm{tr} D \, d\Omega = 0 \tag{4}$$

Trial functions are introduced in the residuals for the interpolated field variables. In contrast to the common velocity-pressure formulation (see Chap. 13), in the hybrid formulation both the velocity (components u_i) and the stress (deviatoric and spherical portions: S_λ and p, respectively) are represented with trial functions [TONG & PIAN 1969; PIAN 1995]:

$$S_\lambda = \sum_{l=1}^{L} \hat{N}_\lambda^l \, s^l \qquad p = \sum_{m=1}^{M} \tilde{N}^m \, P^m \qquad u_i = \sum_{n=1}^{N} N_i^n \, U^n \qquad (\lambda=1,5; \, i=1,3) \tag{5}^{1}$$

where the $\hat{N}^l$, $\tilde{N}^m$, N^n are terms in the respective spatial interpolation functions, (equivalent to the weighting functions Ψ, Υ, and Φ introduced above) and the s^l, P^m, U^n are the respective coefficients, developed from a solution of the *global* system of equations. For the interpolation of stress terms (S_λ and p), piecewise discontinuous functions are specified which satisfy equilibrium within an element a priori [BRATIANU & ATLURI, 1983]. The limits L, M, and N denote the dimensions of the interpolation variables specific to the element discretization used; for the eight-node brick used in this work, they are 17, 1, and 24, respectively.

Each residual is written using the trial functions. For the crystal slip relation, Eq. 2:

$$\sum_{l=1}^{L} H^{ql} s^l - \sum_{n=1}^{N} R^{qn} U^n = 0 \qquad (q=1,\ldots,L) \tag{6}$$

where

$$H^{ql} \equiv \int_{\Omega_e} \hat{N}_\lambda^q \, \mathbb{P}_{\lambda\mu} \, \hat{N}_\mu^l \, d\Omega$$

and

$$R^{nq} \equiv \int_{\Omega_e} B_\lambda^n \, \hat{N}_\lambda^q \, d\Omega$$

The matrix B contains the gradient of the velocity trial functions such that

$$D_\lambda = \sum_{n=1}^{N} B_\lambda^n \, U^n$$

The traction equilibrium statement for each element then becomes:

---

[1] Explicit summation will be used over the terms following from the finite element discretization (italic superscripts).

$$\sum_{l=1}^{L} R^{sl} s^l - \sum_{m=1}^{M} G^{sm} P^m - f^s = 0 \qquad (s = 1, \ldots, N) \tag{7}$$

where

$$f^s \equiv \int_{\Gamma_e} N_i^s t_i \, d\Gamma$$

relates to the resultant force on an element interface Γ_e and (with **h** as the trace operator)

$$G^{sm} \equiv \int_{\Omega_e} h_\lambda \, B_\lambda^s \, \tilde{N}^m d\Omega$$

The nodal stresses can be eliminated by solving Eq. 6 for the s^l and then substituting the result into Eq. 3. This yields the equation to be solved for each element:

$$\sum_{l=1}^{L} \sum_{q=1}^{L} \sum_{n=1}^{N} R^{sl} \, \bar{H}^{ql} \, R^{qn} \, U^n - \sum_{m=1}^{M} G^{sm} P^m - f^s = 0 \qquad (s = 1, N) \tag{8}$$

where $\bar{H}$ is the inverse of the matrix **H**. This discretized traction residual is solved simultaneously with the discretized residual for the incompressibility constraint:

$$\sum_{n=1}^{N} G^{nr} \, U^n = 0 \qquad (r = 1, M) \tag{9}$$

The global system of equations follows from the assembly of the residuals (Eqs. 8 and 9) for each element. Assembly is based on shared nodal degrees of freedom for the velocities. The solution of this system then yields the velocity field for each element corresponding to its current geometry and state.

1.4 Discretized aggregate

A sample of material is constructed from three-dimensional brick elements, each consisting of an individual crystal. For the sample to be truly representative of a real material, the number of crystals must be relatively large. In the examples that follow, 1000 elements form an aggregate (Fig. 2). The mesh used in the simulations, however, has 4096 elements with 16 elements in each of the three coordinate directions. The aggregate is the inner 1000 elements of this mesh, leaving a layer of elements three elements thick around the outer surface. These elements are part of the simulations but not regarded as part of the aggregate. Rather, the exterior layer of elements acts as an effective medium through which the boundary conditions are imposed on the aggregate.

The initialization of the texture is a central issue in the application of polycrystal-based models. This is especially true for small-scale applications as it is necessary to define the spatial relation of crystals. This may be done by randomly assigning orientations to elements; alternatively, with recently developed experimental capability, the elements of a finite element mesh can be associated with specific orientations measured within a material sample. Here, the lattice orientations of all crystals are assigned from an orientation distribution for the material. The orientation is chosen randomly for each element from the distribution. In the examples discussed later, the materials were assumed to have no initial preferred orientations, so a uniform distribution was used for assigning orientations. Neither do the crystal orientations have any initial correlations with the orientations of neighboring crystals. For simulations involving the evolution of texture, the initial distribution changed with straining accord-

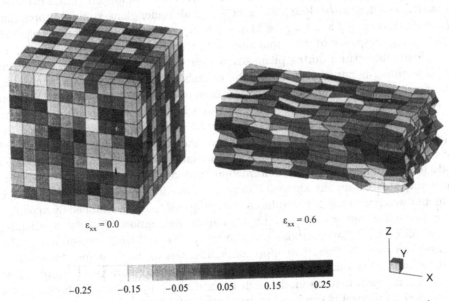

$\varepsilon_{xx} = 0.0$ $\varepsilon_{xx} = 0.6$

−0.25   −0.15   −0.05   0.05   0.15   0.25

Fig. 2. Finite element mesh showing the inner 1000 elements that constitute the aggregate. Shading depicts the value of the strain rate component D_{XY} (whose average is zero), during the first and last steps of a hybrid finite element calculation.

ing to Eq. 83 of Chapter 8 for each crystal. The spin terms needed in the evolution equation are evaluated at the centroid of the element.

2. Response of Polycrystal Aggregates

Plane-strain compression is frequently chosen for detailed study because it is an idealization of flat rolling (in which large strains may be accomplished) and because a great deal is understood concerning the behavior of materials under this loading. Model results can be compared readily either to the extensive body of experimental data for many materials or to the predictions of other models. For models that employ macroscopic variables (e.g. yield surfaces or continuum state variables), the deformation gradient or velocity gradient can be imposed explicitly, circumventing the need to solve the constitutive equations simultaneously with the field equations. When microstructural variables are used, the interaction of macroscopic kinematics and microscopic response involves an additional assumption that enables the macroscopic deformation rate to be partitioned among the individual crystals. With simple assumptions, such as the Taylor hypothesis, again the macroscopic deformation gradient may be applied directly to each crystal, and the stress and texture evolution computed. However, some models require that the partitioning be evaluated on the basis of the straining within each crystal that best satisfies the field equations throughout the volume of the aggregate. Each crystal's deformation depends on both the kinematics of deformation and the balance of tractions at crystal interfaces. In these cases the straining in any one crystal is not known from the macroscopic deformation alone, but rather involves considerations of the response of all crystals in the aggregate simultaneously.

In this chapter we explore the response of materials comprised of face centered cubic (fcc) and hexagonal close packed (hcp) crystals under plane-strain compression. The material is viewed as an aggregate of single crystals whose macroscopic response is the average response of the population of crystals. Of interest is the evolution of crystallographic texture during plane-strain compression, as well as the manner in which straining is partitioned among the crystals of an aggregate. The numerical framework described in previous sections is used to compute the response of such an aggregate in which each crystal coincides with a finite element. By discretizing the microstructure crystal by crystal we can permit individual crystals to respond independently, constrained by local balance laws and compatibility. The anisotropy of the crystal, together with the differences in orientation between neighboring crystals, leads to non-uniformity in the deformation field over the polycrystal. The non-uniformity stems from the kinematic degrees of freedom associated with the finite element discretization. These results can be compared to those obtained by making a more restrictive, but more easily implemented, assumption of uniform straining among all crystals that comprise the polycrystal (*i.e.* a Taylor assumption). The textures computed by these two approaches differ due to the deviations from plane-strain compression that are permitted with the discretized microstructure. This approach has been used to model both fcc and hcp polycrystals. In both cases a discretized polycrystal is constructed; boundary conditions corresponding to plane-strain compression are applied; and the response of the polycrystal is computed.

2.1 Behavior of fcc polycrystals

Crystals in the discretized aggregate were assigned an fcc crystal geometry with the 12 customary $\langle 111 \rangle$ {110} slip systems. The rate dependence, given by m, was set to be 0.05. The slip system strength evolved according to a Voce hardening model in a manner that captured the increase in flow stress observed for tensile tests on aluminum [MATHUR & DAWSON 1989]. The boundary conditions were specified to approximate plane-strain compression in rolling with coordinate axes identified as 'R', 'T', and 'N'. A constant straining direction was simulated using increments of 0.01. The initial and deformed geometry of the elements representing the finite element polycrystal were shown earlier in Fig. 2a of Chapter 8.

Texture evolution

Shown in Fig. 3 are 111 pole figures for 80% cold rolled aluminum as well as the pole figures produced, for an equivalent reduction, by the finite element model and Taylor-type calculations according to the Relaxed and Full Constraints models, respectively. On the right side of the figure, difference 111 pole figures contrasting the hybrid element polycrystal, the relaxed constraints model, and the Taylor model with experimental data for 80% cold rolled commercial purity aluminum are given. Positive densities denote a strengthening of the texture relative to the experimental texture; negative densities indicate the reverse trend. All simulations show a stronger texture development relative to the experimental data. However, the hybrid element polycrystal demonstrates less of this sharpening; maximum deviations are within 2.5 times random (Fig. 3b, right). The relaxed constraints and Taylor simulations (Figs. 3c and 3d, respectively) demonstrate larger regions with greater deviations from experiment than does the hybrid element polycrystal.

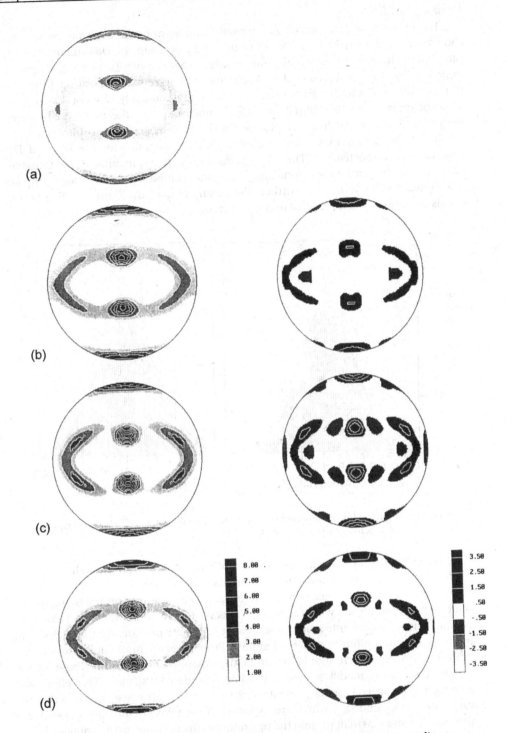

Fig. 3. Pole figures (left) showing experimental data (a) and predictions according to (b) a finite element polycrystal, (c) a Relaxed Constraints, and (c) a Full Constraints model. On the right are the respective difference pole figures against the experiments.

Strain-rate distribution

The instantaneous values of one normal component of the deformation rate are shown for each element, by its shading in Fig. 2a of Chap. 8. Deformation rate is sampled at the element centroid – the location where deviatoric stresses recovered from the constitutive relation (Eq. 1) and the equilibrium interpolation functions (Eq. 6) coincide. Note that the values range over approximately ±25% of the nominal value of unity. The distribution of deformation rate over the polycrystal can be represented in reduced form by plotting in the form of histograms for each component of the deformation rate. Figure 4 shows histograms for the D_{RR} and D_{RT} components, respectively. The histograms are well approximated by Gaussian distributions in both cases, indicating that the component values are normally distributed about the mean. Further, the mean value of the Gaussian distribution equals the nominal value applied to the polycrystal.

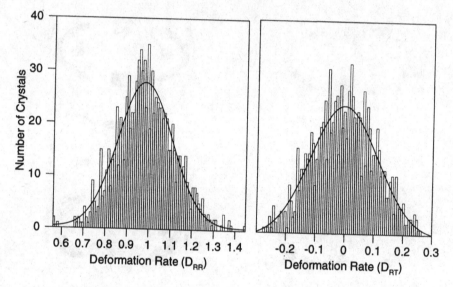

Fig. 4. Histograms for plane-strain compression in a cubic material. D_{RR} and D_{RT} are scaled by the macroscopic average of D_{RR}.

The effect of grain shape on texture evolution is accounted for through the course of deformation by the enforcement of equilibrium between an element and its neighbors. As a consequence, one would expect that the evolution predicted by the finite element polycrystal would reflect the effects of grain aspect ratio. For example, the fully relaxed constraints model described in Chapter 8 provides for the response of infinitely long and infinitesimally flat grains. Proper treatment of equilibrium in the finite element approach should result in deformations approaching those of the relaxed constraints model as grains become flattened in shape. This effect was investigated for several aspect ratios up to 8:1:1/8 (R:T:N). It is seen in Fig. 5 that the variations in D_{NR} and D_{NT}, which are expected in the relaxed constraints model, are present with similar strength from the beginning in the finite element simulation. The variation in D_{RT}, which is forbidden in the relaxed constraints model, and is of about the same strength in the present model in the equiaxed condition, does in fact decrease significantly as the flat grain condition is approached. This component is of

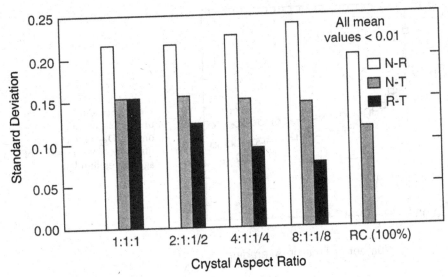

Fig. 5. Influence of grain shape (represented by the aspect ratios) on the standard deviation of the strain-rate components D_{NR}, D_{NT}, and D_{RT}. Plane-strain compression.

particular interest since it is known to favor development of the brass texture component. This indicates that the brass texture is free to evolve when the grain shape is near equiaxed. Continued plane-strain compression leads to increased constraints on this evolution which are in agreement with theoretical assumptions forming the relaxed constraints model.

Neighborhood effects

It is possible to study details of the influence of the local neighborhood on the deviations of a crystal's deformation from the nominal value using the finite element model. For example, the same sample of 1000 orientations can be assigned to elements of the mesh in a different order, thereby giving each crystal a different set of neighbors. The polycrystal is loaded in the same manner as before and the deviations from the nominal value are noted. By repeating this procedure numerous times, the response of a particular orientation to a variety of neighborhoods is generated. This was done for the fcc polycrystal described above for a total of 25 arrangements of the same 1000 orientations. An interesting trend arises: as the number of arrangements taken together increases the standard deviation of the deformation rate for all crystals in the polycrystal decreases (Fig. 6), but does not diminish to zero.

In Fig. 7 are plotted the highest, lowest, and average values of two deformation rate components for all 1000 crystals as a function of each crystal's Taylor factor. For D_{RR} there is a weak correlation of deformation rate with Taylor factor [SARMA & DAWSON 1996]. While crystals with lower Taylor factors exhibit larger values of the deformation rate, the *spread* in values dominates the response. The range in values from highest to lowest for any orientation (Taylor factor) is larger than the deviations seen in the mean values over the full range of Taylor factor. In fact, all orientations exhibit values of deformation ranging from above to below the mean value. For the D_{RT} component, there is no correlation between the deviation from the mean value and the Taylor factor. This latter result is consistent with the fact that the Taylor factor

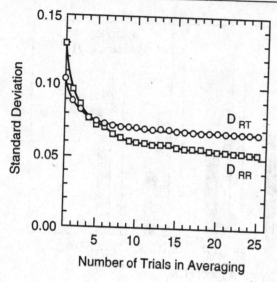

Fig. 6. Standard deviation of the strain-rate components D_{RT} and D_{RR}, as a function of the number of aggregates simulated.

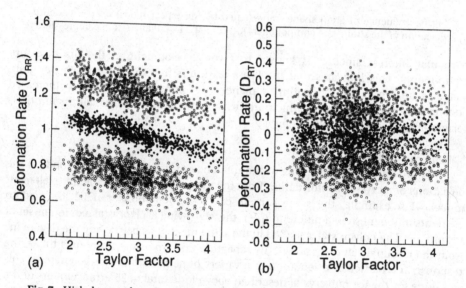

Fig. 7. High, low, and average values of strain rate for all 1000 crystals in 25 arrangements, as a function of the crystal Taylor factor. D_{RR} and D_{RT} are scaled by the average D_{RR}.

is the projection of the stress along the direction of the macroscopic deformation rate. For plane-strain compression, the shear components of the deformation rate are identically zero, so we would not anticipate a correlation similar to that found for the compression or extension components.

From these simulations, we conclude that the effects of the local neighborhood (the particular set of neighboring orientations) plays a very strong role in establishing any deviations from Taylor behavior that arise as a consequence of non-uniform deformations over a polycrystal. From this trend it would appear that the orientation of a crystal is less important in determining the deviation of its deformation rate from the nominal value than is its neighborhood. Several limitations should be emphasized. First, the grain shape is a simple brick-like geometry. The variations in deformation rate may differ from cases where more realistic grain morphology is simulated. Second, the equations are solved only approximately with the finite element method. Of concern is the satisfaction of the constitutive equation in a weak form, rather that exactly point-by-point (strong sense). All of the statistics have been collected at a point corresponding to the element centroids. At these points the constitutive equation are satisfied exactly. However, more extensive study is justified to verify that these results are truly representative of the material behavior.

2.2 Behavior of hcp polycrystals

Hcp crystals often are categorized by the axial ratio ($\frac{c}{a}$ ratio, in the commonly used terminology). Here, we specify an axial ratio of 1.63, corresponding to the ideal value for close packing but not representing a specific metal. Slip in hcp crystals can occur on different families of slip systems. Basal and prismatic systems often exhibit lower critical resolved shear stresses, but by themselves do not possess sufficient modes of deformation to accommodate arbitrary deformations. Pyramidal systems add the missing mode, extension along the crystallographic c-axis, but typically can require higher resolved shear stresses to be active, as discussed in Chapter 11. In addition, twinning exists as a deformation mode for hcp systems, especially when even basal slip is difficult; under some conditions of temperature and stress, it contributes significantly to the net deformation. Here we neglect twinning and postulate pyramidal slip $\{11\bar{2}2\}$ $\langle\bar{1}\bar{1}23\rangle$ systems to close the yield surface and give sufficient slip systems to avoid any inextensible modes. In the simulations performed the strength of these systems relative to the prismatic and basal systems (which were taken as equal) was varied from unity to ten. The pyramidal strength relative to the basal strength has been designated as α^{PYR}. In all cases the rate sensitivity, m, was held fixed at 0.10, a value chosen for numerical convenience though somewhat larger than expected on experimental grounds (see Chapter 8) appropriate to higher temperatures where twinning is not an important contribution. To preserve the relative strength ratios with deformation, no slip system hardening was permitted on any system. This has two effects: the relative strength ratios remain the same in each grain; and differential straining between grains does not cause differential hardening.

The finite element polycrystal was subjected to boundary velocities corresponding to plane-strain compression. The global coordinate directions are associated with the rolling process as: compression – normal (N); extension – rolling (R); and transverse (T). The results give a measure of the variation in deformation rates stemming both from the single-crystal anisotropy and from the influence of neighborhood. The plastic response of hcp crystals has been discussed in some extent in Chapter 11, where a self-consistent polycrystal model was used to describe grain interactions. By assuming a homogeneous effective medium, the latter approach assesses the impact of the grain anisotropy on the deviation of its response with respect to the average.

Stress response

The application of mean field assumptions that impose either equal deformations (upper bound) or equal stresses (lower bound) over all crystals in an aggregate can lead to unrealistic results either in terms of excessively high stresses or severe violations of compatibility. This can be seen in part by examining the fixed-state stress response of a random polycrystal. Upper and lower bounds on the stress differ widely, especially at low rate sensitivity [HUTCHINSON 1977, CHASTEL & DAWSON 1994]. Using the finite element model we compute the effective stress of a polycrystal comprised of hcp crystals for the range of rate sensitivity from 0.01 to one-half. The calculated effective stresses are compared to both upper and lower bound results.

The variation in effective stress with rate sensitivity for plane-strain compression is shown in Fig. 8. Here, the effective stress is normalized by the slip system strength, τ_1^s. This normalized stress diminishes with increasing rate sensitivity. At $m = 0.01$, the normalized effective stress is approximately 5.17, which is lower than that computed from a Taylor assumption [HUTCHINSON 1977] of approximately 6.0. However it is considerably greater than the lower-bound result [PRANTIL &AL. 1995a] of less than 2.0. The effective stresses computed with the finite element model decrease smoothly with m, converging with both the lower-bound and constrained hybrid models to a value of m of 0.3. Thus the finite element model lies closer to the upper bound (Taylor model) at low rate sensitivity and to the lower bound at higher rate sensitivity.

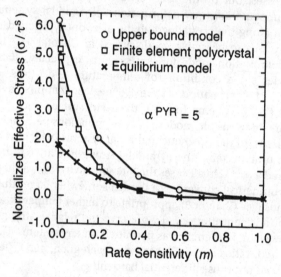

Fig. 8. Stress response of hcp aggregates for various rate sensitivities.

Texture evolution

To examine the texture evolution, 100 equal time steps were imposed to reduce the polycrystal to one half of its original height in plane-strain compression. This produces a true strain of 0.69, which corresponds to a von Mises strain of 0.8. The initial lattice orientation changes with straining for each crystal; the spin terms needed to compute the reorientation are evaluated at the centroid of the element. Mesh distortion became significant after approximately 60 time steps. Consequently, the mesh was repaired after 60 steps by replacing the distorted elements with ones having brick shapes. However, the average aspect ratio of the deformed polycrystal body was preserved. The lattice orientations existing after 60 steps were mapped onto the reconstructed mesh to the same elements. This permitted the simulations to proceed to 100 time steps (50% reduction) in all cases except the highest ratio of pyramidal to prismatic slip system strength. In this case the mesh had to be repaired again after 90 time steps. The finite element polycrystal of the inner 1000 elements at the beginning of the simulation and after 60 time steps (before repair) is shown in Fig. 9 for the case of the pyramidal system strength being five times the basal value. The variability of

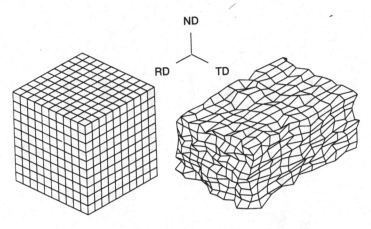

Fig. 9.   Deformed mesh representing an hcp aggregate under plane-strain compression.

straining among the elements/crystals is quite evident, and is appreciably greater than seen for fcc polycrystals.

The ratio of the pyramidal strength to basal/prismatic strength, α^{PYR}, was parametrically varied in the set of simulations performed. Beginning with the rather unrealistic case of α^{PYR} equaling unity, its value was increased to three, five, and finally ten. Two simulations were conducted for each value of α^{PYR}: one being the finite element polycrystal and the other being a Taylor model computation. Figure 10 gives 0001 pole figures for each model for values of the α^{PYR} corresponding to 1, 3, 5, and 10, respectively. The principal features of the texture, especially for α^{PYR} values of 5 and 10, correspond to that reported for cold-rolled, fine-grained zircaloy-2 [LEBENSOHN & TOMÉ 1991a].

Several trends are apparent. In every case the textures derived from the finite element simulations are more diffuse than the corresponding Taylor computation. In the Taylor computations an elliptical 'ring' is centered about the compression (rolling normal) axis, with increasing aspect ratio with increasing α^{PYR}. This feature is mimicked in the finite element simulations, but the ellipse is far less sharp. At high α^{PYR} (especially at α^{PYR} of 10), however, the finite element model does not exhibit this feature to any significant degree, but rather appears more to be a diffuse pattern centered on the compression axis. The increased diffusion of the texture coincides with the greater variability in straining experienced as the α^{PYR} increases. As the α^{PYR} becomes higher and higher, each hcp crystal in essence inherits an inextensible direction normal to the basal plane. This makes it less and less likely that crystal deformations will coincide with the macroscopic value since each crystal becomes incapable of a full range of deformation modes. Thus the qualitative differences between the Taylor and finite element polycrystal predictions emerge. As shown in Fig. 11, the Taylor models continue to utilize the pyramidal systems even when the α^{PYR} is high because those systems offer the only means for extension normal to the basal plane. The finite element polycrystal exhibits greater crystal-to-crystal strain variations as it prefers to shift straining to more favorably oriented crystals than to force the strong pyramidal systems to be active. This is quantitatively substantiated by

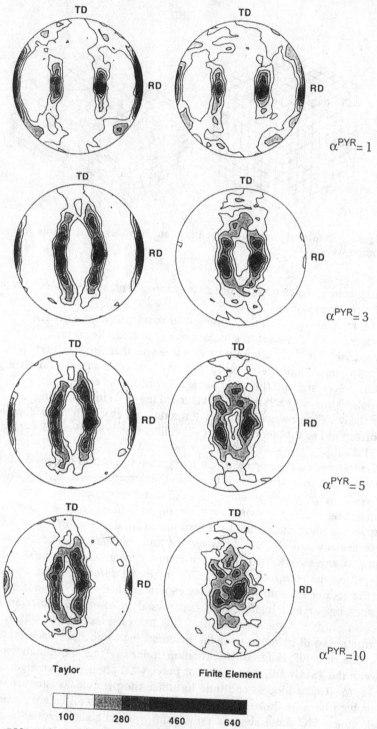

Fig. 10. 0001 pole figures for a range of relative pyramidal strengths α^{PYR}: comparison of predictions according to Taylor and finite element assumptions.

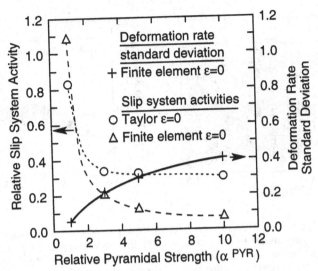

Fig. 11. Slip system activities for a range of relative pyramidal strengths. Also shown is the standard deviation of the deformation rate component D_{RT}.

the increasing value of the standard deviation of the shearing deformation rate component in the plane containing the transverse and rolling directions with increasing α^{PYR} value. This has a very substantial impact on the texture development, as is evident from Fig. 10.

Mean field assumptions other than the Taylor hypothesis, such as self consistent [LEBENSOHN & TOMÉ 1993a], constrained hybrid [PARKS & AHZI 1990] and equilibrium-based [CHASTEL & DAWSON 1994, PRANTIL &AL. 1995a] models, often permit crystal-to-crystal straining variations, but have not shown a similar strong tendency toward quite diffuse textures as the crystals become inextensible. Figure 12a shows the range in deformation rate of crystals in the finite element simulation plotted against the angle that the crystal c-axis makes with the compression direction. The variation in deformation rate is larger, and exhibits two different behaviors when the crystal c-axes are perpendicular to the compression direction. If the c-axis lies in the extension direction the deformation is difficult since the pyramidal systems must be activated. If the c-axis lies in the constrained direction the deformation can be accommodated by the prismatic systems, which requires considerably lower stress.

In Fig.12b, the variations in crystal deformation rates (D_{RR} component) with c-axis orientation are shown for computations performed using the Taylor hypothesis and the Constrained Hybrid model. The Taylor assumption requires that all crystals have identical deformation rate, so there is only the single value of unity for all orientations. For the Constrained Hybrid model, crystals are allowed to activate only basal and prismatic slip, eliminating the possibility of c-axis extension or compression in each crystal. Instead, a projection operator is constructed that distributes the straining among crystals of an aggregate in a manner that does not require any c-axis deformation in any crystal, yet preserves the correct average deformation rate over the full aggregate. Each crystal's deformation rate depends only on its own orientation and the full orientation distribution. Given a specific texture, a crystal has a unique deformation rate. The resulting distribution is shown in Fig. 12b with respect to the

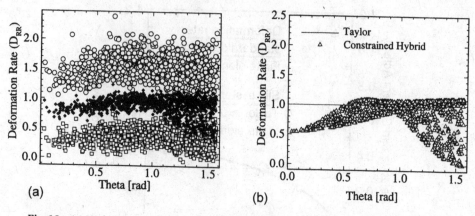

Fig. 12. (a) High, low and average D_{RR}, for all 1000 crystals in 25 arrangements versus the angle the c-axis makes with the compression direction as computed with the finite element formulation. (b) D_{RR}, for 1000 crystals versus the angle the crystal c-axis makes with the compression direction as computed using the constrained hybrid model.

angle the c-axis makes with the compression axis. Crystals with orientations that favor easy activation of basal and prismatic slip demonstrate deformation rates above the average value (1.0), while crystals with orientations that require high stress to activate those systems exhibit deformation rates below the aggregate average. The important point is that the deviation of the deformation rate from the average depends only on the crystal orientation for a given texture. Examples of orientations that are easy or hard to activate can be observed by examining the response of crystals that have the c-axis in the plane perpendicular to the compression axis. Some crystals have the c-axis in the transverse direction, and can be easily deformed, while others have the c-axis oriented close to the extension direction, for which the deformation is close to zero. The magnitude of the deviations in deformation rate for crystals having values above the mean value of 1.0 (peak values of the deformation rate are about 1.2) are considerably smaller those values below it (approaching a value of zero for some orientations).

It is interesting to note that the average crystal responses depicted in Fig. 12a are qualitatively similar to those given by the Constrained Hybrid model. However, the textures predicted by the two are not similar, with the Constrained Hybrid model predicting a very strong c-axis component along the compression axis without the tilt toward the extension axis that is computed for with the finite element model [PRANTIL &AL. 1995a]. The Constrained Hybrid model alleviates the need to activate pyramidal slip and thereby reduces the stress required for plastic flow. However, with the deformation rate depending only on the crystal orientation (and full orientation distribution), the variability in deformation rate that is associated with the local neighborhood is absent and the computed textures do not share the same features with the computations that include this effect.

3. Conclusions

The freedom for crystals of a material to exhibit deformations that differ from that of their neighbors can have an appreciable impact on the evolving texture of a material. Simulation methods can explore this effect by discretizing a material's microstructure explicitly, treating the material as a body with subdomains defined by the crystals and computing its motion by solution of the full set of field equations. Detailed effects of grain interactions may be studied and used to improve models based on simpler mean field assumptions.

Here we have applied a finite element formulation for polycrystal plasticity to the response of fcc and hcp materials. Strain heterogeneity over the volume of the polycrystal leads to significant differences in the crystallographic texture under plane-strain compression. As the single-crystal anisotropy increases, the variation in strain over a polycrystal increases and the deviation between the finite element polycrystal and a corresponding Taylor model increase.

Chapter 13

FINITE ELEMENT SIMULATIONS OF METAL FORMING

FINITE ELEMENT SIMULATIONS OF METAL FORMING

P. R. Dawson and A. J. Beaudoin

Finite element analyses of the deformation of polycrystalline solids require knowledge of material response to perform the elemental integrations. For a simulation of the plastic deformation of a solid, the yield condition and flow law must be specified for the material in a manner that reflects the material's state, its dependencies on strain rate and temperature, and its symmetries. Polycrystal plasticity provides a microstructurally motivated constitutive model for the flow which defines the anisotropic yield condition and flow rule for a material element comprised of an aggregate of crystals. The distribution of lattice orientations, along with the strength of the crystal slip systems, defines the material state. The kinetic relations used for the single-crystal response give the rate dependencies, and the anisotropy is derived from the collective behavior of anisotropic single crystals. Balance laws are not applied to the crystals of the aggregate *per se*, but rather the aggregate is interrogated to define the stress and flow properties.

As a state variable representation for plastic flow, polycrystal plasticity theory has several distinct advantages. Implicit with the use of state variable models is the ability to initialize the state. Orientation distributions are directly accessible via x-ray measurement and well-established methodologies for interpreting those measurements. The crystal hardness distributions are not as direct, but for the restricted assumption of common hardening among all crystals of an aggregate the hardness may be initialized from simple compression testing. Polycrystal theory also provides a direct means for updating the material state via integration of the evolution equations for the crystal lattice orientation and the hardness,

1. Introduction

Large-scale applications are those in which the body, herein called the workpiece, is very much larger than individual crystals (Fig. 1). In these applications, the polycrystal is not a body itself, but rather is a representation of the microstructural state. Care must be exercised in assuring consistency between the macroscopic material element volume and the polycrystal dimensions. It is possible to think of macroscopic deformation gradients within the workpiece that are large, but yet only vary slowly across the dimension of a polycrystal. As such we may consider the polycrystal to be subjected to a uniform deformation locally. 'Locally' here refers to a point on the macroscopic scale, so that we permit only single (tensor) values of stress and velocity gradient. Thus the dimension of a crystal must be small compared to the dimension over which the macroscopic velocity gradient changes appreciably. In turn, the

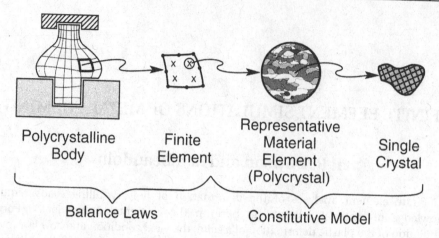

Fig. 1. Relationship of crystals to elements in large-scale simulations.

polycrystal must contain a sufficient number of crystals and inherent appropriate symmetry relations such that the above arguments of homogenization are justified. A representative material element is a volume of material that satisfies these requirements. Constitutive assumptions, such as a Taylor hypothesis or self consistent theories, dictate the manner in which the macroscopic deformation is projected onto a polycrystal, and subsequently, how the resulting responses (stresses) of those crystals are averaged to obtain a macroscopic value.

1.1 Background

Polycrystal plasticity and finite element formulations for the motion of deforming bodies have been used together to simulate a number of forming processes. Both Lagrangian and Eulerian formulations have been reported. In Lagrangian formulations, the mesh that discretizes the body moves with the deforming material, with nodal points becoming tags for material points. In contrast, in Eulerian formulations the mesh discretizes a spatially fixed volume, and material flows through that volume during the deformation process. To evolve the state in either case, streampaths of the material must be computed and evolution equations integrated along these trajectories. For the Lagrangian case, identifying the streampaths is trivial since the nodes in the mesh correspond to the same material points throughout a process. However, in the Eulerian case, computing streampaths is more complex with the material passing through the mesh. Substantial benefit is possible for cases of steady flow since the evolution of properties proceeds through the mesh along a streamline.

The rolling of plates, sheet forming, and forging have been studied with combined polycrystal plasticity and finite element capabilities. A variety of technical issues arise as the focus of these applications. Asymmetric deformations (earing), texture gradients within the workplace (and consequential effects on mechanical properties), macroscopic shear banding, and residual stresses are among the issues examined. For example, the rolling of plate and drawing of wire using an Eulerian formulation were reported by MATHUR & DAWSON [1989] and MATHUR &AL. [1990] using both Taylor and relaxed constraints assumptions for polycrystals comprised of fcc crystals. Here the material was assumed to have an initially uniform orientation distribution, and texture gradients in the products were predicted. VAN BAEL &AL. [1991] have implemented a procedure for computing the anisotropic stiffness from an approximation to the average Taylor factor written using a series expansion of terms involving the strain rate. The expansion parameters may be computed from the texture and thus updated with continued deformation. They have implemented the model in a Lagrangian elasto-plastic formulation and simulated rolling of bcc metals.

KALIDINDI &AL. [1992] reported on the implementation of a polycrystal model based on the Taylor assumption into a commercial elasto-plastic displacement based code. They compared the textures computed to those measured for laboratory micro-forgings of copper. CHASTEL & DAWSON [1991] modeled the rolling of initially textured silicon steels with an Eulerian formulation using a model for pencil glide slip modes. Detailed comparisons of through-thickness texture variations were made with experiment. CHASTEL &AL. [1993] used the same Eulerian visco-plastic formulation to study texturing of rocks in the Earth's mantle. Here, because of low crystal symmetry an equilibrium based model was employed to partition the deformation among crystals of an aggregate. Comparisons were made between computed and measured seismic velocities through the mantle. MANIATTY [1991] extended this Eulerian visco-plastic formulation to include elasticity and examined the influence of initial texture on the residual stress distribution in the product. SMELSER & BECKER [1991] have studied drawing of cups with a commercial code that calls the polycrystal theory through a user defined module, both using single-crystal and polycrystal representations.

One special concern arising with formulations that include elasticity is the effective integration of the stress. This involves the crystal kinematics, especially the decomposition of motion in term of elastic and plastic components. Both explicit and implicit procedures have been devised. PEIRCE &AL. [1982] developed an explicit procedure. Implicit procedures have been reported by KALIDINDI &AL. [1992] and by MANIATTY &AL. [1992], and CUITIÑO & ORTIZ [1992a, 1992b].

Recent progress in parallel computing has been exploited for large-scale applications. Parallel computing hardware, together with rewriting of algorithms for this architecture, have made it possible to simulate 3-D deformations involving several thousand elements each with hundreds of crystals defining every aggregate. For example, BEAUDOIN &AL. [1993] examined the evolution of texture in compression test specimens; hydroforming was studied by BEAUDOIN &AL. [1994] with respect to the formation of ears. Comparison of hydroforming simulation with experiment demonstrated the ability to predict both the location and strength of ears in the initially textured aluminum sheet [BEAUDOIN &AL. 1994].

1.2 Linking crystal behavior with macroscopic response

In applying polycrystal plasticity, it is assumed that the properties at any point in the workpiece are determined from a collection of anisotropic crystals that underlies that point. The modes of the crystal deformation are set by the slip system geometry, given that slip is the dominant deformation mechanism and other mechanisms can be neglected (including twinning, grain boundary sliding, and diffusion). An assumption, or rule, is employed to partition the macroscopic (average) strain among the individual crystals within an aggregate. A variety of rules are possible, some of which constitute bounds. By modeling the linkage between the microscopic (crystal) and macroscopic (continuum) scales with partitioning rules, the need to define each crystal's neighbors is circumvented. Instead, the aggregate of crystals can be thought of as a set of co-existing orientations whose averaged responses define the macroscopic properties. The orientations play the role of state variables, together with the crystal strengths and the grain shape. As state variables, the orientations replace the need to remember the deformation path, but require initialization to begin an analysis.

Different assumptions have been used to link the crystal behavior to the macroscopic deformation and thus to provide a means to partition the deformations and the stresses among the crystals of an aggregate. For crystals with high symmetry, such as fcc or bcc crystals with many slip systems of comparable strength, and nearly equiaxed shape, it often is assumed that each crystal of the aggregate experiences the same rate of deformation. This partitioning rule (referred to as the Taylor assumption) is extended to include the spin, so the crystal velocity gradient is common to all crystals [ASARO & NEEDLEMAN 1985]. However, heavily worked crystals can become flat and elongated. Then, based on the aspect ratio of the grains (as computed using the principal stretches of the macroscopic deformation gradient), some crystals within an aggregate are subjected to relaxed kinematic constraints [HONNEFF & MECKING 1978, KOCKS & CANOVA 1981]. For these crystals the continuity of the tractions and displacements is imposed based on the orientation of the largest interface between crystals. Given that each crystal deforms homogeneously, this leads to a mixed combination of stress and deformation constraints between crystals of an aggregate.

For crystals which have markedly anisotropic yield surfaces, say those with fewer than five independent slip systems, equal partitioning of the deformation among all crystals of an aggregates leads to physically unrealistic conditions. A crystal may be forced to deform in a mode that is not available to it from combinations of the existing slip systems or may exhibit the very high stress levels needed to activate unfavorable slip systems. In these cases, assumptions must be invoked that permit unequal straining. For example an equilibrium assumption may be made (all crystals experience the same stress state), or crystals may be permitted to deform using only favorable slip systems using a constrained hybrid approach such as that offered by PARKS & AHZI [1990].

More general partitioning rules exist that provide some measure both of equilibrium and compatibility. The visco-plastic self consistent theory presented previously (Chapter 11) is one example. This avenue will not be discussed here because of its limited use within finite element formulations, even though it use within finite element simulations has been reported recently [LOGÉ &AL. 1996]. Instead, we focus on the commonly made assumption, the Taylor hypothesis, and a modification of this rule that accounts for elongated crystal shapes, the relaxed constraints assumption.

Deformation rate assumption

Stated in terms of the crystal kinematics, the Taylor assumption requires that the macroscopic deformation gradient equals its microscopic counterpart:

$$\mathbf{D}^c = \mathbf{D}$$

(1)

where $\mathbf{D}$ is the macroscopic deformation rate. In contrast to the Taylor assumption, when relaxed kinematic constraints are imposed only those components of the deviatoric rate of deformation that lie in the plane of the shared flat grain boundary are required to be the same from crystal to crystal and equal to the macroscopic quantity. Letting $\mathbf{e}^1$, $\mathbf{e}^2$, and $\mathbf{e}^3$ be base vectors aligned so that the 3 direction lies perpendicular to this plane, then

$$\mathbf{e}^i \cdot \mathbf{D}^c \cdot \mathbf{e}^j = \mathbf{e}^i \cdot \mathbf{D} \cdot \mathbf{e}^j \qquad\qquad i, j = 1, 2 \qquad\qquad (2)$$

Furthermore, the tractions on this plane are equal, giving:

$$\mathbf{e}^i \cdot \mathbf{S}^c \cdot \mathbf{e}^3 = \mathbf{e}^i \cdot \mathbf{S} \cdot \mathbf{e}^3 \qquad\qquad i = 1, 2 \qquad\qquad (3)$$

where $\mathbf{S}$ is the macroscopic deviatoric Cauchy stress. The deformation rate may vary from crystal to crystal, but must average to the macroscopic value:

$$\mathbf{D} = \frac{1}{N} \sum_c w^c \mathbf{D}^c \qquad\qquad (4)$$

where N is the number of crystals in the aggregate [KOCKS & CANOVA 1981, MATHUR &AL 1990] and w^c is the weight afforded each crystal.

Local spin assumptions

To complete the link between the microscopic and macroscopic scales the material spin within each grain must be related to the macroscopic spin. A Taylor-like hypothesis is that the local spin $\mathbf{W}^g$ is equal to the macroscopic spin $\mathbf{W}$:

$$\mathbf{W}^g = \mathbf{W} \qquad\qquad (5)$$

This is not the only possibility, however. Conceptually, the relaxed constraints model is intended for those instances when crystal are flat and the aggregate can be thought of as layers of thin flat crystals. Throughout the deformation the layers remain parallel. This assumption is accomplished by requiring that the principal axes of the stretch ellipsoids of all crystals under relaxed constraints remain aligned. Based on this requirement, the microscopic spin of a crystal is defined by equating the rate of rotation of the principal axes of the grain's stretch ellipsoid (grain axes), $\mathbf{W}^e$, to that of the surrounding matrix (Chap. 8 Sec. 3.2), $\mathbf{W}^E$:

$$\mathbf{W}^e = \mathbf{W}^E \qquad\qquad (6)$$

where the e and E superscripts refer to the Euler spin. It is possible to apply either of the above rules for determining the grain spin, with the choice being based on the physical attributes of the system. In the limit of equiaxed grains, the two assumptions are equivalent.

2. Rolling Simulations

Flat rolling of metals is an important industrial process for which the evolution of texture is pronounced. Reductions in the material thickness to reduce the stock gauge translate into large strains with commensurate property alterations. The principal driving tractions for the reductions often are the frictional tractions between the workpieces and the rolls, leading to through-thickness gradients in deformations and properties that affect the product quality. Rolling is performed both hot and cold, and the evolution of texture is apparent in both. Here we address the rolling of aluminum [MATHUR &AL. 1990] and steel [SARMA & DAWSON 1996] alloys, showing the effects of

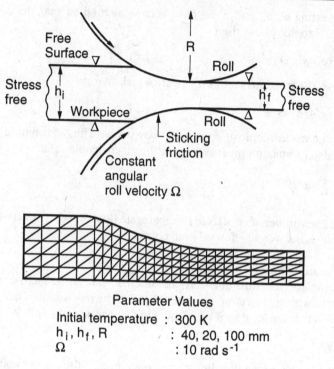

Parameter Values

Initial temperature : 300 K

h_i, h_f, R : 40, 20, 100 mm

Ω : 10 rad s^{-1}

Fig. 2. Boundary conditions and finite element mesh used in simulation of flat rolling.

modeling assumptions (in the case of aluminum) and comparisons to measured data (in the case of steel).

2.1 Rolling of aluminum alloys

The flat rolling of aluminum sheets was simulated to model the evolution of crystallographic texture using Taylor and Relaxed Constraints assumptions. The geometry, the boundary conditions, and the finite element mesh used for simulating the rolling operation are shown schematically in Fig. 2. Three rolling passes each of a nominal thickness reduction of 50% were simulated to give an overall thickness reduction of 87.5%. The macroscopic deformation was assumed to be two-dimensional (plane strain), although slip in the crystals was fully three-dimensional. The thickness of the entering sheet of 1100 aluminum was 40 mm and the sheet temperature at the beginning of each pass was initialized to be 300 K. The 100 mm diameter rolls were assumed rigid during the deformation. With sticking friction conditions at the roll/workpiece interface. The speed of angular rotation was constant at 10 rad s^{-1}, corresponding to a tangential velocity of 1 in s^{-1} at the roll surface. The aluminum sheets were cooled by convective losses to the atmosphere and to the rolls. Material constants typical of 1100 aluminum were used in the rolling simulations.

At every point of the material, its state was represented by an associated polycrystal, which consisted of a collection of 200 grains. The orientation of the grains prior to the start of the deformation was chosen randomly from a uniform distribution. The distribution of the hardness of all the slip systems for each grain prior to the first rolling pass was assumed to be uniform in the polycrystal as well as through

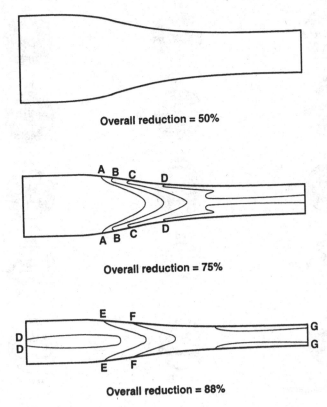

Fig. 3. Distribution of fraction of grains in the aggregate with relaxed constraints (contour A=10%;contour interval is 10%).

the thickness of the sheet. Further, the grains were assumed to be initially equiaxed. For subsequent rolling passes the slip system hardness and the grain orientations at the inlet of the control volume were computed from the predictions of the simulation for the previous pass. The net deformation of each polycrystal was required to be plane strain to coincide with the macroscopic motion.

Due to symmetry about the centerplane only the upper half of the workpiece was discretized. Each finite element mesh was made up of 240 six-node triangular elements with 539 nodes for the velocity and temperature. The velocity and temperature fields were approximated by continuous piecewise quadratic inter-polation functions. A step length which was equivalent to ten percent of the characteristic length of the smallest element in the mesh was used to define and integrate along the streamline. The initial guess for the flow field was obtained from a rolling simulation using an isotropic phenomenological constitutive model [KOCKS 1976]. Three to four further iterations were required to obtain a converged anisotropic steady-state flow field.

Figure 3 shows the distribution of the volume fraction of grains with Relaxed Constraints. The evolution of crystallographic texture is shown in Fig. 4 as the metal deforms through the first three passes. These passes correspond to overall thickness reductions of 50%, 75% and 88%, respectively. The corresponding texture evolution

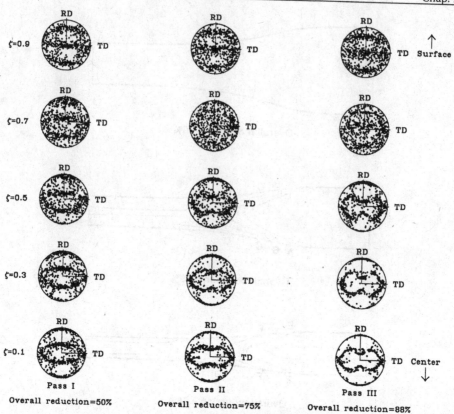

Fig. 4. 111 pole figures of the predicted through-thickness texture variation after three rolling passes using the method of Relaxed Constraints.

predicted by the Taylor model without accounting for the influence of the evolving grain shape on plastic anisotropy is shown in Fig. 5. In both Figs. 4 and 5 the location of material points is given by a normalized distance s, such that $s = 0$ on the center-plane of the workpiece and $s = 1$ on the surface of the workpiece. The volume fraction of grains with Relaxed Constraints remains very small during the first rolling pass. Consequently, the texture predictions after the first rolling pass by the Relaxed Constraints model differs little from the texture predictions obtained using the Taylor assumption. During the next two rolling passes, as the volume fraction of grains with Relaxed Constraints increases, the texture predicted by the two models deviate significantly, especially near the centerplane of the workpiece. The major components of the predicted textures are shifted in Euler space and the minor texture components differ in strength.

Looking first at the behavior near the centerplane of the workpiece, the shear component of the rate of deformation is very small in comparison to the normal components and, consequently, the deformation field is very similar to the 'ideal plane-strain' conditions. The forming model based on Taylor assumptions predicts texture near the centerplane with a fiber from the 'brass' texture component ({011} $\langle 21\bar{1} \rangle$) orientation towards the {4 4 11} $\langle 11\ 11\ \bar{8} \rangle$ stable orientation. This is the well-known 'Taylor' texture component, which is known to differ from the experimental

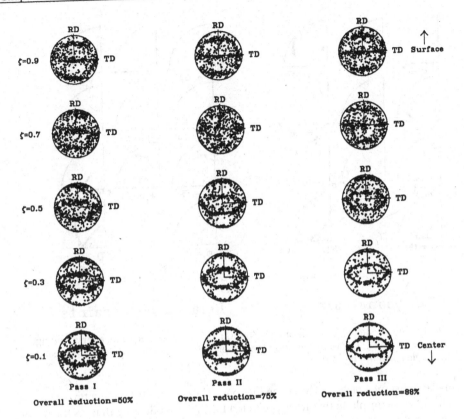

Fig. 5. 111 pole figures of the predicted through-thickness texture variation after three rolling passes using the Taylor hypothesis.

rolling texture. The results which include the effects of grain shape evolution predict the centerplane texture in better agreement with the experimental 'copper' texture component ($\{112\} \langle 11\bar{1}\rangle$). In contrast, near the surface the amount of shear deformation induced by the rolls is comparable to the compressive deformation, especially for large nominal thickness reductions. The final surface texture indicates $\{001\} \langle 110\rangle$ as the preferred orientation by both the models with the texture predicted by the Relaxed Constraints model being sharper.

The main reason for the inhomogeneity of the rolling texture is the varying amounts of shear deformation experienced by different material layers through the thickness of the workpiece. Near the centerplane, the shear component of the deformation always remains small when compared to the normal component. As a result, the deformation-induced texture for material points near the centerplane is progressively sharper with deformation. However, material points near the surface of the workpiece experience a shear component of the deformation which may be comparable with the normal component. Further, the shear component changes direction at the neutral point inside the deformation zone. This characteristic of the shear component of the deformation field continues to affect the development of deformation-induced surface texture even after the third rolling pass. From Figs. 6 and 7 it is clear that a texture similar to the centerplane texture is produced up to a

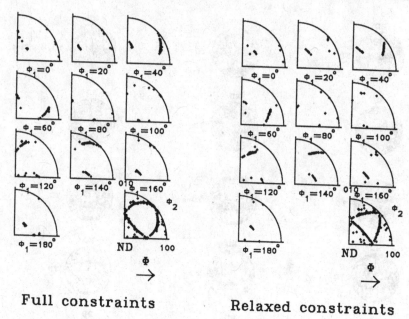

Full constraints   **Relaxed constraints**

Fig. 6. The predicted orientation distribution near the centerplane after a rolling reduction of 88%. The representation is a sample orientation distribution (SOD) using the angle convention of Bunge.

normalized distance of $s \leq 0.3$ after the first pass. After the third pass, a texture that is similar to the centerplane texture is predicted for $s \geq 0.5$ indicating that, as the overall thickness reduction increases, the through-thickness variation of the texture tends to localize in a zone near the surface of the workpiece where the shear deformation is still substantial.

The rolling texture predicted by the two polycrystalline models have been examined in greater detail by comparing the OD functions. Figures 6 and 7 compare the orientation distributions of the grains underlying material points at two points through the thickness of the rolled workpiece. The location of the material points are the same as the location of the material points whose pole figures have been plotted in Figs. 4 and 5. Near the centerplane of the workpiece, the difference in the texture predictions by the two methods is very clear. The full Taylor model reproduces the classical Taylor component while the Relaxed Constraints model which accounts for grain morphology predicts the well-known copper component. Near the surface, the differences in the texture predictions are again significant. Whereas the shear component of the deformation smears the texture for the Taylor model, the texture predicted by the current model is much sharper.

2.2 Rolling of ferritic stainless steels

Often it is desirable to model only certain stages of the rolling, such as the cold rolling that occurs near the end of the complete processing history of a metal. In such cases the material already exhibits pronounced crystallographic texture that it inherits from the prior stages of processing. To simulate only the latter stages, the texture at the start of these stages must be initialized from measurements. The rolling process may then be simulated and changes in texture that occur during the latter stages of

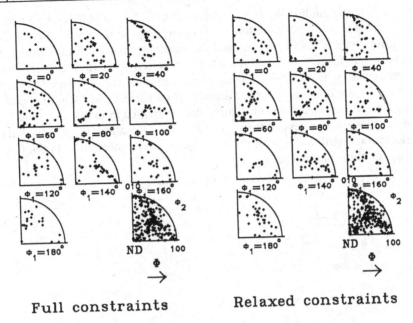

Full constraints   **Relaxed constraints**

Fig. 7. As in Fig. 6, but near the surface of the rolled sheet.

rolling can be predicted. Here we show the comparison of predicted textures for cold rolled chrome steel with measured textures at several through-thickness locations.

Three rolling passes were simulated. The mesh consisted of 180 quadratic triangular elements arranged in five layers, representing one half of the workpiece. The rolling schedule is given in Table I. The roll diameter was 63.5 mm and the starting thickness was 3.61 mm. The density was taken to be 7.715×10^3 kg/M^3 ; the specific heat used in the simulations was 490 J/kg K; and the thermal conductivity was estimated at 25 W/m^{2}K. The kinematics were driven by friction along the roll, using a friction model which determines the friction traction as a constant times the velocity difference between the tool and the workpiece. For the first pass, the friction constant was 1×10^3 N s/m^3; on the second and third passes, it was 6×10^8 N s/m^3. The problem was thermally coupled with heat being generated by internal stresses and being released through convection at the boundary. The convection coefficients were 200 W/m^2 K on the free surfaces and were 2000 W/m^2 K under the roll; the ambient temperature was 300 K. Internal heat was computed assuming that 90% of the mechanical work was converted to heat. Details of the methods used in coupling the material flow and the heat transfer for rolling may be found in DAWSON [1984].

Table I. Rolling schedule

pass	% reduction	roll speed (m/s)	inlet temperature (K)
1	24	1.3	300
2	19.5	2.8	330
3	17	2.8	370

The textures were initialized using popLA [KALLEND &AL. 1991a] on each layer from experimental data. Four sets of starting textures were used, representing the starting textures at the midline ($s = 0.0$), at 40% height ($s = 0.4$), at 80% height ($s = 0.8$), and at the surface ($s = 1.0$). The bottom layer of elements in the mesh received the midline texture; the next two rows received the texture from $s = 0.4$; the fourth row received the texture from $s = 0.8$; and the top row received the surface texture. After the first pass, subsequent textures were initialized using the outlet textures from the previous pass. Figure 8 shows the initial textures at $s = 0.0$, 0.8 and 1.0 in the form of 110 pole figures. The recalculated pole figures also are plotted from the corresponding initial weights files after processing them. The discrete orientations were converted to crystal orientation distributions (CODs). Pole figures were generated from the CODs following a smoothing operation.

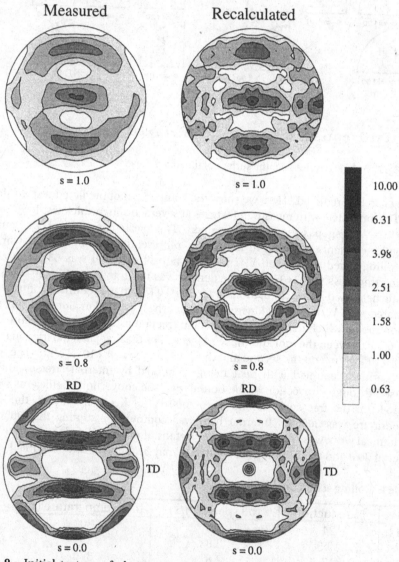

Fig. 8. Initial texture of chrome steel sheet at three positions through the thickness. Experimentally measured pole figures and pole figures recalculated from discretization of the texture (used in the finite element analysis) are shown.

The chrome steel sheet had been subjected to hot rolling prior to the cold rolling operation modeled here. As a consequence of the hot rolling, the material developed a texture which varied through the thickness of the sheet. Examination of the pole figures in Fig. 8 shows that the initial weights files capture the essential features of the through-thickness texture variation in the hot rolled band. The centerplane ($s = 0.0$) shows a rolling texture consequence of the deformation mode being plane-strain compression near the center of the sheet. Near the surface ($s = 1.0$ and $s = 0.8$), the texture shows the influence of the shear deformation which is predominant at the surface.

The finite element simulation of each rolling pass produced texture results at the outlet in the form of weights files. These weights files were used as input textures for the next pass, and for generating pole figures as described above. Texture predictions

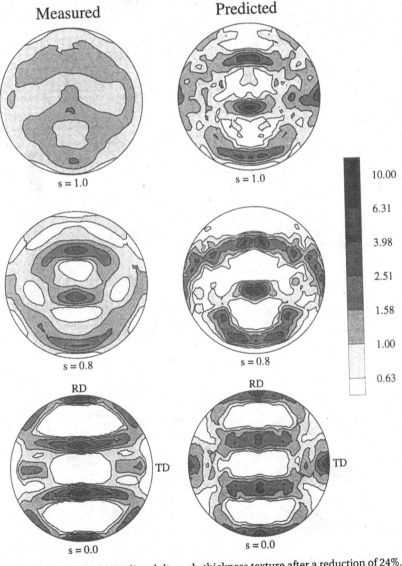

Fig. 9. Measured and predicted through-thickness texture after a reduction of 24%.

after 24%, 39% and 49% reduction are shown as 110 pole figures in Figs. 9–11. Also shown in these figures are the experimentally measured pole figures. The centerplane experimental texture qualitatively shows little change from the initial plane-strain compression texture. The simulations capture the main trends in this evolution. Near the surface, the texture originally has components stemming from shear deformation. The texture becomes more diffuse by 24% reduction, and develops features of a plane-strain compression texture by 39% reduction, which become stronger with further reduction. The simulated textures near the surface show the similar behavior but are much sharper. Errors in the computed textures appear to stem from the inadequate

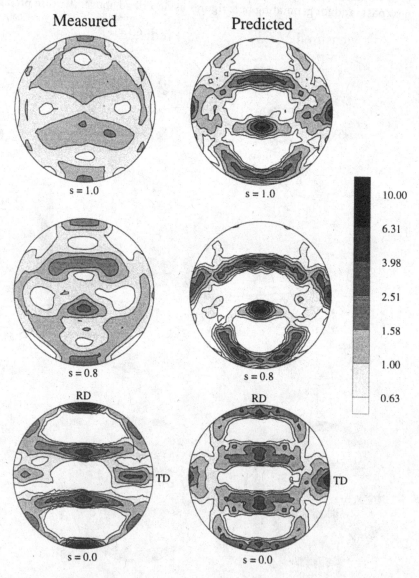

Fig. 10. Measured and predicted through-thickness texture after a reduction of 39%.

discretization of the original texture. While the discretization may approximate the initial texture near the surface, it lacks the detail necessary to simulate the randomization of the texture near the surface. The effects of shearing near the surface from friction are mitigated in the simulation in comparison to the experiment due to a combination of the use of the Taylor hypothesis and the original texture being too sharp. This is also noticeable in the texture at the centerplane, where the recalculated pole figure shows peaks near the circumference along the transverse direction (TD) axis. These peaks remain in the texture and appear in the pole figures at later reductions.

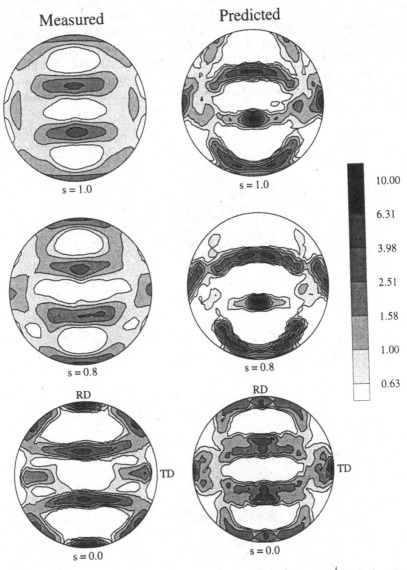

Fig. 11. Measured and predicted through-thickness texture after a reduction of 49%.

3. Sheet Forming Simulations

Sheet forming is a commonly used process for manufacturing a variety of components in automobiles, appliances, containers, and other products. Strength often is a requirement of the component, so forming the component from material that has been processed to provide strength is necessary. Strengthening likely follows from complex thermo-mechanical processing; as a consequence, significant anisotropy may be developed in the sheet. This initial anisotropy may be sufficiently strong to have a significant consequence on the forming operation. The anisotropy may produce 'ears' or may influence the point of instability in which formability is limited by localized thinning. Here we discuss two application of finite element simulations of sheet forming in which polycrystal plasticity is used explicitly to represent the material behavior. The hydroforming application [DAWSON &AL.1992b, BEAUDOIN &AL. 1993] demonstrates the ability of such a formulation to capture the detailed behavior in the deformation mode of a part formed from textured material. The limiting dome height application [BRYANT &AL. 1994] shows the use of this modeling capability to understand the origins of formability differences in materials produced under varying processing practices.

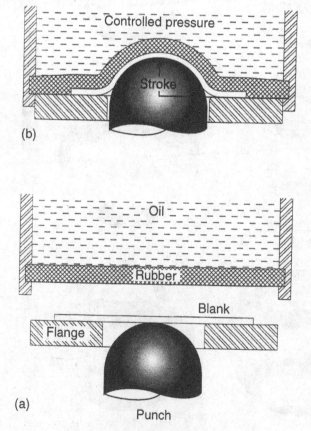

Fig. 12. Schematic of the hydroforming process.

3.1 Hydroforming of an aluminum hemisphere

Hydroforming consists of an advancing tool moving against a metal sheet that is constrained by backup tooling in the form of a rubber pad and pressurized oil bath. The sheet conforms to the tool geometry to produce the desired part, as depicted in Fig. 12 for conditions at the start of the process and after appreciable deformation.

The forming of a thin circular blank of AA5052 aluminum alloy was simulated and benchmarked with experiment. The blank had a radius of 117.5 mm and a thickness of 2.3 mm. The tool was hemispherical with a radius of 71.4 mm; its vertical speed was constant at 12.7 mm/s. In the experiment the tool surface was roughened, so a condition of sticking friction was imposed at the tool-sheet contacting interface. Well lubricated (frictionless) sliding was prescribed along the contact zone between the sheet and the flat flange and rubber pad. Correspondingly, lubrication was applied to the flange surface. The hydrostatic pressure was set at 3 MPa at the start of the process, and was ramped up as the deformation proceeded, reaching a peak value of approximately 34 MPa at the end of the operation.

The initial texture of the blank is shown in Fig. 13. This texture is representative of a recrystallized texture, dominated by the cube component with some retained rolling components. The sheet properties were assumed to be spatially uniform, so all elements were initialized with identical initial textures using weighted orientation files [KOCKS &AL. 1991b]. A total of 256 weighted grains were used to describe each of the textures. Hardening in the aggregate was prescribed so as to match experimental tensile test results for the blank material.

The geometry alone is initially axisymmetric. However, the initial textures present in the sheet were both orthorhombic, so full 3-D analyses were required. The global Z axis was taken normal to the sheet; orthorhombic symmetry in the initial textures implies symmetry planes normal to the X and Y axes. Taking advantage of this symmetry permits the discretization of only the quarter section of the sheet. In the examples presented, 1500 eight-noded brick elements define the workpiece. The deformation was simulated using time steps of 0.05 s to reach a dome height of equal to the tool radius after approximately 6 s.

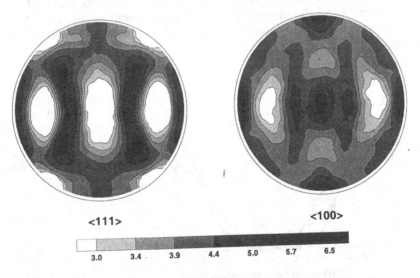

<111>   <100>

| 3.0 | 3.4 | 3.9 | 4.4 | 5.0 | 5.7 | 6.5 |

Fig. 13. Initial texture used in the simulation of hydroforming.

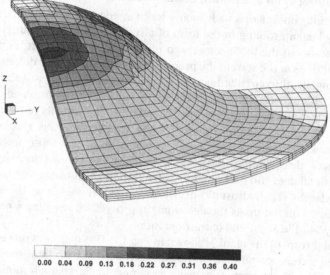

0.00 0.04 0.09 0.13 0.18 0.22 0.27 0.31 0.36 0.40

Fig. 14. Deformed mesh showing accumulated plastic strain.

Figure 14 shows deformed meshes with the process partially completed, corresponding to about 3 s after the process initiated. The shading in the plot depicts the accumulated plastic strain. This variation of plastic strain leads to the formation of ears along the workpiece periphery lying on the flange. As can be seen in the figure, the ears are forming in two different locations – in the vicinity of reduced plastic strain along the X and Y axes. The percentage earing is a common measure of this effect and often is quantified by a relative value, e_ρ:

$$e_\rho = \frac{\rho - \rho_{min}}{\rho_{min}} \qquad (7)$$

where ρ is the radial distance measured from the centerline of the tool to the outer edge of the sheet. Earing for the simulation and experimental workpiece are shown in Fig. 15. Similar to the experimental result, the simulation predicts the development of

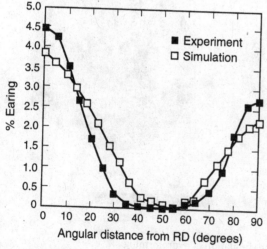

Fig. 15. Comparison of measured and computed earing profiles of sheet deformed by hydroforming.

an ear along the rolling direction (X-axis) and a relatively smaller ear along the transverse direction (Y-axis). The ears are a bit smaller and broader in the simulation as compared to experiment. A region of uniform radius from 40° to 65° – developed in the experiment – is also captured by the simulation.

3.2 Limiting dome height formability test

The limiting dome height test is used widely to assess the formability properties of metals. The test, shown schematically in Fig. 16 consists of stretching a rectangular sheet over a hemispherical punch. The sheet is constrained by a circular drawbead. Because the sheet is rectangular, the extent to which it is constrained by the drawbead is dependent on its width in comparison to the radius of the drawbead. By having a wide sheet, the stress state is near biaxial tension at the punch pole. With a narrow sheet, the stress state is closer to uniaxial.

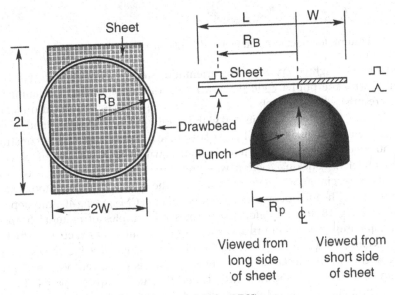

Fig. 16. Schematic of the limiting dome height (LDH) test.

The effect of processing practice on formability has been reported for an aluminum alloy [BRYANT &AL. 1994] in which the measured R-ratio for each rolling practice has been correlated with its measured dome height prior to failure. The various processing practices induce different crystallographic textures in the sheet, which translate to different initial conditions for the limiting dome height test. Pole figures for two of the practices reported in BRYANT &AL. [1994] are shown in Fig. 17. They differ principally in the strength of the Goss component, with greater Goss strength occurring in the sheet with higher measured R-value. The crystallographic texture affects the performance in the limiting dome height test through the effect of anisotropy in plastic flow on the thinning of the sheet.

Simulation of the limiting dome height test was conducted using the large-scale velocity pressure formulation outlined above for initial textures prescribed for the high and low R-value processing practices. The mesh used to discretize a symmetric quarter of the test specimen was comprised of 3600 brick elements, each with an

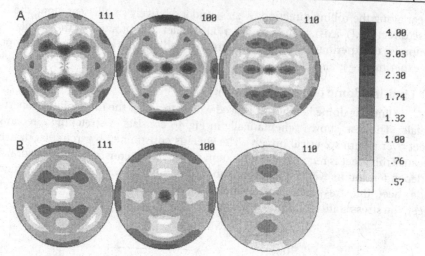

Fig. 17. Textures for Practice A and Practice B. RD up, stereographic projection.

aggregate of 256 weighted crystals. The orientation and weighting of the crystals was determined with the popLA code [KALLEND &AL. 1991a]. Identical hardening parameters were prescribed for each sheet based on tensile tests performed on samples from the sheet. Each sheet was oriented with its rolling direction corresponding to the long axis of the rectangular dome height specimen. The specimen was constrained against movement where it contacted the hold down tooling; the drawbead was not explicitly modeled. Sticking friction was prescribed for the interface between the punch and the specimen, which corresponded to the unlubricated conditions used in the experiments. The full operation was simulated on a CM-5 using 40 time steps.

Shown in Fig. 18 are the deformed meshes for samples of material from each practice, depicting strain contours and showing the contours of specimen thickness. The simulation corresponding to the texture having a high R-value shows a greater tendency to thin as the dome height increases. This is consistent with experiment: greater dome height is associated with lower R-value. Further, part failures are observed experimentally near the locations of greatest strain magnitude shown in these

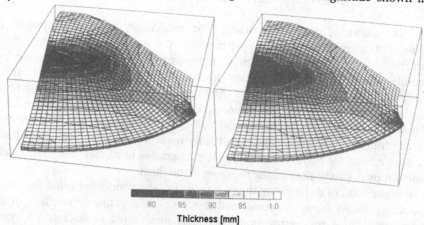

Fig. 18. Deformed mesh corresponding the completed tests showing thickness variations for Practice A (left) and Practice B (right).

4. Numerical Methodologies

Like models for polycrystal plasticity, finite element formulations come in many different forms. Because of this it is not advisable, nor even feasible, to attempt to establish the best wedding of microstructural theory with numerical treatment of a boundary value problem. For example, finite element codes are available based on explicit dynamic formulations as well as implicit static methodologies. Both displacements and velocities have been chosen as the primitive kinematic variables. Some formulations neglect elasticity while others are constructed around its existence. Introduction of a polycrystal plasticity model into a finite element formulation introduces further variants; polycrystal plasticity models have been developed that incorporate an extended Taylor hypothesis while others utilize self consistent premises to allow for richer variety of grain interactions. It not surprising that various combinations of these have been investigated, nor is it surprising that no single combination has demonstrated clearly its overall superiority.

As background to the use of finite element analyses of deformation of polycrystalline materials, it is helpful to review a few basic features of finite element methodology. The finite element method is a technique used principally for solving partial differential equations approximately. The method is not synonymous with any physical theory (system of equations), although its heaviest use probably has been in solid and structural mechanics. While there are a few features common to all finite element formulations, there is no single universal formulation. Rather, there is considerable variety in the implementations of the method, usually motivated by aspects of the systems of equations being solved. For solid mechanics applications, for example, major differences arise between kinematically-based formulations (having displacements or velocities as primary variables) and mixed methods (in which both kinematic and stress quantities are primary variables). Hybrid methods provide additional richness through attention to physically-motivated partitioning of the domain. Analytical reduction of the dimensionality of the problem in many instances yields greater efficiency together with more accurate results. There are two features of finite element methods, however, that are pervasive: locally defined interpolation functions (within elements) and governing equations satisfied in their weak form. These points are summarized in the succeeding sections as they are both central to the finite element method and are important in the merger of finite elements and polycrystal plasticity.

4.1 Interpolation and residuals

The choice of interpolation functions for defining the spatial variations of variables is at the root of why finite elements are called finite elements. Variables are defined piecewise over the domain, with the region described by one of the piecewise functions, or segments of the full field, being an element. While elements in theory can become arbitrarily small, they always remain finite and represent some fraction of the full domain. Continuity requirements on the mappings of elements onto the domain assure that the entire domain is spanned by the collection of elements, and thereby that the collection of piecewise functions fully defines the variable over the complete domain. Because the elements can be made as small as desired, polynomials of low degree (linear or quadratic) can be chosen without limiting the complexity of the complete solution.

Two attributes of the interpolation are critical: the degree of continuity of the interpolation within and across elements, and the completeness of the (polynomial) interpolation functions. Both must be adequately treated to guarantee convergence of the method. The necessary level of inter-element continuity in a given variable depends on the order of it and its derivatives in the weak form of the governing equation (discussed next). The completeness requirement focuses on the ability of arbitrarily small elements to retain an adequate approximation of the integrand of integrals appearing in the weak form. Other attributes or constraints may arise, such as the interrelationship of interpolated quantities. For example, the relation between stress and deformation in the constitutive relations often implies similar types of interpolations should be used for these quantities for consistency. However, the two predominant issues are the continuity and completeness.

Finite element solutions stem from weighted residuals formed on one or more of the governing equations, where by governing equations we mean the Euler equations for the system. Weighted residuals require that the weighted equations are satisfied in some average sense over the full or partitioned domain. The meaning of average depends on the nature of the weighting functions. In finite element formulations it is common, but not universal, to use identical interpolation functions for the primary variables (called the trial functions) and for the weighting functions, resulting in (Bubnov) Galerkin formulations. The important point to remember is that the solutions may not satisfy the governing (differential) equations at any given point. Rather, the solution is the best possible solution within the limitations of the discretization, where discretization incorporates both the resolution of the mesh of elements and the degree of the interpolation functions. In this sense, the solution is 'weaker' than one which could satisfy the differential equation pointwise. In the sections that follow that deal with different implementations of finite elements and polycrystal plasticity, we will emphasize this point. The velocity/pressure formulation imposes the weak form only on the equilibrium equation, while in the hybrid formulation discussed in Chapter 12 residuals are formed on both the equilibrium and the constitutive equations.

4.2 Velocity/pressure formulation

The governing equations which result from balance of momentum and conservation of mass are solved using a finite element discretization of the domain. A weighted residual is formed from the differential form of the equilibrium equation[1] and appropriate weighting functions, Φ:

$$\int_{\Omega} \Phi \cdot \text{div} \sigma \, d\Omega = 0 \qquad (8)$$

By means of standard finite element techniques, Eq. 8 may be rewritten as:

$$\int_{\Omega} \text{grad} \Phi \cdot S \, d\Omega - \int_{\Omega} \text{div} \Phi \, p \, d\Omega = \int_{\Gamma} \Phi \cdot t \, d\Gamma \qquad (9)$$

where the stress has been split into deviatoric (S) and volumetric (p) parts, and t represents the surface traction vector. (The presence of a body force would lead to one additional term.) S is eliminated from Eq. 9 using a visco-plastic constitutive relationship of the form

$$S = \mathbb{Q} : D \qquad (10)$$

where $\mathbb{Q}$ is a fourth-order stiffness tensor obtained by inverting and averaging the single-crystal compliances, $\mathbb{P}$:

$$\mathbb{Q} = \sum_c w^c \mathbb{Q}^c = \sum_c w^c \left(\mathbb{P}^c \right)^{-1} \qquad (11)$$

where w^c is the weighted contribution of crystal c. Each crystal compliance is generated using the single-crystal relationships as outlined in Chapter 8. The result is a matrix that relates the rate of deformation and the deviatoric stress in the crystal and takes the form:

$$\mathbb{P}^c = \sum_s \frac{\dot{\varepsilon}}{\tau_1^s} \left(\frac{m^s : S^c}{\tau_1^s} \right)^{n-1} m^s \otimes m^s \qquad (12)$$

---

[1] Inertial effects and body forces in the momentum balance and elasticity in the material response are neglected here.

Introducing shape functions for the velocity and pressure fields:

$$u_i = \sum_{n=1}^{N} N_i^n U^n \quad \text{and} \quad p = \sum_{m=1}^{M} \tilde{N}^m P^m \qquad (i=1,3) \qquad (13)$$

where u_i is a component of the velocity at an arbitrary point, U^n is a nodal point degree of freedom and N and $\tilde{N}$ are interpolation functions for the velocity and hydrostatic pressure, respectively. For the interpolation of hydrostatic pressure, p, piecewise discontinuous functions are specified – identical to those used in the hybrid method (Chap. 12).

Introducing the trial and weight functions into Eq. 9 gives

$$\sum_{n=1}^{N} K^{sn} U^n - \sum G^{sm} P^m = f^s \qquad (s=1,N) \qquad (14)$$

where

$$K^{sn} \equiv \int_{\Omega} B_\lambda^s \, \mathbb{Q}_{\lambda\mu} \, B_\mu^n \, d\Omega \qquad (15)$$

and

$$G^{sm} \equiv \int_{\Omega_e} h_\lambda \, B_\lambda^s \, \tilde{N}^m \, d\Omega, \qquad (16)$$

In Eqs. 14 and 15, h is the trace operator and the matrix $\mathbf{B}$ contains the gradient of the velocity trial functions such that

$$D_\lambda = \sum_{n=1}^{N} B_\lambda^n U^n$$

The right-hand side of Eq. 14 relates to the resultant force on one surface of the element:

$$f^s \equiv \int_{\Gamma_e} N_i^s t_i \, d\Gamma \qquad (17)$$

Eq. (14) is combined with the incompressibility constraint, which appears in an identical manner to the hybrid formulation:

$$\sum_{n=1}^{N} G^{nr} U^n = 0 \qquad (r=1,M) \qquad (18)$$

to give the global system of equations which may be solved for the velocity field and pressure distribution. The stresses are recovered from the visco-plastic matrix using the computed deformation rates. The resulting matrix equation is nonlinear and is solved with an iterative scheme. This requires information about the mechanical behavior at the integration points for a given estimate for the flow field. For our applications we assume that an aggregate is located within every element of the finite element mesh and that the mechanical behavior can be determined by interrogation of the aggregate. At each iteration on the velocity field, $\mathbb{Q}$ is re-computed based on the macroscopic deformation rate and the rule for partitioning that deformation rate among the crystals within the aggregate. The stiffness $\mathbb{Q}$ and the deformation rate are then used to eliminate the stress tensor from the weak form of equilibrium and to compute a new estimate for the flow field. As in the case of the hybrid formulation, the presence of the incompressibility constraint requires special attention in the conjugate gradient procedure [BEAUDOIN &AL. 1993].

4.3 Updating the texture and hardness

As material deforms, its state changes; these changes are recorded in polycrystal plasticity theory through evolution of the crystallographic texture and the slip system hardnesses. Texture evolves with reorientation of the crystals according to an equation for the reorientation rate, $\dot{R}^*$, of a crystal's lattice that is derived from kinematics and the slip system activity (Chap. 8 Sec. 3.3)

$$\dot{R}^* = (W^g - W^c) \cdot R^* \tag{20}$$

Here W^g is the skew portion of the crystal velocity gradient, and W^c is the spin associated with plastic slip as given by (Chap. 8 Eq. 12):

$$W^c = \sum_s q^s \dot{\gamma}^s \tag{21}$$

q^s is the skew portion of the Schmid tensor for the slip system s.

The hardness τ_1^s evolves with straining according to a simple relationship that embodies a saturation hardness, $\kappa\tau_v$ (Chap. 8 Sec. 1.5):

$$\dot{\tau}_1^s = \Theta_0 \left(\frac{\tau_1^s - \tau_0}{\kappa\tau_v} \right)^{\kappa} \dot{\Gamma} \tag{22}$$

where the saturation hardness may depend on strain rate and temperature, $\dot{\Gamma}$ is shear rate summed over all slip systems within the crystal, and Θ_0, τ_0, and κ, are material parameters. Here we assume equal hardening for all slip systems, neglecting any variations from differing levels of slip system activity. No attempt is made to quantify the dislocation structure in any detail, but rather slip is assumed to harden the crystal uniformly over its volume and equally on all of its slip systems.

The evolution of the material state may be computed for either Eulerian and Lagrangian formulations by appropriately following the material. For Eulerian cases, streamline methods enable the tracking of particle trajectories through a steady flow field. For Lagrangian formulations, particles follow streampaths of the deforming material's motion which usually coincide with the movement of the mesh. In both cases the history of straining may be evaluated and used in the integration of the differential equation along its characteristic.

4.4 Parallel computing strategies

The global system of equations is challenging to solve using a data parallel architecture owing to the poor condition caused by the incompressibility constraint. A pre-conditioned conjugate gradient method is used, with special provisions for handling the constraint [BEAUDOIN &AL. 1993]. Many of the computational tasks may be performed in a manner to exploit parallel computer architectures. For example, the forming of all elemental matrices is done concurrently.

The parallel implementation of the mathematical formulation discussed here consists of three independent data structures: a set of unassembled finite elements; a set of assembled nodal degrees of freedom; and a set of crystals. Any interaction between the three data structures results in communication between the processing nodes of the parallel architecture. These three data-sets must be mapped to the processing nodes of the parallel architecture such that there is arithmetic load balance, the locality of reference is maximized, and the contention for the communication channels in the network interconnecting the processing nodes is kept to a minimum.

The finite elements in the mesh are grouped into sub-meshes for mapping to the processing nodes using a parallel implementation of the recursive spectral bisection algorithm [MATHUR 1995]. No assumptions are made regarding the topology of the mesh. The mapping obtained by this algorithm attempts to maximize the locality of the sub-mesh residing on each processing node by ensuring that the finite elements comprising the sub-meshes are geometrically connected and have a surface area to volume ratio that is close to optimal.

Nodal points that are interior to the sub-mesh residing on a processing node are mapped to that same processing node. Boundary nodes must be assigned to one of the processing nodes with which they are associated, or replicated on all of the processing nodes in which they appear. Only boundary nodes necessitate communication. The implementation reported here assigns the boundary nodes to one of the partitions with which they are associated in a random manner. Randomization minimizes the contention for the communication channels that interconnect the processing nodes.

The finite element formulation outlined earlier permits an arbitrary number of crystals in each element. For the case of one crystal per element, the crystals are mapped to the processing nodes using the same mapping as the finite elements. For the case when the crystal to element ratio is greater than one, the set of crystals is viewed as a two-dimensional data set (a dense matrix). The rows of the matrix represent the crystals within a polycrystal and the columns of the matrix represent the polycrystals. The number of polycrystals (= number of columns) is the same as the number of finite elements (N_e) comprising the mesh. The number of rows is the crystal to element ratio (N_c). Typically all the crystals in one row would be assigned to the processor that holds the associated element.

All the finite element meshes that are used in the simulations reported here are comprised of one type of element only. The arithmetic operation count for each crystal is almost uniform. Consequently, there is near perfect arithmetic load balance for all element-wise computations. The blocked mapping scheme results in an even distribution of the crystals on the processing nodes. This is sufficient for achieving arithmetic load balance without sophisticated data dependence.

Concurrency is exploited at the level of unassembled finite elements, assembled nodal points and crystals. The crystal behavior equations and the nodal stress versus nodal velocity relation are solved/inverted concurrently for all the crystals in the aggregate. The unassembled element-wise matrices are evaluated concurrently for all finite elements in the mesh. The resulting sparse linear system is solved using a conjugate gradient method; no explicit assembly of the sparse coefficient matrix is required. The poor condition of the coefficient matrix caused by the incompressibility constraint, however, requires special treatment.

4.5 Implementation issues

In most finite element implementations, isoparametric elements are employed in which gradients of velocity (or displacement) may exist across an element. Further, elemental integrations are performed using quadrature in which the integrand is evaluated at specific points within an element. One can consider the material in the vicinity of a quadrature point as a region on which the homogenization is performed on the crystal responses of an aggregate. Sufficient crystals must reside in this region for the homogenization to be valid, as discussed in the previous paragraph. Variations in the velocity gradient within an element imply, however, that all of the quadrature points are not experiencing the same strain rate. The material's rate sensitivity, texture, and hardening characteristics become important then because of the influence they exert on the variations in stress in the presence of differences in the strain rate over the domain of the element. For convergence of the finite element solution for the motion, the stress should approach a smooth distribution within an element and across its boundaries to neighboring elements. Smoothness of stress will not be achieved without some smoothness of the properties.

This requires that the number of crystals within an aggregate be sufficiently large to represent an orientation distribution well, and that the distributions within an element not vary greatly. In fact, the use of a single aggregate within each element, placed at its centroid, provides excellent performance in finite element simulations because it restricts the possible variations of stress. For consistency then, as elements become sufficiently small that smooth stresses are achieved, they each must still remain sufficiently large to physically contain an aggregate that is representative of the complete distribution and over which a legitimate homogenization may be performed. In Chapter 12 a hybrid formulation was reviewed that permitted solutions in cases of few (including only one) crystals per element. In the hybrid development, the residual on equilibrium is constructed to seek the best distribution of interelement tractions. Discussion of the relative merits can be found in BEAUDOIN &AL. [1995a].

5. Concluding Remarks

Polycrystal plasticity theory, viewed solely as a constitutive framework motivated by the physics of deformation in the microstructure, provides the finite element analyst benefit as compared to the more common continuum constitutive models. The simulation captures the underlying physics of metal deformation to a greater degree. However, benefits extend beyond the initial gains which are garnered – at an admittedly significant computational expense – through the incorporation of a more detailed material description.

The development of polycrystal plasticity theory was founded predominantly on the prediction of texture evolution for simple, proportional deformation paths. In the course of this development, a rich array of experimental and analytic tools were advanced. A finite element analysis which embraces polycrystal plasticity provides the capability to address problems with gradients in deformation typical of industrial forming operations. The aforementioned tools attain new significance. For example, a material having a deformation history which produced spatial gradients in the texture may be characterized; quantitative texture analysis provides a pathway to describe material state at different points in such a graded material. The resulting description of state may be used to initialize a finite element model for a subsequent forming operation, the simulation itself leading to gradients following from inter-action of the deformation and evolution of the material anisotropy (such as in the examples of rolling chrome steel and hydroforming). At any point in the simulation, the tools of quantitative texture analysis can then be invoked to post-process the finite element results in a likeness reflecting experiment. Results are communicated through graphical descriptions which give a physical feel to the model.

It is not simply then an argument for increased accuracy which prompts the combining of polycrystal plasticity theory with finite element analysis. In adopting the myriad variables which describe the material state as a polycrystal aggregate, one is rewarded by the combined ability to initialize the anisotropic state of the material in a tractable manner, accommodate gradients in deformation which follow from interplay of the material state and boundary conditions, and illustrate the simulation results in identical fashion to allied experiments.

Chapter 14

PLASTICITY MODELING IN MINERALS AND ROCKS

1. **Verification of Plasticity Models with Anisotropic Minerals**
    1.1 Mechanical twinning and heterogeneous axisymmetric deformation: calcite
    1.2 The issue of rate sensitivity: quartz
    1.3 Heterogeneous deformation due to high plastic anisotropy
       Halite. Quartz
    1.4 A deformation-based model of dynamic recrystallization: quartz and ice

2. **Textures as a Diagnostic Tool in Naturally Deformed Rocks**
    2.1 Estimation of overall strain: morphological textures in sheet silicates
    2.2 Strain paths. Thrusting versus crustal thinning
       Calcite. Quartz
    2.3 Texture transitions with metamorphic grade: quartz

3. **Physical Properties: Seismic Anisotropy in the Earth's Mantle**
    3.1 Anisotropy of seismic velocities along the San Andreas fault in California
    3.2 Convection in the Earth's upper mantle

4. **Conclusions**

PLASTICITY MODELING IN MINERALS AND ROCKS

H.-R. Wenk

While the study of preferred orientation in geological materials is old, quantitative modeling of texture evolution is relatively recent. In the past much of the interpretation has been based on intuition rather than physical principles [WENK & CHRISTIE 1991]. Early applications of the Taylor–Bishop–Hill theory to halite [SIEMES 1974] and quartz [LISTER &AL. 1978, LISTER & HOBBS 1980] laid the groundwork for later, more refined models. In this Chapter we would like to present a few examples in which polycrystal plasticity theory has been useful for interpreting the deformation histories of rocks from observations of textures, and in which anisotropic physical properties can be explained on the basis of textures. Also, in some cases minerals have been helpful in understanding material properties and elucidating basic differences between theories. Because of their low crystal symmetry, of the high dependence of flow stress on strain rate, and of the high plastic anisotropy, minerals have been important in advancing such concepts as rate sensitivity, mechanical twinning and heterogeneous deformation. Even when the crystal symmetry is high, as in halite, ionic and covalent crystals with directional bonds are often more plastically anisotropic than cubic metals. The examples will provide some insight why geologists have been pursuing texture analysis for a long time, dating back to D'HALLOY [1833] who was the first to introduce the term 'texture' to describe directional features in polycrystals. The Chapter also raises awareness about the significance of minerals for developing and refining modeling methods. This is not a comprehensive review but rather a set of typical examples.

The Chapter is divided into three parts. The first part deals with complications encountered with plasticity models in mineral systems. The second part uses textures to reveal information about the geological deformation history. The third part relates textures to macroscopic seismic properties. The minerals used in this Chapter emphasize trigonal *calcite*, trigonal *quartz*, cubic *halite*, monoclinic *mica* and orthorhombic *olivine*. Calcite is ductile compared to silicates. Limestone or marble beds, consisting of calcite, often concentrate large deformations in tectonic shear zones. Quartz is of profound geological significance because, due to its common occurrence and its relatively high ductility compared to other rock-forming silicate minerals, its deformation behavior controls much of the rheology of the Earth's crust. Halite intrudes as salt domes into overlying sediments, producing complicated deformation structures with strong texture. Orientation patterns in sheet silicates have been used for a long time to estimate the finite strain which accumulated during deformation in the crust. Olivine is the main constituent of the Earth's upper mantle and critical for deformation processes in the deeper interior and for seismic anisotropy.

1. Verification of Plasticity Models with Anisotropic Minerals

1.1 Mechanical twinning and heterogeneous axisymmetric deformation: calcite

Calcite is a mineral with many slip systems, each family with a different critical resolved shear stress. BAUMHAUER [1879] did the classical experiment to induce mechanical twinning by applying a stress with a knife. Twinning can be modeled in plasticity theories in a similar way to slip, yet there are some significant differences (see Chapters 2, 8 and 11). Whereas slip is more or less homogeneous throughout the grain, twinning occurs in localized regions, in the case of calcite as thin lamellar bands, and comprises a certain volume fraction, which is often small. The twinned fraction has a specific orientation relation with respect to the matrix. Whereas slip can occur in opposite directions, though often with different critical resolved shear stresses, twinning (in most cases) can only occur in a single direction (or 'sense'). In the case of $e^+=\{\bar{1}018\}\langle 40\bar{4}1\rangle$ twinning in calcite, it is in the positive sense (defined as positive if shear causes extension along the c-axis [0001], Fig. 1).

Whenever twinning is initiated it amounts to splitting each grain into two parts with different orientations, which is not practical for most simulation procedures where the number of grains is kept constant. Two schemes are used to circumvent this problem. One consists of abandoning the concept of individual grains which reorient and instead making a volume fraction transfer between fixed cells in Euler space [TOMÉ &AL. 1991a]. The second is a statistical method which either retains the matrix or the twin, depending on their relative volume fractions and the outcome of a random number generator [VAN HOUTTE 1978]. The statistical scheme has been refined to match the accumulated volume fractions of the twin-reoriented grains with that given by the twinning shears, and also only to pick grains for twinning if the accumulated twin fraction has reached a certain threshold. (See Chapter 11 and TOMÉ &AL. [1991a].)

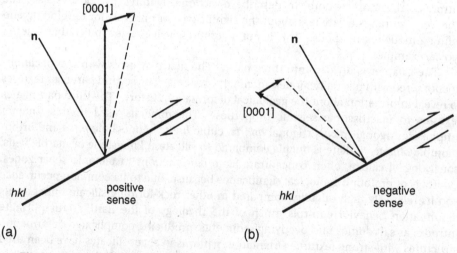

Fig. 1.   Definition used for shear sense of calcite and hexagonal, trigonal and tetragonal crystals in general. Deformation by slip or twinning on a lattice plane *hkl* in the positive sense causes extension along the c-axis [0001] (a) and compression in the negative sense (b) [TURNER &AL. 1954].

Using this method, texture development in calcite in plane-strain compression has been modeled with the Taylor theory. The trend in texture variations observed in experimental pole figures can be explained by differences in activity of available slip systems. Table I compares observed deformation systems of the two geologically important rhombohedral carbonates dolomite and calcite (see also DE BRESSER & SPIERS [1997]). Critical shear stresses are well established through experiments on single crystals. They vary particularly as a function of temperature and to a lesser degree of strain rate. As the importance of mechanical twinning diminishes with increasing temperature, the texture changes from a (0001) maximum near the principal shortening direction to a (0001) double maximum (Chap. 6 Fig. 4).

Table I. Major deformation mechanisms in rhombohedral carbonates. Shear stresses (in MPa) for low temperature (LT, 100°C) and strain rates of about 10^{-5} s^{-1} interpolated and estimated from Table 1 p. 362 of WENK [1985c].

	LT	HT
Calcite CaCO$_3$		
slip:		
$r^-=\{10\bar{1}4\}\langle20\bar{2}\bar{1}\rangle$	50	15
$r^+=\{10\bar{1}4\}\langle\bar{2}021\rangle$	80	20
$f=\{\bar{1}012\}\langle0\bar{2}\bar{2}1\rangle\langle\bar{2}20\bar{1}\rangle$	150	30
$f^*=\{\bar{1}012\}\langle02\bar{2}\bar{1}\rangle\langle2\bar{2}01\rangle$	200	50
twinning:		
$e^+=\{\bar{1}018\}\langle40\bar{4}\bar{1}\rangle$ $\gamma=0.69$	10	7
Dolomite (Mg,Ca) CO$_3$		
slip:		
$c=(0001)\langle2\bar{1}\bar{1}0\rangle$	50	130
$f=\{\bar{1}012\}\langle0\bar{2}\bar{2}\bar{1}\rangle\langle-220\bar{1}\rangle$	170	100
$f^*=\{\bar{1}012\}\langle02\bar{2}\bar{1}\rangle\langle2\bar{2}01\rangle$	250	150
$r^-=\{10\bar{1}4\}\langle\bar{1}2\bar{1}0\rangle$	250	150
twinning:		
$f=\{\bar{1}012\}\langle10\bar{1}1\rangle$ $\gamma=0.59$	90	100

Some examples which interpret texture transitions in calcite based on a topology analysis of the single-crystal yield surface and deformation mode maps have been given in Chapter 11. TOMÉ &AL. [1991b] documented that for plane-strain deformation of calcite different polycrystal plasticity models (Taylor, relaxed constraints and self-consistent theories) provide rather similar texture results (Fig. 2). This is due to a fairly equiaxed single-crystal yield surface, compared to other minerals and particularly in a subspace that contains most plane-strain paths, with many (21) potentially active slip systems. Simulated texture patterns compare favorably with experiments (Chap. 6 Fig. 6) [WAGNER &AL. 1982, TAKESHITA &AL. 1987].

For axial compression this is different and Taylor simulations do not at all agree with experiments. Experimental inverse pole figures show a concentration of compression axes at negative rhombs $\{0h\bar{h}l\}$ with a shoulder towards (0001) (Fig. 3a) whereas Taylor simulations have a concentration at positive rhombs $\{h0\bar{h}l\}$ (Fig. 3b). The experimental pattern for axial compression is qualitatively similar to that obtained experimentally when calcite is deformed in plane-strain compression and the sample is rotated about the compression direction [Fig. 3c, WENK &AL. 1986]. This is reminiscent of the 'curling' effect during compression of fcc metals and extension of bcc metals where individual crystals deform locally in plane strain even though the

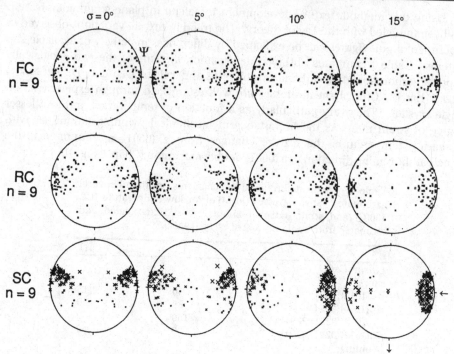

Fig. 2. Orientation distribution diagrams, presented as oblique sections for calcite polycrystal, deformed in pure shear. They illustrate similar textures for simulations with the full constraint Taylor model (FC), the relaxed constraints option (RC), and the visco-plastic self-consistent model (SC). The compression axis is horizontal, the extension axis vertical; 50% equivalent strain, stress exponent $n=9$ [TOMÉ &AL. 1991b].

macroscopic deformation is axisymmetric [HOSFORD 1964]. Such locally hetero-geneous deformation is consistent with the calcite microstructure which shows in a section perpendicular to the compression direction grains with high aspect ratios in different directions.

The observed texture with a maximum near (01$\bar{1}$8) is well reproduced with a relaxed constraints Taylor theory that relaxes strain increment components ($d\varepsilon_{22}$–$d\varepsilon_{11}$) and $d\varepsilon_{12}$ [VAN HOUTTE 1981] (Fig. 3d). Each crystal can freely choose two of its principal strain directions normal to the compression direction and two of the magnitudes along these directions. Contrary to the relaxation for flat grains, which produces minimal compatibility problems, this type of relaxation produces local mismatch between grains that have to be accommodated by bending, grain-boundary sliding and brittle processes, all of which are observed in the microstructures. Grain curling may well occur in calcite, but it is automatically accounted for in self-consistent model calculations, without the need to impose explicit conditions [LEBENSOHN &AL. 1997].

1.2 The issue of rate sensitivity: quartz

Trigonal quartz was the first mineral system on which Taylor texture simulations were performed and the pioneering work of LISTER &AL. [1978] had a great impact by providing a quantitative tool to interpret mineral textures that so far were approached by intuition.

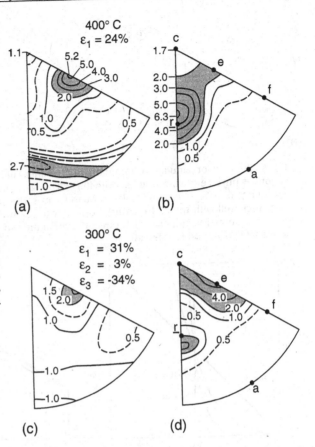

Fig. 3. Deformation of calcite in axial compression, inverse pole figures. (a) Axial compression experiment, (b) Taylor simulations for axial compression, (c) compression axis for plane-strain compression experiment, (d) Taylor simulation for plane-strain compression with curling which averages about the compression direction [WENK &AL. 1986]. Equal-area projection, density regions above 2 m.r.d. are shaded.

Whereas deformation systems and critical shear stresses in calcite and dolomite have been determined in single-crystal experiments, experimental determinations are much more difficult for quartz, in part due to the low ductility at moderate confining pressure, mechanical Dauphiné twinning and the α-β phase transformation. Therefore, quartz deformation mechanisms are much less well known. But it is generally agreed that at low temperature basal slip is dominant; as temperature increases first prismatic and finally pyramidal slip becomes active, similar to observations in hexagonal metals. In models, qualitative critical shear stress ratios for low and high temperature have been assumed (Table II) [LISTER &AL. 1978 and WENK &AL. 1989a].

Table II. Slip systems in quartz and assumed critical-shear-stress ratios [WENK &AL. 1989a]. Model α, β and γ, corresponding to low, medium and high temperature.

slip plane/direction			α	β	γ
(0001)	$\langle 2\bar{1}\bar{1}0\rangle$	basal	1	1	1
$\{10\bar{1}0\}$	$\langle\bar{1}2\bar{1}0\rangle$	prismatic	6	0.4	1
$\{10\bar{1}1\}$	$\langle\bar{1}2\bar{1}0\rangle$	pyramidal	3	3	∞
$\{10\bar{1}1\}$	$\langle\bar{2}113\rangle$	pyramidal	6	6	∞
$\{2\bar{1}\bar{1}1\}$	$\langle\bar{1}\bar{1}23\rangle$	pyramidal	∞	∞	2.5
$\{2\bar{1}\bar{1}1\}$	$\langle\bar{1}2\bar{1}3\rangle$	pyramidal	∞	∞	3

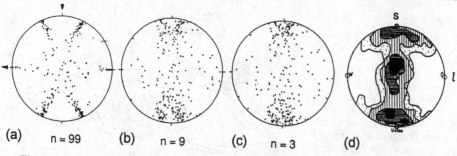

Fig. 4. (a-c) Taylor simulations for quartz in pure shear illustrating the effect of stress exponent (inverse of strain-rate sensitivity) [WENK &AL. 1989a]. The stress exponent n=3 (in $\dot{\gamma}=\tau^n$) which is applicable to quartz, gives the most realistic texture and compares well with the first measured quartz texture on the right side (d) [SCHMIDT 1925] (compression and extension directions are indicated by arrows, s is the normal to the schistosity and l the lineation).

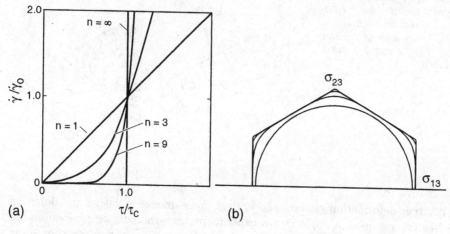

Fig. 5. Visco-plastic deformation of quartz. (a) Graph illustrating the power law to describe strain-rate sensitivity for different stress exponents. (b) Two-dimensional sections through the single-crystal yield surface of quartz according model 'α' (Table II) ,for stress exponents 3, 9 and 99. Stress coordinates σ_{23} - σ_{13} are indicated.

There is not much experimental information about the importance of 'trigonality', e.g. the difference between slip on $\{10\bar{1}1\}$ and $\{01\bar{1}1\}$ which are structurally different planes, and quartz is mostly treated as hexagonal in deformation modeling. Unfortunately, many of the predicted patterns for quartz, such as that in Fig. 4a, had little resemblance to natural and experimentally produced textures (Fig. 4d).

It turned out that an important consideration in quartz is the strain-rate sensitivity of the flow stress. Both Taylor and Sachs assumed a rigid-plastic behavior; implying that no deformation (i.e. dislocation movement) occurs until the critical resolved shear stress is reached, at which point deformation is instantaneous. This is rarely an adequate description (Chapter 8). Especially at intermediate and high temperatures where lattice diffusion and climb become active, there is a rather high rate dependence. Crystals deform even at low stresses, albeit more slowly. As described in Chapter 8, a power law is often used to describe visco-plastic deformation and strain-rate sensitivity (Fig. 5a).

$$(\dot{\gamma}/\dot{\gamma}_0) = (\tau/\tau_0)^n \tag{1}$$

where $\dot{\gamma}$ is the shear strain rate and τ the resolved shear stress. Such a power law is at least applicable in the vicinity of some threshold stress τ_0 coupled to some reference strain rate $\dot{\gamma}_0$. Metallurgists have originally introduced the visco-plastic condition as a way to deal with the choice of slip system combinations at vertices of the single-crystal yield surface in cubic metals where ambiguities exist in the rigid-plastic case [ASARO & NEEDLEMAN 1985]. In low symmetry minerals this ambiguity is generally of no concern but the strain-rate sensitivity is much higher and has an influence on textures. Minerals are somewhere intermediate between a Newtonian viscous behavior ($n=1$) where the strain rate is proportional to the stress and a rigid-plastic behavior ($n=\infty$). In visco-plastic deformation many systems are active in each crystal at all times, yet some only to a small degree. For quartz with a low stress exponent of 3 (high strain-rate sensitivity of 1/3) the single-crystal yield surface is highly rounded (Fig. 5b) and strain is distributed more uniformly over many systems, causing a reduction in anisotropy.

Whereas it was found that strain-rate sensitivity is not very significant for texture development in such systems as calcite, halite and olivine, for quartz the influence on texture is profound (Fig. 4a-c) [WENK &AL. 1989a]. The visco-plastic texture, displayed as 0001 pole figures, is smoother than the rigid-plastic texture, corresponding better to some natural quartz textures such as that in the first published fabric diagram (Fig. 4d, SCHMIDT [1925]). The maximum in the center of pole figure 4d will be discussed later, in the context of recrystallization.

1.3 Heterogeneous deformation due to high plastic anisotropy

Halite

The cubic mineral halite (NaCl) has a typical ionic structure. Since salt mines have gained recognition as potential repository sites for nuclear waste, the long-term deformation behavior of halite has become of particular importance and numerous studies have been devoted to investigate ductility of mono- and polycrystalline halite. Halite is isostructural with periclase (MgO, 'magnesia') which composes much of the lower mantle of the Earth and, together with the perovskite phase MgSiO$_3$, controls its rheology, just as quartz determines the rheology in the crust and olivine in the upper mantle. Other isostructural minerals are galena (PbS, e.g. SIEMES [1970]) and cinnabar (HgS), both important ore minerals.

Table III. Assumed slip systems and resolved-shear-stress ratios for halite [WENK &AL. 1989b, derived from CARTER & HEARD 1970].

slip system	LT	HT
{110}⟨1$\bar{1}$0⟩	1	1
{100}⟨011⟩	5	1

NaCl deforms at low temperature preferentially by slip on {110}⟨1$\bar{1}$0⟩; other systems with a much higher critical resolved shear stress such as {100}⟨011⟩ and {111}⟨1$\bar{1}$0⟩ are subordinate (CARTER & HEARD [1970], Fig. 6 and Table III). In spite of the high cubic symmetry, the preferred {110}⟨1$\bar{1}$0⟩ system has only two independent variants and therefore does not satisfy the von Mises criterion. In order to obtain an

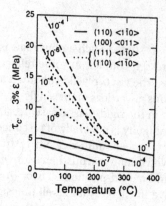

Fig. 6. Critical resolved shear stress of slip systems in halite as a function of temperature at different strain rates [CARTER & HEARD 1970].

Fig. 7. Two hundred randomly distributed orientations represented in the inverse pole figure sector for the cubic crystal. The size of the symbols indicates the effective stress corresponding to the Taylor factor for low-temperature extension or compression of halite. Orientations with the largest symbols are five times harder (36 MPa) than those with the smallest symbols (6.6 MPa).

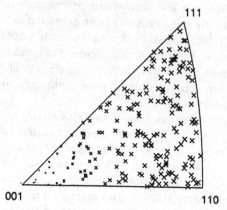

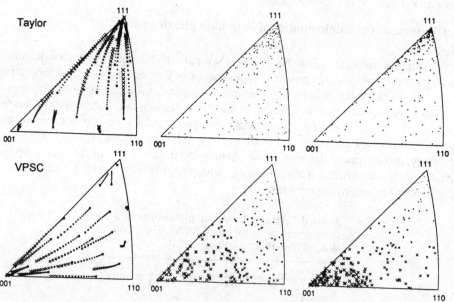

Fig. 8. Texture development during deformation of halite in axial extension, according to the Taylor model (top) and the visco-plastic self-consistent theory (VPSC, bottom) with no work hardening. The first column gives rotation trajectories for 20 representative orientations (in 5% strain increments), the second column gives distribution of 200 grains after 50% strain and the third column after 100% strain. The size of the symbols indicates the relative grain integrated plastic deformation [WENK &AL. 1989b].

arbitrary shape change, harder systems need to be activated. Fig. 7 shows the variation of the Taylor factor in extension and for the low temperature (LT) conditions given in Table III, for different orientations, plotted in the cubic inverse pole figure sector. Orientations with extension axes near (001) (6.6 MPa) are in a 'Taylor sense' almost six times 'softer' (small symbols) than those with axes near (111) and (110) (36MPa) (large symbols). One would expect that crystals which are favorably oriented for weak (110) slip, i.e. those near (001), deform more than others, contrary to assumptions of the Taylor model that all grains maintain the same shape.

Texture development in this system has been approached with the Taylor model [SIEMES 1974] and the visco-plastic self-consistent theory [WENK &AL. 1989b]. Contrary to most other systems investigated so far, results for the two theories are entirely different. The Taylor theory predicts extension axes to rotate towards (111) (Fig. 8, upper row) whereas the self-consistent theory predicts rotations towards (001) (Fig. 8, lower row). The difference in texture can be understood in terms of slip system activities (Fig. 9). The Taylor maximum at (111) is caused by significant $\{111\}\langle1\bar{1}0\rangle$ slip required to maintain compatibility. In fact this hard system accommodates on the average most of the strain. In the self-consistent case practically only $\{110\}\langle1\bar{1}0\rangle$ is active even though other systems were also allowed, resulting in a (001) texture. The self-consistent texture is close to that observed in extrusion experiments [SKROTZKI & WELCH 1983] (Fig. 10).

Predicted texture development is much slower with the self-consistent model than with the Taylor model. What is the reason for this? The salt system has another interesting feature: for each system (110)[1$\bar{1}$0] there is an equivalent system (1$\bar{1}$0)[110] (slip plane normal and slip direction exchanged) with exactly the same Schmid factor;

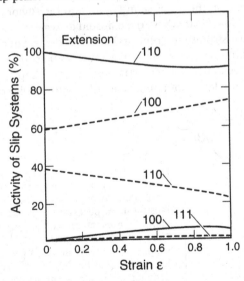

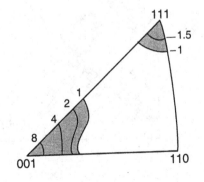

Fig. 9. Slip system activity in halite during deformation in extension as a function of strain, predicted by the Taylor model (dashed lines) and the visco-plastic self-consistent theory (solid lines).

Fig. 10. Inverse pole figure for experimentally extruded halite at 23° C [SKROTZKI & WELCH 1983]. Region above 1 m.r.d. is shaded, contours are labeled. Stereographic projection.

therefore the same activity. The shears on these systems compensate each other's spins. Superficially it may appear that, if only $\{110\}\langle1\bar{1}0\rangle$ is active, no rotations should occur and no texture should develop. In the Taylor model all grains maintain the same shape and since spins on the associated $\{110\}$ $\langle1\bar{1}0\rangle$ and $\{1\bar{1}0\}$ $\langle110\rangle$ systems compensate, all rotations are due to slip on other systems. In the self-consistent simulation, with only $\{110\}$ $\langle1\bar{1}0\rangle$ activity, no texture develops due to shears on that slip system. However, individual grain shapes are another cause for grain rotations. Indeed, at low strain (50%) the self-consistent texture is weak, but as the orientation dependent grain shape distribution evolves, texture develops. With increasing strain some grains become strongly elongated and for those rotation increments are large as is best seen in the rotation trajectories.

In Fig. 8 the symbol size is indicative of the deformation of individual grains as defined by an 'average grain shape parameter' ε_{av}. (The size of the symbols indicates relative average change in grain shape ε_{av}, or grain integrated plastic deformation. It is defined as $\sqrt{(3/2)}$ times the integral over the von Mises strain rate). In Taylor simulations all grains deform by the same amount and at a given deformation step the grain shape of all grains is identical. In self-consistent calculations grains near (001) deform much more (large symbols) than those near (111). At 100% extension, ε for Taylor simulations is 1.2, for self-consistent simulations it ranges from 0.3 to 2.3.

A more detailed assessment of intercrystalline heterogeneity is ratios to evaluate the shapes of the finite strain ellipsoid of individual grains, by investigating grain aspect ratio (Fig. 11). In such diagrams the distance from the origin increases with total deformation and the location determines whether a grain ellipsoid is elongated, flattened or is in plane strain. The macroscopic strain on the aggregate after 100% elongation is indicated by the filled circle on the abscissa in Fig. 10 and for Taylor all grains plot at that location (log b/c=0.65, log a/b=0, where the ellipsoid axes are $a<b<c$). Self-consistent predictions show a wide distribution: some grains barely deform at all, others show very large deformation, many have a plane-strain geometry and some are even oblate. It is clear that for aggregates in which large variations of grain shape are observed after deformation, the Taylor model is not applicable, even with provisions of adding relaxed constraints.

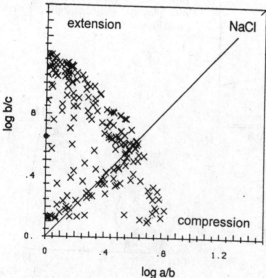

Fig. 11. Deformation simulation of halite with the visco-plastic self-consistent theory in extension to 100% equivalent strain. Diagram illustrating aspect ratios of strain ellipsoid axes of individual grains. Note the wide distribution of shapes. For Taylor all grains plot at log $a/b = 0$ and log $b/c = 0.65$. The distance from the origin indicates the degree of deformation. Even though the macroscopic strain prescribes extension (i.e. constriction), many grains deform in plane strain (diagonal line) and some even in the field of compression (i.e. flattening).

At higher temperature critical shear stresses of all systems are similar (Fig. 6 and Table III) and in this case there is no noticeable difference between Taylor and self-consistent simulations. The example of halite illustrates that there are many parameters which can be used to discriminate between polycrystal plasticity models and some can be experimentally relatively easily determined. One of those is texture, others are identity of active slip systems, number of active slip systems in a grain, total deformation of a grain, shape of deformed grains, relationships between deformed grains and their neighbors. Unfortunately many of these data are not available for halite and most other non-metallic materials. Due to its high plastic anisotropy at low temperature and the particular geometry of slip systems, halite would be an excellent system for a comprehensive experimental and theoretical investigation of polycrystal plasticity.

Quartz

There have been several studies of the orientation-dependent grain shape distribution for quartz. In quartzites one observes often highly deformed and relatively undeformed grains adjacent to each other in the same sample. BOUCHEZ [1977] recorded grain aspect ratios in addition to orientation and noticed that c-axes of weakly deformed grains are at high angles to Y (Fig. 12a), whereas strongly deformed grains cluster around the intermediate fabric direction Y (in the schistosity plane and normal to the lineation, in the center of the pole figure) (Fig. 12b). This has since then been observed in many cases. If texture development of quartz is modeled with the self-consistent theory (n=3) for pure shear conditions, highly deformed grains are indeed predicted in the intermediate strain direction, as indicated by the large symbols in Fig. 12c. Although plastic anisotropy in quartz is reduced by rate sensitivity, there are still large differences in strength between grains of different orientations.

Naturally, if models allow heterogeneous (intragranular) deformation, incompatibility between grains has to be accommodated somehow and this cannot be done within a context that assumes grains to deform homogeneously. As has been described in Chapter 11, an attempt has been made to model a system with high plastic anisotropy causing large intragranular heterogeneity, a composite of quartz

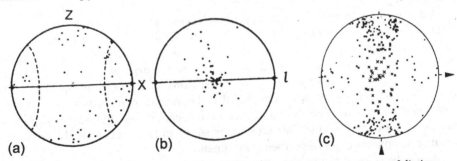

Fig. 12. c-axis pole figures of quartz. (a) Naturally deformed quartzite, foliation horizontal and lineation *l* to the right (fabric coordinates X and Z, used in the geological literature are indicated, Y is in the center). Collection of grains with an aspect ratio <4 (weakly deformed grains). (b) Same sample as (a) but a collection of strongly deformed grains with an aspect ratio >8 [BOUCHEZ 1977]. (c) Visco-plastic self-consistent simulations of pure shear deformation. Symbol sizes are proportional to strain [WENK &AL. 1989a]. Arrows indicate the compression and extension direction respectively.

and rigid mica particles [CANOVA &AL. 1992], by introducing microstructure in the self-consistent model and dividing each grain into domains (small cubes). It was possible to simulate gradients in strain rate and stress in single grains, and variations depend on the neighborhood. Predicted quartz textures of a less anisotropic single-phase quartzite are stronger than those for a more anisotropic quartzite with suspended rigid particles. The latter show a higher dispersion of intracrystalline strain rates around rigid inclusions. These results for polyphase aggregates are still preliminary and deformation of such systems will in the future probably be more adequately modeled with finite element methods.

1.4 A deformation-based model of dynamic recrystallization: quartz and ice

Many geological materials are partially recrystallized and this ought to be incorporated into deformation models that simulate texture evolution. In some cases recrystallized textures are similar to deformation textures, in other cases they are very different. In quartz and halite it was noticed that the recrystallization texture emphasized those orientations for which the self-consistent method predicted large deformation. This is the motivation for exploring a qualitative model for dynamic recrystallization based on an analysis of the deformation behavior [WENK &AL. 1997a], following up on concepts introduced by JESSELL & LISTER [1990].

It is acknowledged that recrystallization is a very complicated process which depends on many different factors [HAESSNER 1978, HUMPHREYS & HATHERLY 1995]. Because of prevailing uncertainties, recrystallization and its effect on textures has been largely omitted in this book. There are microstructural characteristics influencing recrystallization that are distinct for each material, but a common factor is previous deformation. Thus, it seems appropriate to discuss some elements of dynamic recrystallization based on deformation models. As we have emphasized above, in highly anisotropic systems some grains deform much more than others and there is a large variation in the dislocation microstructure, such as dislocation density and subgrain geometry. In dynamic recrystallization, recrystallization occurs as the material deforms [GUILLOPE & POIRIER 1979]. It may involve both nucleation of new grains and replacement of a texture component by another one, while deformation proceeds.

A driving force for recrystallization is the strain energy stored in individual grains. Highly deformed grains (or highly deformed regions within a grain) have a tendency to nucleate new grains, given the availability of significant misorientations, and those with a low stored energy have a tendency to grow at the expense of their neighbors. The stored energy is introduced by accumulations of dislocations during work hardening and hardening is determined by the overall deformation of a grain. Depending on the respective importance of nucleation and boundary migration processes, the recrystallization textures are expected to favor either highly deformed texture components or less deformed components.

As a polycrystal deforms, slip systems are activated. At the microstructural level, slip causes dislocations to multiply and interact. It becomes increasingly difficult for dislocations to propagate, causing an increase in the critical shear stress. Crystals which are favorably oriented relative to the applied stress deform more and harden more than those which are unfavorably oriented. Also, more highly deformed grains have a tendency to develop a cell structure with more highly misoriented subgrains

[HAASEN 1993]. Both processes are inducive for nucleation. These concepts can be empirically introduced into a self-consistent polycrystal deformation model, using a few parameters to characterize growth and nucleation [LEBENSOHN &AL. 1997].

A brief qualitative description of the model for dynamic recrystallization follows [WENK &AL. 1997a, LEBENSOHN &AL. 1997]. Deformation simulations with the self-consistent model [MOLINARI &AL. 1987] provide a population of grains with a variation in deformation and correspondingly in dislocation density. The microstructural hardening of slip systems during deformation provides an incremental strain energy to grains after each deformation step. Grains with a high stored energy are likely to be invaded by their neighbors with a lower stored energy. In the model the stored energy of each grain is compared with the average stored energy of the polycrystal (calculated from the increments in shear stresses). If the stored energy of a grain is lower than the average, it grows; if it is higher, it shrinks. The grain-boundary velocity is proportional to the difference in stored energy. A grain may disappear. After each simulation step grain sizes are renormalized so that the average over the whole aggregate remains constant. A single parameter determines the growth velocity.

So-called 'nucleation' of grains, or generation of new large-angle boundaries, is a difficult and unresolved problem. It is natural to assume it to be easier in more highly deformed grains, even if it is less clear whether stress (hardening) or strain (mis-orientations) are most influential. In the model the probability of nucleation depends exponentially on the deformation of an individual grain and is described by a second empirical parameter. Nucleation is only allowed to take place if the deformation is above a certain threshold (third parameter). This corresponds to the requirement for high-angle subgrain boundaries which become sufficiently mobile and this will occur only after a critical strain is reached.

For simplicity and for lack of other information it is assumed that, if nucleation occurs, a new crystal replaces the old one completely (same size) and that the new crystal is in the same orientation as the old one but the stored energy of the new crystal is set equal to zero by returning to the initial critical resolved shear stress, and the grain is restored to an equiaxed shape. During dynamic recrystallization all grains are evaluated after each deformation step and allowed to grow or shrink or nucleate and the characteristics (deformation, size, orientation, shape, slip systems) of all grains are updated. The development of the microstructure and texture is determined by the balance between nucleation and growth. These processes are described in the model by only three parameters: a mobility parameter, a nucleation parameter and a nucleation threshold. Since the model relies on the difference in deformation for different orientations, it is particularly applicable for materials with a high plastic anisotropy such as some of the systems described in the previous sections.

Of particular interest is *quartz*, because for this mineral deformation models have been unable to explain the widely observed (0001) maximum in the intermediate strain direction (Fig. 13a and compare also Fig. 4d). In pure shear simulations a broad maximum near the shortening axis (vertical) develops (Fig. 13b, after 10 and 14 steps of deformation, compare also Fig. 12c which indicates individual grain deformation). Grains in the intermediate strain direction (center of pole figure) have a low Taylor factor (and are therefore highly deformed) compared to those at the periphery. With properly adjusted recrystallization parameters the pole figure is effectively purged of grains at high angles to the intermediate strain direction (Fig. 13c). Dynamic recryst-allization is allowed after 8 steps, each with a 5% increment in equivalent strain.

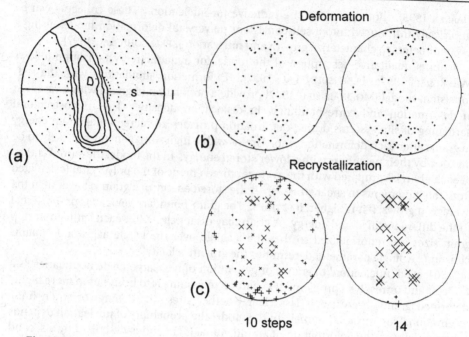

Fig. 13. c-axis pole figures for quartz. (a) Typical texture of a naturally recrystallized quartzite with a (0001) maximum in the 'intermediate' fabric direction, normal to the lineation *l* and in the plane of schistosity *s*. (b) Pure shear deformation texture after 10 and 14 5% steps modeled with the visco-plastic self-consistent theory (compare with Fig. 12c). (c) Dynamic recrystallization after 8 steps. Grains with diagonal crosses (x) have nucleated, those with parallel crosses (+) have not. The symbol size is proportional to the grain size which changes due to boundary migration [WENK &AL. 1997a].

Already after 10 steps most grains near the center have nucleated (symbol x) and become strain-free, retaining their old orientation. They grow and may nucleate again (symbol size is proportional to grain size); those near the periphery, though less deformed, get consumed by the newly nucleated grains. After 14 steps peripheral grains have disappeared completely, also some of the highly deformed grains disappear if nucleation does not keep up with boundary migration. The model is in reasonable agreement with textures in natural quartzites (e.g. Fig. 13a and Chap. 6 Fig. 16a and Fig. 17a) and observations that grains in the intermediate strain direction grow at the expense of their neighbors [JESSELL 1988, LLOYD & FREEMAN 1991]. The model also provides mechanical information. Figure 14 shows a predicted strain energy versus deformation curve with oscillations due to the complicated combination of geometric (texture) hardening/softening, nucleation softening and growth hardening/softening. The oscillations are also present in the stress/strain curve, though attenuated, and diminish with increasing strain just as observed in some metal systems [GLOVER & SELLARS 1973].

Another example is *ice*. Ice deforms mainly by basal slip (0001) $\langle 11\bar{2}0 \rangle$. For axial compression the deformation model predicts a high (0001) concentration near the compression axis [CASTELNAU &AL. 1996, Fig. 15a, center of pole figure] and this agrees with observations from experiments [DUVAL 1981]. It also corresponds to textures observed in polar ice sheets thought to have deformed by thinning [THORSTEINSSON &AL.

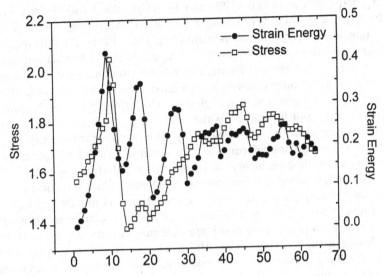

Fig. 14. Development of average strain energy (closed symbols) and stress (open symbols) during dynamic recrystallization of a quartz polycrystal as a function of strain (to 64 5% steps).

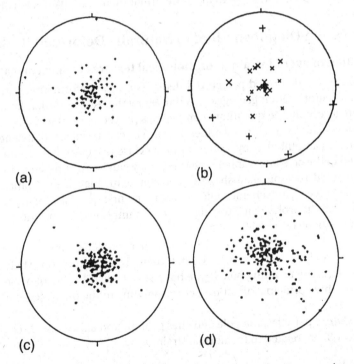

Fig. 15. 0001 pole figures for ice, deformed in compression. Compression axis is in center of pole figure. (a) Deformation simulation with the visco-plastic self-consistent theory to 14 5% steps, (b) dynamic recrystallization after 8 steps; symbols are the same as in Fig. 13c, (c) natural deformation texture of ice from the GRIP drill core in the Greenland ice sheet (2779.15m), (d) natural recrystallization texture that occurs in layers (2795.65m) [THORSTEINSSON &AL. 1997].

1997]. In these ice sheets, such as the GRIP core from Greenland, there is often a rapid transition from a fine grained, very strong deformation texture (Fig. 15c) to a much weaker but quite coarse recrystallization texture (Fig. 15d). In the dynamic recrystallization simulations, intermediate orientations that are highly deformed nucleate (x) and grow (Fig. 15b). Also, some high-angle (hard) orientations grow. However, as grains rotate towards the compression axis, they are deformed and are invaded by less deformed, non-nucleated grains at high angles (+). Most grains near the compression direction disappear, except for the few which were initially exactly aligned with the compression direction and, because of basal slip with the compression axis perpendicular to the slip plane, do not deform at all. The result is a wider orientation spread, a weaker texture, with only a few large grains remaining, similarly to Fig. 15d.

The method which introduces dynamic recrystallization in a statistical way into polycrystal plasticity models is a first-order approach and is mainly applicable for highly anisotropic systems such as minerals. In those the variation in deformation and stored energy is large and may dominate over mechanisms based on neighbor relations or intragranular heterogeneities. The observed textures are very sensitive to the choice of the three parameters. The model does not produce new orientations but, depending on the parameters, either strongly deformed 'soft' orientations, or 'hard' orientations dominate the recrystallization texture. Generally, in systems investigated so far – ice is an exception – intermediate orientations vanish and the recrystallization texture is stronger than the deformation texture [WENK &AL. 1997a].

2. Textures as a Diagnostic Tool in Naturally Deformed Rocks

2.1 Estimation of overall strain: morphological textures in sheet silicates

The texture of a deformed polycrystal depends on the deformation path and on the total deformation. Often geologists need to know the total deformation of a given rock and here a very simple quantitative model has proved to be surprisingly successful. Many rocks contain either platy minerals (mica or clay particles) or needle-shaped minerals (such as amphiboles, pyroxenes or plagioclase feldspar); in both the morphology is linked to the crystallographic structure. They can be considered as rigid strain markers. For rigid rod- or plate-shaped inclusions with high aspect ratios, MARCH [1932] designed an elegant scheme which became the first quantitative application to relate strain and 'morphologic' texture. No assumptions about intracrystalline deformation mechanisms need be made.

The March theory is based on the concept that marker grains are rotated in a homogeneously deforming matrix, and that there is no interaction between the particles (Fig. 16). The resulting pole figure has a simple distribution of orthorhombic symmetry with three principal directions, corresponding to the main axes of the strain ellipsoid.

For *rod-shaped* particles, the pole densities ρ_i of the rod axis in the three principal directions i (i=1,3) are related to the normal strains ε_i by

$$\varepsilon_i = \rho_i^{1/3} - 1 \tag{1}$$

The pole densities, expressed in m.r.d., are determined on a normalized pole-figure.

For rigid *platelets* the relationship is

$$\varepsilon_i = \rho_i^{-1/3} - 1 \tag{2}$$

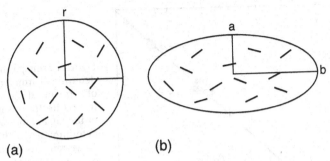

Fig. 16. Schematic illustration of the March theory. (a) randomly oriented lines in a circle, (b) preferred orientation of lines develops after deformation of the circle to an ellipse.

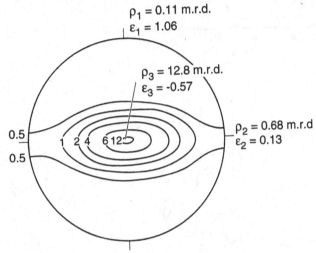

Fig. 17. 001 pole figure of muscovite from Cambrian slates in North Wales [WOOD & OERTEL 1980]. Contours of pole densities are indicated (in m.r.d.). The pole densities ρ are used to determine the March strain ε.

where ρ_i is the pole density of the plate normal (i.e. (001) in the case of mica).

Figure 17 is a 001 pole figure of muscovite from a Cambrian slate in North Wales [WOOD & OERTEL 1980]. Principal densities are 12.8 m.r.d., 0.68 and 0.11. From these one obtains strain components of -0.57, 0.13 and 1.06.

The original March model assumes that the initial particle orientation distribution is random. OWENS [1973] generalized it to any arbitrary starting distribution. Another modification is for texture development during compaction, i.e. with a volume change [OERTEL & CURTIS 1972]. The compaction strain ε_c is related to the maximum platelet pole density ρ_{max}

$$\varepsilon_c = \rho_{max}^{-1/2} - 1 \ . \tag{3}$$

This is relevant for porous, unconsolidated sediments and has also been applied to compaction of powders. The March model is justified for moderate strains. At larger strains grain interactions become significant and textures develop less rapidly than predicted by the model (see Fig. 35 of Chapter 6). The March model, applied to sheet silicates, has been widely and successfully used as a method to determine strain;

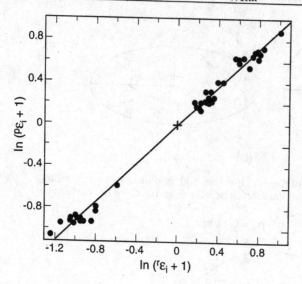

Fig. 18. Principal natural strains determined from measurements of preferred orientation of muscovite (ρ_{ε_i} contained in the ordinate) and independent strain determination from direct measurements of the dimension of reduction spots (r_{ε_i} contained in the abscissa) for slates from 14 localities in North Wales [WOOD & OERTEL 1980].

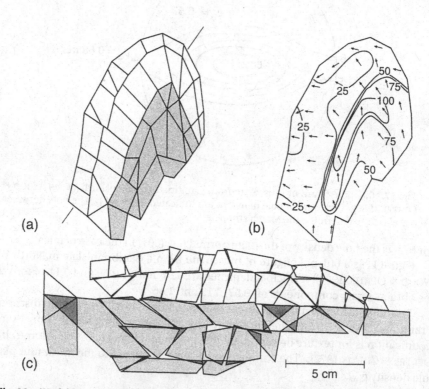

Fig. 19. 'Unfolding' of a fold. (a) A layered fold consisting of silty slate (light) and slate (grey) has been divided into sectors. (b) In each sector the preferred orientation of the phyllosilicate chlorite was measured. Arrows mark the direction of the (001) maximum in the pole figures and contours give the pole densities of the maximum. (c) From the preferred orientation pattern the strain was calculated according to the March theory and then applied to each block in reverse. The result produces two flat layers with minimal misfits [OERTEL & ERNST 1978].

and comprehensive reviews are given by OERTEL [1983] and CHEN & OERTEL [1991]. We illustrate the applicability for two examples in which strain has also been determined independently and results agree with the predictions from the March model.

WOOD & OERTEL [1980] studied preferred orientation in purple Cambrian slates from the slate belt of North Wales. Slates are low-grade metamorphic clay-rich rocks that have been deformed. From the (001) pole densities in measured pole figures, such as that in Fig. 17, they determined natural strains ranging from -1 to 1. The strains determined from the pole figures of muscovite and chlorite correlate very well with independent strain markers (Fig. 18). In that particular case the independent strain markers are so-called 'reduction spots' which are seen as green ellipsoidal spots in the purple matrix. The reduction spots are caused by local removal of hematitic pigment and were originally spherical. Mostly reduction spots are not available and geologists have to rely on phyllosilicate textures to estimate strain.

A second example is a fold of slate beds [OERTEL & ERNST 1978]. The fold consists of two layers (Fig. 19a). The top two beds are composed of a silty slate, the bottom three beds are a mudstone. The specimen was cut perpendicular to the fold axis and divided into sectors. In a meticulous analysis the preferred orientation was measured in each sector by x-ray diffraction in transmission geometry. From pole figures, March strains were derived and then the sector was restored to its presumed original shape by imposing to it the inverse of the determined plastic strain (Fig. 19b).

Starting at the right hinge below the double layer of silty mudstone, each unstrained block was translated, rotated and adjusted in volume so that the bottom bedding plane became horizontal, and so that adjacent portions of the same bed fit in height. The resulting shape for the two silt beds was that of an original lenticular body, probably a ripple formed by water waves. Blocks in the underlying mudstone layers were also strained in the reverse direction, translated and rotated in the same way, but so that bedding planes were rotated to the horizontal. They were successively fitted to the underside of the silt body, block by block and bed by bed. The mutual fit of the unstrained blocks is a measure for the compatibility of the individual measurements. In most parts this fit is very good, except for a misfit at the left end. A silt-filled worm track in that region indicated to the authors that late bioturbation interfered locally with the strain pattern. The remainder of the pattern is explained by a history of variable compaction and folding.

The two examples and many applications in the literature deal with relatively simple deformation histories, mainly compression and compaction. For more general situations, if inclusions do not have nearly infinite aspect ratios, and if laminar flow of the matrix occurs with significant amount of simple shear, then the more basic theory of JEFFERY [1923] needs to be applied, which has been expanded to textures by WILLIS [1977] and GHOSH & RAMBERG [1976].

2.2 Strain paths. Thrusting versus crustal thinning

Calcite

The previous section dealt with the determination of the finite strain. Often the detailed strain history is of concern and textures are significant because they are often sensitive to the path. For example, a structural geologist needs to determine if a tectonic zone has been subject to non-coaxial shearing in a shear zone (Fig. 20, bottom) or coaxial crustal thinning (Fig. 20, top). Both paths can lead to an identical finite strain. Naturally deformed calcite rocks, limestones and marbles, often display

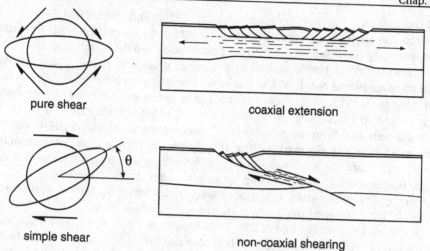

Fig. 20. Cartoon comparing crustal deformation by coaxial thinning (top) and non-coaxial shearing (bottom). Coaxial thinning may occur by pure shear (plane-strain compression), whereas non-coaxial shearing occurs by simple shear. The finite strain (shape of ellipse) may be the same but in simple shear the ellipse is inclined by an angle θ to the shear plane.

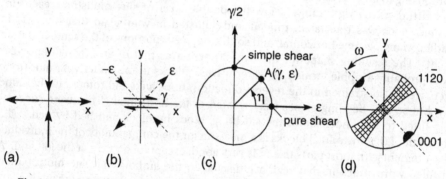

Fig. 21. Definition of coordinate systems used for the texture simulations in pure shear (a) and simple shear (b). (c) is a diagram to define the strain partitioning factor η discussed in the text. (d) is a schematic pole figure in the coordinate system of (a,b) with an oblique (0001) maximum and a (11$\bar{2}$0) girdle. The angle of asymmetry ω between the (0001) maximum and the shear-plane normal (y) can be used to determine the strain partitioning factor by comparing the measured value with results from polycrystal plasticity simulations.

strong preferred orientation. Texture patterns of calcite in deformed marbles have been used to determine the partitioning of deformation into a coaxial deformation component and a non-coaxial (simple shear) component and to infer the deformation history. One of the universal principles of texture interpretation is symmetry and it has been widely applied in geological situations: The texture symmetry cannot be lower than the symmetry of the strain path [PATERSON & WEISS 1961], if it started out uniform. For coaxial deformation one expects orthorhombic pole figures, whereas a non-coaxial path is likely to produce monoclinic pole figures.

It appears appropriate to clarify some of the nomenclature for strain paths, used in mechanics and structural geology. The following terms are strictly equivalent:

Mechanics	Structural Geology
(axial) compression	flattening
(axial) tension	elongation, constriction
simple shear	simple shear (non-coaxial)
plane-strain compression	pure shear (coaxial)

Simple shear and pure shear are plane-strain paths, i.e. there is no deformation perpendicular to a particular plane (e.g. the z-axis). A typical *simple shear* displacement gradient tensor in 2-D with a displacement along x (shear direction) can be decomposed into two parts (normalized):

$$\beta = \begin{pmatrix} 0 & 1 \\ 0 & 0 \end{pmatrix} = \begin{pmatrix} 0 & \frac{1}{2} \\ \frac{1}{2} & 0 \end{pmatrix} + \begin{pmatrix} 0 & \frac{1}{2} \\ -\frac{1}{2} & 0 \end{pmatrix} \tag{4}$$

where the first part is the strain and the second part the rotation (or spin) (Fig. 21b). The velocity gradient tensor L used in previous Chapters is the time derivative of β. To obtain the total deformation from the incremental deformation one has to integrate over the path and one obtains [SHRIVASTAVA &AL. 1982] the 'deformation gradient matrix'

$$F = \begin{pmatrix} 1 & \gamma \\ 0 & 1 \end{pmatrix} \tag{5}$$

This corresponds to a shear $\gamma = \sqrt{3}\, \varepsilon_{eq} = 2 / \tan 2\theta$ where ε_{eq} is the equivalent von Mises strain, and θ the angle between shear plane and the long axis of the strain ellipsoid (Fig. 20, bottom).

A *pure shear* displacement gradient tensor (in the mechanics convention) is

$$\beta = \begin{pmatrix} 0 & 1 \\ 1 & 0 \end{pmatrix} \tag{6}$$

It is equivalent to

$$\beta = \begin{pmatrix} 1 & 0 \\ 0 & -1 \end{pmatrix} \tag{7}$$

except that the coordinate system is rotated 45° around the z-axis (Fig. 21a). This is still referred to as 'pure shear' in structural geology. Figure 20 illustrates with arrows how a circle is deformed to an ellipse. For differential increments the arrow for pure shear is always at 45°. Both Eq. 6 and 7 are symmetric tensors and free of a rotational component. A succession of 'pure shear' increments (but expressed in the coordinate system of Eq. 7) integrates to the deformation gradient tensor

$$F = \begin{pmatrix} e^{\sqrt{\frac{3}{2}}\,\varepsilon_{eq}} & 0 \\ 0 & e^{-\sqrt{\frac{3}{2}}\,\varepsilon_{eq}} \end{pmatrix} . \tag{8}$$

There are *mixed paths* between simple shear and pure shear, still in plane strain, to arrive at the same deformation. As a generalization of Eq. 4, a displacement gradient tensor may be defined (and decomposed into symmetric and anti-symmetric parts):

$$\beta = \begin{pmatrix} \varepsilon & \gamma \\ 0 & -\varepsilon \end{pmatrix} = \begin{pmatrix} \varepsilon & \frac{\gamma}{2} \\ \frac{\gamma}{2} & -\varepsilon \end{pmatrix} + \begin{pmatrix} 0 & \frac{\gamma}{2} \\ -\frac{\gamma}{2} & 0 \end{pmatrix} \tag{9}$$

The components of longitudinal strain (ε) and rotation ($\gamma/2$) can be represented in a 'Mohr-circle' diagram for finite strain (Fig. 21c) and in this we can define $\tan \eta = \gamma/2\varepsilon$ as the ratio of simple shear with respect to pure shear and geologists refer to it as the 'strain partitioning factor'. In this diagram we obtain $\varepsilon = \cos \eta$ and $\gamma/2 = \sin \eta$ and therefore can calculate the displacement gradient tensor for different partitioning factors ($\tan \eta$):

$$\beta = \begin{pmatrix} \cos\eta & \sin\eta \\ \sin\eta & -\cos\eta \end{pmatrix} + \begin{pmatrix} 0 & \sin\eta \\ -\sin\eta & 0 \end{pmatrix} = \begin{pmatrix} \cos\eta & 2\sin\eta \\ 0 & -\cos\eta \end{pmatrix} \tag{10}$$

There are different ways to determine the partitioning factor from textures. One approach would be to compare textures of a naturally deformed calcite rock with textures produced in experiments performed along a known deformation path. Unfortunately, as distinct from metals, most minerals are only ductile at reasonable strain rates if a high confining pressure is applied. This is relatively easy to achieve for axial compression geometry which is the reason why the great majority of experimental data on deformation of minerals are for compression only, even though most geological deformations follow a different path, particularly plane strain during folding and shearing. Plane-strain deformation is very difficult to achieve in the laboratory, under a confining pressure and high temperature [HEARD 1985].

Another approach for obtaining $\tan \eta$ is to compare natural textures with those obtained from polycrystal plasticity models. The philosophy is to first develop a deformation model for axial compression and test it by comparing it with axial compression experiments. Once the model adequately predicts this geometry, one may attempt to predict the behavior along to any other strain paths which cannot be reproduced experimentally. For calcite there are some plane-strain experiments (Fig. 3, Chapter 6 [KERN 1971, KERN & WENK 1983]) and Taylor simulations satisfactorily reproduce these textures (Chap. 6 Fig. 4). Therefore the model appears reliable and has been applied to intermediate cases. Calculated 0001 pole figures for plane strain document a symmetrical (orthorhombic) pure shear pole figure and an asymmetrical (monoclinic) simple shear pole figure and intermediate states (Fig. 22). The relative amount of simple shear can be quantified by measuring the angle of asymmetry ω (defined in Fig. 21d) between the (0001) maximum and the shear-plane normal. From polycrystal plasticity calculations one can construct an empirical determinative diagram to assess the amount of simple shear from the asymmetry of the (0001) texture maximum [WENK &AL. 1987] (Fig. 23a).

In practice, geologists collect oriented rock samples in the field, then measure pole figures in the laboratory relative to geological coordinates, such as schistosity plane and lineation direction which define the shear plane. From the asymmetry ω of the 0001 pole figure maximum relative to the shear plane the sense of shear can be inferred. From the angle of asymmetry and using the determinative diagram in Fig. 23a, the strain partitioning can be estimated. Whereas many marbles in core complexes of the Western United States show almost symmetrical patterns of (0001) axes [ERSKINE &AL. 1993, Fig. 23b] and presumably formed during crustal extension of the Basin and Range province, limestones from the spreading nappes in the Alps have generally highly asymmetric texture patterns attributed to shearing on thrust planes

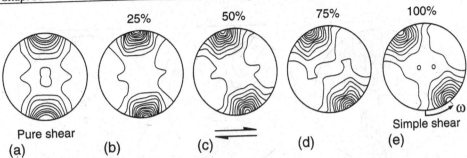

Fig. 22. 0001 pole figures of calcite obtained with the Taylor model for 100% equivalent strain. Pure shear (plane-strain compression) on left, mixed modes in center, and simple shear on right [WENK &AL. 1987]. Note the increasing asymmetry ω of the [0001] maximum relative to the shear plane (horizontal). Sense of shear is indicated.

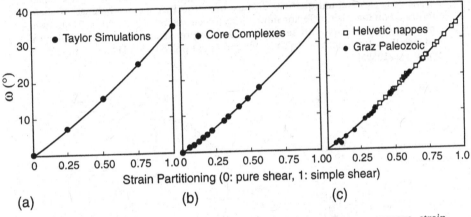

Fig. 23. (a) Determinative diagram with angle of asymmetry ω versus strain partitioning factor (percentage of simple shear deformation) as obtained from Taylor simulations (Fig. 22). Results for (b) marble mylonites from core complexes of the American Cordillera [ERSKINE &AL. 1993] and (c) various limestone textures from Alpine spreading nappes [RATSCHBACHER &AL. 1991].

[DIETRICH & SONG 1984, RATSCHBACHER &AL. 1991, Fig. 23c]. In Alpine spreading nappes the component of simple shear ranges from 0 to 100%.

While metallurgists mostly use plasticity theory to predict future behavior, geologists have applied it to reconstruct the past history.

Quartz

For calcite this method to determine strain partitioning has been fairly successful. For quartz it has been much more controversial. Quartz pole figures often deviate from orthorhombic and even monoclinic symmetry (Chap. 6 Fig. 15) which implies non-coaxial deformation or superposition of different deformation events. Simple shear experiments [DELL' ANGELO & TULLIS 1989, Chap. 6 Fig. 15] and simulations produce indeed monoclinic textures.

LISTER & WILLIAMS [1979] used Taylor simulations for simple shear of quartz to cautiously interpret asymmetric c-axis diagrams of CARRERAS &AL. [1977] (Fig. 24a). In Taylor simulations as well as in self-consistent simulations (Fig. 24c) the c-axis girdle

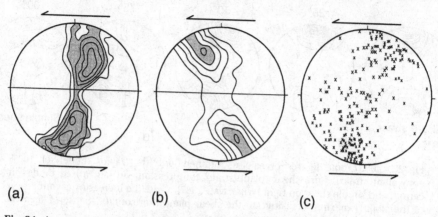

(a)   (b)   (c)

Fig. 24. Asymmetric (monoclinic) quartz 0001 pole figures on which the sense of shear is indicated. (a) Natural quartzite texture from the Pyrenees with sense of shear interpreted on the basis of Taylor simulations [CARRERAS &AL. 1977], (b) quartzite from Moyne thrust, Scotland with sense of shear interpreted on the basis of geological evidence [LAW 1990], (c) visco-plastic self-consistent polycrystal plasticity simulations [WENK &AL. 1989a].

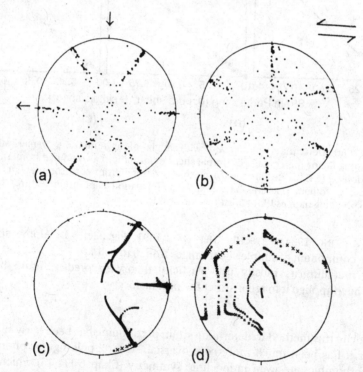

(a)   (b)

(c)   (d)

Fig. 25. Texture development in quartz, model γ (high temperature) and $n=99$ [WENK &AL. 1989a] represented in 0001 pole figures. The pole figure for simple shear (b) is at a first glance the same as a pole figure in pure shear (a) but rotated against the sense of shear. However, the development of texture is entirely different as illustrated with rotation trajectories with 5% strain increments of selected grains (c,d). The size of the symbols increases with strain. The pure shear texture develops stable orientation components, the simple shear texture is a dynamic state.

is inclined against the sense of shear relative to the shear plane, just as in calcite. As with other materials, pole figures of simple shear textures for quartz resemble pure shear textures but are rotated against the sense of shear (e.g. calcite, Fig. 22 and quartz, Fig. 25a,b). In detail there are deviations and the way that individual crystallites arrive at these orientations is entirely different for simple shear and pure shear as demonstrated with rotation trajectories (Fig. 25c,d). For pure shear the texture develops as stable orientations. In simple shear individual crystallites always rotate with the sense of shear and a simple shear texture is a dynamic state, displaying texture maxima where rotation increments are smallest (Chap. 8 Sec. 3.4).

The interpretation of LISTER & WILLIAMS [1979] is not universally accepted. Some researchers suggest an opposite sense of shear for quartz mylonites along the Moyne thrust, Scotland [LAW 1990] (Fig. 24b), based on independent geological evidence. It is possible that large flattened grains ('porpyroclasts') behave differently from a finer-grained and more equiaxed groundmass.

Part of the problem with shear sense may arise from the fact that many natural quartzites are at least partially recrystallized. Concerning shear sense we note that in experiments at low strain with no recrystallization, texture patterns agree with plasticity theory, whereas at high strain, with significant recrystallization, the asymmetry for the large, non-recrystallized grains is reversed [DELL'ANGELO & TULLIS 1989]. These large grains, which dominate the texture, may float like passive markers in a dynamically recrystallizing matrix. In such a case one would expect rigid body rotations *in* the sense of shear. Another reason may be selective nucleation and grain growth during recrystallization, similar to the selection of orientations near the intermediate fabric direction discussed in Sec. 1.4. Indeed, self-consistent simulations indicate that orientations with **c**-axes rotated with the sense of shear from the shear-plane normal are more highly deformed and may therefore be more likely to nucleate and then grow.

Clearly there are still unresolved problems with natural quartz textures: In many cases the shear sense cannot be predicted correctly with polycrystal plasticity theory without taking complicating circumstances into account and some caution against indiscriminate application is warranted. The low triclinic symmetry of pole figures in most metamorphic quartzites remains another puzzling feature. A superposition of different deformation episodes with a combination of deformation and dynamic and static recrystallization are the most likely reasons for low texture symmetry.

2.3 Texture transitions with metamorphic grade: quartz

The relative ease with which certain slip systems operate depends on the physical conditions prevailing during deformation. Temperature is the most important factor that affects critical resolved shear stresses. Since the texture patterns are linked to slip system activity, one should observe changes with temperature. Indeed such changes have been well established experimentally for calcite, dolomite and quartz as described in Chapter 6. These transitions could be reproduced with simulations based on a topology analysis of the single-crystal yield surface for calcite [TOMÉ &AL. 1991b] and quartz [LISTER &AL. 1978] and directly related to changes in critical resolved shear stresses and slip system activity. For calcite, by changing the critical shear stress of the secondary slip system f^- and that of e^+ twinning relative to the primary slip system r^-, a whole range of different texture types is produced, as illustrated effectively in a topology diagram (Chap. 11 Fig. 15).

One would expect to find similar changes in naturally deformed rocks from regions with temperature gradients, referred to by petrologists as 'metamorphic grade'. In the same mountain belt one can study rocks which have been deformed at different temperatures. So far nobody has described texture transitions in naturally deformed calcite and dolomite rocks. All observed textures of carbonates seem to conform with the 'low temperature' type, based on experiments. In the case of quartz there is more variety: BEHR [1964] documented, with increasing metamorphic grade (increasing temperature and pressure), a transition from a crossed girdle (Fig. 26 top) to a small-circle girdle type texture (Fig. 26 bottom) in quartzites and granulites from the Erzgebirge in Saxony. The c-axis pole figures have almost orthorhombic symmetry (Fig. 26 left side) except in the vicinity of an igneous intrusion where they become distinctly triclinic due to geometric constraints (Fig. 26 right side). The difference in textures (Fig. 26 top to bottom) was thought to be due to a change from predominantly basal to prismatic and pyramidal slip in a metamorphic temperature gradient and was interpreted by LISTER & DORNSIEPEN [1982] on the basis of plasticity simulations. But factors such as changes in strain history, e.g. a transition from compressional deformation with small-circle girdles to an extensional or plane-strain deformation with crossed girdles, may have a similar effect as temperature [HOFMANN 1974, TAKESHITA & WENK 1988]. This discussion illustrates that textures of such common minerals as quartz, which have been extensively studied, still leave many unanswered questions.

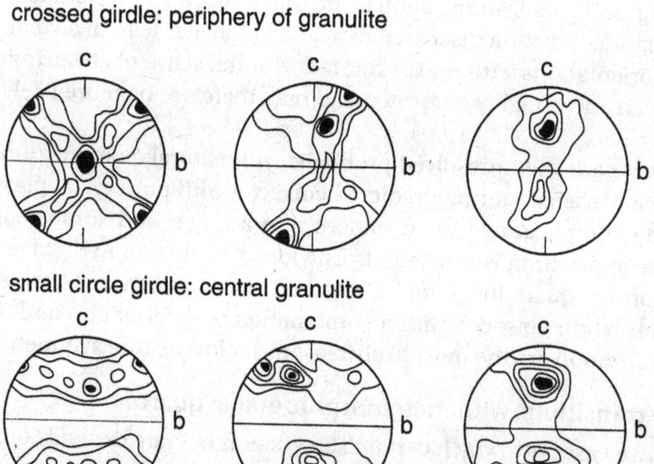

crossed girdle: periphery of granulite

small circle girdle: central granulite

Fig. 26. Quartz 0001 pole figures from the Saxony granulites. Top: crossed girdle from the peripheral northern part of the granulite which was deformed at lower temperature. Bottom: small-circle girdles in the central part of the granulite, at higher temperature. From left to right increasing triclinic distortion is caused by a gabbro intrusion [BEHR 1964]. c is the normal to the foliation and b the lineation direction.

3. Physical Properties: Seismic Anisotropy in the Earth's Mantle

3.1 Anisotropy of seismic velocities along the San Andreas fault in California

Anisotropy of physical properties in polycrystals can have various sources. As discussed in Chapter 1 it may be due to a layered arrangement of components or due to crystallographic preferred orientation. In geological materials the presence of oriented microfractures, presence of partial melt and to a lesser extent the grain shape are also influential. Wave propagation through porous media and media with oriented fractures has been studied by CRAMPIN [1981] who illustrates the importance of fractures in rocks of the upper crust [CRAMPIN 1987]. Velocities through a rock with oriented fractures are much higher parallel to the fractures than across them. With increasing pressure microfractures close and their influence on anisotropy diminishes. KERN & WENK [1990] measured in phyllosilicate rich rocks at 50 MPa a velocity anisotropy of 19%, due to both crystal alignment and oriented microfractures. Above 500 MPa, where microfractures are largely closed, the anisotropy is reduced to 12% and only due to crystallographic preferred orientation.

In recent years seismic anisotropy in the mantle, which is largely due to the texture of olivine, has received much attention [e.g. SILVER 1996]. SAVAGE & SILVER [1993], based on seismic transverse (s) wave data measured in California, proposed a model for the vicinity of the San Andreas fault with an upper layer in which the fast velocity direction is more or less parallel to the San Andreas fault (SE–NW) and a lower layer with a fast direction inclined about 60° to the San Andreas fault (E–W). These results can be interpreted based on polycrystal plasticity modeling.

In the *upper layer* the crust of the western North American plate consists of many belts that are aligned in a SE–NW direction (Fig. 27). These units are heterogeneous in composition. Many are rich in sheet and chain silicates with a similar SE–NW alignment, and these silicates have a very high intrinsic seismic anisotropy. Deformation has been predominantly brittle, as evidenced by numerous SE–NW trending faults

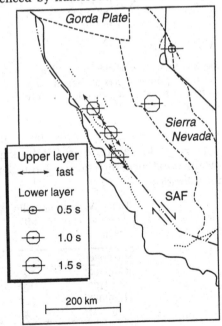

Fig. 27. Seismic velocities along the San Andreas fault (SAF) in California [SAVAGE & SILVER 1993]. Dashed lines separate tectonic units. Dotted lines indicate structural trends. From interpretation of shear-wave splitting measurements they predict for stations along the San Andreas fault an upper layer with a fast direction parallel to the fault and a lower layer with an E–W fast direction.

indicated on Fig. 27, and microscopic and macroscopic fractures are present that enhance anisotropy. One may expect therefore a seismic signal with a fast vibration direction parallel to the San Andreas fault on the basis of structural heterogeneity.

The observed anisotropy in the *lower layer* is very different with a fast direction inclined to the San Andreas fault (E–W). Below 20 km depth, deformation was predominantly ductile and at the high pressure it is unlikely that open fractures could contribute to anisotropy. An imposed shear stress would not result in local fracture but in the development of a ductile shear zone with intercrystalline deformation of component crystals. What if we assume that the same displacements that produced brittle faulting near the surface activated ductile deformation at depth?

The upper mantle of the Earth is largely composed of the orthorhombic mineral olivine. Olivine crystals exhibit about a 25% difference in longitudinal (p) velocities between the slowest [010] and the fastest crystal directions [100] (Fig. 28a). The velocity distribution can be mapped on the sphere and displays a pattern with orthorhombic symmetry, corresponding to the symmetry of the olivine crystal (Fig. 28b) [VERMA 1960]. In an aggregate with preferred orientation of component crystals one also expects a directional dependence of seismic wave propagation.

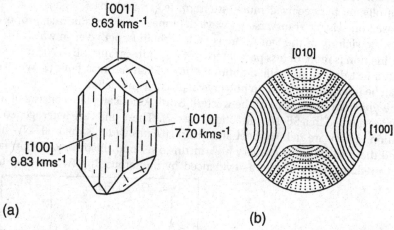

Fig. 28. Elastic properties of an olivine single crystal. Longitudinal p-velocities along the three principal crystal axes are indicated in (a); the contoured p-wave velocity surface in equal-area projection is shown in (b). Contour interval is 0.1 km/s, dotted below 8.5 km/s.

We can use similar arguments about polycrystal deformation, developed in the previous section for calcite, to predict the alignment of olivine in a vertical shear zone with a horizontal shear direction underneath the San Andreas fault and then calculate seismic velocities.

Table IV. Assumed slip systems and resolved shear stresses (in MPa) for Olivine [BAI &AL. 1991].

slip system	LT	HT
(010)[100]	20	15
(001)[100]	20	16
(100)[001]	20	40
(010)[001]	40	35

Olivine has only a few potential slip systems: (010)[100], (001)[100] and (100)[001]. Resolved shear stresses established from single-crystal experiments for lower and higher temperature conditions are listed in Table IV. The slip systems are not sufficient to deform a crystal to an arbitrary shape and therefore the strict Taylor model requiring uniform deformation of all grains cannot be applied. Normal strain components cannot be accommodated by existing slip systems. CHASTEL &AL. [1993] used a lower-bound approach (an extension of the Sachs 1928 model) assuming stress equilibrium and WENK &AL. [1991b] used the visco-plastic self-consistent theory [MOLINARI &AL. 1987]. Both models yielded similar results: in compression one finds an alignment of [100] axes at high angles to the compression direction, in agreement with experiments (Chap. 6 Fig. 19).

In simple shear simulations 500 initially randomly oriented grains are deformed in 100 increments of simple shear to a total equivalent strain of 200% which corresponds to a shear of $\gamma = \varepsilon_{eq} \sqrt{3} = 3.5$. During this deformation the principal axis of the strain ellipse gradually rotates into the shear plane (Fig. 29, top). The orientation patterns are shown side by side for 50%, 100%, 150% and 200% equivalent strain as 100 pole figures in Fig. 29 (center rows) for low and high-temperature conditions. At low temperature (LT) the patterns display weak preferred orientation at 50%, becoming distinct at 100%, increasing moderately to 200%. The [100] maximum is at about 25° to the shear plane, displaced against the sense of shear. At high temperature (HT) the texture pattern is similar, with an asymmetric [100] maximum, but texture development is faster. The example illustrates the sensitivity of texture development on the activity of slip systems and in geological situations the shear stress ratios are not exactly known. In order to have some confidence in interpretations, one needs to explore a range of conditions.

Using methods described in Chapter 7 the elastic tensor of these olivine simple shear textures is calculated with a self-consistent average, based on the stiffnesses of the single crystal and the textures (Table V). As with texture, elastic anisotropy develops much faster for models with HT shear stress ratios.

Table V. Elastic properties of olivine single crystal, an olivine aggregate with no preferred orientation, and polycrystals deformed in simple shear with increasing equivalent strain. Stiffness coefficients in Voigt notation, in GPa. (Compare Fig. 29.)

	C_{11}	C_{12}	C_{13}	C_{16}	C_{22}	C_{23}	C_{26}	C_{33}	C_{36}	C_{44}	C_{45}	C_{55}	C_{66}
Crystal	324	59	79	0	198	78	0	249	0	67	0	81	79
Isotropic	238	78	78	0	238	78	0	238	0	81	0	81	81
LT													
50%	241	78	78	6	236	77	5	239	1	80	2	81	81
100%	247	78	79	8	230	76	6	240	1	79	3	82	81
150%	254	77	79	9	224	76	5	241	1	78	3	84	80
200%	259	77	80	7	220	75	4	243	1	77	3	85	79
HT													
50%	246	79	79	11	233	77	9	235	2	79	4	83	83
100%	260	77	80	17	222	77	10	237	2	76	5	82	85
150%	263	75	80	19	219	77	10	243	1	74	4	82	86
200%	269	74	79	18	216	77	8	244	1	74	4	82	85

Note: for monoclinic sample symmetry: $C_{14} = C_{15} = C_{24} = C_{25} = C_{34} = C_{35} = C_{46} = C_{56} = 0$

Changes in the elastic properties can be visualized in p-wave velocity surfaces shown in Fig. 29 (bottom). These velocity surfaces show a statistical orthorhombic

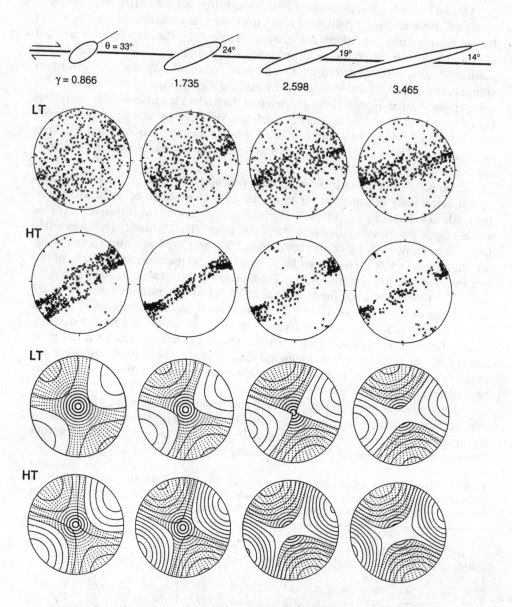

Fig. 29. Deformation of olivine during simple shear. Shear plane is horizontal and sense of shear is indicated. The principal axis of the finite strain ellipsoid gradually rotates into the shear plane (top row). Textures for low-temperature (LT) and high-temperature (HT) conditions (Table IV) predicted with the lower-bound model are represented as 100 pole figures (center rows). From the texture and single-crystal elastic properties, elastic properties for the polycrystal were calculated by self-consistent averaging. From elastic properties p-wave velocity surfaces were obtained and are represented in equal-area projection. Contour interval is 0.1 km/s, dotted below 8.5 km/s.

symmetry but with the principal axes inclined against the shear plane. The velocity maximum for both LT and HT conditions is inclined about 25° to the shear plane which is just about what is observed (Fig. 27). If the movements of the Pacific and the North American plates are driven by similar shear displacements in the upper mantle, then the observed type of anisotropy should develop. From the pattern of seismic anisotropy, and assuming that olivine is the major constituent of the shear zone, one can also interpret the correct right lateral sense of shear.

3.2 Convection in the Earth's upper mantle

The geologic processes which have been discussed so far occur close to the surface and are accessible to direct sampling and detailed measurements. Indirect observations suggest that deformation occurs also in the deeper parts of the Earth and that anisotropy is produced during convection. Very roughly the Earth consists of concentric shells: a crust, an upper mantle, a lower mantle, an outer core and an inner core (Fig. 30). Apart from the liquid outer core, the Earth is mostly a polycrystalline solid and subject to the same laws as polycrystalline metals. Seismological evidence suggests that the upper mantle [NISHIMURA & FORSYTH 1989] and the inner core [WOODHOUSE &AL. 1986, VINNIK &AL. 1994] are anisotropic and that the anisotropy may be produced by deformation.

The upper mantle has been investigated in most detail. Deformation of the upper mantle is directly expressed in processes in the Earth's crust, such as plate tectonics, sea floor spreading, volcanic activity, earthquakes and mountain building. One of these examples is the development of the San Andreas fault, discussed in the previous section. In the mantle large cells of convection are induced by instabilities and driven by temperature gradients [e.g. HAGER & O'CONNELL 1981]. Convection models were generally based on the assumption that the material was a viscous liquid with neither internal structure nor directional properties, even though most of the mantle is crystalline.

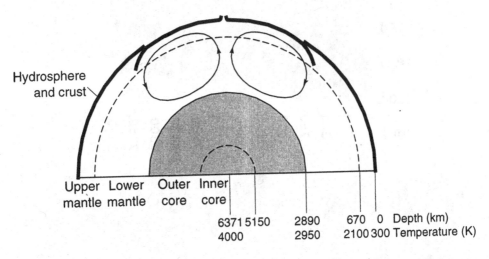

Fig. 30. Very schematic cross section, illustrating the structure of the Earth. Only the outer core is in a largely liquid state. Large convection cells, driven by temperature gradients, exist in the mantle. They produce subduction of the lithosphere at trenches and uprising of lava near ridges. Two convection cells are indicated.

On a microscopic scale, olivine is deformed in the upper mantle by intra-crystalline processes, slip of dislocations in the crystal lattice, and accompanying dynamic recrystallization. It is therefore pertinent to investigate deformation and anisotropy of the upper mantle with methods introduced in this book. A convection cell in the Earth is a very large heterogeneous system. Heterogeneities and anisotropy change as deformation proceeds. At each point, the next deformation increment is controlled by the accumulated previous states and the strength of the preferred orientation may either increase or decrease. In order to correctly predict the flow behavior at any point in the mantle, we need to consider the *microscopic* deformation processes in all single crystals over the whole *macroscopic* deformation history. Micro-macro linking problems have been described in Chapter 13 for predicting anisotropy changes during metal forming. A similar approach can be applied to even larger systems such as the Earth where the deformation extends over thousands of kilometers [CHASTEL &AL. 1993].

A finite element model based on crystal plasticity is used to predict texture development. A rectangular region, 2900 km deep and 3000 km long, is considered to contain the convection cell. The upper and lower boundary temperatures are 1600 and 2900 K respectively and buoyancy associated with temperature gradients drives the convective flow within the cell. The upper and lower mantle have different mineralogical compositions due to phase transformations with increasing pressure. In the model it is assumed that the upper layer consists of olivine and texture evolves. The lower layer, composed mainly of perovskite and periclase type structures, is assumed to be isotropic, in accordance with seismic observations that document

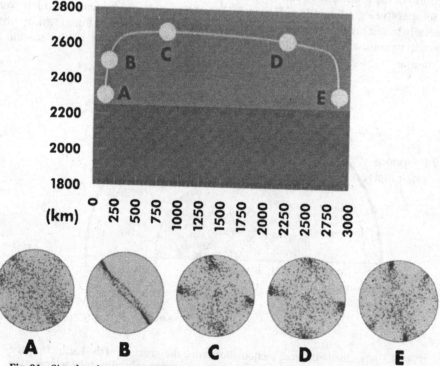

Fig. 31. Simulated texture development during convective flow in the upper mantle. Evolution of texture along a streamline from A to E, represented by olivine 100 pole figures (1000 grains). During upwelling (left) a strong texture develops and changes.

absence of anisotropy and experiments suggesting that superplasticity may be the dominant deformation mechanism [KARATO &AL. 1995]. As material passes through the boundary layer the texture is assumed to be randomized through phase transformations.

The equations that govern such a large system include conservation of energy, balance of linear momentum, kinematic requirements for compatibility, relations between stress and deformation from polycrystal plasticity, and boundary conditions. These equations are solved with the finite element method which gives approximate solutions for the velocity and temperature distribution. An Eulerian frame is specified in which the mesh remains fixed and rocks flow through it. Properties are evaluated at each mesh point after each increment.

A mesh point is assumed to consist of an aggregate of many hundreds of olivine crystals. To obtain mechanical aggregate properties, the plastic stiffness of each crystal is computed with a visco-plastic polycrystal model and then averaged. Since each crystal exhibits an anisotropic behavior, the average depends on texture and changes as convection proceeds. An iterative scheme is used to solve the coupled sets of equations describing the boundary value problems. First the flow field is solved at a fixed state. In a second step the temperature distribution within the cell is calculated. The third step consists in updating the mechanical state as defined by the orientation of the crystals and the slip system hardness. The sequence is repeated, usually over several iterations, until convergence is obtained.

Figure 31 follows texture development in the upper mantle along a streamline in pole figures of olivine [100]-axes. During upwelling a strong texture develops very rapidly (B). The preferred orientation stabilizes during spreading (C,D) and attenuates during subduction (E). The pole figures are distinctly asymmetric due to the component of simple shear. While the finite strain along a streamline increases monotonically, the texture does not. Neither does it completely revert to randomness as the aggregate reenters the lower layer (E).

The texture patterns in selected elements of the two-dimensional convection cell are illustrated, again as pole figures of olivine [100]-axes, in Fig. 32. Only the upper mantle is shown. It has been surprising to find great heterogeneity, with some regions highly textured and others almost isotropic.

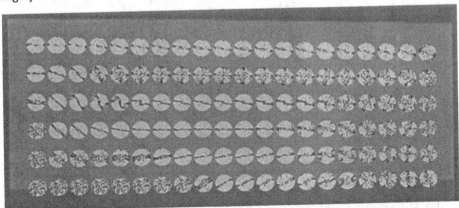

Fig. 32. Preferred orientation patterns of olivine (100 pole figures) in the model upper mantle of the Earth. Rapid texture development during upwelling (left side), little change during spreading, and decrease in texture strength during subduction (right side).

Knowing the orientation of crystallites in a convection cell and the elastic properties of single crystals, one can then average over sectors of the cell to get seismic velocities in different directions. The variations of the p-wave velocities in this simulated textured mantle predict a similar azimuthal anisotropy (Fig. 32a) as that derived from seismic data (MORRIS &AL. [1969], Fig. 32b). Seismologists noticed that seismic waves travel about 5% faster perpendicular to oceanic ridges than parallel to the ridges. Already in 1964, HESS interpreted this as a result of a preferential alignment of crystals with directional properties, and proposed that this alignment was attained during the convection process. Geologists have indeed observed a strong preferred alignment of olivine crystals in rocks that have been brought up from the mantle e.g. as inclusions in oceanic basalts or as large blocks of mantle material that have been juxtaposed with the crust (MERCIER [1985] and Chap. 6 Fig. 20).

From such models it is evident that the mantle is not a structureless viscous medium but shares many properties with a single crystal which displays internal structure and anisotropy. The approach described links processes that occur within crystals on an atomic scale to macroscopic dimensions of the size of continents. By including polycrystal plasticity, geophysical deformation modeling is entering a new state of refinement and future convection modeling should include the complexities of anisotropy and mechanical properties in addition to compositional heterogeneity.

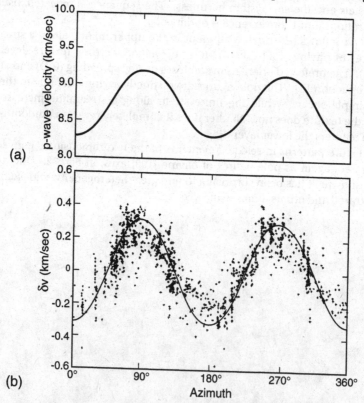

Fig. 33. Azimuthal variation of seismic longitudinal wave velocities (in km/s); (a) obtained by averaging the upper 50 km of the mantle based on texture predictions with an anisotropic convection model [CHASTEL &AL. 1993], (b) observed p-wave velocities in the vicinity of Hawaii (0° is N, 90° is E) [MORRIS &AL. 1969].

4. Conclusions

The previous Chapters have documented that deformation of polycrystalline cubic metals is well understood and a high level of sophistication has been reached in mechanical experiments, texture measurements, quantitative determination of orientation distributions, simulating processes under a wide range of conditions to the extent that many 'experiments' can be reliably and much more efficiently done on the computer than in the laboratory. Predictions of textures and properties may show slight deviations, depending which theories and boundary conditions are applied, and improvements constitute refinements. To provide a comprehensive background and make the reader familiar with these methods and procedures was the central aim of this book.

Clearly 'texture and anisotropy' does not stop here. Common processes such as recrystallization of cubic metals, which produce strong anisotropy, are much less understood and have not been discussed here in any detail. In hexagonal metals one is beginning to be able to evaluate the influence of competing slip and twinning systems. For the yet more complicated systems discussed in this Chapter there are some qualitative and plausible solutions, but many intriguing puzzles are emerging that have yet to be solved. Here the field of texture analysis is still wide open. Even for a relatively simple structure such as halite, there is no agreement on which theories should be applied under which conditions, and experimental data are lacking to support or reject a model. Anisotropic and heterogeneous systems remain a challenge for theoreticians and experimentalists to develop new approaches to better describe polycrystal plasticity in a more general way. It is apparent that large plastic anisotropy of component crystals is one of the major limitations for the classical Taylor model. Complications are not restricted to minerals: many ceramics, almost all polyphase systems and ordered intermetallic materials face similar challenges.

For these same systems new methods need to be developed for texture measurements. Extraction of texture information from continuous diffraction spectra is one direction. Orientation imaging with the SEM and probably with synchrotron radiation will provide a new dimension for evaluating the grain-to-grain interaction and defining many microstructural characteristics.

Taylor models will continue to be valuable for cubic metals, particularly with the various modifications introduced in the last decade. But new theories need to be developed to deal with materials where the behavior of a grain depends on its immediate (not statistical) environment. It is anticipated that FEM methods, briefly introduced in the Chapters 12 and 13, will become ever more important tools for polycrystal plasticity. There has not been much discussion of internal elastic strains producing local residual stresses and their relationship to texture. This is a field where much new progress has been made and which, no doubt, will become an important part of texture analysis in the future.

Also textures produced by processes other than deformation have been barely touched. Preferential growth in different environments can produce very strong textures and highly anisotropic materials with many applications in solid state physics.

If this book has provided a sound basis for such exciting future developments, it has accomplished its goal.

Appendix

THE ELASTIC INCLUSION PROBLEM

Carlos Tomé

The assumption that the grains of a polycrystal can be treated as inclusions embedded in the homogeneous medium represented by the other grains is the cornerstone of all self-consistent models of aggregates. As a consequence, in this Appendix we will review in some detail the equations associated with the elastic inclusion formalism. For a thorough discussion of this problem the reader is referred to the book of MURA [1991].

The same mathematical formalism used to address the elastic problem is applicable to the elasto-plastic, the visco-elastic and the visco-plastic polycrystal. In the latter case, however, a fictitious small compressibility has to be introduced in order to use the results of the elastic inclusion formalism. Otherwise, the formal treatment of visco-elastic media requires us to solve the incompressibility and the stress equilibrium equations simultaneously. The reader is referred to the work of MOLINARI &AL. [1987] for a treatment of the latter problem.

1. Transformation strains and Green function technique

The basic equations describing the response of elastic media are Hooke's law and the stress equilibrium condition

$$\sigma_{ij} = \mathbb{C}_{ijkl}\, u_{k,l} \tag{A1}$$

$$\sigma_{ij,j} = -f_i(\mathbf{x}) \tag{A2}$$

where σ is the Cauchy stress tensor, $u_{k,l}$ the displacement gradient (also called the distortion) tensor, $\mathbb{C}$ is the tensor of elastic stiffness and $\mathbf{f}$ are the volumetric forces acting on an element of the solid. Equations A1, A2 apply locally, and all the above magnitudes may be a function of position. This system can be solved using the standard Green function technique, which is particularly suitable for treating the problem of infinite media and uniform stiffness $\mathbb{C}$.

The elastic Green function of infinite media $G_{km}(\mathbf{x}-\mathbf{x}')$ is defined as the displacement parallel to the k-axis of a material point located at $\mathbf{x}$, induced by a unit force applied at $\mathbf{x}'$ in the direction m. As a consequence of its definition, the Green tensor $\mathbf{G}$ is the solution to the following equilibrium equation

$$\mathbb{C}_{ijkl}\, G_{km,lj}(\mathbf{x}-\mathbf{x}') = -\delta_{im}\, \delta(\mathbf{x}-\mathbf{x}') \tag{A3}$$

It is straightforward to see that the solution of Eqs. A1, A2 is given by the linear superposition of displacements

$$u_k(\mathbf{x}) = \int_V G_{km}(\mathbf{x}-\mathbf{x}')\, f_m(\mathbf{x}')\, d\mathbf{x}' \tag{A4}$$

while the associated distortions are simply given by

$$u_{k,l}(\mathbf{x}) = \int_V G_{km,l}(\mathbf{x}-\mathbf{x}')\, f_m(\mathbf{x}')\, d\mathbf{x}' \tag{A5}$$

Now, consider a homogeneous elastic medium of stiffness $\mathbb{C}$, subjected to uniform loading conditions at infinity. The overall stress $\bar{\sigma}$ and the elastic displacement $\bar{u}$ are uniform and linearly related through Hooke's law:

$$\bar{\sigma}_{ij} = \mathbb{C}_{ijkl}\, \bar{u}_{k,l} \tag{A6}$$

Next, assume that a 'transformation' strain $\varepsilon^t(\mathbf{x})$ (also called 'stress free' strain) takes place inside the homogeneous medium, inducing local variations of stress and strain. For what follows we define the internal stress $\tilde{\sigma} = \sigma - \bar{\sigma}$ and the internal displacement $\tilde{u} = u(\mathbf{x}) - \bar{u}$ that the transformation induces, as the local deviations from the uniform stress and displacement $\bar{\sigma}$ and $\bar{u}$, respectively. Expressing the elastic strain as the difference between the total and the transformation strains, Hooke's law becomes:

$$(\bar{\sigma}+\tilde{\sigma})_{ij} = \mathbb{C}_{ijkl}\left[(\bar{u}+\tilde{u})_{k,l} - \varepsilon^t_{kl}\right] \tag{A7}$$

Observe that it is the *total* and not the *elastic* strain that enters in Eq. (A7). The volume averages of the deviations are zero by definition, and their local variations follow from combining Eqs. A6 and A7:

$$\tilde{\sigma}_{ij} = \mathbb{C}_{ijkl}\left[(\tilde{u}_{k,l} - \varepsilon^t_{kl})\right] \tag{A8}$$

In addition, the equilibrium equation also applies to the stress deviation tensor

$$(\bar{\sigma}+\tilde{\sigma})_{ij,j} = \tilde{\sigma}_{ij,j} = 0 \tag{A9}$$

In what follows we favor formulating the inclusion problem in terms of the deviations of the stress and the *total* strain with respect to an otherwise uniform state. Replacing Eq. A8 in A9 leads to a second order differential equation in the internal displacements:

$$\mathbb{C}_{ijkl}\, \tilde{u}_{k,lj} = -f_i^* \tag{A10}$$

where $f_i^* = -\mathbb{C}_{ijkl}\, \varepsilon^t_{kl,j}(\mathbf{x})$

plays the role of a fictitious distribution of volume forces. This system can be solved using the Green function technique described above. Replacing the fictitious force in Eq. A5 gives for the displacement gradient

$$\tilde{u}_{k,l}(\mathbf{x}) = -\int_V G_{km,l}(\mathbf{x}-\mathbf{x}')\, \mathbb{C}_{mjrs}\, \varepsilon^t_{rs,j}(\mathbf{x}')\, d\mathbf{x}' \tag{A11}$$

After integrating by parts and using Gauss' theorem we obtain a general equation describing the internal displacements induced in a homogeneous elastic medium with stiffness $\mathbb{C}$ by an arbitrary volumetric transformation ε^t:

$$\tilde{u}_{k,1}(\mathbf{x}) = -\int_V G_{km,1j}(\mathbf{x} - \mathbf{x}') \, \mathbb{C}_{mjrs} \, \varepsilon^t_{rs}(\mathbf{x}') \, d\mathbf{x}' \tag{A12}$$

2. The elastic Eshelby inclusion

A particular case of Eq. A12 is when ε^t is uniform within a region Ω and null outside, in which case the integration reduces to the domain Ω and $\tilde{u}_{k,1}$ is a linear function of ε^t. ESHELBY [1957] demonstrates that when the region Ω has an ellipsoidal shape and $\mathbf{x} \in \Omega$, the integral is independent of $\mathbf{x}$. As a consequence, $\tilde{u}_{k,1}$ turns out to be uniform within the domain of the inclusion, and given by the tensor product [MURA 1987, Chapter 3]:

$$\tilde{u}_{i,j} = \Gamma_{ikjl} \, \mathbb{C}_{klmn} \, \varepsilon^t_{mn} \tag{A13}$$

where $\mathbb{C}$ has to be expressed in the system defined by the principal axes of the ellipsoid and:

$$\Gamma_{ikjl} = \frac{1}{4\pi} \int_0^\pi \sin\theta \, d\theta \int_0^{2\pi} \gamma_{ikjl} \, d\phi \tag{A14}$$

is a non-symmetric fourth order tensor, with

$$\gamma_{ikjl} = K_{ik}^{-1}(\xi) \, \xi_j \, \xi_1 \tag{A15}$$

$$K_{ip}(\xi) = \mathbb{C}_{ijpl} \, \xi_j \, \xi_1 \tag{A16}$$

$$\text{and } \xi_1 = \frac{\sin\theta \cos\phi}{a_1} \; ; \; \xi_2 = \frac{\sin\theta \sin\phi}{a_2} \; ; \; \xi_3 = \frac{\cos\theta}{a_3} \tag{A17}$$

Here $0 < \phi < 2\pi$ and $0 < \theta < \pi$ are spherical coordinates that define the direction of the vector ξ with respect to the principal axes of the ellipsoid, of length $2a_1$, $2a_2$, $2a_3$. The elastic strain and rotation tensors induced by the transformation are the symmetric and skew-symmetric components of the displacement gradient (Eq. A13). They are uniform within the domain Ω of the inclusion and given by:

$$\tilde{\varepsilon}_{ij} = \frac{1}{2}\left(\tilde{u}_{i,j} + \tilde{u}_{j,i}\right) = \mathbb{E}_{ijmn} \, \varepsilon^t_{mn} \tag{A18a}$$

where

$$\mathbb{E}_{ijmn} = \frac{1}{4}\left(\Gamma_{ikjl} + \Gamma_{iljk} + \Gamma_{jkil} + \Gamma_{jlik}\right) \mathbb{C}_{klmn} = \Gamma^{sym} : \mathbb{C} \tag{A18b}$$

and

$$\tilde{\omega}_{ij} = \frac{1}{2}\left(\tilde{u}_{i,j} - \tilde{u}_{j,i}\right) = \mathbb{W}_{ijmn} \, \varepsilon^t_{mn} \tag{A19a}$$

where

$$\mathbb{W}_{ijmn} = \frac{1}{4}\left(\Gamma_{ikjl} + \Gamma_{iljk} - \Gamma_{jkil} - \Gamma_{jlik}\right) \mathbb{C}_{klmn} = \Gamma^{skew} : \mathbb{C} \tag{A19b}$$

Here $\mathbb{E}$ is the classical Eshelby tensor and, together with $\mathbb{W}$, it couples the transformation strain with the relaxation displacements induced in the ellipsoidal domain Ω.

Both $\mathbb{E}$ and $\mathbb{W}$ are functions of the anisotropic stiffness moduli $\mathbb{C}_{ijkl}$ and of the aspect ratios of the ellipsoid, and can be integrated numerically in a straightforward manner using the algorithms given by Eqs. A14 to 19. It is evident from the previous equations that the tensors $\mathbb{E}$ and $\mathbb{W}$ are homogeneous functions of degree zero of the elastic constants and of the axes of the ellipsoid. As a consequence they depend on the ratio of the ellipsoid axes and on the relative values of the stiffness tensor components.

It is convenient, for the later treatment of inhomogeneities, to express the previous equations in terms of the strain deviations instead of the transformation strain. Using Eqs. A8 and A18, the internal stress in the inclusion can be written

$$\tilde{\sigma} = - \mathbb{C} : \mathbb{R} : \tilde{\varepsilon}$$

$$(A20a)$$

where

$$\mathbb{R} = (\mathbb{I} - \mathbb{E}) : \mathbb{E}^{-1}$$

$$(A20b)$$

is the *reaction tensor*, so called because it 'modulates' the stress reaction induced by the deviation in strain between the inclusion and the matrix. The reaction tensor admits a straightforward interpretation: according to Eq. A18, the Eshelby tensor $\mathbb{E}$ represents the fraction of the transformation that is accommodated by the matrix and, as a consequence, it ranges from the identity to the null tensor, $\mathbb{O}$. If each extreme is taken as a limit case in Eq. A20, $\mathbb{E} \equiv \mathbb{I}$ corresponds to an infinitely compliant matrix, the reaction tensor is zero and, as a consequence, the stress deviation is zero. Conversely, $\mathbb{E} \equiv \mathbb{O}$ corresponds to an infinitely stiff matrix and an infinite reaction tensor, which requires zero strain deviations in Eq. A20 if the stresses are to remain bounded. As a consequence, the upper and the lower bounds discussed in Section 3.1 of Chapter 7 can be regarded as limit cases of the interaction equation, and are associated with the matrix being infinitely stiff or infinitely compliant, respectively.

Fully analytic forms of the Eshelby tensor $\mathbb{E}$ have been worked out only for some spheroidal inclusions in elastically isotropic media [MURA 1987, Chap. 2]. In particular, for spherical inclusions the components of $\mathbb{E}$ adopt the simple form

$$\mathbb{E}_{iiii} = \frac{7 - 5v}{15(1 - v)} \qquad i = 1,3$$

$$\mathbb{E}_{iijj} = \frac{-(1 - 5v)}{15(1 - v)} \qquad i,j = 1,3 \ ; \ i \neq j \qquad (A21)$$

$$\mathbb{E}_{ijij} = \frac{4 - 5v}{15(1 - v)} \qquad i,j = 1,3 \ ; \ i \neq j$$

where it can be seen that they only depend on Poisson's ratio v. For the more complex case of spheroidal inclusions in media with hexagonal and cubic symmetry, KINOSHITA & MURA [1971] and LIN & MURA [1973] have been able to express the Eshelby tensor as one-dimensional integrals.

The Eqs. A13 to A19, on the other hand, are completely general: they give the components of the Eshelby tensor for an inclusion of arbitrary ellipsoidal shape embedded in a medium with arbitrary elastic symmetry. They are used in Section 3.5 of Chapter 7 to calculate $\mathbb{E}$ and $\mathbb{R}$, for equiaxed and elongated grains of cubic, hexagonal, trigonal and orthotropic symmetry, embedded in an elastically anisotropic polycrystal. They have also been used by LEBENSOHN & TOMÉ [1993b] for analyzing the stability of twin lamelli in cubic, hexagonal and trigonal crystals.

3. The inhomogeneity problem

Although transformation strains are important in several situations, they rarely take place inside a region having the same elastic properties as the surrounding medium. In addition, here we are concerned with the calculation of the overall elastic and thermal properties of a crystalline aggregate, where the local variation in the elastic and thermal properties associated with the distribution of crystal orientations has to be accounted for. As a consequence, it is important to solve the so called inhomogeneity problem. In what follows, we demonstrate that a correspondence can be established between the distortions induced by a stress-free transformation in a homogeneous medium, and the distortions induced by a fluctuation in the elastic properties of the medium (an inhomogeneity) *when the latter is subjected to external stresses*.

Assume that instead of having a homogeneous medium where a volumetric transformation $\varepsilon^t(\mathbf{x})$ takes place, we have an inhomogeneous medium subjected to a stress $\bar{\sigma}$ at infinity. If the local dependence of the elastic constants is $\mathbb{C}(\mathbf{x}) = \bar{\mathbb{C}} + \tilde{\mathbb{C}}(\mathbf{x})$, the constitutive and equilibrium Eqs. A1,A2 adopt the form

$$\sigma_{ij} = \left(\bar{\mathbb{C}} + \tilde{\mathbb{C}}\right)_{ijkl} \left(\bar{u} + \tilde{u}\right)_{k,l} \tag{A22}$$

$$\bar{\mathbb{C}}_{ijkl}\, \tilde{u}_{k,lj} = -\left(\tilde{\mathbb{C}}_{ijkl}\, u_{k,l}\right)_{,j} \tag{A23}$$

The second member can be treated as a fictitious volume force and the associated displacement gradients, given by Eq. A5, are:

$$\tilde{u}_{k,l}(\mathbf{x}) = \int_V G_{km,l}(\mathbf{x} - \mathbf{x}') \left(\tilde{\mathbb{C}}_{mjrs}\, u_{r,s}\right)_{,j}(\mathbf{x}')\, d\mathbf{x}' \tag{A24}$$

which is an integro-differential equation in the displacement gradients. The strain tensor components can be written in a more symmetric form as

$$\tilde{\varepsilon}_{k,l}(\mathbf{x}) = \frac{1}{2}\left(\tilde{u}_{k,l} + \tilde{u}_{l,k}\right) = \int_V \mathbb{G}_{kmlj}(\mathbf{x} - \mathbf{x}') \left(\tilde{\mathbb{C}}_{mjrs}\, u_{r,s}\right)(\mathbf{x}')\, d\mathbf{x}' \tag{A25a}$$

where

$$\mathbb{G}_{kmlj} = \frac{1}{4}\left(G_{km,lj} + G_{kj,lm} + G_{lm,kj} + G_{lj,km}\right) \tag{A25b}$$

is a fourth order symmetric tensor. In the derivation of Eq. A25 we have used integration by parts, Gauss' theorem, and the symmetry $\tilde{\mathbb{C}}_{mjrs} = \tilde{\mathbb{C}}_{jmrs}$. Notice that the integration by parts requires us to take derivatives with respect to the variable $\mathbf{x}'$, while the derivatives of the Green tensor indicated by the subindices after the comma are taken with respect to the argument. As a consequence, the minus sign is accounted for explicitly in Eq. A25.

Equation A25 is used by ZELLER & DEDERICHS [1973] as the starting point for a perturbation scheme aimed at calculating the effective elastic moduli of polycrystals. Also, an approach based on Eq. A25 was used by MOLINARI &AL. [1987] for modeling the visco-plastic response of polycrystals and is described in Chapter 11 of this book.

In what follows we analyze the simplest form of this equation, relevant to the modeling of elastic polycrystals.

4. The inhomogeneous inclusion

A comparison of Eqs. A11 and A24 indicates that if the inhomogeneity is uniform within a domain Ω, then it is equivalent to a uniform transformation strain within the same domain. As a consequence, the result of Section 2 applies and deformation will be uniform within Ω. In what follows, we will concentrate on the particular case in which an inhomogeneous region Ω, having ellipsoidal shape and uniform elastic constants $\mathbb{C}=\overline{\mathbb{C}}+\widetilde{\mathbb{C}}$, is embedded in an infinite medium V with constants $\overline{\mathbb{C}}$, uniformly loaded at infinity with a stress $\overline{\sigma}$. For completeness, we are going to assume that uniform transformation strains $\overline{\varepsilon}^{t}$ and $\varepsilon^{t}=\overline{\varepsilon}^{t}+\widetilde{\varepsilon}^{t}$ take place in the surrounding medium and in the domain Ω, respectively. The stress, given in terms of the total strain, is:

$$\overline{\sigma}+\widetilde{\sigma} = (\overline{\mathbb{C}}+\widetilde{\mathbb{C}}) : [(\overline{\varepsilon}+\widetilde{\varepsilon})-(\overline{\varepsilon}^{t}+\widetilde{\varepsilon}^{t})] \qquad \text{in } \Omega$$

$$\overline{\sigma}+\widetilde{\sigma} = \overline{\mathbb{C}} : (\overline{\varepsilon}+\widetilde{\varepsilon}-\overline{\varepsilon}^{t}) \qquad \text{in } V\!-\!\Omega$$

$$= \overline{\mathbb{C}} : (\overline{\varepsilon}-\overline{\varepsilon}^{t}) \qquad \text{in } \Gamma_{V} \qquad\qquad (A26)$$

where the last equation represents the conditions at the external boundary Γ_{D} of the medium. If the *inhomogeneity* were replaced by an *inclusion* of the same size and shape undergoing a transformation $(\varepsilon^{t}+\varepsilon^{*})$, the stress in the medium would be described by the following equations:

$$\overline{\sigma}+\widetilde{\sigma} = \overline{\mathbb{C}} : (\overline{\varepsilon}+\widetilde{\varepsilon})-(\overline{\varepsilon}^{t}+\widetilde{\varepsilon}^{t})-\varepsilon^{*}] \qquad \text{in } \Omega$$

$$\overline{\sigma}+\widetilde{\sigma} = \overline{\mathbb{C}} : [(\overline{\varepsilon}+\widetilde{\varepsilon}-\overline{\varepsilon}^{t}) \qquad \text{in } V\!-\!\Omega$$

$$\overline{\sigma} = \overline{\mathbb{C}} : (\overline{\varepsilon}-\overline{\varepsilon}^{t}) \qquad \text{in } \Gamma_{V} \qquad\qquad (A27)$$

A comparison of Eqs. A26 and A27 indicates that it is always possible to choose ε^{*} such that the stress in the equivalent inclusion (Eq. A27) is the same as the stress in the inhomogeneous domain Ω (Eq. A26). Since the stress at the inclusion-matrix interface is left unchanged by this procedure, the replacement will go unnoticed outside the domain Ω. The condition that ε^{*} has to fulfill is

$$-\overline{\mathbb{C}} : \varepsilon^{*} = \widetilde{\mathbb{C}} : [(\overline{\varepsilon}+\widetilde{\varepsilon})-(\overline{\varepsilon}^{t}+\widetilde{\varepsilon}^{t})] \qquad\qquad (A28)$$

Observe that $\varepsilon^{*}=0$ if there is no inhomogeneity (i.e. $\widetilde{\mathbb{C}}\equiv\mathbb{O}$). If the last equation in A27 is subtracted from the others, the problem can be formulated, as before, in terms of the local deviations $\widetilde{\sigma}$ and $\widetilde{\varepsilon}$:

$$\widetilde{\sigma} = \overline{\mathbb{C}} : [\widetilde{\varepsilon} - (\widetilde{\varepsilon}^{t}+\varepsilon^{*})] \qquad \text{in } \Omega$$

$$\widetilde{\sigma} = \overline{\mathbb{C}} : \widetilde{\varepsilon} \qquad \text{in } V\!-\!\Omega \qquad\qquad (A29)$$

In addition, the stress deviation has to obey the equilibrium equation

$$\widetilde{\sigma}_{ij,j} = 0 \qquad\qquad (A30)$$

Expressed in this form, the problem of the inhomogeneous inclusion is equivalent to the one of the elastic inclusion presented in Section 2. The strain deviation is uniform in the inclusion and it is related to the *effective* transformation strain through Eq. A18 as:

$$\tilde{\varepsilon} = \mathbb{E} : (\tilde{\varepsilon}^t + \varepsilon^*) \tag{A31}$$

And similarly the deviation in the elastic rotation (antisymmetric component of the displacement gradient) is given by A19 as:

$$\tilde{\omega} = \mathbb{W} : (\tilde{\varepsilon}^t + \varepsilon^*) = \mathbb{W} : \mathbb{E}^{-1} : \tilde{\varepsilon} \tag{A32}$$

Here the Eshelby tensors $\mathbb{E}$ and $\mathbb{W}$ are a function of the elastic constants $\overline{\mathbb{C}}$ of the medium, and the inhomogeneity effect is accounted for by the fictitious transformation strain ε^* given by Eq. A28. In addition, inserting Eq. A31 into A29 shows that the inhomogeneity is described by an interaction equation of the same form as Eq. A20 for the inclusion, namely:

$$\tilde{\sigma} = -\overline{\mathbb{C}} : (\mathbb{I} - \mathbb{E}) : \mathbb{E}^{-1} : \tilde{\varepsilon} \tag{A33}$$

This form of the interaction equation is used in Chapter 7 and most of Chapter 11.

NOTATION

All vector, tensor, matrix, and set quantities are written as **bold-face** symbols. For example, an 'orientation' is called **g**, regardless of whether a set of Euler angles is used to represent it or a rotation matrix. Similarly, the elastic stiffness is symbolically written as $\mathbb{C}$, regardless of whether the components are, in a current application, represented by fourth-rank tensors or by a 6×6 matrix. The difference between vectors and second-rank tensors is not evident from the symbol; but fourth-rank tensors are written in a 'blackboard bold' font (e.g., $\mathbb{C}$). Components are written non-bold, upright, with upright indices. Polynomial coefficients are upright with *italic* indices. Only scalar quantities that are Latin (rather than Greek) are written in *italics*.

Multiplications of tensor quantities of any rank are most unmistakably expressed by index notation, using the summation convention. However, symbolic notation is often advantageous. There are various conventions for symbolic notation; we have adopted the 'tensor' (rather than 'matrix') notation in this book. In the following, we show it in parallel with the index notation on a few examples, which will then be commented on.

$$\beta_{ij} = s_i\, n_j \qquad\qquad\qquad \boldsymbol{\beta} = \mathbf{s} \otimes \mathbf{n} \tag{1}$$

$$K_i = \sigma_{ij}\, n_j \qquad\qquad\qquad \mathbf{K} = \boldsymbol{\sigma} \cdot \mathbf{n} \tag{2}$$

$$F_{ij} = V_{ik}\, R_{kj} \qquad\qquad\qquad \mathbf{F} = \mathbf{V} \cdot \mathbf{R} \tag{3}$$

$$dw = \sigma_{ij}\, d\varepsilon_{ij} \qquad\qquad\qquad dw = \boldsymbol{\sigma} : d\boldsymbol{\varepsilon} \tag{4}$$

$$\sigma_{ij} = \mathbb{C}_{ijkl}\, \varepsilon_{kl} \qquad\qquad\qquad \boldsymbol{\sigma} = \mathbb{C} : \boldsymbol{\varepsilon} \tag{5}$$

$$\sigma_\mu = \mathbb{C}_{\mu\nu}\, \varepsilon_\mu \qquad\qquad\qquad \boldsymbol{\sigma} = \mathbb{C} : \boldsymbol{\varepsilon} \tag{6}$$

$$\sigma_{ii} = \sigma_{ij}\, \delta_{ij} \qquad\qquad\qquad \mathrm{tr}\, \boldsymbol{\sigma} \tag{7}$$

(1) The dyadic product of two vectors yields a second-rank tensor.

(2) Multiplication of a second-rank tensor and a vector yields another vector. A sum over the repeated index j is implied, which is also called a 'single contraction' and represented by a single dot in 'tensor notation'.

(3) 'Matrix multiplication' of two matrices yields another matrix. This corresponds to 'single contraction' in tensor notation.

(4) 'Double contraction' (symbol :) of two second-rank tensors yields a scalar. (Note: other notations use a single dot whenever the result is a scalar.) *Also note that the symbol* d *is used as a general differential operator (not necessarily a <u>total</u> differential.*

(5) Multiplication of a fourth-rank tensor with a second-rank tensor yields another second-rank tensor. A sum over the two repeated indices is implied; thus, this is a 'double contraction', shown by a double dot.

(6) A quantity that is known to be a symmetric second-rank tensor is represented as a 1×6 matrix, which can be obtained from a multiplication of a 6×6 matrix (used to represent a fourth-rank tensor) and another 1×6 matrix (used to represent a second-rank tensor). The symbolic notation does not change. (The Greek indices in (6) go from 1 to 6.)

(7) Occasionally the symbol tr is borrowed from matrix notation for the trace; δ_{ij} is the unit matrix.

Other comments:

(8) For function symbols, we sometimes use upright (non-bold), sometime script font.

(9) An over-caret ('hat') is used in three ways:

- Over a **bold**-face quantity to specify its non-dimensional, 'directional' part. It is multiplied with a (generally dimensioned) scalar to make the whole quantity. (The scale factor is not necessarily a normalization to 1.)

- Over scalars (especially τ and σ) for the maximum value (e.g., at zero absolute temperature: the 'Mechanical Threshold').

- The Carson transform.

(10) An over-dot signifies a time derivative (not necessarily of a state function, see point (4) above). In Chap.3 Eqs. 3B-6B, over-dots relate to symmetry properties.

(11) ∇ is used for the gradient operator.

(12) A prime (′) is sometimes used for the deviatoric part of a tensor (i.e., with trace zero).

(13) The asterisk (*) is sometimes used for the complex conjugate (sometimes just for a special value; and sometimes it is part of a definition; e.g. W^*).

(14) The superscript $^\mathsf{T}$ is uniformly used for the transpose of a matrix.

The following lists have been compiled as a reference tool. The *Chapter number* and (colon) *Equation number* (or *'Fig' number*) are listed on the right. Underlined are the locations of <u>definitions</u>. Not listed are symbols used only in a local context.

Index of Latin Symbols

D	diffusivity tensor (*D* when scalar)	*1:15,22,23*
	plastic strain rate tensor (deviatoric)	*1:13´;7:69,73;8:11,29;11:1;12:1;13:10*
$\hat{D}$	straining direction	*8:15,21*
$\tilde{D}$	local deviation in strain rate	*11:10*
d	grain size	*8:80*
	plane spacings (twinning d_t; lattice d_{hkl})	*1:12e;2:Fig.12;4:1*
$\mathbb{E}$	Eshelby tensor	*7:57,91;11:10c;A18*
E	Young's modulus	*1:24;7:63*
$\mathcal{E}$	evolution function	*8:35*
F	deformation gradient (matrix)	*8:85-87,97;14:5,8*
F_i	diffraction form factor	*3:17*
F_l^v	coefficients of pole distribution (Bunge)	*3:4B*
f	body force, per unit volume	*11:28;A2*
	resulting force on element interface	*12:7-8*
f(**g**)	orientation distribution function (OD)	*3:5,5B*
$\tilde{f}, \tilde{\tilde{f}}$:	even and odd part of f(**g**)	*3:10*
$\mathcal{F}$	function symbol	*8:31,32*
G	symmetric derivative of the Green tensor G	*11:31;A25b*
G	the elastic Green tensor	*A3,4*
ΔG	activation energy (free enthalpy)	*1:23*
g	crystal orientation (in sample frame): set of angles or rotation matrix	*2:6*
$\mathbf{g}^E$	orientation of stretch ellipsoid (grain shape)	*8:89,Fig.30*
d**g**	element in orientation space (Δ**g** when finite)	*2:14(15)*
Δ**g**	misorientation	*2:8,9a*
g (superscript): grain (referring to morphology not orientation)		*8:91,93*
	also: index of cell in orientation space	*(p.479)*

H	symmetry operator (matrix)	*1:6;8:31′*
h, *hkl*	Miller indices	*2:1b,18;3:17*
hst	hardening matrix	<u>*8:39*</u>
h(*t*)	Heaviside step function	*7:74,82*
h	trace operator	*12:7;13:17*
$\mathcal{H}$	hardening function	<u>*8:74*</u>
$\mathbb{I}$	unit tensor of fourth (or arbitrary) rank	*7:57,62a*
I	unit matrix	*8:79*
$\bar{\text{I}}$	inversion operator	*1:8*
I	intensity	*4:2*
I^{bg}	background intensity	*4:5*
I^{norm}	normalized intensity	*4:11*
J_i	tensor invariants	*1:14*
$\mathbb{K},\mathbb{K}^c$	linear creep compliance tensor (polycrystal; single crystal)	*7:69,98*
$\mathbf{K}_1$	habit plane in twinning	*2:12a,2:Fig.20*
$K^{(\lambda)}$	eigenvalues of $\mathbb{K}^c$	*7:104*
k	wave vector	*7:65*
k	Miller index	*2:1b*
	Boltzmann constant	*8:51*
k_y	Hall–Petch constant	*8:80*
$K_l^{\,m}$	coefficients of inverse pole figure (Bunge)	*2:Fig.38*
$k_l^{\,n}$	symmetrized harmonic functions (Bunge)	*3:3B*
$\mathcal{K}$	kinetic function	*8:51′,58,62*
L	velocity gradient (plastic)	<u>*1:12*</u>*;8:13*
l	Miller index	*2:1b*

l	order of harmonic function	3:5,5B
$\mathbb{M};\mathbf{M}^c$	visco-elastic compliance (polycrystal; single crystal)	7:72,70
$\hat{\mathbb{M}}$	Carson transform of $\mathbb{M}$	7:89
$\mathbf{M}$	Mirror plane operator	1:9
$\mathbf{M}^c$	Taylor tensor: stress deviator scaled by yield surface size	8:23
$\mathbf{M}_\varepsilon$	Taylor tensor defined by strains, not stresses	8:42,49
M^c	Taylor factor (normalized plastic work, regardless of model used)	2:24
	– especially for crystal c. (M is often used for the *average*.)	10:Fig.30a
M_i	multiplicity of pole i	3:12
$\mathbf{m}$	Schmid tensor (sym. part of unit distortion for slip, m when scalar)	<u>8:9;5:2,6</u>
m	physical relative rate sensitivity of flow stress	<u>8:52</u>
$\hat{N}^l, \tilde{N}^m, N^n$	interpolation functions in FEM	12:5;13:13
N	upper limit for running index n	7:102;12:5;13:13,14
	number of counts	4:10
$\mathbf{n}$	normal to a surface or to a to slip plane (unit vector $\hat{n}$)	8:8
	axis of rotation in axis/angle description of misorientations	2:9a
n	order of diffraction	4:1
	stress exponent in rate sensitive numerical kinetic relation	8:55
$\mathbb{P}$	plasticity tensor: visco-plastic compliance	8:75,<u>76</u>;11:2–8;12:1;13:12
$\mathbb{P}^{eff}$	effective plastic compliance	8:82
$\mathbf{P}$	pole	2:Fig.3
$\mathbf{p}$	the direction of a line	8:86
	a misorientation in quaternion description	2:11d,e
p	pressure	1:21;8:79;11:28;13:13
P_l^m	Legendre polynomial	3:3
$p_h(\mathbf{y}), p(\alpha,\beta)$	pole distribution function	2:17;3:1,2,3,3B,17

$\mathbb{Q}$	visco-plastic stiffness tensor	11:24;13:11
$\mathbf{Q}$	quaternion (four components)	2:11
Q	activation energy	1:22;8:59
$\mathbf{q}$	heat flow (vector)	1:1
	vector part of quaternion	2:11b
	normal to a plane	8:86'
	skew part of unit distortion tensor for slip	5:3,7;<u>8:9'</u>
q $(\bar{q}, \Delta q)$	contraction ratio (average and span)	<u>10:6-8</u>
q_0	scalar part of quaternion	2:11b
Q_{lm}	coefficients in pole figure series	3:3,6

$\mathbb{R}$	elastic reaction tensor	7:56,<u>A20</u>
$\mathbf{R}$	rotation operator	1:4,11'
$\mathbf{R}$	Rodrigues vector	2:10a
$\mathbf{R}^*$	crystal orientation, identical to $\mathbf{g}$ (represented as matrix)	13:20
R	Lankford coefficient of strain anisotropy in sheet	<u>10:4</u>
r	same on differential basis	<u>10:5</u>
$\mathbf{r}$	direction	2:1a
$r(\alpha,\beta)$	inverse pole figure density	3:9
R_{ln}	harmonic coefficients of inverse pole figure	3:8

$\mathbb{S},\mathbb{S}_{ijkl},\mathbb{S}_{\lambda\mu}$	elastic compliance tensor (components, matrix)	<u>1:18'</u>;7:10,63
$\mathbf{S},\mathbf{S}^c$	stress deviator (c: in crystal)	8:29,79;11:24;12:1;13:10
$\mathbf{S}$	plane of shear in twinning	2:12a
$\mathbf{S}^{(\lambda)}$	eigenvalues of $\mathbb{S}$	7:26,29
s	Carson transform of time	7:78

s (superscript): slip (or twinning) system

T	absolute temperature	*1:1*
t	twinning transformation matrix	*2:12b*
	surface tractions (a vector)	*12:3*
t	time	*7:64*
	thickness	*4:2*
$T_l^{\mu\nu}$	generalized spherical harmonics (Bunge)	*3:5B;7:6*
^{T}	(superscript): transpose	*1:5*
tr	sometimes used for Trace	*12:4*
u	displacement vector (u: scalar displacement)	*7:64;A4*
	velocity vector in FEM	*13:12*
u₁	displacement in twinning	*2:12a*
$u_{k,l}$	displacement gradient tensor	*A1,5*
uvw	direction indices	*2:1a*
V	left stretch tensor	*8:87*
v	elastic wave velocity	*7:68*
W	rotational Eshelby tensor	*11:11;A19*
W	spin tensor (skew part of velocity gradient)	*1:13";8:84,94*
W^g	same at grain level	
$\tilde{\mathbf{W}}$	local deviation in rotation rate	*11:11*
W^c, **W**^{p}	'plastic spin' of crystal (due to slip)	*8:12,92;(13:21)*
W^E (**W**^{e})	Euler spin (of grain)	*8:89-91;13:6*
W*	reorientation rate of crystal axes	*8:93-96*
w	work per unit volume (only used differentially or as rate)	*8:17,68*
w^c	statistical weight of a crystal	*7:2,3;13:4,11*
W$_{lmn}$	generalized harmonic coefficients (Roe)	*3:5;7:6*

Index of Greek Symbols

$\dot{\gamma}_0$	reference strain rate	*11:1*
γ_t	twinning shear	*2:12e;8:2*
δ	virtual variation	*8:1,2*
	azimuth angle in oblique section (Matthies)	*2:5*
	angle in Debye ring	*4:13*
δ_{ij}	Kronecker symbol (unit tensor)	
Δ	difference operator	*2:15*
ε	elastic strain tensor (*plastic* only as dε)	*7:9;(8:15′)*
$\tilde{\varepsilon}$	local deviation in ε	*7:56*
$\hat{\varepsilon}(s)$	Carson transform of $\varepsilon(t)$	*7:88*
ε^t	transformation strain tensor	*A7*
ε^*	fictitious transformation strain	*A27*
$\dot{\varepsilon}$	plastic strain rate (scalar)	*8:15,51,67′*
η_1	twinning direction	*2:12a,2:Fig.20*
η	azimuth of axis in crystal coordinates	*3:6*
θ	(one-half of the) Bragg angle	*4:1*
θ	crystal hardening rate (sub-zero: initial, sub-one: for single slip)	*8:38,40*
	angle of twist	*10:2*
Θ	Euler angle (distance between z and Z poles)	*2:2,6,2:Fig.3,4*
	polycrystal hardening rate	*8:74*
κ, κ_{ij}	thermal conductivity (tensor, components)	*1:1*
ς	elastic compressibility	*1:21;7:49*
	parameter in work-hardening relation	*8:37′;13:22*

λ_i	eigenvalues	1:14;8:88
λ	wavelength of radiation	4:1
	diffusive jump distance	1:23
μ	shear modulus (relevant combination for dislocation problems)	8:58,62
	x-ray absorption coefficient	4:2
	Euler angle in oblique section	2:4
ν	Euler angle in oblique section	2:3
ν	Poisson's ratio	10:9
Ξ	radial distance of axis from Z in sample coordinate system	3:8
ξ	radial distance of pole from z in crystal coordinate system	3:6
ρ	material density	4:3;7:64
	dislocation density	8:35
	fractional radial distance	2:2;13:7
	angular density of rod axes	14:1-3
$\sigma, \sigma_{ij}, \sigma_\mu$	Cauchy stress	1:9;7:98,106;A1
$\bar{\sigma}, \tilde{\sigma}$	average and local deviation of σ	7:56;A7
$\hat{\sigma}(s)$	Carson transform of $\sigma(t)$	7:88
σ_f	flow stress (a scalar measure of the size of the yield surface)	8:65,70
σ_{vM}	equivalent stress according to von Mises	8:68;10:10
σ	Euler angle in oblique sections, according to Matthies	2:5
	standard deviation	4:10
Σ	sum	
τ	shear stress	10:1

τ	relaxation time	*7:100*
τ^c	scalar parameter of strength (measure of crystal yield surface size)	*8:26*
τ^s	critical resolved shear stress on system s (with hat: at 0 K)	*8:18,51;11:1*
τ_ε	part of flow stress due to strain hardening	*8:34*
τ_v	scaling stress in strain-hardening law ('Voce stress')	*8:35*
Υ	weighting function in FEM (a scalar)	*12:4*
Φ	weighting function in FEM (a 1-D array)	*12:3;13:8,9*
ϕ	tilting angle in goniometer	*4:Fig.1*
	Euler angle (azimuth in inverse pole figure)	*2:Fig.4*
φ_1	Euler angle (azimuth in pole figure)	*2:Table I*
φ_2	Euler angle (azimuth in inverse pole figure)	*2:Table I*
φ^+, φ^-	Euler angles in oblique section (Bunge)	*2:5,2:Fig.15*
Φ	Euler angle (Bunge: distance between z and Z poles)	*2:Table I*
	(Roe: azimuth in inverse pole figure)	*2:Table I*
	plastic potential	*8:21*
χ	goniometer azimuth	*4:Fig.1*
Ψ	weighting function in FEM (a 2-D array)	*12:2*
Ψ	Euler angle (azimuth in pole figure)	*2:Fig.3*
ω	skew part of distortion tensor (used only differentially)	*8:5*
$\tilde{\omega}$	deviation in elastic rotation tensor	*A19*
ω	angle of rotation in axis/angle description	*2:9b*
	angle of rotation in goniometer (for rocking curves)	*4:12a*
	wave frequency	*7:65*
Ω	volume in orientation space (or other space)	*2:13,19;12:3;13:14*

REFERENCES

Proceedings ICOTOM – International Conference on Textures of Materials:

Textures in Research and Practice. Proc. Int. Symp. Clausthal-Zellerfeld,
J. Grewen & G. Wassermann, eds. (Berlin: Springer, 1969).

Quantitative Analysis of Textures. Proc. Int. Seminar Cracow 1971,
J. Karp & al., eds. (The Society of Polish Metallurgical Engineers).

*3ème Colloque Européen sur les Textures de Déformation et de Recristallisation des
Métaux et leur Application Industrielle.* Proc. Conf. at Pont-à-Mousson 1973, R.
Penelle, ed. (Société Française de Métallurgie).

Texture and the Properties of Materials. Proc. 4th Int. Conf. on Textures, Cambridge
1975, G.J. Davies &al., eds. (London: The Metals Society).

Textures of Materials. Proc. 5th Int. Conf. on Textures, Aachen 1978, G. Gottstein & K.
Lücke, eds. (Berlin: Springer), 2 volumes.

Sixth Int. Conf. on Textures of Materials, 1981 (Tokyo: The Iron and Steel Institute of
Japan).

Seventh Int. Conf. on Textures of Materials, C.M. Brakman, P. Jongenburger & E.J.
Mittemeijer, eds. (Netherlands Soc. for Materials Science).

Eighth Int. Conf. on Textures of Materials, J.S. Kallend & G. Gottstein, eds. (Warrendale,
PA: The Metallurgical Society, 1988).

Ninth Int. Conf. on Textures of Materials. Special Issue, R. Penelle & C. Esling, eds., of
Textures Microstruct. **14-18** (1991).

Textures of Materials ICOTOM-10, H.J. Bunge, ed. (Switzerland: Trans. Tech. Pubs.,
1994). [This is also listed as *Materials Forum* **157-162** (1994).]

Texture of Materials ICOTOM-11, Z. Liang, L. Zuo & Y. Chu, eds. (Beijing: International
Academic Publishers, 1996).

REFERENCES

ADAMS, B.L. (1986), Description of the intercrystalline structure distribution in poly-crystalline metals. *Metall. Trans.* **17A**, 2199-2207. [2]

ADAMS, B.L. & FIELD, D.P. (1991), A statistical theory of creep in polycrystalline materials. *Acta Metall. Mater.* **39**, 2405-2417. [8]

ADAMS, B.L., ZHAO, J. & GRIMMER, H. (1990), Discussion of the representation of inter-crystalline misorientation in cubic materials. *Acta Crystallogr.* **A46**, 620-622. [2]

ADAMS, B.L., WRIGHT, S.I. & KUNZE, K. (1993), Orientation imaging: The emergence of a new microscopy. *Metall. Trans.* **24A**, 819-831. [4]

ADCOCK, F. (1922), The internal mechanism of cold-work and recrystallization in cupro-nickel. *J. Inst. Met.* **2**, 73. [5]

AERNOUDT, E. (1978), Calculation of deformation textures according to the Taylor model. In *Textures of Materials*, G. Gottstein & K. Lücke, eds. (Berlin: Springer) 45-66. [8]

AERNOUDT, E. & STÜWE, H.-P. (1970), Die Endlagen der Verformungstextur, insbesondere bei kfz. Metallen. *Z. Metallk.* **61**, 128-136. [8,9]

AERNOUDT, E., GIL-SEVILLANO, J. & VAN HOUTTE, P. (1987), Structural background of yield and flow. In *Constitutive Relations and their Physical Basis*, S.I. Andersen &al., eds. (Roskilde, Denmark: Risø Nat. Lab.) 1-38. [10]

AERNOUDT, E., VAN HOUTTE, P & LEFFERS, T. (1993), Deformation and textures of metals at large strain. In *Materials Science and Technology* (Weinheim, Germany: VCH) 6, 89-136. [8,9]

AHLBORN, H. & WASSERMANN, G. (1962), Über die doppelte Fasertextur in gezogenen Drähten kubisch flächenzentrierter Metalle. *Z. Metallk.* **53**, 422-427. [9]

AHLBORN, H. & WASSERMANN, G. (1963), Einfluß von Verformungsgrad und -temperatur auf die Textur von Silberdrähten. *Z. Metallk.* **54**, 1-6. [9]

AKHTAR, A. (1973a), Basal slip in zirconium. *Acta Metall.* **21**, 1-11. [11]

AKHTAR, A. (1973b), Compression of zirconium single crystals parallel to the c-axis. *J. Nucl. Mater.* **47**, 79-86. [11]

AKHTAR, A. (1975), Prismatic slip in zirconium single crystals at elevated temperatures. *Metall. Trans.* **6A**, 1217-1222. [11]

ALAM, M.N., BLACKMAN, M. & PASHLEY, D.W. (1954), High-angle Kikuchi patterns. *Proc. R. Soc. Lond.* **A221**, 224-242. [4]

ALEKSANDROV, K.S. & AISENBERG, L.A. (1966), Method of calculating the physical constants of polycrystalline materials. *Soviet Physics – Doklady* **11**, 323-325. [7]

ALTMANN, S.I. (1986). *Rotations, Quaternions and Double Groups* (Oxford: Clarendon Press). [2]

ANAND, L. & KOTHARI, M. (1996), A computational procedure for rate-independent crystal plasticity. *J. Mech. Phys. Sol.* **44**, 525-558. [8]

ANTONIADIS, A., BERRUYER, J. & FILHOL, A. (1990), Maximum-likelihood methods in powder diffraction refinements. *Acta Crystallogr.* **A46**, 692-711. [4]

ARGON, A.S. (1973), Stability of plastic deformation. In *The Inhomogeneity of Plastic Deformation*, R.E. Reed-Hill, ed. (Metals Park OH: Amer. Soc. Metals) 161-189. [8]

ARMINJON, M. & IMBAULT, D. (1996), Variational micro-macro model and deformation textures predicted for steels. *Textures Microstruct.* **26-27**, 191-220. [8,9]

ASARO, R.J. (1979), Geometrical effects in the inhomogeneous deformation of ductile single crystals. *Acta Metall.* **27**, 445-453. *[10]*

ASARO, R.J. & NEEDLEMAN, A.(1985), Texture development and strain hardening in rate dependent polycrystals. *Acta Metall.* **33**, 923-953. *[8,10,12,13,14]*

ASARO, R.J., AHZI, Ś., BLUMENTHAL, W. & DIGIOVANNI, A. (1992), Mechanical processing of high J_c BSCCO superconductors. *Philos. Mag.* **A66**, 517-538. *[6]*

ASBECK, H. & MECKING, H. (1978), Influence of friction and geometry of deformation on texture inhomogeneities during rolling of Cu single crystals as an example. *Mater. Sci. Eng.* **34**, 111-119. *[5]*

ASHBY, M.F. (1970), The deformation of plastically non-homogeneous materials. *Philos. Mag.* **21**, 399-424. *[8,9]*

ATLURI, S.N. (1975), On 'hybrid' finite element methods in solid mechanics. *Advances in Computer Methods for Partial Differential Equations*, R. Vichinevetsky, ed. (New Brunswick, NJ: Rutgers Univ. Press) 346-355. *[13]*

AVÉ LALLEMANT, H. & CARTER, N.L. (1970), Syntectonic recrystallization of olivine and modes of flow in the upper mantle. *Geol. Soc. Am. Bull.* **81**, 2203-2220. *[6]*

AZAROFF, L.V. (1968), *Elements of X-ray Crystallography.* (New York: McGraw-Hill) 610pp. *[4]*

AZRIN, M. & BACKOFEN, W.A. (1970), The deformation and failure of a biaxially stretched sheet. *Metall. Trans. A* **1**, 2857-2865. *[10]*

BACKOFEN, W.A. (1950), The torsion texture of copper. *Trans. Metall. Soc. AIME* **188**, 1454-1459. *[5]*

BACKOFEN, W.A. (1972), *Deformation Processing* (Reading MA: Addison Wesley) *[9,10]*

BACKSTRØM, S.P., RIEKEL, C., ABEL, S., LEHR, H. & WENK, H.-R. (1996), Micro-texture analysis by synchrotron x-ray diffraction of nickel-iron alloys prepared by microelectroplating. *J. Appl. Crystallogr.* **29**, 118-124. *[4]*

BACON, G.E. (1975), *Neutron Diffraction.* (Oxford: Clarendon Press) 436 pp. *[4]*

BACROIX, B. (1986), *Prediction of high temperature deformation textures in fcc metals.* Ph.D. Thesis (Montreal: McGill University). *[9]*

BACROIX, B. & JONAS, J.J. (1988), The influence of non-octahedral slip on texture development in fcc metals. *Textures Microstruct.* **8–9**, 267-311. *[9]*

BACZMANSKI, A., WIERZBANOWSKI, K., JURA J., HAIJE, W.G., HELMHOLDT, R.B. & MANIAWSKI, F. (1993), Calculation of the rotation rate field on the basis of experimental texture data. *Philos. Mag.* **A67**, 155-171. *[8]*

BAI, Q., MACKWELL, S.J. & KOHLSTEDT, D.L. (1991), High temperature creep of olivine single crystals. I. Mechanical results for buffered samples. *J. Geophys. Res.* **96**, 2441-2463. *[6,14]*

BAKER, D.W. & WENK, H.-R. (1972), Preferred orientation in a low symmetry quartz mylonite. *J. Geol.* **80**, 81-105. *[3]*

BAKER, D.W., WENK, H.-R. & CHRISTIE, J.M. (1969), X-ray analysis of preferred orientation in fine-grained quartz aggregates. *J. Geol.* **77**, 143-172. *[4]*

BALLINGER, R.G. (1979). *The anisotropic behavior of Zircaloy-2.* (New York: Garland Publishing). *[11]*

BARBER, D.J. & WENK, H.-R. (1979), Deformation twinning in calcite, dolomite, and other rhombohedral carbonates. *Phys. Chem. Mineral.* **5**, 141-165. *[2]*

BARBER, D.J., HEARD, H.C. & WENK, H.-R. (1981), Deformation of dolomite single crystals from 20-800°C. *Phys. Chem. Mineral.* **7**, 271-286. *[6]*

BARBER, D.J., WENK, H.-R. & HEARD, H.C.(1994), The plastic deformation of poly-crystalline dolomite: Comparison of experimental results with theoretical predictions. *Mater. Sci. Eng.* **A175**, 83-104. *[4,6,11]*

BARLAT, F. & RICHMOND, O. (1987), Prediction of tricomponent plane stress yield surfaces and associated flow and failure behavior of strongly textured f.c.c. polycrystalline sheets. *Mater. Sci. Eng.* **95**, 15-29. *[10]*

BARRETT, C.S. & LEVENSON, L.H. (1940), The structure of aluminum after compression. *Trans. Metall. Soc. AIME* **137**, 112-127. *[5]*

BARRETT, C.S. & MASSALSKI, T.B. (1980), *Structure of Metals*, 3rd Revised Ed. (Oxford: Pergamon) 654pp. *[1,2,4]*

BATE, P.S. (1992), Texture inhomogeneity and limit strains in aluminium sheet. *Scripta Metall.* **27**, 515-520. *[10]*

BATE, P.S. (1993), The effects of combined strain-path and strain-rate changes in aluminum. *Metall. Trans.* **A24**, 2679-2689. *[10]*

BAUMHAUER, H. (1879), Ueber künstliche Kalkspath Zwillinge nach -½R. *Z. Kristallogr.* **3**, 588-591. *[14]*

BAY, B., HANSEN, N., HUGHES, D.A. & KUHLMANN-WILSDORF, D. (1992), Overview No. 96: evolution of fcc deformation structures in polyslip. *Acta Metall. Mater.* **40**, 205-219. *[5,9]*

BEAUDOIN, A.J., MATHUR, K.K., DAWSON, P.R. & JOHNSON, G.C. (1993), Three-dimen-sional deformation process simulation with explicit use of polycrystalline plasticity models. *Int. J. Plast.* **9**, 833-860. *[12,13]*

BEAUDOIN, A.J., DAWSON, P.R., MATHUR, K.K., KOCKS, U.F. & KORZEKWA, D.A. (1994), Application of polycrystalline plasticity to sheet forming. *Comp. Meth. Appl. Mech. Eng.* **117**, 49-70. *[13]*

BEAUDOIN, A.J., DAWSON, P.R., MATHUR, K.K. & KOCKS, U.F. (1995a), A hybrid finite element formulation for polycrystal plasticity with consideration of macro-structural and microstructural linking. *Int. J. Plast.* **11**, 501-521. *[9,12,13]*

BEAUDOIN, A.J., MECKING, H. & KOCKS, U.F. (1995b). Development of local shear bands and orientation gradients in fcc polycrystals. In *Simulation of Materials Processing: Theory, Methods and Applications.* Shan-Fu Shen & P. Dawson, eds. (Rotterdam: Balkema) 225-230. *[5]*

BEAUDOIN, A.J., MECKING, H. & KOCKS, U.F. (1996), Development of localized orientation gradients in fcc polycrystals. *Philos. Mag.* **A73**, 1503-1517. *[5,8,9]*

BECKER, P., HEIZMANN, J.J. & BARO, R. (1977), Relations topotaxiques entre des cristaux naturels d'hématite et la magnétite qui en est issue par réduction a basse température. *J. Appl. Crystallogr.* **10**, 77-78. *[6]*

BECKER, R. (1991a), Analysis of texture evolution in channel die compression - 1. Effects of grain interaction. *Acta Metall. Mater.* **39**, 1211-1230. *[13]*

BECKER, R. (1991b), Applications of crystal plasticity constitutive models. In *Modeling the Deformation of Crystalline Solids*, T.C. Lowe &al., eds. (Warrendale PA: The Minerals, Metals & Materials Soc.) 249-269. *[13]*

BECKER, R. (1995), Pencil glide formulation for polycrystal modelling. *Scripta Metall. Mater.* **32**, 2051-2054. *[9]*

BECKER, R. & LALLI, L.A. (1991), Analysis of texture evolution in channel die com-pression Part 11: effects of grains which shear. *Textures Microstruct.* **14-18**, 145-150. *[13]*

BECKER, R. & PANCHANADEESWARAN, S. (1989), Crystal rotations represented as Rodrigues vectors. *Textures Microstruct.* **10**, 167-194 [2]

BECKER, R. & PANCHANADEESWARAN, S. (1995), Effect of grain interactions on deformation and local texture in polycrystals. *Acta Metall. Mater.* **43**, 2701-2719. [5]

BEHR, H.-J. (1964), Die Korngefügefazies der Zweigürteltektonite im kristallinen Grundgebirge Sachsens. *Abh. Deutsch. Akad. Wiss. Berlin, Kl. Bergbau*, **1**, 1-46. [14]

BELLIER, S. & DOHERTY, R.D. (1977), The structure of deformed aluminium and its recrystallization - investigations with transmission Kossel diffraction, *Acta Metall.* **25**, 521-538. [5]

BENNETT, K., WENK, H.-R., DURHAM, W.B. & STERN, L.A. (1997), Preferred crystallographic orientation in the ice I→II transformation and the flow of ice II. *Philos. Mag.* **A76**, 413-435. [4,6]

BERGMANN, H.W., FROMMEYER, G. & WASSERMANN, G. (1978), The dependence of the texture and microstructure in two-phase composites on the yield stresses of the components. In *Textures of Materials*, G. Gottstein & K. Lücke, eds. (Berlin: Springer) 371-377. [11]

BERTHELOOT, L., VAN HOUTTE, P. & AERNOUDT, E. (1976). Texture development in tube drawing. In *Fourth Int. Conf. on Texture and the Properties of Materials (ICOTOM-4).* (Cambridge: The Metals Society) 64-73. [5]

BERVEILLER, M. & ZAOUI, A. (1978), An extension of the self-consistent scheme to plastically flowing polycrystals. *J. Mech. Phys. Solids.* **26**, 325-344. [8]

BESSIÈRES, J., HEIZMANN, J.J. & EBERHARDT, A. (1991), XRD quantitative phase analysis of textured materials by using a curved P.S.D. *Textures Microstruct.* **14-18**, 157-162. [4]

BEVIS, M. & CROCKER, A.G. (1969), Twinning modes in lattices. *Proc. R. Soc. Lond.* **A313**, 509-529. [2]

BIELER, T.R., NOEBE, R.D., WHITTENBERGER, J.D. & LUTON, M.J. (1992), Extrusion textures in NiAl and reaction milled NiAl/AlN composites. *Materials Research Society Symposium Proceedings*, **273**, 165-170. [5]

BILBY, B.A. & CROCKER, A.G. (1965), The theory of the crystallography of deformation twinning. *Proc. R. Soc. Lond.* **A288**, 240-255 [2]

BIOT, M.A. (1954), Theory of stress-strain relations in anisotropic viscoelasticity and relaxation phenomena. *J. Appl. Phys.* **25**, 1385-1391. [7]

BISELLI, C. & MORRIS, D.G. (1996), Microstructure and strength of Cu-Fe *in situ* composites after very high drawing strains. *Acta Mater.* **44**, 493-504. [5]

BISHOP, J.F.W. (1953), A theoretical examination of the plastic deformation of crystals by glide. *Philos. Mag.* **44**, 51-64. [8]

BISHOP, J.F.W. (1954), A theory of the tensile and compressive textures of face-centered cubic metals. *J. Mech. Phys. Solids* **3**, 130-142. [9]

BISHOP, J.F.W. & HILL, R. (1951a), A theory of plastic distortion of a polycrystalline aggregate under combined stresses. *Philos. Mag.* **42**, 414-427. [8]

BISHOP, J.F.W. & HILL, R. (1951b), A theoretical derivation of the plastic properties of a polycrystalline face-centred metal. *Philos. Mag.* **42**, 1298-1307. [8,9,10]

BISHOP, J.R. (1981), Piezoelectric effects in quartz-rich rocks. *Tectonophysics* **77**, 297-321. [6]

BITTER, F. (1937). *Introduction to Ferromagnetism* (New York: McGraw-Hill). [2]

BLACIC, J.D. (1975), Plastic deformation mechanisms in quartz: The effect of water. *Tectonophysics* **27**, 271-294. [6]

BLANDFORD, P., SZPUNAR, J.A., DANIEL, D. & JONAS, J. (1991), Textural through-thickness inhomogeneity and its influence on the prediction of plastic anisotropy. *Textures Microstruct.* **14-18**, 525-530. [5]

BLICHARSKI, M., NOURBAKHSH, S. & NUTTING, J. (1979), Structure and properties of plastically deformed alpha Ti. *Met. Sci.* **13**, 516-522. [5]

BOAS, W. & HARGREAVES, M.E. (1948), On the inhomogeneity of plastic deformation in the crystals of an aggregate. *Proc. R. Soc. Lond.* **A193**, 89-97. [8]

BOAS, W. & MACKENZIE, J.K. (1950), Anisotropy in metals. *Prog. Metal Physics* (Pergamon) **2**, 90-120. [1]

BOAS, W. & OGILVIE, G.J. (1954), The plastic deformation of a crystal in a polycrystalline aggregate. *Acta Metall.* **2**, 655-659. [8]

BÖCKER, A., BROKMEIER, H.G. & BUNGE, H.J. (1991), Description of preferred orientation in Al_2O_3 ceramics. *J. Europ. Ceram. Soc.* **7**, 187-194. [6]

BOLMARO, R.E. & KOCKS, U.F. (1992), A comparison of the texture development in pure and simple shear and during strain path changes. *Scripta Metall. Mater.* **27**, 1717-1722. [2,8]

BOLMARO, R.E., GUERRA, F.M., KOCKS, U.F., BROWNING, R.V., DAWSON, P.R., EMBURY, J.D. & POOLE, W.J. (1993), On plastic strain distribution and texture development in fiber composites. *Acta Metall. Mater.* **41**, 1893-1905. [5,8,11]

BOLMARO, R.E., BROWNING, R.V., GUERRA, F.M. & ROLLETT, A.D. (1994), Texture development in Ag-Ni powder composites. *Mater. Sci. Eng.* **A175**, 113-124. [11]

BONARSKI, J.T., WCISLAK, L. & BUNGE, H.J. (1994), Investigation of inhomogeneous textures of coatings and near-surface layers. In *Textures of Materials ICOTOM-10*, H.J. Bunge, ed. (Switzerland: Trans. Tech. Pubs.) 111-118. [4]

BOUCHEZ, J.L. (1977), Plastic deformation of quartzites at low temperature in an area of natural strain gradient. *Tectonophysics* **39**, 25-50. [14]

BOWEN, A.W. (1990), Texture development in high strength aluminium alloys. *Mater. Sci. Technol.* **6**, 1058-1071. [5]

BOWMAN, K.J. (1991), Texture from domain switching of tetragonal zirconias. *J. Am. Ceram. Soc.* **74**, 2690-2692. [6]

BOWMAN, K.J. & CHEN, I.-W. (1993), Transformation textures in zirconia. *J. Am. Ceram. Soc.* **76**, 113-122. [6]

BOWMAN, K.J., REYES-MOREL, P.E. & CHEN, I.W. (1988), Texture from deformation of zirconia-containing ceramics. In *Eighth Int. Conf. on Textures of Materials*. J.S. Kallend & G. Gottstein, eds. (Warrendale, PA: The Metallurgical Soc.) 811-816. [6]

BRANDMÜLLER, J. & WINTER, F.X. (1985), Influence of symmetry on the static and dynamic properties of crystals. *Z. Kristallogr.* **172**, 191-231. [1]

BRATIANU, C. & ATLURI, S.N. (1983), A hybrid finite element method for Stokes flow: Part I - formulation and numerical studies. *Comp. Meth. Appl. Mech. Eng.* **36**, 23-37. [12]

BROCKHOUSE, B.N. (1953), The initial magnetization of nickel under tension. *Canad. J. Phys.* **31**, 339-355. [4]

BRODESSER, S., CHEN, S. & GOTTSTEIN, G. (1990), Microstructure & microtexture development during high temperature low cycle fatigue. *Textures Microstruct.* **14-18**, 1179-1184. [4]

BRONKHORST, C.A., KALIDINDI, S.R. & ANAND, L. (1991), An experimental and analytical study of the evolution of crystallographic texturing in fcc materials. *Textures Microstruct.* **14-18**, 1031-1036. *[8,12]*

BRONKHORST, C.A., KALIDINDI, S.R. & ANAND, L. (1992), Polycrystal plasticity and the evolution of crystallographic texture in face centered cubic metals. *Phil. Trans. R. Soc. Lond.* **A341**, 443-477. *[12]*

BROWN, G.M. (1970), A self-consistent polycrystalline model for creep under combined stress states. *J. Mech. Phys. Solids* **18**, 367-381. *[7]*

BRUNET, F., GERMI, P., PERNET, M. & WENK, H.-R. (1996), Effect of boron doping on the texture and microstructure of diamond films. In *Textures of Materials ICOTOM-11*, Z. Liang, L. Zuo & Y. Chu, eds. (Beijing: International Academic Publishers) 1105-1110.

BRYANT, J.D., BEAUDOIN, A.J. & VANDYKE, R.T. (1994), The effect of crystallographic texture on the formability of AA2036 autobody sheet. *SAE Techn. Paper Series*, 940161. *[10,13]*

BUDIANSKY, B. & WU, T.T. (1962), Theoretical prediction of plastic strains of poly-crystals. *Proc. Fourth U.S. National Congr. Appl. Mech.* (ASME) 1175. *[7,8]*

BUERGER, M.J. (1956), *Elementary Crystallography* (New York: Wiley). *[1]*

BUISKOOL TOXOPEUS, J.M.A. (1977), Fabric development of olivine in peridotite mylonite. *Tectonophysics* **39**, 55-72. *[6]*

BUNGE, H.-J. (1965), Zur Darstellung allgemeiner Texturen. *Z. Metallk.* **56**, 872-874. *[2,3]*

BUNGE, H.-J. (1969), *Mathematische Methoden der Texturanalyse* (Berlin: Akademie-Verlag). *[3]*

BUNGE, H.-J. (1982), *Texture Analysis in Materials Science – Mathematical Methods* (London: Butterworths). *[1,2,3,5]*

BUNGE, H.-J. (1985), Textures in multiphase alloys. *Z. Metallk.* **76**, 92-101. *[4]*

BUNGE, H.-J. (1986). *Experimental Techniques of Texture Analysis* (Oberursel, Germany: Deutsche Gesell. Metallkunde). *[4]*

BUNGE, H.-J. (1988), Calculation and representation of complete ODF. In *Eighth Int. Conf. on Textures of Materials.* J.S. Kallend & G. Gottstein, eds. (Warrendale, PA: The Metallurgical Society) 69-78. *[2]*

BUNGE, H.-J. (1991), Textures in non-metallic materials. *Textures Microstruct.* **14-18**, 283-326. *[6]*

BUNGE, H.-J. & ESLING, C. (1979), Determination of the odd part of the ODF. *J. Phys. Lett.* **40**, 627-628. *[3]*

BUNGE, H.-J. & WENK, H.-R. (1977), Three dimensional texture analysis of three quartzites (trigonal crystal & triclinic specimen symmetry). *Tectonophysics* **40**, 257-285. *[6]*

BUNGE, H.-J., ESLING, C. & MULLER, J. (1980), The role of an inversion center in texture analysis. *J. Appl. Crystallogr.* **13**, 544-554. *[1]*

BUNGE, H.-J., ESLING, C. & MÜLLER, J. (1981), The influence of crystal and sample symmetries on the orientation distribution function. *Acta Crystallogr.* **A37**, 889-899. *[2]*

BUNGE, H.-J., WENK, H.-R. & PANNETIER, J. (1982), Neutron diffraction texture analysis using a position sensitive detector. *Textures Microstruct.* **5**, 153-170. *[4]*

CAHN, R.W. (1953), Plastic deformation of alpha-uranium; twinning and slip. *Acta Metall.* **1**, 49-70. *[2]*

CAHN, R.W. (1991), Measurement and control of texture. In *Materials Science and Technology*, R.W. Cahn, P. Haasen & E.J. Kramer, eds. (Weinheim, Germany: VCH) **15**, 430-479. [5]

CANOVA, G.R. & KOCKS, U.F. (1984), The development of deformation textures and resulting properties of fcc metals. In *Seventh Int. Conf. on Textures of Materials*. C.M. Brakman, P. Jongenburger & E.J. Mittemeijer, eds. (Netherlands: Soc. for Materials Science) 573-579. [8]

CANOVA, G.R., SHRIVASTAVA, S., JONAS, J.J. & G'SELL, C. (1982), The use of torsion testing to assess material formability. In *Formability of Metallic Materials – 2000 A.D.*, J.R. Newby & B.A. Niemeier, eds. (American Society for Testing and Materials) ASTM-STP **753**, 189-210. [10]

CANOVA, G.R., KOCKS, U.F. & JONAS, J.J. (1984a), Theory of torsion texture development. *Acta Metall.* **32**, 211-226. [2,5,9]

CANOVA, G.R., KOCKS, U.F. & STOUT, M.G. (1984b), On the origin of shear bands in textured polycrystals. *Scripta Metall.* **18**, 437-442. [5,10]

CANOVA, G.R., KOCKS, U.F., TOMÉ, C.N. & JONAS, J.J. (1985), The yield surface of textured polycrystals. *J. Mech. Phys. Solids* **33**, 371-397. [1,8,10]

CANOVA, G.R., FRESSENGEAS, C., MOLINARI, A. & KOCKS, U.F. (1988), Effect of rate sensitivity on slip system activity and lattice rotation. *Acta Metall.* **36**, 1961-1970. [8,9]

CANOVA, G.R., WENK, H.-R. & MOLINARI, A. (1992), Deformation modelling of multiphase polycrystals: case of a quartz-mica aggregate. *Acta Metall. Mater.* **40**, 1519-1530. [11,14]

CARRARD, M. & MARTIN, J. (1987), A study of (001) glide in [112] aluminium single crystals– I. Creep characteristics. *Philos. Mag.* **56**, 391-405. [5]

CARRERAS, J., ESTRADA, A. & WHITE, S. (1977), The effects of folding on the c-axis fabrics of a quartz mylonite. *Tectonophysics* **47**, 15-42. [14]

CARREROT, H., RIEU, J., THOLLET, G. & ESNOUF, C. (1991), Texture analysis in alumina plasma sprayed coatings. *Textures Microstruct.* **14-18**, 383-388. [6]

CARTER, N.L. & HANSEN, F.D. (1983), Creep of rocksalt. *Tectonophysics* **92**, 275-333. [6]

CARTER, N.L. & HEARD, H.C. (1970), Temperature and rate dependent deformation of halite. *Am. J. Sci.* **269**, 193-249. [6,14]

CASTELNAU, O., DUVAL, P., LEBENSOHN, R.A. & CANOVA, G.R. (1996), Viscoplastic modeling of texture development in polycrystalline ice with a self-consistent approach; comparison with bound estimates. *J. Geophys. Res.* **101**, 13851-13868. [14]

CHAKI, T.K. & LI, J.C.M. (1986), Latent hardening in polycrystalline copper and comparison with high density polyethylene. *J. Mater. Sci.* **21**, 3038-3042. [10]

CHARQUET, D., ALHERITIERE, E. & BLANC, G. (1987), Cold-rolled and annealed textures of zircaloy-4 thin strips. In *Proc. 7th Int. Symp. on 'Zirconium in the Nuclear Industry'*, ASTM-STP 939, R.B. Adamson & L.F.P. Van Swam, eds. (Philadelphia: ASTM) 663-672. [11]

CHASTEL, Y.B. (1993). *Modeling Anisotropic Deformations in Polycrystalline Materials*. Ph.D. Thesis (Ithaca, NY: Cornell University). [5]

CHASTEL, Y.B. & DAWSON, P.R. (1991), Comparison of polycrystalline rolling simulations with experiments. In *Modeling the Deformation of Crystalline Solids*, Lowe &al., eds. (Warrendale, PA: TMS) 225-238. [13]

CHASTEL, Y.B. & DAWSON, P.R. (1994). An equilibrium-based model for anisotropic deformations of polycrystalline materials. In *Textures of Materials ICOTOM-10*, H.J. Bunge, ed. (Switzerland: Trans. Tech. Pubs.) 1747-1752. [8,12,13]

CHASTEL, Y.B., DAWSON, P.R., WENK, H.-R. & BENNETT, K. (1993), Anisotropic convection with implications for the upper mantle. *J. Geophys. Res.* **B98**, 17757-17771. *[13,14]*

CHATEIGNER, D., GERMI, P. & PERNET, M. (1992), Texture analysis by the Schulz reflection method: Defocalization corrections for thin films. *J. Appl. Crystallogr.* **25**, 766-769. *[4]*

CHATEIGNER, D.J., WENK, H.-R. & PERNET, M. (1997a), Orientation analysis of bulk YBCO from incomplete neutron diffraction data. *J. Appl. Crystallogr.* **30**, 43-48. *[4,6]*

CHATEIGNER, D., WENK, H.-R., BARBER, D.J. & PATEL, A. (1997b), Analysis of preferred orientations in PST and PZT thin films on various substrates. *Ferroelectrics* (in press). *[6]*

CHEN, C., RYDER, D.F. & SPURGEON, W.A. (1989), Synthesis and microstructure of highly oriented lead titanate thin films prepared by a sol-gel method. *J. Am. Ceram. Soc.* **72**, 1495-1498. *[6]*

CHEN, I.W. & XUE, L.A. (1990), Development of superplastic structural ceramics. *J. Am. Ceram. Soc.* **73**, 2585-2609. *[6]*

CHEN, I.W., WU, X., KEATING, S.J., KEATING, C.Y., JOHNSON, P.A. & TIEN, T.Y. (1987), Texture development in $YBa_2Cu_3O_x$ by hot extrusion and hot-pressing. *J. Am. Ceram. Soc.* **70**, C388-C390. *[6]*

CHEN, N., BIONDO, A.C., DORRIS, S.E., GORETTA, K.C., LANAGAN, M.T., YOUNGDAHL, C.A. & POEPPEL, R.B. (1993), Sinter-forged $(Bi,Pb)_2Sr_2Ca_3Cu_2O_x$ superconductors. *Supercond. Sci. Technol.* **6**, 674-677. *[6]*

CHEN, R.T. & OERTEL, G. (1991), Determination of March strain from phyllosilicate preferred orientation: A semi-numerical method. *Tectonophysics* **200**, 173-185. *[4,6,14]*

CHEN, S.R. & KOCKS, U.F. (1991a), High-temperature plasticity in copper polycrystals. In *High Temperature Constitutive Modeling*, A.D. Freed & K.P. Walker, eds. (New York: The Amer. Soc. Mechanical Engineers) 1-12. *[10]*

CHEN, S.R. & KOCKS, U.F. (1991b), Texture and microstructure development in Al-2%Mg during high-temperature deformation. In *Hot Deformation of Aluminum Alloys*. T.G. Langdon &al. eds. (Warrendale, PA: The Minerals, Metals & Materials Society) 89-104. *[9,10]*

CHIN, G.Y. & MAMMEL, W.L. (1969), Generalization and equivalence of the minimum work (Taylor) and maximum work (Bishop-Hill) principles for crystal plasticity. *Trans. Metall. Soc. AIME* **245**, 1211-1214. *[8]*

CHIN, G.Y., THURSTON, R.N. & NESBITT, E.A. (1966), Finite plastic deformation due to crystallographic slip. *Trans. Metall. Soc. AIME* **236**, 69-75. *[8]*

CHIN, G.Y., HOSFORD, W.F. & MENDORF, D.R. (1969), Accommodation of constrained deformation in f.c.c. metals by slip and twinning. *Proc. R. Soc. Lond.* **A309**, 433-456. *[8]*

CHOI, C.S., PRASK, H.J. & OSTERTAG, C.P. (1989), Texture study of magnetically aligned $YBa_2Cu_3O_7$-type materials by neutron diffraction. *J. Appl. Crystallogr.* **22**, 465-469. *[6]*

CHOI, C.S., BAKER, E.F. & OROSZ, J. (1993), Application of ODF to the Rietveld profile refinement of polycrystalline solid. *Adv. in X-ray Analysis* **37**, 49-57. (Plenum Press, N.Y.) *[4]*

CHRISTIAN, J.W. (1975). *The Theory of Transformations in Metals and Alloys*, 2nd Edition (Pergamon Press). *[2]*

CHRISTIAN, J.W. & MAHAJAN, S. (1995), Deformation twinning. *Progress in Materials Science* **39**, 1-157. *[8]*

CHRISTODOULOU, N., CAUSEY, A.R., WOO, C.H., TOMÉ, C.N., KLASSEN, R.J. & HOLT, R.A. (1993), Modelling the effect of texture and dislocation structure on irradiation creep of zirconium alloys. *In Effects of Radiation on Materials: 16th Int. Symp. ASTM STP 1175.* A.S. Kumar &al., eds. (Philadelphia: Amer. Soc. for Testing and Materials). *[7]*

CLARK, J.B., GARRETT Jr., R.K., JUNGLING, T.L., VANDERMEER, R.A. & VOLD, C.L. (1991), Effect of processing variables on texture and texture gradients in tantalum. *Metall. Trans.* **22A**, 2039-2048. *[5]*

CMSSL for CM-Fortran: CM-5 Version, Version 3.1. (Cambridge, MA: Thinking Machines Corporation, 1994). *[13]*

COTTRELL, A.H. (1953). Dislocations and Plastic Flow in Crystals (Oxford University Press). *[8]*

CRAMPIN, S. (1981), A review of wave motion in anisotropic and cracked elastic media. *Wave Motion* **3**, 242-391. *[14]*

CRAMPIN, S. (1987), Geological and industrial applications of extensive-dilatancy anisotropy. *Nature* **328**, 491-496. *[14]*

CROCKER, A.G., HECKSCHER, F., BEVIS, M. & GUYONCOURT, D.M.M. (1966), A single surface analysis of deformation twins in crystalline mercury. *Philos. Mag.* **13**, 1191-1205. *[2]*

CUITIÑO, A.M. & ORTIZ, M. (1992a), A material independent method for extending stress update algorithms from small strain plasticity to finite plasticity kinematics. *Eng. Comp.* **9**, 437-451. *[13]*

CUITIÑO, A.M. & ORTIZ, M. (1992b), Computational modeling of single crystals. *Modelling and Simulation in Mater. Sci. Eng.* **1**, 255-263. *[13]*

CULLITY, B.D. (1978), *Elements of X-ray Diffraction (2nd edition)* (Addison-Wesley, Reading), 555 pp. *[4]*

CURIE, P. (1884), Sur la symétrie. *Bull. Soc. Minéral. France* **7**, 418-457. *[1]*

DAHMS, M. & BUNGE, H.-J. (1979), The use of the positivity condition in pole figure inversion. In *Eighth Int. Conf. on Textures of Materials*, J.S. Kallend & G. Gottstein, eds. (Warrendale, PA: The Metallurgical Society) 79-85. *[3]*

DAHMS, M., WU, Y. & BUNGE, H.J. (1988). Texture development in tetragonal materials. In *Eighth Int. Conf. on Textures of Materials*, J.S. Kallend & G. Gottstein, eds. (Warrendale, PA: The Metallurgical Society) 685-690. *[5]*

DAMJANOVIC, A. (1965), On the mechanism of metal electrocrystallization. *Plating* **52**, 1017-1026. *[5]*

DANDEKAR, D.P. (1968), Pressure dependence of the elastic constants of calcite. *Phys. Rev.* **172**, 873-877. *[7]*

DANIEL, D. & JONAS, J.J. (1990), Measurement and prediction of plastic anisotropy in deep-drawing steels. *Metall. Trans.* **21A**, 331-343. *[10]*

DAVIES, G.J. & KALLEND, J.S. (1971), The texture transition in f.c.c. metals. In *Quantitative Analysis of Textures* (Cracow: Academy of Mining and Metallurgy), 345-353. *[9]*

DAVIES, G.J., KALLEND, J.S. & RUBERG, T. (1975), Quantitative textural measurement on cube-texture copper. *Met. Sci.* **9**, 421-424. *[10]*

DAWSON, P.R. (1984), A model for the hot or warm forming of metals with special use of deformation mechanism maps. *Int. J. Mech. Sci.* **26**, 227-244. *[13]*

DAWSON, P.R., BEAUDOIN, A.J. & MATHUR, K.K. (1992a), Simulating deformation-induced texture in metal forming. *Numerical Methods of Industrial Forming Processes*, NUMIFORM '92, (Rotterdam: Balkema), 25-33. *[13]*

DAWSON, P.R., BEAUDOIN, A.J., MATHUR, K.K., KOCKS, U.F. & KORZEKWA, D.A. (1992b), Crystallographic texture effects in hydroforming of aluminum sheet. In *Numerical Methods for Simulation of Industrial Metal Forming Processes*, M.J. Saran &al. eds. (New York: ASME) AMD-**156**, 1-10. *[13]*

DAWSON, P.R., CHASTEL, Y.B., SARMA, G.B. & BEAUDOIN, A.J. (1993), Simulating texture evolution in Fe-3%Si with polycrystal plasticity theory. In *Proc. Int. Conf. Modeling Metal Rolling Processes*. (London: The Institute of Materials) 264-273. *[13]*

DAWSON, P.R., BEAUDOIN, A.J. & MATHUR, K.K. (1994), Finite element modeling simulations of polycrystals. *Computational Material Modeling*, A.K. Noor & A. Needleman, eds. (Amer. Soc. Mech. Eng. AD) **42**, 37-52. *[13]*

DE BRESSER, J.H.P. & SPIERS, C.J. (1997), Strength characteristics of the r, f and c slip systems in calcite. *Tectonophysics* **272**, 1-23. *[14]*

DE RANGO, P., LEES, M., LEJAY, P., SULPICE, A., TOURNIER, R., INGOLD, M., GERMI, P. & PERNET, M. (1991), Texturing of magnetic materials at high temperature by solidification in a magnetic field. *Nature* **349**, 770-772. *[6]*

DECKER, B.F., ASP, E.T. & HARKER, D. (1948), Preferred orientation determination using a Geiger counter x-ray diffraction goniometer. *J. Appl. Phys.* **19**, 388-392. *[4]*

DELL'ANGELO, L.N. & TULLIS, J. (1989), Fabric development in experimentally sheared quartzites. *Tectonophysics* **169**, 1-21. *[4,6]*

DEN BROK, S.W.J. & SPIERS, C.J. (1991), Experimental evidence for water weakening of quartzite by microcracking plus solution-precipitation creep. *J. Geol. Soc. Lond.* **14**, 541-548. *[6]*

DETWILER, R.J., MONTEIRO, P.M., WENK, H.-R. & ZHONG, Z. (1988), Texture of calcium hydroxide near the cement paste-aggregate interface. *Cement and Concrete Res.* **18**, 823-829. *[6]*

DEZILLIE, L., VAN HOUTTE, P. & AERNOUDT, E. (1988), Simulation of the rolling texture of a 3004 aluminium alloy taking account of the initial texture. In *Eighth Int. Conf. on Textures of Materials*, J.S. Kallend & G. Gottstein, eds. (Warrendale, PA: The Metallurgical Society) 357-368. *[8,9]*

D'HALLOY, O.J.J. (1833), *Introduction à la Géologie* (Paris: Levrault). *[14]*

DIENES, J. (1987), Theory of Deformation. *Los Alamos National Laboratory Report* LA-11063-MS, Vol. I. *[1]*

DIETRICH, D. & SONG, H. (1984), Calcite fabrics in a natural shear environment; the Helvetic nappes of western Switzerland. *J. Struct. Geol.* **6**, 19-32. *[14]*

DIETZ, P. & GIELESSEN, H. (1995), Texture development and anisotropic photoelastic properties in rolled silver chloride. *Textures Microstruct.* **24**, 105-119. *[6]*

DILLAMORE, I.L. & KATOH, H. (1971), Polycrystalline plasticity and texture development in cubic metals. In *Quantitative Analysis of Textures* (Cracow: Academy of Mining and Metallurgy) 315-344. *[9]*

DILLAMORE, I.L. & KATOH, H. (1974), The mechanisms of recrystallization in cubic metals with particular reference to their orientation dependence. *Metal Sci.* **8**: 73-83. *[5]*

DILLAMORE, I.L. & ROBERTS, W.T. (1965), Preferred orientation in wrought and annealed metals. *Metall. Rev.* **10**, 271-380. *[5,9]*

DILLAMORE, I.L., MORRIS, P. L., SMITH, C.J.E. & HUTCHINSON, W. B. (1972), Transition bands and recrystallization in metals. *Proc. R. Soc. Lond.* **A329**, 405-420. *[5]*

DINGLEY, D.J. (1981), A comparison of diffraction techniques in the SEM. *Scanning Electron Microscopy* **4**, 273-286. *[4]*

DINGLEY, D.J. & RANDLE, V. (1992), Review. Microtexture determination by electron-backscatter diffraction. *J. Mater. Sci.* **27**, 4545-4566. *[4]*

DIOT, C. & ARNAULT, V. (1991), Orientation anisotropy in SiC matrix of unidirectional SiC/SiC composite. *Textures Microstruct.* **14-18**, 389-395. *[6]*

DOHERTY, R.D. (1978), Nucleation. In *Recrystallization of Metallic Materials*, F. Haeßner, ed. (Stuttgart: Dr. Riederer Verlag) 23-62. *[5]*

DOHERTY, R.D., GOTTSTEIN, G., HIRSCH, J., HUTCHINSON, W.B., LÜCKE, K., NES, E. & WILBRANDT, P.J. (1988). Report of panel on recrystallization textures: mechanisms and experiments. In *Eighth Int. Conf. on Textures of Materials*, J.S. Kallend & G. Gottstein, eds. (Warrendale, PA: The Metallurgical Society) 563-572. *[5]*

DOHERTY, R.D., HUGHES, D., HUMPHREYS, F.J., JONAS, J.J., JUUL JENSEN, D., KASSNER, M.E., McNELLEY, T.R. & ROLLETT, A.D. (1997), Current issues in recrystallization. *Mater. Sci. Eng.* (in press). *[5]*

DOLLAR, M., DYMEK, S., HWANG, S.J. & NASH, P. (1993), The role of microstructure on strength and ductility of hot-extruded mechanically alloyed NiAl. *Metall. Trans.* **24A**, 1993-1999. *[5]*

DOLLASE, W.A. (1986), Correction of intensities for preferred orientation in powder diffractometry: Application of the March model. *J. Appl. Crystallogr.* **19**, 267-272. *[4]*

DORNBUSCH, H.J., SKROTZKI, W. & WEBER, K. (1994), Development of microstructure and texture in high-temperature mylonites from the Ivrea Zone. In *Textures of Geological Materials*. H.J. Bunge &al., eds. (Oberursel, Germany: Deutsch. Gesell. Metallkunde) 187-202. *[6]*

DROSD, R. & WASHBURN, J.(1982), Some observations on the amorphous to crystalline transformation in silicon. *J. Appl. Phys.* **53**, 397-403. *[6]*

DUFFY, J., CAMPBELL, J.D. & HAWLEY, R.H. (1971), On the use of a torsional split Hopkinson bar to study strain rate effects in 1100-O aluminum. *J. Appl. Mech.* **38**, *Trans. ASME*, **93**, Series E, 83-91. *[10]*

DUGGAN, B.J., LÜCKE, K., KÖHLHOFF, G. & LEE, C.S. (1993), On the origin of cube texture in copper. *Acta Metall. Mater.* **41**, 1921-1927. *[5]*

DUNCAN, J.L. & JOHNSON, W. (1968), The ultimate strength of rectangular anisotropic diaphragms. *Int. J. Mech. Sci.* **10**, 143-155. *[10]*

DUNST, D. & MECKING, H. (1996), Analysis of experimental and theoretical rolling textures of 2-phase titanium alloys. *Z. Metallk.* **87**, 498-507. *[M,11]*

DUNST, D., DENDIEVEL, R. & MECKING, H. (1994), The development of deformation textures as a function of the phase fractions in two-phase titanium-based alloys. In *Textures of Materials ICOTOM-10*, H.J. Bunge, ed. (Switzerland: Trans. Tech. Pubs.), 665-672. *[11]*

DURHAM, W.B., KIRBY, S.H., HEARD, H.C., STERN, L.A. & BORO, C.O. (1988), Water ice phases II, II, V: plastic deformation and phase relationships. *J. Geophys. Res.* **93**, 10191-10208. *[6]*

DUVAL, P. (1979), Creep and recrystallization of polycrystalline ice. *Bull. Minéral.* **102**, 80-85. *[6]*

DUVAL, P. (1981), Creep and fabrics of polycrystalline ice under shear and compression. *J. Glaciol.* **27**, 129-140. *[14]*

DUVAL, P., ASHBY, M.F. & ANDERMAN, I. (1983), Rate-controlling processes in the creep of polycrystalline ice. *J. Phys. Chem.* **87**, 4066-4074. *[6]*

EICHELKRAUT, H., ABBRUZZESE, G. & LÜCKE, K. (1988), A theory of texture controlled grain growth- II. numerical and analytical treatment of grain growth in the presence of two texture components. *Acta Metall.* **36**, 55. *[5]*

ENGLER, O. (1996), Nucleation and growth during recrystallisation of aluminium alloys investigated by local texture analysis. *Mater. Sci. Technol.* **12**, 859-872. *[5,9]*

ENGLER, O., HIRSCH, J. & LÜCKE, K. (1989), Texture development in Al-1.8%Cu depending on the precipitation state – Part I: Rolling Textures. *Acta Metall.* **37**, 2743-2753. *[5]*

ENGLER, O., WAGNER, P., SAVOIE, J., PONGE, D. & GOTTSTEIN, G. (1993), Strain rate sensitivity of flow stress and its effect on hot rolling texture development. *Scripta Metall. Mater.* **28**, 1317-1322. *[9]*

ENGLER, O., GOTTSTEIN, G., POSPIECH, J. & JURA, J. (1994a), Statistics, evaluation, and representation of single grain orientation measurements. In *Textures of Materials ICOTOM-10*, H.J. Bunge, ed. (Switzerland: Trans. Tech. Pubs.) 259-274. *[2]*

ENGLER, O., HIRSCH, J., KARHAUSEN, K. & GOTTSTEIN, G. (1994b), Influence of rolling temperature on the texture gradient in an Al-Mg-Si alloy. In *Textures of Materials ICOTOM-10*, H.J. Bunge, ed. (Switzerland: Trans. Tech. Pubs.) 673-678. *[5]*

ENGLER, O., PITHAN, C. & LÜCKE, K. (1994c), Rolling texture development in Cu-Mn alloys. In *Textures of Materials ICOTOM-10*, H.J. Bunge, ed. (Switzerland: Trans. Tech. Pubs.) 679-683 .*[5,9]*

ENGLER, O., YANG, P., GOTTSTEIN, G., JURA, J.& POSPIECH, J. (1994d), Behaviour of statistical texture parameters applied to single grain orientation measurements in recrystallized Al-Mn. In *Textures of Materials ICOTOM-10*, H.J. Bunge, ed. (Switzerland: Trans. Tech. Pubs.) 933-938. *[4]*

ENGLER, O., ESCHER, C. & GOTTSTEIN, G. (1996a), Single grain orientation measurements applied to the formation and growth of recrystallization nuclei. *Textures Microstruct.* **26-27**, 337-359. *[4]*

ENGLER, O., SACHOT, E., EHRSTRÖM, J.C., REEVES, A. & SHAHANI, R. (1996b), Recrystallisation and texture in hot deformed aluminium alloy 7010 thick plates. *Mater. Sci. Technol.* **12**, 717-729. *[4,5]*

ENGLISH, A.T. & CHIN, G.Y. (1965), On the variation of wire texture with stacking fault energy in f.c.c. metals and alloys. *Acta Metall.* **13**, 1013-1016. *[9]*

ERSKINE, B.G., HEIDELBACH, F. & WENK, H.-R. (1993), Lattice preferred orientations and microstructures of deformed Cordilleran marbles: correlation of shear indicators and determination of strain path. *J. Struct. Geol.* **15**, 1189-1205. *[M,14]*

ESHELBY, J. (1957), The determination of the elastic field of an ellipsoidal inclusion, and related problems. *Proc. R. Soc. Lond.* **A241**, 376-396. *[8,A]*

EULER, L. (1775), Formulae generales. *Nov. Comm. Acad. Sci. Imp. Petrop.* **20**, 189-207. *[2]*

FAN, L.S. & MULLER, R.S. (1988), As-deposited low-strain LPCVD polysilicon. *Proc. 1988 IEEE Solid State Sensors & Actuators Workshop*, Hilton Head, S.C., 55-58. *[6]*

FELDMANN, K. (1989), Texture investigation by neutron time-of-flight diffraction. *Textures Microstruct.* **10**, 309-323. *[4]*

FELDMANN, K, BETZL, M., KLEINSTEUBER, W. & WALTHER, K. (1991), Neutron time-of-flight texture analysis. *Textures Microstruct.* **14-18**, 59-64. *[4]*

FENG, Y.C., LAUGHLIN, D.E. & LAMBETH, D.N. (1994), Formation of crystallographic texture in rf sputter-deposited Cr thin films. *J. Appl. Phys.* **76**, 7311-7316. *[5]*

FERRARI, M. & LUTTEROTTI, L. (1994), Method of simultaneous determination of anisotropic residual stresses and texture by x-ray diffraction. *J. Appl. Phys.* **76**, 7246-7255. *[4]*

FIELD, D.P. (1995), On the symmetric domain of cubic misorientations. *Scripta Metall. Mater.* **32**, 67-70. *[2]*

FIELDS, D.S. & BACKOFEN, W.A. (1957), Determination of strain hardening characteristics by torsion testing. *Proc. Amer. Soc. for Testing and Materials* **57**, 1259-1272. *[10]*

FISHER, E.S. (1966), Temperature dependence of the elastic moduli in alpha-uranium single crystals, part IV (298K-923K). *J. Nucl. Mater.* **18**, 39-54. *[7]*

FLEISCHER, R.L. (1987), The number of active slip systems in polycrystalline brass: implications for ductility in other structures. *Acta Metall.* **35**, 2129-2136. *[8]*

FOLTYN, S.R., ARENDT, P.N., WU, X.D., BLUMENTHAL, W.R., COTTON, J.D., COULTER, J.Y., HULTS, W.L., SAFAR, H.F. SMITH, J.L. & PETERSON, D.E. (1995), Pulsed laser deposition of thick $YBa_2Cu_3O_{7-8}$ films on flexible substrates. In *Proc. Int. Workshop on Superconductivity, Maui, Hawaii* (Pittsburgh, PA: Mater. Res. Soc.) 105-108. *[M]*

FORTUNIER, R. & DRIVER, J.H. (1987), Grain reorientations in rolled aluminum sheet: comparison with predictions of continuous constraints model. *Acta Metall.* **35**, 1355-1366. *[13]*

FORTUNIER, R. & HIRSCH, J. (1987), Computer simulation of the rolling texture development. In *Theoretical Methods of Texture Analysis*, H.-J. Bunge, ed. (Oberursel: Deutsche Gesellschaft f. Metallkunde) 231-239. *[2]*

FRANCIOSI, P. & STOUT, M.G. (1988), Hardening anisotropy evolution of FCC polycrystals in channel die testing. In *Strength of Metals and Alloys (ICSMA 8)*, P.O. Kettunen, T.K. Lepistö & M.E. Lehtonen, eds. (Oxford: Pergamon) 325-330. *[10]*

FRANCIOSI, P., STOUT, M.G., O'ROURKE, J., ERSKINE, B. & KOCKS, U.F. (1987), Channel die tests on Al and Cu polycrystals: study of the prestrain history effects on further large strain texture. *Acta Metall.* **35**, 2115-2128. *[10]*

FRANK, F.C. (1953), A note on twinning in alpha-uranium. *Acta Metall.* **1**, 71-74. *[2]*

FRANK, F.C. (1965), On Miller-Bravais indices and four-dimensional vectors. *Acta Crystallogr.* **18**, 862-866. *[2]*

FRANK, F.C. (1987), Orientation mapping. *Metall. Trans.* **A19**, 403-408. *[2]*

FRANSSEN, R.C.M.W. & SPIERS, C.J. (1990), Deformation of polycrystalline salt in compression and in shear at 250-350°C. In *Deformation Mechanisms. Rheology and Tectonics*, R.J. Knipe & E.H. Rutter, eds. (London: Geological Soc., Special Publ.) **45**, 201-213. *[6]*

FREDA, A. & CULLITY, B.D. (1959), Quantitative deformation textures of aluminum, copper, silver, and iron wires. *Trans. AIME* **215**, 530-535. *[9]*

FREDERICK, S.F. & LENNING, G.A. (1975), The influence of prior texture on the cold rolled texture Ti-6Al-4V. *Metall. Trans.* **6A**, 1467-1468. *[5]*

GALE, B. & GRIFFITHS, D. (1960), Influence of instrumental aberrations on the Schultz technique for the measurement of pole figures. *Brit. J. Appl. Phys.* **11**, 96-102. *[4]*

GANDIN, C.-A., RAPPAZ, M., WEST, D. & ADAMS, B.L., (1995), Grain texture evolution during the columnar growth of dendritic alloys. *Metall. Mater. Trans.* **26A**, 1543-1551. *[5]*

GANGLI, P., ROOT, J. & FONG, R. (1995), Investigation of textures and interfaces in a Zr-2.5Nb alloy with zirconium hydrides. *Canadian Metall. Quarterly* **34**, 211-218. *[11]*

GERLING, R., BARTELS, A., CLEMENS, H., OEHRING, M. & SCHIMANSKY, F.-P. (1997), Properties of two-phase intermetallic $(Ti,Nb),(Al,Si)+(Ti,Nb)_5(Si.Al)_3$ p/m bulk and sheet material. *Acta Mater.* **45**, 4057-4066. *[M]*

GHOMSHEI, M.M. & TEMPLETON, T.L. (1989), Piezoelectric and a-axis fabric along a quartz vein. *Phys. Earth Planet. Int.* **55**, 374-386. *[6]*

GHOMSHEI, M.M., NAROD, B.B., TEMPLETON, T.L., ARROTT, A.S. & RUSSELL, R.D. (1988), Piezoelectric pole figure of a vein quartz sample. *Textures Microstruct.* **7**, 303-316. *[3]*

GHOSH, A.K. & BACKOFEN, W.A. (1973), Strain hardening and instability in biaxially stretched sheets. *Metall. Trans.* **A4**, 1113-1123. *[10]*

GHOSH, S.K. & RAMBERG, H. (1976), Reorientation of inclusions by combination of pure shear and simple shear. *Tectonophysics* **34**, 1-70. *[14]*

GIL SEVILLANO, J. (1993), Flow stress and work hardening. In *Materials Science and Technology* (Weinheim, Germany: VCH) **6**, 19-88. *[8]*

GIL SEVILLANO, J., VAN HOUTTE, P. & AERNOUDT, E. (1975), Deutung der Schertexturen mit Hilfe der Taylor-Analyse. *Z. Metallk.* **66**, 367-373. *[9,10]*

GIL SEVILLANO, J., VAN HOUTTE, P. & AERNOUDT, E. (1977), The contribution of macroscopic shear bands to the rolling texture of fcc metals. *Scripta Metall.* **11**, 581-585. *[2,9]*

GILORMINI, P., BACROIX, B. & JONAS, J.J. (1988a), A new approach to the prediction of deformation textures in bcc polycrystals on the basis of ⟨111⟩ pencil glide. In *Eighth Int. Conf. on Textures of Materials*, J.S. Kallend & G. Gottstein, eds. (Warrendale, PA: The Metallurgical Society) 381-386. *[9]*

GILORMINI, P., BACROIX, B. & JONAS, J.J. (1988b), Theoretical analyses of ⟨111⟩ pencil glide in b.c.c. crystals. *Acta Metall.* **36**, 231-256. *[9]*

GLASSER, F.P. & SAGOE-CRENTSIL, K.K. (1989), Steel in concrete: Part II – Electron microscopy analysis. *Mag. Concrete Res.* **41**, 213-220. *[6]*

GLEASON, G.C., TULLIS, J. & HEIDELBACH, F. (1993), The role of dynamic recrystallization in the development of lattice preferred orientations in experimentally deformed quartz aggregates. *J. Struct. Geol.* **15**, 1145-1168. *[6]*

GLOVER, G. & SELLARS, C.M. (1973), Recovery and recrystallization during high temperature deformation of α-iron. *Metall. Trans.* **4**, 765-775. *[14]*

GOEMAN, U. & SCHUMANN, H. (1977), Zur Deutung des petrographischen Gefüges einer natürlichen Halitprobe. *Z. Kali- und Steinsalz* **7**, 171-173. *[6]*

GOODMAN, D.J., FROST, H.J. & ASHBY, M.F. (1981), The plasticity of polycrystalline ice. *Philos. Mag.* **43**, 665-695. *[6]*

GOTOH, M. (1978), A finite element formulation for large elastic-plastic deformation analysis of polycrystals and some numerical considerations on polycrystalline plasticity. *Int. J. Numer. Meth. Eng.* **12**, 101-114. *[13]*

GOTTSTEIN, G. (1988), Automatic microtexture determination with synchrotron radiation. In *Eighth Int. Conf. on Textures of Materials*, J.S. Kallend & G. Gottstein, eds. (Warrendale, PA: The Metallurgical Society) 195-202. *[4]*

GRASSO, G., PERIN, A., HENSEL, B. & FLUKIGER, R. (1993), Pressed and cold-rolled Ag-sheathed Bi (2223) tapes. A comparison. *Physica* C **217**, 335-341. *[6]*

GREEN, H.W., GRIGGS, D.T. & CHRISTIE, J.M. (1970), Syntectonic and annealing recrystallization of fine-grained quartz aggregates. In *Experimental & Natural Rock Deformation*, P. Paulitsch, ed. (Berlin: Springer Verlag) 272-335. *[6]*

GREWEN, J. (1973). Textures of hexagonal metals and alloys and their influence on industrial application. In *3ème Colloque Européen sur les Textures de Déformation et de Recristallisation des Métaux et leur Application Industrielle*. Proc. of the Conference at Pont-à-Mousson 1973, R. Penelle, ed. (Société Française de Métalurgie). *[5]*

GREWEN, J. & WASSERMANN, G. (1955), Über die idealen Orientierungen einer Walz-textur. *Acta Metall.* **3**, 354-360. *[2]*

GRIGGS, D.T. (1967), Hydrolytic weakening of quartz and other silicates. *Geophys. J. R. Astron. Soc.* **14**, 19-31. *[6]*

GRIGGS, D.T. & BLACIC, J.D. (1965), Quartz: anomalous weakness of synthetic crystals. *Science* **147**, 292-295. *[6]*

GROVES, G.W. & KELLY, A. (1963), Independent slip systems in crystals. *Philos. Mag.* **8**, 877-887. *[8]*

GROVES, G.W. & KELLY, A. (1969), Change of shape due to dislocation climb. *Philos. Mag.* **19**, 977-986. *[8]*

GUIDI, M., ADAMS, B.L. & ONAT, E.T.(1992), Tensorial representation of the orientation distribution function in cubic polycrystals. *Textures Microstruct.* **19**, 147-167. *[2]*

GUILLOPE, M. & POIRIER, J.-P.(1979), Dynamic recrystallization during creep of single-crystalline halite: An experimental study. *J. Geophys. Res.* **84**, 5557-5567. *[14]*

GUINIER, A. & GRAF, R. (1952), Dispositif supprimant l'asymétrie des diagrammes due à la hauteur du faisceau issu d'un monochromateur. *Acta Crystallogr.* **5**, 150. *[4]*

HAASEN, P. (1993), How are new orientations generated during primary recrystallization? *Metall. Trans.* **24A**, 1001-1015. *[14]*

HAESSNER, F. (1978), *Recrystallization of Metallic Materials*. (Stuttgart: Riederer). *[14]*

HAESSNER, F., POSPIECH, J. & SZTWIERTNIA, K. (1983), Spatial arrangement of orientations in rolled copper. *Mater. Sci. Eng.* **57**, 1-14. *[2,4]*

HAGER, B.H. & O'CONNELL, R. J. (1981), A simple global model of plate dynamics and mantle convection. *J. Geophys. Res.* **86**, 4843-4867. *[14]*

HALDAR, P., HOEHN, J.G., RICE, J.A. & MOTOWIDLO, L.R. (1992), Enhancement in critical current density of Bi-Pb-Sr-Ca-Cu-O tapes by thermomechanical processing: Cold rolling versus uniaxial pressing. *Appl. Phys. Lett.* **60**, 495-497. *[6]*

HANSEN, J. & MECKING, H. (1976), Influence of the geometry of deformation on the rolling texture of fcc metals. In *Fourth Int. Conf. on Texture and the Properties of Materials (ICOTOM-4)*, (Cambridge: The Metals Society) 34-37. *[5]*

HANSEN, J., POSPIECH, J. & LÜCKE, K. (1978), *Tables for Texture Analysis of Cubic Crystals* (Berlin: Springer). *[1,2]*

HANSEN, N. (1992), Deformation microstructures. *Scripta Metall. Mater.* **27**, 1447-1452. *[5]*

HARBEKE, G., KRAUSBAUER, L., STEIGMEIER, E.F., WIDMER, A.E., KAPPERT, H.F. & NEU-GEBAUER, G. (1983), LPCVD polycrystalline silicon: growth and physical properties of in-situ phosphorus doped und undoped films. *RCA Review* **44**, 287-312. *[6]*

HARREN, S.V. & ASARO, R.J. (1989), Nonuniform deformations in polycrystals and aspects of the validity of the Taylor model. *J. Mech. Phys. Solids* **37**, 191-232. *[13]*

HARREN, S.V., LOWE, T.C., ASARO, R.J. & NEEDLEMAN, A. (1989), Analysis of large-strain shear in rate-dependent face-centred cubic polycrystals: correlation of micro- and macromechanics. *Phil. Trans. R. Soc. Lond.* **A328**, 443-500. *[9,10]*

HART, E.W. (1976), Constitutive relations for the nonelastic deformation of metals. *J. Eng. Mater. Tech.* **98**, 193-202. *[8]*

HASHIN, Z. & SHTRIKMAN, S. (1962), On some variational principles in anisotropic and non-homogeneous elasticity. *J. Mech. Phys. Solids* **10**, 335-343. *[7]*

HATHERLY, M. & HUTCHINSON, W.B. (1979). *An Introduction to Textures in Metals*, Monograph No.5 (London: The Institution of Metallurgists), 76pp. *[5]*

HAVLICEK, F., TOKUDA, M., HINO, S. & KRATOCHVIL, J. (1992), Finite element analysis of micro-macro transition in polycrystalline plasticity. *Int. J. Plast.* **8**, 477-499. *[13]*

HAVNER, K. (1992). *Finite Deformation of Crystalline Solids* (New York: Cambridge University Press). *[8]*

HEARD, H.C. (1985), Chapter 23, Experimental determination of mechanical properties. In *Preferred Orientation in Deformed Metals and Rocks: An Introduction to Modern Texture Analysis*, H.-R. Wenk, ed. (Orlando, FL: Academic Press) 485-506. *[14]*

HEARD, H.C. & RALEIGH, C.B. (1972), Steady state flow in marble at 500–800°C. *Geol. Soc. Am. Bull.* **83**, 935-956. *[11]*

HECKER, S.S. (1972), Experimental investigation of corners in the yield surface. *Acta Mech.* **13**, 69-86. *[10]*

HECKER, S.S. (1976), Experimental studies of yield phenomena in biaxially loaded metals. In *Constitutive Equations in Viscoplasticity* (New York: Amer. Soc. Mech. Eng.) 1-33. *[10]*

HECKER, S.S. & STOUT, M.G. (1984), Strain hardening of heavily cold worked metals. In *Deformation, Processing and Structure*, George Krauss, ed. (Metals Park, OH: American Society for Metals) 1-46. *[10]*

HEDEGAARD, C. & WENK, H.-R. (1997), Microstructure and the texture patterns of mollusc shells. *J. Mollusc Studies* (in press). *[6]*

HEIDELBACH, F., WENK, H.-R., MUENCHAUSEN, R.E., FOLTYN, S., NOGAR, N. & ROLLETT, A.D. (1992), Textures of laser ablated thin films of $YBa_2Cu_3O_{7-d}$ as a function of deposition temperature. *J. Mater. Res.* **7**, 549-557. *[4,6]*

HEIDELBACH, F., WENK, H.-R., CHEN, S.R., POSPIECH, J. & WRIGHT, S. (1996), Orientation and misorientation characteristics of annealed, rolled and recrystallized copper. *Mater. Sci. Eng.* **A215**, 39-49. *[2,4]*

HEILMANN, P.T., CLARK, W.A.T. & RIGNEY, D.A. (1982), Computerized method to determine crystal orientations from Kikuchi patterns. *Ultramicroscopy* **9**, 365-372. *[4]*

HEINZ, A. & NEUMANN, P. (1991), Representation of orientation and disorientation data for cubic, hexagonal, and orthorhombic crystals. *Acta Crystallogr.* **A47**, 780-789. *[2]*

HEIZMANN, J.J. & LARUELLE, C. (1986), Simultaneous measurement of several x-ray pole figures. *J. Appl. Crystallogr.* **19**, 467-572. *[4]*

HEIZMANN, J.J., EL-ABDOUNI, H., VADON, A. & LARUELLE, C. (1988a), Evolution of the texture of iron oxides during their reduction. In *Eighth Int. Conf. on Textures of Materials*, J.S. Kallend & G. Gottstein, eds. (Warrendale, PA: The Metallurgical Soc.), 795-800. *[6]*

HEIZMANN, J.J., LARUELLE, C., VADON, A., ROUSSELET, C., CIOSMAK, D. & BERTRAND, G. (1988b), Texture of tantalum and niobium oxyde scales. In *Eighth Int. Conf. on Textures of Materials*, J.S. Kallend & G. Gottstein, eds. (Warrendale, PA: The Metallurgical Soc.) 783-787. *[6]*

HELLING, D.E., MILLER, A.K. & STOUT, M.G. (1986), An experimental investigation of the yield loci of 110-O aluminum, 70:30 brass and an overaged aluminum alloy after various prestrains. *J. Eng. Mater. Techn.* **108**, 313-320. *[8,10]*

HELMING, K. (1991), Minimal pole figure ranges for quantitative texture analysis. *Textures Microstruct.* **14-18**, 187-192. *[2,4]*

HELMING, K. (1992), Minimum pole figure ranges for quantitative texture analysis. *Textures Microstruct.* **19**, 45-54. *[3]*

HELMING, K., MATTHIES, S. & VINEL, G.W. (1988), ODF representation by means of σ-sections. In *Eighth Int. Conf. on Textures of Materials*, J.S. Kallend & G. Gottstein, eds. (Warrendale, PA: The Metallurgical Society) 55-60. *[2]*

HELMING, K., WENK, H.-R., CHOI, C.S. & SCHÄFER, W. (1994), Description of quartz textures by components. Examples from metamorphic rocks. In *Textures of Geological Materials*, H.J. Bunge &al.,eds. (Oberursel, Germany: Deutsch. Gesell. Metallkunde) 303-325. *[6]*

HELMING, K., GEIER, S., SCHRECK, M., HESSMER, R., STRITZKER, B. & RAUSCHENBACH, B. (1995), Texture analysis of vapor deposited diamond films on silicon by the component method. *J. Appl. Phys.* **77**, 4765-4770. *[6]*

HERSHEY, A.V. (1954), The plasticity of an isotropic aggregate of anisotropic f.c.c single crystals. *J. Appl. Mech.* **21**, 241-249. *[8]*

HESS, H.H. (1964), Seismic anisotropy of the uppermost mantle under oceans, *Nature*, **203**, 629. *[14]*

HEUER, A.H. (1966), Deformation twinning in corundum. *Philos. Mag.* **13**, 379-393. *[2]*

HEUER, A.H. (1970), Plastic deformation in polycrystalline alumina. *Proc. Brit. Ceram. Soc.* **10**, 173-184. *[6]*

HEUER, A.H., SELLERS, D.J. & RHODES, W.H. (1969), Hot-working of aluminum oxide: I, Primary recrystallization and texture. *J. Am. Ceram. Soc.* **52**, 468-474. *[6]*

HEUER, A.H., TIGHE, N.J. & CANNON, R.M. (1980), Plastic deformation of fine-grained alumina (Al_2O_3): II Basal slip and nonaccomodated grain-boundary sliding. *J. Am. Ceram. Soc.* **63**, 53-57. *[6]*

HILL, R. (1948), A theory of yielding and plastic flow of anisotropic metals. *Proc. R. Soc. Lond.* **193A**, 281-297. *[10]*

HILL, R. (1950), *The Mathematical Theory of Plasticity* (Oxford University Press). *[8,10]*

HILL, R. (1952), The elastic behavior of a crystalline aggregate. *Proc. Phys. Soc.* **A65**, 349-354. *[7]*

HILL, R. (1965), Continuum micromechanics of elasto-plastic polycrystals. *J. Mech. Phys. Solids* **13**, 89-101. *[7,8]*

HILL, R. (1967), The essential structure of constitutive laws for metal composites and polycrystals. *J. Mech. Phys. Solids* **15**, 79-95. *[7,8,9,11]*

HILL, R. (1978), Aspects of invariance in solid mechanics. *Adv. Appl. Mech.* **18**, 1-76. *[8]*

HILL, R. (1979), Theoretical plasticity of textured aggregates. *Math. Proc. Camb. Phil. Soc.* **75**, 179-191. *[10]*

HILL, R. & HUTCHINSON, J.W. (1992), Differential hardening in sheet metal under biaxial loading: a theoretical framework. *J. Appl. Mech.* **59**, S1-S9. *[10]*

HILL, R., HECKER, S.S. & STOUT, M.G. (1994), An investigation of plastic flow and differential work hardening on orthotropic brass tubes under fluid pressure and axial load. *Int. J. Solids Struct.* **31**, 2999-3021. *[10]*

HIRSCH, J. (1984), *Rolling textures in fcc metals.* Ph.D. Thesis, RWTH Aachen, West Germany. *[5]*

HIRSCH, J. (1990), Correlation of deformation texture and microstructure. *Mater. Sci. Technol.* **6**, 1048-1057. *[9]*

HIRSCH, J. & LÜCKE, K. (1988a), Mechanism of deformation and development of rolling textures in polycrystalline f.c.c. metals – I. Description of rolling texture development in Cu-Zn alloys. *Acta Metall.* **36**, 2863-2882. *[5,9]*

HIRSCH, J. & LÜCKE, K. (1988b), Mechanism of deformation and development of rolling textures in polycrystalline f.c.c. metals – II. Simulation and interpretation of experiments on the basis of Taylor-type theories. *Acta Metall.* **36**, 2883-2904. *[9]*

HIRSCH, J., LÜCKE, K. & MECKING, H. (1984). Comparison of experimental and theoretical rolling textures of fcc metals. In *Seventh Int. Conf. on Textures of Materials*, C.M. Brakman, P. Jongenburger & E.J. Mittemeijer, eds. (Netherlands Soc. for Materials Science) 83-88. *[5,9]*

HIRSCH, J., LÜCKE, K. & HATHERLY, M. (1988), Mechanism of deformation and development of rolling textures in polycrystalline f.c.c. metals – III. The influence of slip inhomogeneities and twinning. *Acta Metall.* **36**, 2905-2927. *[9]*

HIRSCH, P., HOWIE, A., NICHOLSON, R.B., PASHLEY, D.W. & WHELAN, M.J. (1977), *Electron Microscopy of Thin Crystals.* Krieger Publishing Co., Malabar, Florida, 2nd ed., 563 pp. *[4]*

HIRTH, G. & TULLIS, J. (1992), Dislocation creep regimes in quartz aggregates. *J. Struct. Geol.* **14**, 145-159. *[6]*

HJELEN, J., ØRSUND, R., HOEL, E., RUNDE, P., FURU, T. AND NES, E. (1993), EBSP, Progress in technique and applications. *Textures Microstruct.* **20**, 29-40. *[4]*

HOCKETT, J.E. (1959). The rolling pressures of uranium sheet and plate, LA-2233 (Los Alamos National Laboratory). *[5]*

HOFMANN, J. (1974), Die Quartzteilgefüge von Metamorphiten und Anatexiten, dargestellt am Beispiel des Osterzgebirges (DDR). *Freiberger Forsch.* **C297**, 107 pp. *[14]*

HØIER, R., BENTDAL, J., DAALAND, O. & NES, E. (1994), A high resolution transmission electron diffraction method for on-line texture analysis. In *Textures of Materials ICOTOM-10*, H.J. Bunge, ed. (Switzerland: Trans. Tech. Pubs.) 143-148. *[4]*

HONEYCOMBE, R.W.K. (1984), *The plastic deformation of metals*, 2nd ed., (London: Edward Arnold). *[13]*

HONNEFF, H. & MECKING, H. (1978), A method for the determination of the active slip systems and orientation changes during single crystal deformation. In *Textures of Materials*, G. Gottstein & K. Lücke, eds. (Berlin: Springer) 265-275. *[9,13]*

HONNEFF, H. & MECKING, H. (1981), Analysis of the deformation texture at different rolling conditions. In *Sixth Int. Conf. on Textures of Materials* (Tokyo: The Iron and Steel Institute of Japan) 347-352. *[9]*

HORNBOGEN, E. (1979), Combined reactions. *Metall. Trans.* **10A**, 947-972. *[5]*

HOSFORD, W.F. (1964), Microstructural changes during deformation of [110]-fiber textured metals. *Trans. Metall. Soc. AIME* **230**, 12-15. *[5,9,14]*

HOSFORD, W.F. (1965), Axially symmetric flow of aluminum single crystals. *Trans. Metall. Soc. AIME* **233**, 329-333. *[8]*

HOSFORD, W.F. (1979a), Incorporating work hardening into yield loci calculations. In *Strength of Metals and Alloys*, P. Haasen, V. Gerold & G. Kostorz, eds. (Oxford: Pergamon) 775-780. *[10]*

HOSFORD, W.F. (1979b), On yield loci of anisotropic cubic metals. *Proc. 7th North American Metal Working Research Conf.* (Dearborn MI: Soc. of Manufacturing Engineers) 191-197. *[10]*

HOSFORD, W.F (1993). *The Mechanics of Crystals and Textured Polycrystals* (Oxford University Press) 248 pp. *[8,10]*

HOSFORD, W.F. & BACKOFEN, W.A. (1964), Strength and plasticity of textured metals. In *Fundamentals of Deformation Processing* (Syracuse NY: Syracuse University Press) 259-298. *[10]*

HOSFORD, W.F. & CADDELL, R.M. (1983). *Metal Forming–Mechanics and Metallurgy*. (Englewood Cliffs NJ: Prentice-Hall). 320 pp. *[10]*

HOWE, R.T. (1988), Surface micromachining for microsensors and microactuators. *J. Vacuum Sci. Tech.* **B6**, 1809-1813. *[6]*

HU, H. &CLINE, R.S. (1961), Temperature dependence of rolling textures in high-purity silver. *J. Appl. Phys.* **32**, 760-763. *[5]*

HU, H. & GOODMAN, S.R. (1963), Texture transition in copper. *Trans. Metall. Soc. AIME* **227**, 627-639. *[5,9]*

HU, H., CLINE, R.S. & GOODMAN, S.R. (1961), Texture transition in high-purity silver and its correlation with stacking fault frequency. *J. Appl. Phys.* **32**, 1392-1399. *[5]*

HUANG, J., KRULEVITCH, P., JOHNSON, G.C., HOWE, R.T. & WENK, H.-R. (1990), Investigation of texture and stress in undoped polysilicon films. In *Polysilicon Thin Films and Interfaces*, T. Kamins, C.V. Thompson & B. Raicu, eds. (Mater. Res. Soc. Symp. Proc. **182**) 201-206. *[6]*

HUANG, S.C., LAFORCE, R.P., RITTER, A.M. & GOEHNER, R.P. (1985), Rapid solidification characterisics in melt spinning a Ni-base superalloy. *Metall. Trans.* **16A**, 1773-1779. *[5]*

HUGHES, D.A. (1986), *Strain Hardening of f.c.c. Metals and Alloys at Large Strains*. Ph.D. Thesis (Stanford, CA: Stanford University). *[10]*

HUGHES, D.A. & HANSEN, N. (1991), Microstructural evolution in nickel during rolling and torsion. *Mater. Sci. Technol.* **7**, 544-553. *[9]*

HUGHES, D.A. & KUMAR, A. (1996), The effect of deformation mode on local orientations and high angle boundaries. In *Textures of Materials ICOTOM-11*, Z. Liang, L. Zuo & Y. Chu, eds. (Beijing: International Academic Publishers) 1345-1350. *[8,9]*

HUGHES, D.A. & WENK, H.R. (1988), The effect of stacking fault energy on the texture of nickel-cobalt solid solutions at large strains. In *Eighth Int. Conf. on Textures of Materials*, J.S. Kallend & G. Gottstein, eds. (Warrendale, PA: The Metallurgical Society) 455-460. *[5,9]*

HUIJSER-GERITS, E.M.C. & RIECK, G.D. (1974), Defocusing effects in the reflexion technique for the determination of preferred orientation. *J. Appl. Crystallogr.* **7**, 286-290. *[4]*

HUMPHREYS, F.J. (1977), The nucleation of recrystallization at second phase particles in deformed aluminum. *Acta Metall.* **25**, 1323-1344. *[5]*

HUMPHREYS, F.J. (1983), The determination of crystallographic textures from selected areas of a specimen by electron diffraction. *Textures Microstruct.* **6**, 45-62. *[4]*

HUMPHREYS, F.J. (1988), Experimental techniques for microtexture determination. In *Eighth Int. Conf. on Textures of Materials*, J.S. Kallend & G. Gottstein, eds. (Warrendale, PA: The Metallurgical Society) 171-182. *[4]*

HUMPHREYS, F.J. & HATHERLY, M. (1995), *Recrystallization and Related Annealing Phenomena.* (Oxford University Press). *[5,14]*

HUMPHREYS, F.J. & KALU, P.N. (1990), The development of microstructure and texture during the deformation & annealing of particle-containing polycrystals. *Textures Microstruct.* **14-18**, 703-708. *[4]*

HUSSIEN, S.A., MAHMOOD, S.T. & MURTY, K.L. (1988). Texture and mechanical anisotropy gradients in recrystallized zircaloy TREX. In *Eighth Int. Conf. on Textures of Materials*, J.S. Kallend & G. Gottstein, eds. (Warrendale, PA: The Metallurgical Society) 843-848. *[5]*

HUTCHINGS, M.T. & KRAWITZ, A.D., eds. (1993). *Measurement of Residual and Applied Stress using Neutron Diffraction* (Netherlands: Kluwer Acad. Publ.). *[4]*

HUTCHINSON, J.W. (1964a), Plastic stress-strain relations of fcc polycrystalline metals hardening according to the Taylor rule. *J. Mech. Phys. Solids* **12**, 11-24. *[8]*

HUTCHINSON, J.W. (1964b), Plastic deformation of b.c.c. polycrystals. *J. Mech. Phys. Solids* **12**, 25-33. *[8]*

HUTCHINSON, J.W. (1970), Elastic-plastic behaviour of polycrystalline metals and composites. *Proc. R. Soc. Lond.* **A319**, 247-272. *[8]*

HUTCHINSON, J.W. (1976), Bounds and self-consistent estimates for creep of polycrystalline materials. *Proc. R. Soc. Lond.* **A348**, 101-127. *[7,11]*

HUTCHINSON, J.W. (1977), Creep and plasticity of hexagonal polycrystals as related to single crystal slip. *Metall. Trans.* **8A**, 1465-1469. *[12,13]*

HUTCHINSON, W.B. (1994), Practical aspects of texture control in low carbon steels. In *Textures of Materials ICOTOM-10*, H.J. Bunge, ed. (Switzerland: Trans. Tech. Pubs.) 1917-1928. *[5]*

HUTCHINSON, W.B., DUGGAN, B.J. & HATHERLY, M. (1979), Development of deformation texture and microstructure in cold- rolled Cu-30Zn. *Metals Tech.* **6**, 398-403. *[5]*

HUTCHINSON, W.B., RYDE, L., BATE, P.S. & BACROIX, B. (1996), On the description of misorientations and interpretation of recrystallization textures. *Scripta Mater.* **35**, 579-582. *[2]*

IBE, G. & LÜCKE, K. (1972), Description of orientation distributions of cubic crystals by means of 3D rotation coordinates. *Texture* **1**, 87-98. *[2]*

ILLINGWORTH, J. & KITTLER, J. (1988), A survey of the Hough transform. *Computer Vision, Graphics and Image Processing* **44**, 87-116. *[4]*

IMHOF, J. (1977), Determination of an approximation of the orientation distribution function using only one pole figure. *Z. Metallk.* **68**, 38- 43. *[3]*

INOKUTI, Y. & DOHERTY, R.D. (1978), Transmission Kossel study of the structure of compressed iron and its recrystallization behavior. *Acta Metall.* **26**, 61-80. *[5]*

International Critical Tables (New York: McGraw Hill, 1929). *[7]*

International Tables for Crystallography (1983): Vol. A, T. Hahn, ed. (Reidel) *[1,2]*

International Tables for X-ray Crystallography (1962), Vol. III. Physical and Chemical Tables (Birmingham, England: The Kynoch Press). *[4]*

IOSIPESCU, N. (1967), New accurate procedure for single shear testing of metals. *J. Mater.* **2**, 537-566. *[10]*

ITO, K., MUSICK, R. & LÜCKE, K. (1983), The influence of iron content and annealing temperature on the recrystallization textures of high-purity aluminum-iron alloys. *Acta Metall.* **31**, 2137-2149. *[5]*

IVANKINA, T.I., NIKITIN, A.N., VOITUS, W. & WALTHER, K. (1991), Texture analysis and investigation of piezoelectric properties of natural quartz. *Textures Microstruct.* **14-18**, 421-429. *[6]*

JACKSON, P.J. & BASINSKI, Z.S. (1967), Latent hardening and the flow stress in copper single crystals. *Can. J. Physics* **45**, 707-735. *[9,10]*

JASIENSKI, Z., POSPIECH, J., PIATOWSKI, A., KUSNIERZ, J., LITWORA, A., PAWLIK, K. & PAUL, H. (1994), Influence of shear banding on the texture in rolled and channel-die compressed polycrystalline copper. In *Textures of Materials ICOTOM-10*, H.J. Bunge, ed. (Switzerland: Trans. Tech. Pubs.) 1231-1236. *[5]*

JAYNES, E.T. (1957), Information theory and statistical mechanics. *Phys. Rev.* **106**, 620-630. *[3]*

JEFFERY, G.B. (1923), The motion of ellipsoidal particles immersed in a viscous fluid. *Proc. R. Soc. Lond.* **A102**, 161-179. *[6,14]*

JENSEN, D.J. & LEFFERS, T. (1989), Fast texture measurements using position sensitive detector. *Textures Microstruct.* **10**, 361-374 *[4]*

JESSEL, M.W. (1988), Simulation of fabric development in recrystallizing aggregates – I. Description of the model. *J. Struct. Geol.* **10**, 771-778. *[14]*

JESSEL, M.W. & LISTER, G.S. (1990), A simulation of the temperature dependence of quartz fabrics. In *Deformation Mechanisms, Rheology and Tectonics*, R.J. Knipe & E.H. Rutter, eds. (London: Geol. Soc. Spec. Publ.) **54**, 353-362. *[14]*

JETTER, L.K. & BORIE, B.S. (1953), A method for the quantitative determination of preferred orientation. *J. Appl. Phys.* **24**, 532-535. *[4]*

JETTER, L.K. & MCHARGUE, C.J. (1957), Preferred orientation in extruded uranium rod. *Trans. Metall. Soc. AIME* **209**, 291-292. *[5]*

JI, S. & MAINPRICE, D. (1988), Natural deformation fabrics of plagioclase: Implications for slip systems and seismic anisotropy. *Tectonophysics* **147**, 145-163. *[6]*

JOHNSON, G.C. & FERRARI, M. (1988), A demonstration of the significance of the odd-order coefficients in the harmonic method. In *Eighth Int. Conf. on Textures of Materials*, J.S. Kallend & G. Gottstein, eds. (Warrendale, PA: The Metallurgical Society) 115-121. *[2,3,6]*

JONAS, J.J., CANOVA, G.R., SHRIVASTAVA, S.C. & CHRISTODOULOU, N. (1982), Sources of the discrepancy between the flow curves determined in torsion and in axi-symmetric tension and compression testing. In *Plasticity of Metals at Finite Strain: Theory, Computation and Experiment*, E.H. Lee & R.L. Mallett, eds. (Palo Alto, CA: Quicksilver Printing) 206-222. *[10]*

JOO, Y.-C. & THOMPSON, C.V. (1994). Electromigration lifetimes of single-crystal aluminum lines with different crystallographic orientations. In *Materials Reliability in Microelectronics*, P. Borgesen &al., eds.(Pittsburgh PA: MRS) 319-324. *[5]*

JORDAN, P.G. (1987), The deformation behavior of bimineralic limestone-halite aggregates. *Tectonophysics* **135**, 185-197. *[6]*

JOUBERT, P., LOISEL, B., CHOUAN, Y. & HAJI, L. (1987), The effect of low pressure on the structure of LPVCO polycrystalline silicon films. *J. Electrochem. Soc.* **134**, 2541-2545. *[6]*

JOY, D.C., NEWBURY, D.E. & DAVIDSON, D.L. (1982), Electron channeling patterns in the scanning electron microscope. *J. Appl. Phys.* **55**, R81-122. *[4]*

JURA, J. (1991), Evaluation of the parameters of texture components on the basis of a discrete form of orientation distribution. *Textures Microstruct.* **14-18**, 193-197. *[2]*

JUUL JENSEN, D. & HANSEN, N. (1990), Flow stress anisotropy in aluminum. *Acta Metall. Mater.* **38**, 1369-1380. *[8,9]*

JUUL JENSEN, D. & KJEMS, J.K. (1983), Apparatus for dynamical texture measurements by neutron diffraction using a position sensitive detector. *Textures Microstruct.* **5**, 239-251. *[4]*

KALIDINDI, S.R. & ANAND, L. (1992), An approximate procedure for the evolution of crystallographic texture in bulk deformation processing of fcc metals. *Int. J. Mech. Sci.* **34**, 309-329. *[13]*

KALIDINDI, S.R., BRONKHORST, C.A. & ANAND, L. (1992), Crystallographic texture evolution in bulk deformation processing of fcc metals. *J. Mech. Phys. Solids* **40**, 537-569. *[10,13]*

KALLEND, J.S. & DAVIES, G.J. (1969), Crystallite distribution about {110}⟨1̄12⟩ in cold rolled α-brass. *J. Inst. Metals* **97**, 350-352. *[2]*

KALLEND, J.S. & DAVIES, G.J. (1971a), A simulation of texture development in f.c.c. metals. *Philos. Mag.* **18**, 471-490. *[9]*

KALLEND, J.S. & DAVIES, G.J. (1971b), The elastic and plastic anisotropy of cold-rolled sheets of copper, gilding metal, and α-brass. *J. Inst. Metals* **99**, 257-260. *[10]*

KALLEND, J.S., DAVIES, G.J. & MORRIS, P.P., (1976), Texture transformations: the misorientation distribution function. *Acta Metall.* **24**, 361-368. *[3]*

KALLEND, J.S., KOCKS, U.F., ROLLETT, A.D. & WENK, H.-R. (1991a), Operational texture analysis. *Mater. Sci. Eng.* **A132**, 1-11. [Obtain CORRECTED VERSION from U.F. Kocks] *[2,3,9,13]*

KALLEND, J.S., SCHWARZ, R.B. & ROLLETT, A.D. (1991b). Resolution of superimposed diffraction peaks in texture analysis of a $YBa_2Cu_3O_7$ polycrystal. *Textures Microstruct.* **13**, 189-197. *[3,6]*

KAMB, B. (1972), Experimental recrystallization of ice under stress. In *Flow and Fracture of Rocks* (H.C. Heard, I.Y. Borg, N.C. Carter & C.B. Raleigh, eds.). *Geophys. Monogr.* **16**, 211-241 (Washington, DC: Am. Geoph. U.). *[6]*

KAMB, W.B. (1959), Ice petrofabric observations from Blue Glacier, Washington, in relation to theory and experiment. *J. Geophys. Res.* **64**, 1891-1909. *[6]*

KAMINS, T. (1988), *Polycrystalline Silicon for Integrated Circuit Applications.* (Boston: Kluwer Academic Publishers). 290pp. *[6]*

KAMINS, T., MANOLIN, J. & TUCKER, R.N. (1972), Diffusion of impurities in poly-crystalline silicon. *J. Appl. Phys.* **43**, 83-91. *[6]*

KÄMPF, H., ERTEL, A., BANKWITZ, P., BETZL, M. & ZÄNKER, G. (1986), Texturanalyse an Evaporiten der Lagerstätte Zielitz. Nachweis einer Fasertextur in Halititen. *Z. angew. Geol.* **33**, 104-107. *[6]*

KANETAKE, N., TOZAWA, Y. & KATO, T. (1981), Yield loci of aluminum sheets and the use of texture data to predict them. In *Proc. Sixth Int. Conf. on Strength of Metals and Alloys*, S. Nagashima, ed. (Iron and Steel Institute of Japan) **2**, 1101-1110. *[10]*

KARATO, S.-I. (1987), Seismic anisotropy due to lattice preferred orientation of minerals: Kinematic or dynamic? In *High-Pressure Research in Mineral Physics*, M.H. Manghnani & Y. Syono, eds. (Orlando, FL: Academic Press) 455-471. *[6]*

KARATO, S.-I. (1988),The role of recrystallization in the preferred orientation of olivine. *Phys. Earth Planet. Inter.* **51**, 107-122. *[6]*

KARATO, S.-I., ZHANG, S. & WENK, H.-R. (1995), Superplasticity in Earth's lower mantle: evidence from seismic anisotropy and rock physics. *Science* **270**, 458-461. *[6,14]*

KEELER, S.P. & BACKOFEN, W.A. (1963), Plastic instability and fracture in sheets stretched over rigid punches. *Trans. ASM* **56**, 25-48. *[10]*

KERN, H. (1971), Dreiaxiale Verformungen an Solnhofen Kalkstein im Temperatur-bereich von 20°C-650°C. Röntgenographische Gefügeuntersuchungen mit dem Texturgoniometer. *Contrib. Mineral. Petrol.* **31**, 39-66. *[6,14]*

KERN, H. & BRAUN, G. (1973), Deformation und Gefügeregelung von Steinsalz im Temperaturbereich 20-200°C. *Contrib. Mineral. Petrol.* **40**, 169-181. *[6]*

KERN, H. & RICHTER, A. (1985), Microstructures and textures in evaporites. In *Preferred Orientation in Deformed Metals and Rocks: An Introduction to Modern Texture Analysis*, H.-R. Wenk, ed. (Orlando, FL: Academic Press) 317-333. *[6]*

KERN, H. & WENK, H.-R. (1983), Calcite texture development in experimentally induced ductile shear zones. *Contrib. Mineral. Petrol.* **83**, 231-236. *[6,14]*

KERN, H. & WENK, H.-R. (1990), Fabric-related velocity anisotropy and shear wave splitting in rocks from the Santa Rosa mylonite zone, California. *J. Geophys. Res.* **95**, 11213-11223. *[14]*

KIM, C.J., YOON, D.S., LEE, J.S., CHOI, C.G., LEE, W.J. & NO, K. (1994), Electrical characteristics of (100),(111), and randomly aligned lead zirconate titanate thin films. *J. Appl. Phys.* **76**, 7478-7482. *[6]*

KINOSHITA, N. & MURA, T. (1971), Elastic fields of inclusions in anisotropic media. *Phys. Stat. Sol.* (a) **5**, 759-768. *[A]*

KLEIN, H., ESLING, C. & BUNGE, H.J. (1988), Model calculations of deformation textures on the basis of orientation flow fields. In *Eighth Int. Conf. on Textures of Materials*, J.S. Kallend & G. Gottstein, eds. (Warrendale, PA: The Metallurgical Society) 307-312. *[8]*

KNORR, D.B. (1993). The role of texture on the reliability of aluminum-based inter-connects. In *Mater. Res. Soc. Symp., Materials Reliability in Microelectronics III*, (Pittsburgh, PA: MRS,) 75-86. *[5]*

KOCHENDÖRFER, A. (1941), *Plastische Eigenschaften von Kristallen und metallischen Werkstoffen* (Berlin: Springer). *[8]*

KOCKS, U.F. (1958), Polyslip in polycrystals. *Acta Metall.* **6**, 85-94. *[9,10]*

KOCKS, U.F. (1964a), Independent slip systems in crystals. *Philos. Mag.* **10**, 187-193. *[8]*

KOCKS, U.F. (1964b), Latent hardening and secondary slip in aluminum and silver. *Trans. Metall. Soc. AIME* **230**, 1160-1167. *[8,9,10]*

KOCKS, U.F. (1967). In a session discussion in *Can. J. Phys.* **45**, 1134. *[9]*

KOCKS, U.F. (1970), The relation between polycrystal deformation and single crystal deformation. *Metall. Trans.* **1**, 1121-1143. *[1,8,9,10,13]*

KOCKS, U.F. (1975), Constitutive relations for slip. In *Constitutive Equations in Plasticity*, A.S. Argon, ed. (Cambridge, MA: MIT Press) 81-115. *[8]*

KOCKS, U.F. (1976), Laws for work-hardening and low-temperature creep. *J. Eng. Mater. Technol.* **98**, 76-85. *[8,13]*

KOCKS, U.F. (1985), Dislocation interactions: flow stress and strain hardening. In *Dislocations and Properties of Real Materials* (London: Inst. Metals) 125-143. *[8]*

KOCKS, U.F. (1987), Constitutive behavior based on crystal plasticity. In *Unified Constitutive Equations for Creep and Plasticity*, A.K. Miller, ed. (London: Elsevier Appl. Sci.) 1-88. [8,10]

KOCKS, U.F. (1988), A symmetric set of Euler angles and oblique orientation space sections. In *Eighth Int. Conf. on Textures of Materials*, J.S. Kallend & G. Gottstein, eds. (Warrendale, PA: The Metallurgical Society) 31-36. [2]

KOCKS, U.F. (1991), Reliable modeling of complex behavior. In *Modeling the Deformation of Crystalline Solids*, T.C. Lowe &al., eds. (Warrendale, PA: TMS) 175-188. [8]

KOCKS, U.F. (1994), Mechanisms and models for large-strain heterogeneous plasticity. *Mater. Sci. Eng.* **A175**, 49-54. [8]

KOCKS, U.F. & CANOVA, G.R. (1981), How many slip systems and which? In *Deformation of Polycrystals*, N. Hansen &al., eds. (Roskilde, Denmark: Risø National Lab.) 35-44. [8,9,13]

KOCKS, U.F. & CHANDRA, H. (1982), Slip geometry in partially constrained deformation. *Acta Metall.* **30**, 695-709. [8,9]

KOCKS, U.F. & CHEN, S.R. (1993), Constitutive laws for deformation and dynamic recrystallization in cubic metals. In *Aspects of High Temperature Deformation and Fracture in Crystalline Materials* (JIMIS-7), Y. Hosoi &al., eds. (Tokyo: Jap. Inst. Metals) 593-600. [8]

KOCKS, U.F. & NECKER, C.T. (1994), Polycrystal models to fit experiments. In *Numerical Prediction of Deformation Processes and the Behavior of Real Materials*, S.I Andersen &al., eds. (Roskilde, DK: Risø Nat. Lab.) 45-58. [9,10]

KOCKS, U.F. & WESTLAKE, D.G. (1967), The importance of twinning for the ductility of cph polycrystals. *Trans. Metall. Soc. AIME* **239**, 1107-1109. [2,8]

KOCKS, U.F., ARGON, A.S. & ASHBY, M.F. (1975), Thermodynamics and kinetics of slip. *Prog. Mater. Sci.* **19**, 1-288. [1,8,9]

KOCKS, U.F., CANOVA, G.R. & JONAS, J.J. (1983), Yield vectors in f.c.c. crystals. *Acta Metall.* **31**, 1243-1252. [9]

KOCKS, U.F., TOMÉ, C. & CANOVA, G.R. (1986), Effective-cluster simulation of polycrystal plasticity. In *Large Deformation of Solids: physical basis and mathematical modeling.* J. Gittus, J. Zarka & S. Nemat-Nasser, eds. (London: Elsevier) 99-106. [9]

KOCKS, U.F., CANOVA, G.R., TOMÉ, C.N., ROLLETT, A.D. & WRIGHT, S.I. (1988a), *Computer Code LA-CC-88-6* (Los Alamos, NM: Los Alamos National Laboratory). [9]

KOCKS, U.F., STOUT, M.G. & ROLLETT, A.D. (1988b), The influence of texture on strain hardening, In *Strength of Metals and Alloys (ICSMA-8)*, P.O. Kettunen, T.K. Lepistö & M.E. Lehtonen, eds. (Oxford: Pergamon) 25-34. [10]

KOCKS, U.F., FRANCIOSI, P. & KAWAI, M. (1991a), A forest model of latent hardening and its application to polycrystal deformation. In *Ninth Int. Conf. on Textures of Materials.* Special Issue, R. Penelle & C. Esling, eds., of *Textures Microstruct.* **14-18**, 1103-1114. [8,9]

KOCKS, U.F., KALLEND, J.S. & BIONDO, A.C. (1991b), Accurate representations of general textures by a set of weighted grains. *Textures Microstruct.* **14-18**, 199-204. [2,5,13]

KOCKS, U.F., CHEN, S.R. & DAWSON, P.R. (1994a), Failures to model the development of a cube texture during high-temperature compression of Al-Mg alloys. In *Textures of Materials ICOTOM-10*, H.J. Bunge, ed. (Switzerland: Trans. Tech. Pubs.) 1797-1802. [9]

KOCKS, U.F., DAWSON, P.R. & FRESSENGEAS, C. (1994b), Kinematics of plasticity related to the state and evolution of the material microstructure. *J. Mech. Behaviour of Materials* **5**(2), 107-128. *[8,11]*

KOCKS, U.F., WRIGHT, S.I. & BEAUDOIN, A.J. (1996), The sensitivity of yield surface predictions to the details of a texture. In *Textures of Materials ICOTOM-11.* Z. Liang, L. Zuo & Y. Chu, eds. (Beijing: Int. Academic Publishers) 763-768. *[10]*

KOLLIA, C. & SPYRELLIS, N. (1993), Textural modifications in nickel electrodeposition under pulse reversed current. *Surface Coatings Technology* **57**, 71-75. *[5]*

KORZEKWA, D.R., JACOBSON, L.A. & HIRT, C.W. (1992). Modeling planar flow casting with FLOW-3D. In *Melt-Spinning and Strip Casting*, E.F. Matthys, ed. (Warrendale, PA: TMS) 107-122. *[5]*

KOUDDANE, R., MOLINARI, A. & CANOVA, G.R. (1993), Self-consistent modelling of heterogeneous visco-elastic and elasto-visco-plastic materials. In *Large Plastic Deformations. Fundamentals and Applications to Metal Forming*, C. Teodosiu, J.L. Raphanel & F. Sidoroff, eds. (Rotterdam: Balkema) 129-141. *[7]*

KOUDDANE, R., ZOUHAL, N., MOLINARI, A. & CANOVA, G.R. (1994), Complex loading of visco-plastic materials: micro-macro modelling. *Mater. Sci. Eng.* **A175**, 31-36. *[7]*

KRATKY, K.O. (1930), Ein Röntgengoniometer für die Polykristalluntersuchung. *Z. Kristallogr.* **72**, 529-540. *[4]*

KRIEGER LASSEN, N.C. & BILDE-SØRENSEN, J.B. (1993), Calibration of an electron back scattering pattern set-up. *J. Microscopy* **170**, 125-129. *[4]*

KRIEGER LASSEN, N.C., JUUL JENSEN, D. & CONRADSEN, K. (1992), Image processing procedures for analysis of electron back scattering patterns. *Scanning Microscopy* **6**, 115-121. *[4]*

KRÖNER, E. (1961), Zur plastischen Verformung des Vielkristalls. *Acta Metall.* **9**, 155-161. *[7,8]*

KRULEVITCH, P., NGUYEN, Tai.D., JOHNSON, G.C., WENK, H.-R. & GRONSKY, R. (1991), LPCVD polycrystalline silicon thin films: The evolution of structure, texture and stress. In *Evolution of Thin-Film & Surface Microstructure*, C.V. Thompson, J.Y. Tsao & D.J. Srolovitz, eds. (Pittsburgh, PA: Mat. Res. Soc.) **202**, 167-172. *[6]*

KRUSHEV, L., PANGAROVA, V. & PANGAROV, N. (1968), The effect of crystal orientation on the strength of copper deposited at high current densities. *Plating* **55**, 841-842. *[5]*

KUMAKURA, H., TOGANO, K., MAEDA, H., KASE, J. & MORIMOTO, T. (1991), Anisotropy of critical current density in textured $Bi_2Sr_2Ca_1Cu_2O_x$ tapes. *Appl. Phys. Lett.* **58**, 2830-2832. *[6]*

KUMAR, A. & DAWSON, P.R. (1995), Polycrystal plasticity modeling with finite elements over orientation space. *Comp. Mech.* **17**, 10-25. *[8,13]*.

KUNZE, K., WRIGHT, S.I., ADAMS, B.L. & DINGLEY, D.J. (1993), Advances in automatic EBSP single orientation measurements. *Textures Microstruct.* **20**, 41-54. *[4]*

KUNZE, K., ADAMS, B.L., HEIDELBACH, F. & WENK, H.-R. (1994), Local microstructural investigations in recrystallized quartzite using orientation imaging microscopy. In *Textures of Materials ICOTOM-10*, H.J. Bunge, ed. (Switzerland: Trans. Tech. Pubs.) **157-162**, 1243-1250. *[4,6]*

LANDAU, L.D. & LIFTSHITZ, E.M. (1959). *Theory of Elasticity.* (Oxford: Pergamon). *[7]*

LANGE, F.F. (1973), Relation between strength, fracture energy and microstructure of hot-pressed Si_3N_4. *J. Am. Ceram. Soc.* **56**, 518-522. *[6]*

LANGOUCHE, F., AERNOUDT, E. & VAN HOUTTE, P. (1989), Quantitative texture measurements on thin wires. *J. Appl. Crystallogr.* **22**, 533-538. *[4]*

LANKFORD, W.F., SNIDER, S.C. & BAUSCH, J.A. (1950), New criteria for predicting the press performance of deep drawing steels. *Trans. Amer. Soc. Metals* **42**, 1197-1232. *[10]*

LARSON, D.C. & KOCKS, U.F. (1963), Discussion. In *Recovery and Recrystallization of Metals*, L. Himmel, ed. (New York: Interscience) 239-240. *[9]*

LAW, R.D. (1990), Crystallographic fabrics: a selective review of their applications to research in structural geology. In *Deformation Mechanisms, Rheology and Tectonics*, R.J. Knipe & E.H. Rutter, eds. (London: Geol. Soc. Spec. Publ.) **54**, 335-352. *[14]*

LAWS, N. & MCLAUGHLIN, R. (1978), Self-consistent estimates for the viscoelastic creep compliances of composite materials. *Proc. R. Soc. Lond.* **A359**, 251-273. *[7]*

LEBENSOHN, R. & TOMÉ, C. (1991), Modelling twinning in texture development codes. *Textures Microstruct.* **14-18**, 959-964. *[5,11]*

LEBENSOHN, R.A. & TOMÉ, C.N. (1993a), A self-consistent anisotropic approach for the simulation of plastic deformation and texture development of polycrystals – Application to zirconium alloys. *Acta Metall. Mater.* **41**, 2611-2624. *[10,11]*

LEBENSOHN, R.A. & TOMÉ, C.N. (1993b), A study of the stress state associated with twin nucleation and propagation in anisotropic materials. *Philos. Mag.* **A67**, 187-206. *[8,A]*

LEBENSOHN, R.A. & TOMÉ, C.N. (1994), A self-consistent visco-plastic model: prediction of rolling textures of anisotropic polycrystals. *Mater. Sci. Eng.* **A175**, 71-82. *[9,11]*

LEBENSOHN, R.A. & TOMÉ, C.N. (1996), Yield loci calculation of hexagonal materials using a self-consistent polycrystalline model. *Textures Microstruct.* **26-27**, 513-529. *[11]*

LEBENSOHN, R.A., SANCHEZ, P.V. & POCHETTINO, A.A. (1994), Modelling texture development of zirconium alloys at high temperatures. *Scripta Metall. Mater.* **30**, 481-486. *[11,13]*

LEBENSOHN, R.A., GONZALEZ, M.I., TOMÉ, C.N. & POCHETTINO, A.A. (1996), Measurement and prediction of texture development during a rolling sequence of zircaloy-4 tubes. *J. Nucl. Mater.* **229**, 57-64. *[11]*

LEBENSOHN, R.A., WENK, H.-R. & TOMÉ, C.N. (1997), Modelling deformation and recrystallization textures in calcite. *Acta Mater.*, in press. *[14]*

LEBER, S. (1961), Cylindrical textures in tungsten and other body-centered cubic metals. *Trans. Amer. Soc. Metals* **53**, 697-713. *[4]*

LEE, C.S., DUGGAN, B.J. & SMALLMAN, R.E. (1993), Deformation banding in copper. *Philos. Mag. Lett.* **68**, 185-190. *[5]*

LEE, E.H. (1969), Elastic-plastic deformation at finite strains. *J. Appl. Mech.* **36**, 1-14. *[8]*

LEE, K. & BOWMAN, K.J. (1992), Texture and anisotropy in silicon nitride. *J. Am. Ceram. Soc.* **75**, 1748-1755. *[6]*

LEE, K., SANDLIN, M.S. & BOWMAN, K.J. (1993), Toughness anisotropy in textured ceramic composites. *J. Am. Ceram. Soc.* **76**, 1793-8000. *[6]*

LEFFERS, T. (1979), A modified Sachs approach to the plastic deformation of polycrystals as a realistic alternative to the Taylor model. In *Strength of Metals and Alloys*, P. Haasen, V. Gerold & G. Kostorz, eds. (Oxford: Pergamon) 769-774. *[8,9]*

LEFFERS, T. (1988), Deformation textures: simulation principles (Panelist's contribution). In *Eighth Int. Conf. on Textures of Materials*, J.S. Kallend & G. Gottstein, eds. (Warrendale, PA: The Metallurgical Society) 273-284. *[9]*

LEFFERS, T. (1996), The brass-type texture once again. In *Texture of Materials ICOTOM-11*, Z. Liang, L. Zuo & Y. Chu, eds. (Beijing: International Academic Publishers) 299-306. *[9]*

LEFFERS, T. & BILDE-SØRENSEN, J.B. (1990), Intra- and intergranular heterogeneities in the plastic deformation of brass during rolling. *Acta Metall. Mater.* **38**, 1917-1926. *[9]*

LEFFERS, T. & JUUL JENSEN, D. (1986), Evaluation of the effect of initial texture on the development of deformation texture. *Textures Microstruct.* **6**, 231-254. *[9]*

LEFFERS, T. & JUUL JENSEN, D. (1991), The relation between texture and microstructure in rolled fcc metals. *Textures Microstruct.* **14-18**, 933-952. *[9]*

LEFFERS, T., ASARO, R.J., DRIVER, J.H., KOCKS, U.F., MECKING, H., TOMÉ, C. & VAN HOUTTE, P. (1988), Deformation textures: simulation principles. In *Eighth Int. Conf. on Textures of Materials*, J.S. Kallend & G. Gottstein, eds. (Warrendale, PA: The Metallurgical Society) 265-272. *[5,9,13]*

LEHAZIF, R., DORIZZI, P. & POIRIER, J.P. (1973), Glissement {110}⟨110⟩ dans les métaux de structure cubique faces centrées. *Acta Metall.* **21**, 903-911. *[9]*

LEIBFRIED, G. & BREUER, N. (1978). *Point Defects in Metals. I: Introduction to the Theory.* (Berlin: Springer). *[7]*

LEISS, B., HELMING, K., SIEGESMUND, S. & WEBER, K. (1994), Quantitative texture analysis of naturally deformed polyphase dolomite marbles and its kinematic significance. In *Textures of Materials ICOTOM-10*, H.J. Bunge, ed. (Switzerland: Trans. Tech. Pubs.) 789-794. *[6]*

LEONHARDT, A., SCHLÄFER, D., SEIDLER, M., SELBMANN, D. & SCHÖNHERR, M. (1982), Microhardness and texture of TiCx layers on cemented carbides. *J. Less-Common Met.* **87**, 63-69. *[6]*

LEQUEU, P., GILORMINI, P., MONTHEILLET, F., BACROIX, B. & JONAS, J.J. (1987a), Yield surfaces for textured polycrystals – I. Crystallographic approach. *Acta Metall.* **35**, 439-451. *[8]*

LEQUEU, P., GILORMINI, P., MONTHEILLET, F., BACROIX, B. & JONAS, J.J. (1987b), Yield surfaces for textured polycrystals – II. Analytical approach. *Acta Metall.* **35**, 1159-1174. *[10]*

LESLIE, W.C. (1981), *The Physical Metallurgy of Steels* (New York: McGraw-Hill). *[8]*

LI, B.S., CHERNG, J.S., BOWMAN, K.J. & CHEN, I.W. (1988), Domain switching as a toughening mechanism in tetragonal zirconia. *J. Am. Ceram. Soc.* **71**, C362-C364. *[6]*

LI, F. & BATE, P.S. (1991), Strain path change effects in cube textured aluminum sheet. *Acta Metall. Mater.* **39**, 2639-2650. *[10]*

LI, W.-Y. & SORENSEN, O.T. (1995), Texture formed by platelet alignment in ceramic platelet ceramic matrix composites. *Textures Microstruct.* **24**, 53-65. *[6]*

LIANG, Z., WANG, F. & XU, J. (1988), Inverse pole figure determination according to the maximum entropy method. In *Eighth Int. Conf. on Textures of Materials*, J.S. Kallend & G. Gottstein, eds. (Warrendale, PA: The Metallurgical Society) 111-114. *[3]*

LILE, R.C. (1978), The effect of anisotropy on the creep of polycrystalline ice. *J. Glac.* **21**, 475-479. *[6]*

LIN, S.C. & MURA, T. (1973), Elastic fields of inclusions in anisotropic media (II). *Phys. Stat. Sol.* (a) **15**, 281-285. *[A]*

LINDHOLM, U.S., NAGY, A., JOHNSON, G.R. & HOEGFELDT, J.M. (1981), Large strain, high strain rate testing of copper. *J. Eng. Mater. Technol.* **102**, 376-381. *[10]*

LINSSEN, G., MENGELBERG, H.-D. & STÜWE, H.-P. (1964), Zyklische Texturen von Drähten kubisch flächenzentrierter Metalle. *Z. Metallk.* **55**, 600-604. *[1,5]*

LIPKIN, J. & LOWE, T.C. (1989), Axial effects during reversed torsional deformation. In *Advances in Plasticity*, A.S. Khan & M. Tokuda, eds. (Pergamon Press) 625-628. *[10]*

LIPKIN, J., CHIESA, M.L. & BAMMANN, D.J. (1988), Thermal softening of 304L stainless steel: experimental results and numerical simulations. In *IMPACT '87: Int. Conf. on Impact Loading and Dynamic Behavior of Materials*, C. Chiem, H.D. Kunze & L.W. Meyer, eds. (Bremen: Informationgesellschaft, Verlag) 687-694. *[10]*

LISTER, G.S. & DORNSIEPEN, V.F. (1982), Fabric transitions in the Saxony granulite terrain. *J. Struct. Geol.* **4**, 81-92. *[14]*

LISTER, G.S. & HOBBS, B.E. (1980), The simulation of fabric development during plastic deformation and its application to quartzite: The influence of deformation history. *J. Struct. Geol.* **2**, 355-370. *[14]*

LISTER, G.S. & WILLIAMS, P.F. (1979), Fabric development in shear zones: Theoretical controls and observed phenomena. *J. Struct. Geol.* **1**, 283-297. *[14]*

LISTER, G.S., PATERSON, M.S. & HOBBS, B.E. (1978), The simulation of fabric development during plastic deformation and its application to quartzite: The model. *Tectonophysics* **45**, 107-158. *[14]*

LLOYD, G.E. (1987), Backscattered electron techniques. *Mineral. Mag.* **51**, 3-19. *[4]*

LLOYD, G.E. & FREEMAN, B. (1991), SEM electron channelling analysis of dynamic recrystallization in a quartz grain. *J. Struct. Geol.* **13**, 945-953. *[14]*

LLOYD, L.T. & BARRETT, C.S. (1966), Thermal expansion of alpha-uranium. *J. Nucl. Mater.* **18**, 55-59. *[7]*

LLOYD, L.T. & CHISWIK, H.H. (1955), Deformation mechanisms of alpha-uranium single crystals. *Trans. AIME* **203**, 1206-1214. *[2]*

LOGAN, R.W. & HOSFORD, W.F. (1980), Upper-bound anisotropic yield locus calculations assuming ⟨111⟩-pencil glide. *J. Mech. Phys. Solids.* **22**, 419-430. *[10]*

LOGÉ, R., CHASTEL, Y., DUMAS, G., LAMY, V., GEX, D. & LEBENSOHN, R. (1996), Texture induced anisotropy of Zy4 during compression: modelling and experimental validation. In *Texture of Materials ICOTOM-11*, Z. Liang, L. Zuo & Y. Chu, eds. (Beijing: International Academic Publishers) 818-823. *[13]*

LOVATO, M.L. & STOUT, M.G. (1992), Compression testing techniques to determine the stress/strain behavior of metals subject to finite deformation. *Metall. Trans.* **23A**, 935-951. *[10]*

LOVETT, D.R. (1989). *Tensor Properties of Crystals.* (Bristol: Adam Hilger) 139 pp.

LOWE, T.C., HARREN, S., ASARO, R.J. & NEEDLEMAN, A. (1987), Analysis of the evolution of texture and axial stresses in fcc polycrystals subject to large strain shear. In *UTAM Symposium on Yielding, Damage, and Failure of Anisotropic Solids*, J.P. Boehler, ed. (London: Mechanical Engineering Publications) 335-358. *[9]*

LÜCKE, K., POSPIECH, J., JURA, J. & HIRSCH, J. (1986), On the presentation of orientation distribution functions by model functions. *Z. Metallk.* **77**, 312-321. *[3]*

LÜCKE, K., POSPIECH, J., VIRNICH, K.H. & JURA, J. (1981), On the problem of the reproduction of the true orientation distribution from pole figures. *Acta Metall.* **29**, 167-185. *[2]*

LUTTEROTTI, L., POLONIOLI, P., ORSINI, P.G. & FERRARI, M. (1994), Stress and texture analysis of zirconia coatings by XRD total pattern fitting. *Mater. and Design Techn. ASME* **62**, 15-20. *[4,6]*

LUTTEROTTI, L., MATTHIES, S., WENK, H.-R., SCHULTZ, A.J. & RICHARDSON JR., J.W. (1997), Combined texture and structure analysis of deformed limestone from time-of-flight diffraction spectra. *J. Appl. Phys.* **81**, 594-600. *[4]*

MA, Y. & BOWMAN, K.J. (1991), Texture in hot-pressed or forged alumina. *J. Am. Ceram. Soc.* **74**, 2941-2944. *[6]*

MA, Y., LEE, Y. & BOWMAN, K.J. (1994), Deformation textures in hot-worked polyphase ceramics. *Mater. Sci. Eng.* **A175**, 167-176. *[6]*

MACEWEN, S.R., CHRISTODOULOU, N., TOMÉ, C., JACKMAN, J., HOLDEN, T.M., FABER, J. JR. & HITTERMAN, R.L. (1988), The evolution of texture and residual stress on zircaloy-2. In *Eighth Int. Conf. on Textures of Materials*, J.S. Kallend & G. Gottstein, eds. (Warrenville, PA: The Metallurgical Soc.) 825-836. *[11]*

MACEWEN, S.R., TOMÉ, C.N. & FABER, J. JR. (1989), Residual stresses in annealed Zircaloy. *Acta Metall.* **37**, 979-989. *[7,11]*

MACKENZIE, J.K. (1958), Second paper on statistics associated with the random disorientation of cubes. *Biometrica* **45**, 229-240. *[2]*

MAKINDE, A., THIBODEAU, L. & NEALE, K.W. (1992), Development of an apparatus for biaxial testing using cruciform specimens. *Exp. Mech.* **32**, 138-144. *[10]*

MANIATTY, A.M. (1991). *Eulerian Elast-viscoplastic Formulation for Modeling Steady-state Deformations with Strain-induced Anisotropy*. Ph.D. Thesis (Ithaca, N.Y.: Cornell Univ.). *[13]*

MANIATTY, A.M., DAWSON, P.R. & LEE, Y.S. (1992), A time integration algorithm for elasto-viscoplastic cubic crystals applied to modeling polycrystalline deformation. *Int. J. Num. Meth. Engr.* **35**, 1565-1588. *[13]*

MANSOUR, S.A. & VEST, R.W. (1992), The dependence of ferroelectric and fatigue behaviors of PZT films on microstructure and orientation. *Integr. Ferroelectrics* **1**, 57-69. *[6]*

MARCH, A. (1932), Mathematische Theorie der Regelung nach der Korngestalt bei affiner Deformation. *Z. Kristallogr.* **81**, 285-297. *[4,6,14]*

MARCINIAK, Z. & KUCZYŃSKI, K. (1967) Limit strains in the processes of stretch-forming sheet metal. *Int. J. Solids Struct.* **9**, 609-620. *[10]*

MARGEVICIUS, R.W. & COTTON, J.D. (1995), Study of the brittle-to-ductile transition in NiAl by texture analysis. *Acta Metall. Mater.* **43**, 645-655. *[5,8]*

MARGEVICIUS, R.W. & LEWANDOWSKI, J.J. (1993), Deformation texture of hydrostatically extruded polycrystalline NiAl. *Scripta Metall. Mater.* **29**, 1651-1654. *[5]*

MARTIN, J.L. (1993), Non-compact slip in close-packed metals. In *Aspects of High Temperature Deformation and Fracture in Crystalline Materials*, Y. Hosoi &al., eds. (Nagoya: The Japan Inst. Metals) 3-10. *[9]*

MATHUR, K.K. (1995), Parallel algorithms for large scale simulations in materials processing. In *Simulation of Materials Processing*, S-F. Shen & P. Dawson, eds. (Rotterdam: Balkema) 109-114. *[13]*

MATHUR, K.K. & DAWSON, P.R. (1989), On modeling the development of crystallographic texture in bulk forming processes. *Int. J. Plast.* **5**, 67-94. *[12,13]*

MATHUR, K.K. & DAWSON, P.R. (1990), Texture development during wire drawing. *J. Eng. Mater. Technol.* **112**, 292-297. *[13]*

MATHUR, K.K., DAWSON, P.R. & KOCKS, U.F. (1990), On modeling anisotropy in deformation processes involving textured polycrystals with distorted grain shape. *Mech. Mater.* **10**, 183-202. *[9,13]*

MATTHIES, S. (1979), On the reproducibility of the orientation distribution function of texture samples from pole figures (ghost phenomena). *Phys. Stat. Sol. (b)* **92**, K135-138. *[3,6]*

MATTHIES, S. (1988), On the basic elements of and practical experience with the WIMV algorithm, an ODF reproduction method with conditional ghost correction. In *Eighth Int. Conf. on Textures of Materials*, J.S. Kallend & G. Gottstein, eds. (Warrendale, PA: The Metallurgical Society) 37-48. *[3]*

MATTHIES, S. & HUMBERT, M. (1993), The realization of the concept of the geometric mean for calculating physical constants of polycrystalline materials. *Phys. Stat. Sol. (b)* **177**, K47-K50. *[7]*

MATTHIES, S. & VINEL, G.W. (1982), On the reproduction of the orientation distribution function of textured samples from reduced pole figures using the concept of conditional ghost correction. *Phys. Stat. Sol. (b)* **112**, K111-114. *[2,3]*

MATTHIES, S. & VINEL, G.W. (1994), On some methodical development concerning calculations performed directly in the orientation space. In *Textures of Materials ICOTOM-10*, H.J. Bunge, ed. (Switzerland: Trans. Tech. Pubs.) 1641-1646. *[2]*

MATTHIES, S. & WAGNER, F. (1996), On a 1/n law in texture related single orientation analysis. *Phys. Stat. Sol.(b)* **196**, K11. *[4]*

MATTHIES, S. & WENK, H.-R. (1992), Optimization of texture measurements by pole figure coverage with hexagonal grids. *Phys. Stat. Sol. (a)* **133**, 253-257. *[2,4]*

MATTHIES, S., VINEL, G.W. & HELMING, K. (1987), *Standard Distributions in Texture Analysis – Maps for the Case of Cubic-Orthorhombic Symmetry*, Vol.1: (Berlin: Akademie-Verlag). *[1,2]*

MATTHIES, S., WENK, H.-R. & VINEL, G.W. (1988), Some basic concepts of texture analysis and comparison of three methods to calculate orientation distributions from pole figures. *J. Appl. Crystallogr.* **21**, 285-304. *[2,3]*

MATTHIES, S., HELMING, K. & KUNZE, K. (1990a), On the representation of orientation distributions by σ-sections – I. General properties of σ-sections. *Phys. Stat. Sol. (b)* **157**, 71-83. *[2]*

MATTHIES, S., HELMING, K. & KUNZE, K. (1990b), On the representation of orientation distributions by σ-sections – II. Consideration of crystal and sample symmetry, examples. *Phys. Stat. Sol. (b)* **157**, 489-507. *[2]*

MATTHIES, S., LUTTEROTTI, L. & WENK, H.-R. (1997), Advances in texture analysis from diffraction spectra. *J. Appl. Crystallogr.* **30**, 31-42. *[4]*

MAURICE, C & DRIVER, J.H. (1993), High-temperature plane-strain compression of cube oriented aluminum crystals. *Acta Metall. Mater.* **41**, 1653-1664. *[9]*

MAURICE, C., DRIVER, J.H. & TÓTH, L.S. (1992), Modeling high-temperature rolling textures of fcc metals. *Textures Microstr.* **19**, 211-227. *[9]*

MCCLINTOCK, F.A. & ARGON, A.S. (1966), *Mechanical Behavior of Materials* (Reading, MA: Addison-Wesley). *[1]*

MCHARGUE, C.J., JETTER, L.K. & OGLE, J.C. (1959), Preferred orientation in extruded aluminum rod. *Trans. Metall. Soc. AIME* **215**, 831-837. *[5]*

MCHUGH, P.E., ASARO, R.J. & SHIH, C.F. (1993), Computational modeling of metal matrix composite materials - 1. Isothermal deformation patterns in ideal microstructures. *Acta Metall.* **41**, 1461-1476. *[13]*

MCLAREN, A.C., COOK, R.F., HYDE, S.T. & TOBIN, R.C. (1983), The mechanisms of the formation and growth of water bubbles and associated dislocation loops in synthetic quartz. *Phys. Chem. Mineral.* **9**, 79-94. *[6]*

MECIF, A., BACROIX, B. & FRANCIOSI, P. (1997), Temperature and orientation-dependent plasticity features of Cu and Al single crystals under axial compression. *Acta Mater.* **45**, 371-381. *[10]*

MECKING, H. (1977), Description of hardening curves of fcc single and polycrystals. In *Work Hardening in Tension and Fatigue*, A.W. Thompson, ed. (Warrendale, PA: The Metallurgical Soc.of AIME) 67-87. [8]

MECKING, H. (1980), Deformation of polycrystals. In *Strength of Metals and Alloys*, P. Haasen, V. Gerold & G. Kostorz., eds. (Oxford: Pergamon) 1573-1594. *[8]*

MECKING, H. (1981a), Computer simulation of texture development. In *Sixth Int. Conf. on Textures of Materials* (Tokyo: The Iron and Steel Institute of Japan) 53-66. *[8,9]*

MECKING, H. (1981b), Low temperature deformation of polycrystals. In *Deformation of Polycrystals*, N. Hansen &al., eds. (Roskilde, Denmark: Risø National Lab.) 73-86. *[8,9]*

MECKING, H. (1985), Textures in metals. In *Preferred Orientation in Deformed Metals and Rocks: An Introduction to Modern Texture Analysis*, H.-R. Wenk, ed. (Orlando, FL: Academic Press) 267-315. *[9]*

MECKING, H. & KOCKS, U.F. (1981), Kinetics of flow and strain-hardening. *Acta Metall.* **29**, 1865-1875. *[8]*

MECKING, H. & LÜCKE, K. (1969), Die Verfestigung von Silber-Einkristallen zwischen 77 und 1200°K. *Z. Metallk.* **60**, 185-195. *[8]*

MECKING, H., HARTIG, C. & KOCKS, U.F. (1996a), Deformation modes in γ-TiAl as derived from the single crystal yield surface. *Acta Mater.* **44**, 1309-1321. *[5,8]*

MECKING, H., KOCKS, U.F. & HARTIG, C. (1996b), Taylor factors in materials with many deformation modes. *Scripta Mater.* **35**, 465-471. [8]

MELLOR, P.B. (1956), Stretch forming under fluid pressure. *J. Mech. Phys. Solids* **5**, 41-56. *[10]*

MERCIER, J.C.C. (1985), Olivine and pyroxenes. In *Preferred Orientation in Deformed Metals and Rocks: An Introduction to Modern Texture Analysis*, H.-R. Wenk, ed., (Orlando, FL: Academic Press) 407-430. *[6,14]*

MICHALUK, C., BINGERT, J. & CHOI, C.S. (1993), The effects of texture and strain on the r-value of heavy gauge tantalum plate. In *Textures of Materials ICOTOM-10*, H.J. Bunge, ed. (Switzerland: Trans. Tech. Pubs.) 1653-1662. *[5]*

MICHNO, M.J. & FINDLEY, W.N. (1976), An historical perspective of yield surface investigations for metals. *Int. J. Non-linear Mechanics* **11**, 59-82. *[10]*

MILLER, M.P., DAWSON, P.R. & BAMMANN, D.J. (1995), Reflecting microstructural evolution in hardening models for polycrystalline metals. In *Simulation of Materials Processing*, S-F. Shen & P. Dawson, eds. (Rotterdam: Balkema) 295-300. *[10]*

MISES, R. VON (1913), Mechanik der festen Körper im plastisch-deformablen Zustand. *Nachrichten Könogl. Gesellsch. Wissenschaften Göttingen, Math.-Phys. Klasse*, 582-592. [8,10]

MISES, R. VON (1928), Mechanik der plastischen Formänderung von Kristallen. *Z.Angew. Math.& Mech.* **8**, 161-185. *[8,9]*

MISHRA, S., DÄRMANN, C. & LÜCKE, K. (1984), On the development of the Goss texture in iron–3% silicon. *Acta Metall.* **32**, 2185-2201. *[5]*

MIYAJI, H. & FURUBAYASHI, E.-I. (1991), Effect of stress on the variant selection in martensitic transformation. *Textures Microstruct.* **14-18**, 561-566. *[5]*

MOAN, G.D. & EMBURY, J.D. (1979), A study of the Bauschinger effect in Al-Cu alloys. *Acta Metall.* **27**, 903-914. *[10]*

MODARESSI, A., AHRA, E.H. & HEIZMANN, J.J. (1991), Texture of magnetite observed during the phase transformation hematite-magnetite. *Textures Microstruct.* **14-18**, 437-442. *[6]*

MOLINARI, A. & TÓTH, L.S. (1994a), Tuning a self-consistent viscoplastic model by finite element results. Part I: Modeling. *Acta Metall. Mater.* **42**, 2453-2458. *[11]*

MOLINARI, A. & TÓTH, L.S. (1994b), Tuning a self-consistent viscoplastic model by finite element results. Part II: Application to torsion textures. *Acta Metall. Mater.* **42**, 2459-2466. *[11]*

MOLINARI, A., CANOVA, G.R. & AHZI, S. (1987), A self-consistent approach of the large deformation polycrystal viscoplasticity. *Acta Metall.* **35**, 2983-2994. *[8,11,14,A]*

MOLS, K., VAN PREAT, K. & VAN HOUTTE, P. (1984), A generalized yield locus calculation from texture data. In *Seventh Int. Conf. on Textures of Materials*, C.M. Brakman, P. Jongenburger & E.J. Mittemeijer, eds. (Netherlands Soc. for Materials Science) 651-656. *[10]*

MOMENT, R.L. (1972), Rolling texture of Pu-Ga alloys as a function of temperature. *Metall. Trans.* **3**, 1639-1644. *[5]*

MONTESIN, T. & HEIZMANN, J.J. (1991a), Evolution of crystallographic texture in thin wires. *J. Appl. Crystallogr.* **25**, 665-673. *[4]*

MONTESIN, T. & HEIZMANN, J.J. (1991b), Evolution of the texture in steelcord. *Textures Microstr.* **14-18**, 573-578. *[4]*

MONTESIN, T., HEIZMANN, J.J. & VADON, A. (1991), Absorption corrections for x-ray texture measurement of any shape sample. *Textures Microstr.* **14-18**, 567-572. *[4]*

MONTHEILLET, F., COHEN, M. & JONAS, J.J. (1984), Axial stresses and texture development during the torsion testing of Al, Cu, and α-Fe. *Acta Metall.* **32**, 2077-2089. *[2,5,10]*

MORAWIEC, A. (1989), Calculation of polycrystal elastic constants from single crystal data. *Phys. Stat. Sol. (b)* **154**, 535-541. *[7]*

MORAWIEC, A. (1990), The rotation rate field and geometry of orientation space. *J. Appl. Cryst.* **23**, 374-377. *[8]*

MORAWIEC, A. (1992), A method of texture inhomogeneity estimation from reflection pole figures. *J. Phys.: Cond. Matter* **4**, 339-349. *[4]*

MORAWIEC, A. (1995), Misorientation-angle distribution of randomly oriented symmetric objects. *J. Appl. Cryst.* **28**, 289-293. *[2]*

MORAWIEC, A. & FIELD, D.P. (1996), Rodrigues parametrization for orientation and misorientation distributions. *Philos. Mag.* **A73**, 1113-1130. *[2]*

MORAWIEC, A. & POSPIECH, J. (1989a), Some information on quaternions useful in texture calculations. *Textures Microstruct.* **10**, 211-242. *[2]*

MORAWIEC, A. & POSPIECH, J. (1989b), Properties of projection lines in the space of the orientation distribution function. *Textures Microstruct.* **10**, 243-264. *[2]*

MORII, K., MECKING, H. & NAKAYAMA, Y. (1985), Development of shear bands in fcc single crystals. *Acta Metall.* **33**, 379-386. *[9]*

MORII, K., HARTIG, C., MECKING, H., NAKAYAMA. Y. & LÜTJERING, G. (1988), Deformation modes and texture formation in two phase Ti-6Al-4V alloys. In *Eighth Int. Conf. on Textures of Materials*, J.S. Kallend & G. Gottstein, eds. (Warrendale, PA: The Metallurgical Society) 991-996. *[11]*

MORRIS, G.B., RAITT, R.W. & SHOR, G.G. (1969). Velocity anisotropy and delay time maps of the mantle near Hawaii. *J. Geophys. Res.* **74**, 4300-4316. *[14]*

MORRIS, P.R. (1959), Reducing the effects of nonuniform pole distribution in inverse pole figure studies *J. Appl. Phys.* **30**, 595-596. *[4]*

MORRIS, P.R. (1970), Elastic constants of polycrystals. *Int. J. Eng. Sci.* **8**, 49-61. *[7]*

MORRIS, P.R. (1971). Crystallite orientation analysis for materials with tetragonal, hexagonal and orthorhombic crystal symmetries. In *Quantitative Analysis of Textures*. Proc. Int. Seminar Cracow 1971, J. Karp & al., eds. (The Society of Polish Metallurgical Engineers). 87. *[5]*

MORRIS, P.R. & HECKLER, A.T. (1968), Crystallite orientation analysis for rolled cubic metals. *Adv. X-Ray Analysis* **11**, 454-472. *[2]*

MOTOWIDLO, L.R., GALINSKI, G., OZERYANSKY, G., ZHANG, W. & HELLSTROM, E.E. (1994), Dependence of critical current density on filament diameter in round multifilament Ag-sheathed $Bi_2Sr_2CaCu_2O_x$ wires processed in O_2. *Appl. Phys. Lett.* **65**, 2731-2733. *[6]*

MÜCKLICH, A., MATTHIES, S. & HENNIG, K. (1981), Fiber texture studies and ghost phenomena (Experiments on fir-tree texture). *Sixth Int. Conf. on Textures of Materials* (The Iron and Steel Institute of Japan) 1266-1273. *[1]*

MÜCKLICH, A., HENNIG, K., BOUILLOT, J. & MATTHIES, S. (1984) Study of an Fe-Ni alloy by means of Mossbauer spectroscopy and neutron diffraction. In *Proc. ICOTOM 7*, C.M. Brakman, P. Jongenburger & E.J. Mittemeijer, eds. (Zwijndrecht, NL: Netherlands Soc. Mater. Sci.) 657-662. *[4]*

MUDDLE, B.C. & HANNINK, R.H.J. (1986), Crystallography of the tetragonal to monoclinic transformation in MgO-partially-stabilized zirconia. *J. Am. Ceram. Soc.* **69**, 547-555. *[6]*

MUELLER, M., RICHARDSON, J., SCHULTZ, A., ROSS, F. & REICHEL, D. (1988). Assessment of grain size and preferred orientation by neutron diffraction on IPNS zircaloy-clad uranium target material. In *Eighth Int. Conf. on Textures of Materials*, J.S. Kallend & G. Gottstein, eds. (Warrendale, PA: The Metallurgical Society) 209-214. *[5]*

MURA, T. (1991). *Micromechanics of Defects in Solids* (2nd ed., Dordrecht: Martinus-Nijhoff). *[7,A]*

NADAI, A. (1950). *Theory of Flow and Fracture in Solids*, 2nd ed. (New York: McGraw-Hill) p. 349. *[10]*

NARUTANI, T. & TAKAMURA, J. (1991), Grain size strengthening in terms of dislocation density measured by resistivity. *Acta Metall. Mater.* **39**, 2037-2049. *[8]*

NAUER-GERHARDT, C.U. & BUNGE, H.J. (1986), Orientation determination by optical methods. In *Experimental Techniques of Texture Analysis*, H.J. Bunge, ed. (Oberursel, Germany: Deutsche Gesell. Metallkunde) 125-145. *[4]*

NEALE, K.W. & CHATER, E. (1980), Limit strain predictions for strain-rate sensitive anisotropic sheets. *Int. J. Mech. Sci.* **22**, 563-574. *[10]*

NEALE, K.W. & ZHOU, Y. (1991), Simulation of the behavior of fcc polycrystals. In *Modeling the Deformation of Crystalline Solids*, T.C. Lowe &al., eds. (Warrendale PA: The Minerals, Metals and Materials Society) 239-246. *[8]*

NECKER, C.T. (1997), *Recrystallization Texture in Cold Rolled Copper*. Ph.D. Thesis (Philadelphia PA: Drexel University). *[5,9]*

NECKER, C.T., DOHERTY, R.D. & ROLLETT, A.D. (1993), Quantitative measurement of the development of recrystallization texture in OFE copper. In *Textures of Materials ICOTOM-10*, H.J. Bunge, ed. (Switzerland: Trans. Tech. Pubs.) 1021-1026. *[2]*

NEEDLEMAN, A., ASARO, R.J., LEMOND, J. & PEIRCE, D. (1985), Finite element analysis of crystalline solids. *Comp. Meth. Applied Mech. Eng.* **52**, 689-708. *[13]*

NES, E., HIRSCH, J. & LÜCKE, K. (1984). On the origin of the cube recrystallization texture in directionally solidified aluminium. In *Seventh Int. Conf. on Textures of Materials*, C.M. Brakman, P. Jongenburger & E.J. Mittemeijer, eds. (Netherlands Soc. for Materials Science) 663-674. *[5]*

NEUMANN, P. (1991), Representation of orientations of symmetrical objects by Rodrigues vectors. *Textures Microstruct.* **14-18**, 53-58. *[2]*

NEUMANN, P. (1992), The role of geodesic and stereographic projections for the visualization of directions, rotations, and textures. *Phys. Stat. Sol.(a)* **131**, 555-567. *[2]*

NICOLAS, A., BOUDIER, F. & BOULLIER, A.-M. (1973), Mechanisms of flow in naturally and experimentally deformed peridotites. *Am. J. Sci.* **273**, 853-876. *[6]*

NISHIMURA, C. & FORSYTH, D. (1989), The anisotropic structure of the upper mantle in the Pacific. *Geophys. J.* **96**, 203-229. *[14]*

NORTON, J.T. (1948), A technique for quantitative determination of texture of sheet metals. *J. Appl. Phys.* **19**, 1176-1178. *[4]*

NOUDEM, J.G., BEILLE, J., BEAUGNON, E., BOURGAULT, D., CHATEIGNER, D., GERMI, P., PERNET, M., SULPICE, A. & TOURNIER, R. (1995), Magnetic melt texturing combined with hot pressing applied to superconducting (2223) Bi-Pb-Sr-Ca-Cu-ceramics. *Supercond. Sci. Technol.* **8**, 558-563. *[6]*

NYE, J.F. (1957), *Physical Properties of Crystals; their representation by tensors and matrices* (Oxford University Press). [Has been re-issued many times, with same page nos.] *[1,7]*

O'BRIEN, D.K., WENK, H.-R., RATSCHBACHER, L. & YOU, Z. (1987), Preferred orientation of phyllosilicates in phyllonites and ultramylonites. *J. Struct. Geol.* **9**, 719-730. *[6]*

OERTEL, G. (1970), Deformation of a slaty, lapillar tuff in the Lake District, England. *Geol. Soc. Am. Bull.* **81**, 1173-1188. *[6]*

OERTEL, G. (1983), The relationship of strain and preferred orientation of phyllosilicate grains in rocks - review. *Tectonophysics* **100**, 413-447. *[6,14]*

OERTEL, G. & CURTIS, C.D. (1972), Clay-ironstone concretion preserving fabrics due to progressive compaction. *Geol. Soc. Am. Bull.* **83**, 2597-2606. *[14]*

OERTEL, G. & ERNST, W.G. (1978), Strain and rotation in a multilayered fold. *Tectonophysics* **48**, 77-106. *[14]*

OERTEL, G., ENGELDER, T. & EVANS, K. (1989), A comparison of the strain of crinoid columnals with that of their enclosing silty and shaly matrix on the Appalachian Plateau, New York. *J. Struct. Geol.* **11**, 975-993. *[6]*

OHASHI, Y., OHNO, N. & KAWAI, M. (1982), Evaluation of creep constitutive equations for type 304 stainless steel under repeated multiaxial loading. *J. Eng. Mater. Technol.* **104**, 159-164. *[10]*

OHASHI, Y., KAWAI, M. & MOMOSE, T. (1986), Effects of prior plasticity on subsequent creep of type 316 stainless steel at elevated temperature. *J. Eng. Mater. Technol.* **108**, 68-74. *[10]*

ONO, N. & SATO, A. (1988), Plastic deformation governed by the stress induced martensite transformation in polycrystals. Trans. *Jap. Inst. Metals* **29**, 267-273. *[8]*

ORD, A. & HOBBS, N.E. (1986), Experimental control of the water-weakening effect in quartz. *Amer. Geophys. U. Geophys. Monogr.* **36**, 51-72. *[6]*

OROWAN, E. (1934), Zur Kristallplastizität III. *Z. Phys.* **89**, 634-659 (Fig.9). *[8]*

OTTE, H.M. & CROCKER, A.G. (1965), Crystallographic formulae for hexagonal lattices. *Phys. Stat. Sol.* **9**, 441-450. *[2]*

OWENS, W.H. (1973), Strain modification of angular density distributions. *Tectonophysics* **16**, 249-261. *[14]*

PAN, J. & RICE, J.R. (1983), Rate sensitivity of plastic flow and implication for yield-surface vertices. *Int. J. Solids Struct.* **19**, 973-987. *[10]*

PANCHANADEESWARAN, S., BECKER, R., DOHERTY, R.D. & KUNZE, K. (1994), Direct observation of orientation changes following channel die compression of polycrystalline aluminum: comparison between experiment and model. In *Textures of Materials ICOTOM-10*, H.J. Bunge, ed. (Switzerland: Trans. Tech. Pubs.) 1277-1282. *[5]*

PANGAROV, N.A. (1965), Preferred orientations in electro-deposited metals. *J. Electroanalytical Chemistry* **9**, 70-85. *[5]*

PARKHOMENKO, E.I. (1971), *Electrification Phenomena in Rocks*. (Translation by G. V. Keller. New York: Plenum Press) 285 pp. *[6]*

PARKS, D.M. & AHZI, S. (1990), Polycrystalline plastic deformation and texture evolution for crystals lacking five independent slip systems. *J. Mech. Phys. Solids* **38**, 701-724. *[8,11,12,13]*

PATERSON, M.S. (1989), The interaction of water with quartz and its influence in dislocation flow - an overview. In *Rheology of Solids and of the Earth*, S.I. Karato & M. Toriumi, eds. (Oxford Univ. Press.) 107-142. *[6,14]*

PATERSON, M.S. & WEISS, L.E. (1961), Symmetry concepts in the structural analysis of deformed rocks. *Geol. Soc. Amer. Bull.* **72**, 841-882. *[1,14]*

PAWLIK, K., POSPIECH, J. & LÜCKE, K. (1991), The ODF approximation from pole figures with the aid of the ADC method. *Textures Microstruct.* **14-18**, 25-30. *[3]*

PEIRCE, D., ASARO, R.J. & NEEDLEMAN, A. (1982), An analysis of nonuniform and localized deformation in ductile single crystals. *Acta Metall.* **30**, 1087-1119. *[8,13]*

PELLERIN, N., ODIER, P., SIMON, P. & CHATEIGNER, D. (1994), Nucleation and growth mechanisms of textures YBaCuO and the influence of Y_2BaCuO_5. *Physica C* **222**, 133-148. *[6]*

PENTECOST, J.L. & WRIGHT, C.H. (1964), Preferred orientation in ceramic materials due to forming techniques. In *Advances in X-Ray Analysis Vol. 7*, M. Mueller, G.R. Mallett & M. Fay, eds. (New York: Plenum Press) 174-181. *[6]*

PEROVIĆ, B. & KARASTOJKOVIĆ, Z. (1980), Variation of plastic anisotropy in relation to texture of copper and brass sheet. *Metals Technology* **7**, 79-82. *[10]*

PERUTZ, M.F. & SELIGMON, G. (1939), A crystallographic investigation of glacier structure and the mechanism of glacier flow. *Proc. R. Soc. Lond.* A **172**, 335-360. *[6]*

PEYRAC, C. & PENELLE, R. (1988), Induced texture in tension and in deep drawing. In *Eighth Int. Conf. on Textures of Materials*, J.S. Kallend & G. Gottstein, eds. (Warrendale, PA: The Metallurgical Society) 977-983. *[10]*

PHILIPPE, M.J., (1994), Texture formation in hexagonal materials. In *Textures of Materials ICOTOM-10*, H.J. Bunge, ed. (Switzerland: Trans. Tech. Pubs.) 1337-1350. [5]

PHILLIPS, F.C. (1963), *An Introduction to Crystallography* (New York: Wiley). [1]

PIAN, T.H.H. (1995), Evolution of hybrid stress finite element method and an example analysis of finite strain deformation of a rigid-viscoplastic solid. In *Simulation of Materials Processing*, S-F. Shen & P. Dawson, eds. (Rotterdam: Balkema) 23-29. [12]

PIEHLER, H.R. & BACKOFEN, W.A. (1971), A theoretical examination of the plastic properties of bcc crystals deforming by ⟨111⟩ pencil glide. *Metall. Trans.* **2**, 249-255. [9,10]

PIERRON, F., VAUTRIN, A. & HARRIS, B. (1995), The Iosipescu in-plane shear test: validation on an isotropic material. *Exp. Mech.* **35**, 130-136. [10]

PIROUZ, P., LAWLOR, B.F., GEIPEL, T., BILDE-SØRENSEN, J.B., HEUER, A.H. & LAGERLÖF, K.P.D. (1996), On basal slip and basal twinning in sapphire (α-Al_2O_3): 2. A new model of basal twinning. *Acta Mater.* **44**, 2153-2164. [2]

PLAYER, M.A., MARR, G.V., GU, E., SAVALONI, H., ÖNCAN, N. & MUNRO, I.H. (1992), Preferred orientation in erbium thin films observed using synchrotron radiation. *J. Appl. Crystallogr.* **25**, 770-777. [4]

POCHETTINO, A.A., GANNIO, N., VIAL EDWARDS, C. & PENELLE, R. (1992), Texture and pyramidal slip in Ti, Zr and their alloys. *Scripta Metall. Mater.* **27**, 1859-1863. [11]

POCHETTINO, A.A., SANCHEZ, P., LEBENSOHN, R.A. & TOMÉ, C.N. (1994), Temperature effects on rolling texture formation in zirconium alloys. In *Textures of Materials ICOTOM-10*, H.J. Bunge, ed. (Switzerland: Trans. Tech. Pubs.) 783-788. [11]

POOLE, W.J., EMBURY, J.D., KOCKS, U.F. & BOLMARO, R.E. (1991). Texture development in Cu-W composites. In *Metal Matrix Composites – Processing, Microstructure and Properties*, N. Hansen & al., eds. (Roskilde, Denmark: Risø National Laboratory) 587-593. [5]

POOLE, W.J., EMBURY, J.D., MACEWEN, S.R. & KOCKS, U.F. (1994), Large strain deformation of a copper tungsten composite system (I & II). *Philos. Mag.* **A69**, 645-665 & 667-687. [8,11]

POSPIECH, J. (1972), Die Parameter der Drehung und die Orientierungsverteilungs-funktion. *Kristall und Technik* **7**, 1057-1072. [2]

POSPIECH, J. & LÜCKE, K. (1975), The rolling textures of copper and α-brasses discussed in terms of the orientation distribution function. *Acta Metall.* **23**, 997-1007. [2]

POSPIECH, J., SZTWIERTNIA, J. & HAESSNER, F. (1986), The misorientation distribution function. *Textures Microstruct.* **6**, 201-215. [2]

POSPIECH, J., LÜCKE, K. & SZTWIERTNIA, J. (1993), Orientation distributions and orientation correlation functions for description of microstructures. *Acta Metall. Mater.* **41**, 305-321. [2,4]

POUDENS, A., BACROIX, B. & BRETHEAU, T. (1995), Influence of microstructures and particle concentrations on the development of extrusion textures in metal matrix composites. *Mater. Sci. Eng.* **A196**, 219-228. [11]

POWELL, R.W. & GRIFFITHS, E. (1937), The variation with temperature of the thermal conductivity and the x-ray structure of some micas. *Proc. R. Soc. Lond.* **A163**, 189-198. [1]

PRAGER, W. (1959), *An Introduction to Plasticity* (Reading, MA: Addison-Wesley). [8]

PRANTIL, V.C., JENKINS, J.T. & DAWSON, P.R. (1993), An analysis of texture and plastic spin for planar polycrystals. *J. Mech. Phys. Solids* **41**, 1357-1382. *[2]*

PRANTIL, V.C., DAWSON, P.R. & CHASTEL, Y.B. (1995a), Comparison of equilibrium based plasticity models and a Taylor-like hybrid formulation for deformations of constrained crystal systems. *Modeling and Simulation in Materials Science and Engineering* **3**, 215-234. *[12]*

PRANTIL, V.C., JENKINS, J.T. & DAWSON, P.R. (1995b), Modeling deformation induced textures in titanium using analytical solutions for constrained single crystal response. *J. Mech. Phys. Solids* **43**, 1283-1302. *[8]*

PRIESMEYER, H.G., LARSEN, J. & MEGGERS, K. (1994), Neutron diffraction for non-destructive strain/stress measurements in industrial devices. *J. Neutron Res.* **2**, 31-52. *[4]*

PRZYSTUPA, M.A., VASUDEVAN, A.K. & ROLLETT, A.D. (1994), Crystallographic texture gradients in the aluminum 8090 matrix alloy and 8090 particulate composites. *Mater. Sci. Eng.* **A186**, 35-44. *[5]*

PURSEY, H. & COX, H.L. (1954), The correction of elasticity measurements on slightly anisotropic materials. *Philos. Mag.* **45**, 295-302. *[3]*

RAABE, D. (1992). *Texturen kubisch-raumzentrierter Überhangsmetalle.* Ph.D. Thesis (Aachen, Germany: Rheinisch-Westfälische Techn. Hochschule). *[9]*

RAABE, D. & LÜCKE, K. (1994), Rolling and annealing textures of bcc metals. In *Textures of Materials ICOTOM-10*, H.J. Bunge, ed. (Switzerland: Trans. Tech. Pubs.) 597-610. *[5,9]*

RAABE, D., HERINGHAUS, F., HANGEN, U. & GOTTSTEIN, G. (1995), Investigation of a Cu-20 mass% Nb in situ composite. *Z. Metallk.* **86**, 405-415. *[5]*

RAMAN, V., WATANABE, T. & LANGDON, T.G. (1989), A determination of the structural dependance of cyclic migration in polycrystalline aluminum using electron channeling pattern analysis. *Acta Metall.* **37**, 705-714. *[4]*

RANDLE, V. (1992). *Microstructure Determination and its Applications* (London: Inst. Mater. Sci.) *[2]*

RASTEGAEV, M.V. (1940), A new method of homogeneous compression of specimens for determining flow stress and the coefficient of internal friction. *Zadovsk Lab.* **6**, 345. *[10]*

RATSCHBACHER, L., WENK, H-R. & SINTUBIN, M. (1991), Calcite textures: examples from nappes with strain-path partitioning. *J. Struct. Geol.* **13**, 369-384. *[M,14]*

RAY, R.K., JONAS, J.J. & HOOK, R.E. (1993), Cold rolling and annealing textures in low carbon and extra low carbon steels. *Int. Mater. Reviews.* **39**, 129-172. *[5]*

RAY, R.K., JONAS, J.J., BUTRON-GUILLÉN, M.P. & SAVOIE, J. (1994), Transformation textures in steels. *Iron & Steel Inst. Jap.* **34**, 927-942. *[5]*

READ, W.T. (1950), Effect of stress-free edges in plane shear of a flat body. *J. Appl. Mech.* **17**, 349-352. *[10]*

REGAZZONI, G., KOCKS, U.F. & FOLLANSBEE, P.S. (1987), Dislocation kinetics at high strain rates. *Acta Metall.* **35**, 2865-2875. *[8]*

REGENET, P.J. & STÜWE, H.-P. (1963), Zur Entstehung von Oberfächentexturen beim Walzen kubisch-flächenzentrieter Metalle. *Z. Metallk.* **54**, 273-278. *[5]*

REIDINGER, F. & WHALEN, P.J. (1987), Texture on ground, fractured and aged Y-TZP surfaces. *Mat. Res. Soc. Symp. Proc.* **78**, 25-33. *[6]*

REIMER, L. (1985), *Scanning electron microscopy.* (Berlin: Springer) 457 pp. *[4]*

RENOUARD, M. & WINTENBERGER, M. (1976), Déformation homogène par glissements de dislocations de monocristaux de structure cubique face centrées sous l'effet de contraintes et de déplacements imposés. *Comptes Rendues Acad. Sci. Paris* **283** (B), 237-240. *[9]*

RENOUARD, M. & WINTENBERGER, M. (1981), Détermination de l'amplitude des glissements dans la déformation plastique homogène d'un monocristal sous l'effet de contraintes et de déplacements imposés. *Comptes Rendues Acad. Sci. Paris* **291** (II), 385-388. *[8,9]*

REYES-MOREL, P.E. & CHEN, I.W. (1988), Transformation plasticity of CeO_2-stabilized tetragonal zirconia polycrystals: I, Stress assistance and autocatalysis. *J. Am. Ceram. Soc.* **71**, 343-353. *[6]*

RHYS-JONES, T.N. (1990), Thermally sprayed coating system for surface protection and clearance control applications in aero engines. *Surf. and Coat. Technol.* **43/44**, 402-415. *[6]*

RICE, J.R. (1970), On the structure of stress-strain relationships for time-dependent plastic deformation of metals. *J. Appl. Mech.* **37**, 728-737. *[8,9]*

RIDHA, A. & HUTCHINSON, W. (1982), Recrystallization mechanisms and the origin of cube texture in Copper. *Acta Metall.* **30**, 1929-1939. *[5]*

RIETVELD, H.M. (1969), A profile refinement method for nuclear and magnetic structures. *J. Appl. Crystallogr.* **2**, 65-71. *[4]*

RODRIGUES, O. (1840), Des lois géométriques qui régissent les déplacements d'un système solide dans l'espace, et de la variation des coordonnées provenant de ces déplacements considérés independamment des causes qui peuvent les produire. *J. Mathématique Pures et Appliquées* **5**, 380-440. *[2]*

ROE, R.-J. (1965), Description of crystallite orientation in polycrystalline materials III, general solution to pole figure inversion. *J. Appl. Phys.* **36**, 2024-2031. *[2,3]*

ROEDER, R.K., BOWMAN, K. & TRUMBLE, K.P. (1995), Texture and microstructure development in Al_2O_3-platelet reinforced $Ce-ZrO_2/Al_2O_3$ laminates produced by centrifugal consolidation. *Textures Microstruct.* **24**, 43-52. *[6]*

ROLLETT, A.D. (1991). Comparison of experimental and theoretical texture development in alpha-uranium. In *Modeling the Deformation of Crystalline Solids*, T.C. Lowe &al., eds. (Warrendale PA: The Minerals, Metals and Materials Society) 361-368. *[5]*

ROLLETT, A.D. & KOCKS, U.F. (1988), Computer simulation of pencil glide in b.c.c. metals. In *Eighth Int. Conf. on Textures of Materials*, J.S. Kallend & G. Gottstein, eds. (Warrendale, PA: The Metallurgical Society) 375-380. *[9]*

ROLLETT, A.D., CANOVA, G.R. & KOCKS, U.F. (1987a), The effect of the cube texture component on the earing behavior of rolled f.c.c. metals. In *Formability and Metallurgical Structure*, A.K. Sachdev & J.D. Embury, eds. (Warrendale PA: The Metallurgical Society) 147-157. *[10]*

ROLLETT, A.D., KOCKS, U.F. & DOHERTY, R.D. (1987b), Stage IV work hardening in cubic metals. In *Formability and Metallurgical Structure*, A.K. Sachdev & J.D. Embury, eds. (Warrendale PA: The Metallurgical Society) 211-255. *[8]*

ROLLETT, A.D., LOWE, T., KOCKS, U.F. & STOUT, M.G. (1988). The microstructure and texture of torsion-reverse torsion experiments. In *Eighth Int. Conf. on Textures of Materials*, J.S. Kallend & G. Gottstein, eds. (Warrendale, PA: The Metallurgical Society) 473-478. *[5]*

ROLLETT, A.D., JUUL JENSEN, D. & STOUT, M.G. (1992), Modelling the effect of micro-
structure on yield anisotropy. In *Modelling of Plastic Deformation and its
Enginering Applications*, S.I. Andersen &al., eds. (Roskilde, Denmark: Risø Nat.
Lab.) 93-109. *[8,9,10]*

ROUGIER, Y., STOLZ, C. & ZAOUI, A. (1993), Représentation spectrale en visco-élasticité
linéaire des matériaux hétérogènes. *Comptes Rendus Acad. Sci. Paris Serie II* **316**,
1517-1522. *[7]*

ROUGIER, Y., STOLZ, C. & ZAOUI, A. (1994), Self-consistent modelling of elastic-
viscoplastic polycrystals. *Comptes Rendus Acad. Sci. Paris Serie II* **318**, 145-151. *[7]*

RUER, D. & BARO, R. (1977), A new method for the determination of the texture of
materials of cubic structure from incomplete reflection pole figures. *Adv. X-Ray
Analysis* **20**, 187-200. *[3]*

RUNDE, M., ROUTBORT, J.L., MUNDY, J.N., ROTHMAN, S.J., WILEY, C.L. & XU, X. (1992),
Diffusion of ^{18}O in Bi_2Sr_2CuO single crystals. *Phys. Rev.* **B46**, 3142-3144. *[1]*

SACHS, G. (1928), Zur Ableitung einer Fliessbedingung. *Z. Ver. Dtsch. Ing.* **72**, 734-736.
[8,14]

SALEM, A., SHANNON, J.L. & BRADT, R.C. (1989), Crack growth resistance of textured
alumina. *J. Am. Ceram. Soc.* **72**, 20-27. *[6]*

SALINAS RODRÍGUEZ, A. (1995), Grain size effects on the texture evolution of α-Zr. *Acta
Metall. Mater.* **43**, 485-498. *[11]*

SALINAS RODRÍGUEZ, A.S. & JONAS, J.J. (1992), Evolution of textures in zirconium alloys
deformed uniaxially at elevated temperatures. *Metall. Trans.* **23A**, 271-293. *[5,11]*

SALINAS RODRÍGUEZ, A. & ROOT, J.H. (1991), Texture evolution in Zr-2.5%Nb deformed
in uniaxial compression. *Textures Microstruct.* **14-18**, 1239-1244. *[11]*

SANDER, B. (1950), *Einführung in die Gefügekunde der Geologischen Körper*, Vol. **2**.
(Wien: Springer). *[4,6]*

SANDHAGE, K.H., RILEY Jr., G.N. & CARTER, W.L.(1991), Critical issues in the OPIT
processing of high-J_c BSCCO superconductors. *JOM* **43**, 21-25 (The Min. Met. &
Mater. Soc.). *[6]*

SANDLIN, M.S. & BOWMAN, K.J. (1992), Green body processing effects on SiC whisker
textures in alumina matrix composites. *Ceram. Eng. Sci. Proc.* **13**, 661-668. *[6]*

SANDLIN, M.S., LEE, F. & BOWMAN, K.J. (1992), Simple geometric model for assessing
whisker orientation in axisymmetric SiC-whisker-reinforced composites. *J. Am.
Ceram. Soc.* **75**, 1522-1528. *[6]*

SANDLIN, M.S., PETERSON, C.R. & BOWMAN, K.J. (1994), Texture measurements on
materials containing platelets using stereology. *J. Am. Ceram. Soc.* **77**, 2127-2131.
[4,6]

SARMA, G.B. & DAWSON, P.R. (1996), Effects of interaction among crystals on the
inhomogeneous deformations of polycrystals. *Acta Mater.* **44**, 1937-1953. *[8,12,13]*

SAVAGE, M.K. & SILVER, P.G. (1993), Mantle deformation and tectonics: constraints
from seismic anisotropy in the western United States. *Phys. Earth Planet. Interiors.*
78, 207-227. *[14]*

SCHAEBEN, H. (1988), Entropy optimization in texture goniometry. *Phys. Stat. Sol. (b)*
148, 63-72. *[3]*

SCHAEBEN, H. (1990), Parameterizations and probability distributions of orientations.
Textures Microstruct. **13**, 51-54. *[2]*

SCHAEBEN, H. (1993), Numerical determination of the variation width of feasible ODFS.
Textures Microstruct. **21**, 55-62. *[3]*

SCHÄFER, W., MERZ, P., JANSEN, E. & WILL, G. (1991), Neutron diffraction texture analysis of multiphase and low-symmetry materials using the position-sensitive detector Julios and peak deconvolution methods. *Textures Microstruct.* **14-18**, 65-71. *[4]*

SCHAPERY, R.H. (1962), Approximate methods of transform inversion for viscoelastic stress analysis. *Proc. Fourth U.S. Nat. Congr. Appl. Mech.* (ASME) 1075-1085. *[7]*

SCHAPERY, R.H. (1974), Viscoelastic behavior and analysis of composite materials. In *Mechanics of Composite Materials, Chap. 4*, G.P. Sendeckyj, ed. (Orlando, FL: Academic Press) 85-168. *[7]*

SCHEFFZÜK, C. (1996), Neutron diffraction texture analysis of natural and experimentally deformed halite samples of the Merkers mine. *Z. Geol. Wiss.* **24**, 403-410. *[6]*

SCHMID, E. (1924), Zn – normal stress law. *Proc. Int. Congr. Appl. Mech.* (Delft), 342. *[8]*

SCHMID, E. & WASSERMANN, G. (1927), Über die Texturen hartgezogener Drähte. *Z. Phys.* **42**, 779-794. *[9]*

SCHMID, S.M. & CASEY, M. (1986), Complete fabric analysis of some commonly observed quartz c-axis patterns. *Am. Geophys. U. Geophys. Monogr.* **36**, 263-286. *[6]*

SCHMID, S.M., BOLAND J.N. & PATERSON, M.S. (1977), Superplastic flow in fine grained limestone. *Tectonophysics* **43**, 257-291. *[6]*

SCHMID, S.M., CASEY, M. & STARKEY, J. (1981), An illustration of the advantages of a complete texture analysis described by the orientation distribution function (ODF) using quartz pole figure data. *Tectonophysics* **78**, 101-117. *[6]*

SCHMIDT, N.H. & OLESEN, N.Ø. (1989), Computer-aided determination of crystal lattice orientation from electron channeling patterns in the SEM. *Canadian Mineralogist* **27**, 15-22. *[4]*

SCHMIDT, W. (1925), Gefügestatistik. *Tschermaks Mineral. Petrog. Mitt.* **38**, 392-423. *[14]*

SCHMIDT, W. (1932), *Tektonik und Verformungslehre.* (Berlin: Bornträger). *[6]*

SCHMITT, J.H. (1986). *Contribution à l'étude de la micro-plasticité des aciers.* Ph.D. Thesis, Inst. Nat. Polytechnique, Grenoble. *[10]*

SCHRECK, M., HESSMER, R., GEIER, S., RAUSCHENBACH, B. & STRITZKER, B. (1994), Structural characterization of diamond films grown epitaxially on silicon. *Diamond and Related Materials* **3**, 510-514. *[6]*

SCHUBNIKOW, A. & ZINSERLING, K. (1932), Über die Schlag- und Druckfiguren und über die mechanischen Quarzzwillinge. *Z. Kristallogr.* **83**, 243-264. *[2]*

SCHULZ, L.G. (1949a), A direct method of determining preferred orientation of a flat transmission sample using a Geiger counter x-ray spectrometer. *J. Appl. Phys.* **20**, 1030-1033. *[4]*

SCHULZ, L.G. (1949b), Determination of preferred orientation in flat transmission samples using a Geiger counter x-ray spectrometer. *J. Appl. Phys.* **20**, 1033-1036. *[4]*

SCHWARZER, R.A. & WEILAND, H. (1988), Texture analysis by the measurement of individual grain orientations. – Electron microscopical methods and application to dual phase steel. *Textures Microstruct.* **8-9**, 443-456. *[4]*

SEKINE, K., VAN HOUTTE, P., GIL SEVILLANO, J. & AERNOUDT, E. (1981), The transition of torsional deformation textures in fcc metals. In *Sixth Int. Conf. on Textures of Materials* (Tokyo: The Iron and Steel Institute of Japan) 396-407. *[5]*

SELVAMANICKAM, V. & SALAMA, K. (1990), Anisotropy and intergrain current density in oriented grained bulk YBa$_2$Cu$_3$O$_x$ superconductor. *Appl. Phys. Lett.* **57**, 1575-1577. *[6]*

SHINDO, T. & FURUKAWA, T. (1981). Colony structure and its texture in a ferritic stainless steel. In *Sixth Int. Conf. on Textures of Materials* (Tokyo: The Iron and Steel Institute of Japan) 846-855. *[5]*

SHIRATORI, E., IKEGAMI, K. & YOSHIDA, F. (1979), Analysis of stress strain relations by use of an anisotropic hardening plastic potential. *J. Mech. Phys. Sol.* **27**, 213-229. *[10]*

SHRIVASTAVA, S.C., JONAS, J.J. & CANOVA, G. (1982), Equivalent strain in large deformation torsion testing: Theoretical and practical considerations. *J. Mech. Phys. Solids* **30**, 75-90. *[14]*

SIEGESMUND, S., HELMING, K. & KRUSE, R. (1994), Complete texture analysis of a deformed amphibolite: comparison between neutron diffraction and U-stage data. *J. Struct. Geol.* **16**, 131-142. *[6]*

SIEMES, H. (1970), Experimental deformation of galena ores. In *Experimental and Natural Rock Deformation*, P. Paulitsch, ed. (Berlin: Springer) 165-208. *[14]*

SIEMES, H. (1974), Anwendung der Taylor-Theorie auf die Regelung von kubischen Mineralen. *Contrib. Mineral. Petrol.* **43**, 149-157. *[14]*

SIEMES, H. & HENNIG-MICHAELI, C. (1985), Ore minerals. In *Preferred Orientation in Deformed Metals and Rocks: An Introduction to Modern Texture Analysis*, H.-R. Wenk, ed. (Orlando, FL: Academic Press) 335-360. *[6]*

SILVER, P.G. (1996), Seismic anisotropy beneath the continents: Probing the depths of geology. *Ann. Rev. Earth Planet. Sci.* **24**, 385-432. *[14]*

SIMMONS, G. & WANG, H. (1971), *Single Crystal Elastic Constants and Calculated Aggregate Average Properties – A Handbook* (Cambridge, MA: MIT Press). *[1,7]*

SINTUBIN, M. (1994), Texture types in shales and slates. In *Textures of Geological Materials*, H.-J. Bunge &al. eds. (Oberursel, Germany: DMG Informationsgesellschaft) 221-229. *[6]*

SINTUBIN, M., WENK, H.-R. & PHILLIPS, D. (1995), Texture development in platy materials: comparison of Bi2223 aggregates with phyllosilicate fabrics. *Mater. Sci. Eng.* **A202**, 157-171. *[6]*

SKROTZKI, W. & HAASEN, P. (1981), Hardening mechanisms of ionic crystals on {110} and {100} slip planes. *J. Phys. Colloque C3* **42**, 119-148 (Supplement 6). *[6]*

SKROTZKI, W. & WELCH, P. (1983), Development of texture and microstructure in extruded ionic polycrystalline aggregates. *Tectonophysics* **99**, 47-62. *[6,14]*

SKROTZKI, W., WEDEL, A., WEBER, K. & MÜLLER, W.F. (1990), Microstructure and texture in lherzolites of the Balmuccia massif and their significance regarding the thermomechanical history. *Tectonophysics* **179**, 227-251. *[6]*

SKROTZKI, W., HELMING, K., BROKMEIER, H.-G., DORNBUSCH, H.-J. & WELCH, P. (1995), Textures in pure shear deformed rock salt. *Textures Microstruct.* **24**, 133-141. *[6]*

SMELSER, R.E. & BECKER, R. (1991), Earing in cup drawing of a (100) single crystal. In *ABACUS User's Group Conf. Proc., Oxford*, 457-471. *[13]*

SOWERBY, R., VIANA, C.S.D. & DAVIES, G.J. (1980), The influence of texture on the mechanical response of commercial purity sheet in some simple forming processes. *Mater. Sci. Eng.* **46**, 23-51. *[10]*

SPIERS, C.J. (1979), Fabric development in calcite polycrystals deformed at 400 C. *Bull. Minéral.* **102**, 282-289. *[6]*

SREE HARSHA, K.S. & CULLITY, B.D. (1962), Effect of initial orientation on the deformation texture and tensile and torsional properties of copper and aluminum wires. *Trans. Metall. Soc. AIME* **224**, 1189-1193. *[9]*

STARKEY, J. (1964), An x-ray method for determining the orientation of selected crystal planes in polycrystalline aggregates. *Am. J. Sci.* **262**, 735-752. *[4]*

STEINLAGE, G., ROEDER, R., TRUMBLE, K., BOWMAN, K., LI, S. & MCELRESH, M. (1994), Preferred orientation of BSCCO via centrifugal slip casting. *J. Mater. Res.* **9**, 833-836. *[6]*

STEPHENS, A.W. (1968). *Texture and mechanical anisotropy in the copper-zinc system.* Ph.D. Thesis, University of Arizona. *[5,10]*

STÖREN, S. & RICE, J.R. (1975), Localized necking in thin sheets. *J. Mech. Phys. Solids.* **23**, 421-441. *[10]*

STOUT, M.G. & HECKER, S.S. (1983), Role of geometry in plastic instability and fracture of tubes and sheet. *Mech. Mater.* **2**, 23-31. *[10]*

STOUT, M.G., MARTIN, P.L., HELLING, D.E. & CANOVA, G.R. (1985), Multiaxial yield behavior of 1100 aluminum following various magnitudes of prestrain. *Int. J. Plast.* **1**, 163-174. *[10]*

STOUT, M.G., FOLLANSBEE, P.S. & LOVATO, M.L. (1987), An evaluation of incremental testing to large strain in OFE copper. *Mech. Mater.* **6**, 1-10. *[10]*

STOUT, M.G., KALLEND, J.S., KOCKS, U.F., PRZYSTUPA, M.A. & ROLLETT, A.D. (1988), Material dependence of deformation texture development in various deformation modes. In *Eighth Int. Conf. on Textures of Materials*, J.S. Kallend & G. Gottstein, eds. (Warrendale, PA: The Metallurgical Society) 479-484. *[5,9]*

STÜWE, H.P. (1961), Texturbildung bei der Primärrekristallisation. *Z. Metallk.* **52**, 34-44. *[4]*

SUDO, M., HASHIMOTO, S. & TSUKATANI, I. (1981). Recrystallization texture of cold rolled high strength silicon containing steel sheets. In *Sixth Int. Conf. on Textures of Materials* (Tokyo: The Iron and Steel Institute of Japan) 1076-1085. *[5]*

SUE, J.A. & TROUE, H.H. (1987), Effect of crystallographic orientation on erosion characteristics of arc evaporation titanium nitride coating. *Surface & Coatings Tech.* **33**, 169-181. *[6]*

SUITS, J.C. & CHALMERS, B. (1961), Plastic microstrain in silicon-iron. *Acta Metall.* **9**, 854-860. *[8]*

SURI, A.K., NIMMAGADDA, R. & BUNSHAH, R.F. (1980), Synthesis of titanium nitrides by activated reactive evaporation. *Thin Solid Films* **72**, 529-533. *[6]*

SWIFT, H.W. (1947), Length changes in metals under torsional overstrain. *Engineering (London)* **163**, 253-257. *[10]*

SZPUNAR, J.A. (1990), Energy dispersive diffractometry for quantitative texture studies. *Textures Microstruct.* **12**, 243-248. *[4]*

SZPUNAR, J.A. (1996), Textures in thin films. In *Texture of Materials ICOTOM-11*, Z. Liang, L. Zuo & Y. Chu, eds. (Beijing: International Academic Publishers) 1095-1104. *[6]*

SZTWIERTNIA, K. & HAESSNER, F. (1990), The orientation characteristics of different recrystallization stages in copper. *Textures Microstruct.* **14-18**, 641-646. *[4]*

SZTWIERTNIA, K. & HAESSNER, F. (1994), Orientation aspects of the morphological elements of the microstructure in highly cold rolled pure copper and phosphorus copper. In *Textures of Materials ICOTOM-10*, H.J. Bunge, ed. (Switzerland: Trans. Tech. Pubs.) 1291-1298. *[5]*

TAKESHITA, T. & WENK, H.-R. (1988), Plastic anisotropy and geometric hardening in quartzites. *Tectonophysics* **149**, 345-361. *[14]*

TAKESHITA, T., TOMÉ, C.N., WENK, H.-R. & KOCKS, U.F. (1987), Single crystal yield surface for trigonal lattices: Application to texture transitions in calcite polycrystals. *J. Geophys. Res.* **B92**, 12917-12920. *[6,8,11,14]*

TAKESHITA, T., KOCKS, U.F. & WENK, H.-R. (1989), Strain path dependence of texture development in aluminum. *Acta Metall.* **37**, 2595-2611. *[2]*

TAKESHITA, T., WENK, H.-R., MOLINARI, A. & CANOVA, G. (1990), Simulation of dislocation assisted plastic deformation in olivine polycrystals. In *Deformation Processes in Minerals, Ceramics and Rocks*, D.J. Barber & P.G. Meredith, eds. (London: Unwin Hyman) 365-377. *[6,11]*

TAYLOR, G.I. (1938a), Plastic strain in metals. *J. Inst. Met.* **62**, 307-324. *[8,9,12,13]*

TAYLOR, G.I. (1938b), Analysis of plastic strain in a cubic crystal. In *Contributions to the Mechanics of Solids* (New York: Macmillan) 218-224. *[8,9]*

TAYLOR, G.I. (1956), Strains in crystalline aggregate. In *Deformation and Flow of Solids*, R. Grammel, ed. (Berlin: Springer) 3-12. *[8,9,10]*

TENCKHOFF, E. (1970a), Defocusing for the Schulz technique of determining preferred orientation. *J. Appl. Phys.* **41**, 3944-3948. *[4]*

TENCKHOFF, E. (1970b), Der Einfluß der Reduktionswerte auf die Entwicklung von Textur und Texturgradienten in Zircaloy Rohren. *Z. Metallk.* **61**, 64-71. *[5]*

TENCKHOFF, E. (1974), Operable deformation systems and mechanical behavior of textured zircaloy tubing. In *Zirconium in Nuclear Applications ASTM STP 551.* 179-200. *[11]*

TENCKHOFF, E. (1978), The development of the deformation texture in zirconium during rolling in sequential passes. *Metall. Trans.* **A9**, 1401-1412. *[5,11]*

TENCKHOFF, E. (1988). *Deformation Mechanisms, Texture, and Anisotropy in Zirconium and Zircaloy* (Philadelphia: ASTM). *[5]*

TEWARI, S.N. (1988), Effect of melt spinning on grain size and texure in Ni-Mo alloys. *Metall. Trans.* **19A**, 1711-1720. *[5]*

THORSTEINSSON, T., KIPFSTUHL, J. & MILLER, H. (1997), Textures and fabrics in the GRIP ice core. *J. Geophys. Res.* (in press) *[14]*

TIEM, S., BERVEILLER, M. & CANOVA, G.R. (1986), Grain shape effects on the slip system activity and on the lattice rotations. *Acta Metall.* **34**, 2139-2149. *[7,8,11]*

TIETZ, L.A., CARTER, C.B., LATHROP, D.K., RUSSEK, S.E., BUHRMAN, R.A. & MICHAEL, J.R.J. (1989), Crystallography of $YBa_2Cu_3O_{6+x}$ thin film-substrate interfaces. *J. Mater. Res.* **4**, 1072-1081. *[6]*

TOMÉ, C. & KOCKS, U.F. (1985), The yield surface of hcp single crystals. *Acta Metall.* **33**, 603-621. *[8,10,11,12]*

TOMÉ, C., CANOVA, G.R., KOCKS, U.F., CHRISTODOULOU, N. & JONAS, J.J. (1984), The relation between macroscopic and microscopic strain hardening in f.c.c. polycrystals. *Acta Metall.* **32**, 1637-1653. *[8,9,10,13]*

TOMÉ, C., POCHETTINO, A. & PENELLE, R. (1988), Analysis of plastic anisotropy of rolled zircaloy-4. In *Eighth Int. Conf. on Textures of Materials*, J.S. Kallend & G. Gottstein, eds. (Warrendale, PA: The Metallurgical Society) 985-990. *[11]*

TOMÉ, C.N., LEBENSOHN, R.A. & KOCKS, U.F. (1991a), A model for texture development dominated by deformation twinning: application to zirconium alloys. *Acta Metall. Mater.* **39**, 2667-2680. *[8,11,12,14]*

TOMÉ, C.N., WENK, H.-R., CANOVA, G.R. & KOCKS, U.F. (1991b), Simulations of texture development in calcite: comparison of polycrystal plasticity theories. *J. Geophys. Res.* **96**, 11865-11875. *[11,14]*

TOMÉ, C.N., SO, C.B., WOO, C.H. (1993), Self-consistent calculation of steady-state creep and growth in textured zirconium. *Philos. Mag.* **A67**, 917-930. *[7]*

TOMÉ, C.N., CHRISTODOULOU, N., TURNER, P.A., MILLER, M.A., WOO, C.H., ROOT, J. & HOLDEN, T.M. (1996), Role of internal stresses in the transient of irradiation growth of Zircaloy-2. *J. Nucl. Mater.* **227**, 237-250. *[7]*

TOMILIN, M.G. (1990), New physical method to study characteristics of the surfaces coated by NLC. *Mol. Cryst. Liq. Cryst.* **193**, 7-10. *[6]*

TONG, P. & PIAN, T.H.H. (1969), A variational principle and convergence of a finite element method based on assumed stress distribution. *Int. J. Solids Struct.* **5**, 463. *[12,13]*

TÓTH, L.E. (1971), *Transition Metal Carbides and Nitrides.* (New York: Academic Press). *[6]*

TÓTH, L.S. & VAN HOUTTE, P. (1992), Discretization techniques for orientation distribution functions. *Textures Microstruct.* **19**, 229-244. *[2]*

TÓTH, L.S., GILORMINI, P & JONAS, J.J. (1988), Effect of rate sensitivity on the stability of torsion textures. *Acta Metall.* **36**, 3077-3091. *[8,9]*

TÓTH, L.S., NEALE, K.W. & JONAS, J.J. (1989), Stress response and persistence characteristics of the ideal orientations of shear textures. *Acta Metall.* **37**, 2197-2210. *[8]*

TOZAWA, Y. (1978), Plastic deformation behavior. In *Mechanics of Sheet Metal Forming: Material Behavior and Deformation Analysis*, D.P. Koistinen & N-M. Wang, eds. (Plenum Press) 81-109. *[10]*

TRESCA, H. (1864), Mémoire sur l'écoulement des corps solides soumis à des fortes pressions. *Comptes Rendus Hébdom. Acad. Sci. Paris* **59**, 754-758. *[8,10]*

TRUSZKOWSKI, W., KRÓL, J. & MAJOR, B. (1980), Inhomogeneity of rolling texture in fcc metals. *Metall. Trans.* **11A**, 749-758. *[5]*

TUCKER, G.E.G. (1961), Texture and earing in deep drawing of aluminium. *Acta Metall.* **9**, 275-286. *[10]*

TULLIS, J. & TULLIS, T. (1972), Preferred orientation of quartz produced by mechanical Dauphiné twinning: thermodynamics and axial experiments. *Am. Geophys. U. Geophys. Monogr.* **16**, 67-82. *[6]*

TULLIS, J. & WENK, H.-R. (1994), The effect of muscovite on the strength and lattice preferred orientations of experimentally deformed quartz aggregates. *Mater. Sci. Eng.* **A175**, 209-220. *[6]*

TULLIS, J., CHRISTIE, J.M. & GRIGGS, D.T. (1973), Microstructure and preferred orientations of experimentally deformed quartzites. *Geol. Soc. Am. Bull.* **84**, 297-314. *[6]*

TULLIS, T.E. (1976), Experiments on the origin of slaty cleavage and schistosity. *Geol. Soc. Am. Bull.* **87**, 745-753. *[6]*

TULLIS, T.E. & WOOD, D.S. (1975), Correlation of finite strain from both reduction bodies and preferred orientation of mica in slate from Wales. *Geol. Soc. Am. Bull.* **86**, 632-638. *[6]*

TUOMINEN, M., GOLDMAN, A.M., CHANG, Y.Z. & JIANG, P.Z. (1990), Magnetic anisotropy of high-T_c superconductors. *Phys. Rev. B* **42**, 412-419. *[6]*

TURNER, F.J., GRIGGS, D.T. & HEARD, H.C. (1954), Experimental deformation of calcite crystals. *Bull. Geol. Soc. Am.* **65**, 883-934. *[14]*

TURNER, F.J., GRIGGS, D.T., CLARK, R.H. & DIXON, R.H. (1956), Deformation of Yule marble, part VII: development of oriented fabrics at 300°C-400°C. *Geol. Soc. Am. Bull.* **67**, 1259-1294. *[6]*

TURNER, P.A. & TOMÉ, C.N. (1993), Self-consistent modeling of visco-elastic polycrystals: application to irradiation creep and growth. *J. Mech. Phys. Solids* **41**, 1191-1211. *[7]*

TURNER, P.A. & TOMÉ, C.N. (1994), A study of residual stresses in zircaloy-2 with rod texture. *Acta Metall. Mater.* **42**, 4143-4153. *[8]*

TURNER, P.A., TOMÉ, C.N. & WOO, C.H. (1994), Self-consistent modeling of non-linear visco-elastic polycrystals: an approximate scheme. *Philos. Mag.* **A70**, 689-711. *[7]*

ULLEMEYER, K., HELMING, K. & SIEGESMUND, S. (1994), Quantitative texture analysis of plagioclase. In *Textures of Geological Materials*, H.-J. Bunge &al., eds. (Oberursel, Germany: Deutsch. Gesell. Metallkunde) 93-108. *[6]*

VADON, A., RUER, D. & BARO, R. (1980), The generalization and refinement of the vector method for the texture analysis of polycrystalline materials. *Adv. X-Ray Analysis* **23**, 349-360. *[2]*

VALIANT, L. (1982), A scheme for fast parallel communication. *SIAM J. Comp.*, **11**, 350-361. *[13]*

VAN ACKER, K., DE BUYSER, L., CELIS, J.P. & VAN HOUTTE, P. (1994), Characterization of thin nickel electrocoatings by the low-incident-beam-angle diffraction method. *J. Appl. Crystallogr.* **27**, 56-66. *[4]*

VAN BAEL, A., VAN HOUTTE, P., AERNOUDT, E., HALL, F.R., PILLINGER, I., HARTLEY, P. & STURGESS, C.E.N. (1991), Anisotropic finite-element analysis of plastic metal forming processes. *Textures Microstr.* **14-18**, 1007-1012. *[13]*

VAN HOUTTE, P. (1978), Simulation of the rolling and shear texture of brass by the Taylor theory adapted for mechanical twinning. *Acta Metall.* **26**, 591-604. *[2,14]*

VAN HOUTTE, P. (1981), Adaptation of the Taylor theory to the typical substructure of cold rolled FCC metals. In *Sixth Int. Conf. on Textures of Materials* (Tokyo: The Iron and Steel Institute of Japan) 428-437. *[9,14]*

VAN HOUTTE, P. (1983), The use of a quadratic form for the determination of non-negative texture functions. *Textures Microstruct.* **6**, 1-20. *[3]*

VAN HOUTTE, P. (1984), Some recent developments in the theories for deformation texture prediction. In *Seventh Int. Conf. on Textures of Materials*, C.M. Brakman, P. Jongenburger & E.J. Mittemeijer, eds. (Netherlands Soc. for Materials Science) 7-23. *[9]*

VAN HOUTTE, P. (1986), The effect of strain path on texture: theoretical and experimental considerations. In *Strength of Metals and Alloys*, H.J. McQueen &al., eds. (Oxford: Pergamon) 1701-1725. *[9]*

VAN HOUTTE, P. (1987), Calculation of the yield locus of textured polycrystals using the Taylor and the relaxed Taylor theory. *Textures Microstruct.* **7**, 29-72. *[10]*

VAN HOUTTE, P. (1991), A method for the generation of various ghost correction algorithms – the example of the positivity method and the exponential method. *Textures Microstruct.* **13**, 199-212. *[3]*

VAN HOUTTE, P. (1994), Application of plastic potentials on strain rate sensitive and insensitive anisotropic materials. *Int. J. Plast.* **10**, 719-748. *[8]*

VAN HOUTTE, P. (1995), Heterogeneity of plastic strain around an ellipsoidal inclusion in an ideal plastic matrix. *Acta Metall. Mater.* **43**, 2859-2879. *[8,9]*

VAN HOUTTE, P. (1996), Microscopic strain heterogeneity and deformation texture prediction. In *Texture of Materials ICOTOM-11*, Z. Liang, L. Zuo & Y. Chu, eds. (Beijing: International Academic Publishers) 236-247. *[9]*

VAN HOUTTE, P. & AERNOUDT, E. (1976), Considerations on the crystal and the strain symmetry in the calculation of deformation textures with the Taylor theory. *Mater. Sci. Eng.* **23**, 11-22. *[2]*

VAN HOUTTE, P. & DE BUYSER, L. (1993), The influence of crystallographic texture on diffraction measurements of residual stress. *Acta Metall. Mater.* **41**, 323-336. *[4]*

VAN HOUTTE, P., GIL SEVILLANO, J. & AERNOUDT, E. (1979), Models for shear band formation in rolling and extrusion. *Z. Metallk.* **70**, 426-432, 503-508. *[9]*

VAN HOUTTE, P., AERNOUDT, E. & SEKINE, K. (1981), Orientation distribution function measurements of copper and brass torsion textures. In *Sixth Int. Conf. on Textures of Materials* (Tokyo: The Iron and Steel Institute of Japan) 337-346. *[5]*

VASUDEVAN, A.K., FRICKE, W.G., MALCOLM, R.C., BUCCI, R.J., PRZYSTUPA, M.A. & BARLAT, F. (1988), On through thickness crystallographic texture gradient in Al-Li-Cu-Zr alloy. *Metall. Trans.* **19A**, 731-732. *[5]*

VATNE, H.E., SHAHANI, R. & NES, E. (1996), Deformation of cube oriented grains and formation of recrystallized cube grains in a hot deformed commercial AlMgMn aluminum alloy. *Acta Mater.* **44**, 4447-4462. *[5]*

VENABLES, J.A. & HARLAND, C.J. (1973), Electron back-scattering patterns – A new technique for obtaining crystallographic information in the scanning electron microscope. *Philos. Mag.* **27**, 1193-1200. *[4]*

VERHOEVEN, J.D., CHUMBLEY, L.S., LAABS, F.C. & SPITZIG (1991), Measurement of filament spacing in deformation processed Cu-Nb alloys. *Acta Metall.* **39**, 2825-2834. *[5]*

VERMA, R.K. (1960). Elasticity of some high-density crystals. *J. Geophys. Res.* **65**, 757-766. *[14]*

VIANA, C.S. DA C., KALLEND, J.S. & DAVIES, G.J. (1979), The use of texture data to predict the yield locus of metal sheets. *Int. J. Mech. Sci.* **21**, 355-371. *[10]*

VIGLIN, A.S. (1960), A quantitative measure of the texture of a polycrystalline material – texture function. *Fiz .Tverd. Tela* **2**, 2463-2476. *[2,3]*

VINNIK, L., ROMANOWICZ, B. & BREGER, L. (1994), Anisotropy in the center of the inner core. *Geophys. Res. Lett.* **21**, 1671-1674. *[14]*

VOIGT, W. (1928), *Lehrbuch der Kristallphysik* (Leipzig: Teubner). *[1]*

VON DREELE, R.B. (1997), Quantitative texture analysis by Rietveld refinement. *J. Appl. Cryst.* **30**, 517-525. *[4]*

WAGNER, F., WENK, H.-R., ESLING, C.E. & BUNGE, H.-J. (1981), Importance of odd coefficients in texture calculations for trigonal triclinic symmetries. *Phys. Status Solidi (a)* **67**, 269-285. *[6]*

WAGNER, F., WENK, H.-R., KERN, H., VAN HOUTTE, P. & ESLING, C. (1982), Development of preferred orientation in plane strain deformed limestone: Experiment and theory. *Contrib. Mineral. Petrol.* **80**, 132-139. *[14]*

WAGNER, P., ENGLER, O. & LÜCKE, K. (1995), Formation of Cu-type shear bands and their influence on deformation and texture of rolled (112)[111] single crystals. *Acta Metall. Mater.* **43**, 3799-3812. *[9]*

WAGONER, R.H. & WANG, N-M. (1979), An experimental and analytical investigation of in-plane deformation of 2036-T4 aluminum sheet. *Int. J. Mech. Sci.* **21**, 255-264. *[10]*

WALKER, T., MATTERN, N. & HERRMANN, M. (1995), Texture and microstructure in hot-pressed Si_3N_4. *Textures Microstruct.* 24, 75-91. [6]

WALTER, J.L. & DUNN, C.G. (1960), Growth of (110)[001]-oriented grains in high-purity silicon iron – a unique form of secondary recrystallization. *Trans. Met. Soc. AIME* 218, 1033-1038. [5]

WALTER, J. L. & DUNN, C.G. (1959), Tertiary recrystallization in silicon iron. *Trans. Met. Soc. AIME* 215, 465-471. [5]

WALTHER, K., ULLEMEYER, K., HEINITZ, J., BETZL, M. & WENK, H.-R. (1995), TOF texture analysis of limestone standard: Dubna results. *J. Appl. Crystallogr.* 28, 503-507. [4]

WASSERMANN, G. (1963), Der Einfluß mechanischer Zwillingsbildung auf die Entstehung der Walztexturen kubisch flächenzentrierter Metalle. *Z. Metallk.* 54, 61-65. [9]

WASSERMANN, G. & GREWEN, J. (1962). *Texturen Metallischer Werkstoffe* (Berlin: Springer). [9]

WASSERMANN, G., BERGMANN, H.W. & FROMMEYER, G. (1978), Deformation textures in two-phase systems. In *Textures of Materials*, G. Gottstein & K. Lücke, eds. (Berlin: Springer) 37-46. [5,11]

WATTS, A.B. & FORD, H. (1955), On the basic yield stress curve for a metal. *Proc. Inst. Mech. Engrs.* 169, 1141-1156. [10]

WEBER, K. (1981), Kinematic and metamorphic aspects of cleavage-formation in very low-grade metamorphic slates. *Tectonophysics* 78, 291-306. [6]

WEILAND, H. & PANCHANADEESWARAN, S. (1993), Local pole figures in the TEM – Measurement and ODF determination. *Textures Microstruct.* 20, 67-86. [4]

WEILAND, H., LIU, J. & NES, E. (1990), Substructure and local orientation evolution of aluminum single crystals during deformation. *Textures Microstruct.* 14-18, 109-114. [4]

WEISS, L.E. & WENK, H.-R. (1985), Symmetry of pole figures and textures. In *Preferred Orientation in Deformed Metals and Rocks: An Introduction to Modern Texture Analysis*, H.-R. Wenk, ed. (Orlando, FL: Academic Press) 49-72. [2]

WEISSENBERG, K. (1922), Statistische Anisotropie in kristallinen Medien. *Annalen d. Physik* 69, 409-435. [1,2]

WEISSENBERG, K. (1924), Zur Systematik und Theorie der Wachstums- und Deformationsstrukturen. *Z Kristallogr.* 61, 58-74. [1]

WEISSENBERG, K. (1935), Le mécanique des corps déformables. *Arch. des Sciences physique et naturelles* (Geneva, Switzerland) 17, 44-106, 130-171. [1]

WELCH, P.I., RATKE, L. & BUNGE, H.-J. (1983), Consideration of anisotropy parameters in polycrystalline metals. *Z. Metallk.* 74, 233-237. [10]

WENK, H.-R. (1963), Eine Gefüge-Röntgenkamera, *Schweiz. mineral. petrog. Mitt.* 43, 707-719. [4]

WENK, H.-R. (1965),. Eine photographische Röntgengefügeanalyse, *Schweiz. mineral. petrog. Mitt.* 45, 518-550. [4,6]

WENK, H.-R., ed. (1985a), *Preferred Orientation in Deformed Metals and Rocks: An Introduction to Modern Texture Analysis* (Orlando, FL: Academic Press) 610 pp. [1,6,7]

WENK, H.-R. (1985b), Chapter 2, Measurement of pole figures. In *Preferred Orientation in Deformed Metals and Rocks: An Introduction to Modern Texture Analysis*, H.-R. Wenk, ed. (Orlando, FL: Academic Press) 11-47. [4]

WENK, H.-R. (1985c), Chapter 17, Carbonates. In *Preferred Orientation in Deformed Metals and Rocks: An Introduction to Modern Texture Analysis*, H.-R. Wenk, ed. (Orlando, FL: Academic Press) 361-384. *[14]*

WENK, H.-R. (1991), Standard project for pole figure determination by neutron diffraction. *J. Appl. Crystallogr.* 24, 920-927. *[4]*

WENK, H.-R. (1992), Advantages of monochromatic radiation for texture determination of superconducting thin films. *J. Appl. Crystallogr.* 25, 524-530. *[4]*

WENK, H.-R. (1994a), Texture analysis with TOF neutrons. In *Time-of-Flight-Diffraction at Pulsed Neutron Sources*, J.D. Jorgensen & A.J. Schultz, eds. (*Trans. Am. Cryst. Ass.* 29) 95-108. *[4]*

WENK, H.-R. (1994b), Preferred orientation patterns in deformed quartzites. In *Reviews in Mineralogy Vol.* 29, *Silica Physical Behavior, Geochemistry and Materials Applications*, P.J., C.T. Prewitt & G.V. Gibbs, eds. (Washington, DC: Mineral. Soc. Amer.) 177-208. *[6]*

WENK, H.-R. & CHRISTIE, J.M. (1991), Review paper: Comments on the interpretation of deformation textures in rocks. *J. Struct. Geol.* 13, 1091-1110. *[6,14]*

WENK, H.R. & KOCKS, U.F. (1987), The representation of orientation distributions. *Metall. Trans.* 18A, 1083-1092. *[2]*

WENK, H.-R. & PANNETIER, J. (1990), Texture development in deformed granodiorites from the Santa Rosa mylonite zone, southern California. *J. Struct. Geol.* 12, 177-184. *[4,6]*

WENK, H.-R. & PHILLIPS, D. (1992), Highly textured Bi-2223 aggregates produced by cold pressing of powders. *Physica C* 200, 105-112. *[6]*

WENK, H.-R. & SHORE, J. (1975), Preferred orientation in experimentally deformed dolomite. *Contrib. Mineral. Petrol.* 52, 115-126. *[6]*

WENK, H.-R. & TROMMSDORFF, V. (1965), Koordinatentransformation, mittelbare Orientierung, Nachbarwinkelstatistik. *Beitr. Mineral. Petrogr.* 11, 559-585. *[4]*

WENK, H.-R., VENKITASUBRAMANYAN, C.S., BAKER, D.W. & TURNER, F.J. (1973), Preferred orientation in experimentally deformed limestone. *Contrib. Mineral. Petrol.* 38, 81-114. *[6]*

WENK, H-R., KERN, H., SCHÄFER, W. & WILL, G. (1984), Comparison of x-ray and neutron diffraction in texture analysis of carbonate rocks. *J. Struct. Geol.* 6, 687-692. *[4]*

WENK, H.-R., BUNGE, H.-J., JANSEN, E. & PANNETIER, J. (1985), Preferred orientation of plagioclase – neutron diffraction and U-stage data. *Tectonophysics* 126, 271-284. *[4]*

WENK, H.-R., KERN, H., VAN HOUTTE, P. & WAGNER, F. (1986), Heterogeneous strain in axial deformation of limestone, textural evidence. In *Mineral & Rock Deformation: Laboratory Studies*, B.E. Hobbs & H.C. Heard, eds., *Am. Geoph. U. Geophys. Monogr.* 36, 287-295. *[6,14]*

WENK, H.-R., TAKESHITA, T., BECHLER, E., ERSKINE, B.G. & MATTHIES, S. (1987), Pure shear and simple shear calcite textures. Comparison of experimental, theoretical and natural data. *J. Struct. Geol.* 9, 731-745. *[6,14]*

WENK, H.-R., VERGAMINI, P.J. & LARSON, A.C. (1988), Texture analysis by TOF measurements of spallation neutrons with a 2d position sensitive detector. *Textures Microstruct.* 8-9, 443-456. *[4]*

WENK, H.-R., CANOVA, G., MOLINARI, A. & KOCKS, U.F. (1989a), Viscoplastic modeling of texture development in quartzite. *J. Geophys. Res.* 94, 17895-17906. *[11,14]*

WENK, H.-R., CANOVA, G., MOLINARI, A. & MECKING, H. (1989b), Texture development in halite: Comparison of Taylor model and self-consistent theory. *Acta Metall.* **37**, 2017-2029. *[11,14]*

WENK, H.-R., PANNETIER, J., BUSSOD, G. & PECHENIK, A. (1989c), Preferred orientation in experimentally deformed $YBa_2Cu_3O_{6+x}$. *J. Appl. Phys.* **65**, 4070-4073. *[6]*

WENK, H.-R., SINTUBIN, M., HUANG, J., JOHNSON, G.C. & HOWE, R.T. (1990), Texture analysis of polycrystalline silicon films. *J. Appl. Phys.* **67**, 572-574. *[6]*

WENK, H.-R., LARSON, A.C., VERGAMINI, P.J. & SCHULTZ, A.J. (1991a), TOF of pulsed neutrons and 2d detectors for texture analysis of deformed polycrystals. *J. Appl.. Phys.* **70**, 2035-2040. *[4]*

WENK, H.-R., BENNETT, K., CANOVA, G. & MOLINARI, A. (1991b), Modelling plastic deformation of peridotite with the self-consistent theory. *J. Geophys. Res.* **96B**, 8337-8349. *[6,11,14]*

WENK, H.-R., MATTHIES, S. & LUTTEROTTI, L. (1994a), Texture analysis from diffraction spectra. In *Textures of Materials ICOTOM-10*, H.J. Bunge, ed. (Switzerland: Trans. Tech. Pubs.) **157-162**, 473-479. *[4]*

WENK, H.-R., PAWLIK, K., POSPIECH, J. & KALLEND, J.S. (1994b), Deconvolution of superposed pole figures by discrete ODF methods: Comparison of ADC and WIMV for quartz and calcite with trigonal crystal and triclinic specimen symmetry. *Textures Microstruct.* **22**, 233-260. *[3,4]*

WENK, H.-R., HEIDELBACH, F., CHATEIGNER, D. & ZONTONE, F. (1996a), Laue orientation imaging. *J. Synchr. Rad.* **4**, 95-101. *[4]*

WENK, H.-R., CHAKMOURADYAN, A. & FOLTYN, S. (1996b), Quantitative texture measurements of YBCO films on various substrates. *Mater. Sci. Eng.* **A205**, 9-22. *[6]*

WENK, H.-R., CHATEIGNER, D., PERNET, M., BINGERT, J., HELLSTROM, E. & OULADDIAF, B. (1996c). Texture analysis of Bi 2212 and 2223 tapes and wires using neutron diffraction. *Physica C* **272**, 1-12. *[6]*

WENK, H.-R., CANOVA, G.C., BRECHET, Y. & FLANDIN, L. (1997a), A deformation-based model for recrystallization of anisotropic materials. *Acta Mater.* **45**, 3283-3296. *[11,14]*

WENK, H.-R., MATTHIES, S., DONOVAN, J. & CHATEIGNER, D. (1997b), BEARTEX, a Windows-based program system for quantitative texture analysis. *J. Appl. Cryst.* **31** (in press). *[4]*

WESTON, J.E. (1980), Origin of strength anisotropy in hot pressed silicon nitride. *J. Mat. Sci.* **15**, 1568-1576. *[6]*

WEVER, F. (1924), Über die Walzstruktur kubisch kristallisierender Metalle. *Z. Phys.* **28**, 69-90. *[4]*

WILKINSON, A.J. & DINGLEY, D.J. (1991), Quantitative deformation studies using electron backscatter patterns. *Acta Metall. Mater.* **39**, 3047-3055. *[4]*

WILL, G., SCHÄFER, W. & MERZ, P (1989), Texture analysis by neutron diffraction using a linear position sensitive detector. *Textures Microstruct.* **10**, 375-387. *[4]*

WILLIAMS, R.O. (1962), Shear textures in copper, brass, aluminum, iron, and zirconium. *Trans. Metall. Soc. AIME* **224**, 129-139. *[5]*

WILLIAMS, R.O. (1967), Analytical solutions for representing fiber textures as axial pole figures. *J. Appl. Phys.* **38**, 4029-4033. *[2]*

WILLIAMS, R.O. (1968a), Analytical methods for representing complex textures by biaxial pole figures. *J. Appl. Phys.* **39**, 4329- 4335. *[3]*

WILLIAMS, R.O. (1968b), The representation of the textures of rolled copper, brass, and aluminum by biaxial pole figures. *Trans. Metall. Soc. AIME* **242**, 104-115. *[2]*

WILLIS, D.G. (1977), A kinematic model of preferred orientation. *Geol. Soc. Am.. Bull.* **88**, 883-894. *[14]*

WILLKENS, C.A., CORBIN, N.D., PUJARI, V.K., YECKLEY, R.L. & MANGAUDIS, M.J. (1988), The influence of microstructure orientation on the fracture toughness of Si_3N_4 based materials. *Ceram. Eng. Sci. Proc.* **9**, 1367-1370. *[6]*

WILSON, D.V. (1969), Controlled directionality of mechanical properties in sheet metals. *Metall. Reviews* **14**, 175-188. *[10]*

WILSON, C.J.L. & RUSSELL-HEAD, D.S. (1982), Steady-state preferred orientation of ice deformed in plane strain at -10°C. *J. Glac.* **28**, 145-160. *[6]*

WITHERS, R. & BURSILL, L. (1980), Higher order structural relationships between hematite and magnetite. *J. Appl. Crystallogr.* **13**, 346-353. *[6]*

WOO, C.H. (1985), Polycrystalline effects on irradiation creep and growth in textured zirconium. *J. Nucl. Mater.* **131**, 105-117. *[7]*

WOO, C.H. (1987). *Materials for Nuclear Reactor Core Applications.* (London: British Nuclear Eng. Soc.) 65. *[7]*

WOOD, D.S. & OERTEL, G. (1980), Deformation in the Cambrian slate belt of Wales. *J. Geol.* **88**, 309-326. *[14]*

WOODHOUSE, J.H., GIARDINI, D. & LI, X.-D. (1986), Evidence for inner core anisotropy from free oscillations. *Geophys. Res. Lett.* **13**, 1549-1552. *[14]*

WOODTHORPE, J. & PEARCE, R. (1970), The anomalous behavior of aluminum sheet under biaxial tension. *Int. J. Mech. Sci.* **12**, 341-347. *[10]*

WOOLLEY, R.L. (1953), The Bauschinger effect in some face-centered and body-centered cubic metals. *Philos. Mag.* **44**, 597-618. *[10]*

WRIGHT, S.I. (1993), A review of automated orientation imaging microscopy (OIM). *J. Computer-Assisted Microscopy* **5**, 207-221. *[4]*

WRIGHT, S.I. & ADAMS, B.L. (1992), Automatic analysis of electron backscatter diffraction patterns. *Metall. Trans.* **A23**, 759-767. *[4]*

WRIGHT, S.I. & KOCKS, U.F. (1996), A comparison of different texture analysis techniques. In *Texture of Materials ICOTOM-11*, Z. Liang, L. Zuo & Y. Chu, eds. (Beijing: International Academic Publishers) 53-62. *[4]*

WRIGHT, S.I., ADAMS, B.L. & KUNZE, K. (1993a), Application of a new automatic lattice orientation measurement technique to polycrystalline aluminum. *Mater. Sci. Eng.* **A160**, 229-240. *[4]*

WRIGHT, S.I., ROLLETT, A.D. & STOUT, M.G. (1993b), In plane anisotropy of as-rolled copper sheet. *Scripta Metall. Mater.* **28**, 985-990. *[7]*

WRIGHT, S.I., BEAUDOIN, A.J. & GRAY, G.T. (1994a), Texture gradient effects in tantalum. In *Textures of Materials ICOTOM-10*, H.J. Bunge, ed. (Switzerland: Trans. Tech. Pubs.) 1695-1700. *[5]*

WRIGHT, S.I., GRAY, III, G.T. & ROLLETT, A.D. (1994c), Textural and microstructural gradient effects on the mechanical behavior of a tantalum plate. *Metall. Mater. Trans.* **25A**, 1025-1031. *[5]*

WRIGHT, S.I., BINGERT, J.F. & ZERNOW, L. (1996), Microtextural zones in a copper shaped charge particle. *Mater. Sci. Eng.* **A207**, 224-227. *[2]*

WU, T-Y., BASSANI. J.L. & LAIRD, C. (1991), Latent hardening in single crystals. *Proc. R. Soc. Lond.* **435**, 1-19. *[8]*

XIONG FULIN, CHANG, R.P.H., HAGERMAN, M.E., KOZHEVNIKOV, V.L., POEPPELMEIER, K.R., ZHOU HAITAIN, WONG, G.K., KETTERSON, J.B. & WHITE, C.W. (1994), Pulsed excimer laser deposition of potassium titanyl phosphate films. *Appl. Phys. Lett.* **64**, 161-163. [6]

YAO, Z. & WAGONER, R.H. (1993), Active slip in aluminum multicrystals. *Acta Metall.* **41**, n. 2, 451-468. [13]

YEUNG, W.Y., HIRSCH, J. & HATHERLY, M. (1988). Rolling texture of fine grained 70:30 brass. In *Eighth Int. Conf. on Textures of Materials*, J.S. Kallend & G. Gottstein, eds. (Warrendale, PA: The Metallurgical Society) 467-472. [5]

YOO, M.H. (1981), Slip, twinning, and fracture in hexagonal close-packed metals. *Metall. Trans.* **12A**, 409-418. [2]

YOO, M.H. & LOH, B.T.M. (1970), Structural and elastic properties of zonal twin dislocations in anisotropic crystals. In *Fundamental Aspects of Dislocation Theory*, J.A. Simmins, R. deWit & R. Bullough, eds. (Nat. Bur. Stand. Spec. Pub. 317) 479-493. [2]

YOUNG, C.T. & LYTTON, J.L. (1972), Computer generation and identification of Kikuchi projections. *J. Appl. Phys.* **43**, 1408-1417. [4]

ZAEFFERER, S. & SCHWARZER, R.A. (1994), Automated measurement of single grain orientations in the TEM. *Z. Metallk.* **85**, 585-591. [4]

ZELLER, R. & DEDERICHS, P.H. (1973), Elastic constants of polycrystals. *Phys. Stat. Sol.* (b) **55**, 831-842. [A]

ZHANG, S. & KARATO, S.-Y. (1995), Lattice preferred orientation of olivine aggregates deformed in simple shear. *Nature* **375**, 774-777. [6]

ZHANG, Y. & JENKINS, J.T. (1993), The evolution of anisotropy of a polycrystalline aggregate. *J. Mech. Phys. Solids* **41**, 1213-1243. [8]

ZHAO, J. & ADAMS, B.L. (1988), Definition of an asymmetric domain for intercrystalline misorientation in cubic materials in the space of Euler angles. *Acta Crystallogr.* **A44**, 326-336. [2]

ZHAO, J., ADAMS, B.L. & MORRIS, P.R. (1988), A comparison of measured and texture-estimated misorientations in type 304 stainless steel tubing. *Textures Microstruct.* **8 & 9**, 493-508. [2]

ZINSERLING, K. & SHUBNIKOV, A. (1933), Über die Plastizität des Quarzes. *Z. Kristallogr.* **85**, 454-461. [6]

SUBJECT INDEX

The camera-ready copy for this book was designed and produced by U. F. Kocks, using Word for Windows95 (Version 7.0) by Microsoft Corp., MathType by Design Science Inc., the Utopia font set by Adobe Systems Inc., and the LaserJet 4V printer by Hewlett Packard Co.

rinted in the United States
y Bookmasters